Methods of Molecular Quantum Mechanics

THEORETICAL CHEMISTRY

A Series of Monographs

T. E. PEACOCK: Electronic Properties of Aromatic and Heterocyclic Molecules

R. McWEENY and B. T. SUTCLIFFE: Methods of Molecular Quantum Mechanics

J. LINDBERG and Y. ÖHRN: Propagators in Quantum Chemistry

M. S. CHILD: Molecular Collision Theory

C. D. H. CHISHOLM: Group Theoretical Techniques in Quantum Chemistry

E. R. DAVISON: Reduced Density Matrices in Quantum Chemistry

A. C. HURLEY: Electron Correlation in Small Molecules

D. P. CRAIG and T. THIRUNAMACHANDRAN: Molecular Quantum Electrodynamics

GAD FISCHER: Vibronic Coupling

R. McWEENY: Methods of Molecular Quantum Mechanics. Second Edition

R. H. TIPPING and J. F. OGILVIE: The Vibrational–Rotational Spectroscopy of Diatomic Molecules

N. H. MARCH: Electron Density Theory of Atoms and Molecules

Methods of Molecular Quantum Mechanics

R. McWEENY

University of Pisa,
Pisa, Italy

SECOND EDITION

ACADEMIC PRESS
Harcourt Brace & Company Publishers
London San Diego New York
Boston Sydney Tokyo Toronto

ACADEMIC PRESS LIMITED
24/28 Oval Road
London NW1 7DX

United States Edition published by
ACADEMIC PRESS INC.
San Diego, CA 92101

Second printing 1996

British Library Cataloguing in Publication Data

McWeeny, R.
Methods of molecular quantum mechanics.—(Theoretical chemistry).—2nd ed.
1. Chemistry. Quantum theory
I. Title II. Series

541.2′8

ISBN 0-12-486551-8
ISBN 0-12-486552-6 PBK

FIRST EDITION (R. McWeeny and B. T. Sutcliffe) 1969
Reprinted with SI units and revisions 1976
Reprinted 1978

Typeset in Northern Ireland by The Universities Press (Belfast) Ltd.
Printed and bound in Great Britain by Hartnolls Ltd. Bodmin, Cornwall.

Preface to the Second Edition

The last twenty years have seen remarkable advances in molecular quantum mechanics. The traditional methods expounded in the first edition of this book have been implemented on a grand scale, and the *ab initio* calculation of molecular electronic structure and properties, using freely available programs and increasingly powerful computers, has become commonplace. Indeed "computational chemistry" now has its own books and journals and plays a respected role in the study of molecular processes of all kinds—whether in interstellar space or in the chemical laboratory or in biological systems.

Were that the only development, the first edition would have required no more than minor revisions. In fact, however, the underlying theoretical fabric of quantum chemistry has also been the subject of considerable change, largely as a result of the incorporation of methods originating in theoretical physics. Such methods are still unfamiliar in chemistry and, because of their complexity and sophistication, their use has been widely resisted; but their power and generality is now beyond doubt and no theoretical chemist can afford to be ignorant of them—their inclusion was therefore inevitable.

In preparing this second edition I have retained as far as possible the structure and spirit of the original; and again the aim has been to produce "a book that would stand by itself without the need for constant reference to other books and journals", taking the reader—typically a beginning graduate student in chemistry or physics—up to the research frontiers. The frontiers are now more distant, and the new edition consequently larger, but the same deductive and step-by-step approach is used and the demands made on the reader should be no more severe. This is a "teaching book" rather than a guide to the research literature, and the problems at the end of each chapter should provide a valuable supplement to the text.

The earlier chapters, still essentially introductory, are not greatly changed, but subsequent chapters contain an increasing amount of new material, some being completely new. Chapter 2 includes more on vector spaces, matrices and the variation method; while second-quantization ideas are introduced in Chapter 3, following the discussion of the Slater method, so that both approaches may be used and compared in later chapters. The treatment of spin eigenfunctions and their connection with permutation symmetry now occupies a whole new chapter (Chapter 4). Electron distribution functions (Chapter 5) retain their importance and new material on spin, current and kinetic-energy densities has been added.

Chapter 6 on molecular orbital theory has been rewritten and expanded to cover more general open-shell states and ensemble averages; the following chapter on valence bond theory now includes several new sections covering recent progress on the *ab initio* implementation of the VB approach. Chapter 8 is new and is devoted mainly to the optimization of multiconfiguration wavefunctions that contain either localized or delocalized orbitals; it therefore represents the logical sequel to the two preceding chapters. Traditional and second-quantization methods are developed side by side to demonstrate their equivalence and to familiarize the reader with the newer techniques.

Two more major advances are introduced in Chapters 9 and 10. Both chapters are quite new and are concerned with the calculation of high-precision wavefunctions involving many thousands of electron configurations. They include a first-principles development of many-body perturbation theory, and its diagrammatic implementation, followed by an elementary exposition (again from first principles) of the so-called "unitary group approach" now used in much of the current research literature.

The discussion of electric and magnetic properties of molecules retains its importance, especially with the growth of new and sensitive experimental techniques in spectroscopy and magnetic resonance. Chapter 11, on static properties, is a rewritten and expanded version of the corresponding chapter in the first edition and covers effects due to the presence of small (relativistic) terms in the Hamiltonian and to the application of external fields. Nowadays, however, *dynamic* effects are also of growing importance, and a new Chapter 12 is devoted to time-dependent perturbations and the general theory of linear response.

From the study of linear response it is but a short step to propagators and Green's functions. Chapter 13 develops all the basic concepts in this comparatively new and rapidly expanding field, along with some of the methodology for actually calculating propagators. The approach is carefully related both the the time-dependent perturbation theory of the previous chapter and to the so-called EOM (equation-of-motion) methods, treated in a final section, which now enjoy great popularity in the study of excitation, ionization and attachment processes.

The final chapter looks beyond individual molecules and their properties—to *interactions*: the "non-bonded" interactions between different molecules, or between different groups in a large molecule, are fundamental in determining not only the behaviour of matter in bulk (from imperfect gases to liquids and molecular crystals) but also the course of chemical reactions and, to give only one example, the secondary structure of protein chains and other polymers. These interactions form the interface between molecular theory and the rest of physics and chemistry: they are considered from a unified and general standpoint in Chapter 14.

I have tried to be reasonably comprehensive, but only within a well-defined area, which excludes, apart from the study of nuclear motion, all computational details (a subject now large enough to require separate monographs—e.g. Wilson, 1986) and all topics that explicitly involve interaction with a radiation field, for which reference may be made to the admirable treatment by Craig and Thirunamachandran (1984).

Again, the book may be used in various ways. An introductory graduate course would involve selected parts of the first six chapters and Appendices 1–3; a course on electronic properties of molecules might be based on Chapters 11 and 12; and an advanced course could start from second-quantization (Section 3.6), following with its applications in Chapter 8, and passing to a selection of topics from the rest of the book (e.g. spin eigenfunctions, VB theory and unitary group methods; or diagrammatic perturbation theory and propagators).

Thanks are due to Dr B. T. Sutcliffe, coauthor of the first edition, for many useful discussions of what the present volume should contain and for his continuing interest during the last few years. Although he was unable to participate in the writing of the new edition, and responsibility for revisions, new material and (no doubt) mistakes must therefore be mine, his help in critically reading all chapters and in suggesting some of the problems and references has been greatly appreciated. A companion volume, in which he intends to deal with nuclear-motion effects and with molecules moving in the presence of time-dependent fields, is planned.

Finally, I wish to record my thanks to Academic Press for their patient cooperation.

R. McWeeny

Contents

Units

SI conventions are assumed throughout this book and recommended symbols‡ are employed throughout, with two small exceptions. For typographcial convenience, $4\pi\epsilon_0$ (ϵ_0 being the permittivity of free space) is often denoted by the single symbol κ_0; and β has been retained for the Bohr magneton (recommended symbol μ_B). Instead of $\mu_0/4\pi$ (μ_0 being the permeability of free space), we have used the equivalent combination $1/\kappa_0 c^2$, c being the velocity of light.

The SI base units are the kilogramme (kg), metre (m), second (s) and ampère (A). Some commonly occurring derived units are given below.

Quantity	Unit	Symbol	Dimensional definition(s)
Energy	joule	J	$kg\,m^2\,s^{-2}$
Force	newton	N	$kg\,m\,s^{-2}$ $(J\,m^{-1})$
Power	watt	W	$kg\,m^2\,s^{-3}$ $(J\,s^{-1})$
Electric charge	coulomb	C	A s
Electric potential difference	volt	V	$kg\,m^2\,s^{-3}\,A^{-1}$ $(J\,A^{-1}\,s^{-1})$
Electric field strength	volt/metre	—	$kg\,m\,s^{-3}\,A^{-1}$ $(V\,m^{-1})$
Electric capacitance	farad	F	$kg^{-1}\,m^{-2}\,s^4\,A^2$ $(A\,s\,V^{-1})$
Magnetic flux density	tesla	T	$kg\,s^{-2}\,A^{-1}$ $(V\,s\,m^{-2})$
Frequency	hertz	Hz	s^{-1} (cycle per second)

Multiples and fractions of the base units are indicated by a prefix (e.g. centi $\equiv 10^{-2}$) as follows:

prefix	giga	mega	kilo	milli	micro	nano	pico	atto
multiple	10^9	10^6	10^3	10^{-3}	10^{-6}	10^{-9}	10^{-12}	10^{-15}
symbol	G	M	k	m	μ	n	p	a

The electron volt (1.6021×10^{-19} J) and the gauss (10^{-4} T) are still permitted, and commonly used, units. For further details the reader should consult the reference already given.

‡ See e.g. *Quantities, Units and Symbols*, 2nd edn. Royal Society, London, 1975.

Atomic units‡ are also employed. In atomic units, e, $\hbar$, m and κ_0 ($=4\pi\epsilon_0$) all take unit values and may thus be dropped from all equations provided the symbols occurring are reinterpreted as the numerical (i.e. dimensionless) measures of the quantities they represent. For example, Z is the numerical measure, in atomic units, of charge on the nucleus (Ze) and $[l(l+1)]^{1/2}$ is the numerical measure of an angular momentum $[l(l+1)]^{1/2}\hbar$. At the end of a calculation the results will of course appear in atomic units; if an energy is computed as 6.2 this means $E=6.2E_h$, where E_h is the atomic unit of energy. It is therefore necessary only to know (i) the atomic units of common physical quantities, in terms of e, $\hbar$, m, κ_0, and (ii) how these units are related to the corresponding SI units. In practice, it is convenient to introduce the "bohr" and the "hartree"

$$a_0=\hbar^2\kappa_0/me^2, \qquad E_h=me^4/\kappa_0^2\hbar^2$$

and to express other quantitites in terms of the most convenient combinations of e, $\hbar$, m, a_0 and E_h. Typical results are indicated below.

Quantity	Atomic unit	SI equivalent
Mass	m	9.1091×10^{-31} kg
Charge	e	1.6021×10^{-19} C
Action	$\hbar$	1.0545×10^{-34} J s
Permittivity	κ_0	$4\pi\times8.8542\times10^{-12}$ F m^{-1}
Length	a_0	$5.2916\ 10^{-11}$ m
Energy	E_h	4.3594×10^{-18} J (27.211 eV)
Time	$\hbar E_h^{-1}$	2.4189×10^{-17} s
Velocity	$E_h e\hbar^{-1}$	2.1876×10^{6} m s^{-1}
Force	$E_h a_0^{-1}$	8.2383×10^{-8} N
Power	$E_h^2\hbar^{-1}$	1.8023×10^{-1} W
Electric potential difference	$E_h e^{-1}$	2.7211×10^{1} V
Electric field strength	$E_h e^{-1}a_0^{-1}$	5.1422×10^{11} V m^{-1}
Electric dipole moment	ea_0	8.4778×10^{-30} C m
Magnetic flux density	$\hbar e^{-1}a_0^{-2}$	2.3506×10^{5} T
Magnetic dipole moment	$e\hbar m^{-1}$	1.8546×10^{-23} J T^{-1}

It should be noted that the Bohr magneton ($e\hbar/2m$), the magnetic moment associated with one unit ($\hbar$) of orbital angular momentum, takes

‡ Strictly speaking, these are *quantities,* not units; for example, the charge on the electron is *measured* in terms of the conventionally defined SI unit (the coulomb). The term "unit" is nevertheless adopted, in conformity with everyday usage.

the value $\frac{1}{2}$ in atomic units. The fine-structure constant is

$$\alpha = e^2(\kappa_0 \hbar c) = 1/137.030,$$

and hence the velocity of light in atomic units has the numerical value 137.030. When equations are expressed in atomic units, the orders of magnitude of small terms in the Hamiltonian (Chapter 11) are conveniently indicated by powers of $1/c$.

1 Introductory Survey

1.1 SCHRÖDINGER'S EQUATION

The electronic structure and properties of any molecule, in any of its available stationary states, may be determined in principle by solution of Schrödinger's (time-independent) equation. For a system of N electrons, moving in the potential field due to the nuclei, this takes the form

$$\mathsf{H}\Psi(\boldsymbol{x}_1, \boldsymbol{x}_2, \ldots, \boldsymbol{x}_N) = E\Psi(\boldsymbol{x}_1, \boldsymbol{x}_2, \ldots, \boldsymbol{x}_N), \qquad (1.1.1)$$

where H is the Hamiltonian operator‡

$$\mathsf{H} = \sum_i \mathsf{h}(i) + \tfrac{1}{2}\sum_{i,j}{}' g(i, j). \qquad (1.1.2a)$$

Here

$$\mathsf{h}(i) = -\frac{\hbar^2}{2m}\nabla^2(i) + V(i), \qquad (1.1.2b)$$

a Hamiltonian operator for electron i, while (SI units, see page xiii)

$$g(i, j) = e^2/\kappa_0 r_{ij} \qquad (1.1.2c)$$

is the electrostatic interaction between electrons i and j. The operator $\mathsf{h}(i)$ consists of two parts. The first is the kinetic-energy operator, which in cartesian coordinates becomes

$$-\frac{\hbar^2}{2m}\nabla^2(i) = -\frac{\hbar^2}{2m}\left(\frac{\partial^2}{\partial x_i^2} + \frac{\partial^2}{\partial y_i^2} + \frac{\partial^2}{\partial z_i^2}\right), \qquad (1.1.3)$$

while the second is simply the potential energy of electron i in the field of the nuclei. The notation $\mathsf{h}(i)$, for example, indicates that the operator works on (e.g. differentiates with respect to) the variables x_i, y_i, z_i

‡ A sans serif typeface (Univers) will be used throughout to denote operators. An exception is made when the operator is merely a numerical multiplier (i.e. does not involve differentiation etc.), in which case no typographical distinction is necessary. Also, summation ranges are omitted if clear from the context: thus i, j in (1.1.2a) run over all electrons $1, 2, \ldots, N$.

defining the position of particle i. The function Ψ depends upon all such variables; we use the bold letter $\mathbf{x}_i$ to symbolize collectively all the variables needed in referring to particle i, noting that the set must later be extended to include a "spin variable". It should be noted that m and $-e$ are the mass and charge of the electron while $\hbar$ is Planck's constant h divided by 2π. The prime on the second summation in (1.1.2a) simply indicates that the term $i=j$ should not be counted, and the factor $\frac{1}{2}$ ensures that the interaction of each pair of particles is counted once only.

We assume that the nuclei in the molecule are fixed in space so that the potential energy of electron i is given by

$$V(i) = -\sum_n \frac{Z_n e^2}{\kappa_0 r_{ni}} \qquad (r_{ni} = |\mathbf{r}_i - \mathbf{R}_n|), \tag{1.1.4}$$

where $\mathbf{R}_n$ and eZ_n are the position vector and charge of nucleus n while r_{ni} is the magnitude of the position vector‡ of electron i relative to nucleus n. In rectangular cartesian coordinates, r_{ni} is given by

$$r_{ni} = [(x_i - X_n)^2 + (y_i - Y_n)^2 + (z_i - Z_n)^2]^{1/2}. \tag{1.1.5}$$

The potential energy arising from mutual repulsion of electrons i and j is given by the term $g(i, j)$, in which r_{ij} is just the separation $|\mathbf{r}_j - \mathbf{r}_i|$.

We frequently express the operators in (1.1.2) in Hartree atomic units (see page xiv), the units of charge§, mass and "action" then being e, m and $\hbar$ respectively. The unit of length in this system is the Bohr radius $a_0 = 5.292 \times 10^{-11}$ m and the unit of energy is the Hartree $E_{\mathrm{h}} = 27.21$ eV. More precise values of these and many other constants have been compiled by Cohen and DuMond (1965). In atomic units,

$$\left.\begin{aligned} \mathsf{h}(i) &= -\tfrac{1}{2}\nabla^2(i) - \sum_n \frac{Z_n}{r_{ni}}, \\ g(i, j) &= \frac{1}{r_{ij}}. \end{aligned}\right\} \tag{1.1.6}$$

The *wavefunction* (or *state function*) $\Psi(\mathbf{x}_1, \mathbf{x}_2, \ldots, \mathbf{x}_N)$, which describes the electronic state of the molecule, must possess certain mathematical properties, in particular that of "quadratic integrability". If Φ is any solution of (1.1.1), this requirement takes the form

$$\int \Phi^* \Phi \, \mathrm{d}\mathbf{x} = \int |\Phi|^2 \, \mathrm{d}\mathbf{x} = K \quad (K \text{ finite}), \tag{1.1.7}$$

‡ The vector itself is $\mathbf{r}_{ni} = \mathbf{r}_i - \mathbf{R}_n$, pointing from n to i.

§ Note that the actual charge of the electron is $-e$ or -1 atomic unit.

where $\mathrm{d}\boldsymbol{x}$ is used to denote the general volume element $\mathrm{d}\boldsymbol{x} = \mathrm{d}\boldsymbol{x}_1\,\mathrm{d}\boldsymbol{x}_2 \ldots$ and the integration is over all space (i.e. the full range of all variables). But since $\Psi = c\Phi$ also satisfies (1.1.1), we are at liberty to choose the constant c so that

$$\int |\Psi|^2\,\mathrm{d}\boldsymbol{x} = |c|^2 \int |\Phi|^2\,\mathrm{d}\boldsymbol{x} = |c|^2 K = 1.$$

A function Ψ possessing this property, namely

$$\int |\Psi|^2\,\mathrm{d}\boldsymbol{x} = 1, \tag{1.1.8}$$

is said to be *normalized.* We usually assume that eigenfunctions are normalized, and in this case $|\Psi|^2$ has the physical interpretation that

$$|\Psi|^2\,\mathrm{d}\boldsymbol{x}_1\,\mathrm{d}\boldsymbol{x}_2 \ldots \mathrm{d}\boldsymbol{x}_N = \begin{pmatrix} \text{probability of finding electrons} \\ 1, 2, \ldots, N \text{ simultaneously in} \\ \text{volume elements } \mathrm{d}\boldsymbol{x}_1, \mathrm{d}\boldsymbol{x}_2, \ldots, \mathrm{d}\boldsymbol{x}_N \end{pmatrix}. \tag{1.1.9}$$

Full discussions of the mathematical properties that a many-electron wavefunction must possess are available elsewhere (Kato, 1957; Kemble, 1937). A function possessing the required properties (quadratic integrability etc.) is said to belong to the "class L^2" (or class Q).

The *electronic energy* E is the energy of the N electrons moving in the field provided by the nuclei. It contains the coordinates of the nuclei, assumed fixed, as parameters and may be indicated explicitly as a function of these coordinates: $E = E(\boldsymbol{R}_1, \boldsymbol{R}_2, \ldots)$. For a system with fixed nuclei the total energy of the molecule E_{tot} is just the sum of the electronic energy and the nuclear repulsion energy:

$$E_{\text{tot}} = E + \tfrac{1}{2} \sum_{n,n'}{}' \frac{Z_n Z_{n'}}{R_{nn'}}. \tag{1.1.10}$$

E_{tot} must of course be negative for molecular binding and it must also be less than the sum of the energies of the separated atoms if the molecule is to be stable against dissociation into atoms.

Equation (1.1.1) is an *eigenvalue equation* and is characterized by possessing solutions of class L^2 only for certain special values of the *eigenvalue parameter E.* These solutions Ψ_K are the *eigenfunctions* of the operator H, and the corresponding parameter values E_K are the eigenvalues, which in this context are the quantized energies of the allowed stationary states of the electronic system. We note that H is a *Hermitian* operator whose properties are discussed later. In particular, any two of its eigenfunctions that correspond to different energy values E_K, E_L

possess the *orthogonality property*

$$\int \Psi_K^* \Psi_L \, d\mathbf{x} = 0. \tag{1.1.11}$$

When two or more eigenvalues happen to be identical, corresponding to a degenerate state of the system, the different eigenfunctions may still be assumed orthogonal without loss of generality. It is normally assumed that the full set of normalized eigenfunctions of equation (1.1.1) forms a *complete orthonormal system*. This means that an arbitrary function of the same class may be expressed as a linear combination of the Ψs with any required accuracy provided that a sufficient number of terms is included, an idea that we discuss in more detail in Chapter 2.

If (1.1.1) provides an accurate statement of the electronic problem to be considered then the solutions $\Psi_1, \Psi_2, \ldots$ describe the available stationary states of the molecule as completely as possible, and all electronic properties of the molecule in state Ψ_K can be determined as expectation values of appropriate Hermitian operators. Before continuing, however, a little more must be said about (1.1.1) and the relation of its solutions to the *experimentally observed* molecular electronic properties.

Any experimental observation is performed not on a single molecule but on a macroscopic aggregate of molecules; and, secondly, any observation involves some kind of *interaction* with the system. We have taken no account of these facts in setting up the Schrödinger equation (1.1.1) or in the Hamiltonian defined in (1.1.2a); but both must be allowed for when we try to interpret physical and chemical observations using results reached by solving (1.1.1). Very often the environment of a molecule has little effect on some given property of interest; thus, in spectroscopy, the radiation field merely provides a weak perturbation, inducing transitions from one state to another, and properties of the individual molecular states can be accurately inferred from observation. In chemical reactions, however, more drastic changes occur as a result of collision and scattering processes; and since a whole phase (gas, liquid or solid) is involved, the methods of statistical mechanics must also be invoked. Even at the molecular level such processes are time-dependent and their analysis requires the time-dependent Schrödinger equation

$$\mathsf{H}\Psi = i\hbar \frac{\partial \Psi}{\partial t}. \tag{1.1.12}$$

The detailed study of transition and collision processes lies outside the scope of this book, which is concerned almost wholly with the electronic structure and properties of single molecules.

Equation (1.1.1) provides a basis for the discussion of the electronic structure and properties of molecules in their stationary states. In the next few chapters we shall deal almost exclusively with stationary states; but in later chapters we develop theoretical procedures for calculating the effects of time-dependent perturbations, and we then have to start from (1.1.12). This latter development will lead to the study of "propagator" and "equation-of-motion" methods, which now play an important role in molecular quantum mechanics.

In returning to the time-independent equation (1.1.1), it must be stressed that the Hamiltonian (1.1.2a) is still somewhat idealized, even for an isolated system. In writing it down, we have assumed that the nuclei are fixed in space and we have neglected all interactions between particles other than those which are purely electrostatic in origin. We shall consider the inclusion of terms corresponding to more general electromagnetic interactions and relativistic effects in Chapter 11; here we comment only on the assumption of fixed nuclei.

If we wish to include the effects of nuclear motion, the Hamiltonian H (in the approximation of electrostatic forces only) becomes

$$\mathsf{H} = \mathsf{H}_{\mathrm{e}} + \mathsf{H}_{\mathrm{n}} + \mathsf{H}_{\mathrm{en}}, \tag{1.1.13}$$

where the electronic, nuclear and interaction terms are (in atomic units)

$$\mathsf{H}_{\mathrm{e}} = -\tfrac{1}{2}\sum_{i} \nabla^2(i) + \tfrac{1}{2}\sum_{i,j}{}' g(i, j), \tag{1.1.14a}$$

$$\mathsf{H}_{\mathrm{n}} = -\tfrac{1}{2}\sum_{n} \frac{1}{m_n} \nabla^2(n) + \tfrac{1}{2}\sum_{n,n'}{}' Z_n Z_{n'} g(n, n'), \tag{1.1.14b}$$

$$\mathsf{H}_{\mathrm{en}} = -\sum_{n}\sum_{i} \frac{Z_n}{r_{ni}}. \tag{1.1.14c}$$

The corresponding wavefunction must then include both electronic and nuclear coordinates. The validity of the fixed-nucleus model discussed so far was first established by Born and Oppenheimer (1927) (see also Born and Huang, 1954), who expanded the total molecular wavefunction in terms of products of electronic and nuclear wavefunctions, and showed that in good approximation a single product was usually appropriate; the electronic wavefunction is then a solution of (1.1.1), while the nuclear wavefunction is derived from a nuclear eigenvalue equation in which E_{tot} obtained in (1.1.10), as a function of nuclear positions (via solution of (1.1.1)), is used as a potential function. It is because of the validity of this separation, which depends on the large ratio between electronic and nuclear masses, that we may confine our attention initially to a purely *electronic* problem.

Although we are not concerned with nuclear motion in this book, the nuclear-motion Schrödinger equation plays a central role in molecular spectroscopy (Herzberg, 1945, 1950, 1966; Wilson *et al.*, 1955). It must also be borne in mind that motion of the nuclei can sometimes profoundly affect even those properties of a molecule that are generally considered electronic—examples being the Jahn–Teller effect (removal of electronic degeneracy by spontaneous distortion of a molecule) and the Renner effect (modification of spectra by interaction of electronic and vibrational motions). The full discussion of effects arising from nuclear motion (especially in the presence of external fields), even in a moderately rigorous way, leads to complications and difficulties that have yet to be fully overcome: such considerations are taken up elsewhere (Fischer, 1984; Lefebvre-Brion and Field 1986). A useful introduction to this field, which is of great importance in molecular spectroscopy, is contained in the review by Longuet-Higgins (1961).

To summarize: (1.1.1)–(1.1.12) provide a mathematical basis for molecular quantum mechanics in a fixed-nucleus approximation, but the limitations of this theoretical model (which by now should be apparent) must not be forgotten.

Before approaching the main problems, it will be useful to review certain basic topics and to introduce definitions and notation within the context of two simple examples.

1.2 EXAMPLE: THE HELIUM ATOM

We first consider the helium-like atom, with a nucleus of charge Z fixed at the origin. The electronic Hamiltonian defined in Section 1.1 takes the form

$$\mathsf{H}(1,2) = \mathsf{h}(1) + \mathsf{h}(2) + g(1,2), \tag{1.2.1}$$

where‡, in atomic units,

$$\mathsf{h}(i) = -\tfrac{1}{2}\nabla^2(i) - \frac{Z}{r_i}, \tag{1.2.2a}$$

$$g(1,2) = \frac{1}{r_{12}}. \tag{1.2.2b}$$

‡ With only one nucleus the label n in (1.1.4) may be omitted: r_i is the distance of electron i from the nucleus (here $i = 1, 2$) and r_{12} is the interelectronic distance.

First approximation

The solutions of the eigenvalue equation (1.2.1) cannot be obtained in simple closed form owing to the presence of the term $g(1, 2)$. To obtain a very crude description of the system, we may negect this term altogether, i.e. consider a "model" system with non-interacting electrons and the Hamiltonian

$$\mathsf{H}_0(1, 2) = \mathsf{h}(1) + \mathsf{h}(2). \tag{1.2.3}$$

Solutions of the eigenvalue equation

$$\mathsf{H}_0 \Psi = E\Psi \tag{1.2.4}$$

are then easily obtained by the method of separation of variables. We look for a solution of the form

$$\Phi(\boldsymbol{r}_1, \boldsymbol{r}_2) = \phi_1(\boldsymbol{r}_1)\phi_2(\boldsymbol{r}_2), \tag{1.2.5}$$

where each factor contains the variables of only one electron. We denote these by $\boldsymbol{r}_1$, $\boldsymbol{r}_2$ respectively (position vectors) reserving $\boldsymbol{x}_1$, $\boldsymbol{x}_2$ for later use when we include the spin variables. Whether such solutions exist is confirmed by substituting in (1.2.4) and asking what conditions must be satisfied by the two factors. Substitution, followed by division throughout by Φ, yields

$$\frac{\mathsf{h}(1)\phi_1(\boldsymbol{r}_1)}{\phi_1(\boldsymbol{r}_1)} + \frac{\mathsf{h}(2)\phi_2(\boldsymbol{r}_2)}{\phi_2(\boldsymbol{r}_2)} = E. \tag{1.2.6}$$

Now the two terms on the left are entirely independent of each other, being functions of different sets of variables, and their sum can be a constant, for all values of the variables, only if each term is itself a constant. Thus, denoting the two constants by ϵ_1 and ϵ_2, the product (1.2.5) will satisfy the eigenvalue equation (1.2.4) provided that

$$\mathsf{h}(1)\phi_1(\boldsymbol{r}_1) = \epsilon_1\phi_1(\boldsymbol{r}_1), \tag{1.2.7a}$$

$$\mathsf{h}(2)\phi_2(\boldsymbol{r}_2) = \epsilon_2\phi_2(\boldsymbol{r}_2), \tag{1.2.7b}$$

and the total energy will then be

$$E = \epsilon_1 + \epsilon_2. \tag{1.2.8}$$

The names of the variables in (1.2.7) are irrelevant: ϕ_1 and ϕ_2 are functions of position of a point in space and the operator h works on the variables (e.g. cartesian coordinates) defining the point. It follows that ϕ_1 and ϕ_2 may be any solutions of the *one-electron* eigenvalue problem:

$$\mathsf{h}\phi = \epsilon\phi. \tag{1.2.9}$$

The corresponding eigenvalues, or "orbital energies", are ϵ_1 and ϵ_2, and in the state $\Phi(\boldsymbol{r}_1, \boldsymbol{r}_2) = \phi_1(\boldsymbol{r}_1)\phi_2(\boldsymbol{r}_2)$ each electron has its own definite energy: for $\Phi(\boldsymbol{r}_1, \boldsymbol{r}_2)$ is a *simultaneous eigenstate* of the operators $\mathsf{h}(1)$ and $\mathsf{h}(2)$ associated with the energies of the individual particles. Thus

$$\mathsf{h}(1)\Phi(\boldsymbol{r}_1, \boldsymbol{r}_2) = \epsilon_1\Phi(\boldsymbol{r}_1, \boldsymbol{r}_2), \qquad (1.2.10a)$$

$$\mathsf{h}(2)\Phi(\boldsymbol{r}_1, \boldsymbol{r}_2) = \epsilon_2\Phi(\boldsymbol{r}_1, \boldsymbol{r}_2). \qquad (1.2.10b)$$

The fact that the total energy is the sum of the energies of the separate electrons is not unexpected in view of the omission of the interaction term $g(1, 2)$.

These conclusions are easily generalized for any number of electrons. Thus if electrons were non-interacting we should never have to solve anything harder than a 1-electron eigenvalue equation, and by taking a product of 1-electron eigenfunctions, one factor for each electron, we could obtain an exact solution of the central problem of quantum chemistry! When electron interaction is admitted, the problem becomes vastly more difficult, but product functions such as (1.2.5) continue to play an important role, particularly as a means of describing and classifying the possible electronic states of a complicated many-electron system, this classification seldom being upset by the details of electron interaction. The "independent-particle model" (IPM) may indeed be refined, in such a way that this interaction is included to some extent, by using an "average potential" in the definition of h instead of the potential due to the nucleus alone. This development is taken up in Chapter 6.

The 1-electron 1-centre eigenfunctions, which satisfy an equation of the form (1.2.9), are *atomic orbitals* (AOs): they are "hydrogen-like" AOs if we use the Hamiltonian (1.2.2a), corresponding to one electron in the field of the bare nucleus, but may have a somewhat different functional form if they are eigenfunctions of a more general Hamiltonian (1.1.2b) in which the potential V may allow partly for the presence of other electrons. In any case the AOs are essentially those of a *central field,* in which V depends only on the radial distance of the electron from the nucleus, and have characteristic forms which are well described in elementary textbooks on valency. For completeness, the forms of the 1s, 2s, 2p, . . . , 4f orbitals and the significance of the classification are reviewed in Appendix 1.

A product function, such as (1.2.5), or more generally

$$\Phi(\boldsymbol{r}_1, \boldsymbol{r}_2, \ldots, \boldsymbol{r}_N) = \phi_i(\boldsymbol{r}_1)\phi_j(\boldsymbol{r}_2) \ldots \phi_p(\boldsymbol{r}_N), \qquad (1.2.11)$$

provides a specification of the orbitals $\phi_i, \phi_j, \ldots, \phi_p$ to which the N electrons are assigned. Generally the solutions of (1.2.9) form an infinite set $\phi_1, \phi_2, \ldots$, and the orbitals appearing in (1.2.11) are said to be "occupied", all others being "unoccupied", in the many-electron state described by the product. In the helium atom, with no electron interaction, the occupied AOs might, for example, both be 1s, or one may be 1s the other 2s, the corresponding wavefunctions and energies being

$$\Phi_1(\boldsymbol{r}_1, \boldsymbol{r}_2) = 1\mathrm{s}(\boldsymbol{r}_1)\, 1\mathrm{s}(\boldsymbol{r}_2), \qquad E_1 = 2\epsilon_{1\mathrm{s}}, \tag{1.2.12a}$$

$$\Phi_2(\boldsymbol{r}_1, \boldsymbol{r}_2) = 1\mathrm{s}(\boldsymbol{r}_1)\, 2\mathrm{s}(\boldsymbol{r}_2), \qquad E_2 = \epsilon_{1\mathrm{s}} + \epsilon_{2\mathrm{s}}. \tag{1.2.12b}$$

Specification of the set of occupied orbitals defines what we shall call an *orbital configuration,* and this is frequently abbreviated by symbols such as $1\mathrm{s}^2$, 1s 2s, etc. There is, of course, an ambiguity since there is no way of detecting *which* electrons have been assigned to the different orbitals; the resolution of this ambiguity leads to the Pauli principle, which we discuss presently.

Symmetry of the wavefunction

The fact that electrons are indistinguishable places important restrictions on the form of the wavefunction. Suppose we evaluate $\Psi(\boldsymbol{r}_1, \boldsymbol{r}_2) = \Phi_2(\boldsymbol{r}_1, \boldsymbol{r}_2) = 1\mathrm{s}(\boldsymbol{r}_1)\, 2\mathrm{s}(\boldsymbol{r}_2)$ for particular values of the variables indicated by $\boldsymbol{r}_1, \boldsymbol{r}_2$ (defining points 1 and 2); then $|\Psi(\boldsymbol{r}_1, \boldsymbol{r}_2)|^2\, d\boldsymbol{r}_1\, d\boldsymbol{r}_2$ is the probability of finding the two electrons simultaneously in volume elements $d\boldsymbol{r}_1$, $d\boldsymbol{r}_2$ around these points.‡ But now let us switch the electrons, so that the first electron is at point 2 and the second at point 1; the wavefunction evaluated for this new situation is $\Psi(\boldsymbol{r}_2, \boldsymbol{r}_1) = 1\mathrm{s}(\boldsymbol{r}_2)\, 2\mathrm{s}(\boldsymbol{r}_1)$ and the corresponding probability is $|\Psi(\boldsymbol{r}_2, \boldsymbol{r}_1)|^2\, d\boldsymbol{r}_1\, d\boldsymbol{r}_2$. Since the two situations referred to are physically indistinguishable, they should occur with the same probability, and since this is true for any choice of the two points, 1 and 2, the result should hold for all values of the variables:

$$|\Psi(\boldsymbol{r}_2, \boldsymbol{r}_1)|^2 = |\Psi(\boldsymbol{r}_1, \boldsymbol{r}_2)|^2. \tag{1.2.13}$$

The product function defined above clearly violates this requirement, and although acceptable as a solution of (1.2.4) it is not acceptable as a wavefunction defining the state of two indistinguishable particles. The functions $\Psi(\boldsymbol{r}_2, \boldsymbol{r}_1)$ and $\Psi(\boldsymbol{r}_1, \boldsymbol{r}_2)$ should clearly differ at most by a unimodular complex number $e^{i\theta}$, and it can be shown from group theory that the only essentially distinct possibilities that need be considered are

‡ At the moment we are proceeding as if electrons were spinless particles, i.e. as if the spatial variables ($\boldsymbol{r}_i$) alone comprise the full set $\boldsymbol{x}_i$ referred to in Section 1.1.

± 1. The symmetry principle thus implies two possibilities:

$$\Psi(\mathbf{r}_2, \mathbf{r}_1) = \Psi(\mathbf{r}_1, \mathbf{r}_2) \quad \text{(symmetric wavefunction)},$$

$$\Psi(\mathbf{r}_2, \mathbf{r}_1) = -\Psi(\mathbf{r}_1, \mathbf{r}_2) \quad \text{(antisymmetric wavefunction)}.$$

Similar considerations apply to the interchange of any two electrons in a many-electron system, and since any permutation may be regarded as a sequence of interchanges, the more general principle would appear to be

$$\left.\begin{aligned} &\mathsf{P}\Psi(\mathbf{r}_1, \mathbf{r}_2, \ldots, \mathbf{r}_N) = \Psi(\mathbf{r}_1, \mathbf{r}_2, \ldots, \mathbf{r}_N) \quad \text{(symmetric)},\\ &\text{or}\\ &\mathsf{P}\Psi(\mathbf{r}_1, \mathbf{r}_2, \ldots, \mathbf{r}_N) = \epsilon_{\mathsf{P}}\Psi(\mathbf{r}_1, \mathbf{r}_2, \ldots, \mathbf{r}_N) \quad \text{(antisymmetric)}, \end{aligned}\right\} \tag{1.2.14}$$

where P effects any permutation of the arguments $\mathbf{r}_1, \mathbf{r}_2, \ldots, \mathbf{r}_N$, and ϵ_{P} is ± 1 according as the permutation is equivalent to an even or odd number of interchanges (even or odd *parity*†). This generalization is modified when electron spin is recognized, as will be seen presently.

We observe that the helium-atom product function $\Phi_1(\mathbf{r}_1, \mathbf{r}_2)$ set up in (1.2.12a) is acceptable as a wavefunction, being symmetric besides satisfying the eigenvalue equation (1.2.4). On the other hand, neither $\Phi_2(\mathbf{r}_1, \mathbf{r}_2)$ nor $\Phi_2(\mathbf{r}_2, \mathbf{r}_1)$ is acceptable, although each satisfies (1.2.4). Since the eigenvalue equation is linear, however, the two latter functions may be combined with arbitrary coefficients to yield another solution with the same eigenvalue $\epsilon_{1s} + \epsilon_{2s}$. We then observe that the following are eligible as wavefunctions for three states of a helium-like system:

configuration $1s^2$

$$\Psi_1(\mathbf{r}_1, \mathbf{r}_2) = \Phi_1(\mathbf{r}_1, \mathbf{r}_2) = 1s(\mathbf{r}_1)\, 1s(\mathbf{r}_2); \tag{1.2.15a}$$

configuration 1s 2s

$$\begin{aligned}\Psi_2(\mathbf{r}_1, \mathbf{r}_2) &= [\Phi_2(\mathbf{r}_1, \mathbf{r}_2) + \Phi_2(\mathbf{r}_2, \mathbf{r}_1)]/\sqrt{2}\\ &= [1s(\mathbf{r}_1)\, 2s(\mathbf{r}_2) + 1s(\mathbf{r}_2)\, 2s(\mathbf{r}_1)]/\sqrt{2},\end{aligned} \tag{1.2.15b}$$

$$\begin{aligned}\Psi_3(\mathbf{r}_1, \mathbf{r}_2) &= [\Phi_2(\mathbf{r}_1, \mathbf{r}_2) - \Phi_2(\mathbf{r}_2, \mathbf{r}_1)]/\sqrt{2}\\ &= [1s(\mathbf{r}_1)\, 2s(\mathbf{r}_2) - 1s(\mathbf{r}_2)\, 2s(\mathbf{r}_1)]/\sqrt{2}.\end{aligned} \tag{1.2.15c}$$

These functions have been normalized and are mutually orthogonal, as follows easily on integrating their products. Wavefunctions for other states may be set up in a similar way starting from other orbital configurations, e.g. $1s\, 2p_0$: we have merely selected the ground state and the first excited configuration (i.e. next in ascending energy order) of *spherical symmetry*. Such states are frequently degenerate, in the

† ϵ as a parity factor will not be confused with an orbital energy.

independent-particle model, owing to degeneracies among the AOs themselves. When electron interaction is admitted, the energies of the independent-particle states are modified, and degeneracies may be resolved, but the states are still conveniently classified in terms of electron configurations and IPM ideas still play a fundamental role in the theory of atomic spectra (Condon and Shortley, 1935; Slater, 1960). We return to the calculation of energies in Chapter 3, noting only that these may be estimated roughly as expectation values of the full Hamiltonian (1.2.1). At this point, however, we ask how the classification is affected when electron spin is recognized.

Effect of electron spin

It will be recalled that a single electron is not completely characterized by its wavefunction, $\phi(\boldsymbol{r})$ say. Even in a state with no orbital angular momentum, application of a magnetic field reveals that the state is really a degenerate pair or "doublet", the small splitting of the energy levels (Zeeman effect) suggesting an *intrinsic* magnetic moment whose component along the field direction can take only one of two possible values. The splitting arises from small field-dependent terms in the Hamiltonian, which we have so far ignored. The intrinsic angular momentum with which the magnetic moment may be associated is called the "spin", and the permitted values of the spin component are found to be $\pm\frac{1}{2}\hbar$. The direction along which the component is measured is conventionally chosen as the z axis. To describe this situation mathematically, we introduce in addition to the position variable $\boldsymbol{r}$ a "spin variable" s; and we associate operators S_x, S_y, S_z with the three components of spin angular momentum. The spinless Hamiltonian (1.2.1) then works on the "external" variable $\boldsymbol{r}$, while the spin operators work on the "internal" variable s. There is then a close formal analogy between this situation and that considered in the last section, which may be indicated schematically as in Table 1.1.

Just as the product function $\Phi(\boldsymbol{r}_1, \boldsymbol{r}_2)$ is a simultaneous eigenfunction of the operators referring to the two parts of the system, so is the *spin-orbital* $\psi(\boldsymbol{x}) = \phi(\boldsymbol{r})\eta(\mathrm{s})$, where (cf. p. 7) we now adopt $\boldsymbol{x}$ for space–spin variables, i.e. $\boldsymbol{r}$ (space) and s (spin) together. For if we are interested in states with a definite value of S_z, say, then $\eta(\mathrm{s})$, which describes the spin state, will be a solution of

$$\mathsf{S}_z\eta = \lambda\eta, \tag{1.2.16}$$

and if the orbital state is of energy ϵ (neglecting spin terms in the

Table 1.1

	Two electrons (no spin)	One electron (with spin)
Variables (two sets)	$\boldsymbol{r}_1$ (position, electron 1) $\boldsymbol{r}_2$ (position, electron 2)	$\boldsymbol{r}$ (position) s (spin)
Two-part Hamiltonian (interaction neglected)	$\mathsf{H} = \mathsf{h}(1) + \mathsf{h}(2)$	$\mathsf{H} = \mathsf{h} + \mathsf{S}$ (S a spin operator)
States of independent parts	$\phi_1(\boldsymbol{r}_1)$ eigenfunction of $\mathsf{h}(1)$ $\phi_2(\boldsymbol{r}_2)$ eigenfunction of $\mathsf{h}(2)$	$\phi(\boldsymbol{r})$ eigenfunction of h $\eta(s)$ eigenfunction of S
Complete description	$\Phi(\boldsymbol{r}_1, \boldsymbol{r}_2) = \phi_1(\boldsymbol{r}_1)\phi_2(\boldsymbol{r}_2)$	$\psi(\boldsymbol{r}, s) = \phi(\boldsymbol{r})\eta(s)$

Hamiltonian) then $\phi(r)$ satisfies

$$\mathsf{h}\phi = \epsilon\phi. \tag{1.2.17}$$

Consequently, since each operator works only on its own variables,

$$\mathsf{h}\psi = \epsilon\psi, \tag{1.2.18a}$$

$$\mathsf{S}_z\psi = \lambda\psi, \tag{1.2.18b}$$

and ψ is a state in which the particle simultaneously has definite energy in its orbital motion and also a definite z component of spin. Observation shows that (1.2.16) has only two solutions: these are normally written α and β and satisfy

$$\boxed{\mathsf{S}_z\alpha = \tfrac{1}{2}\alpha,} \tag{1.2.19a}$$

$$\boxed{\mathsf{S}_z\beta = -\tfrac{1}{2}\beta.} \tag{1.2.19b}$$

We shall use dimensionless angular-momentum operators throughout in order to avoid the repeated appearance of $\hbar$, which is the natural unit. This convention means that $\hbar\mathsf{S}_z$ etc. are the actual angular momentum operators, and is therefore numerically equivalent to expressing all angular momenta in atomic units. Corresponding to the two choices of spin component, every orbital ϕ yields two possible spin-orbitals, $\psi = \phi\alpha$ and $\bar{\psi} = \phi\beta$.

The observed spin components behave like components of a vector, on rotating the coordinate axes, and from the general postulates of quantum mechanics it is inferred that the spin operators satisfy commutation relations exactly like the orbital angular-momentum operators (see

Appendix 2), namely

$$\boxed{\begin{aligned}\mathsf{S}_x\mathsf{S}_y - \mathsf{S}_y\mathsf{S}_x &= i\mathsf{S}_z,\\ \mathsf{S}_y\mathsf{S}_z - \mathsf{S}_z\mathsf{S}_y &= i\mathsf{S}_x,\\ \mathsf{S}_z\mathsf{S}_x - \mathsf{S}_x\mathsf{S}_z &= i\mathsf{S}_y.\end{aligned}} \tag{1.2.20}$$

It is clear that a spin-orbital ψ may be an exact eigenfunction even *without* neglect of spin effects, provided the spin term in the Hamiltonian (Table 1.1) is just S_z itself or some function of S_z; and this gives a very simple interpretation of the observed Zeeman splitting for a one-electron system with only spin angular momentum. The classical Hamiltonian for a particle with magnetic moment $\boldsymbol{\mu}$ and with a magnetic field B along the z axis would contain the term $-B\mu_z$ in addition to the "orbital" kinetic and potential energy terms. On inserting the observed proportionality factor, $\mu_z = -g\beta S_z$, where β is the Bohr magneton $(e\hbar/2m)$ (relating the magnetic moment to the component of angular momentum $\hbar S_z$) while $g = 2.0023$ is the electronic g factor. The Hamiltonian may thus be written, neglecting any spin–orbit interaction terms

$$\mathsf{H} = \mathsf{h} + g\beta B\mathsf{S}_z \tag{1.2.21}$$

and the states represented by $\psi = \phi\alpha$ and $\bar{\psi} = \phi\beta$ will remain energy eigenstates even when $B \neq 0$. For if ϵ is the energy without magnetic field, ϕ and ϵ satisfying (1.2.17), then for $B \neq 0$

$$\begin{aligned}\mathsf{H}\psi &= \alpha(\mathsf{h}\phi) + g\beta B\phi(\mathsf{S}_z\alpha)\\ &= (\epsilon + \tfrac{1}{2}g\beta B)\psi.\end{aligned} \tag{1.2.22}$$

Thus, in the presence of the field, the energy of the ψ state is increased by $\frac{1}{2}g\beta B$. Similarly, that of the $\bar{\psi}$ state is decreased by $\frac{1}{2}g\beta B$, and the corresponding Zeeman splitting of the doublet is thus $g\beta B$. Such splittings are directly observable in electron spin resonance (ESR) experiments, from which very accurate values of g may be obtained. The theoretical origins of the anomalous (i.e. non-classical) value $g \simeq 2$ are referred to in Chapter 11.

It should be noted that the introduction of a "spin variable" s is not at all necessary and may be regarded simply as a convenient formal device; states are described in quantum mechanics by solutions of eigenvalue equations such as (1.1.1), and, although the operators involved may sometimes be differential operators, working on functions of certain variables, the more general formulation of quantum mechanics is concerned only with symbolic statements involving operators and operands, and especially with operator relationships such as (1.2.20) which are independent of the particular "language" or "representation" employed. The reader who likes to have a more concrete interpretation of the spin variable may regard s as the value of the spin component S_z, and may visualize the spin functions α and β as "spikes" at the points $S_z = +\frac{1}{2}$ and $S_z = -\frac{1}{2}$ respectively. Since $\alpha(s)$ is supposed to describe an electron "with spin $+\frac{1}{2}$" it must vanish for $s \neq \frac{1}{2}$, and the normalization condition then implies that the spike is the limiting form of a function of small width but large height, the limit being approached in such a way that the integral converges. This is essentially the "delta function" (Dirac, 1958). The normalization and orthogonality properties of eigenfunctions representing quantum-mechanical states become, in the case of spin eigenfunctions,

$$\int \alpha^*(s)\alpha(s)\,\mathrm{d}s = \int \beta^*(s)\beta(s)\,\mathrm{d}s = 1, \tag{1.2.23a}$$

$$\int \alpha^*(s)\beta(s)\,\mathrm{d}s = \int \beta^*(s)\alpha(s)\,\mathrm{d}s = 0, \tag{1.2.23b}$$

where the integration sign is used in a formal way, whose meaning will become clear in Chapter 2.

Classification of electronic states

After this digression, we return to the helium-like system under consideration. When spin is included, the electrons must each be assigned to *spin*-orbitals, the first few of which are the products

$$1\mathrm{s}\alpha, \quad 1\mathrm{s}\beta, \quad 2\mathrm{s}\alpha, \quad 2\mathrm{s}\beta, \ldots \quad .$$

For the configuration $1\mathrm{s}^2$ both electrons may occupy the 1s orbital with the following choices of spin factors

$$\left.\begin{aligned}
\Phi_1(\boldsymbol{x}_1, \boldsymbol{x}_2) &= 1\mathrm{s}(\boldsymbol{r}_1)\,\alpha(s_1)\,1\mathrm{s}(\boldsymbol{r}_2)\,\alpha(s_2),\\
\Phi_2(\boldsymbol{x}_1, \boldsymbol{x}_2) &= 1\mathrm{s}(\boldsymbol{r}_1)\,\alpha(s_1)\,1\mathrm{s}(\boldsymbol{r}_2)\,\beta(s_2),\\
\Phi_3(\boldsymbol{x}_1, \boldsymbol{x}_2) &= 1\mathrm{s}(\boldsymbol{r}_1)\,\beta(s_1)\,1\mathrm{s}(\boldsymbol{r}_2)\,\alpha(s_2),\\
\Phi_4(\boldsymbol{x}_1, \boldsymbol{x}_2) &= 1\mathrm{s}(\boldsymbol{r}_1)\,\beta(s_1)\,1\mathrm{s}(\boldsymbol{r}_2)\,\beta(s_2).
\end{aligned}\right\} \tag{1.2.24}$$

It is evident that Φ_1 and Φ_4 are symmetric under interchange of $\boldsymbol{x}_1$ and $\boldsymbol{x}_2$, and hence apparently acceptable, while Φ_2 and Φ_3 are not: all functions, however, have the same eigenvalue of the model operator H_0, namely $E = 2\epsilon_{1\mathrm{s}}$, and so may be mixed to give eigenfunctions conforming to the symmetry conditions analous to (1.2.14) but in which space and spin description are included on the same footing, each $\boldsymbol{r}_i$ being replaced by an $\boldsymbol{x}_i$. The following appear to be acceptable state functions:

$$\left.\begin{aligned}
\Psi_1^{\mathrm{a}}(\boldsymbol{x}_1,\boldsymbol{x}_2) &= [\Phi_2(\boldsymbol{x}_1,\boldsymbol{x}_2) - \Phi_3(\boldsymbol{x}_1,\boldsymbol{x}_2)]/\sqrt{2}\\
&= 1\mathrm{s}(\boldsymbol{r}_1)\,1\mathrm{s}(\boldsymbol{r}_2)[\alpha(s_1)\beta(s_2) - \beta(s_1)\alpha(s_2)]/\sqrt{2},\\
\Psi_1^{\mathrm{s}}(\boldsymbol{x}_1,\boldsymbol{x}_2) &= \Phi_1(\boldsymbol{x}_1,\boldsymbol{x}_2) = 1\mathrm{s}(\boldsymbol{r}_1)\,1\mathrm{s}(\boldsymbol{r}_2)\alpha(s_1)\alpha(s_2),\\
\Psi_2^{\mathrm{s}}(\boldsymbol{x}_1,\boldsymbol{x}_2) &= [\Phi_2(\boldsymbol{x}_1,\boldsymbol{x}_2) + \Phi_3(\boldsymbol{x}_1,\boldsymbol{x}_2)]/\sqrt{2}\\
&= 1\mathrm{s}(\boldsymbol{r}_1)\,1\mathrm{s}(\boldsymbol{r}_2)[\alpha(s_1)\beta(s_2) + \beta(s_1)\alpha(s_2)]/\sqrt{2},\\
\Psi_3^{\mathrm{s}}(\boldsymbol{x}_1,\boldsymbol{x}_2) &= \Phi_4(\boldsymbol{x}_1,\boldsymbol{x}_2) = 1\mathrm{s}(\boldsymbol{r}_1)\,1\mathrm{s}(\boldsymbol{r}_2)\beta(s_1)\beta(s_2)
\end{aligned}\right\}\qquad(1.2.25)$$

and it is evident that each is a space–spin product, the space factor being symmetric under electron interchange, the spin factor symmetric (three cases) or antisymmetric (once). The superscript indicates the *overall* symmetry or antisymmetry.

Similar considerations apply to the configuration 1s 2s, suggesting the following as valid state functions:

$$\left.\begin{aligned}
\Psi_{2,3,4}^{\mathrm{a}} &= \frac{1}{\sqrt{2}}[1\mathrm{s}(\boldsymbol{r}_1)\,2\mathrm{s}(\boldsymbol{r}_2) - 2\mathrm{s}(\boldsymbol{r}_1)\,1\mathrm{s}(\boldsymbol{r}_2)]\\
&\quad\times\begin{cases}\alpha(s_1)\,\alpha(s_2)\\ [\alpha(s_1)\,\beta(s_2) + \beta(s_1)\,\alpha(s_2)]/\sqrt{2} \quad \text{(antisymmetric)},\\ \beta(s_1)\,\beta(s_2).\end{cases}\\
\Psi_4^{\mathrm{s}} &= \frac{1}{\sqrt{2}}[1\mathrm{s}(\boldsymbol{r}_1)\,2\mathrm{s}(\boldsymbol{r}_2) - 2\mathrm{s}(\boldsymbol{r}_1)\,1\mathrm{s}(\boldsymbol{r}_2)]\\
&\quad\times[\alpha(s_1)\,\beta(s_2) - \beta(s_1)\,\alpha(s_2)]/\sqrt{2} \qquad \text{(symmetric)},\\
\Psi_5^{\mathrm{a}} &= \frac{1}{\sqrt{2}}[1\mathrm{s}(\boldsymbol{r}_1)\,2\mathrm{s}(\boldsymbol{r}_2) + 2\mathrm{s}(\boldsymbol{r}_1)\,1\mathrm{s}(\boldsymbol{r}_2)]\\
&\quad\times[\alpha(s_1)\,\beta(s_2) - \beta(s_1)\,\alpha(s_2)]/\sqrt{2} \qquad \text{(antisymmetric)},\\
\Psi_{5,6,7}^{\mathrm{s}} &= \frac{1}{\sqrt{2}}[1\mathrm{s}(\boldsymbol{r}_1)\,2\mathrm{s}(\boldsymbol{r}_2) + 2\mathrm{s}(\boldsymbol{r}_1)\,1\mathrm{s}(\boldsymbol{r}_2)]\\
&\quad\times\begin{cases}\alpha(s_1)\,\alpha(s_2)\\ [\alpha(s_1)\,\beta(s_2) + \beta(s_1)\,\alpha(s_2)]/\sqrt{2} \quad \text{(symmetric)}.\\ \beta(s_1)\,\beta(s_2).\end{cases}
\end{aligned}\right\}\qquad(1.2.26)$$

For non-interacting particles, the states belonging to each configuration would be rigorously degenerate; but inclusion of the interaction $(1/r_{12})$ is expected to resolve any two states differing in the spatial behaviour of the electrons. Thus, on estimating the energies as expectation values of the

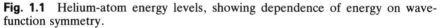

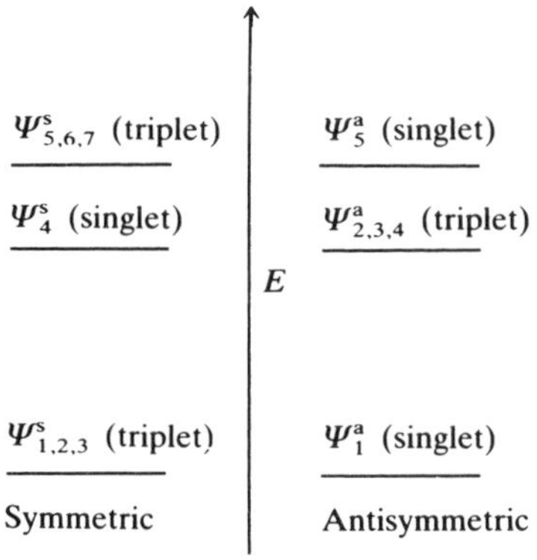

Fig. 1.1 Helium-atom energy levels, showing dependence of energy on wavefunction symmetry.

Hamiltonian (1.2.1), the states $\Psi^a_{2,3,4}$ and Ψ^s_4 should differ in energy from Ψ^a_5 and $\Psi^s_{5,6,7}$; anticipating a later section (3.2), the former should be lower in energy than the latter. The first few energy levels‡, with electron interaction tentatively included but spin effects still ignored (except in so far as they relate to symmetry), may then be indicated as in Fig. 1.1. On applying a magnetic field, the various triplet states would each be resolved (as in the case of a single electron) but not into three components, and may therefore be identified spectroscopically. It will be remembered that spectroscopy verifies conclusively the existence of the states on the right in Fig. 1.1 and the non-existence of those on the left, implying that when spin is included the choice between symmetry and antisymmetry is not open. The choice is determined once and for all by the fact that the particles concerned are *electrons*. Such conclusions are stated as an *antisymmetry principle*:

> The wavefunction $\Psi(\boldsymbol{x}_1, \boldsymbol{x}_2, \ldots, \boldsymbol{x}_N)$ describing any state of an N-electron system is antisymmetric under any permutation of the electrons:
>
> $$\mathsf{P}\Psi(\boldsymbol{x}_1, \boldsymbol{x}_2, \ldots, \boldsymbol{x}_N) = \epsilon_{\mathsf{P}}\Psi(\boldsymbol{x}_1, \boldsymbol{x}_2, \ldots, \boldsymbol{x}_N)$$
>
> where $\epsilon_{\mathsf{P}} = \pm 1$ for permutations of even or odd parity, respectively.

(1.2.27)

‡ For states of spherical symmetry (S states).

This is essentially the quantum-mechanical generalization of Pauli's exclusion principle. The connection is easily made in the case $N = 2$, where the wavefunction may be written as a product of space and spin factors: for if two electrons are put into the same orbital with the same spins (i.e. into the same spin-orbital) the wavefunction can only be *symmetric* (cf. Ψ_1^s and Ψ_3^s above), in violation of the antisymmetry requirement. Two electrons cannot therefore (in an IPM description) occupy the same state or—in Pauli's statement—possess identical sets of quantum numbers. The generality of the principle, which applies for any number of electrons in any kind of system and even when interaction is admitted, will be discussed further in Chapter 3. For more than two electrons the wavefunction has no simple symmetry for interchange of space or spin variables *separately*; exchange of particles implies exchange of space and spin variables together and (1.2.27) applies to this case only.

1.3 EXAMPLE: THE HYDROGEN MOLECULE

In principle, it would appear that the approach outlined in Section 1.2 could be applied in discussing any many-electron system, atomic or molecular: the model Hamiltonian H_0 (cf. (1.2.3)) would describe N non-interacting electrons moving in the field of a number of fixed nuclei; functions of product from (cf. (1.2.5)) would be exact eigenfunctions of H_0, provided that their factors were eigenfunctions of the *one*-electron operator h; and when symmetry considerations were satisfied the resultant functions might still give a fair description of the various electronic states of the actual system in which electron interactions are admitted. There are, however, many complications. In passing from atoms to molecules, the first unpleasant feature we meet is that even the *one*-electron eigenfunctions are not obtainable in closed form when there are several nuclei instead of just one; instead of the 1s, 2s, 2p, . . . , AOs, with their simple analytical forms, we shall need polycentric *molecular orbitals* (MOs) that describe the states of a single electron moving over the whole nuclear "framework". It is indeed possible to proceed in this way, and to build up wavefunctions for a 2-electron *molecule* in exactly the same way as for the helium atom; but the one-electron factors in the product functions are no longer available as ready-made units and their determination is a central problem of MO theory, developed in detail in Chapter 6.

First we briefly discuss the simplest possible 1-electron 2-centre system, the hydrogen molecule ion H_2^+, in order to illustrate the nature of the problem. The one-electron eigenvalue equation is now (using lower-case letters for 1-electron quantities, as in (1.2.9))

$$\mathsf{h}\phi = \epsilon\phi, \tag{1.3.1}$$

where (using atomic units)

$$\mathsf{h} = -\tfrac{1}{2}\nabla^2 - \left(\frac{1}{r_a} + \frac{1}{r_b}\right). \tag{1.3.2}$$

Here r_a and r_b are the distances of the electron from the two nuclei a and b. Equation (1.3.1) was first solved, with high accuracy, by Burrau (1927), who transformed to confocal elliptic coordinates in which the equation is separable, the solution becoming a product of three factors that may be obtained by solution of three separate differential equations. Solutions in this way is not generally possible, and we pass directly to the construction of simpler approximations.

We note that if the electron is close to nucleus a then the $1/r_a$ term in (1.3.2) will be so large that $1/r_b$ may be neglected by comparison; the Hamiltonian operator is then quite close to that for one electron near nucleus a *alone*, i.e. for a single hydrogen atom. If we use $a(\boldsymbol{r})$ for the 1s orbital centred on nucleus a and $b(\boldsymbol{r})$ for that centred on nucleus b, it is therefore reasonable to expect that

$$\phi(\boldsymbol{r}) \simeq c_a a(\boldsymbol{r}) \quad \text{(electron close to a)},$$

$$\phi(\boldsymbol{r}) \simeq c_b b(\boldsymbol{r}) \quad \text{(electron close to b)},$$

where c_a and c_b are at present arbitrary numerical factors (since a solution valid in some given region may be multiplied by any constant and will still satisfy the same differential equation). Since the 1s functions fall rather rapidly from their peak values at the nuclei, a trial function with suitable behaviour near both nuclei (i.e. an appropriate MO) would be

$$\phi(\boldsymbol{r}) = c_a a(\boldsymbol{r}) + c_b b(\boldsymbol{r}). \tag{1.3.3}$$

Also, since the two nuclei are identical, we expect that $|\phi(\boldsymbol{r})|^2$ will be completely symmetrical across a plane cutting the system into two identical halves; there is no reason to expect a higher probability of finding the electron at one end of the molecule than the other. But the required symmetry of

$$|\phi(\boldsymbol{r})|^2 = c_a^2 a^2(\boldsymbol{r}) + c_b^2 b^2(\boldsymbol{r}) + 2c_a c_b a(\boldsymbol{r}) b(\boldsymbol{r}) \tag{1.3.4}$$

implies that $c_b = \pm c_a$, and it therefore appears that two approximate MOs can be constructed from the 1s AOs on the two centres. We denote these by

$$\left.\begin{aligned} A(\boldsymbol{r}) &= m_A[a(\boldsymbol{r}) + b(\boldsymbol{r})], \\ B(\boldsymbol{r}) &= m_B[a(\boldsymbol{r}) - b(\boldsymbol{r})], \end{aligned}\right\} \tag{1.3.5}$$

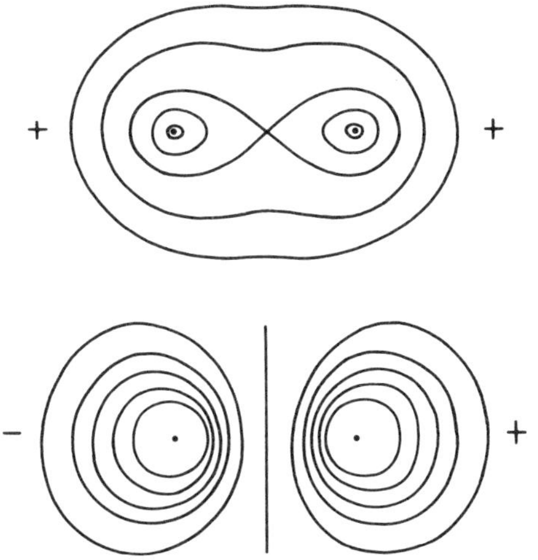

Fig. 1.2 Bonding and antibonding molecular orbitals for H_2^+ (upper and lower respectively). The contours correspond to $|\Psi| = 0.6$, 0.4, 0.3, 0.2, reading outwards from the nuclei.

where the remaining numerical factors in the two cases have been called m_A and m_B, and must be chosen to *normalize* the two approximate wavefunctions in accordance with (1.1.8). The two functions are illustrated schematically in Fig. 1.2 by indicating contours on which $A(\boldsymbol{r})$ (and $B(\boldsymbol{r})$) take a constant numerical value. Approximate values of the energies associated with these alternative states are readily calculated (Pauling and Wilson, 1935) as functions of the separation R of the two nuclei. When the energy of nuclear repulsion is added (as in (1.1.10)) the resultant *total* energies behave as in Fig. 1.3. Evidently the MO corresponding to formation of a stable ion H_2^+ is A; the predicted bond length ($R = 1.32$ Å) and binding energy (1.77 eV) are rather poor by comparison with those obtained by Burrau, but the general form of Burrau's wavefunction is quite well reproduced. The function $B(\boldsymbol{r})$ is also a fair approximation to the wavefunction of the first excited state. MOs built up in this way, in linear-combination-of-atomic-orbitals (LCAO) form, are of great importance throughout molecular quantum mechanics.

The normal molecule H_2, in which *two* electrons move in the field of the two nuclei, may now be discussed in essentially the same way as the helium atom in Section 1.2. In the absence of electron interaction, an approximate spatial wavefunction for the state of lowest energy would be

$$\Phi(\boldsymbol{r}_1, \boldsymbol{r}_2) = A(\boldsymbol{r}_1)\,A(\boldsymbol{r}_2),$$

giving an energy expectation value $E = 2\epsilon_A$. This is compatible only with

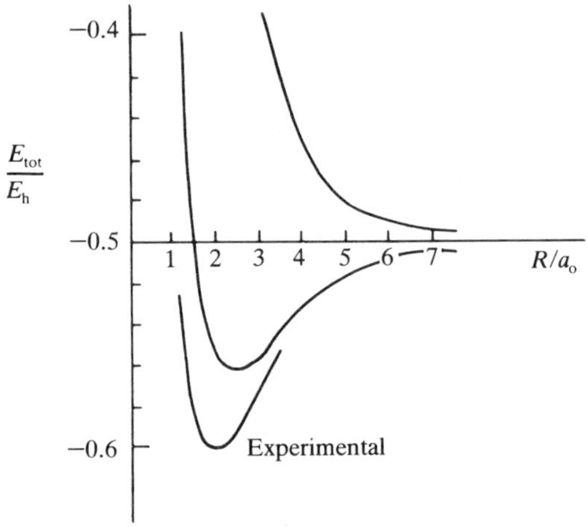

Fig. 1.3 Approximate energies of two states of the H_2^+ ion as functions of internuclear distance, compared with the observed ground-state curve.

the antisymmetric spin factor, giving an acceptable wavefunction

$$\Psi(\boldsymbol{x}_1, \boldsymbol{x}_2) = A(\boldsymbol{r}_1)\, A(\boldsymbol{r}_2)[\alpha(s_1)\, \beta(s_2) - \beta(s_1)\, \alpha(s_2)]/\sqrt{2}, \qquad (1.3.6)$$

addition of the spin factor having no effect on the energy. When electron interaction is recognized, by using the full Hamiltonian H, an approximate ground-state energy may be calculated from this function. The total energy may again be plotted as a function of internuclear distance and (cf. Fig. 1.3) this time gives a minimum at $R = 0.16$ nm and a binding energy (relative to two hydrogen atoms) of 2.65 eV. This simple "MO approximation", though not of high accuracy, accounts for the main features of the simplest chemical bond.

It will be recalled that the first quantum-mechanical explanation of the chemical bond is usually attributed to Heitler and London (1927). Their discussion of the hydrogen molecule, however, was not based on IPM concepts, with each electron assigned to a *molecular* orbital, but rather on an independent-*atom* approach in which the electrons were assigned to *atomic* orbitals. The Heitler–London approach was the basis for *valence bond* (VB) theory, which was important in the early days of quantum chemistry but later fell into disuse. There has been some revival of interest in VB theory now that more powerful computing facilities are available: a full discussion of this method is deferred until Chapter 7, although a preliminary account will appear in Chapter 3.

PROBLEMS 1

1.1 Show that the time-independent equation (1.1.1) follows from (1.1.12) on requiring the solution to have the special form $\Psi_t(\boldsymbol{x}, t) = \Psi(\boldsymbol{x})f(t)$. Obtain the form of the time factor and show that $\Psi_t\Psi_t^*$ then reduces to $\Psi\Psi^*$, which is time-independent: Ψ describes a *stationary state*. [*Hint*: Substitute the assumed form in (1.1.12), consider the effect of the differential operators, divide throughout by Ψ_t, and obtain two independent terms whose sum must vanish. This is the method of "separation of variables".]

1.2 Set up the Hamiltonian operator for an N-electron diatomic molecule whose atoms have atomic numbers Z_1, Z_2 and masses M_1, M_2. Write down also the "clamped-nucleus" Hamiltonian, supposing both nuclei to be at rest.

1.3 Formulate the eigenvalue equation that determines the energy levels of an electron in the field of a (fixed) nucleus of charge Ze. Verify that a wavefunction of the form $\phi = e^{-ar}$ will satisfy the equation provided that a is suitably chosen, and that the energy will then be $E = -\frac{1}{2}Z^2E_h$. Extend the analysis to a function of the form $\phi = (a + br)e^{-cr}$ with parameters a, b, c all non-zero. The two solutions that you find will be hydrogenlike 1s and 2s orbitals; use them to calculate the most-probable distance of the electron from the nucleus in both cases. [*Hint*: Show that for functions depending only on r, ∇^2 takes the form $\mathrm{d}^2/\mathrm{d}r^2 + (2/r)\,\mathrm{d}/\mathrm{d}r$; do the differentiations, cancel an exponential factor, and compare coefficients of each power of r on the two sides of the equation.]

1.4 The operators in (1.2.20) are *angular-momentum* operators for the electron spin. Establish similar commutation properties for the angular-momentum operators associated with orbital motion of an electron, starting from the classical definitions $M_x = yp_z - zp_y$ etc., and show why it is convenient to introduce L_x, L_y, L_z such that $\mathsf{M}_x = \hbar\mathsf{L}_x$ etc. [*Hint*: First study the linear-momentum operators p_x, p_y, p_z, showing for example that $x\mathsf{p}_x - \mathsf{p}_x x = i\hbar$, and then use your results to obtain $\mathsf{M}_x\mathsf{M}_y - \mathsf{M}_y\mathsf{M}_x$ etc.; remember that operators always work on an operand, ϕ say.]

1.5 Show that the operator L_z (Problem 1.4) (and similarly for L_y, L_z) commutes with the potential energy function $V(r)$ for a central-field system. [*Hint*: Think of $V(r)$ as a function of $r^2 = x^2 + y^2 + z^2$ and the operators as *differential* operators.]

1.6 Show that an s-type AO describes an electron with zero values of L_z and L^2 $(= L_x^2 + L_y^2 + L_z^2)$. [*Hint*: Apply the operators, using $\partial\phi/\partial x = (\mathrm{d}\phi/\mathrm{d}r)(\partial r/\partial x)$ etc. for ϕ depending only on r.]

1.7 Show that a p-type function $p_x = xf(r)$ is *not* an eigenfunction of L_z, and therefore does not describe a state of definite L_z, but *is* an eigenfunction of L_x. What would be the expectation value $\langle \mathsf{L}_z \rangle$ in such a state? [*Hint*: Remember that the "expectation value" of a quantity A in state ϕ (normalized) is given by $\langle \mathsf{A} \rangle = \int \phi^*\mathsf{A}\phi\,\mathrm{d}\boldsymbol{x}$, where A is the operator associated with A.]

1.8 Show that $(x \pm iy)f(r)$ is an eigenfunction of L_z with eigenvalue ± 1, and hence that p-type AOs of the form $p_x \pm ip_y$ correspond to states with one unit of angular momentum around the z axis. Discuss the connection between the real and complex AOs in Appendix 1 in the light of this result.

1.9 Consider the diatomic molecule whose Hamiltonian you wrote down in Problem 1.2, with $N = 2$, and let a, b be normalized AOs on a, b. What product functions would you use to describe

(i) electrons 1 and 2 in orbital a, with up-spin and down-spin, respectively;
(ii) electron 1 in a with up-spin and 2 in b with down-spin;
(iii) one electron in each of a, b, with the same spin, the wavefunction being antisymmetric (two possibilities),
(iv) one electron in each of a, b, with opposite spins, the wavefunction being antisymmetric (two possibilities)?

Which of these functions might legitimately be used to describe electronic states?

1.10 Use the Hamiltonian of Problem 1.2 (with $N = 2$) and the various wavefunctions of Problem 1.9 to derive expressions for the energy expectation values

$$E = \langle \mathsf{H} \rangle = \int \Psi^*(\mathbf{x}_1, \mathbf{x}_2)\mathsf{H}\,\Psi(\mathbf{x}_1, \mathbf{x}_2)\,\mathrm{d}\mathbf{x}_1\,\mathrm{d}\mathbf{x}_2.$$

Note that H includes electron interaction and that the derived expressions will *not* yield eigenvalues of H; they can provide only approximate energy levels. [*Hint*: Perform the integrations and express the results in terms of the "overlap", "1-electron" and "2-electron" integrals such as

$$\langle a \mid b \rangle = \int a^*(\mathbf{r})b(\mathbf{r})\,\mathrm{d}\mathbf{r}, \qquad \langle a|\,\mathsf{h}\,|b\rangle = \int a^*(\mathbf{r})\mathsf{h}b(\mathbf{r})\,\mathrm{d}\mathbf{r},$$

$$\langle ab|\,g\,|ab\rangle = \int a^*(\mathbf{r}_1)b^*(\mathbf{r}_2)g(1, 2)a(\mathbf{r}_1)b(\mathbf{r}_2)\,\mathrm{d}\mathbf{r}_1\,\mathrm{d}\mathbf{r}_2,$$

noting that the names ($\mathbf{r}$, $\mathbf{r}_1$, $\mathbf{r}_2$) of the integration variables are arbitrary.]

1.11 Discuss (with the same assumptions as in Problem 1.10) the ground states of the systems H_2 and HeH^+, using an MO approximation in which the electrons are assigned to MOs of the type (1.3.3). Which coefficient will be the larger in HeH^+ and why? Derive expressions for the total electronic energy E, assuming equal coefficients for H_2 and one twice the other for HeH^+, in terms of quantities of the kind used in Problem 1.10.

1.12 Set up excited-state wavefunctions for H_2, assuming one electron in each of the MOs in (1.3.5) with spins opposed and the spin factor (i) symmetric and (ii) antisymmetric. Show that the energy expectation value will be unchanged on identifying the two spins in case (i), indicating a triplet state.

1.13 Obtain energy expressions for the two states in Problem 1.12 in terms of the usual integrals (cf. Problem 1.10), but with MOs A, B in place of AOs a, b; and express your results in terms of the AO integrals. What is the energy separation of the two states, and which do you expect to be the lower? [*Hint*:

Regard each 2-electron integral as the interaction energy of two charge distributions.]

1.14 Compare the energy expression for the H_2 ground state, obtained in Problem 1.10, with that which results (see Problem 1.10) on using the AO-type wavefunction (iv) of Problem 1.9. Try to interpret qualitatively the signs and relative magnitudes of the 1- and 2-electron integrals. Would the alternative energy estimates both behave correctly in the dissociation limit, giving twice the energy of a single hydrogen atom?

1.15 Write down the Hamiltonian for two like particles moving in a field such that $V(i)$ in (1.1.2b) takes the form $V(i) = \frac{1}{2}kr_i^2$, while $g(1, 2) = -\alpha r_{12}^2$. Show that separation into two independent parts may be achieved (cf. Problem 1.1) on introducing new variables

$$R = (r_1 + r_2)/\sqrt{2}, \qquad r = (r_1 - r_2)/\sqrt{2}$$

and that eigenfunctions of product form $\Psi(R, r) = \Phi(R)\phi(r)$ may then be obtained. What does this mean physically?

1.16 Show that the full N-particle Hamiltonian in (1.1.13) can be rewritten as

$$\mathsf{H} = \mathsf{H}_{\text{cm}}(\boldsymbol{R}) + \mathsf{H}_{\text{rel}}(\bar{\boldsymbol{r}})$$

by changing the variables to $\boldsymbol{R}$, for the centre of mass, and $\bar{\boldsymbol{r}}$ $(= \bar{\boldsymbol{r}}_2, \ldots, \bar{\boldsymbol{r}}_N)$, which represent $N-1$ "reduced" variables. Find the explicit forms of H_{cm} and H_{rel}. How do you interpret the separation and what form would the total energy take? [*Hint*: $\boldsymbol{R}$ is defined as $M^{-1}\sum_i m_i\boldsymbol{r}_i$, M being the *total* mass of the system. Only $N-1$ of the variables $\bar{\boldsymbol{r}}_i = \boldsymbol{r}_i - \boldsymbol{R}$ are independent. Use cartesian coordinates, evaluate $\partial^2\Psi/\partial x_1^2$ and $\partial^2\Psi/\partial x_i^2$ (for $i = 2, \ldots, N$) in terms of the new variables, and sum over all particles.]

1.17 How is the derivation in Problem 1.16 changed when $\boldsymbol{R}$ is taken to be the centre of mass of the *nuclei* alone? Show that the part of the resultant Hamiltonian that refers only to the electrons becomes

$$\mathsf{H}_{\text{el}} = -\frac{\hbar^2}{2\mu}\sum_i \nabla^2(i) - \frac{\hbar^2}{2M}\sum_{i,j}{}' \boldsymbol{\nabla}(i)\cdot\boldsymbol{\nabla}(j) + \sum_i V(i) + \frac{e^2}{2(4\pi\epsilon_0)}\sum_{i,j}{}' g(i, j),$$

where operators and potential terms all refer to the new variables, summations are over electrons only, and μ is a "reduced mass" given by $\mu = mM/(m + M)$, where M is the total nuclear mass. The second sum in H_{el} is sometimes called the "mass-polarization" term.

2 Mathematical Methods

2.1 COMPLETE SET EXPANSIONS

Before considering in more detail the approximation methods used in quantum mechanics, we review some of the mathematical tools available.‡ One of the most powerful of these concerns the representation of a function as a mixture, with carefully chosen coefficients, of a number of more elementary functions:

$$f(x) = c_1\phi_1(x) + c_2\phi_2(x) + \ldots \quad . \tag{2.1.1}$$

Familiar examples are the Fourier analysis of, say, the electron density in a crystal into sine and cosine components; and the approximation of the wave function for an electron in a molecule by a linear combination of atomic orbitals (LCAO) on the various atomic centres. Discussion of the validity and accuracy of such expansions introduces the idea of *completeness* of a set of functions.

Consider an arbitrary function of one variable $f(x)$ with $a \leqslant x \leqslant b$; we say the function is defined in the interval (a, b). Consider also some set of functions $\phi_1(x)$, $\phi_2(x)$, ... defined in the same interval. We might attempt to approximate $f(x)$ by using only the first n members of the set:

$$f(x) \simeq f_n(x) = c_1\phi_1(x) + c_2\phi_2(x) + \ldots + c_n\phi_n(x). \tag{2.1.2}$$

In order to get the best approximation of this kind, it is natural to determine the coefficients so that:

$$D = \int_a^b |f(x) - f_n(x)|^2 \, dx = \text{minimum}. \tag{2.1.3}$$

That is, we minimize the mean-square deviation of f_n from f. To do this, we differentiate D with respect to each of the coefficients in turn and set

‡ This chapter is intended to be self-contained and adequate for most purposes, but obviously cannot be comprehensive. Group theory has been omitted because most of what follows can be understood on the basis of a few ideas and results connected with *symmetry*. These results are collected separately in Appendix 3.

the results equal to zero. This gives a set of equations that determine the optimum coefficients, and it may be verified by examining the signs of the second derivatives that D is then a true minimum and not just a stationary value.

In order to improve the approximation, we may add an extra function $\phi_{n+1}(x)$ and again determine the coefficients. At this point we find that it is convenient if the ϕ_i form an *orthonormal set* in the sense

$$\langle \phi_i \mid \phi_j \rangle = \int_a^b \phi_i^*(x)\phi_j(x)\,\mathrm{d}x = \delta_{ij} = \begin{cases} 1 & (i=j), \\ 0 & (i \neq j). \end{cases} \tag{2.1.4}$$

This quantity, for which we employ the "bra–ket" ($\langle\ |, |\ \rangle$) notation due to Dirac (1958), is called the "Hermitian scalar product" of the two functions ϕ_i and ϕ_j (for reasons that will be clear presently), and with orthonormal functions it is then easy to show that the best fit of the function f results when

$$c_i = \langle \phi_i \mid f \rangle = \int_a^b \phi_i^*(x) f(x)\,\mathrm{d}x. \tag{2.1.5}$$

The coefficients thus determined have the "property of finality" in that, no matter how many functions are added in getting a better approximation, the earlier coefficients are unchanged—the coefficient of ϕ_i depending on no other member of the set.

If $D \to 0$ as $n \to \infty$ for any function f defined in the interval (a, b) then the set of functions ϕ_i is said to be complete, and we write

$$f = \sum_{i=1}^{\infty} c_i \phi_i. \tag{2.1.6}$$

It is important to note that our discussion shows that this equality holds only "in the mean": for it to hold "point to point" certain restrictions must be placed on f and ϕ_i. What precisely these restrictions are is a matter of some mathematical nicety. It is well known, however, that if $f(x)$ exhibits a discontinuity at some value of x then $f_n(x)$ frequently includes a narrow "spike" at that point; this may become more pronounced as $n \to \infty$ so long as its contribution to the integral D tends to zero. It is therefore clear that we could find a function that was in the mean quite a good approximate wavefunction, but that behaved badly at certain points. Consequently, considerable care must be exercised in calculating point properties (e.g. electron and spin density at a nucleus) from an approximate wavefunction. It is also important to note that all the functions involved in (2.1.1) should belong to the same class (e.g. class L^2 in the given interval).

The considerations advanced so far apply to functions of a single variable x. To apply them to wavefunctions that may be functions of three variables (x, y, z) or several $(x_1, y_1, z_1, x_2, y_2, \ldots, z_N)$, some generalization is necessary. The extension is suggested by the examples of Section 1.2, where 2-electron wavefunctions were constructed as mixtures of simple products, in the form

$$f(\mathbf{r}_1, \mathbf{r}_2) = \sum_{i,j} c_{ij}\phi_i(\mathbf{r}_1)\phi_j(\mathbf{r}_2) = \sum_{\kappa} c_\kappa \Phi_\kappa(\mathbf{r}_1, \mathbf{r}_2),$$

where $\kappa = (i, j)$ denotes the pair of indices specifying an orbital configuration. Such an expansion is formally similar to (2.1.1) except that every function involves several variables instead of just one. Would it be possible to write an *arbitrary* 2-electron function (including an exact wavefunction) in such a form, admitting as in the case of one electron that the expansion might have to be infinite?

The basis for the belief that the set of all products *is* complete, so that such an expansion is valid, may be indicated in a non-rigorous way by considering a function of just two variables (x_1, x_2). If $\{\phi_i(x_1)\}$‡ is complete for functions of x_1 in a given interval (a, b) and $\{\phi_j'(x_2)\}$ is complete for functions of x_2 in the interval (a', b') then we may at once assert that

$$f(x_1, x_2) = c_1\phi_1(x_1) + c_2\phi_2(x_1) + \ldots$$

for any given value of x_2, the coefficients depending on the choice of x_2. But if the coefficients are functions of x_2, then we may write

$$c_i = c_{i1}\phi_1'(x_2) + c_{i2}\phi_2'(x_2) + \ldots,$$

and insertion in the previous equation gives

$$f(x_1, x_2) = c_{11}\phi_1(x_1)\phi_1'(x_2) + c_{12}\phi_1(x_1)\phi_2'(x_2) \ldots + c_{ij}\phi_i(x_1)\phi_j'(x_2) + \ldots, \tag{2.1.7}$$

which is the required result. The extension to many variables is formally straightforward, and we assume that a wavefunction involving the coordinates of N electrons may be expanded in the form

$$\Psi(\mathbf{r}_1, \mathbf{r}_2, \ldots, \mathbf{r}_N) = \sum c_{ij\ldots p}\phi_i(\mathbf{r}_1)\phi_j(\mathbf{r}_2) \ldots \phi_p(\mathbf{r}_N), \tag{2.1.8}$$

provided that $i, j, \ldots, p$ may run over all possible choices of n orbitals from any set $\phi_1, \phi_2, \ldots, \phi_i, \ldots$ that is itself complete. Here the

‡ A typical function, enclosed in curly brackets, is commonly used to denote a whole set. Thus $\{\phi_i\}$ stands for $\phi_1, \phi_2, \ldots, \phi_i, \ldots$.

variables for each electron, symbolized collectively by $\boldsymbol{r}_k$, are identical in character and range, and the same complete set has been used for each one: an example in which different sets are used for different variables is provided by the spherical-harmonic expansion of a function $\phi(r, \theta, \phi)$ of the polar coordinates r, θ and ϕ in terms of functions of r, times functions of θ (Legendre polynomials), times functions of ϕ (Fourier terms, $e^{\pm im\phi}$).

We must not forget, of course, that electron spin must be taken into account. A slight modification is necessary in order to accommodate the results of Section 1.2. On admitting a spin variable s for each particle along with its space variables $\boldsymbol{r}$ $(=x, y, z)$, the complete set for expanding its wavefunction would comprise all space–spin products of the form $\phi_i(\boldsymbol{r})\eta_j(s)$, where there are only two η_j, namely α and β. If the ϕ_i form a complete set of orbitals, the complete set of space–spin products for one particle may thus be denoted by

$$\phi_1(\boldsymbol{r})\alpha(s), \quad \phi_1(\boldsymbol{r})\beta(s), \quad \phi_2(\boldsymbol{r})\alpha(s), \quad \phi_2(\boldsymbol{r})\beta(s), \quad \text{etc.}$$

These are the *spin-orbitals* introduced in Section 1.2. When it is not necessary to indicate the spin factor explicitly we shall frequently denote a spin-orbital by a single letter such as ψ_i, at the same time using $\boldsymbol{x}$ to denote both space and spin variables $(\boldsymbol{r}, s)$ as in Section 1.2. With spin included, (2.1.8) then requires no formal change:

$$\Psi(\boldsymbol{x}_1, \boldsymbol{x}_2, \ldots, \boldsymbol{x}_N) = \sum c_{ij\ldots p}\psi_i(\boldsymbol{x}_1)\psi_j(\boldsymbol{x}_2)\ldots\psi_p(\boldsymbol{x}_N). \quad (2.1.9)$$

Nearly all of molecular quantum theory since the early work of Slater and others has been based on the approximation of wavefunctions by suitable linear combinations of spin-orbital products. All the most accurate calculations made so far support the belief that complete sets of this kind may be constructed, and most of this book is concerned with their use.

One last point must be mentioned. So far we have assumed that the complete sets have been discrete; the expansions have been written in terms of functions ψ_i, each with a coefficient c_i, where i is an *integer* (or set of integers) enumerating the chosen functions and their coefficients. It is important to realize that not all complete sets are discrete and that not all infinite discrete sets are complete. The wavefunctions of the bound (negative-energy) states of the hydrogen atom form an infinite discrete set, but the set is not complete. The complete set includes the "continuum" of wavefunctions that describe a free electron (positive energy) scattered by the nucleus. If we wish to expand a function in terms of the hydrogen-atom solutions $\phi_i(\boldsymbol{r})$ then we must write the expansion as

$$f(\boldsymbol{r}) = \sum_i c_i\phi_i(\boldsymbol{r}) + \int c(\boldsymbol{\alpha})\phi(\boldsymbol{r}, \boldsymbol{\alpha})\,\mathrm{d}\boldsymbol{\alpha}, \tag{2.1.10}$$

where the integral is over the continuum of states enumerated by the continuous parameter $\boldsymbol{\alpha}$ and the sum is over the discrete states enumerated by the integer i. It is, however, most convenient to use discrete sets; in particular, they allow us to transcribe from operator equations to matrix equations, which are often more convenient both in formal discussions and in actual computation.

2.2 VECTOR SPACES AND MATRICES

It is often useful to stress the analogy between expansions in terms of a set of functions and the representation of a vector in an n-dimensional vector space in terms of basis vectors (see e.g. Morse and Feshbach, 1953; McWeeny, 1963). In terms of basis vectors e_i we write‡

$$\mathsf{v} = \sum_i v_i\mathsf{e}_i, \tag{2.2.1}$$

and this clearly resembles (2.1.6). If $\{\mathsf{e}_i\}$ is a set of orthogonal unit vectors then the components may be expressed as scalar products $v_i = \mathsf{e}_i \cdot \mathsf{v}$. This is why the expression (2.1.5) for the expansion coefficient $c_i = \langle \phi_i \,|\, f \rangle$ is referred to as a (Hermitian) scalar product. In this sense it indicates the "component" of f along ϕ_i. It follows from (2.1.3), (2.1.4) and (2.1.5) that D vanishes when and only when

$$\int_a^b f^*f\,\mathrm{d}x = \sum_i |c_i|^2. \tag{2.2.2}$$

This relation is often used as the condition for completeness. The integral is analogous to the squared length of a vector, and a normalized function therefore has "unit length". In such cases the sum of the squares of the expansion coefficients must approach unity as more terms are added. Generally the scalar product of two functions

$$f = \sum_i c_i\phi_i \quad \text{and} \quad f' = \sum_j c_j'\phi_j$$

‡ Vectors and operators are denoted by sans serif (Univers) letters; matrices (i.e. *arrays*) are denoted by upright bold letters.

is defined by

$$\langle f \mid f' \rangle = \int_a^b f^* f' \, dx = \sum_{i,j} c_i^* c_j' \int_a^b \phi_i^* \phi_j \, dx,$$

and therefore, using matrix notation,

$$\langle f \mid f' \rangle = \mathbf{c}^\dagger \mathbf{S} \mathbf{c}' \tag{2.2.3}$$

Here the square matrix **S**, often called the "overlap matrix" of the basis functions, has for its elements the scalar products

$$S_{ij} = \langle \phi_i \mid \phi_j \rangle = \int_a^b \phi_i^* \phi_j \, dx = \langle \phi_j \mid \phi_i \rangle^* = S_{ji}^*, \tag{2.2.4}$$

which define the *metric* of the function space. For an orthonormal set, **S** reduces to the unit matrix. The matrices **c** and **c**′ are columns of expansion coefficients while $\mathbf{c}^\dagger$ is obtained by transposing **c** and replacing each element by its complex conjugate. More generally, for an arbitrary matrix **M**, the matrix $\mathbf{M}^\dagger$ obtained in a similar way by defining

$$(\mathbf{M}^\dagger)_{ij} = M_{ji}^* \tag{2.2.5}$$

is the Hermitian transpose‡ of **M**. Components will always be collected into a *column* matrix **c**, and $\mathbf{c}^\dagger$ is consequently a *row*. We note that, from the definition of the scalar product in (2.2.4), **S** has the property

$$\mathbf{S}^\dagger = \mathbf{S}. \tag{2.2.6}$$

Such a matrix is said to be *Hermitian-symmetric*. The concise expression on the right-hand side of (2.2.3) is evidently a row times a square matrix times a column (all conformable), yielding a single number (a 1×1 matrix). The space defined by an infinite set of functions $\{\phi_i\}$ with a metric defined by (2.2.4), and with further properties to be described, is called a *Hilbert space*.

We now consider in a general way the transcription of operator equations into matrix language. Suppose that some operator H (for example the Hamiltonian operator for some system) is allowed to act on a function ψ (for example a wavefunction of the system), yielding some new function ψ':

$$\psi' = \mathsf{H}\psi. \tag{2.2.7}$$

If each function is expanded in terms of a complete orthonormal set ϕ_i,

‡ Also called Hermitian conjugate, adjoint, etc.

this becomes

$$\sum_i c_i' \phi_i = \sum_i c_i \mathsf{H} \phi_i.$$

On taking the scalar product with ϕ_j (i.e. multiplying by ϕ_j^* and integrating), and remembering that $\langle \phi_j \mid \phi_i \rangle = 0$ unless $i = j$, we obtain

$$c_j' = \sum_i \langle \phi_j \mid \mathsf{H} \phi_i \rangle c_i \quad \text{(all } j\text{)}. \tag{2.2.8}$$

It is usual to write the scalar product on the right in the more symmetrical form $\langle \phi_j | \mathsf{H} | \phi_i \rangle$ and to call it a "matrix element of H"—relative to the ϕs. The reason is clear, since (2.2.8) identifies element by element the two matrices (columns) $\mathbf{c}'$ and $\mathbf{Hc}$, where $\mathbf{H}$ is the square matrix with elements

$$H_{ji} = \langle \phi_j | \mathsf{H} | \phi_i \rangle. \tag{2.2.9}$$

Thus the operator equation (2.2.7) has a matrix analogue

$$\mathbf{c}' = \mathbf{Hc}, \tag{2.2.10}$$

in which the functions are replaced by columns of expansion coefficients and the operator by a suitably defined square matrix. Strictly, the dimensions of the matrix is infinite and convergence requires discussion: in practice, expansions are made over a limited or "truncated" set and the equivalence of the operator and matrix statements is only approximate. In the same way it can be shown that the operator equality

$$\mathsf{AB} = \mathsf{C} \tag{2.2.11}$$

(i.e. B followed by A, working on any function ψ, gives the same result as C), is equivalent to the matrix equality

$$\mathbf{AB} = \mathbf{C}, \tag{2.2.12}$$

provided that the matrices are infinite, with elements defined exactly as in (2.2.9) but that the finite matrices defined for a limited set need not fulfill this relationship exactly. The matrices $\mathbf{A}, \mathbf{B}, \mathbf{C}, \ldots$ that combine with each other or "have the same multiplication table" as a set of operators $\mathsf{A}, \mathsf{B}, \mathsf{C}, \ldots$ are said to provide a *representation* of the operators; and the representation is *faithful* when two statements such as (2.2.11) and (2.2.12) are exactly equivalent, each implying the truth of the other. Fortunately, many of the applications of quantum mechanics, depending essentially on solution of the Schrödinger equation, are not entirely dependent on the use of faithful representations and hence of infinite matrices: quite precise results may be derived, as we shall find in the next section, even though these may involve incomplete sets and approximate wavefunctions.

It must be stressed that the representation property embodied in (2.2.11) and (2.2.12), with the matrix elements defined as in (2.2.9), depends on the use of an orthonormal basis $\{\phi_i\}$. If the basis is non-orthonormal, with a metric $\mathbf{S}$ defined in (2.2.4), then the "metrically defined" matrices ($\mathbf{A}^S$, $\mathbf{B}^S$, . . . , say) do not reflect the properties of the operators; when $\mathsf{AB} = \mathsf{C}$ it is not generally true that $\mathbf{A}^S\mathbf{B}^S = \mathbf{C}^S$. Since many kinds of basis are used in molecular quantum mechanics, and it is often necessary to pass from one to another, this point needs further discussion.

Any set of basis functions is conveniently collected into a row matrix

$$\boldsymbol{\phi} = (\phi_1 \ \ \phi_2 \ \ \ldots \ \ \phi_i \ \ \ldots),$$

and with this convention an arbitrary function appears as a row–column product

$$\psi = \sum_i \phi_i c_i = (\phi_1 \ \ \phi_2 \ \ \ldots)\begin{pmatrix} c_1 \\ c_2 \\ \vdots \end{pmatrix} = \boldsymbol{\phi}\mathbf{c}. \qquad (2.2.13)$$

The effect of an operator such as H may be regarded as a rotation or "mapping" in which every ϕ_i is sent into its "image" $\phi_i' = \mathsf{H}\phi_i = \sum_j \phi_j \mathsf{H}_{ji}$; this is concisely expressed as

$$\mathsf{H}\boldsymbol{\phi} = (\mathsf{H}\phi_1 \ \ \mathsf{H}\phi_2 \ \ \ldots) = (\phi_1 \ \ \phi_2 \ \ \ldots)\mathbf{H} = \boldsymbol{\phi}\mathbf{H}, \qquad (2.2.14)$$

where the ith column of the square matrix $\mathbf{H}$ contains the expansion coefficients H_{ji} of the new function $\phi_i' = \mathsf{H}\phi_i$. It is the matrices defined in this way that "have the same multiplication table", in the sense of (2.2.12), as the operators themselves. If instead we use $\mathbf{H}^S$ to denote a matrix defined in terms of scalar products, according to (2.2.9), then it follows readily from (2.2.14) that

$$\mathbf{H}^S = \mathbf{SH}, \qquad \mathbf{H} = \mathbf{S}^{-1}\mathbf{H}^S, \qquad (2.2.15)$$

where $\mathbf{S}$ is the overlap matrix defined in (2.2.4). To pass from the metrically defined quantities to the true representation matrices, it is thus necessary to multiply by an inverse overlap matrix. Clearly, there are great advantages in using orthonormal bases, such that $\mathbf{S} = \mathbf{1}$ and the distinction between $\mathbf{H}$ and $\mathbf{H}^S$ in (2.1.15) disappears. The choice of basis is thus a matter of some importance.

The connection between two different bases, $\boldsymbol{\phi}$ and $\bar{\boldsymbol{\phi}}$ may be described as a transformation, or mapping, in which $\bar{\phi}_i$ is the image associated with ϕ_i. With the same conventions as in (2.2.14),

$$\bar{\boldsymbol{\phi}} = \boldsymbol{\phi}\mathbf{T}, \tag{2.2.16}$$

where the kth column of $\mathbf{T}$ contains the expansion coefficients of $\bar{\phi}_k$ in terms of the original set. When the same number of independent functions is used in each basis the relationship may be reversed by finding the *inverse matrix* $\mathbf{T}^{-1}$ defined by

$$\mathbf{T}^{-1}\mathbf{T} = \mathbf{T}\mathbf{T}^{-1} = \mathbf{1},$$

where $\mathbf{1}$ is the *unit matrix* (1s on the diagonal, zeros elsewhere) which, in multiplication, leaves any other matrix unchanged. Multiplication of (2.2.16) by $\mathbf{T}^{-1}$ on the right then gives

$$\boldsymbol{\phi} = \bar{\boldsymbol{\phi}}\mathbf{T}^{-1}. \tag{2.2.17}$$

The basis change (2.2.16) naturally gives a new representation of a given function as in (2.2.13) but with new coefficients. Thus, from (2.2.13) and (2.2.17), $\psi = \bar{\boldsymbol{\phi}}\mathbf{T}^{-1}\mathbf{c}$, and the new coefficients are contained in the column

$$\bar{\mathbf{c}} = \mathbf{T}^{-1}\mathbf{c}. \tag{2.2.18}$$

Similarly, the relationship between the matrices $\mathbf{A}, \mathbf{B}, \ldots$ and $\bar{\mathbf{A}}, \bar{\mathbf{B}}, \ldots$, which represent the same operators $\mathsf{A}, \mathsf{B}, \ldots$ in the different bases $\boldsymbol{\phi}$ and $\bar{\boldsymbol{\phi}}$, is then easily found to be

$$\bar{\mathbf{R}} = \mathbf{T}^{-1}\mathbf{R}\mathbf{T}, \tag{2.2.19}$$

for any operator R. This *similarity transformation* preserves exactly the properties of the matrices: if $\mathbf{AB} = \mathbf{C}$ then clearly $\bar{\mathbf{A}}\bar{\mathbf{B}} = \bar{\mathbf{C}}$ and the alternative representations are *equivalent.*

An important application of a basis change occurs when a given set of functions $\{\phi_i\}$ is not orthonormal; for a new basis set $\{\bar{\phi}_i\}$ that *is* orthonormal may evidently be constructed according to (2.2.16) in an infinite variety of ways. One common choice is to take $\mathbf{T}$ in upper-triangular form so that

$$\bar{\phi}_1 = \phi_1 T_{11}, \qquad \bar{\phi}_2 = \phi_1 T_{12} + \phi_2 T_{22}, \quad \text{etc.},$$

with the numerical coefficients chosen to ensure orthonormality of the new functions; this is usually described as "Schmidt orthonormalization". Another procedure, with certain advantages, follows on noting that after a basis change (2.2.6) the new metric is

$$\bar{\mathbf{S}} = \mathbf{T}^{\dagger}\mathbf{S}\mathbf{T}. \tag{2.2.20}$$

Consequently, by choosing $\mathbf{T} = \mathbf{S}^{-1/2}$ and noting (2.2.6), we ensure that

$\bar{\mathbf{S}} = \mathbf{S}^{-1/2}\mathbf{S}\mathbf{S}^{-1/2} = \mathbf{1}$. This transformation (Löwdin, 1950) leads to a symmetrically orthonormalized basis

$$\bar{\boldsymbol{\phi}} = \boldsymbol{\phi}\mathbf{S}^{-1/2} \tag{2.2.21}$$

in which (again momentarily using $\mathbf{H}^S$ for the matrix defined in (2.2.9)) the matrix associated with the operator H is

$$\mathbf{H} = \mathbf{S}^{1/2}\mathbf{H}^S\mathbf{S}^{-1/2}. \tag{2.2.22}$$

Since $\bar{\phi}$ is an orthonormal basis, such matrices do provide a representation in the sense of (2.2.11) and (2.2.12); moreover, the matrices associated in this way with Hermitian operators possess Hermitian symmetry ($\mathbf{H}^\dagger = \mathbf{H}$) themselves, unlike the representation matrices defined through (2.2.15).

Finally, we note that according to (2.2.20) any transformation leading from one orthonormal basis to another is described by a unitary matrix such that (putting $\mathbf{S} = \bar{\mathbf{S}} = \mathbf{1}$)

$$\mathbf{T}^\dagger\mathbf{T} = \mathbf{1}, \qquad \mathbf{T}^\dagger = \mathbf{T}^{-1}. \tag{2.2.23}$$

Denoting such a matrix by $\mathbf{U}$, the similarity transformation (2.2.19) becomes a unitary transformation

$$\bar{\mathbf{R}} = \mathbf{U}^\dagger\mathbf{R}\mathbf{U}, \tag{2.2.24}$$

while the relationship between the two bases is "reversible" without the need to calculate an inverse matrix:

$$\bar{\boldsymbol{\phi}} = \boldsymbol{\phi}\mathbf{U}, \qquad \boldsymbol{\phi} = \bar{\boldsymbol{\phi}}\mathbf{U}^\dagger. \tag{2.2.25}$$

Unitary representations of the operators that describe mappings in a vector space are particularly convenient (for example in group theory, Appendix 3) and will be used frequently in later chapters.

2.3 THE EIGENVALUE EQUATION

We now discuss the eigenvalue equation

$$\mathsf{H}\Psi = E\Psi, \tag{2.3.1}$$

where Ψ may be a general many-electron wavefunction, which we expand in terms of a suitable complete set $\Phi_1, \Phi_2, \ldots, \Phi_\kappa, \ldots$, as for example in (2.1.9) where the Φs are spin-orbital products:

$$\Psi = \sum c_\kappa \Phi_\kappa. \tag{2.3.2}$$

We use Greek capital letters almost exclusively to indicate *many*-electron functions, reserving lower-case letters (ϕ, ψ, etc.) for orbitals and spin-orbitals. The analysis of Section 2.2 may be taken at once, the matrix equivalent of (2.3.1) becoming

$$\mathbf{Hc} = E\mathbf{c}, \tag{2.3.3}$$

provided that the Φs are orthonormal. We note that the use of non-orthonormal Φs does not, in this instance, raise any real difficulties. The modified equation involves the matrix **M**, defined as in (2.2.4) but in terms of many-electron functions, and is

$$\mathbf{Hc} = E\mathbf{Mc}, \tag{2.3.4}$$

where $M_{\kappa\lambda} = \langle \Phi_\kappa \mid \Phi_\lambda \rangle$. This reduces to (2.3.3) when **M** is the unit matrix.

Before considering the matrix equations in detail, we recall certain properties of the operator H and its eigenfunctions $\Psi_1, \Psi_2, \ldots$. With any operator F we may define the (Hermitian) adjoint operator $\mathsf{F}^\dagger$ by the condition

$$\langle \Phi \mid \mathsf{F}^\dagger \Phi \rangle = \langle \Phi \mid \mathsf{F}\Phi \rangle^* \quad (\text{all } \Phi), \tag{2.3.5}$$

"all Φ" meaning for all functions of the class within which F operates (e.g. class L^2 for 1-electron, or for N-electron functions, etc.). The Hamiltonian operator has the property of being *self-adjoint* or *Hermitian*, $\mathsf{H}^\dagger = \mathsf{H}$, and this is clearly an essential *physical* requirement, ensuring that the expectation value of the energy in any state Φ must be a *real* quantity. The relationship (2.3.5) may be put in a different form by substituting two alternative choices of Φ, namely (i) $\Phi_1 + i\Phi_2$, and (ii) $\Phi_1 - i\Phi_2$ (Φ_1 and Φ_2 arbitrary), and subtracting one of the resultant equations from the other: the result is

$$\langle \Phi_1 \mid \mathsf{F}^\dagger \Phi_2 \rangle = \langle \mathsf{F}\Phi_1 \mid \Phi_2 \rangle \tag{2.3.6}$$

and is sometimes referred to as the "turnover rule"—F operating on the left-hand functions may be taken over to the right if we add the dagger. The scalar product on the right may also be written the other way round if we take its complex conjugate and use (2.2.4) with $\phi_i = \mathsf{F}\Phi_1$ and $\phi_j = \Phi_2$; using the more symmetrical notation for matrix elements of an operator we obtain

$$\langle \Phi_1 | \mathsf{F}^\dagger | \Phi_2 \rangle = \langle \Phi_2 | \mathsf{F} | \Phi_1 \rangle^* \tag{2.3.7}$$

and this tells us that the matrix representing $\mathsf{F}^\dagger$, in any basis, is obtained by transposing and taking the complex conjugate of that representing F. The use of the dagger in defining an adjoint operator is thus completely parallel to its use in indicating the Hermitian transpose (or adjoint) of a *matrix*: if F is represented by $\mathbf{F}$ then $\mathsf{F}^\dagger$ is represented by $\mathbf{F}^\dagger$. The characteristic property of a Hermitian, or self-adjoint operator such as H is thus (since Φ_1 and Φ_2 may be *any* two functions of the basis)

$$H_{\kappa\lambda} = \langle \Phi_\kappa | \mathsf{H} | \Phi_\lambda \rangle = \langle \Phi_\lambda | \mathsf{H} | \Phi_\kappa \rangle^* = H^*_{\lambda\kappa}. \qquad (2.3.8)$$

The matrix representing such an operator is thus Hermitian in the *matrix* sense (2.2.5), $\mathbf{H} = \mathbf{H}^\dagger$. A useful consequence is that the adjoint of a *non*-Hermitian operator such as $\mathsf{A} + i\mathsf{B}$ (A and B Hermitian) is

$$(\mathsf{A} + i\mathsf{B})^\dagger = \mathsf{A} - i\mathsf{B}, \qquad (2.3.9)$$

and is thus obtained simply by reversing the sign of i.

The most important property of the Hamiltonian operator, for our purposes, is that its eigenfunctions $\Psi_1, \Psi_2, \ldots$, may be assumed to form a complete orthogonal set (Kato, 1951): if they also belong to class L^2 we may normalize in the usual way and write

$$\langle \Psi_K | \Psi_L \rangle = \delta_{KL}. \qquad (2.3.10)$$

In the degenerate case $E_K = E_L$ it is not *necessary* that Ψ_K and Ψ_L are orthogonal, but from them we can always find new linear combinations that are still eigenfunctions with the same eigenvalue $E = E_K = E_L$ but that *are* orthogonal. Similar considerations apply to multiple degeneracies, and (2.3.10) may thus be assumed without loss of generality. Proofs of these statements are given in any textbook on quantum mechanics (e.g. Eyring *et al.*, 1944, Chap. 3).

We now return to the matrix equation (2.3.3), which in practice is applied in the finite form that arises when only, say, n members of the complete set are taken into account. What is the significance of the solutions of the resultant "truncated" equation? Written out in full, the matrix equation then becomes

$$\boxed{\begin{aligned} H_{11}c_1 + H_{12}c_2 + \ldots + H_{1n}c_n &= Ec_1, \\ H_{21}c_1 + H_{22}c_2 + \ldots + H_{2n}c_n &= Ec_2, \\ \vdots \qquad\qquad & \quad \vdots \end{aligned}} \qquad (2.3.11)$$

—a system of linear simultaneous equations usually called the "secular equations". For an arbitrary value of E these cannot all be satisfied; for since each may be divided by, say, c_1 the first $n-1$ equations would determine the $n-1$ remaining unknowns (c_κ/c_1 for $\kappa = 2, 3, \ldots, n$) and these would not necessarily satisfy the remaining nth equation. It is only for n *special* values of E, let us call them $E_1^{(n)}, E_2^{(n)}, \ldots, E_n^{(n)}$, that the last equation is compatible with the first $n-1$. The compatibility condition may be expressed in a convenient and general way by taking the terms on the right in (2.3.11) over to the left and considering the determinant of the coefficients of $c_1, c_2, \ldots, c_n$; the necessary and sufficient condition for compatibility is then (Margenau and Murphy, 1956)

$$\Delta^{(n)}(E) = \begin{vmatrix} H_{11} - E & H_{12} & \ldots & H_{1n} \\ H_{21} & H_{22} - E & \ldots & H_{2n} \\ \vdots & \vdots & & \vdots \\ H_{n1} & H_{n2} & \ldots & H_{nn} - E \end{vmatrix} = 0.$$

The "secular determinant", which is a function of the unknown E, is often abbreviated by indicating just the general element, and the condition is then written as

$$\Delta^{(n)}(E) = \det |H_{\kappa\lambda} - E\delta_{\kappa\lambda}| = 0, \qquad (2.3.12)$$

where $\delta_{\kappa\lambda}$ shows that E appears only on the diagonal. It may be noted that the corresponding condition for (2.3.4) is

$$\det |H_{\kappa\lambda} - EM_{\kappa\lambda}| = 0 \qquad (2.3.13)$$

in which E also appears *off* the diagonal. We shall have little occasion to use the theory of determinants, since modern methods of high-speed computation invariably solve the equations (2.3.11) more directly. We need only note that $\Delta^{(n)}(E)$, on expansion, yields a polynomial of the nth degree in E and that the required solutions are its roots. Each E value $E_K^{(n)}$ may then be substituted back into (2.3.11), which can be solved for the coefficients in order to determine a corresponding $\Psi_K^{(n)}$.

To investigate the significance of the solutions, we consider equation (2.3.3), first with a truncation to the first n functions $\Phi_1, \Phi_2, \ldots, \Phi_n$ and secondly with a truncation to the first $n+1$ functions, $\Phi_1, \Phi_2, \ldots, \Phi_{n+1}$. In other words, we consider the two matrix equations, first

$$\mathbf{H}^{(n)}\mathbf{c}^{(n)} = E^{(n)}\mathbf{c}^{(n)} \qquad (2.3.3a)$$

and then

$$\mathbf{H}^{(n+1)}\mathbf{c}^{(n+1)} = E^{(n+1)}\mathbf{c}^{(n+1)}, \qquad (2.3.3b)$$

to see how the solutions converge when the basis is extended. From (2.3.3a) we find n solutions

$$(\Psi_1^{(n)}, E_1^{(n)}), \quad (\Psi_2^{(n)}, E_2^{(n)}), \quad \ldots, \quad (\Psi_n^{(n)}, E_n^{(n)}),$$

while if we solve (2.3.3b) we find $n+1$ solutions

$$(\Psi_1^{(n+1)}, E_1^{(n+1)}), \quad (\Psi_2^{(n+1)}, E_2^{(n+1)}), \quad \ldots, \quad (\Psi_{n+1}^{(n+1)}, E_{n+1}^{(n+1)}).$$

Now it can be shown that the secular determinants $\Delta^{(n)}(E)$ and $\Delta^{(n+1)}(E)$

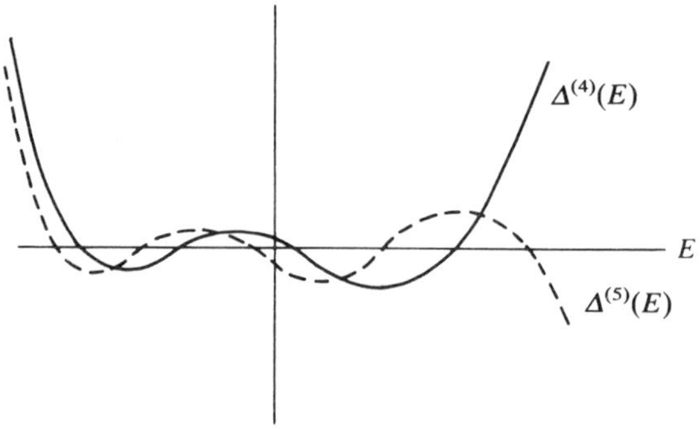

Fig. 2.1 Behaviour of secular polynomials for $n = 4$, 5.

must behave as in Fig. 2.1. Consequently the relative disposition of the roots (which approximate the energy levels $E_1, E_2, \ldots$) must be as in Fig. 2.2, the levels in any approximation ($n + 1$ functions) *separating* those of the previous approximation (n functions). In other words, by extending the basis every approximate energy level is depressed, and the exact levels, which would be obtained using a complete basis, must be approached monotonically from above. This is the "separation theorem" due to Hylleraas and Undheim (1930) and MacDonald (1933), and also discussed extensively by Epstein (1974).

The preceding results provide a firm foundation for most of the approximation methods used in molecular quantum mechanics. Whenever an approximate wavefunction (of given symmetry) is written as a one-to-one correspondence between the approximate energy levels,

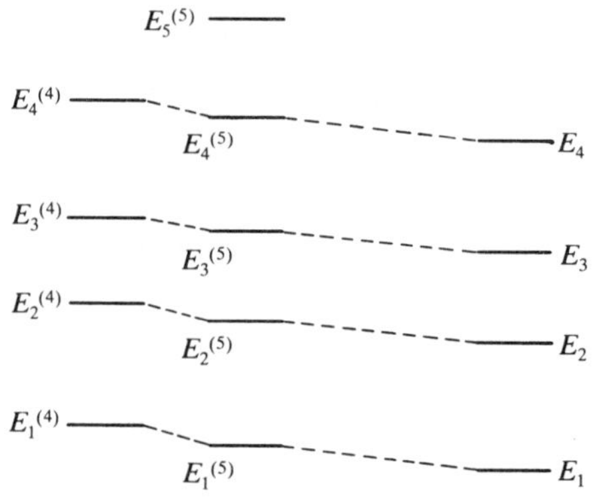

Fig. 2.2 Convergence of approximate energy values.

however poor they may be, and the exact levels (for states of the given symmetry), the Kth approximate level being an upper bound to the Kth exact level. The reason for the proviso "of given symmetry" (Appendix 3) is simply that all matrix elements connecting functions Φ_κ and Φ_λ of different symmetry (e.g. the symmetric and antisymmetric functions in (1.3.5)) vanish identically, in which case the determinant factorizes into separate blocks yielding different sets of entirely independent solutions; the theorem assumes that the $(n+1)$th function is connected with the others by at least one *non*-vanishing matrix element and therefore applies only *within* a set of functions of given symmetry type. These results are also valid when the basis employed is non-orthogonal and (2.3.4) is appropriate, and are not affected by degeneracies among the eigenvalues (Davies, 1960).

We now collect some important results relating to the solution of the matrix eigenvalue equation (2.3.4). First we note that the solutions have orthogonality properties analogous to those of the solutions of the operator equation (2.3.1). Thus two sets of coefficients, $\mathbf{c}_K$ and $\mathbf{c}_L$, corresponding to eigenvalues E_K and E_L, satisfy

$$\mathbf{H}\mathbf{c}_K = E_K\mathbf{M}\mathbf{c}_K, \qquad \mathbf{H}\mathbf{c}_L = E_L\mathbf{M}\mathbf{c}_L,$$

and on multiplying from the left by $\mathbf{c}_K^\dagger$ and $\mathbf{c}_L^\dagger$ respectively, taking the Hermitian transpose of the second equation, and subtracting, we obtain

$$(E_K - E_L^*)\mathbf{c}_L^\dagger\mathbf{M}\mathbf{c}_K = \mathbf{0}. \qquad (2.3.14)$$

This equation is analogous to one encountered in proving the eigenfunction orthogonality property (2.3.10). The usual arguments show that all eigenvalues are real and that, even when degeneracies occur, the eigenvectors may be chosen orthornormal in the sense

$$\mathbf{c}_K^\dagger\mathbf{M}\mathbf{c}_L = \delta_{KL}, \qquad (2.3.15)$$

corresponding to the orthonormality of the (exact or approximate) eigenfunctions that they represent. The associated energies are then given by

$$E_K = \mathbf{c}_K^\dagger\mathbf{H}\mathbf{c}_K. \qquad (2.3.16)$$

It is possible to combine these statements by collecting the columns $\mathbf{c}_1, \mathbf{c}_2, \ldots$ into a single square matrix

$$\mathbf{C} = (\mathbf{c}_1 \mid \mathbf{c}_2 \mid \ldots). \qquad (2.3.17)$$

For if $\mathbf{E} = \mathrm{diag}\,(E_1 \;\; E_2 \;\; \ldots)$ is the diagonal matrix of eigenvalues then it follows readily that

$$\mathbf{E} = \mathbf{C}^\dagger\mathbf{H}\mathbf{C}, \qquad \mathbf{C}^\dagger\mathbf{M}\mathbf{C} = \mathbf{1}. \qquad (2.3.18)$$

C is then referred to as the matrix that "brings to diagonal form" both **M** and **H**: it describes a new basis in which the operator H has vanishing matrix elements between different functions. When only orthonormal sets are admitted, $\mathbf{M} = \mathbf{1}$, and **C** is the matrix of a unitary transformation as defined in (2.2.24) such that $\mathbf{C}^\dagger = \mathbf{C}^{-1}$. In the notation used previously

$$\boldsymbol{\Psi} = \boldsymbol{\Phi}\mathbf{C}, \tag{2.3.19}$$

where, since $\mathbf{c}_K$ is the Kth column of **C**,

$$\Psi_K = \boldsymbol{\Phi}\mathbf{c}_K = \sum_\kappa \Phi_\kappa c_\kappa^{(K)} = \sum_\kappa \Phi_\kappa C_{\kappa K}.$$

It should be noticed that the full set of equations of the type (2.3.4), one for each eigenvalue and eigenvector, may be collected in the form

$$\mathbf{HC} = \mathbf{MCE}, \tag{2.3.20}$$

in which all the matrices are square.

2.4 VARIATION THEORY

The familiar variation theorem may be regarded as a particular instance of the inequalities referred to in the last section. Thus putting $n = 1$ in the n-term expansion (2.3.2) corresponds to using a single function Φ_1 (not necessarily normalized) as an approximate eigenfunction. The determinant (2.3.12) then reduces to a single element $H_{11} - EM_{11}$, and on using $\Psi = \Phi_1$,

$$\boxed{E = \frac{\langle \Psi | \mathsf{H} | \Psi \rangle}{\langle \Psi | \Psi \rangle} \geqslant E_1,} \tag{2.4.1}$$

where E_1 is the energy of the first state Ψ_1, in ascending energy order (subject to any constraints associated with symmetry). It is for this reason that the simple approximate wavefunctions for the helium atom and hydrogen molecule, used in Chapter 1, yield upper bounds on the energies of the states considered.

The variation method utilizes the result (2.4.1) by incorporating variable parameters into any suitable trial function Ψ and adjusting them until E reaches its minimum value, thus obtaining "best" approximations to the ground-state energy and wavefunction. When Ψ in (2.4.1) is freely variable in the sense that $\Psi \to \Psi + \delta\Psi$, where the infinitesimal function $\delta\Psi$ is completely unconstrained (not arising from variation of a finite number of numerical parameters in a function of given form), the energy

E is said to be a functional of Ψ; the functional $E = E[\Psi]$ takes a value that depends on the functional form of Ψ, not merely on particular numerical values of a set of independent variables (adjustable parameters). The analysis of the behaviour of $E[\Psi]$ with respect to variation of Ψ is "functional analysis" and parallels the study of the behaviour of $f(x)$ with respect to variation of x (i.e. analysis in the usual sense): thus, for example, it is possible to discuss variations in terms of "functional derivatives" $\delta E/\delta\Phi$. We shall, however, make little use of such concepts, in spite of their formal value (see e.g. Feynman and Hibbs, 1965), since variation functions are most commonly defined in terms of a finite number of parameters, to which the ordinary rules of differential calculus can be applied. The other problems we meet can also be solved by elementary methods.

It is easily verified that if Ψ in (2.4.1) is taken to be a linear variation function of the form $c_1\Phi_1 + c_2\Phi_2 + \ldots + c_n\Phi_n$ then the condition that E be stationary against variation of $c_1, c_2, \ldots, c_n$ leads back to the secular equation (2.3.4) in the truncated form corresponding to an n-term expansion. In other words, "best approximate" solutions at any level may be obtained by seeking stationary values of the corresponding energy functional. Thus the finite-basis form of (2.4.1) is

$$E = \sum_{\kappa,\lambda} c_\kappa^* c_\lambda \langle \Phi_\kappa | \mathsf{H} | \Phi_\lambda \rangle \Big/ \sum_{\kappa,\lambda} c_\kappa^* c_\lambda \langle \Phi_\kappa \mid \Phi_\lambda \rangle, \tag{2.4.2a}$$

or in matrix form, admitting non-orthogonal functions for generality,

$$E = \frac{\mathbf{c}^\dagger \mathbf{H} \mathbf{c}}{\mathbf{c}^\dagger \mathbf{M} \mathbf{c}}. \tag{2.4.2b}$$

Let us now consider a variation in which the coefficients change ($\mathbf{c} \to \mathbf{c} + \delta\mathbf{c}$) to give a new value of E:

$$E + \delta E = \frac{(\mathbf{c} + \delta\mathbf{c})^\dagger \mathbf{H} (\mathbf{c} + \delta\mathbf{c})}{(\mathbf{c} + \delta\mathbf{c})^\dagger \mathbf{M} (\mathbf{c} + \delta\mathbf{c})}.$$

On expanding the denominator using the binomial theorem and keeping only terms of first order in small quantities, we obtain easily

$$\delta E = \delta\mathbf{c}^\dagger(\mathbf{H}\mathbf{c} - E\mathbf{M}\mathbf{c}) + (\mathbf{c}^\dagger\mathbf{H} - E\mathbf{c}^\dagger\mathbf{M})\delta\mathbf{c}.$$

The two terms are seen to be complex conjugate‡, and the condition that E be stationary for arbitrary $\delta\mathbf{c}$ is thus that the coefficient of $\delta\mathbf{c}$ (and

‡ Take the Hermitian transpose of the second term, remembering that $(\mathbf{AB})^\dagger = \mathbf{B}^\dagger\mathbf{A}^\dagger$, $\mathbf{H}^\dagger = \mathbf{H}$, $\mathbf{M}^\dagger = \mathbf{M}$, and that for a single element the asterisk and dagger are equivalent.

likewise of $\delta\mathbf{c}^\dagger$) should vanish. Thus

$$\mathbf{Hc} = E\mathbf{Mc}, \tag{2.4.3}$$

which is the condition already stated in (2.3.4). The eigenvalue equation in matrix form, referred to either a complete or a truncated set, orthonormal or otherwise, is thus entirely equivalent to a stationary-value condition

$$E = \frac{\mathbf{c}^\dagger\mathbf{Hc}}{\mathbf{c}^\dagger\mathbf{Mc}} = \text{stationary value.} \tag{2.4.4}$$

If we denote the Kth solution of (2.4.3) by $(\mathbf{c}_K, E_K)$‡ it is clear on substituting and multiplying from the left by $\mathbf{c}_K$ that the stationary values are the eigenvalues themselves.

Similar considerations apply to the eigenvalue equation in operator form,

$$\mathsf{H}\Psi = E\Psi, \tag{2.4.5}$$

which has the variational equivalent

$$E = \frac{\langle\Psi|\,\mathsf{H}\,|\Psi\rangle}{\langle\Psi\,|\,\Psi\rangle} = \text{stationary value} \tag{2.4.6}$$

for any *functional* variation $\Psi \rightarrow \Psi + \delta\Psi$. The matrix forms, with a *complete* set, simply restate these conditions; with a truncated set they provide the link between exact and approximate results.

Now that we have established the equivalence between an eigenvalue equation (operator form or matrix form) and a stationary-value condition (for functional variation or linear parameter variation) we turn to the general problem of finding the stationary values of the energy functional when all types of parameter, linear and non-linear, are admitted. A very convenient machinery for this purpose has been developed by Moccia (1974), whose approach we now adopt.

We denote the wavefunction, which may contain any number of parameters $p_1, p_2, \ldots, p_i, \ldots$ (assumed independent), by

$$\Psi = \Psi(p_1, p_2, \ldots \,|\, \boldsymbol{x}_1, \boldsymbol{x}_2, \ldots, \boldsymbol{x}_N) = \Psi(\mathbf{p}; \boldsymbol{x}) \tag{2.4.7}$$

and write for the corresponding energy functional

$$E = H(p_1, p_2, \ldots) = \frac{\langle\Psi|\,\mathsf{H}\,|\Psi\rangle}{\langle\Psi\,|\,\Psi\rangle}. \tag{2.4.8}$$

The particular values $E_1, E_2, \ldots$ for which this functional becomes stationary against infinitesimal variations $\Psi \rightarrow \Psi + \delta\Psi$ (including those resulting from parameter variations $p_j \rightarrow p_j + \delta p_j$) are then, in the usual

‡ Note that $\mathbf{c}_K$ denotes the Kth eigenvector (i.e. column matrix)—not a single component.

sense, approximations to the exact eigenvalues of the many-electron Hamiltonian; while the corresponding variation functions will be parametrically determined approximations to the exact eigenfunctions.

The conditions for a stationary value of expression (2.4.8) are of course

$$\frac{\partial H}{\partial p_1} = \frac{\partial H}{\partial p_2} = \ldots = 0, \tag{2.4.9}$$

while the differential $\mathrm{d}H$ is easily found to be

$$\mathrm{d}H = \sum_j \left\{ \frac{\langle \Psi | \mathsf{H} | \partial \Psi / \partial p_j \rangle}{\langle \Psi | \Psi \rangle} - \frac{\langle \Psi | \mathsf{H} | \Psi \rangle \langle \Psi | \partial \Psi / \partial p_j \rangle}{\langle \Psi | \Psi \rangle^2} \right\} \mathrm{d}p_j + \text{c.c.}, \tag{2.4.10}$$

where c.c. indicates a complex-conjugate term arising from variation of the $\langle \Psi |$ factor in the scalar products. Now the quantity in curly brackets is simply the j-component $\partial H / \partial p_j$ of a "gradient vector" on the energy surface $E = H(p_1, p_2, \ldots)$, each component indicating a rate of change of E in the "direction" corresponding to variation of one parameter. In general, therefore, we introduce a gradient component

$$(H\nabla)_j = \frac{\langle \Psi | (\mathsf{H} - H) | \partial \Psi / \partial p_j \rangle}{\langle \Psi | \Psi \rangle} \tag{2.4.11a}$$

and the complex-conjugate component

$$(\nabla H)_j = \frac{\langle \partial \Psi / \partial p_j | (\mathsf{H} - H) | \Psi \rangle}{\langle \Psi | \Psi \rangle}. \tag{2.4.11b}$$

The notation $H\nabla$, or ∇H, corresponds to differentiation of the function on the right of the operator (i.e. in the ket), or on the left (i.e. in the bra) respectively. For a stationary point the gradient must vanish in all directions, and hence

$$(\nabla H)_j = (H\nabla)_j = 0 \quad (\text{all } j). \tag{2.4.12}$$

These are the general conditions for a stationary point.

Higher derivatives of the energy may be expressed in a similar way. If we evaluate the general second derivative, writing for brevity $\partial \Psi / \partial p_j = \Psi^j$ etc., we encounter the quantities

$$(\nabla\nabla H)_{jk} = \frac{\langle \Psi^{jk} | (\mathsf{H} - H) | \Psi \rangle - (\nabla H)_j \langle \Psi^k | \Psi \rangle - \langle \Psi^j | \Psi \rangle (\nabla H)_k}{\langle \Psi | \Psi \rangle}, \tag{2.4.13}$$

$$(\nabla H \nabla)_{jk} = \frac{\langle \Psi^j | (\mathsf{H} - H) | \Psi^k \rangle - (\nabla H)_j \langle \Psi | \Psi^k \rangle - \langle \Psi^j | \Psi \rangle (H\nabla)_k}{\langle \Psi | \Psi \rangle}. \tag{2.4.14}$$

A third quantity, $(H\nabla\nabla)_{jk}$, is simply the complex conjugate of (2.4.13), both differentiations being applied to the righthand Ψ.

All of the above expressions are conveniently regarded as "tensor components" of rank 1 and 2. Thus $(\nabla H)_j$ is the j-component of a first-rank tensor that we denote by $\boldsymbol{\nabla} H$, while $(\nabla H\nabla)_{jk}$ is the jk-component of a second-rank tensor $\boldsymbol{\nabla} H \boldsymbol{\nabla}$. These tensor components possess the following symmetry properties

$$\left.\begin{aligned}(\nabla H)_j = (H\nabla)_j^*, \qquad (\nabla H\nabla)_{jk} = (\nabla H\nabla)_{kj}^*, \\ (\nabla\nabla H)_{jk} = (\nabla\nabla H)_{kj} = (H\nabla\nabla)_{jk}^* = (H\nabla\nabla)_{kj}^*,\end{aligned}\right\} \tag{2.4.15}$$

which reflect the Hermitian symmetry of the operator and the assumption that Ψ is a well-behaved function of the parameters.

The variation of (2.4.8) around a point $\mathbf{p}_0 = (p_1^{(0)}, p_2^{(0)}, \ldots)$, at which $E = H_0$, is readily expressed in terms of the tensor components. If we put $p_j = p_j^{(0)} + d_j$ and expand up to terms quadratic in the "parameter displacements" d_j then

$$\begin{aligned}\delta E = &\sum_j [d_j^*(\nabla H)_j + (H\nabla)_j d_j] \\ &+ \sum_{j,k} [\tfrac{1}{2} d_j^* d_k^* (\nabla\nabla H)_{jk} + d_j^* (\nabla H\nabla)_{jk} d_k + \tfrac{1}{2}(H\nabla\nabla)_{jk} d_j d_k] + \ldots,\end{aligned} \tag{2.4.16}$$

where the two sums represent the first- and second-order energy changes and all derivatives are evaluated at the point $\mathbf{p}_0$.

When $\mathbf{p}_0$ is a stationary point (2.4.12) holds and the linear term vanishes; but the point will be a local minimum only if the quadratic term is always positive. This latter condition may be expressed in terms of the matrices with elements

$$M_{jk} = (\nabla H\nabla)_{jk}, \qquad Q_{jk} = (\nabla\nabla H)_{jk}. \tag{2.4.17}$$

It is necessary that the sum and difference $\mathbf{M} \pm \mathbf{Q}$ both be positive-definite, and this is often called a "stability condition". Stability is an important criterion, especially in the variational calculation of electronic properties (Hall, 1961), and has received considerable attention (see e.g. Čížek and Paldus, 1967).

It is often useful to write (2.4.16) in matrix form, using the definitions in (2.4.17). Thus, on collecting the parameter variations and the gradient components into column vectors $\mathbf{d}$ and $\mathbf{a}$ respectively, it follows that

$$E = H_0 + \begin{pmatrix}\mathbf{d} \\ \mathbf{d}^*\end{pmatrix}^\dagger \begin{pmatrix}\mathbf{a} \\ \mathbf{a}^*\end{pmatrix} + \frac{1}{2}\begin{pmatrix}\mathbf{d} \\ \mathbf{d}^*\end{pmatrix}^\dagger \begin{pmatrix}\mathbf{M} & \mathbf{Q} \\ \mathbf{Q}^* & \mathbf{M}^*\end{pmatrix}\begin{pmatrix}\mathbf{d} \\ \mathbf{d}^*\end{pmatrix} + \ldots, \tag{2.4.18}$$

where in the partitioned columns the elements of **a** and **d** are followed by their complex conjugates. More briefly, and in an obvious notation, this may be written

$$E = H_0 + \mathbf{D}^\dagger\mathbf{A} + \tfrac{1}{2}\mathbf{D}^\dagger\mathbf{B}\mathbf{D}, \tag{2.4.19}$$

and the symmetry relations (2.4.15) show that **B** is Hermitian-symmetric. Evidently, **A** contains the first derivatives (gradient components) while **B** (the Hessian) contains the second derivatives.

The preceding equations are valid for all types of parameter variation and lead to efficient methods of optimizing a variational wavefunction. Thus, if the parameter values $\mathbf{p}_0$ do not give a stationary point in parameter space, but one sufficiently close for a quadratic approximation to be good, we may seek such a point ($\mathbf{p} = \mathbf{p}_0 + \mathbf{d}$, say) by inserting $\mathbf{D} + \delta\mathbf{D}$ in (2.4.19) and requiring $\delta E = 0$ for the variation $\delta\mathbf{D}$ around the new point $\mathbf{p}$. This variation, to first order, is

$$\delta E = \delta\mathbf{D}^\dagger\mathbf{A} + \tfrac{1}{2}\delta\mathbf{D}^\dagger\mathbf{B}\mathbf{D} + \tfrac{1}{2}\mathbf{D}^\dagger\mathbf{B}\delta\mathbf{D},$$

and, on adding the complex conjugate (remembering that δE must be real and that $\mathbf{B}^\dagger = \mathbf{B}$), the stationary condition will become

$$2\delta E = \delta\mathbf{D}^\dagger(\mathbf{A} + \mathbf{B}\mathbf{D}) + (\mathbf{A} + \mathbf{B}\mathbf{D})^\dagger\delta\mathbf{D} = 0.$$

Since the two terms are complex-conjugate, and hence linearly independent, each must separately vanish. In the partitioned form corresponding to (2.4.18), the parameter values for a stationary point will then follow (subject to the approximation of a quadratic "basin") by solution of

$$\begin{pmatrix} \mathbf{a} \\ \mathbf{a}^* \end{pmatrix} + \begin{pmatrix} \mathbf{M} & \mathbf{Q} \\ \mathbf{Q}^* & \mathbf{M}^* \end{pmatrix}\begin{pmatrix} \mathbf{d} \\ \mathbf{d}^* \end{pmatrix} = \mathbf{0}. \tag{2.4.20}$$

Formally, the solution involves the inverse of the Hessian matrix, which may be rather large; but in practice it is usually more convenient to solve the linear system (2.4.20) directly.

When all quantities are real a considerable reduction occurs, (2.4.20) yielding immediately (twice over) the equation

$$\mathbf{a} + (\mathbf{M} + \mathbf{Q})\mathbf{d} = \mathbf{0}, \tag{2.4.21}$$

whose dimensions are only half those of (2.4.20).

2.5 PARTITIONING AND PERTURBATION METHODS

Variational techniques, either in their general form (Section 2.4) or in the form corresponding to use of a linear variation function (Section 2.3), are central to many of the approximation methods currently in use: they are flexible and general and lead to upper bounds on the energy levels that may be improved by systematic variation of parameters.

In certain situations, however, it is expedient to use some form of perturbation theory in which the Hamiltonian H is assumed to differ only slightly from an operator H_0, whose eigenvalues and eigenfunctions are "known". We write $\mathsf{H} = \mathsf{H}_0 + \mathsf{H}'$ and regard H' as a small "perturbation" of the "unperturbed Hamiltonian" H_0. For example, the electron interaction term in the helium-atom Hamiltonian (Section 1.2) could be regarded as a perturbation of H_0 defined in (1.2.3), whose exact eigenfunctions are orbital products of the form (1.2.5). This approach is fully discussed in most books on quantum mechanics, but, although of great formal value, it has various limitations; in particular, it may converge poorly (or not at all), and a complete set of eigenfunctions, even of the unperturbed Hamiltonian, is not generally available.

For present purposes it is more useful to concentrate on other approaches, which start from the finite-basis form of the linear variation method. In many forms of "variation–perturbation" theory, exact unperturbed eigenfunctions are not required and the partitioning of the Hamiltonian into two terms is secondary to a partitioning of the basis. At the same time, as we shall see, it is possible to retrieve the equations of conventional perturbation theory by making an appropriate choice of basis.

To develop the variation–perturbation approach we may suppose that Φ_1 is an approximation to the required eigenfunction and that we wish to improve this approximation by adding functions $\Phi_2, \Phi_3, \ldots$ and solving the secular equations (2.3.3). More generally, let us suppose a given orthonormal basis $\{\Phi_\kappa\}$ divided into two subsets, A and B say, which contain n_A and n_B functions respectively, and ask how the results obtained using n_A functions are improved by adding the n_B extra functions. The matrices $\mathbf{H}$ and $\mathbf{c}$ may be partitioned accordingly (Löwdin, 1951, 1962), and the secular equations become

$$\begin{pmatrix} \mathbf{H}^{AA} & \mathbf{H}^{AB} \\ \mathbf{H}^{BA} & \mathbf{H}^{BB} \end{pmatrix} \begin{pmatrix} \mathbf{c}^{A} \\ \mathbf{c}^{B} \end{pmatrix} = E \begin{pmatrix} \mathbf{c}^{A} \\ \mathbf{c}^{B} \end{pmatrix}, \tag{2.5.1}$$

where $\mathbf{c}^a$ is composed of n_A coefficients and $\mathbf{c}^B$ of n_B. $\mathbf{H}^{AA}$ contains the matrix elements $H_{11}, H_{12}, \ldots, H_{1n_A}, H_{21}, H_{22}, \ldots, H_{2n_A}$, etc., and the other blocks are defined similarly. This equation is equivalent to the two equations

$$\mathbf{H}^{AA}\mathbf{c}^{A} + \mathbf{H}^{AB}\mathbf{c}^{B} = E\mathbf{c}^{A},$$
$$\mathbf{H}^{BA}\mathbf{c}^{A} + \mathbf{H}^{BB}\mathbf{c}^{B} = E\mathbf{c}^{B},$$

which may be solved formally by obtaining $\mathbf{c}^B$ from the second equation, assuming E given, and inserting the result in the first. If this is done, it follows that

$$\boxed{\mathbf{H}_{\text{eff}}\mathbf{c}^A = E\mathbf{c}^A,} \tag{2.5.2}$$

in which, using $\mathbf{1}_{BB}$ for the $n_B \times n_B$ unit matrix,

$$\boxed{\mathbf{H}_{\text{eff}} = \mathbf{H}^{AA} + \mathbf{H}^{AB}(E\mathbf{1}_{BB} - \mathbf{H}^{BB})^{-1}\mathbf{H}^{BA}.} \tag{2.5.3}$$

Equation (2.5.2) is exactly equivalent to (2.5.1), but we have reduced the secular problem from one of n dimensions to one of n_A dimensions by defining an "effective Hamiltonian" to allow for the other functions.

Since $\mathbf{H}_{\text{eff}}$ depends on E, which is not yet known, the solution must generally be obtained by iteration. Let us consider first, for example, the case $n_A = 1$, which will be appropriate if Φ_1 is non-degenerate. We may then choose $c_1 = 1$ since it is not necessary at this stage to normalize the eigenfunction. The effective Hamiltonian (2.5.3) then reduces to a single element (1×1 matrix), and (2.5.2) gives

$$E = \mathbf{H}_{\text{eff}} = f(E).$$

To obtain an explicit form for the energy, we insert the first approximation $E = H_{11}$ in the second term of (2.5.3) and expand the inverse matrix using the binomial theorem (assuming that the expansion converges). The result, to second order in the off-diagonal elements of $\mathbf{H}$, is easily found to be

$$E = H_{11} + \sum_{\kappa>1} \frac{H_{1\kappa}H_{\kappa 1}}{H_{11} - H_{\kappa\kappa}}. \tag{2.5.4}$$

This estimate for E may be used in a further iteration; but to second order the results are unchanged. To obtain the corresponding wavefunction, we note that the unknown expansion coefficients are contained in $\mathbf{c}^B$, which is given by

$$\mathbf{c}^B = (E\mathbf{1}_{BB} - \mathbf{H}^{BB})^{-1}\mathbf{H}^{BA}\mathbf{c}^A. \tag{2.5.5}$$

In the present instance ($n_A = 1$) the binomial expansion yields, to first order in off-diagonal elements,

$$c_\kappa = \frac{H_{\kappa 1}}{H_{11} - H_{\kappa\kappa}} \quad (\kappa > 1). \tag{2.5.6}$$

In (2.5.4) and (2.5.6) the superscripts are no longer needed, the matrix elements being indicated explicitly.

The above equations resemble those of the ordinary Rayleigh–Schrödinger (RS) perturbation theory (see e.g. Dalgarno, 1961), to which they reduce as a special case, but in some ways they are more general. The basis is finite and quite arbitrary; it is not necessary to separate the Hamiltonian into an unperturbed part and a perturbation; and the availability of a complete set of exact eigenfunctions is not assumed. The similarity between these results and those of conventional perturbation theory is sometimes misleading, and comparisons must be made with great care.

The connection with RS perturbation theory becomes clear on writing

$$\mathsf{H} = \mathsf{H}_0 + \lambda \mathsf{H}' \quad (\lambda \to 1), \tag{2.5.7}$$

and choosing, in place of the arbitrary basis, the eigenfunctions of H_0 that satisfy

$$\mathsf{H}_0 \Phi_K = E_K^{(0)} \Phi_K, \tag{2.5.8}$$

where the state labels $K, L, \ldots$ will now be used instead of the arbitrary indices κ, λ. The "perturbation parameter" λ is introduced for convenience in classifying orders of magnitude, and the perturbation is "switched on" by allowing λ to go from 0 to 1. Again, assuming the A set to contain only one function Φ_1, we may employ the same argument used in establishing (2.5.4), the other functions Φ_L ($L \neq 1$) being mixed with Φ_1 under the influence of the perturbation.

The matrix elements that appear in (2.5.4) are now, remembering the orthonormality of the Φ_K in (2.5.8),

$$H_{11} = \langle \Phi_1 | (\mathsf{H}_0 + \lambda \mathsf{H}') | \Phi_1 \rangle = E_1^{(0)} + \lambda \langle \Phi_1 | \mathsf{H}' | \Phi_1 \rangle,$$

$$H_{1L} = \langle \Phi_1 | (\mathsf{H}_0 + \lambda \mathsf{H}') | \Phi_L \rangle = \lambda \langle \Phi_1 | \mathsf{H}' | \Phi_L \rangle.$$

On inserting these results in (2.5.4) and (2.5.6) and replacing subscript 1 by K, since the choice of *which* function constitutes the A set is entirely arbitrary, we obtain the familiar expressions for the energy (up to second order) and the wavefunction (to first order) of the Kth state of the perturbed system:

$$E_K = E_K^{(0)} + \langle \Phi_K | \mathsf{H}' | \Phi_K \rangle + \sum_{L(\neq K)} \frac{\langle \Phi_K | \mathsf{H}' | \Phi_L \rangle \langle \Phi_L | \mathsf{H}' | \Phi_K \rangle}{E_K^{(0)} - E_L^{(0)}}, \tag{2.5.9}$$

$$\Psi_K = \Phi_K - \sum_{L(\neq K)} \Phi_L \frac{\langle \Phi_L | \mathsf{H}' | \Phi_K \rangle}{E_K^{(0)} - E_L^{(0)}}. \tag{2.5.10}$$

Here the coefficients obtained from (2.5.6) have been used in the wavefunction expansion and λ has been suppressed. We note that the wavefunction is normalized only with neglect of second-order terms, and that (2.5.9) does not give an upper bound to the exact energy, even for the ground state. This is easily remedied: we simply calculate

$$E'_K = \frac{\langle \Psi_K | \mathsf{H} | \Psi_K \rangle}{\langle \Psi_K | \Psi_K \rangle} \tag{2.5.11}$$

and obtain an energy that is correct to *third* order. For the lowest state of a given symmetry, E'_K is an upper bound to the exact energy. It is often stated that a wavefunction correct to first order yields an energy correct to second order, without stating in what sense: in fact a function correct to first order determines the energy to third order, according to (2.5.11), but the expectation value of H in state Ψ_K does not give the energy correctly even to second order unless renormalization is taken into account by including the denominator in (2.5.11). We return to this point in a later section.

Higher-order perturbation theory, as well as other applications of the partitioning technique, will be taken up in later chapters: here we mention only two generalizations. First, various forms of degenerate perturbation theory may be obtained by using the degenerate states of interest as the A set; the equations already given may indeed be used as they stand with $n_A \neq 1$, the only new feature being the solution of the A-set eigenvalue problem (2.5.2). Secondly, the partitioning approach may also be applied with a non-orthogonal basis by partitioning both $\mathbf{M}$ and $\mathbf{H}$ in (2.5.1). In this case (2.5.2) is replaced by

$$\mathbf{H}_{\text{eff}}\mathbf{c}^A = E\mathbf{M}^{AA}\mathbf{c}^A, \tag{2.5.12}$$

where the effective Hamiltonian is

$$\mathbf{H}_{\text{eff}} = \mathbf{H}^{AA} + (\mathbf{H}^{AB} - E\mathbf{M}^{AB})(E\mathbf{M}^{BB} - \mathbf{H}^{BB})^{-1}(\mathbf{H}^{BA} - E\mathbf{M}^{BA}). \tag{2.5.13}$$

We assume normalized basis functions, consider again the case $\mathbf{c}^A = 1$, and expand the inverse matrix in powers of its off-diagonal part: the result is, instead of (2.5.4),

$$E = H_{11} + \sum_{\kappa>1} \frac{(H_{1\kappa} - H_{11}M_{1\kappa})(H_{\kappa 1} - H_{11}M_{\kappa 1})}{H_{11} - H_{\kappa\kappa}}. \tag{2.5.14}$$

The formula for the coefficients becomes in this case, to first order,

$$c_\kappa = \frac{H_{\kappa 1} - H_{11}M_{\kappa 1}}{H_{11} - H_{\kappa\kappa}} \quad (\kappa \neq 1), \tag{2.5.15}$$

and higher orders may again be obtained by iteration.

PROBLEMS 2

2.1 Show that the desired "property of finality" (p. 26) requires the orthogonality of the functions $\{\phi_r\}$. [*Hint:* Write down the sets of conditions $\partial D/\partial c_r = 0$, first for n functions and then for $n+1$; assume that the c_r satisfy the first set and require that the *same* coefficients, with c_{n+1} added, satisfy the second set.]

2.2 Generalize (2.1.3) by including a "weight factor" $w(x)$, real and positive in the interval (a, b), in the integrand and find how (2.1.4) and (2.1.5) must be changed. Note that the functions $\{\phi_r\}$ are then "orthonormal with weight factor w" and that the set $\{w^{1/2}\phi_r\}$ is orthonormal in the usual sense.

2.3 Use the functions $\phi_n(x) = x^n$ $(n = 0, 1, 2, \ldots)$ to construct the first few members of a set of orthogonal polynomials, defined in the interval $(-1, +1)$, by means of the Schmidt process (p. 33). These functions are the Legendre polynomials $P_n(x)$. Show that the set $\{P_n(\cos\theta)\}$ is orthogonal in the interval $(0, \pi)$ with an appropriate weight factor $w(\theta)$.

2.4 Use again the functions x^n to construct (i) polynomials orthonormal in the interval $(0, \infty)$ with weight factor e^{-x} and (ii) polynomials orthonormal in $(-\infty, +\infty)$ with weight factor e^{-x^2}. These are the polynomials of Laguerre and Hermite respectively. Now define the Laguerre and Hermite *functions,* which are orthonormal in the usual sense but for the *infinite* intervals.

2.5 The wavefunctions for a "particle in a box", defined in $(0, L)$, are $\phi_n(x) = (2/L)^{1/2}\sin(n\pi x/L)$. Calculate n-term approximations to the function $f(x)$ that has the value $(2/L)^{1/2}$ in the interval $(0, \frac{1}{2}L)$ but is zero in $(\frac{1}{2}L, L)$, for $n = 2$, 6, 10, and verify (by plotting the results) that, while the approximation appears to converge *in the mean,* it does not do so at the point of discontinuity $x = \frac{1}{2}L$. Prove that as $D \to 0$ the sum of the squares of the expansion coefficients tends to 1, and give a vector-space interpretation of this result (cf. p. 29).

2.6 If e_1, e_2 are unit vectors along the x and y axes, a rotation R_ϕ through angle ϕ about the z axis produces images $\mathsf{e}_1' = \mathsf{R}_\phi\mathsf{e}_1$, $\mathsf{e}_2' = \mathsf{R}_\phi\mathsf{e}_2$. By expressing the images in terms of e_1, e_2, using the notation of (2.2.14), obtain the *matrix* $\mathbf{R}_\phi$ associated with the operator R_ϕ. Verify that when $\mathsf{R}_\phi = \mathsf{R}_{\phi_2}\mathsf{R}_{\phi_1}$ $(\phi = \phi_1 + \phi_2$ being the rotation produced by R_{ϕ_1} followed by $\mathsf{R}_{\phi_2})$, $\mathbf{R}_\phi = \mathbf{R}_{\phi_2}\mathbf{R}_{\phi_1}$, where the matrices are combined by matrix multiplication. Now read pp. 30–31; the matrices provide a 2-dimensional *representation* of the *group* consisting of all rotations around the z axis.

2.7 Instead of e_1, e_2 in Problem 2.6, use $\bar{\mathsf{e}}_1 = \mathsf{e}_1$, $\bar{\mathsf{e}}_2 = \frac{1}{2}\sqrt{3}\mathsf{e}_2 - \frac{1}{2}\mathsf{e}_1$. Obtain the matrix $\bar{\mathbf{R}}$ for the rotation $\phi = 60°$ and compare it with the original matrix $\mathbf{R}$. Write down the transformation matrix $\mathbf{T}$ and verify that $\bar{\mathbf{R}} = \mathbf{T}^{-1}\mathbf{R}\mathbf{T}$. [*Hint:* See (2.2.16) and (2.2.19); obtain $\mathbf{T}^{-1}$ by expressing e_1, e_2 in terms of $\bar{\mathsf{e}}_1$, $\bar{\mathsf{e}}_2$.)

2.8 Obtain the matrix $\mathbf{S}$ of scalar products for $\bar{\mathsf{e}}_1$, $\bar{\mathsf{e}}_2$ in Problem 2.7, with the usual definition $\mathsf{v}_1 \cdot \mathsf{v}_2 = v_1 v_2 \cos\theta_{12}$. Form the matrix $\bar{\mathbf{R}}^S$, with elements

$\bar{e}_i \cdot (R\bar{e}_j) = \langle \bar{e}_i | R | \bar{e}_j \rangle$, and verify the relationship (2.2.15). Note the special advantages of using an orthonormal basis.

2.9 Show that when (2.2.14) is adopted as the *definition* of the matrix $\mathbf{H}$ associated with a general operator H the representation property embodied in (2.2.11) and (2.2.12) is independent of whether or not the basis functions (or vectors) are orthonormal. [*Hint:* Apply B to the set $\boldsymbol{\phi}$, as in (2.2.14), and then apply A to $\boldsymbol{\phi}' = B\boldsymbol{\phi} = \boldsymbol{\phi}\mathbf{B}$. The final result must be equivalent to using the single operator AB (= C). Note that scalar products are nowhere involved.]

2.10 When using an independent-particle model (IPM), an MO ϕ is often expressed in LCAO form, i.e. as a linear combination of atomic orbitals ($\chi_1, \chi_2, \ldots, \chi_m$, say), and the expansion coefficients are determined from secular equations as in Section 2.3: the Hamiltonian is the *one*-electron operator h and the "orbital energy" ϵ. Find the MOs and orbital energies for a linear H_3 molecule, taking χ_1, χ_2, χ_3 to be 1s AOs on the three centres and including only nearest-neighbour matrix elements, with (Hückel, 1931)

$$\langle \chi_i | h | \chi_i \rangle = \alpha \quad (i = 1, 2, 3), \qquad \langle \chi_1 | h | \chi_2 \rangle = \langle \chi_2 | h | \chi_3 \rangle = \beta, \qquad \langle \chi_1 | h | \chi_3 \rangle = 0$$

and with neglect of overlap. [*Hint:* Divide all equations by β and put $(\alpha - \epsilon)/\beta = -x$; the parameters (both negative) are thus eliminated and x represents the energy in units of β relative to the "reference level" α.]

2.11 Reconsider Problem 2.10, introducing a new basis in which $\bar{\chi}_1 = (\chi_1 + \chi_3)/\sqrt{2}$, $\bar{\chi}_2 = \chi_2$, $\bar{\chi}_3 = (\chi_1 - \chi_3)/\sqrt{2}$. Show that the matrix $\mathbf{H}$ is transformed to "diagonal block form", the secular equations separating into two independent sets. Make a table to show how your basis functions and MOs behave under the symmetry operation of reflection across the plane bisecting the molecule. Express your results in matrix form, as in Problem 2.7. [*Hint:* Express the new matrix elements $\langle \bar{\chi}_i | h | \bar{\chi}_j \rangle$ in terms of α and β and formulate secular equations to determine the new mixing coefficients $\bar{c}_i$.]

2.12 The basis functions introduced in Problem 2.11 are said to be "symmetry-adapted" or to be "symmetry functions". Use a similar procedure to obtain MOs and orbital energies for (i) linear H_4 and (ii) square-planar H_4. [*Hint:* Orbitals that are exchanged (more generally permuted) by a symmetry operation are said to be "equivalent"; they can be combined to give symmetry functions. In case (ii) you may use *two* reflections, classifying the functions as $\left(\frac{+ | +}{+ | +}\right)$ $\left(\frac{+ | -}{+ | -}\right)$ etc.]

2.13 Show how the results of the last two problems may be generalized to (i) a straight chain or (ii) a ring of N equispaced hydrogen atoms, by direct solution of the secular equations without the use of symmetry. Obtain the AO coefficients $c_n^{(K)}$ in each MO ϕ_K for the chain with $N = 6$ and indicate the MO forms by drawing circles around the N centres with radii proportional to corresponding coefficients (with signs added); compare these with the "box" wavefunctions in Problem 2.5. Show that for the *ring* with $N = 6$ there are two pairs of degenerate MOs, which, by taking suitable linear combinations, may be written in real form. [*Hint:* The N secular equations have the common form $\beta c_{n-1} + \alpha c_n + \beta c_{n-1} = \epsilon c_n$

except when $n = 1$ or N. Try for a solution $c_n = e^{in\theta}$, relating ϵ to θ and noting a degeneracy, which suggests a general solution $c_n = Ae^{in\theta} + Be^{-in\theta}$. Eliminate the constants by using *boundary conditions* $c_0 = c_{N+1}$ (fictitious atoms!) for the chain, or $c_{N+n} = c_n$ (periodicity) for the ring, and by normalizing.]

2.14 Derive the "turnover rule" (p. 35) from the definition (2.3.5) of the adjoint operator. [*Hint:* See text, just before (2.3.6).]

2.15 Show that any real positive-definite matrix $\mathbf{M}$ can be written as $\mathbf{M} = \mathbf{U}^\dagger\mathbf{U}$, where $\mathbf{U}$ is real and *upper-triangular,* thus formalizing the Schmidt process (p. 33). Show also that in this case the calculation of the inverse matrix, needed in relating orthogonal and non-orthogonal bases, is particularly simple.

2.16 Show how the general matrix-eigenvalue equation (2.3.4) may be reduced to the simpler form $\bar{\mathbf{H}}\bar{\mathbf{c}} = E\bar{\mathbf{c}}$ by changing to a new basis $\bar{\mathbf{\Phi}} = \mathbf{\Phi}\mathbf{M}^{-1/2}$, this being the many-electron analogue of the orbital transformation (2.2.21). How is $\bar{\mathbf{H}}$ related to $\mathbf{H}$ and, for any given approximate eigenfunction $\Psi_K = \mathbf{\Phi}\mathbf{c}_K$, how are the new expansion coefficients $\bar{\mathbf{c}}_K$ related to the old $\mathbf{c}_K$?

2.17 Show that any $m \times m$ Hermitian matrix $\mathbf{H}$ can be written in terms of its eigenvalues and eigenvectors in the form $\mathbf{H} = \sum_K E_K\mathbf{c}_K\mathbf{c}_K^\dagger = \sum_K E_K\boldsymbol{\rho}_K$ and that the matrices $\boldsymbol{\rho}_K$ describe "projection operators" in the vector space spanned by the eigenvectors. [*Hint:* An arbitrary vector is expressible as a linear combination, $\mathbf{c} = \sum_K \alpha_K\mathbf{c}_K$. The effect of an operator is obtained on multiplying by its associated matrix.]

2.18 A function $f(\mathbf{M})$ of a matrix $\mathbf{M}$ may be defined as the matrix with the same eigenvectors $\mathbf{c}_K$ as $\mathbf{M}$ but with eigen*values* $f(m_K)$ instead of m_K. Give a procedure for obtaining such a matrix and relate your result to that in Problem 2.17.

2.19 A non-singular square matrix $\mathbf{A}$ may be partitioned into blocks, as in (2.5.1). Show that, with a similar partitioning, the inverse matrix is given by

$$\mathbf{A}^{-1} = \begin{pmatrix} \mathbf{A}_{11}^{-1}(\mathbf{1}_{11} + \mathbf{A}_{12}\mathbf{B}^{-1}\mathbf{A}_{21}\mathbf{A}_{11}^{-1}) & -\mathbf{A}_{11}^{-1}\mathbf{A}_{12}\mathbf{B}^{-1} \\ -\mathbf{B}^{-1}\mathbf{A}_{21}\mathbf{A}_{11}^{-1} & \mathbf{B}^{-1} \end{pmatrix},$$

where $\mathbf{B} = \mathbf{A}_{22} - \mathbf{A}_{21}\mathbf{A}_{11}^{-1}\mathbf{A}_{12}$. Use this result to devise a recursive method of calculating an inverse. [*Hint:* Assume the inverse of an $n \times n$ block to be known and add a row and a column; use the above result. Start from the 1×1 block.]

2.20 Prove the "separation theorem" (p. 38).

2.21 Use the separation theorem to show how the discussion of the ground and first excited states of the helium atom (p. 15) could be modified in order to obtain upper bounds on the energies, still using orbital products as expansion functions.

2.22 Use a variation function $\phi = e^{-\alpha r^2}$ to find an upper bound on the ground-state energy of the hydrogen atom. Show that for atomic number $Z \neq 1$ corresponding results are obtained on replacing r by Zr: what does this mean? Use a more general function $(1 + ar)e^{-\alpha r^2}$ to obtain energy-gradient and second-derivative tensors (pp. 43–44) and evaluate these quantities at the point where $a = 0$ and α has the value already found. Set up an equation of the form (2.4.21) to obtain better parameter values and calculate the improved energy. (The process may be repeated if desired.)

2.23 Take a linear combination $\exp(-\alpha_1 r^2) + c \exp(-\alpha_2 r^2)$ instead of the functions used in Problem 2.22. Start from the point $\alpha_1 = \alpha_2 = \alpha$ (value from Problem 2.22), $c = 0$ and find improved parameter values in a similar way. Look for a second stationary point to obtain a solution of 2s type. [*Hint:* If c is chosen to satisfy a 2×2 secular equation then the uppper root will be an upper bound to the 2s energy.]

2.24 Use an expansion in unperturbed AOs to formulate the Rayleigh–Schrödinger perturbation equations for a hydrogen atom in an electric field F along the z axis. Give a 2-term approximation to the perturbed 1s orbital, noting the general effect of the field, and estimate the electric polarizability α from the result $E = E_0 - \frac{1}{2}\alpha F^2$. [*Hint:* You will need the result $\langle 2\mathrm{p}_z | z | 1\mathrm{s} \rangle = 2^8/3^5\sqrt{2}\, a_0$.]

2.25 Reconsider Problem 2.24, but with a variation function $\phi = (1 + az)e^{-\zeta r}$. Formulate variational equations to determine *all* parameters (cf. Problem 2.22), noting that r really means $|\boldsymbol{r} - \boldsymbol{r}_0|$ ($\boldsymbol{r}_0$ being the position vector of the nucleus) and that the exponential function should be allowed to 'float' away from the nucleus. [*Hint:* Put $r = [x^2 + y^2 + (z - z_0)^2]^{1/2}$ and find the derivatives with respect to a, c, z_0, showing that they can all be expressed in terms of integrals over Slater-type orbitals (Appendix 1) with a variety of n, l, m quantum numbers.]

2.26 The first three electronic states Φ_1, Φ_2, Φ_3 of a given system have energies (in suitable units) $E_1 = E_2 = 0$, $E_3 = 10.0$. An applied perturbation is described by the matrix

$$\mathbf{H}' = \begin{pmatrix} 1.0 & 0.0 & 1.0 \\ 0.0 & 2.0 & 2.0 \\ 1.0 & 2.0 & 3.0 \end{pmatrix}.$$

Show how the splitting of the degenerate levels can be obtained by solving a 2×2 matrix-eigenvalue equation with an "effective" Hamiltonian; obtain the shifted levels to second order. How could you improve your results?

2.27 Generalize (2.5.14) and (2.5.15) to admit functions that are neither orthogonal nor normalized. (The results will be needed in Chapter 14.)

3 Many-Electron Wavefunctions

3.1 ANTISYMMETRY AND THE SLATER METHOD

In discussing the helium atom (Section 1.2) the antisymmetry requirement on the electronic wavefunction was easily satisfied; for with only two electrons the function would be written as a product of space and spin factors, one of which had to be antisymmetric, the other symmetric. This is possible even for an exact eigenfunction of the Hamiltonian (1.2.1), as well as for an orbital product. The construction of an antisymmetric *many*-electron function is less easy. We have seen in Section 1.2 that for a general permutation (involving both space and spin variables) an antisymmetric function has the property

$$\mathsf{P}\Psi = \epsilon_{\mathsf{P}}\Psi, \tag{3.1.1}$$

where $\epsilon_{\mathsf{P}} = \pm 1$ according as P is equivalent to an even or odd number of interchanges. For more than two electrons, however, it is not possible to find a single space–spin product with this property. In general, a large number of different products must be suitably combined to give a fully antisymmetrical wavefunction, and the choice of such a combination is a fairly advanced exercise in group theory. These difficulties are avoided in Slater's approach (Slater, 1929, 1931), which we consider first: at a later stage we examine the connection with group theoretical methods.

From the assumption of completeness of the set of all products of N spin-orbitals, each factor being drawn from an orthonormal complete set of spin-orbitals, we may write any wavefunction

$$\Psi(\boldsymbol{x}_1, \boldsymbol{x}_2, \ldots, \boldsymbol{x}_N) = \sum_{A,B,\ldots,X} c_{AB\ldots X}\psi_A(\boldsymbol{x}_1)\psi_B(\boldsymbol{x}_2)\ldots\psi_X(\boldsymbol{x}_N), \tag{3.1.2}$$

and the coefficients are then in principle determined (cf. (2.1.5)) by

$$c_{AB\ldots X} = \int \psi_A^*(\boldsymbol{x}_1)\psi_B^*(\boldsymbol{x}_1)\ldots\psi_X^*(\boldsymbol{x}_N) \times \Psi(\boldsymbol{x}_1, \boldsymbol{x}_2, \ldots, \boldsymbol{x}_N)\,\mathrm{d}\boldsymbol{x}_1\,\mathrm{d}\boldsymbol{x}_2\ldots\mathrm{d}\boldsymbol{x}_N. \tag{3.1.3}$$

Now consider the coefficient of the product in which the spin-orbitals have been selected in a different order, obtained by applying a permutation P to the labels. $c_{\mathrm{P}(AB\ldots X)}$ will be expressed as in (3.1.3) except for the order of the subscripts $A, B, \ldots, X$. But now let us permute the variables, *throughout* the integrand, in such a way that $\boldsymbol{x}_1$ comes back into ψ_A^*, $\boldsymbol{x}_2$ into ψ_B^*, etc., remembering that such a change has no effect on the integral because the names of the variables are arbitrary. The expression for $c_{\mathrm{P}(AB\ldots X)}$ will then differ from that for $c_{AB\ldots X}$ only by a permutation of variables in the Ψ factor, and since Ψ is antisymmetric this factor must, according to (3.1.1), become $\epsilon_{\mathrm{P}}\Psi$. Consequently

$$c_{\mathrm{P}(AB\ldots X)} = \epsilon_{\mathrm{P}} c_{AB\ldots X}.$$

The coefficients of all spin-orbital products involving the same *selection* of spin-orbitals (differing only in their order), can therefore differ only in sign. If we use $c_{AB\ldots X}$ for the coefficient of the products in which the spin-orbitals occur in a standard order (e.g. "dictionary" order) and abbreviate this particular ordered set to κ, it follows that

$$\Psi(\boldsymbol{x}_1, \boldsymbol{x}_2, \boldsymbol{x}_N) = \sum_{\kappa} c_{\kappa} \Phi_{\kappa}(\boldsymbol{x}_1, \boldsymbol{x}_2, \ldots, \boldsymbol{x}_N). \tag{3.1.4}$$

It will be convenient to include a normalizing factor M_k in the function

$$\Phi_{\kappa}(\boldsymbol{x}_1, \boldsymbol{x}_2, \ldots, \boldsymbol{x}_N) = M_{\kappa} \sum_{\mathrm{P}} \epsilon_{\mathrm{P}} \mathrm{P} \psi_A(\boldsymbol{x}_1)\psi_B(\boldsymbol{x}_2) \ldots \psi_X(\boldsymbol{x}_N), \tag{3.1.5}$$

which is an *antisymmetrized spin-orbital product.* The summation in (3.1.4) is now only over distinct ordered configurations of spin-orbitals, and with each such configuration is associated the antisymmetrized function (3.1.5), formed by permuting all the spin-orbitals, or equivalently all the variables, and summing with the appropriate + or − signs. This sum is, of course, the expansion of a (Slater) determinant

$$\Phi_{\kappa}(\boldsymbol{x}_1, \boldsymbol{x}_2, \ldots, \boldsymbol{x}_N) = M_{\kappa} \begin{vmatrix} \psi_A(\boldsymbol{x}_1) & \psi_B(\boldsymbol{x}_1) & \ldots & \psi_X(\boldsymbol{x}_1) \\ \psi_A(\boldsymbol{x}_2) & \psi_B(\boldsymbol{x}_2) & \ldots & \psi_X(\boldsymbol{x}_2) \\ \vdots & \vdots & & \vdots \\ \psi_A(\boldsymbol{x}_N) & \psi_B(\boldsymbol{x}_N) & \ldots & \psi_X(\boldsymbol{x}_N) \end{vmatrix}$$

$$= M_{\kappa} \det |\psi_A(\boldsymbol{x}_1)\ \psi_B(\boldsymbol{x}_2)\ \ldots\ \psi_X(\boldsymbol{x}_N)|, \tag{3.1.6}$$

where the abbreviated form displays only the elements on the diagonal.

It is interesting to note that the antisymmetric function (3.1.5) is, in effect, generated from the single spin-orbital product

$$\psi_A(\boldsymbol{x}_1)\psi_B(\boldsymbol{x}_2) \ldots \psi_X(\boldsymbol{x}_N)$$

by applying the operator $\sum_{\mathsf{P}} \epsilon_{\mathsf{P}}\mathsf{P}$, i.e. "make each permutation of the variables, add a coefficient ϵ_{P}, and sum the results". This operator could have been introduced from group theory (Appendix 3): for the permutations form a group and the numbers ± 1 form a one-dimensional representation; an antisymmetric function belongs to this representation, according to (3.1.1), and such a function may be generated by the prescription (A3.25). It is in fact useful to introduce the "antisymmetrizer"

$$\mathsf{A} = \frac{1}{N!} \sum_{\mathsf{P}} \epsilon_{\mathsf{P}} \mathsf{P}, \tag{3.1.7}$$

so normalized that operating on an already antisymmetric function it leaves the function completely unchanged. This operator, the simplest example of a *projection operator,* has the characteristic property of *idempotency* $\mathsf{AA}\Psi = \mathsf{A}\Psi$ or $\mathsf{A}^2 = \mathsf{A}$ and is useful in later sections.

The approximate wavefunctions for helium, obtained in Section 1.2, must evidently be expressible in terms of determinants. Thus, for the configuration $(1\text{s})^2$, the determinant $M_\kappa \det |1\text{s}(\boldsymbol{r}_1)\alpha(s_1)\ 1\text{s}(\boldsymbol{r}_2)\beta(s_2)|$ expands to give, on normalizing,

$$\Psi_1 = \frac{1}{\sqrt{2}} 1\text{s}(\boldsymbol{r}_1)\, 1\text{s}(\boldsymbol{r}_2)\, [\alpha(s_1)\beta(s_2) - \beta(s_1)\alpha(s_2)],$$

which coincides with the approximate ground-state function Ψ_1^{a} of Section 1.2. For the configuration 1s 2s, on the other hand, we may write down antisymmetrized products such as

$$\begin{aligned} &\det |1\text{s}(\boldsymbol{r}_1)\alpha(s_1)\ 2\text{s}(\boldsymbol{r}_2)\alpha(s_2)| \\ &\qquad = 1\text{s}(\boldsymbol{r}_1)\alpha(s_1)\, 2s(\boldsymbol{r}_2)\alpha(s_2) - 1\text{s}(\boldsymbol{r}_2)\alpha(s_2)\, 2\text{s}(\mathbf{r}_1)\alpha(s_1), \end{aligned}$$

and simple rearrangement then shows that the antisymmetric wavefunctions of the excited states, corresponding respectively to Ψ_2^{a}, Ψ_3^{a}, Ψ_4^{a}, Ψ_5^{a}, obtained in Section 1.2 may be written as ‡

$$\begin{aligned} \Psi_2 &= \frac{1}{\sqrt{2}} \det |1\text{s}\alpha\ \ 2\text{s}\alpha|, \\ \Psi_3 &= \tfrac{1}{2}(\det |1\text{s}\alpha\ \ 2\text{s}\beta| + \det |1\text{s}\beta\ \ 2\text{s}\alpha|), \\ \Psi_4 &= \frac{1}{\sqrt{2}} \det |1\text{s}\beta\ \ 2\text{s}\beta|, \\ \Psi_5 &= \tfrac{1}{2}(\det |1\text{s}\alpha\ \ 2\text{s}\beta| - \det |1\text{s}\beta\ \ 2\text{s}\alpha|). \end{aligned}$$

‡ For ease of writing, we no longer indicate the variables explicitly. Thus $\det |1\text{s}\alpha\ \ 2\text{s}\beta|$ stands for $\det |1\text{s}(\boldsymbol{r}_1)\alpha(s_1)\ \ 2\text{s}(\boldsymbol{r}_2)\beta(s_2)|$ etc. The left–right order of the variables will always be $1, 2, 3, \ldots$.

The *symmetric* functions are automatically rejected by using Slater determinants, and in this way the quantum-mechanical form of the Pauli principle is imposed at the outset. Pauli's original form of the principle applies only to a one-determinant approximation and follows immediately from the fact that a determinant such as (3.1.6) vanishes if any two spin-orbitals are identical (this leading to two identical columns); it is therefore impossible to find an antisymmetric wavefunction describing two electrons in the same spin-orbital in an IPM description.

The above considerations suggest that fairly good approximate wavefunctions, when written in terms of determinants based on an intelligent choice of orbitals, may yield expansions of the form (3.1.4) containing only a small number of terms—though an exact function would require an infinite number. We consider the theory of such expansions in detail in later sections and here simply recall that the coefficients c_κ that yield best approximations to the electronic states and their energies may always be determined, in principle, by solving the secular problem of Section 2.3, namely

$$\mathbf{Hc} = E\mathbf{Mc}, \tag{3.1.8}$$

where the matrices **H** and **M** have elements

$$H_{\kappa\lambda} = \langle \Phi_\kappa | \mathsf{H} | \Phi_\lambda \rangle, \qquad M_{\kappa\lambda} = \langle \Phi_\kappa \mid \Phi_\lambda \rangle, \tag{3.1.9}$$

and the functions Φ_κ, in the simplest form of the expansion, are Slater determinants. This is the essence of Slater's method.

3.2 CALCULATION OF THE ENERGY: HELIUM ATOM

To anticipate some of the problems involved in energy calculations, we use the helium atom functions already derived to obtain the actual energy levels. The 2-electron Hamiltonian (neglecting spin terms) is

$$\mathsf{H} = \mathsf{h}(1) + \mathsf{h}(2) + \frac{1}{r_{12}},$$

and we first assume that the functions Ψ_K $(K = 1, \ldots, 5)$ discussed in Section 3.1 are valid approximate wavefunctions for the first five states (of spherical symmetry). The corresponding energies may then be estimated as the expectation values $E_K = \langle \Psi_K | \mathsf{H} | \Psi_K \rangle$. It is of course unnecessary to employ normalized functions from the outset, for we may always replace Ψ_K by $M_K \Psi_K$, where M_K is a normalizing factor, and use $E_K = |M_K|^2 \langle \Psi_K | \mathsf{H} | \Psi_K \rangle$, which gives, on using the normalizing condition $\langle M_K\Psi_K \mid M_K\Psi_K\rangle = 1$ (cf. (2.4.1)),

$$E_K = \frac{\langle \Psi_K | \mathsf{H} | \Psi_K \rangle}{\langle \Psi_K \mid \Psi_K \rangle}, \tag{3.2.1}$$

and it remains to evaluate the quantities $H_{KK} = \langle \Psi_K | \mathsf{H} | \Psi_K \rangle$, $M_{KK} = \langle \Psi_K \mid \Psi_K \rangle$. The energies estimated in this way will not in general be *upper bounds* on the corresponding exact energies. To get upper bounds, we should regard $\Psi_1, \ldots, \Psi_5$ as expansion functions (like the Φs in Section 2.3), from which better approximations to the actual state functions may be constructed by linear combination: the mixing coefficients and *upper-bound* energies would then follow from solution of the secular equations (3.1.8). We note that this choice of expansion functions corresponds to using linear combinations of determinants for the Φs in (3.1.4) instead of the individual determinants—a point to which we return in Section 3.3. It is clear from (2.3.13) that the estimates in (3.2.1) would arise on neglecting off-diagonal elements $H_{KL} - EM_{KL}$; for in that case the roots would be approximated by the values of E that would give zero for any one of the diagonal elements $H_{KK} - EM_{KK}$, yielding the result (3.2.1). To obtain upper bounds to the energies, we therefore need to evaluate both diagonal and off-diagonal terms:

$$\left.\begin{aligned} H_{KL} &= \langle \Psi_K | \mathsf{H} | \Psi_L \rangle = \int \Psi_K^*(\boldsymbol{x}_1, \boldsymbol{x}_2) \mathsf{H} \Psi_L(\boldsymbol{x}_1, \boldsymbol{x}_2)\, d\boldsymbol{x}_1\, d\boldsymbol{x}_2, \\ M_{KL} &= \langle \Psi_K \mid \Psi_L \rangle = \int \Psi_K^*(\boldsymbol{x}_1, \boldsymbol{x}_2) \Psi_L(\boldsymbol{x}_1, \boldsymbol{x}_2)\, d\boldsymbol{x}_1\, d\boldsymbol{x}_2. \end{aligned}\right\} \tag{3.2.2}$$

Before proceeding, we note that functions with different spin factors, when written in the factorized form used in Section 1.2, must give $H_{KL} = M_{KL} = 0$. For example,

$$\begin{aligned} H_{12} = \tfrac{1}{2} \int & 1s^*(\boldsymbol{r}_1)\, 1s^*(\boldsymbol{r}_2)\, [\alpha(s_1)\beta(s_2) - \beta(s_1)\alpha(s_2)]^* \mathsf{H} \\ & \times [1s(\boldsymbol{r}_1)\, 2s(\boldsymbol{r}_2) - 2s(\boldsymbol{r}_1)\, 1s(\boldsymbol{r}_2)][\alpha(s_1)\alpha(s_2)]\, d\boldsymbol{r}_1\, ds_1\, d\boldsymbol{r}_2\, ds_2 \end{aligned}$$

and, since H does not operate on spins, the spin integrations may be performed immediately, yielding a factor

$$\int [\alpha(s_1)\beta(s_2) - \beta(s_1)\alpha(s_2)]^* \alpha(s_1)\alpha(s_2)\, ds_1\, ds_2.$$

But this will vanish, owing to orthogonality of the spin functions (1.2.23), and hence $H_{12} = 0$. The only non-zero off-diagonal element is in fact H_{15}, which we evaluate for illustration. The spin factors in Ψ_1 and Ψ_5 are both

$[\alpha(s_1)\beta(s_2) - \beta(s_1)\alpha(s_2)]/\sqrt{2}$, and spin integration therefore yields

$$\tfrac{1}{2}\int [\alpha(s_1)\beta(s_2) - \beta(s_1)\alpha(s_2)]^*[\alpha(s_1)\beta(s_2) - \beta(s_1)\alpha(s_2)]\, ds_1\, ds_2.$$

If we preserve the order of the subscripts 1, 2 throughout, this may be expanded and written in the abbreviated form

$$\tfrac{1}{2}(\langle \alpha\beta \mid \alpha\beta\rangle - \langle \alpha\beta \mid \beta\alpha\rangle - \langle \beta\alpha \mid \alpha\beta\rangle + \langle \beta\alpha \mid \beta\alpha\rangle).$$

The first and last terms are each unity, the others zero, and spin integration therefore gives only a factor unity, verifying the normalization of the spin function. Thus

$$H_{15} = \frac{1}{\sqrt{2}}\int 1s^*(\boldsymbol{r}_1)\, 1s^*(\boldsymbol{r}_2)\, \mathsf{H}[1s(\boldsymbol{r}_1)\, 2s(\boldsymbol{r}_2) + 2s(\boldsymbol{r}_1)\, 1s(\boldsymbol{r}_2)]\, d\boldsymbol{r}_1\, d\boldsymbol{r}_2.$$

This reduces easily, the three terms in H giving contributions (respectively)

(i) $\langle 1s|\, \mathsf{h}\, |1s\rangle\langle 1s \mid 2s\rangle + \langle 1s|\, \mathsf{h}\, |2s\rangle\langle 1s \mid 1s\rangle$,
(ii) $\langle 1s \mid 1s\rangle\langle 1s|\, \mathsf{h}\, |2s\rangle + \langle 1s \mid 2s\rangle\langle 1s|\, \mathsf{h}\, |1s\rangle$,
(iii) $\langle 1s\, 1s|\, g\, |1s\, 2s\rangle$,

where, for example, remembering that the names of integration variables are immaterial

$$\langle 1s|\, \mathsf{h}\, |2s\rangle = \int 1s^*(\boldsymbol{r}_1)\, \mathsf{h}(1)\, 2s(\boldsymbol{r}_1)\, d\boldsymbol{r}_1 = \int 1s^*(\boldsymbol{r}_2)\, \mathsf{h}(2)\, 2s(\boldsymbol{r}_2)\, d\boldsymbol{r}_2.$$

The 2-electron integral cannot be factorized. When the AOs are assumed orthogonal, the terms listed above together give

$$H_{15} = [2\langle 1s|\, \mathsf{h}\, |2s\rangle + \langle 1s\, 1s|\, g\, |1s\, 2s\rangle]/\sqrt{2}. \tag{3.2.3}$$

It may be verified, in a similar way, that the diagonal matrix elements are

$$\left.\begin{aligned} H_{11} &= 2\langle 1s|\, \mathsf{h}\, |1s\rangle + \langle 1s\, 1s|\, g\, |1s\, 1s\rangle, \\ H_{55} &= \langle 1s|\, \mathsf{h}\, |1s\rangle + \langle 2s|\, \mathsf{h}\, |2s\rangle + \langle 1s\, 2s|\, g\, |1s\, 2s\rangle \\ &\quad + \langle 1s\, 2s|\, g\, |2s\, 1s\rangle, \\ H_{22} = H_{33} = H_{44} &= \langle 1s|\, \mathsf{h}\, |1s\rangle + \langle 2s|\, \mathsf{h}\, |2s\rangle \\ &\quad + \langle 1s\, 2s|\, g\, |1s\, 2s\rangle - \langle 1s\, 2s|\, g\, |2s\, 1s\rangle, \end{aligned}\right\} \tag{3.2.4}$$

and that all others are zero, while $M_{11} = M_{22} = \ldots = M_{55} = 1$. The actual integrals are easily evaluated and first approximations to the energies follow from (3.2.1). In terms of one-electron energies $\epsilon_{1s} = \langle 1s|\, \mathsf{h}\, |1s\rangle$

and $\epsilon_{2s} = \langle 2s|\, \mathsf{h}\, |2s\rangle$, we obtain in particular

$$E_{2,5} = \epsilon_{1s} + \epsilon_{2s} + J \mp K, \tag{3.2.5}$$

where

$$\left.\begin{aligned} J &= \langle 1s\,2s|\, g\, |1s\,2s\rangle = \int 1s^*(\boldsymbol{r}_1)\, 1s(\boldsymbol{r}_1)\left(\frac{1}{r_{12}}\right) 2s^*(\boldsymbol{r}_2)\, 2s(\boldsymbol{r}_2)\, d\boldsymbol{r}_1\, d\boldsymbol{r}_2, \\ K &= \langle 1s\,2s|\, g\, |2s\,1s\rangle = \int 1s^*(\boldsymbol{r}_1)\, 2s(\boldsymbol{r}_1)\left(\frac{1}{r_{12}}\right) 2s^*(\boldsymbol{r}_2)\, 1s(\boldsymbol{r}_2)\, d\boldsymbol{r}_1\, d\boldsymbol{r}_2 \end{aligned}\right\} \tag{3.2.6}$$

are both essentially *positive*, representing electrostatic repulsion energies between two smeared-out charge distributions—with densities $1s^2$, $2s^2$ in J, but both with density 1s 2s in K. The order of the levels in Fig. 1.1 in Chapter 1 (singlet above triplet) thus appears to be essentially correct. More rigorously, the non-zero H_{15} induces a slight mixing between Ψ_1 and Ψ_5, and the energies are not $E_1 = H_{11}$, $E_5 = H_{55}$ but appear instead as the roots of the 2×2 block of the secular determinant, namely

$$\begin{vmatrix} H_{11} - E & H_{15} \\ H_{51} & H_{55} - E \end{vmatrix} = 0. \tag{3.2.7}$$

The effect of including H_{15} is to slightly increase the separation of E_1 and E_5, the former being depressed below H_{11} and the latter raised above H_{55}, but the effect is much too small to change the order of the levels.

The analysis of a many-electron problem using direct expansion of matrix elements would obviously be excessively tedious; the calculation must be systematized by deriving general rules for the reduction to 1- and 2-electron integrals. In Slater's approach the wavefunction is written in terms of determinants and the general matrix element is evaluated for any pair of determinants.

3.3 MATRIX ELEMENTS BETWEEN ANTISYMMETRIZED PRODUCTS

We consider first a single determinant, written according to (3.1.5) as an antisymmetrized product

$$\Phi(\boldsymbol{x}_1, \boldsymbol{x}_2, \ldots, \boldsymbol{x}_N) = M \sum_{\mathsf{P}} \epsilon_{\mathsf{P}} \mathsf{P} \psi_A(\boldsymbol{x}_1) \psi_B(\boldsymbol{x}_2) \ldots \psi_X(\boldsymbol{x}_N) \tag{3.3.1}$$

and assume orthonormal spin orbitals $\psi_A, \psi_B, \ldots, \psi_X$. Because the demonstration is relevant later, we indicate briefly how the matrix elements are found; further details may be found in most books on quantum mechanics. First let us consider the normalization integral $\langle \Phi | \Phi \rangle$. It is clear that when expanded Φ consists of $N!$ products

$$\Phi = M[\psi_A(\boldsymbol{x}_1)\psi_B(\boldsymbol{x}_2) \ldots \psi_X(\boldsymbol{x}_N) - \psi_A(\boldsymbol{x}_2)\psi_B(\boldsymbol{x}_1) \ldots \psi_X(\boldsymbol{x}_N) + \ldots \text{etc.}].$$

The integrand $\Phi^*\Phi$ will thus contain $(N!)^2$ products, each of the form

$$\pm [\psi_A(\boldsymbol{x}_i)\psi_B(\boldsymbol{x}_j) \ldots, \psi_X(\boldsymbol{x}_k)][\psi_A(\boldsymbol{x}_{i'})\psi_B(\boldsymbol{x}_{j'}) \ldots \psi_X(\boldsymbol{x}_{k'})]$$

where $\boldsymbol{x}_i, \boldsymbol{x}_j, \boldsymbol{x}_k, \ldots$ and $\boldsymbol{x}_{i'}, \boldsymbol{x}_{j'}, \boldsymbol{x}_{k'}, \ldots$ indicate the coordinates of electrons 1, 2, 3, ..., N in some order, not necessarily the same in both halves of the product. It is evident, however, that unless $i = i'$, $j = j'$, etc. in a particular product, then on performing the integration, the product will give no contribution to the result, because of spin-orbital orthogonality. Thus there can only be $N!$ non-vanishing contributions, and each contribution, with normalized spin-orbitals, must be 1, giving

$$\langle \Phi | \Phi \rangle = M^2 N!. \tag{3.3.2}$$

The normalizing factor M therefore has the value $(N!)^{-1/2}$, and in future we may write, instead of (3.3.1),

$$\Phi(\boldsymbol{x}_1, \boldsymbol{x}_2, \ldots, \boldsymbol{x}_N) = \left(\frac{1}{N!}\right)^{1/2} \sum_{\mathsf{P}} \epsilon_{\mathsf{P}} \mathsf{P} \psi_A(\boldsymbol{x}_1)\psi_B(\boldsymbol{x}_2) \ldots \psi_X(\boldsymbol{x}_N). \tag{3.3.3}$$

The expectation value of the 1-electron part of the Hamiltonian, namely $\sum_{i=1}^{N} \mathsf{h}(i)$, will reduce to the sum of N identical terms, since the coordinates of each electron appear symmetrically in the corresponding integrand, and by the same kind of arguments used above it is easy to show that

$$\begin{aligned}\left\langle \Phi \middle| \sum_i \mathsf{h}(i) \middle| \Phi \right\rangle &= \frac{1}{N!} N(N-1)!(\langle \psi_A | \mathsf{h} | \psi_A \rangle + \langle \psi_B | \mathsf{h} | \psi_B \rangle + \ldots) \\ &= \sum_R \langle \psi_R | \mathsf{h} | \psi_R \rangle \end{aligned} \tag{3.3.4}$$

The expectation value of $\sum'^{N}_{i,j=1} g(i, j)$ involves slightly more subtle considerations, since contributions may arise from terms which differ by an interchange of two electrons; but an essentially similar argument shows that

$$\left\langle \Phi \middle| \sum_{i,j}{}' g(i, j) \middle| \Phi \right\rangle = \sum_{R,S}{}' (\langle \psi_R \psi_S | g | \psi_R \psi_S \rangle - \langle \psi_R \psi_S | g | \psi_S \psi_R \rangle), \tag{3.3.5}$$

where, in general

$$\langle \psi_R \psi_S | g | \psi_T \psi_U \rangle = \int \psi_R^*(\mathbf{x}_1)\psi_S^*(\mathbf{x}_2)\left(\frac{1}{r_{12}}\right)\psi_T(\mathbf{x}_1)\psi_U(\mathbf{x}_2)\, d\mathbf{x}_1\, d\mathbf{x}_2. \quad (3.3.6)$$

Thus the expectation value of the energy (i.e. the diagonal matrix element of H) with a normalized determinant (3.3.3) is

$$\langle \Phi | \mathsf{H} | \Phi \rangle = \sum_R \langle \psi_R | \mathsf{h} | \psi_R \rangle + \tfrac{1}{2} \sum_{R,S} (\langle \psi_R \psi_S | g | \psi_R \psi_S \rangle - \langle \psi_R \psi_S | g | \psi_S \psi_R \rangle), \quad (3.3.7a)$$

where the prime in the second sum is omitted because the term for $R = S$ vanishes identically. The first term in the double sum is often called the *coulomb integral*, the second the *exchange integral*.

We shall also be concerned with off-diagonal matrix elements, between pairs such as

$$\Phi(\mathbf{x}_1, \mathbf{x}_2, \ldots, \mathbf{x}_N) = (N!)^{-1/2} \sum_{\mathsf{P}} \epsilon_{\mathsf{P}} \mathsf{P} \psi_A(\mathbf{x}_1)\psi_B(\mathbf{x}_2) \ldots \psi_X(\mathbf{x}_N), \quad (3.3.8a)$$

$$\Phi'(\mathbf{x}_1, \mathbf{x}_2, \ldots, \mathbf{x}_N) = (N!)^{-1/2} \sum_{\mathsf{P}} \epsilon_{\mathsf{P}} \mathsf{P} \psi'_A(\mathbf{x}_1)\psi'_B(\mathbf{x}_1) \ldots \psi'_X(\mathbf{x}_N) \quad (3.3.8b)$$

in which the primed orbitals may differ from the unprimed. Fortunately, there are non-vanishing results in only two cases, which we list below for reference:

one spin-orbital difference $(\psi'_R \neq \psi_R)$

$$\langle \Phi' | \mathsf{H} | \Phi \rangle = \langle \psi'_R | h | \psi_R \rangle + \sum_{S(\neq R)} (\langle \psi'_R \psi_S | g | \psi_R \psi_S \rangle - \langle \psi'_R \psi_S | g | \psi_S \psi_R \rangle); \quad (3.3.7b)$$

two spin-orbital differences $(\psi'_R \neq \psi_R,\ \psi'_S \neq \psi_S)$

$$\langle \Phi' | \mathsf{H} | \Phi \rangle = \langle \psi'_R \psi'_S | g | \psi_R \psi_S \rangle - \langle \psi'_R \psi'_S | g | \psi_S \psi_R \rangle. \quad (3.3.7c)$$

The results (3.3.7) are called "Slater's rules" after their original formulator (Slater, 1929, 1931; see also Condon, 1930).

The assumption that the spin-orbitals are orthonormal is central to the analysis indicated above. If that assumption had not been made, we should have had to consider all the possible $(N!)^2$ terms in the expansion of a matrix element. It is, however, possible to establish the general result even for non-orthogonal spin-orbitals. The results are still formally quite simple and were first obtained by Löwdin (1955).‡

‡ The rest of this section is not essential for continuity and may be omitted until required.

We consider a pair of determinants of spin-orbitals and, for convenience of exposition, use the temporaty notation

$$\Phi_a(\boldsymbol{x}_1, \boldsymbol{x}_2, \ldots, \boldsymbol{x}_N) = M_a(N!)^{-1/2} \det |a_1(\boldsymbol{x}_1)\ a_2(\boldsymbol{x}_2)\ \ldots\ a_N(\boldsymbol{x}_N)|, \quad (3.3.9a)$$

$$\Phi_b(\boldsymbol{x}_1, \boldsymbol{x}_2, \ldots, \boldsymbol{x}_N) = M_b(N!)^{-1/2} \det |b_1(\boldsymbol{x}_1)\ b_2(\boldsymbol{x}_2)\ \ldots\ b_N(\boldsymbol{x}_N)|, \quad (3.3.9b)$$

in which a typical pair of spin-orbitals has an overlap integral $\langle a_i \mid b_j \rangle$. The integral $\langle \Phi_a \mid \Phi_b \rangle$ then reduces to $N!$ identical terms, cancelling the factor $1/N!$, and thus to

$$\langle \Phi_a \mid \Phi_b \rangle = M_a M_b \int a_1^*(\boldsymbol{x}_1) a_2^*(\boldsymbol{x}_2) \ldots a_N^*(\boldsymbol{x}_N)$$
$$\times \sum_{\mathsf{P}} \epsilon_{\mathsf{P}} \mathsf{P} b_1(\boldsymbol{x}_1) b_2(\boldsymbol{x}_2) \ldots b_N(\boldsymbol{x}_N)\, \mathrm{d}\boldsymbol{x}_1\, \mathrm{d}\boldsymbol{x}_2 \ldots \mathrm{d}\boldsymbol{x}_N.$$

This expression consists of $N!$ terms, each one giving on integration a result

$$\pm \langle a_1 \mid b_i \rangle \langle a_2 \mid b_j \rangle \ldots \langle a_N \mid b_p \rangle,$$

where $i, j, \ldots, p$ label the spin-orbitals into which $\boldsymbol{x}_1, \boldsymbol{x}_2, \ldots \boldsymbol{x}_N$ have been sent by the permutation P and the $\pm$ sign is the ϵ_{P} of this permutation.‡ Clearly the $N!$ terms comprise the expansion of a determinant, which we shall call D_{ab}. Thus $\langle \Phi_a \mid \Phi_b \rangle = M_a M_b D_{ab}$, where

$$D_{ab} = \det |\langle a_1 \mid b_1 \rangle\ \langle a_2 \mid b_2 \rangle\ \ldots\ \langle a_N \mid b_N \rangle|. \quad (3.3.10)$$

To normalize the functions (case $a = b$), we must therefore choose $M_a = D_{aa}^{-1/2}$, $M_b = D_{bb}^{-1/2}$, and obtain

$$\langle \Phi_a \mid \Phi_b \rangle = (D_{aa} D_{bb})^{-1/2} D_{ab}. \quad (3.3.11)$$

The matrix elements of 1- and 2-electron terms may be evaluated by noting that each term of $\sum_{i=1}^N \mathsf{h}(i)$ gives an identical contribution to the matrix element $\langle \Phi_a | \sum_i \mathsf{h}(i) | \Phi_b \rangle$ and each term of $\sum'^{N}_{i,j=1} g(i, j)$ gives an identical contribution to $\langle \Phi_a | \sum'_{i,j} g(i, j) | \Phi_b \rangle$ (the product $\Phi_a^* \Phi_b$ being symmetrical in electron labels). Thus

$$\left\langle \Phi_a \middle| \sum_i \mathsf{h}(i) \middle| \Phi_b \right\rangle = N \langle \Phi_a | \mathsf{h}(1) | \Phi_b \rangle, \quad (3.3.12)$$

$$\left\langle \Phi_a \middle| \sum_{i,j}{}' g(i, j) \middle| \Phi_b \right\rangle = N(N-1) \langle \Phi_a | g(1, 2) | \Phi_b \rangle. \quad (3.3.13)$$

‡ Strictly, the ϵ_{P} of the *inverse* permutation P^{-1}, since a permutation of variables is equivalent to the inverse permutation of orbital labels; but P and P^{-1} have the same parity (both even or both odd), so the distinction is not important here.

To perform the integrations, we pick out the terms involving $\boldsymbol{x}_1$ and $\boldsymbol{x}_2$: thus

$$\Phi_a(\boldsymbol{x}_1, \boldsymbol{x}_2, \ldots, \boldsymbol{x}_N) = (N!D_{aa})^{-1/2} \sum_{i=1}^{N} a_i(\boldsymbol{x}_1)(-1)^{1+i} \times \det |a_1(\boldsymbol{x}_i)\ a_2(\boldsymbol{x}_2)\ \ldots\ a_N(\boldsymbol{x}_N)|_{(i)}, \tag{3.3.14}$$

where the determinant on the right is the sum of $(N-1)!$ permutations of the variables $\boldsymbol{x}_2, \ldots, \boldsymbol{x}_N$ among the $N-1$ spin-orbitals that remain when $\boldsymbol{x}_1$ is exchanged with $\boldsymbol{x}_i$ and the factor $a_i(\boldsymbol{x}_1)$ is withdrawn. This determinant is a *minor* of that which appears in (3.3.9a), obtained by striking out the row and column containing $a_i(\boldsymbol{x}_1)$, while the minor with the appropriate sign $(-1)^{1+i}$ is the *cofactor* of $a_i(\boldsymbol{x}_1)$. Similarly, we may extract the factors containing both variables, $\boldsymbol{x}_1$ and $\boldsymbol{x}_2$, grouping together the remaining products into minors obtained by striking out *two* rows and columns of the original determinant:

$$\Phi_a(\boldsymbol{x}_1, \boldsymbol{x}_2, \ldots, \boldsymbol{x}_N) = (N!D_{aa})^{-1/2} \sum_{i,k=1}^{N} \epsilon_{ik} a_i(\boldsymbol{x}_1) a_k(\boldsymbol{x}_2)(-1)^{1+i+k} \times \det |a_1(\boldsymbol{x}_i)\ a_2(\boldsymbol{x}_k)\ a_3(\boldsymbol{x}_3)\ \ldots\ a_N(\boldsymbol{x}_N)|_{(ik)}, \tag{3.3.15}$$

where $\epsilon_{ik} = +1$ for i, k in the same order as in the original determinant, -1 for i, k reversed, or

$$\epsilon_{ik} = \begin{cases} +1 & (i<k), \\ -1 & (i>k), \end{cases}$$

and the subscript (ik) indicates the appropriate minor. The sum may also be written in the ordered form $(i<k)$ as

$$\Phi_a(\boldsymbol{x}_1, \boldsymbol{x}_2, \ldots, \boldsymbol{x}_N) = (N!D_{aa})^{-1/2} \sum_{i<k}^{N} [a_i(\boldsymbol{x}_1) a_k(\boldsymbol{x}_2) - a_k(\boldsymbol{x}_1) a_i(\boldsymbol{x}_2)] \times (-1)^{1+i+k} \det |a_1(\boldsymbol{x}_i)\ a_2(\boldsymbol{x}_k)\ a_3(\boldsymbol{x}_3)\ \ldots a_N(\boldsymbol{x}_N)|_{(ik)}. \tag{3.3.16}$$

On inserting (3.3.14) and a similar expression for Φ_b into (3.3.12), we obtain a sum of terms; in each term the integrand factorizes into a part referring to $\boldsymbol{x}_1$ and a part referring to $\boldsymbol{x}_2, \boldsymbol{x}_3, \ldots, \boldsymbol{x}_N$; the latter yields an overlap integral between $(N-1)$-electron functions and may be treated in the same way as $\langle \Phi_a \mid \Phi_b \rangle$, while the $\boldsymbol{x}_1$ factor gives a one-electron integral $\langle a_i|\,\mathsf{h}\,|b_j\rangle$. In this way we obtain

$$\left\langle \Phi_a \middle| \sum_i \mathsf{h}(i) \middle| \Phi_b \right\rangle = (D_{aa}D_{bb})^{-1/2} \sum_{i,j=1}^{N} \langle a_i|\,\mathsf{h}\,|b_j\rangle\, D_{ab}(a_ib_j), \quad (3.3.17)$$

where $D_{ab}(a_ib_j)$ denotes the cofactor of the element $\langle a_i \mid b_j\rangle$ in the determinant D_{ab} of (3.3.10), i.e. $(-1)^{i+j}$ times the minor obtained by removing the row and column that contains it. In a similar way, putting (3.3.15) and a similar expression for Φ_b into (3.3.13), we may derive

$$\left\langle \Phi_a \middle| \sum_{i,j}{}' g(i,j) \middle| \Phi_b \right\rangle = (D_{aa}D_{bb})^{-1/2} \sum_{i,k=1}^{N}{}' \sum_{j,l=1}^{N}{}' \epsilon_{ik}\epsilon_{jl} \times \langle a_ia_k|\, g\, |b_jb_l\rangle\, D_{ab}(a_ia_kb_jb_l), \quad (3.3.18)$$

where $D_{ab}(a_ia_kb_jb_l)$ is the cofactor of D_{ab} defined by deleting the rows and columns containing $\langle a_i \mid b_j\rangle$, $\langle a_k \mid b_l\rangle$ and attaching a factor $(-1)^{i+j+k+l}$ to the resultant minor. The ϵs have the same significance as in (3.3.15).

Although expressions (3.3.17) and (3.3.18) are quite general, they are awkward to use because a minor has to be evaluated for every term on the right-hand side. Fortunately, this can be avoided whenever the matrix of overlap integrals is non-singular. Thus, if $\mathbf{S}_{ab}$ is the matrix whose determinant appears in (3.3.10), and $\mathbf{S}_{ab}^{-1}$ is its inverse, then Jacobi's ratio theorem (Archbold, 1961) shows that

$$\left.\begin{aligned} D_{ab}(a_ib_j) &= (\mathbf{S}_{ab}^{-1})_{ji}D_{ab}, \\ \epsilon_{ik}\epsilon_{jl}D_{ab}(a_ia_kb_jb_l) &= [(\mathbf{S}_{ab}^{-1})_{ji}(\mathbf{S}_{ab}^{-1})_{lk} - (\mathbf{S}_{ab}^{-1})_{li}(\mathbf{S}_{ab}^{-1})_{jk}]D_{ab}. \end{aligned}\right\} \quad (3.3.19)$$

All the necessary coefficients are then determined by a single matrix inversion.

3.4 CONFIGURATION INTERACTION

Once we have the expressions for matrix elements of the Hamiltonian operator H between determinants $\Phi_1, \Phi_2, \ldots, \Phi_\kappa, \ldots$, it is, in principle, straightforward to use the expansion (3.1.4) to obtain a wavefunction approximation. Since each term is defined by stating a "configuration" of occupied spin-orbitals, this expansion procedure is often referred to as the method of "configuration interaction" (CI). The term "configuration", however, is frequently used with a variety of more specialized meanings, and to avoid ambiguity we review the terminology to be used. A *spin-orbital configuration* is a complete specification of all the occupied spin-orbitals defining one antisymmetrized product (determinant). An *orbital configuration,* on the other hand, is defined by a statement of the occupied orbitals—to which spin factors may be added in many different ways. A considerable number of determinants, differing in spin allocations, may therefore "belong" to one orbital configuration; with a spinless Hamiltonian, all these are degenerate in energy. Finally, the term *configuration,* without qualification, was introduced in atomic theory by Condon and Shortley (1935) to mean a specification of *types* of occupied orbitals, each type including all the members of a degenerate set of symmetry orbitals (Appendix 3) with either choice of spin factor. All the functions belonging to such a configuration are degenerate in energy for an IPM Hamiltonian (cf. Section 1.2) in which both spin and electron interactions are omitted. For example, we might speak of the configuration $1s^2\,2s^2\,2p^4$, implying only that (apart from the filled 1s and 2s orbitals) there are four 2p electrons, each orbital having any one of the values $m = 0, \ \pm 1$ and taking either spin factor. The different determinants compatible with this specification may then be mixed, essentially as in Section 3.2, to yield a number of states whose degeneracy is wholly or partially resolved when electron interaction is admitted. A complete account of this procedure, which leans heavily on the theory of angular momentum, is available in the standard works on atomic structure (Condon and Shortley, 1935; Eyring *et al.*, 1944, Chap. 9; Slater, 1960).

Here we are not concerned with the special reductions that apply to free atoms, and emphasize those general features of the theory that are equally applicable to systems of lower symmetry. We note that in molecular theory the term "CI calculation" is often applied indiscriminately to describe the mixing of *spin-orbital* configurations *within* a given configuration or orbital configuration; this is not, of course, CI in the more precise sense of Condon and Shortley.

It is clear that the expansion (3.1.4) will generally contain a very large number of terms, even when we admit Φs formed from a relatively small orbital basis. For example, if we consider a 5-electron system and use 10 basis orbitals then each spin-orbital configuration is defined by selecting 5 out of 20 spin-orbitals, and the full expansion will consequently contain 20!/15!5! or 15 564 terms. Even in this simple case, solution of the secular equations (3.1.8) requires a large computer and special techniques; and the dimensions of the problem rise astronomically with further increase in the numbers of electrons and basis orbitals.

Fortunately, however, it may turn out that many of the Φs my have symmetry properties different from that of the state being approximated, and according to a general theorem (A3.22) may be discarded; and it may also turn out that many orbital configurations are of negligible importance (their inclusion giving no significant improvement in the energy) and may thus be omitted at the outset. The considerations in Section 3.2 support this conjecture, particularly for the ground state which appears to be well represented even by a one-term expansion (a single determinant).

First we consider the effect of symmetry, in particular of the "spin symmetry" that is present even for a molecule with no spatial symmetry. If we assume in first approximation a spin-free Hamiltonian, it is clear that H and the total spin operators

$$\mathsf{S}^2 = \mathsf{S}_x^2 + \mathsf{S}_y^2 + \mathsf{S}_z^2, \tag{3.4.1}$$

$$\mathsf{S}_z = \sum_{i=1}^{N} \mathsf{S}_z(i) \tag{3.4.2}$$

(S_x, S_y being sums analogous to S_z) will commute, and a well-known theorem (Eyring *et al.*, 1944, p. 37) then states that the matrix elements in (3.1.9) will vanish if Φ_κ and Φ_λ are spin eigenfunctions corresponding to different spin eigenvalues (S and/or M).‡ This result is a consequence of the "symmetry" of the Hamiltonian, which is invariant under rotation of the axis of spin quantization, and whose eigenfunctions therefore have characteristic transformation properties under rotations (cf. Appendix 3, p. 539). It suggests immediately that if any individual Φ is *not* a spin eigenfunction then we should set up new linear combinations that *are* spin eigenfunctions, in order to eliminate as many matrix elements as possible. If we use the new basis functions, grouping together those with identical pairs of spin eigenvalues, then the matrices **H** and **M** will assume block form, and the secular equations will break into separate sets, one for each choice of S, M. Consequently, in looking for a state of given (S, M) it is necessary to consider only an expansion over Φs of the same (S, M). Moreover, with a spinless Hamiltonian, every state of given S is $(2S+1)$-fold degenerate (i.e. the individual terms of a spin multiplet are not resolved) and it is therefore necessary to consider only one, mathematically convenient, value of M (e.g. $M = S$). The number of terms in (3.1.4) corresponding to any given choice of orbital configuration may thus be very greatly reduced, at the price of using specific *combinations* of determinants for the Φs instead of individual determinants. Such combinations were in fact very effectively used in the example of Section 3.2. The general problem of constructing the spin eigenfunctions is taken up in Section 4.2.

‡ The eigenvalues of S^2 and S_z are $S(S+1)$ and M respectively, provided that we use the dimensionless operators of Section 1.2. See also Appendix 2.

Spatial symmetry may be utilized in a similar way. According to the theorems in Appendix 3 (p. 541), the expansion of any wavefunction Ψ of given symmetry species contains only symmetry functions of the *same* species. The situation is precisely analogous to that which arises in the case of spin; for the eigenvalues (S, M) are in fact the labels that define the different basis functions $(M = S,\ S-1, \ldots -S)$ of a $(2S+1)$-dimensional representation D_S of the group of rotations in spin space, and therefore correspond to the species labels (α, i) used in Appendix 3. Functions of pure symmetry species, with respect to *spatial* symmetry operations, may again be built up by linear combination of the basic determinants; for molecules, this is easily accomplished by the methods of Appendix 3, and adequate illustrations appear in later sections.

To summarize: it is often convenient to assume that the basis functions in an expansion such as (3.1.4) are all of the same pure symmetry species with respect to both space and spin symmetry operations. The "symmetry-adapted" functions are no longer necessarily single Slater determinants, and may often be linear combinations of a considerable number. Such functions, constructed within the individual configurations, are often called "configurational functions" (CFs), a terminology that we frequently adopt.‡

It is usually expedient to set up the symmetry-adapted Φs at the outset of any calculation, using the methods to be discussed presently, and the number of terms in the CI expansion (3.1.4) will then be drastically reduced. It must be stressed that, in principle, there is no *need* to work with symmetry-adapted functions, and that the solutions of the full secular problem would automatically come out with correct symmetry properties; in practice, however, the secular equations are so big that there are compelling reasons for seeking every possible reduction.

When we consider the possibility of shortening the CI expansion further, by discarding configurations that seem unlikely to contribute significantly, the situation becomes much less clear-cut. The "length" of the expansion (i.e. the number of terms whose coefficients are substantial) is found to be critically dependent on the choice of spin-orbitals $(\psi_A, \psi_B, \ldots)$ from which the Φs are constructed. The "best" orbitals will be those that yield accurate CI functions of moderate length, corresponding to *rapidly convergent* expansions. Since the use of CI expansions (explicitly or otherwise) underlies most approximation methods, the determination of such orbitals is one of the central problems of computational quantum chemistry. We shall take up this problem in detail

‡ The term configurational *state* functions (CSFs) is also widely used; but, except when constructed from IPM orbitals, such functions have little connection with actual states.

at a later stage; here we shall merely indicate the two traditionally established ways of choosing the set of orbitals, deferring a full discussion of the methods to which they lead until Chapters 6 and 7.

The first, and oldest, way of choosing the set is to adopt *atomic orbitals* (AOs) centred on the various nuclei in the molecule; this leads to the valence bond (VB) method. The second is to use *molecular orbitals,* as in the example of Section 1.3, which generally extend over the molecule as a whole; this leads to the molecular orbital (MO) method. The MOs are normally of restricted form, again, as in Section 1.3, being constructed as linear combinations of the AOs (the LCAO approximation). Historically, the two methods have come to be regarded as in some sense representing two "extreme" approaches to the problem of determining molecular wavefunctions. In fact, in a very real sense, they are simply two ways of implementing the same CI method, leading to alternative CI expansions with different convergence properties. The example of 5 electrons and 10 AOs (p. 67) clarifies their relationship: from the 10 AOs we form 20 *atomic* spin-orbitals and 15 564 AO determinants in (3.1.4), corresponding to the available spin-orbital configurations; on the other hand, we could construct 10 linearly independent MOs from the 10 AOs, 20 *molecular* spin-orbitals, and 15 564 MO determinants; if we solved the secular problem in each case, we should arrive at exactly the same approximate energies and wavefunctions, though the expansions themselves would be entirely different in appearance. The reason for this identity is simply that each MO determinant may be expanded, on using the LCAO forms of the MOs, in terms of AO determinants. In general, an MO-product expands into a combination of *all* AO products, with certain numerical coefficients, and consequently (on adding spin factors and antisymmetrizing) the expanded MO determinants, even from one MO configuration, may contain *all possible* AO determinants. A general MO function may therefore be expressed in terms of the *full set* of all AO determinants that can be formed using the original restricted basis of AOs; and a given many-determinant function can thus be written equivalently in MO or VB form. The fact that the MO determinants may be written as linear combinations of the full set of AO determinants, and vice versa, means that the many-electron basis functions (Φs) in each case define the same subspace of the many-electron Hilbert space, the "full CI" subspace determined by the choice of the AOs themselves; the basic determinants used in the MO and VB methods are in fact related by an equation $\bar{\mathbf{\Phi}} = \mathbf{\Phi}\mathbf{T}$ (cf. (2.2.16) that relates alternative *orbital* bases in *one*-electron space) and the expansion coefficients according to (2.2.18) by $\bar{\mathbf{c}} = \mathbf{T}^{-1}\mathbf{c}$, so that

$$\Psi = \mathbf{\Phi}\mathbf{c} = \bar{\mathbf{\Phi}}\bar{\mathbf{c}}.$$

This is true generally and remains true when restrictions are imposed on the coefficients so as to fulfil appropriate symmetry requirements. It must be emphasized, however, that a *subset* of MO configurations (for example those with only singly occupied orbitals) is *not* equivalent to a corresponding subset of AO configurations (as may be verified by expansion); the equivalence holds only at the full CI level. Clearly, if we take the full sets, the choice of which particular orbitals or employ (AOs or LCAO MOs) is irrelevant; but if we take only a few configurations then we shall want to ensure that these are effective (i.e. give good convergence) by carefully choosing the orbitals. Historically, MO theory has evolved from a *one-determinant* first approximation, the orbitals being MOs; VB theory from a many-determinant approximation, the orbitals being AOs. In this sense, the methods start from opposite extremes; but when each is carried to its logical conclusion, within the framework of the CI approach, there must be convergence to the same final results. There are of course many possibilities between these two extremes, with orbitals chosen to describe, for example, localized chemical bonds, chemical groups or atomic inner shells, but the considerations of this section are quite general.

The possible approaches described so far are illustrated in Section 3.5, where we give a detailed discussion of the hydrogen molecule. Before doing so, however, we note an important difference between the MOs and the AOs. The MOs that give the best possible 1-determinant approximation to a wavefunction arise (as we shall find in Chapter 6) from a 1-electron eigenvalue equation and are therefore *orthogonal*, according to the discussion of Section 2.3. On the other hand, the AOs on a number of different centres are generally non-orthogonal. Consequently, while the matrix-element expressions (3.3.7) are appropriate for MO theory, VB theory (unless we are prepared to make many further approximations) requires the more cumbersome expressions whose 1- and 2-electron parts are given in (3.3.17) and (3.3.18). The difficulty of handling the VB wavefunctions has led to a somewhat uneven development of the two methods, which will be apparent in later chapters. Mathematically, the MO method (even with extensive CI) is easiest to work with; but the VB approach has played an important role in the conceptual development of molecular quantum mechanics and is intimately related to the group-theoretical methods that we discuss later. What is most important at this stage is the underlying unity of the CI approach, which should not be lost sight of amidst the technical details of its implementation.

3.5 AN EXAMPLE: THE HYDROGEN MOLECULE

The electronic structure of the H_2 molecule was considered from an elementary standpoint in Section 1.3, where the ground state was described by assigning both electrons to a "bonding" molecular orbital $A = m_A(a+b)$ composed of 1s AOs a and b on the two centres. Here we start instead from the AOs themselves, illustrating the procedures associated with complete set expansions, and then explore the connection between the CI expansion and the simpler forms of MO and VB theory.

From the 1s orbitals a and b we obtain four spin-orbitals

$$a\alpha, \quad b\alpha, \quad a\beta, \quad b\beta.$$

There are six possible spin-orbital configurations, the corresponding determinants being (omitting the variables, cf. footnote on p. 57)

$$\left.\begin{aligned} D_1 &= \det|a\alpha\ b\beta|, & D_2 &= \det|a\beta\ b\alpha|, \\ D_3 &= \det|a\alpha\ a\beta|, & D_4 &= \det|b\alpha\ b\beta|, \\ D_5 &= \det|a\alpha\ b\alpha|, & D_6 &= \det|a\beta\ b\beta|, \end{aligned}\right\} \qquad (3.5.1)$$

of which D_1, D_2, D_5, D_6 belong to the orbital configuration ab, D_3 to a^2 and D_4 to b^2. The considerations of Sections 1.2 and 3.2 suggest that these should first be combined, where necessary, so as to factorize into space and spin parts: the resulting functions are

$$\left.\begin{aligned} \Phi_1 &= D_1 - D_2, & \Phi_2 &= D_3, & \Phi_3 &= D_4, \\ \Phi_4 &= D_1 + D_2, & \Phi_5 &= D_5, & \Phi_6 &= D_6. \end{aligned}\right\} \qquad (3.5.2)$$

These are in fact spin eigenfunctions, as will become apparent in Chapter 4, with the following quantum numbers (S, M) respectively:

$$(0,0), \quad (0,0), \quad (0,0), \quad (1,0), \quad (1,1), \quad (1,-1).$$

Consequently, the singlet ground state may be approximated by a mixture of the first three terms only, while the remaining functions must describe three distinct and non-mixing excited states.

To incorporate *spatial* symmetry requirements, we note that all functions are invariant under spatial rotations‡ about the bond axis, so that we need consider only how they behave under the inversion i, which is the only non-trivial operation of the appropriate symmetry subgroup $\mathrm{C_s} = \{\mathrm{E}, \mathrm{i}\}$. The effect of i on the orbitals is simply to interchange a and b, and hence (remembering that interchange of the columns changes the sign of a determinant) we obtain

D	D_1	D_2	D_3	D_4	D_5	D_6
$\mathrm{i}D$	$-D_2$	$-D_1$	D_4	D_3	$-D_5$	$-D_6$

‡ Rotations applied to the orbital factors alone. For group theory see Appendix 3.

and, for the spin eigenfunctions,

Φ	Φ_1	Φ_2	Φ_3	Φ_4	Φ_5	Φ_6
$\mathsf{i}\Phi$	Φ_1	Φ_3	Φ_2	$-\Phi_4$	$-\Phi_5$	$-\Phi_6$

Now the group C_s has only two irreducible representations, both unidimensional: with their conventional names they are

	E	i
A_g	1	1
A_u	1	−1

This means (Appendix 3) that functions of species A_g are "even" (gerade) under inversion, while A_u functions (changing sign) are "odd" (ungerade). The combinations of Φs that belong to the two species may be obtained either by inspection or by use of a "projection operator" (A3.23). For example, the projector for species A_u is $\mathsf{E}-\mathsf{i}$, and, referring to the transformation properties listed above, $(\mathsf{E}-\mathsf{i})\Phi_1=0$ (no A_u function can be projected from Φ_1) while $(\mathsf{E}-\mathsf{i})\Phi_2=\Phi_2-\Phi_3$ (a function of A_u symmetry). Thus we find (indicating S, M values in parentheses):

$$\left.\begin{aligned}
&\textit{even}: \\
&\Phi_1^g=\Phi_1=D_1-D_2 \quad (0,0),\\
&\Phi_2^g=\Phi_2+\Phi_3=D_3+D_4 \quad (0,0);\\
&\textit{odd}: \\
&\Phi_3^u=\Phi_2-\Phi_3=D_3-D_4 \quad (0,0),\\
&\Phi_4^u=\Phi_4=D_1+D_2 \quad (1,0),\\
&\Phi_5^u=\Phi_5=D_5 \quad (1,1),\\
&\Phi_6^u=\Phi_6=D_6 \quad (1,-1).
\end{aligned}\right\} \tag{3.5.3}$$

Since mixing occurs only between functions of the same space and spin symmetry species, the odd functions already provide the only 2-orbital approximations to four of the electronic states, while best approximations to the even states (both spin singlets) will arise on solving a 2×2 secular equation for Φ_1^g and Φ_2^g. In other words, by using a basis of symmetry functions, instead of the single Slater determinants, the secular problem is reduced to trivial dimensions. The change of basis evidently induces a similarity transformation of the form (2.2.19) under which the matrices **H** and **M** are brought to block form. This procedure, which is nearly always followed in configuration interaction studies, is summarized in Table 3.1.

Because we started from AO determinants, the resultant wavefunctions are in VB form. We now consider the nature of their constituent

Table 3.1 Steps in a configuration interaction calculation.

Slater determinants	D_1	D_2	D_3	D_4	D_5	D_6
(S, M)-classified spin eigenfunctions[a]	Φ_1	(0, 0) Φ_2	Φ_3	(1, 0) Φ_4	(1, 1) Φ_5	(1, −1) Φ_6
Symmetry functions (spatial symmetry A_g or A_u)	Φ_1^g	(A_g) Φ_2^g	(A_u) Φ_3^u	(A_u) Φ_4^u	(A_u) Φ_5^u	(A_u) Φ_6^u
Approximate state functions arising by linear combination	Ψ_1^g	Ψ_2^g	Ψ_3^u	Ψ_4^u	Ψ_5^u	Ψ_6^u

[a] Matrix elements vanish (and hence there is no mixing) between functions separated by vertical lines.

spin eigenfunctions, taking for illustration the ground-state function:

$$\Psi_1^g = c_1\Phi_1 + c_2\Phi_2 + c_3\Phi_3 \quad (c_2 = c_3). \tag{3.5.4}$$

Values of the coefficients follow from the lowest-energy solution of the 2×2 secular problem, the required matrix elements being easily evaluated in terms of those for basic determinants, using (3.3.17) and (3.3.18), or from first principles, along the lines of Section 3.2. The spin-eigenfunctions Φ_1, Φ_2 and Φ_3, defined in terms of AOs, are usually referred to as VB "structures" and are depicted as in Fig. 3.1. The function Φ_1 is said to represent a "covalent structure" in which one electron is associated with each AO, spins coupled to give a singlet function. The functions Φ_2 and Φ_3 on the other hand are said to describe "ionic" or "polar" situations in which one orbital is doubly occupied and the other empty, corresponding to a shift of charge to one end of the molecule or the other. The optimal ground-state function is therefore a certain mixture of covalent and polar VB structures. If we expand further, it may of course be written in terms of the basic determinants as

$$\Psi_1^g = d_1D_1 + d_2D_2 + \ldots + d_6D_6, \tag{3.5.5}$$

although in this special case only the first four coefficients are non-zero.

In "simple" VB theory it is customary to neglect the "ionic" structures because they describe energetically "unlikely" situations, and the corresponding function

$$\Psi_{HL} = c\Phi_1 = c(D_1 - D_2) \tag{3.5.6}$$

Fig. 3.1 Non-polar and polar VB structures. Hydrogen molecule.

(where c is a normalizing factor) is the one used by Heitler and London (1927) in the first successful discussion of the bond in the hydrogen molecule.

We now turn to the MO approach and proceed in a similar way. The MOs of Section 1.3 are

$$A = m_A(a+b), \qquad B = m_B(a-b)$$

and from the spin-orbitals $A\alpha$, $A\beta$, $B\alpha$, $B\beta$ we build determinants

$$\left.\begin{aligned} D_1' &= \det|A\alpha\ B\beta|, & D_2' &= \det|A\beta\ B\alpha|, \\ D_3' &= \det|A\alpha\ A\beta|, & D_4' &= \det|B\alpha\ B\beta|, \\ D_5' &= \det|A\alpha\ B\alpha|, & D_6' &= \det|A\beta\ B\beta|. \end{aligned}\right\} \tag{3.5.7}$$

These are analogous to (3.5.1) and may be combined into spin-eigenfunctions analogous to (3.5.2), namely

$$\left.\begin{aligned} \Phi_1' &= D_1' - D_2', & \Phi_2' &= D_3', & \Phi_3' &= D_4', \\ \Phi_4' &= D_1' + D_2', & \Phi_5' &= D_5', & \Phi_6' &= D_6', \end{aligned}\right\} \tag{3.5.8}$$

and it is clear that these functions are already of g or u type (unlike the VB structures), owing to the simple symmetry properties of the MOs. Thus, under inversion i, $A \to A$ and $B \to -B$, and each determinant is thus multiplied by ± 1. The secular problem may again be formulated and solved, giving a ground-state function

$$\Psi_1^g = c_2'\Phi_2' + c_3'\Phi_3' \tag{3.5.9}$$

identical with (3.5.4), which may of course be written in terms of MO determinants in the form

$$\Psi_1^g = d_1'D_1' + d_2'D_2' + \ldots + d_6'D_6', \tag{3.5.10}$$

which is comparable to (3.5.5). It turns out that in these MO expansions the dominant term is Φ_2' $(= D_3')$, corresponding to both electrons in the bonding MO. This is the one-determinant wavefunction of "simple" MO theory.

The resemblance between the CI expansions of MO and VB theory is clearly no more than formal. In MO theory the available configurations may be indicated in an energy-level diagram (Fig. 3.2), and in this case

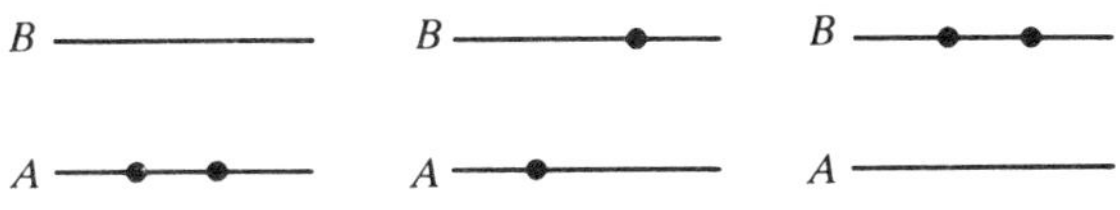

Fig. 3.2 MO configurations. Hydrogen molecule.

(using the IPM terminology of Section 1.2) yield (a) the "ground state" with both electrons in the bonding MO, (b) singly-excited states (with four possible spin couplings), and (c) a doubly-excited state, with both electrons promoted to the antibonding MO. The analogues of the AO-determinants (3.5.1) are MO-determinants with an entirely different physical meaning. Thus $D_3' = \det|A\alpha\, A\beta|$ is a good MO approximation to the ground state whereas $D_3 = \det|a\alpha\, a\beta|$ described a hypothetical ionic situation (Fig. 3.1b) with both electrons on the same nucleus. Similarly, the spin eigenfunction $\Phi_1' = D_1' - D_2'$ describes an excited singlet state (one electron promoted to the antibonding MO), whereas Φ_1 $(=\Psi_{\mathrm{HL}})$ was a good AO approximation to the *ground* state.

In comparing MO and VB wavefunctions, it is often said that MO theory gives undue weight to the ionic structures (Figs 3.1b, c) while VB theory in its simple form leaves them out altogether. To see what this means, it is only necessary to expand the simple configuration MO function in terms of AO determinants. The result is naturally of the general form (3.5.5):

$$\begin{aligned}\Phi_3' &= \det|a\alpha\ a\beta| + \det|a\alpha\ b\beta| + \det|b\alpha\ a\beta| + \det|b\alpha\ b\beta|\\ &= \Phi_1 + \Phi_2 + \Phi_3,\end{aligned}$$

where the spin eigenfunctions are the VB structures depicted in Fig. 3.1. The simplest VB approximation, namely $\Psi_{\mathrm{HL}} = \Phi_1$, is obtained by discarding the ionic structures Φ_2 and Φ_3. The simplest MO approximation apparently includes these structures with the same coefficients as the covalent structure Φ_1. The best CI function, obtained by optimizing the coefficients in (3.5.1), is in fact approached somewhat more closely by the Heitler–London function (Φ_1) than by the MO function (Φ_2'). In obtaining the "best" function, using all determinants, it is quite immaterial whether we employ VB structures ($\Phi_1, \Phi_2, \ldots$), both covalent and ionic, or the MO functions ($\Phi_1', \Phi_2', \ldots$), including all "excitations". In fact we may construct the Φs from *any* set of linearly independent combinations of the original AOs, chosen simply for computational convenience. We investigate the convergence of such calculations in later chapters.

3.6 AN ALTERNATIVE APPROACH: FOCK SPACE

Although the CI method, introduced in Section 3.4 and illustrated in the last section, is in principle extremely simple, requiring nothing more than the use of Slater's rules (p. 63) and enough computing power to solve very large systems of secular equations, it is sometimes useful to adopt a more sophisticated formalism in which the actual handling of determinants is replaced by purely algebraic manipulations. The formalism we now introduce was first developed in physics, where its use is very widespread, but since the mid-nineteensixties similar techniques have also gained wide acceptance in quantum chemistry (see e.g. Jørgensen and Simons, 1981). In this section we present the basic formalism; its value and power will become clear in later chapters.

A wavefunction $\Phi_{ij\ldots p}(\mathbf{x}_1, \mathbf{x}_2, \ldots, \mathbf{x}_N)$, for an independent-particle model in which spin-orbitals $\psi_i, \psi_j, \ldots, \psi_p$ are occupied by electrons, is the Schrodinger representation of a *state vector* or *ket,* denoted by $|ij\ldots p\rangle$, where the labels simply indicate the spin-orbitals to which the electrons are assigned. The wavefunction itself is a *function* of the electronic variables, and we indicate the one-to-one correspondence between state vector and wavefunction by

$$|ij\ldots p\rangle \leftrightarrow \Phi_{ij\ldots p}(\mathbf{x}_1, \mathbf{x}_2, \ldots, \mathbf{x}_N).$$

Instead of the spin-orbital labels, we could alternatively just give the occupation numbers (0, 1) of the whole ordered set of available spin-orbitals: for example, $|010010\ldots\rangle$ would stand for the 2-electron state with electrons in ψ_2 and ψ_5, and the corresponding Schrödinger function would be

$$\Phi_{2,5}(\mathbf{x}_1, \mathbf{x}_2) = 2^{-1/2} \det |\psi_2(\mathbf{x}_1) \;\; \psi_5(\mathbf{x}_2)|.$$

When a state vector is represented by a linear combination of such kets, we speak of the "occupation-number representation": the corresponding vector space, whose basis vectors comprise all possible kets—corresponding to *any number of electrons*—is called the *Fock space.* The Fock space thus focuses attention on the state vectors themselves, rather than the wavefunctions that represent them, and extends the space to include state vectors for any number of electrons. This approach originated in field theory, where the "particles" are photons that are "created" or "annihilated" in emission or absorption processes. The use of Fock space is most commonly described as the method of "second quantization".

Slater's rules (Section 3.3) give us, using the Schrödinger representation, a prescription for setting up matrix elements of the Hamiltonian H between state vectors associated with different spin-orbital configurations (e.g. $\langle ij\ldots r'\ldots p|\,\mathsf{H}\,|ij\ldots r\ldots p\rangle$ in the case of one spin-orbital difference). In Fock space we can get an alternative prescription by regarding the 1- and 2-electron integrals as parameters (calculated once and for all in terms of the chosen spin-orbitals) and introducing *Fock-space operators* that "look at" the occupation numbers and then tell us what combination of parameters will give the correct matrix element.

To this end, we define operators that "connect" the different basis vectors in Fock space. We use the standard representation property that if a vector Φ is represented by a function $\Phi(\boldsymbol{x})$ then the vector $\alpha\Phi$ is represented by $\alpha\Phi(\boldsymbol{x})$, α being any complex number: in symbols, $\Phi \leftrightarrow \Phi(\boldsymbol{x})$ requires $\alpha\Phi \leftrightarrow \alpha\Phi(\boldsymbol{x})$, where $A \leftrightarrow B$ means "with A we associate B and vice versa". Examples of this association are, remembering that the state vectors are represented by determinants,

$$|i\rangle \leftrightarrow \psi_i(\boldsymbol{x}_1), \qquad |ij\rangle \leftrightarrow \frac{1}{\sqrt{2}} \begin{vmatrix} \psi_i(\boldsymbol{x}_1) & \psi_j(\boldsymbol{x}_1) \\ \psi_i(\boldsymbol{x}_2) & \psi_j(\boldsymbol{x}_2) \end{vmatrix},$$

and, from above,

$$|ji\rangle \leftrightarrow \frac{1}{\sqrt{2}} \begin{vmatrix} \psi_j(\boldsymbol{x}_1) & \psi_i(\boldsymbol{x}_1) \\ \psi_j(\boldsymbol{x}_2) & \psi_i(\boldsymbol{x}_2) \end{vmatrix} = -\frac{1}{\sqrt{2}} \begin{vmatrix} \psi_i(\boldsymbol{x}_1) & \psi_j(\boldsymbol{x}_1) \\ \psi_i(\boldsymbol{x}_2) & \psi_j(\boldsymbol{x}_2) \end{vmatrix} \leftrightarrow -|ij\rangle.$$

The order in which the spin-orbitals appear is thus important (as a consequence of the antisymmetry principle), and we adopt the convention that in a ket $|ij\ldots p\rangle$ the associated determinant is

$$\Phi_{ij\ldots p}(\boldsymbol{x}_1, \boldsymbol{x}_2, \ldots, \boldsymbol{x}_N) = (N!)^{-1/2} \det |\psi_i(\boldsymbol{x}_1) \;\; \psi_j(\boldsymbol{x}_2) \;\; \ldots \;\; \psi_p(\boldsymbol{x}_N)|.$$

It is of course unnecessary to allow all permutations of labels, since, for example, $|ji\rangle = -|ij\rangle$ and we want only a single linearly independent set of kets; we normally adopt the set in which the labels occur in dictionary order. One vector of the Fock space is that which represents the "vacuum", with no particles present; thus typical vectors are

$$|\text{vac}\rangle, \quad |i\rangle, \quad |ij\rangle, \quad \ldots, \quad |ij\ldots p\rangle, \quad \ldots,$$

and any vector is multiplied by ϵ_{P} $(=\pm 1)$ under a permutation P of labels.

To connect vectors representing different numbers of electrons, we introduce a *creation operator* a_r^+ such that

$$\mathsf{a}_r^+ \, |ij\ldots p\rangle = \begin{cases} |ij\ldots pr\rangle & (r \not\subset (ij\ldots p)), \\ 0 & (r \subset (ij\ldots p)). \end{cases} \tag{3.6.1}$$

Thus if $ij\ldots p$ *does not* include r then an $(N+1)$th particle is added in ψ_r; if $ij\ldots p$ *does* include r then the associated determinant would vanish by antisymmetry (two identical columns) and would represent the zero vector. Putting the new spin-orbital at the end of the list merely fixes the phase. For example,

$$\mathsf{a}_j^+ \, |i\rangle = |ij\rangle, \qquad \mathsf{a}_i^+ \, |j\rangle = |ji\rangle = -\,|ij\rangle.$$

Clearly, all state vectors may be produced from the vacuum. Thus

$$\mathsf{a}_i^+ \, |\text{vac}\rangle = |i\rangle, \qquad \mathsf{a}_j^+ \mathsf{a}_i^+ \, |\text{vac}\rangle = |ij\rangle,$$
$$\mathsf{a}_p^+ \ldots \mathsf{a}_j^+ \mathsf{a}_i^+ \, |\text{vac}\rangle = |ij \ldots p\rangle;$$

the first operator to be applied creates a particle in the "first" spin-orbital, the last creates one in the "last" spin-orbital. Also

$$\mathsf{a}_i^+ \mathsf{a}_j^+ \, |\text{vac}\rangle = |ji\rangle = -|ij\rangle = -\mathsf{a}_j^+ \mathsf{a}_i^+ \, |\text{vac}\rangle,$$

while $\mathsf{a}_i^+ \mathsf{a}_i^+$ must produce a vector with a vanishing associated determinant, i.e. the zero vector. The creation operators thus have the anticommutation property

$$\mathsf{a}_r^+ \mathsf{a}_s^+ + \mathsf{a}_s^+ \mathsf{a}_r^+ = 0, \tag{3.6.2}$$

which reflects the antisymmetry property of the basis vectors, and it is readily verified that this result is valid no matter how many operators are applied to $|\text{vac}\rangle$, and hence for all vectors of Fock space.

In the same way, we may define an *annihilation* operator a_r^- that "undoes" the effect of a_r^+:

$$\left.\begin{aligned} \mathsf{a}_r^- \, |ij \ldots pr\rangle &= |ij \ldots p\rangle \quad && (r \subset (ij \ldots p)), \\ \mathsf{a}_r^- \, |ij \ldots p\rangle &= 0 && (r \not\subset (ij \ldots p)), \end{aligned}\right\} \tag{3.6.3}$$

where the second part of the definition reflects the impossibility of striking out a spin-orbital already struck out, and implies that $\mathsf{a}_r^- \mathsf{a}_r^- = 0$. Thus a_r^+ and a_r^- are analogous to step-up and step-down operators. If r is contained *within* the set $ij \ldots p$ then the antisymmetry property allows us to rewrite (3.6.3) in the form

$$\mathsf{a}_r^- \, |ij \ldots p\rangle = \begin{cases} (-1)^{\nu_r} \, |ij \ldots \not{r} \ldots p\rangle & (r \subset (ij \ldots p)), \\ 0 & (r \not\subset (ij \ldots p)), \end{cases} \tag{3.6.4}$$

where ν_r interchanges are necessary in moving r to the end of the list. The properties of the operators a_r^- are analogous to those of a_r^+; thus

$$\mathsf{a}_r^- \mathsf{a}_s^- + \mathsf{a}_s^- \mathsf{a}_r^- = 0, \tag{3.6.5}$$

as may easily be verified.

Finally, we note that the operator $\mathsf{a}_s^+ \mathsf{a}_r^-$ has an effect corresponding to changing an occupied spin-orbital from ψ_r to ψ_s, for (assuming r contained in $ij \ldots p$ and s absent)

$$\begin{aligned} \mathsf{a}_s^+ \mathsf{a}_r^- \, |ij \ldots r \ldots p\rangle &= \mathsf{a}_s^+ (-1)^{\nu_r} \, |ij \ldots \not{r} \ldots p\rangle \\ &= (-1)^{\nu_r} \, |ij \ldots \not{r} \ldots ps\rangle \\ &= |ij \ldots s \ldots p\rangle, \end{aligned}$$

since a second phase factor must be added in moving the final s to the position initially occupied by r, and $(-1)^{2\nu_r}=1$. If $r \not\subset (ij\ldots p)$ or $s \subset (ij\ldots p)$ then the result must be zero. Similarly, assuming r in the list but s absent,

$$\begin{aligned}\mathsf{a}_r^- \mathsf{a}_s^+ \,|ij\ldots p\rangle &= \mathsf{a}_r^- \,|ij\ldots ps\rangle \\ &= (-1)^{\nu_r+1}\,|ij\ldots \not{r}\ldots ps\rangle \\ &= -\,|ij\ldots s\ldots p\rangle,\end{aligned}$$

where again ν_r is the number of interchanges needed to take r to the end of the list $ij\ldots p$. Again, if $r \not\subset (ij\ldots p)$ or $s \subset (ij\ldots p)$ then the result must vanish. In the special case $s=r$,

$$\mathsf{a}_r^+ \mathsf{a}_r^- \,|ij\ldots p\rangle = \begin{cases}|ij\ldots p\rangle & (r \subset (ij\ldots p)),\\ 0 & \text{otherwise},\end{cases}$$

$$\mathsf{a}_r^- \mathsf{a}_r^+ \,|ij\ldots p\rangle = \begin{cases}|ij\ldots p\rangle & (r \not\subset (ij\ldots p)),\\ 0 & \text{otherwise},\end{cases}$$

so $\mathsf{a}_r^+\mathsf{a}_r^- + \mathsf{a}_r^-\mathsf{a}_r^+$ is equivalent to the *unit operator* for any ket. These relations may be summarized in the form

$$\mathsf{a}_r^+\mathsf{a}_s^- + \mathsf{a}_s^-\mathsf{a}_r^+ = \delta_{rs}. \tag{3.6.6}$$

The properties of the creation and annihilation operators are completely specified by (3.6.2), (3.6.5) and (3.6.6).

The requirement that the Fock space should be a *metric space*, allowing us to define scalar products as in Schrödinger space, introduces a slight simplification. If, for the moment, we use $\Phi_{ij\ldots p}$ for a vector in Fock space, the normalization condition for all basis vectors requires that

$$\langle \mathsf{a}_r^+\Phi_{ij\ldots p} \mid \mathsf{a}_r^+\Phi_{ij\ldots p}\rangle = \langle \Phi_{ij\ldots p} \mid (\mathsf{a}_r^+)^\dagger \mathsf{a}_r^+\Phi_{ij\ldots p}\rangle = 1 \quad (r \not\subset (ij\ldots p)),$$

where the adjoint operator, indicated by the dagger, is defined in (2.3.5). Thus $(\mathsf{a}_r^+)^\dagger\mathsf{a}_r^+\,|ij\ldots p\rangle = |ij\ldots p\rangle$; but $\mathsf{a}_r^+\,|ij\ldots p\rangle = |ij\ldots pr\rangle$, and hence $(\mathsf{a}_r^+)^\dagger\,|ij\ldots pr\rangle = |ij\ldots p\rangle$. Thus $(\mathsf{a}_r^+)^\dagger$ has the property that defines a_r^-; and a similar argument shows that $(\mathsf{a}_r^-)^\dagger$ has the same effect as a_r^+. We thus conclude that a_r^+ coincides with the adjoint of a_r^- and vice versa. It is therefore unnecessary to define two distinct operators: it is customary to use

$$\begin{aligned}\mathsf{a}_r^\dagger \ (= \mathsf{a}_r^+) &= \text{creation operator},\\ \mathsf{a}_r \ (= \mathsf{a}_r^-) &= \text{annihilation operator}.\end{aligned}$$

The basic properties of the operators are then, from (3.6.2), (3.6.5) and

(3.6.6),

$$\boxed{\begin{aligned} \mathsf{a}_r^\dagger \mathsf{a}_s^\dagger + \mathsf{a}_s^\dagger \mathsf{a}_r^\dagger &= 0, \\ \mathsf{a}_r \mathsf{a}_s + \mathsf{a}_s \mathsf{a}_r &= 0, \\ \mathsf{a}_r \mathsf{a}_s^\dagger + \mathsf{a}_s^\dagger \mathsf{a}_r &= \delta_{rs}. \end{aligned}} \tag{3.6.7}$$

These *anti*commutation rules are often written $[a_r^\dagger, a_s^\dagger]_+ = 0$, etc.

It remains only to express in terms of the a_rs the Fock-space operator A^{F} equivalent to the Schrödinger operator A^{S}. We consider first the one-electron operator (e.g. a term $\mathsf{h}(i)$ in the Hamiltonian)

$$\mathsf{A}^{\mathrm{S}}(i)\psi_k(\boldsymbol{x}_i) = \sum_{k'} \psi_{k'}(\boldsymbol{x}_i) A_{k'k},$$

where $A_{k'k} = \langle k' | A^{\mathrm{S}} | k \rangle$ is a matrix element‡ of the operator, and note that for a many-electron system there occurs always the symmetric sum

$$\mathsf{A}^{\mathrm{S}} = \sum_{i=1}^{N} \mathsf{A}^{\mathrm{S}}(i).$$

Thus, in Schrödinger space,

$$\begin{aligned} \mathsf{A}^{\mathrm{S}} \det |\psi_{k_1}(\boldsymbol{x}_1) \;\; \psi_{k_2}(\boldsymbol{x}_2) \;\; \ldots \;\; \psi_{k_i}(\boldsymbol{x}_i) \;\; \ldots| \\ = \sum_{k_i, k_i'} \det |\psi_{k_1}(\boldsymbol{x}_1) \;\; \ldots \;\; \psi_{k_i'}(\boldsymbol{x}_i) \;\; \ldots| \, A_{k_i' k_i}, \end{aligned}$$

where the determinant on the right has the substitution $\psi_{k_i} \rightarrow \psi_{k_i'}$. The corresponding statement in Fock space is

$$\mathsf{A}^{\mathrm{F}} |k_1 k_2 \ldots k_i \ldots \rangle = \sum_{k_i, k_i'} |k_1 k_2 \ldots k_i' \ldots \rangle A_{k_i' k_i}.$$

But the ket with $k_i \rightarrow k_i'$ can be written $\mathsf{a}_{k_i'}^\dagger \mathsf{a}_{k_i} |k_1 k_2 \ldots k_N\rangle$ and we can remove the restriction that k_i be occupied because if k_i is *not* in the list then the ket is destroyed. Hence, renaming the spin-orbitals and noting that the result applies for *any ket,*

$$\mathsf{A}^{\mathrm{F}} = \sum_{r,s} A_{rs} \mathsf{a}_r^\dagger \mathsf{a}_s. \tag{3.6.8}$$

The beauty of this result is that it applies irrespective of how many electrons the system contains—the number is implied by the number of spin-orbitals in the list, and need not be indicated in the operator.

Exactly similar reasoning applies to a two-electron operator, and the

‡ It is often sufficient to show only the orbital indices in the bra and the ket, a notation that from now on we use frequently.

full Hamiltonian becomes

$$\boxed{\mathsf{H} = \sum_{r,s} \langle r|\,\mathsf{h}\,|s\rangle \mathsf{a}_r^\dagger \mathsf{a}_s + \tfrac{1}{2} \sum_{r,s,t,u} \langle rs|\,g\,|tu\rangle \mathsf{a}_r^\dagger \mathsf{a}_s^\dagger \mathsf{a}_u \mathsf{a}_t} \tag{3.6.9}$$

where the second sum is the Fock-space equivalent of the electron-interaction operator $\frac{1}{2}\sum_{i,j} g(i, j)$.

The expection value of the energy, corresponding to any vector $|\ \rangle$ of Fock space, is now seen to be

$$E = \langle \mathsf{H} \rangle = \sum_{r,s} \langle \mathsf{a}_r^\dagger \mathsf{a}_s \rangle \langle r|\,\mathsf{h}\,|s\rangle + \tfrac{1}{2} \sum_{r,s,t,u} \langle \mathsf{a}_r^\dagger \mathsf{a}_s^\dagger \mathsf{a}_u \mathsf{a}_t \rangle \langle rs|\,g\,|tu\rangle, \tag{3.6.10}$$

whatever the number of electrons.

In later chapters we shall sometimes formulate general arguments using the algebraic properties of the Fock-space operators; sometimes we shall employ the more traditional methods; and in some chapters both approaches will be used side by side.

PROBLEMS 3

3.1 Verify that the antisymmetrizer (3.1.7) has the property $\mathsf{A}^2 = \mathsf{A}$ (idempotency). What factor must be attached in order to obtain, from a product of N orthonormal spin-orbitals, a *normalized* determinant?

3.2 Obtain the matrix element H_{15} in (3.2.7) and, assuming it to be a small quantity, estimate its first-order effect on the approximate energies (E_1, E_5) and wavefunctions (Ψ_1, Ψ_5). This is a simple example of CI (p. 66) involving the configurations [$1s^2$] and [1s 2s]: now extend the CI to include [$2s^2$].

3.3 Use Slater's rules to obtain a ground-state energy expression for the configuration $[\phi_A^2\phi_B]$ of a 3-electron system, first in terms of integrals over spin-orbitals and then after eliminating spin.

3.4 The function $\Psi(\boldsymbol{x}_1, \boldsymbol{x}_2, \ldots, \boldsymbol{x}_N) = \det|\psi_1(\boldsymbol{x}_1)\ \psi_2(\boldsymbol{x}_2)\ \ldots\ \psi_N(\boldsymbol{x}_N)|$ is the determinant of the matrix $\boldsymbol{\Psi}$ whose ith row is $\boldsymbol{\psi}(\boldsymbol{x}_i) = \psi_1(\boldsymbol{x}_i) \ldots \psi_N(\boldsymbol{x}_i)$. Show that under the transformation $\boldsymbol{\psi} \rightarrow \boldsymbol{\psi}' = \boldsymbol{\psi}\mathbf{U}$, where $\mathbf{U}$ is any non-singular $N \times N$ matrix, $\Psi(\boldsymbol{x}_1, \boldsymbol{x}_2, \ldots, \boldsymbol{x}_N)$ is merely multiplied by a constant, and hence that the energy expectation value for a one-determinant wavefunction is invariant against general linear mixing of the occupied spin-orbitals. Why is it usual to choose *orthonormal* mixtures?

3.5 The "outer" or "Krönecker" product of two matrices ($\mathbf{C} = \mathbf{A} \times \mathbf{B}$) is defined by $C_{ik,jl} = A_{ij}B_{kl}$ (double-index labelling of rows and columns). By writing a set of spin-orbitals in the form $\boldsymbol{\psi} = \boldsymbol{\phi} \times \boldsymbol{\sigma}$, where $\boldsymbol{\phi}$ is a row of orbitals and $\boldsymbol{\sigma} = (\alpha\ \beta)$, extend the argument of Problem 3.4 to the case where orbitals and spin factors are *separately* transformed. [*Hint*: The matrix product of two outer products is the outer product of two matrix products, i.e. $(\mathbf{A}\times\mathbf{B})(\mathbf{C}\times\mathbf{D})=(\mathbf{AC})\times(\mathbf{BD})$. Note also that the numbers of occupied α- and β-type orbitals in the determinant must be equal.]

3.6 Suppose that the N spin-orbitals in Problem 3.4 are expressed in terms of a spin-orbital *basis* $\boldsymbol{\chi}=(\chi_1\ \chi_2\ \dots\ \chi_m)$, according to $\boldsymbol{\psi}=\boldsymbol{\chi}\mathbf{T}$, where $\mathbf{T}$ is an $m\times N$ matrix; the matrix $\boldsymbol{\Psi}$ in Problem 3.4 may then be written as $\boldsymbol{\Psi}=\mathbf{XT}$ (the ik-element of $\mathbf{X}$ being $\chi_k(\boldsymbol{x}_i)$). Show that $\boldsymbol{\Psi}$ can then be expressed as

$$\Psi(\boldsymbol{x}_1,\boldsymbol{x}_2,\dots,\boldsymbol{x}_N)=\sum_{k_1\leqslant k_2\leqslant\dots\leqslant k_N} X\begin{pmatrix}1 & 2 & \dots & N\\ k_1 & k_2 & \dots & k_N\end{pmatrix} T\begin{pmatrix}k_1 & k_2 & \dots & k_N\\ 1 & 2 & \dots & N\end{pmatrix},$$

where the X factor is a determinant containing N columns $(k_1,k_2,\dots,k_N)$ selected from the matrix $\mathbf{X}$ and the summation runs over all selections. Obtain the T factor and find the number of independent terms in the expansion. (This latter is the dimension of the "full CI" space.) [*Hint*: Note that $\Psi=N!\mathsf{A}[\Psi_{11}\Psi_{33}\dots\Psi_{NN}]$, where the antisymmetrizer works on the first indices in each pair and $\Psi_{ii}=\psi_i(\boldsymbol{x}_i)=\sum_{k_i}X_{ik_i}T_{k_ii}$.]

3.7 Show that if $F(\boldsymbol{x}_1,\boldsymbol{x}_2,\dots,\boldsymbol{x}_N)$ is a linear combination of spin-orbital products *with a common orbital factor* and is an eigenfunction of the spin operators (3.4.1) and (3.4.2) then $\mathsf{P}F(\boldsymbol{x}_1,\boldsymbol{x}_2,\dots,\boldsymbol{x}_N)$ (P permuting both space and spin variables) has the same property. Hence show that an *antisymmetric* space–spin eigenfunction can be set up by constructing the spin eigenfunction from *spin products* alone, attaching any desired orbital factor, and then finally antisymmetrizing to get a linear combination of determinants. [*Hint*: Apply a permutation to S^2F and S_zF and note the symmetry of the operators.]

3.8 Set up singlet and triplet wavefunctions, with $M=0$, for a configuration $[\phi_A^2\phi_B^2\dots\phi_R^2\phi_P\phi_S]$, corresponding to removal of an electron from ϕ_P of a closed-shell ground state and "promotion" into an empty orbital ϕ_S. [*Hint*: Use the approach of Problem 3.7, with a spin product $\alpha\beta\alpha\dots\beta$ for the doubly occupied orbitals, followed by $(\alpha\beta\pm\beta\alpha)$, verifying also the spin-eigenfunction property.]

3.9 Calculate (in terms of 1- and 2-electron integrals) first approximations to the excitation energies from the closed-shell ground state to the singlet and triplet states in Problem 3.8. Note the simple form of the singlet–triplet splitting. [*Hint*: Obtain matrix elements for the two spin-orbital determinants and hence singlet and triplet expectation energies; subtract the energy of the closed-shell configuration $[\phi_A^2\phi_B^2\dots\phi_P^2]$, noting cancellations. It is convenient to use ψ_R, $\bar{\psi}_R$ for $\phi_R\alpha$, $\phi_R\beta$, eliminating spins in a final step.]

3.10 In the ground state of the carbon atom, the 4 valence electrons are assigned to the 2s and 2p orbitals (s_0, p_{-1}, p_0, p_{+1}, say) with the angular dependence indicated in Table A1.1 in Appendix 1. Set up wavefunctions for the ^{1}S, ^{3}P and ^{1}D states of the configuration [$2s^2\,2p^2$] and express their energies in terms of 1- and 2-electron integrals. Why is there no ^{3}D function? [*Hint*: You will need to know something about angular momentum (Appendix 2). Note that orbital angular-momentum operators working on s orbitals give zero, so only the p electrons (1 and 2, say) need be considered. Start from $p_{+1}(\mathbf{r}_1)p_{+1}(\mathbf{r}_2)$ and use the step-down procedure (p. 528) to find eigenfunctions of L_z and L^2. Add the s orbital and appropriate spin factors (singlet or triplet), antisymmetrize, and use Slater's rules on the resultant determinants.]

3.11 How would you generalize the procedure used in Problem 3.10 for an atom such as Si[(Ne) $3s^2\,3p^2$] whose neon-like core contains orbitals of non-zero angular momentum? [*Hint*: Write out the full spin-orbital product for (Ne) $3s^2$ followed by $3p_{+1}\alpha 3p_{+1}\beta$: write L^2 in the form (A2.5) and show that, on antisymmetrizing, L^2 working on the product will give no contributions from the first 12 spin-orbitals—since repeated spin-orbitals will always be present.]

3.12 The water molecule H_1—O—H_2 has the symmetry of the letter V, the symmetry operations being C_2 (rotate through π around the vertical axis), σ (make the reflection that interchanges H_1 and H_2), and σ' (reflect across the molecular plane)—to which must be added E (do nothing). The operations are elements of a *group* (Appendix 3) called C_{2v}. Construct a group *multiplication table* in which the entry in the R row and S column, for any two elements R, S, is the *product* RS (=T), i.e. the single operation that has the same effect as S followed by R.

3.13 Verify that the following four rows of numbers (named according to convention)

	E	C_2	σ	σ'
A_1	1	1	1	1
A_2	1	1	−1	−1
B_1	1	−1	1	−1
B_2	1	−1	−1	1

provide *representations* of the group C_{2v} (Problem 3.12) in the usual sense (p. 31). [*Hint*: Use the multiplication table from Problem 3.12. Regard the numbers (above) as 1 × 1 matrices, **D**(R) being the one associated with R, and examine the matrix products **D**(R) **D**(S).]

3.14 A simple MO wavefunction for H_2O may be set up by assigning the 10 electrons to the following MOs:

ϕ_1 oxygen inner-shell orbital (~1s);
ϕ_2 lone pair (~2p normal to plane);
ϕ_3 lone pair (~2p along symmetry axis);
ϕ_4 bonding MO (totally symmetric);
ϕ_5 bonding MO (antisymmetric under σ).

Sketch the probable forms of the MOs and assign them to the representations of Problem 3.13. Write down a 1-determinant wavefunction for the ground state Φ_0: to which representation does it belong? Consider singlet and triplet excited functions formed by promoting an electron from ϕ_5 to (i) a vacant MO (ϕ_6) of A_2 symmetry, and (ii) a vacant MO of B_1 symmetry: what symmetries will they have? [*Hint*: Consider the effect of each operation, $\phi \rightarrow \phi' = \mathsf{R}\phi = (?)$. In this case each "image" ϕ' is just a multiple of ϕ, and each orbital "carries a 1-dimensional representation" of C_{2v}. Examine the behaviour of the orbital products used in constructing the ground and excited states, noting that spin factors may be ignored in discussing *spatial* symmetry.]

3.15 Describe how you would set up a limited CI calculation on the water molecule, organizing your work along the lines of Table 3.1 (p. 74) and indicating which elements of the CI matrix will be zero.

3.16 Return to Problem 2.12 and use simple group theory to classify the symmetry orbitals according to their behaviour under the operations of the group C_{4v}. Put atom 1 on the x axis, the z axis being the principal axis of symmetry, and name the representations according to the following conventions: (i) A, B indicate 1-dimensional representations in which ± 1 is associated with the generator (C_4 of the rotations); (ii) E, T indicate 2- and 3-dimensional representations; (iii) representations of the same A or B type are distinguished by subscripts 1, 2 according as ± 1 is associated with the generator of the reflections (σ_1, say, across the (x, z) plane). Examine the behaviour of a set of d orbitals (p. 522) centred on the origin, making a similar classification.

3.17 Construct explicitly some representations of the group C_{4v} in Problem 3.16 by considering the behaviour under symmetry operations of the unit vectors $(\mathsf{e}_1, \mathsf{e}_2, \mathsf{e}_3)$ along the axes. Set up corresponding Wigner operators (A3.22) and use them to obtain symmetry orbitals, starting from s and p orbitals on *one* of the 4 atoms. How would you mix the symmetry orbitals to obtain a pair of MOs of E symmetry? [*Hint*: Refer to Problem 2.6. If you have any difficulty, all matrices, for the irreps (see Appendix 3) of the 32 most common point groups, are listed elsewhere (McWeeny, 1963).]

3.18 Use the results obtained in Problems 2.12, 3.16 and 3.17 to set up *many*-electron symmetry functions for a simple CI calculation on the ground state of square-planar H_4. Is this likely to be a singlet or a triplet? Note that C_{4v} is only a subgroup of the full point group (D_{4h}) but that the symmetry classification is easily modified by observing the behaviour of the various functions with respect to the new generating operation (i, the inversion). [*Hint*: Set up orbital products corresponding to the two lowest-energy configurations in IPM approximation; make a table to show how they behave under the symmetry operations; construct (for example by projection) linear combinations of various symmetry species; add spin factors and antisymmetrize to obtain determinants; functions of the same species will mix.]

3.19 Use the second-quantization form of the Hamiltonian (3.6.9) and the anticommutation properties (3.6.7) to derive the expectation-value expression (3.3.7a). Then modify the argument to obtain the *off*-diagonal elements of H. [*Hint*: Take $\Phi = |12\ldots N\rangle = \mathsf{a}_N^\dagger \ldots \mathsf{a}_2^\dagger \mathsf{a}_1^\dagger |\text{vac}\rangle$ as the N-electron state vector and use (3.6.7) to show that the general 1-electron term is $\langle \psi_r | \mathsf{h} | \psi_s \rangle$ $(\langle 12\ldots N | 12\ldots N\rangle \delta_{rs} - \langle 12\ldots Ns | 12\ldots Nr\rangle)$—which gives $\delta_{rs}\langle \psi_r | \mathsf{h} | \psi_s \rangle$ for $r, s \in 1, 2, \ldots, N$, but zero otherwise. Do likewise for the 2-electron terms. Note that a spin-orbital change ($\psi_k \rightarrow \psi_{k'}$) can be produced by applying $\mathsf{a}_{k'}^\dagger \mathsf{a}_k$ to Φ.]

3.20 Show that with *non*-orthonormal orbitals the anticommutation rules (3.6.7) must be replaced by $[a_p, a_q]_+ = [a_p^\dagger, a_q^\dagger]_+ = 0$, and $[a_p^\dagger, a_q]_+ = S_{pq}$, where $S_{pq} = \langle \psi_p \mid \psi_q \rangle$. Hence rederive the results given in (3.3.17) and (3.3.18). [*Hint*: For a symmetrically orthonormalized basis $\bar{\psi} = \psi \mathbf{S}^{-1/2}$ the corresponding operators $\bar{a}_i = \sum_j a_j(\mathbf{S}^{-1/2})_{ji}$ etc. will satisfy (3.6.7). Use $a_q = \sum_j \bar{a}_j(\mathbf{S}^{1/2})_{jq}$ etc. to obtain $[a_p^\dagger, a_q]_+$ in terms of $[\bar{a}_i^\dagger, \bar{a}_j]_+$.]

3.21 What combinations of creation and annihilation operators would produce, when acting on the closed-shell ground state, the singlet and triplet excited functions in Problem 3.8? Use second-quantization methods to obtain the singlet and triplet excitation energies (cf. Problem 3.9). [*Hint*: Use $a_{p\alpha}$, $a_{p\beta}$, etc. to indicate explicitly the operators that create or annihilate an electron in ϕ_p with given spin.]

4 Spin and Permutation Symmetry

4.1 SPIN EIGENFUNCTIONS

For a 2-electron system, like the helium atom considered in Section 1.2, there is no difficulty in setting up electronic wavefunctions with the characteristic spin factors that appear in (1.2.25) and (1.2.26). Even when, as in Section 3.5, the two-electron functions are expressed in terms of Slater determinants, it is easy to show that the appropriate linear combinations can be expressed as products of space and spin factors, each with symmetry or antisymmetry under exchange of space or spin variables respectively. In fact, as will be clear presently, the spin functions that appear are *eigenfunctions* of the total spin operators (Appendix 2)

$$\mathsf{S}^2 = \sum_i \mathsf{S}^2(i), \qquad \mathsf{S}_z = \sum_i \mathsf{S}_z(i) \tag{4.1.1}$$

(with $N = 2$). The symmetric factors have eigenvalues with $S = 1$, $M = 0$, ± 1, corresponding to the three possible components of a triplet state; while the antisymmetric factor has $S = M = 0$ and corresponds to a singlet state. We say the spins have been "vector-coupled" to give a resultant spin angular momentum with $S = 1$ or $S = 0$, corresponding to "parallel" or "antiparallel" coupling of the two spins with $S = \frac{1}{2}$.

For a many-electron system, however, it is a matter of some difficulty to construct all the linearly independent spin eigenfunctions arising from a given orbital configuration. Suppose, for example, we consider 5 electrons occupying 5 specific (different) orbitals; by attaching spin factors, α or β for each orbital, and antisymmetrizing, we obtain 32 ($= 2^5$) determinants; and from these determinants, by linear combination, we can set up 32 "vector-coupled" functions with various eigenvalues of total spin. One possible eigenfunction has $S = M = \frac{1}{2}$, but there are no less than *five* independent combinations yielding the same eigenvalues; these are not unique and may be mixed freely in setting up a general spin eigenfunction with $S = M = \frac{1}{2}$; different choices of a linearly independent set arise from different "coupling schemes", which we must now consider.

The various spin eigenfunctions have very specific symmetry properties, not only with respect to rotation of the axes of spin quantization (to be discussed later) but also in their response to a *permutation* of spin variables; and since the Pauli principle is expressed most generally in terms of permutations, according to (1.2.27), the behaviour of the spin eigenfunctions is clearly of fundamental importance. In fact, the problem of constructing antisymmetric wavefunctions was first approached, *before* Slater's introduction of determinants, by considering separately the effect of permutations on orbital factors and spin factors as in the elementary example in Section 1.2. This alternative approach, which leans heavily on group theory, has now re-established itself as a powerful means of handling CI calculations, as well as providing deeper insight into the properties of many-electron wavefunctions. In this chapter we therefore emphasize the fundamental role of the spin functions, even though we may often refer to the parallel approach in terms of Slater determinants.

First we note that the problem of constructing spin eigenfunctions can be isolated from any considerations of the *orbital* form of the wavefunction and also from the overall space–spin antisymmetry requirement (1.2.27). This is because the spin operators S^2 and S_z are symmetric in the particles and so, for example, if $\Theta^{(S,M)}$ is a pure spin function with eigenvalues S, M then the equations

$$\mathsf{S}^2\Theta^{(S,M)}(s_1, s_2, \ldots, s_N) = S(S+1)\Theta^{(S,M)}(s_1, s_2, \ldots, s_N), \quad (4.1.2)$$

$$\mathsf{S}_z\Theta^{(S,M)}(s_1, s_2, \ldots, s_N) = M\Theta^{(S,M)}(s_1, s_2, \ldots, s_N) \quad (4.1.3)$$

imply also (see Problem 3.7) that

$$\mathsf{S}^2\Theta^{(S,M)}(s_2, s_1, \ldots, s_N) = S(S+1)\Theta^{(S,M)}(s_2, s_1, \ldots, s_N),$$

$$\mathsf{S}_z\Theta^{(S,M)}(s_2, s_1, \ldots, s_N) = M\Theta^{(S,M)}(s_2, s_1, \ldots, s_N),$$

since the latter equations may be reached from the former merely by exchanging two variables and noting that S^2 and S_z, being sums of contributions from the N electrons, are independent of the change. The same is clearly true of any permutation; when we permute the variables, $\Theta^{(S,M)}$ remains a spin eigenfunction with the same eigenvalues. If, then, we take any orbital product F and attach the spin factor $\Theta^{(S,M)}$, we can antisymmetrize with respect to both space and spin by applying the operator A defined in (3.1.7): the result will be a properly antisymmetric spin eigenfunction, eligible as one of the expansion functions Φ_κ in the CI method. Since $\Theta^{(S,M)}$ must be some linear combination of spin products, the antisymmetrizer will turn every term in $F\Theta^{(S,M)}$ into a Slater determinant based on the given *orbital* product with some corresponding allocation of spin factors. For illustration, let us consider the product

$$[\alpha(s_1)\beta(s_2)-\beta(s_1)\alpha(s_2)][\alpha(s_3)\beta(s_4)-\beta(s_3)\alpha(s_4)],$$

which we find later to be a spin eigenfunction with $S=M=0$. If we attach this spin factor to an orbital product $A(\boldsymbol{r}_1)B(\boldsymbol{r}_2)C(\boldsymbol{r}_3)D(\boldsymbol{r}_4)$ and multiply out, we shall obtain a combination of spin-orbital products

$$\begin{aligned}&A(\boldsymbol{r}_1)\alpha(s_1)B(\boldsymbol{r}_2)\beta(s_2)C(\boldsymbol{r}_3)\alpha(s_3)D(\boldsymbol{r}_4)\beta(s_4)\\&\quad -A(\boldsymbol{r}_1)\beta(s_1)B(\boldsymbol{r}_2)\alpha(s_2)C(\boldsymbol{r}_3)\alpha(s_3)D(\boldsymbol{r}_4)\beta(s_4)\\&\quad -A(\boldsymbol{r}_1)\alpha(s_1)B(\boldsymbol{r}_2)\beta(s_2)C(\boldsymbol{r}_3)\beta(s_3)D(\boldsymbol{r}_4)\alpha(s_4)\\&\quad +A(\boldsymbol{r}_1)\beta(s_1)B(\boldsymbol{r}_2)\alpha(s_2)C(\boldsymbol{r}_3)\beta(s_3)D(\boldsymbol{r}_4)\alpha(s_4)\end{aligned}$$

in which corresponding pairs of orbitals take spin factors α, β or β, α (reversed order) in all possible ways, a minus sign being added for each reversal. When such a sum of products is antisymmetrized, we obtain, with the usual abbreviation (p. 57), a sum of Slater determinants

$$\begin{aligned}&\det|A\alpha\ B\beta\ C\alpha\ D\beta|-\det|A\beta\ B\alpha\ C\alpha\ D\beta|\\&\qquad\qquad -\det|A\alpha\ B\beta\ C\beta\ D\alpha|+\det|A\beta\ B\alpha\ C\beta\ D\alpha|.\end{aligned}$$

We note in passing that for a configuration of doubly occupied orbitals, with $B=A$, $D=C$, etc., all such determinants become identical since, for example, $\det|A\beta\ A\alpha\ C\alpha\ C\beta|=-\det|A\alpha\ A\beta\ C\alpha\ C\beta|$, differing only by interchange of two columns; and a single determinant of *doubly* occupied orbitals is therefore automatically a spin eigenfunction with $S=M=0$. More generally, the question of how to combine the Slater determinants into spin eigenfunctions now clearly reduces to the determination of pure spin functions satisfying (4.1.2) and (4.1.3).

We need several properties of the spin operators. The commutation rules (1.2.20) for the spin operators of *one* electron, imply similar rules for the *total* spin operators (cf. Appendix 2):

$$\mathsf{S}_\alpha\mathsf{S}_\beta-\mathsf{S}_\beta\mathsf{S}_\alpha=i\mathsf{S}_\gamma\quad(\alpha\beta\gamma=xyz,\ yzx,\ zxy),\tag{4.1.4}$$

where

$$\mathsf{S}_\alpha=\sum_{i=1}^{N}\mathsf{S}_\alpha(i)\quad(\alpha=x,\ y,\ z),\tag{4.1.5}$$

and we continue to use the dimensionless operators introduced in Section 1.2, the actual angular-momentum operators being $\hbar\mathsf{S}_\alpha$. It is also convenient to introduce "step-up" and "step-down" operators, S^+ and S^-, defined by

$$\mathsf{S}^\pm=\mathsf{S}_x\pm i\mathsf{S}_y.\tag{4.1.6}$$

If $\Theta^{(S,M)}$ is a normalized spin eigenfunction satisfying (4.1.2) and (4.1.3) then the effect of these operators is

$$\boxed{\begin{aligned} \mathsf{S}^+\Theta^{(S,M)} &= [(S+M+1)(S-M)]^{1/2}\Theta^{(S,M+1)},\\ \mathsf{S}^-\Theta^{(S,M)} &= [(S-M+1)(S+M)]^{1/2}\Theta^{(S,M-1)},\end{aligned}} \tag{4.1.7}$$

i.e. to produce another spin eigenfunction with the value of M stepped up or down by one unit, or to destroy the function if the resultant M would go outside the range $+S$ to $-S$. For a one-electron system, (4.1.7) gives

$$\mathsf{S}^+\alpha = 0, \qquad \mathsf{S}^+\beta = \alpha, \qquad \mathsf{S}^-\alpha = \beta, \qquad \mathsf{S}^-\beta = 0.$$

More generally, (4.1.7) provides a way of constructing all the $2S+1$ spin eigenfunctions belonging to a given family with fixed S and with $M = S, S-1, \ldots, -S$ from just one member; thus, for example, we may focus attention on the construction of $\Theta^{(S,S)}(M = S)$ and may obtain all other functions, if we need them, by operating repeatedly with the step-down operator S^-. In practice, it is usually sufficient to consider just one choice of M; since with a spinless Hamiltonian all choices lead to the same energy (and expectation values of other spinless operators), while matrix elements of spin-dependent operators may be dealt with more easily by other methods (Chapter 11). We shall therefore consider mainly the case $M = S$. We note also that the total spin operator $\mathsf{S}^2 = \mathsf{S}_x^2 + \mathsf{S}_y^2 + \mathsf{S}_z^2$ may be written in alternative forms by introducing S^+ and S^-:

$$\mathsf{S}^2 = \mathsf{S}^+\mathsf{S}^- - \mathsf{S}_z + \mathsf{S}_z^2 \tag{4.1.8a}$$
$$= \mathsf{S}^-\mathsf{S}^+ + \mathsf{S}_z + \mathsf{S}_z^2, \tag{4.1.8b}$$

and the properties of S^+ and S^- are somewhat more convenient than those of S_x, S_y.

We must now consider a few basic techniques for the construction of the spin eigenfunctions that will be used extensively in later chapters.

4.2 METHODS FOR CONSTRUCTING SPIN EIGENFUNCTIONS

The methods that we consider fall into two types: *synthetic* (see e.g. Kotani *et al.*, 1955), in which the full set of linearly independent eigenfunctions of given S and M is built up by some systematic procedure; and *analytic* (Löwdin, 1955, 1956a), in which a spin eigenfunction with the required values of S and M is *extracted* from an arbitrary function (i.e. a mixture of spin eigenfunctions) by means of a suitable projection operator. There are many possible procedures of both types, all exhaustively treated in the book by Pauncz (1979).

We take the synthetic approach first, starting with the observation that the full set of spin functions (however it may be constructed) has an important group-theoretical significance. The connection with group theory, developed and used in later sections, rests upon the permutation symmetry of the spin operators (p. 88), which ensures that any spin eigenfunction $\Theta^{(S,M)}$ remains a spin eigenfunction, with the same eigenvalues, after permutation of the spin variables. Thus if there are f_S^N linearly independent functions of given S, M, namely

$$\Theta_1^{(S,M)}, \quad \Theta_2^{(S,M)}, \quad \ldots, \quad \Theta_{f_S^N}^{(S,M)},$$

then we may assert that, for any permutation of spin variables,

$$\mathsf{P}\Theta_j^{(S,M)} = \sum_{i=1}^{f_S^N} \Theta_i^{(S,M)} P_{ij}, \tag{4.2.1}$$

since the result must by definition be some linear combination of the full set. This tells us (Appendix 3) that the Θs provide a basis for a *representation* of the "symmetric group" S_N of $N!$ permutations, and that with each permutation P we may associate the matrix $\mathbf{P}$ whose ij element is the expansion coefficient P_{ij}. Such a representation is, in fact, irreducible, and the dimension of this "irrep" is (see Wigner, 1959),

$$\boxed{f_S^N = \binom{N}{\frac{1}{2}N - S} - \binom{N}{\frac{1}{2}N - S - 1} = \frac{(2S+1)N!}{(\frac{1}{2}N + S + 1)!(\frac{1}{2}N - S)!},} \tag{4.2.2}$$

depending only on the value of S, not on M. If we already possess a complete set of spin eigenfunctions, we can therefore use (4.2.1) to set up irreducible representations of the symmetric group: and conversely, if we know from the theory of group representations the actual matrices $\mathbf{P}$ in some standard irreducible representation then we can use the group-theoretical projection operators (A3.23) to generate the spin eigenfunctions.

We now discuss briefly two key methods of coupling the spins, for use in later chapters, leaving the details to Appendix 2.

The spin-pairing method

We start from the two types of coupling for any pair of electrons, corresponding to normalized spin functions

$$\left.\begin{array}{ll} [\alpha(s_1)\beta(s_2) - \beta(s_1)\alpha(s_2)]/\sqrt{2} & (S = M = 0), \\ \alpha(s_1)\alpha(s_2) & (S = M = 1). \end{array}\right\} \tag{4.2.3}$$

It is a trivial matter to verify these spin eigenvalues by operating with S^2 and S_z, and we may easily extend the second result to show that parallel coupling of n spins may also be described by a single product:

$$\alpha(s_1)\alpha(s_2)\ldots\alpha(s_n) \quad (S = M = \tfrac{1}{2}n). \tag{4.2.4}$$

Thus, each term $\mathsf{S}_z(i)$ multiplies the corresponding factor $\alpha(s_i)$ by $\frac{1}{2}$, and summation over all n terms shows that the effect of S_z is to multiply by $\frac{1}{2}n$, i.e. $M = \frac{1}{2}n$; and, using (4.1.8b), each term $\mathsf{S}^+(i)$ annihilates a factor $\alpha(s_i)$, and the only non-zero result from S^2 comes from $\mathsf{S}_z + \mathsf{S}_z^2$, which together gives the original product multiplied by $M(M+1)$.

Consider now two different groups of electrons, described by spin eigenfunctions $\Theta_1^{(S_1M_1)}(s_1, \ldots, s_{N_1})$ and $\Theta_2^{(S_2,M_2)}(s_{N_1+1}, \ldots, s_{N_1+N_2})$. It is shown in Appendix 2 that the set of $(2S_1+1)(2S_2+1)$ possible products (fixed S_1, S_2, all M_1, M_2) may be used to construct spin eigenfunctions (S, M) with $S = S_1 + S_2, S_1 + S_2 - 1, \ldots, |S_1 - S_2|$ and with all allowed values of M, this being a standard exercise in angular-momentum-coupling theory. We need here only two trivial cases of the general result: (i) when $S_1 = S_2 = 0$ the single product

$$\Theta_1^{(0,0)}\Theta_2^{(0,0)}$$

is an eigenfunction with $S = M = 0$, and (ii) when $S_1 = 0$, $S_2 \neq 0$ the single product

$$\Theta_1^{(0,0)}\Theta_2^{(S_2,M_2)}$$

is an eigenfunction with $S = S_2$, $M = M_2$. It follows that any number of paired-spin factors may be combined with the product (4.2.4) to yield, with n_2 parallel-coupled spins, an eigenfunction with $S = M = \frac{1}{2}n_2$. Thus, if we want a spin eigenfunction with total spin S then we shall need to parallel-couple $2S$ spins, and if there are N electrons then we shall consequently need $g = \frac{1}{2}(N - 2S)$ spin pairs. A typical spin eigenfunction would be

$$\begin{aligned}\Theta_\kappa &= 2^{-1/2}[\alpha(s_i)\beta(s_j) - \beta(s_i)\alpha(s_j)] \\ &\quad \times 2^{-1/2}[\alpha(s_k)\beta(s_l) - \beta(s_k)\alpha(s_l)]\ldots\alpha(s_p)\alpha(s_q)\ldots\alpha(s_N),\end{aligned} \tag{4.2.5}$$

where the subscript κ labels the "coupling scheme" associated with this particular selection of spin pairings (i, j), (k, l), etc. while the superscript (S, M) is suppressed.

Although this is a very simple way of constructing spin eigenfunctions, first used many years ago by Heitler, London, Rumer, Hund, Weyl and others, the number of alternative pairing schemes in general exceeds the value f_S^N given by (4.2.2), and this means that linear dependences exist among the functions. This is not a serious disadvantage, however, since linearly independent sets may be selected in various ways. In one well-known diagrammatic method‡, we set down the numbers $1, 2, \ldots, N$ in a ring, representing each factor $[\alpha(s_i)\beta(s_j) - \beta(s_i)\alpha(s_j)]\sqrt{2}$ in (4.2.5) by an arrow $i \rightarrow j$. Any functions with crossed arrows are then easily shown to be expressible in terms of those in which no crossings occur, and may consequently be discarded. To give a concrete example, consider the two (triplet-state) spin functions

$$\Theta_1 = \tfrac{1}{2}[\alpha(s_1)\beta(s_2) - \beta(s_1)\alpha(s_2)][\alpha(s_3)\beta(s_4) - \beta(s_3)\alpha(s_4)]\alpha(s_5)\alpha(s_6),$$

$$\Theta_2 = \tfrac{1}{2}[\alpha(s_1)\beta(s_4) - \beta(s_1)\alpha(s_4)][\alpha(s_2)\beta(s_3) - \beta(s_2)\alpha(s_3)]\alpha(s_5)\alpha(s_6),$$

with which we may associate the Rumer diagrams (respectively)

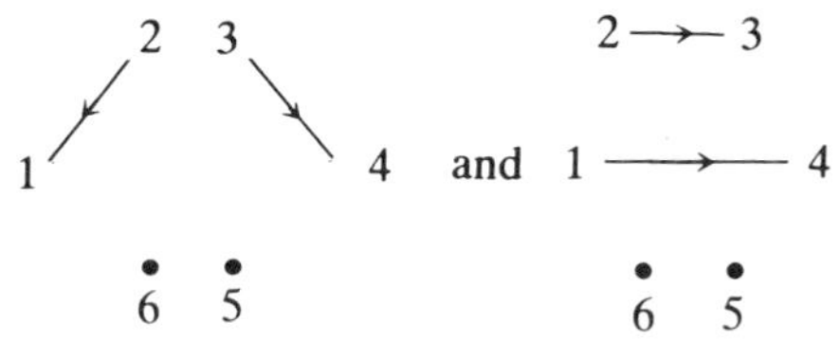

A third function Θ_3, corresponding to a diagram containing the crossed links $1 \rightarrow 3$ and $2 \rightarrow 4$, is then easily seen on multiplying out the spin factors to be expressible as $\Theta_3 = \Theta_1 + \Theta_2$. Any crossed arrows may be eliminated in this way, but, except in the case $S = 0$, more restricted dependences may still remain: an alternative method will be given presently.

Another slight disadvantage of the present construction is the non-orthogonality of the spin functions. Again, this is not serious provided we can easily obtain the non-orthogonality integrals; this turns out to be an extremely simple matter, using the Rumer diagrams, which we take up in Chapter 7 in developing the VB theory.

The branching-diagram method

In the spin-pairing method there appeared to be no systematic way of selecting the pairs and of defining a linearly independent set. An alternative procedure is to start from the known spin states for one or two electrons and add further electrons, one at a time, always coupling the Nth electron spin so as to obtain the N-electron spin eigenfunctions

‡ See e.g. Rumer (1932). The extension to $S \neq 0$ is less well known.

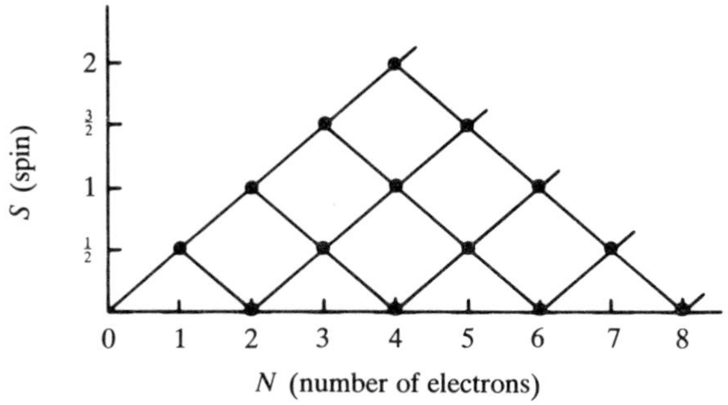

Fig. 4.1 The branching diagram.

in terms of those for $N-1$ electrons. This is a systematic procedure leading to a unique linearly independent set of orthogonal spin eigenfunctions and therefore has certain theoretical advantages. The functions to which it leads are conveniently classified using a "branching diagram" (Fig. 4.1). From the 1-electron states with $S=\frac{1}{2}$, we can obtain 2-electron states with $S=\frac{1}{2}+\frac{1}{2}=1$ or $S=\frac{1}{2}-\frac{1}{2}=0$ by parallel or antiparallel coupling; from the 2-electron states we can obtain similarly 3-electron states with $S=1+\frac{1}{2}=\frac{3}{2}$ (parallel coupling), $S=1-\frac{1}{2}=\frac{1}{2}$ (antiparallel), or $S=0+\frac{1}{2}=\frac{1}{2}$. Each intersection in the branching diagram indicates a state of given S, a line going upwards to the right leading to a state of spin $S+\frac{1}{2}$, one going downwards to a state of spin $S-\frac{1}{2}$.

The actual coupling formulae, which lead from $\Theta_N^{(S,M)}$ for N electrons to $\Theta_{N+1}^{(S\pm 1/2,\, M\pm 1/2)}$ on adding an $(N+1)$th electron, are given in Appendix 2 and may be used (Kotani *et al.*, 1955) to systematically generate all the spin states indicated in the branching diagram, though the explicit forms soon become very cumbersome.‡

It is clear that for every intersection (N, S) in the diagram there will be *several* functions, with the same S, M, constructed by following different routes in the diagram; these constitute the required linearly independent set, and each member is therefore identified by a particular "genealogy" or "parentage". Thus for 5 electrons the spin eigenfunctions with $S=M=\frac{1}{2}$ may be built up by following the five distinct sequences of couplings indicated below:

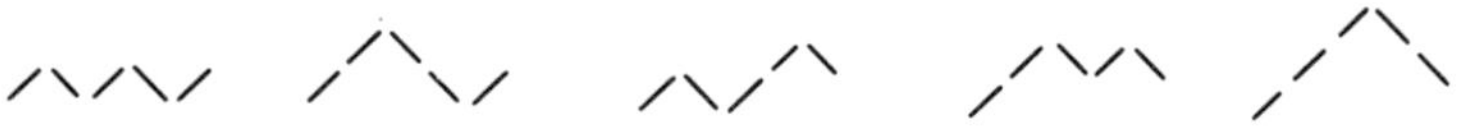

‡ Pauncz (1975) has given an alternative construction which is non-recursive.

Reference to Wigner's formula (4.2.2) indicates that this is the full number expected; and in general the number of distinct routes leading from the origin to the intersection associated with spin S agrees with that given by (4.2.2).

The value of the branching-diagram method is that it leads to spin eigenfunctions that provide *standard* irreducible representations (*irreps*) of the group S_N when the permutations are applied to spin variables, as in (4.2.1); and for many purposes it is possible to exploit the group-theoretical properties of the basis functions *without* ever explicitly constructing them. This approach is developed in Sections 4.3–4.6 and in later chapters.

Projection-operator methods

In the "analytic" methods of constructing spin eigenfunctions we start from an *arbitrary* spin function Θ (e.g. one of the 2^N possible products of α and β factors), which may be expressed in terms of the complete set of $\Theta_\kappa^{(S,M)}$ (with all possible choices of coupling scheme, and hence of S, M, κ) as

$$\Theta = \sum_{S,M} C_\kappa^{(S,M)} \Theta_\kappa^{(S,M)} \tag{4.2.6}$$

and "analyse" the function into its component parts $\Theta_\kappa^{(S,M)}$. The analysis is usually performed by applying a projection operator (cf. p. 73), of which there are many forms. The Wigner operators defined in Appendix 3 provide perhaps the most important example: thus, for the group S_N, (A3.23) gives

$$\rho_{\kappa\lambda}^{(\alpha)} \Theta = \frac{d_\alpha}{N!} \sum_{\mathsf{P}} D_\alpha(\mathsf{P})_{\kappa\lambda}^* \mathsf{P} \Theta = \Theta_\kappa^{(\alpha)} \tag{4.2.7}$$

—a function transforming like the κth basis function of the irrep D_α. Since, in the present instance, α corresponds to a particular choice of S, and κ to a particular coupling scheme‡ (while M is fixed by the numbers of α and β factors in Θ), it is clear that $\Theta_\kappa^{(\alpha)}$ may be identified with a unique term in (4.2.6). Unfortunately, the assumption that the matrices $\mathbf{D}_\alpha(\mathsf{P})$ are available is equivalent to assuming the availability of a full set of spin eigenfunctions—which is what we wish to find. Consequently (4.2.7) does not offer a useful *ab initio* method of generating the $\Theta_\kappa^{(\alpha)}$. A more useful variant of (4.2.7) is obtained by using the characters $\chi_\alpha(\mathsf{P})$ as coefficients instead of the $D_\alpha(\mathsf{P})_{\kappa\lambda}$ in accordance with (A3.25), the

‡ Different irreps are in fact distinguished by the values of S alone, those that differ only in M may be chosen to be identical: κ classifies different basis functions *within* an irrep.

characters of the irreducible representations being well known. Let us take as an example the function

$$\Theta = \alpha\beta\alpha$$

and note that the characters of the permutations in the 2-dimensional representation corresponding to $S = \frac{1}{2}$ are $\chi(\mathsf{P}) = 2$ for the identity, $\chi(\mathsf{P}) = 0$ for single interchanges and $\chi(\mathsf{P}) = -1$ for cyclic permutations. The character operator is then‡ $[2 - (123) - (321)]$ and an eigenfunction with $S = M = \frac{1}{2}$ is

$$2\alpha(s_1)\beta(s_2)\alpha(s_3) - \alpha(s_2)\beta(s_3)\alpha(s_1) - \alpha(s_3)\beta(s_1)\alpha(s_2).$$

Thus (assuming arguments in the standard order)

$$\begin{aligned}\Theta^{(1/2,1/2)} &= 2\alpha\beta\alpha - \alpha\alpha\beta - \beta\alpha\alpha \\ &= (\alpha\beta - \beta\alpha)\alpha - \alpha(\alpha\beta - \beta\alpha),\end{aligned}$$

which is a *mixture* of two linearly independent eigenfunctions with the same values of (S, M). The character operator thus projects from an arbitrary Θ not the standard spin functions but rather some unknown linear combination of them. This is not a serious objection. On the other hand, the need to consider the effect of all permutations makes the method unwieldy in dealing with large numbers of electrons.

Perhaps the most direct way of obtaining the (S, M) component of an arbitrary Θ is that given by Löwdin (1956b). This is based on the successive *annihilation* of each *unwanted* component in the expansion

$$\Theta = \sum_{S,M} c^{(S,M)}\Theta^{(S,M)} \tag{4.2.8}$$

(in which the terms of different κ in (4.2.6) have been summed). Since $\mathsf{S}^2\Theta^{(S,M)} = S(S+1)\Theta^{(S,M)}$, it follows that

$$[\mathsf{S}^2 - S(S+1)]\Theta^{(S,M)} = 0. \tag{4.2.9}$$

The operator on the left is thus an annihilator for the component with total spin S. If we wish to remove all components except that with $S = k$, we may therefore employ the operator

$$\mathsf{O}_k = \prod_{j(\neq k)} \frac{\mathsf{S}^2 - j(j+1)}{k(k+1) - j(j+1)}, \tag{4.2.10}$$

where the denominator is usually added merely to ensure that the desired component is left unchanged.

‡ The notation $(ijk \ldots p)$ means "replace i by j, j by k, . . . , p by i".

To give an example of the use of this operator, we first note that the effect of S^2 on any spin product Θ is

$$\mathsf{S}^2\Theta = (M^2 + \tfrac{1}{2}N)\Theta + \sum_{i,j} \Theta_{ij}, \tag{4.2.11}$$

where Θ_{ij} is derived from Θ by interchanging the α and β spin factors with spin variables s_i and s_j *whenever the factors differ,* and the summation is over all such pairs. This result follows readily by adding (a) and (b) in (4.1.8) and noting that Θ is already an eigenfunction of S_z with eigenvalue $M = \frac{1}{2}(n_\alpha - n_\beta)$. If we now take the example used above, we should get a doublet spin eigenfunction by annihilating the quartet component, i.e. by operating with $(\mathsf{S}^2 - \frac{3}{2}\frac{5}{2})$. The result is

$$(\mathsf{S}^2 - \tfrac{15}{4})\alpha\beta\alpha = [(\tfrac{1}{4} + \tfrac{3}{2}) - \tfrac{15}{4}]\alpha\beta\alpha + (\beta\alpha\alpha + \alpha\alpha\beta) = -2\alpha\beta\alpha + \beta\alpha\alpha + \alpha\alpha\beta,$$

which agrees (apart from an arbitrary factor -1) with the function previously obtained. The construction is easily extended to many-electron functions and has been widely used (e.g. Pauncz, 1979).

4.3 PERMUTATION SYMMETRY AND ITS IMPLICATIONS

In the development of the Slater method (Section 3.1) it was noted that the Pauli principle in the form (1.2.27) could always be satisfied by constructing the electronic wavefunction from determinants (i.e. antisymmetrized products) of spin-orbitals. In an earlier section, however, it was shown that for a two-electron system the antisymmetry principle could also be satisfied by writing the wavefunction as a product of *individually* symmetric or antisymmetric factors—one for spatial variables and the other for spin variables. Since, in the usual first approximation the Hamiltonian does not contain spin variables, it is natural to enquire whether a corresponding "exact" N-electron wavefunction might be written as a "space–spin" product in which the spatial factor is an exact eigenfunction of the spinless Hamiltonian (1.2.1). To investigate this possibility, we need a few basic ideas from group theory (Appendix 3).

In the last section it was noted that a full set of linearly independent eigenfunctions $\{\Theta_\kappa^{(S,M)}\}$ of the spin operators S^2 and S_z provides a basis for (or "carries") a representation of the group (S_N) of $N!$ permutation operators applied to the spin variables $s_1, s_2, \ldots, s_N$. This is a general property (Appendix 3): when an operator possesses an "invariance group" (in this case, for example, S^2 is invariant under all permutations of spin labels) every set of degenerate eigenfunctions carries a representation of the group (as, for example, in (4.2.1)); and such a representation is normally irreducible (i.e. it is an irrep) in the sense that the whole set, not just a subset, is needed in describing the effect of the operators.

Exactly parallel considerations apply to the spatial wavefunctions, which are eigenfunctions of a Hamiltonian operator that is invariant under all permutations of *spatial* variables, and a g-fold degenerate set of (spinless) wavefunctions should likewise carry an irrep of S_N. The fundamental question, even for *exact* solutions of the Schrödinger equation, will therefore be how to combine space and spin functions in order to describe a state of given energy and spin, which at the same time satisfies the Pauli principle; the space–spin function, no matter how it be constructed, must be antisymmetric for any permutation applied simultaneously to space and spin variables.

To answer this question, let us first suppose that a given set of spin eigenfunctions carries a representation, which we for denote by D_α, such that (with standard group-theoretical notation) (4.2.1) becomes‡

$$\mathsf{P}\Theta_i^{(\alpha)} = \sum_j \Theta_j^{(\alpha)} D_\alpha(\mathsf{P})_{ji} \qquad (i, j = 1, 2, \ldots, g_\alpha) \tag{4.3.1}$$

and that we also possess a degenerate set of g_β exact solutions $F_k^{(\beta)}$ of the (spinless) Schrödinger equation (of common energy E), so that

$$\mathsf{P}F_k^{(\beta)} = \sum_l F_l^{(\beta)} D_\beta(\mathsf{P})_{lk} \qquad (k, l = 1, 2, \ldots, g_\beta). \tag{4.3.2}$$

Any linear combination of the space–spin products $F_k^{(\beta)}\Theta_i^{(\alpha)}$ will then be a general exact wavefunction, of energy E, with spin included: it may be written accordingly as

$$\Psi = \sum_{i,k} C_{ik} F_k^{(\beta)} \Theta_i^{(\alpha)}, \tag{4.3.3}$$

and the problem is thus how to choose the coefficients C_{ik} so that Ψ will be *antisymmetric* under simultaneous permutations of space and spin variables. We must admit the eventuality that this *may not be possible* for some sets of degenerate solutions of the Schrödinger equation and that such solutions, although mathematically acceptable, must then be rejected as physically unsatisfactory (violating the Pauli principle).

It is in fact possible (see Wigner, 1959) to construct an antisymmetric

‡ A label (e.g. r or s) is often added to a permutation operator P to indicate which variables it affects; when the intention is clear from the the context we shall avoid such extra labels. We revert to Latin indices for basis functions to avoid confusion with the Greek indices (α, β) labelling the irreps.

Ψ if and only if the irreps D_α and D_β are of the same dimension and are related in a special way; they must be "dual" or "associate", in which case we write $D_\beta = D_{\tilde{\alpha}}$ (and conversely $D_\alpha = D_{\tilde{\beta}}$). The matrices of two such irreps are related according to

$$\mathbf{D}_{\tilde{\alpha}}(\mathsf{P}) = \begin{cases} \epsilon_{\mathsf{P}} \tilde{\mathbf{D}}_\alpha(\mathsf{P}^{-1}) & \text{(general case)}, \\ \epsilon_{\mathsf{P}} \mathbf{D}_\alpha(\mathsf{P})^* & \text{(unitary matrices)}, \\ \epsilon_{\mathsf{P}} \mathbf{D}_\alpha(\mathsf{P}) & \text{(real orthogonal matrices)}, \end{cases} \tag{4.3.4}$$

where the tilde on a matrix indicates the transpose. When the two bases are dual the resultant antisymmetric Ψ is given by

$$\Psi = \sum_i F_i^{(\tilde{\alpha})} \Theta_i^{(\alpha)}, \tag{4.3.5}$$

and is thus a sum over the full set of degenerate spatial wavefunctions multiplied by their spin-eigenfunction "partners". The two-electron functions (1.2.26), though not exact eigenfunctions, illustrate this result. For $M = 0$ there are two spin eigenfunctions: $\Theta_1^{(1)}$ (for $S = 1$) carrying a 1-dimensional irrep D_1 with "matrices" 1, 1; and $\Theta_1^{(2)}$ (for $S = 0$) carrying a 1-dimensional irrep D_2 with "matrices" 1, -1. These irreps are dual, $D_2 = D_{\tilde{1}}$ and $D_1 = D_{\tilde{2}}$, and the $S = 1$ spin function (symmetric) can only be combined with a spatial function that carries $D_{\tilde{1}}$ $(= D_2)$, being antisymmetric; but the $S = 0$ spin function, carrying D_2, can only be combined with a spatial function that carries $D_{\tilde{2}}$ $(= D_1)$, being symmetric. For $M = +1$ there are spin functions that provide two further bases for D_1; and for each of these the only way to construct an antisymmetric space–spin function is to multiply by a spatial wavefunction belonging to D_2. Evidently (4.3.5) provides the generalization, for many-electron systems, of the simple procedure used in Section 1.2.

As the two-electron example suggests, the result (4.3.5) may also be used in constructing *approximate* wavefunctions of orbital form. Thus, starting from an orbital product $\phi_1(\mathbf{r}_1)\phi_2(\mathbf{r}_2) \ldots \phi_N(\mathbf{r}_N)$, it will be possible to construct (for example, using the Wigner operators of Appendix 3) a number of independent bases for an irrep $D_{\tilde{\alpha}}$, the dual of an irrep D_α carried by any desired set of spin eigenfunctions of given S, M: by combining the space and spin partners according to (4.3.5), for each spatial basis, we shall then obtain a number of properly antisymmetric space–spin functions, of given S, M and within the orbital configuration $[\phi_1\phi_2 \ldots \phi_N]$. This means we can construct CFs (cf. p. 69) without any reference to Slater determinants. We shall return to this possibility in a more detailed discussion of the CI method (Chapter 10).

4.4 YOUNG TABLEAUX

It was remarked in Section 4.2 that the sets of *branching-diagram* functions carried certain *standard* irreps of S_N, as used by mathematicians, and we shall need to have some acquaintance with these particular representations. The general theory of the irreps of S_N was first developed by Alfred Young, whose approach has been fully discussed by Rutherford (1948). Here we give only a brief descriptive account (see also Coleman, 1968) in order to make the approach intelligible. Further details are available in textbooks on group theory (e.g. Hamermesh, 1962) and in the book by Pauncz (1979).

A typical element P of the group S_N may be regarded as a permutation of N numbered objects (e.g. labelled variables) among N "cells" or "boxes". A particular type of P may permute λ_1 objects of a certain subset among themselves, λ_2 of a second subset among themselves, and so on; always with no mixing of objects in *different* subsets. A set $[\lambda_1, \lambda_2, \ldots]$ with

$$\lambda_1 \geqslant \lambda_2 \geqslant \lambda_3 \geqslant \ldots \quad (\lambda_1 + \lambda_2 + \lambda_3 + \ldots = N) \tag{4.4.1}$$

is a partition of N, and may be indicated by a *Young shape* (or "diagram" or "pattern"), also denoted by $[\lambda_1, \lambda_2, \ldots]$ or more briefly $[\lambda]$, of the form

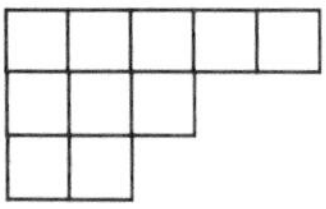

$$[\lambda] = [\lambda_1, \lambda_2, \lambda_3]$$

$$\lambda_1 = 5, \quad \lambda_2 = 3, \quad \lambda_3 = 2$$

with λ_1 cells in the first row, λ_2 in the second, and so on. When the numbers $1, 2, \ldots, N$ are allocated to the cells, in any order, we obtain a *Young tableau*; when the numbers are written sequentially, line by line as in a book, we obtain the *fundamental tableau* $T_1^{[\lambda]}$ (Fig. 4.2); and when, reading along rows and down columns, the numbers always appear in ascending order, we obtain a series of *standard tableaux*. The standard tableaux for S_5, for the partition $[\lambda] = [\lambda_1, \lambda_2] = [3, 2]$, are collected in Fig. 4.2: they appear in a particular order, known as "last-letter

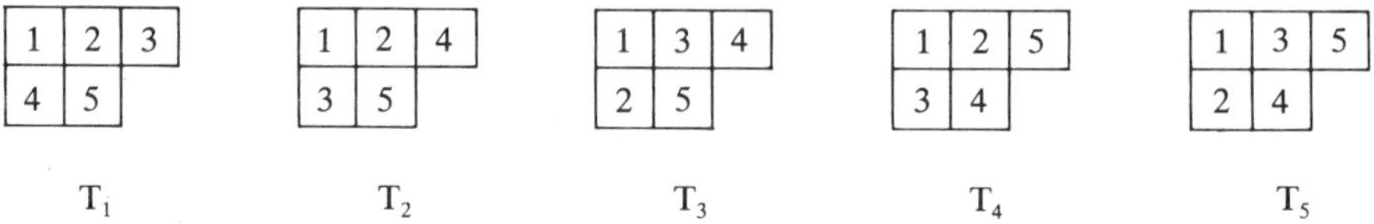

Fig. 4.2 Standard tableaux for $N = 5$ and partition $[3, 2]$.

sequence", in which tableaux with the "last letter" (5) in a given row come before those in which it is in an earlier row, those with 5 in the same row are similarly ordered with respect to the position of the 4, and so on.

The tableaux are important because (i) the shape of a tableau is in one-to-one correspondence with an *irrep* of S_N; (ii) for a given shape, the standard tableaux are in one-to-one correspondence with *basis functions* carrying the irrep, their number thus giving the dimension of the irrep; and (iii) from every set of standard tableaux $\{T_r^{[\lambda]}\}$ it is possible to set up operators (equivalent to linear combinations of permutations) that, when acting upon an arbitrary function of numbered variables, will actually *generate* a set of basis functions that will carry the irrep.

First we define the Young operators. For any tableau $T_r^{[\lambda]}$ it is possible to define a *Young operator* $\mathsf{E}_r^{[\lambda]}$ that, like the antisymmetrizer A in (3.1.7), is a linear combination of permutations (the group elements) and is thus an element of the "group algebra". In dealing with a particular tableau, the labels $[\lambda]$ and r may be suppressed and E may be expressed as

$$\mathsf{E} = \mathsf{NP}, \tag{4.4.2}$$

where the operators P and N are

$$\mathsf{P} = \mathsf{P}_1\mathsf{P}_2 \ldots, \qquad \mathsf{N} = \mathsf{N}_1\mathsf{N}_2 \ldots, \tag{4.4.3}$$

in which P_1 is a sum of all the permutations that move only numbers in row 1, P_2 is a similar sum for row 2, etc., and $\mathsf{N}_1, \mathsf{N}_2, \ldots$ are defined similarly except that the permutations refer to numbers in the *columns* and they are summed with a coefficient ± 1 according to parity, as in (3.1.7). Thus, with the first tableau in Fig. 4.2 we associate an operator

$$\mathsf{E}_1 = [1-(14)][1-(25)][1+(12)+(13)+(23)+(123)+(321)][1+(45)], \tag{4.4.4}$$

where, with the notation used on p. 96, (ijk) for example is the *cyclic* permutation that replaces every number by the one that succeeds it (i by j, j by k, and k by i). The effect of such an operator on a function $\Phi(1, 2, 3, \ldots, N)$ is to permute the indices $1, 2, \ldots, N$ in various ways, summing the results; the resultant function will have certain specific symmetry properties that, as Young was able to show, make it eligible as a basis function for a corresponding irrep of S_N.

Let us construct a spin eigenfunction, which must be some linear combination of the set of standard branching-diagram functions that provide a basis for the irrep $D_{[\lambda]}$ with $[\lambda] = [3, 2]$, by applying the operator E_1 to a "primitive" spin function

$$\theta_1 = \alpha(1)\alpha(2)\alpha(3)\beta(4)\beta(5). \tag{4.4.5}$$

This particular choice clearly simplifies the calculation since θ_1 is already symmetric in the variables 1, 2, 3 and 4, 5, and the P-factors in (4.4.3) are thus redundant, merely multiplying by 12 (for the 6×2 identical terms). The result is

$$\Theta_1 = \mathsf{E}_1\theta_1 = 12[\alpha(1)\beta(4) - \beta(1)\alpha(4)][\alpha(2)\beta(5) - \beta(2)\alpha(5)]\alpha(3), \tag{4.4.6}$$

and this is evidently one of the spin-paired (Rumer) functions of the type (4.2.5); it is not one of the linearly independent set most commonly adopted (in which no crossed arrows appear), but is nevertheless perfectly acceptable as a basis function. As a second function, it might have been expected that $\mathsf{E}_2\theta_1$ (i.e. an alternative projection of the same primitive function) would be suitable. In general, however, this is not so; for an arbitrary primitive f, $\mathsf{E}_1 f$ and $\mathsf{E}_2 f$ are not necessarily partners in the *same basis* for the irrep (see Appendix 3). Let us consider, however, an alternative primitive function

$$\theta_2 = \sigma_{21}\theta_1 = \alpha(1)\alpha(2)\alpha(4)\beta(3)\beta(5),$$

in which σ_{21} is the permutation which leads from tableau T_1 to tableau T_2 ($T_2 = \sigma_{21}T_1$), and operate on θ_2 with the second Young operator E_2. We then find

$$\begin{aligned}\Theta_2 &= \mathsf{E}_2\theta_2 = \mathsf{E}_2\sigma_{21}\theta_1 \\ &= 12[\alpha(1)\beta(3) - \beta(1)\alpha(3)][\alpha(2)\beta(5) - \beta(2)\alpha(5)]\alpha(4),\end{aligned}$$

which is a second spin-paired function. By proceeding in this way, we find five linearly independent spin eigenfunctions, with Rumer diagrams shown in Fig. 4.3,

$$\Theta_1 = \mathsf{E}_1\theta_1, \quad \Theta_2 = \mathsf{E}_{21}\theta_1, \quad \Theta_3 = \mathsf{E}_{31}\theta_1, \quad \Theta_4 = \mathsf{E}_{41}\theta_1, \quad \Theta_5 = \mathsf{E}_{51}\theta_1, \tag{4.4.7}$$

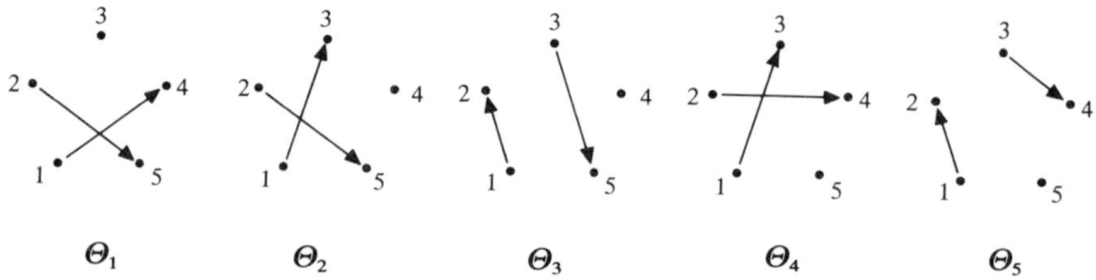

Fig. 4.3 Rumer diagrams corresponding to standard tableaux functions.

where we have introduced (putting $s = 1$)

$$\mathsf{E}_{rs} = \mathsf{E}_r \sigma_{rs}, \tag{4.4.8}$$

which is a generalization, referring to *two* tableaux (T_r, T_s) of the operator defined in (4.4.2). As in the example above, σ_{rs} is the permutation that turns tableau T_s into T_r:

$$\mathrm{T}_r = \sigma_{rs}\mathrm{T}_s. \tag{4.4.9}$$

Young's great achievement was to establish the properties of the operators (4.4.8), to show how they could be used to construct the irreps of S_N, and to devise a simple algorithm which leads to explicit forms of the representation matrices.

4.5 THE STANDARD IRREPS OF S_N

At this point it is already clear that the spin functions listed in (4.4.7) provide a basis for an irrep of S_5; for a general permutation

$$\sigma = \begin{pmatrix} 1 & 2 & 3 & 4 & 5 \\ 1' & 2' & 3' & 4' & 5' \end{pmatrix}$$

applied to Θ_1, say, sends its Rumer diagram (Fig. 4.3) into another diagram with arrows $1' \rightarrow 4'$, $2' \rightarrow 5'$, and point $3'$ free—and these new numbers are simply a permutation of $1, 2, \ldots, 5$. A Rumer diagram with arrows and points permuted in this way, being still a spin eigenfunction with $S = M = \frac{1}{2}$, must be expressible as a linear combination of $\Theta_1, \Theta_2, \ldots, \Theta_5$ (the full linearly independent set), and the expansion coefficients will furnish the first column of the matrix $\mathbf{D}(\sigma)$; and by applying σ to all the basis functions in turn the full matrix can be constructed.

The above observations show that, given a primitive spin product $\theta_s^{[\lambda]}$ in which the variables whose indices stand in the first row of the two-row tableau $\mathrm{T}_s^{[\lambda]}$ are allotted α factors, the rest being given β factors, the "projected" functions

$$\Theta_r^{[\lambda]} = \mathsf{E}_{rs}^{[\lambda]}\theta_s^{[\lambda]} \qquad (s \text{ arbitrary, fixed; } r = 1, 2, \ldots, d_{[\lambda]}) \tag{4.5.1}$$

will provide an irrep $\mathrm{D}_{[\lambda]}$ of S_N. The dimensions of $\mathrm{D}_{[\lambda]}$ are

$$d_{[\lambda]} = f_S^N, \qquad S = \tfrac{1}{2}(\lambda_1 - \lambda_2), \tag{4.5.2}$$

where f_S^N is given in (4.2.2) and the total spin S is just half the excess of α-factors over β-factors. We note that *only two-row tableaux are permitted,* since with more than two rows the antisymmetrizers

$\mathsf{N}_1, \mathsf{N}_2, \ldots$ would annihilate any spin product (at least two αs or βs occurring among the three or more numbers in any column). It is also evident that under the tableau permutations σ_{rs} the basis function $\Theta_r^{[\lambda]}$ in (4.5.1) behave in a particularly simple way:

$$\Theta_r^{[\lambda]} = \sigma_{rs}\Theta_s^{[\lambda]}, \tag{4.5.3}$$

i.e. the basis functions are permuted by the same operations as their corresponding tableaux.

The above properties are not all completely general. In particular, the properties of the $\mathsf{E}_{rs}^{[\lambda]}$ are not identical with those of the group-theoretical projectors defined in Appendix 3, and for many-row tableaux and arbitrary primitive functions they do not in general‡ produce, according to (4.5.1), a set of partners carrying an irrep $D_{[\lambda]}$. Young was able to construct, in terms of the $\mathsf{E}_{rs}^{[\lambda]}$, three types of operator ρ_{rs}, which generate respectively the so-called "natural", "seminormal" and "orthogonal" irreps (all of which are equivalent). We shall be concerned only with the last of these, in which all the matrices are orthogonal, and state Young's general results without proof. In Section 4.6 we show that, for two-row tableaux, the orthogonal irreps can be found from the spin representations based on Rumer functions.

The orthogonal irreps, which from now on are referred to as "standard", may in principle be constructed inductively, starting from the irreps of S_2; constructing those of S_3; and then, assuming all irreps of S_N known, finding a general recipe for passing to those of S_{N+1}. Thus for S_2 there are two irreps, corresponding to partitions $[\lambda]=[2]$ (one row), $[\lambda]=[1, 1]$ (one column), each with one standard tableau:

$$D_{[2]}: \ \begin{array}{|c|c|}\hline 1 & 2 \\ \hline\end{array}\,, \qquad D_{[1,1]}: \ \begin{array}{|c|}\hline 1 \\ \hline 2 \\ \hline\end{array} \tag{4.5.4}$$

Basis functions carrying these two irreps may be denoted by§ $\Phi\left(\begin{array}{|c|c|}\hline 1 & 2 \\ \hline\end{array}\right)$ and $\Phi\left(\begin{array}{|c|}\hline 1 \\ \hline 2 \\ \hline\end{array}\right)$; they are respectively symmetric and antisymmetric under the transposition (12). To obtain irreps of S_3 we add a third cell, obtaining possible shapes

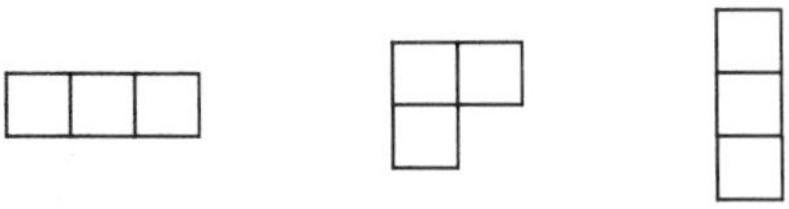

‡ They do, however, serve this purpose for $N \leqslant 5$.

§ Here we use Φ for functions constructed from an *arbitrary* primitive, ϕ, not necessarily a spin function.

Corresponding to these three shapes, there will be three irreps: the first and last are unidimensional, with symmetric and antisymmetric basis functions that we may denote by

$$\mathrm{D}_{[3]}:\quad \Phi_1^{(3)} = \Phi\left(\begin{array}{|c|c|c|}\hline 1 & 2 & 3 \\ \hline\end{array}\right), \qquad \mathrm{D}_{[1^3]}:\quad \Phi_1^{[1^3]} = \Phi\left(\begin{array}{|c|}\hline 1 \\ \hline 2 \\ \hline 3 \\ \hline\end{array}\right).$$

But the second shape admits two standard tableaux, and thus yields a *two*-dimensional irrep with basis functions

$$\mathrm{D}_{[2,1]}:\quad \Phi_1^{[2,1]} = \Phi\left(\begin{array}{|c|c|}\hline 1 & 2 \\ \hline 3 & \multicolumn{1}{c}{} \\ \cline{1-1}\end{array}\right), \qquad \Phi_2^{[2,1]} = \Phi\left(\begin{array}{|c|c|}\hline 1 & 3 \\ \hline 2 & \multicolumn{1}{c}{} \\ \cline{1-1}\end{array}\right).$$

It is easily verified that, using an arbitrary primitive $\phi(1, 2, 3)$, application of the operators $\mathsf{E}_1^{[2,1]}$ and $\mathsf{E}_2^{[2,1]}$ does in this case lead to a pair of functions (Φ_1, Φ_2) that carry the irrep, since the matrices associated with the *generators* of the group, (12) and (23), can be set up and matrix multiplication then yields the rest. This representation is not the one we want, because the matrices are not orthogonal; but by means of a basis change (Section 2.2) we can indeed find an orthogonal irrep. To make the transformation unique (apart from arbitrary normalization and phase factors), it is sufficient to require that (i) when attention is confined to the permutations of S_2 the new functions will provide a standard basis for the irreps of S_2; and (ii) the matrix of the new permutation (23), which introduces the "new" number (3), will appear in orthogonal form. In this example, the matrices of (12) and (23) in the orthogonal irrep $\mathrm{D}_{[2,1]}$ of S_3 are found to be

$$\mathrm{D}_{[2,1]}\quad :\quad (12)\rightarrow\begin{pmatrix}1 & 0\\ 0 & -1\end{pmatrix}, \qquad (23)\rightarrow\begin{pmatrix}-\frac{1}{2} & -\frac{1}{2}\sqrt{3}\\ \frac{1}{2}\sqrt{3} & \frac{1}{2}\end{pmatrix}.$$

On confining attention to permutations of the subgroup S_2 (i.e. considering only (12) and the identity), the basis functions $\Phi_1^{[2,1]}$ and $\Phi_2^{[2,1]}$ behave exactly like the basis functions $\Phi_1^{[2]}$ and $\Phi_1^{[1^2]}$ respectively that carry standard irreps of S_2; and the corresponding matrices appear in block form, the diagonal blocks being the (1-element) "matrices" already known for S_2.

The generalization of the above procedure would be to assume, as an inductive hypothesis, that all irreps of S_{N-1} were known; that any irrep of S_N would assume block form on considering only permutations of $1, 2, \ldots, N-1$, the blocks reproducing the irreps of S_{N-1}; and that the matrix of the "new" permutation $(N-1, N)$ could be found in orthogonal form. All the other new matrices, for permutations involving the number N, could then be found by matrix multiplication. Such an inelegant approach was completely avoided by Young, who constructed irreps purely in terms of the elements E_{rs} of the group algebra, without ever using basis functions.

Young was able to formulate an extremely simple algorithm, which we now present, for writing down the few non-zero elements of the matrix associated with the transposition $(N-1, N)$, thus providing the means of passing from the irreps of S_{N-1} to those of S_N. To this end, it is convenient to consider the process of *reduction,* assuming an irrep $D_{[\lambda]}$ of S_N to be known and asking what happens to the matrices on confining attention to the operations of the subgroup S_{N-1}. Again, an example is helpful. We take the irrep $D_{[3,2]}$ of S_5, whose standard tableaux appear in Fig. 4.2, and ask how the corresponding basis functions must behave under the permutations of S_4; this simply means ignoring or "scratching out" the cells containing 5. The situation is then as shown in Fig. 4.4. Within the subgroup S_4, the functions $\Phi(T_1)$, $\Phi(T_2)$, $\Phi(T_3)$ carry (by construction) the same irrep $D_{[3,1]}$ as the standard basis of functions $\Phi(T_1'), \Phi(T_2')$, $\Phi(T_3')$ with $N=4$; while $\Phi(T_4)$, $\Phi(T_5)$ carry a second irrep $D_{[2,2]}$, which again must coincide with that carried by tableau functions $\Phi(T_1'')$, $\Phi(T_2'')$ for $N=4$. Thus, for any permutation belonging to S_4, the

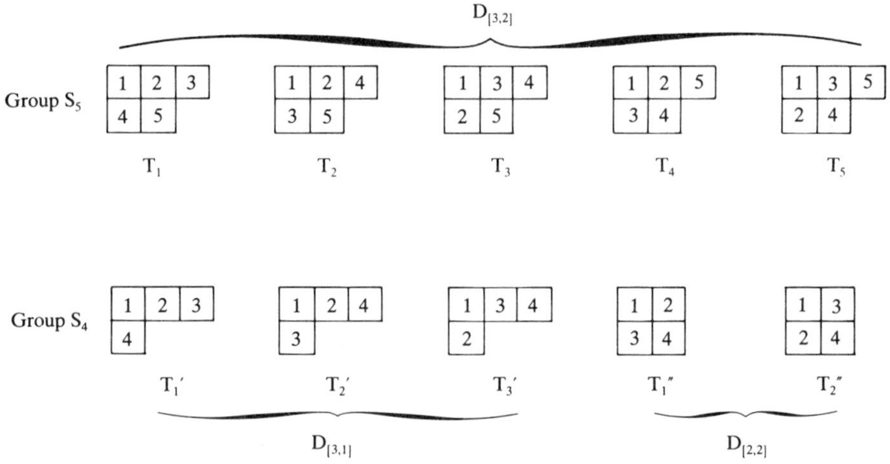

Fig. 4.4 Subduction of a representation. On passing from S_5 to the subgroup S_4, the irrep $D_{[3,2]}$ reduces to the direct sum of $D_{[3,1]}$ and $D_{[2,2]}$.

matrices must assume the block form

$$\mathbf{D}_{[3,2]}(\mathsf{P}) = \begin{array}{c} \\ \mathrm{T}_1 \\ \mathrm{T}_2 \\ \mathrm{T}_3 \\ \mathrm{T}_4 \\ \mathrm{T}_5 \end{array} \begin{array}{c|c} \mathrm{T}_1\ \mathrm{T}_2\ \mathrm{T}_3 & \mathrm{T}_4\ \mathrm{T}_5 \\ \hline \mathbf{D}_{[3,1]}(\mathsf{P}) & \mathbf{0} \\ \hline \mathbf{0} & \mathbf{D}_{[2,2]}(\mathsf{P}) \end{array} = \mathbf{D}_{[3,1]}(\mathsf{P}) \oplus \mathbf{D}_{[2,2]}(\mathsf{P}), \tag{4.5.5}$$

where $\oplus$ indicates the "direct sum" in which the component matrices appear as diagonal blocks of the larger matrix (Appendix 3). The permutations of S_4 mix only functions that correspond to tableaux of the same shape, and the off-diagonal blocks are therefore filled with zeros. In other words, $D_{[3,2]}$ is *reduced* (Appendix 3) on confining attention to the subgroup S_4 to the direct sum of D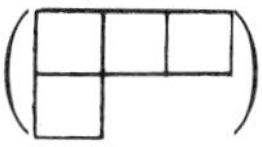 and D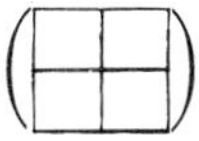. This process is known as *subduction*. It should be noted that the last-letter sequence is always preserved, within each block, when the last letter (5) is scratched out. The order of the basis functions, within each block, is consequently always standard.

For permutations P that do *not* fall within S_4, all of which may be generated by combining permutations of S_4 with the transposition‡ P_{45}, the off-diagonal blocks in (4.5.5) may contain non-zero elements. For the generator (45), however, the whole matrix is extremely sparse. In the general case, the non-zero elements $D_{[\lambda]}(\mathsf{P}_{n-1,n})_{rs}$, between tableau functions $\Phi(\mathrm{T}_r)$ and $\Phi(\mathrm{T}_s)$, are determined by the simple pictorial rules given in Table 4.1. In this table d is the so-called "axial distance" between the cells that contain $n-1$ and n, defined as the number of unit steps along the rectangular paths indicated, and ρ is simply $1/d$; all numbers not shown must match exactly. All other matrix elements are zero.

The rules in Table 4.1 are very easily applied. Thus, referring to Fig. 4.4, the non-zero matrix elements in $D_{[3,2]}$ of the transposition P_{45} will be (with an obvious abbreviation)

$$D_1 = 1, \qquad D_{22} = D_{33} = -\tfrac{1}{2}, \qquad D_{44} = D_{55} = \tfrac{1}{2}, \qquad D_{24} = D_{35} = (\tfrac{3}{4})^{1/2}.$$

Given the diagonal blocks in (4.5.5), i.e. the component irreps of S_4, it is then possible to construct all the matrices of the irrep $D_{[3,2]}$ of S_5. To

‡ To avoid multiple parentheses, (45) is replaced by P_{45}, (ijk) by P_{ijk}, etc.

Table 4.1 Young's algorithm for the matrix elements of the transposition $(N-1, N)$ between tableau functions $\Phi(\mathrm{T}_r)$, $\Phi(\mathrm{T}_s)$.

		Tableau T_r	rr-element
Diagonal elements:	(i) contains	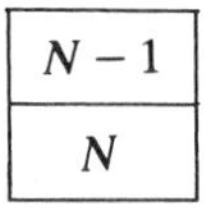	-1
	(ii) contains	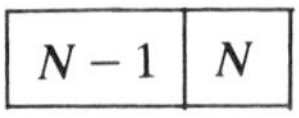	$+1$
	(iii) contains	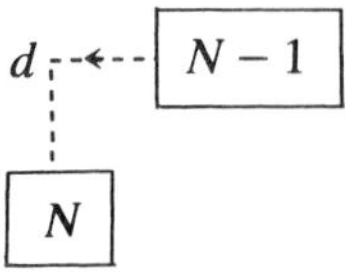	$-\rho$
	(iv) contains	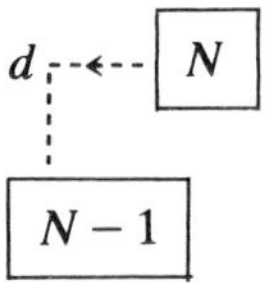	$+\rho$

	Tableau T_r	Tableau T_s	rs-element
	contains	contains	
Off-diagonal elements	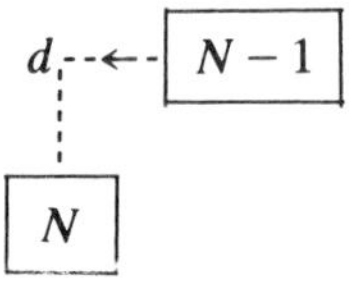	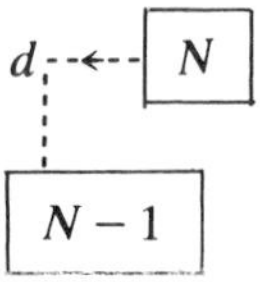	$(1-\rho^2)^{1/2}$

obtain the matrix for transposition (35), for example, it is only necessary to use

$$(jk) = (ji)(ik)(ji),$$

in which i is replaced by j within the cycle (ik), to obtain the corresponding matrix equation

$$\mathbf{D}(\mathsf{P}_{35}) = \mathbf{D}(\mathsf{P}_{34})\mathbf{D}(\mathsf{P}_{45})\mathbf{D}(\mathsf{P}_{34}),$$

in which all matrices refer to generators of the form $(n-1, n)$, as determined by Young's algorithm. The matrices of all transpositions (ij)

may be constructed in this way for any S_N (see e.g. Rettrup (1986) for an efficient computational scheme), and, since a general permutation may always be expressed as a sequence of transpositions, it is in principle possible to generate explicitly all the matrices of any irrep. For some applications, as will be seen later, it is sufficient to calculate only the matrices of the transpositions. If *all* matrices are required then there are indeed practical difficulties (for $N=10$ there are 3 628 800 matrices in each irrep); but the matrices of any $D_{[\lambda]}$ may now be assumed "known", at least in principle.

The above results are of great importance for two main reasons: they allow us to regard the Wigner operators

$$\rho_{rs}^{[\lambda]} = \frac{d_{[\lambda]}}{N!} \sum_{\mathsf{P}} D_{[\lambda]}(\mathsf{P})_{rs} \mathsf{P} \tag{4.5.7}$$

as "known" objects, which prove to be valuable in CI calculations (Chapter 10), where they lead on to the study of unitary-group techniques; and they allow us to classify functions of given permutational symmetry according to the "chains" of irreps in which they appear during the subduction $S_N \to S_{N-1} \to \ldots \to S_1$—a "genealogical" classification, which is again valuable in CI calculations, where vast numbers of CFs need to be sorted and processed.

As a simple example of a subduction chain it is sufficient to consider again the group S_5. Thus $\Phi(T_3)$, a standard basis function associated with tableau T_3 in Fig. 4.4, remains a basis function in each irrep of the sequence

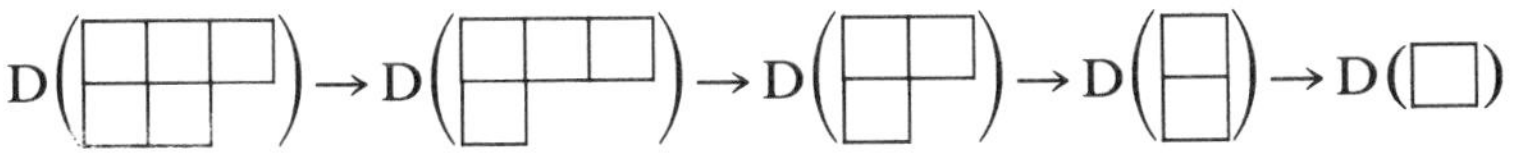

on specializing, successively, to permutations of the first 4, 3, 2, 1 variables; but $\Phi(T_4)$ belongs to the chain

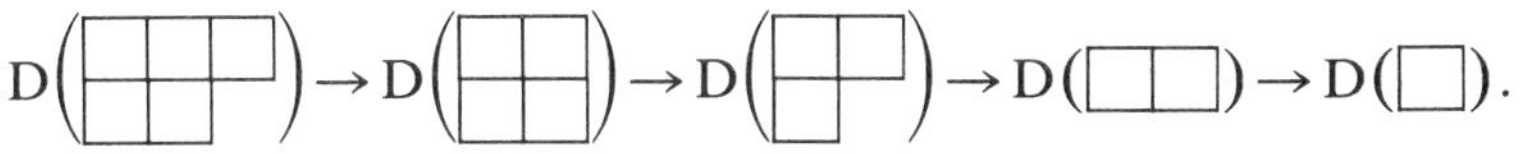

Consequently it is easy to see, for instance, that $\Phi(T_3)$ is antisymmetric under the transposition (12), while $\Phi(T_4)$ is symmetric. If the irreps are numbered then each basis function will be completely specificed by giving a string of integers that indicate its genealogy—a very neat labelling scheme for computational purposes.

4.6 CONNECTIONS BETWEEN DIFFERENT BASES

We conclude this chapter with two final observations. First, the irrep that is the dual of any given $D_{[\lambda]}$ is obtained immediately from (4.3.4); and, since the standard irreps are orthogonal, it is only necessary to multiply the matrices $\mathbf{D}_{[\lambda]}(\mathsf{P})$ by ± 1, according to the parity of P, so that

$$\mathbf{D}_{[\tilde{\lambda}]}(\mathsf{P}) = \epsilon_{\mathsf{P}} \mathbf{D}_{[\lambda]}(\mathsf{P}). \tag{4.6.1}$$

It may be verified that these matrices are obtained also, using Young's algorithm, by associating with $D_{[\tilde{\lambda}]}$ the tableaux that follow on interchanging the rows and columns of the tableaux for $D_{[\lambda]}$. Thus, for S_5, the tableaux in Fig. 4.4 yield a dual set as in Fig. 4.5. It should be noted that these tableaux appear in *inverse* last-letter sequence; thus, when we use (4.3.5) to construct an antisymmetric wavefunction, the spin functions that carry a two-row irrep $D_{[\lambda]}$ being combined with their partners ($\Phi_i^{[\tilde{\lambda}]}$), the latter must be read in reverse order. It is also clear that the only acceptable *spatial* wavefunctions must possess the symmetry properties associated with the basis functions derived from two-*column* tableaux. Since, with a spinless Hamiltonian, the energy of a system is determined by any one spatial factor in (4.3.5), it is also possible to calculate an energy expectation value (with a wavefunction of the form (4.3.5)) *without any reference to spin,* the Θ factors disappearing in the spin integrations. In this sense, and in the approximation of a spinless Hamiltonian, the spin itself has no bearing on the energy of a system—it is enough to specify the appropriate *symmetries* of the spatial wavefunction (corresponding to a two-column tableau), and the spin, as Van Vleck and Sherman (1935) first remarked, serves only as an "indicator" of permissible symmetries. This observation leads to the "spin-free quantum chemistry" of Matsen (1964), to which we return in later chapters.

Secondly, it was stated earlier (p. 95) that the branching diagram spin functions also carried the standard (two-row) irreps of S_N; and that the Rumer functions also carried irreps, although of non-standard form. The

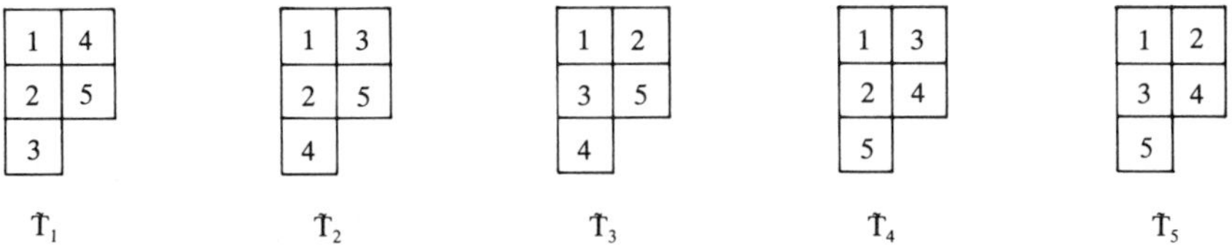

Fig. 4.5 Standard tableaux for $N = 5$ and partition [2, 2, 1]. These tableaux are presented as the *duals* of those in Fig. 4.4, and accordingly appear in *inverse* last-letter sequence.

demonstration (see e.g. Pauncz, 1967) that branching-diagram functions with $S=\frac{1}{2}(\lambda_1-\lambda_2)$ carry the same irreps as the tableau basis functions for $\mathrm{D}_{[\lambda_1\lambda_2]}$ depends on the identification of $\mathbf{D}_S(\mathsf{P}_{N,N-1})$ with $\mathbf{D}_{[\lambda_1\lambda_2]}(\mathsf{P}_{N,N-1})$, the former being obtained by reference to the spin eigenfunctions with $S=\frac{1}{2}(\lambda_1-\lambda_2)$, the latter by reference to Young's algorithm. For $N=2$, 3 the matrices coincide, and the rest follows by induction.

To establish the connection with the Rumer functions, we first note that the *final* function in the last-letter sequence (Fig. 4.2), produced by applying to a suitably "matched" spin product θ_f the Young operator E_f for the final tableau, coincides with the branching diagram function for the "bottom path" (/\/\/) in Fig. 4.1; inspection of the sequence of spin couplings shows that this is generally true. The remaining Rumer functions may all be generated from Θ_f, according to (4.5.3), by applying tableau permutations σ_{rf}. Since the full set spans the standard irrep carried by the branching-diagram functions ($\bar{\Theta}_s$, say), we may express the Rumer functions in the form

$$\Theta_r = \mathsf{E}_{rf}\theta_f = \sum_s C_s^{(r)}\bar{\Theta}_s \tag{4.6.2}$$

and then to try to determine the coefficients.

It may in fact be shown that the sum in (4.6.2) contains no terms corresponding to tableaux that stand to the *left* of T_r in the last-letter sequence. One proof (following from Pauncz, 1979, Section 7.5) depends on adopting the result as an inductive hypothesis, obviously satisfied by $\Theta_r=\Theta_f$, and showing that the permutation σ_{tr} that carries Θ_r into Θ_t produces a function with the same property. (Other proofs, based on the algebraic properties of the E_{rs} (Rutherford, 1948), are more intricate.) The relationship between the Rumer and the standard bases is thus found to be "triangular", provided that the basis functions are written in standard order: in symbols

$$\boldsymbol{\Theta}=\bar{\boldsymbol{\Theta}}\mathbf{V}, \qquad \bar{\boldsymbol{\Theta}}=\boldsymbol{\Theta}\mathbf{U} \qquad (\mathbf{V},\ \mathbf{U}\ \text{upper-triangular}) \tag{4.6.3}$$

It is then easy to identify the expansion coefficients. Thus

$$\bar{\Theta}_f=\Theta_f, \qquad \Theta_{f-1}=a\Theta_{f-1}+b\Theta_f, \qquad \text{etc.,}$$

and, on introducing the usual metric

$$\langle\Theta_r\mid\Theta_s\rangle=\int\Theta_r^*(s_1,\ldots s_N)\Theta_s(s_1,\ldots s_N)\,\mathrm{d}s_1\ldots\mathrm{d}s_N$$

and requiring that the basis $\bar{\boldsymbol{\Theta}}$ be orthonormal, we recognize that the determination of the coefficients is accomplished by the Schmidt process (p. 33). The one-to-one correspondence of basis functions and tableaux, and the fact that the determination of the next coefficient is unique at every stage, ensure that the orthonormal basis generated is indeed the standard basis.

PROBLEMS 4

4.1 Establish the alternative forms (4.1.8a,b) of the operator S^2, starting from the commutation rules (1.2.20).

4.2 Verify that the spin functions in (4.2.3) and (4.2.4) are eigenfunctions of S^2 and S_z, determining their eigenvalues.

4.3 Use (4.1.8) to show that $\mathsf{S}_x\alpha = \frac{1}{2}\beta$, $\mathsf{S}_x\beta = \frac{1}{2}\alpha$, $\mathsf{S}_y\alpha = \frac{1}{2}i\beta$, $\mathsf{S}_y\beta = -\frac{1}{2}i\alpha$, and hence obtain a matrix representation of the operators relative to the basis $\boldsymbol{\eta} = (\alpha\ \beta)$. These matrices, together with that for S_z, are essentially the famous "Pauli matrices".

4.4 Show that exchange of the spin variables s_i, s_j, in any spin function whatever, may be achieved by applying the operator $\mathsf{P}_{ij} = \frac{1}{2}[1 + 4\mathbf{S}(i) \cdot \mathbf{S}(j)]$—a result known as "Dirac's spin-exchange identity". [*Hint:* Any spin product will contain one of the pairs $\alpha(s_i)\alpha(s_j)$, $\alpha(s_i)\beta(s_j)$, $\beta(s_i)\alpha(s_j)$ or $\beta(s_i)\alpha(s_j)$. Express the spin scalar product in terms of components and use the results of Problem 4.3].

4.5 Use the matrix representation in Problem 4.3 to obtain eigenvalues and eigenvectors of the operator S_x. What is the physical significance of your results?

4.6 From experiment it is inferred that $\mathsf{S}_z\alpha = \frac{1}{2}\alpha$, $\mathsf{S}_z\beta = -\frac{1}{2}\beta$ (Stern–Gerlach experiment, field along the z axis). Show, assuming only that space is isotropic (i.e. all directions are equivalent), that the spin operators for a spin-$\frac{1}{2}$ particle must anticommute, i.e. $\mathsf{S}_x\mathsf{S}_y + \mathsf{S}_y\mathsf{S}_x = 0$ etc. Then use simple algebra to *derive* the results of Problem 4.3 without any other assumptions. Hence obtain the commutation relations (1.2.20) as a consequence of the isotropy of space. [*Hint:* Note that $\mathsf{S}_z^2 = \frac{1}{4}\mathsf{I}$ (I being the unit operator) and that on choosing an alternative z axis, z' say, components (and hence associated operators) must change accordingly: $\mathsf{S}_z \rightarrow \mathsf{S}_{z'} = l\mathsf{S}_x + m\mathsf{S}_y + n\mathsf{S}_z$, l, m, n being direction cosines of the new axis relative to the old axes. $\mathsf{S}_{z'}$ must have the same properties as S_z. A choice of sign is open at one stage; it fixes whether the coordinate system is left-handed or right-handed.]

4.7 Follow the prescription for obtaining branching-diagram functions (given in Appendix 3, p. 530) to construct the two linearly independent singlet functions for a 4-electron system. Show that the "bottom path" leads to a Rumer function, and the "top path" to a linear combination of two Rumer functions with non-crossing links.

4.8 Use the 3-electron doublet functions for $S = M = \frac{1}{2}$ (available from Problem

4.7) to obtain a matrix representation of the set of 3! permutations of spin variables. (This is an irreducible representation or 'irrep' of the permutation group S_3.) [*Hint:* Verify that the 6 permutations can all be expressed in terms of two *generators* P_{12} and P_{23}; obtain, referring to (4.3.1), the corresponding matrices $\mathbf{D}_S(P_{ij})$ ($S = \frac{1}{2}$), and obtain the rest by matrix multiplication.]

4.9 Use the irrep set up in Problem 4.8 to define the Wigner operators $\rho_{\kappa\lambda}$ (κ, $\lambda = 1, 2$) that appear in (4.2.7). Verify that on applying them to the spin product $\theta = \alpha\beta\alpha$ they reproduce the branching-diagram functions.

4.10 Confirm (4.2.15) *et seq.*, either by the route suggested or by using the result obtained in Problem 4.4.

4.11 Set up the irrep $D_{\tilde{S}}$, dual to that obtained in Problem 4.8, and use it to form Wigner operators as in Problem 4.9. Apply a suitable pair of operators to a product $a(\mathbf{r}_1)b(\mathbf{r}_2)c(\mathbf{r}_3)$ to generate spatial functions that behave, under permutations of spatial variables, like the two basis vectors. Construct the function (4.3.5) and show that it is indeed antisymmetric under space–spin permutations P_{12}, P_{23}, and hence satisfies the Pauli principle.

4.12 To what linear combination of Slater determinants is the function obtained in Problem 4.11 equivalent? Show that a *second* function may be obtained, by using an alternative pair of Wigner operators, and hence that there are *two* linearly independent doublet functions (with $S = M = \frac{1}{2}$) for the configuration $[abc]$.

4.13 Use the Young tableaux in Fig. 4.2 and the matrix element rules in Table 4.1 to obtain the matrices of the generators P_{12}, P_{23}, P_{34}, P_{45} in the irrep D_λ with $\lambda = [3, 2]$.

4.14 By excluding the generators in Problem 4.13 that contain the number 5, then 4, and so on, verify the successive reductions of D_λ to diagonal block form. Indicate the subduction chains for all the basis functions of the 5-dimensional irrep with $\lambda = [3, 2]$.

4.15 Confirm, by applying the rules in Table 4.1 to the 2-*column* tableaux in Fig. 4.5, that the matrices of the generators of $D_{\tilde{\lambda}}$ with $\tilde{\lambda} = [2, 2, 1]$ are related to those of D_λ (obtained in Problem 4.13) according to (4.6.1).

4.16 Obtain a variational energy expression for a wavefunction in which 3 electrons are assigned to 3 orbitals (different and non-orthogonal) in the order $a(\mathbf{r}_1)b(\mathbf{r}_2)c(\mathbf{r}_3)$, with spins coupled "up–down–up", by using the spin-free approach (p. 110) with a corresponding Wigner operator ρ. [*Hint:* Use diagonal elements of the matrices in Problem 4.11 to generate a function of species corresponding to the first tableau in Fig. 4.5; then reduce the expectation value to $\langle \Omega | \mathsf{H} | \rho\Omega \rangle$, which contains only 1- and 2-electron integrals multiplied by chains of overlaps.]

4.17 The result obtained in Problem 4.16 is not the most general energy expression for the configuration [*abc*], since (see Problem 4.12) there are *two* functions of each symmetry species. Show that a second function is obtained on applying ρ_{12} to $a(\boldsymbol{r}_1)b(\boldsymbol{r}_2)c(\boldsymbol{r}_3)$ *or* by applying the original ρ_{11} to the new product $a(\boldsymbol{r}_1)c(\boldsymbol{r}_2)b(\boldsymbol{r}_3)$. Set up equations to optimize the mixing coefficients in the two-term representation of the ground state. [*Hint:* Proceed as in Problem 4.16, testing the projected functions carefully for possible non-orthogonality.]

5 Digression: The Electron Distribution

5.1 ELECTRON DENSITY FUNCTIONS

We interrupt the logical development at this point in order to discuss the link between the formal mathematics and the physical situation that we are trying to describe. Wavefunctions for molecules become more and more elaborate as computational facilities improve. They involve the coordinates and spins of all electrons, and usually consist of large numbers of determinants constructed from orbitals that may in turn be linear combinations of basis functions containing various numerical parameters. Consequently, the wavefunction itself gives us no clear and simple picture of the electron distribution and how it determines physical and chemical properties. If wavefunctions are to give any understanding of electronic structure, and any general means of relating different properties and comparing the merits of one description with those of another, we must try to extract in some way information about the physically essential features of the electron distribution. This information is in fact contained in a small number of density functions that are comparatively easy to visualize: an electron density, for example, depends on the coordinates of a single point in space, whereas Ψ itself depends on N sets of ccordinates and spins in a many-dimensional configuration space. In this chapter we introduce the density functions in a general way (McWeeny, 1954a, 1955b, 1960; Löwdin, 1955), and discuss in more detail those that are most familiar—the electron density, the spin density and the pair function.

First we consider a single electron in orbital ϕ_A with spin $+\frac{1}{2}$. The wavefunction is then the spin-orbital $\psi_A(\boldsymbol{x}) = \phi_A(\boldsymbol{r})\alpha(s)$, where $\boldsymbol{x}$ as usual stands for the space ($\boldsymbol{r}$) and spin (s) variables collectively, and the statistical interpretation is simple: $|\psi_A|^2\,\mathrm{d}\boldsymbol{x}$ is the probability of finding the electron in (space–spin) volume element $\mathrm{d}\boldsymbol{x}$. In other words, the probability of finding the electron in volume element $\mathrm{d}\boldsymbol{r}$ and with spin between s and $s + ds$ is determined by the *density function*

$$\rho(\boldsymbol{x}) = |\psi_A(\boldsymbol{x})|^2. \tag{5.1.1}$$

More fully, for the spin-orbital $\psi_A(\boldsymbol{x}) = \phi_A(\boldsymbol{r})\alpha(s)$,

$$\begin{pmatrix}\text{probability of}\\ \text{electron in } d\boldsymbol{x}\end{pmatrix} = \rho(\boldsymbol{x})\, d\boldsymbol{x} = |\phi_A(\boldsymbol{r})|^2\, |\alpha(s)|^2\, d\boldsymbol{r}\, ds. \tag{5.1.2}$$

This quantity is zero unless $s = +\frac{1}{2}$, since this is an "up-spin" electron and the spin factor $\alpha(s)$ is interpreted as in Section 1.2. If we are not interested in spin, but only in where the electron is, we can sum over all spin possibilities (i.e. integrate over spin) and obtain the probability of the electron being found in $d\boldsymbol{r}$ with *any* spin. We write the result in the form

$$P(\boldsymbol{r})\, d\boldsymbol{r} = d\boldsymbol{r} \int \rho(\boldsymbol{x})\, ds = |\phi_A(\boldsymbol{r})|^2\, d\boldsymbol{r}$$

(just as if the electron had no spin and was simply put into orbital ϕ_A). The function $\rho(\boldsymbol{x})$ is a probability density, including the spin description, while

$$P(\boldsymbol{r}) = \int \rho(\boldsymbol{x})\, ds \tag{5.1.3}$$

is a probability density without reference to spin, and is obtained by summing (integrating) over spin. These are the prototypes of functions that can be defined for a many-electron system. P is particularly useful because for many purposes it is possible to regard the electron as actually "smeared out" with density P.

We now generalize to the case of many electrons, where the wavefunction is $\Psi(\boldsymbol{x}_1, \boldsymbol{x}_2, \ldots, \boldsymbol{x}_N)$ and has the interpretation

$$\Psi(\boldsymbol{x}_1, \boldsymbol{x}_2, \ldots, \boldsymbol{x}_N)\Psi^*(\boldsymbol{x}_1, \boldsymbol{x}_2, \ldots, \boldsymbol{x}_N)\, d\boldsymbol{x}_1\, d\boldsymbol{x}_2 \ldots d\boldsymbol{x}_N = \begin{pmatrix}\text{probability of electron 1 in } d\boldsymbol{x}_1, \text{ electron 2}\\ \text{simultaneously in } d\boldsymbol{x}_2, \text{ etc.}\end{pmatrix}. \tag{5.1.4}$$

The probability of electron 1 in $d\boldsymbol{x}_1$, other electrons *anywhere* is then

$$d\boldsymbol{x}_1 \int \Psi(\boldsymbol{x}_1, \boldsymbol{x}_2, \ldots, \boldsymbol{x}_N)\Psi^*(\boldsymbol{x}_1, \boldsymbol{x}_2, \ldots, \boldsymbol{x}_N)\, d\boldsymbol{x}_2 \ldots d\boldsymbol{x}_N,$$

and the probability of finding *any* of the N electrons in $d\boldsymbol{x}_1$ is N times this, since the product $\Psi\Psi^*$ is completely symmetrical in the variables of the N electrons and each one therefore has the same probability of being in a given volume element. We write this probability as $\rho_1(\boldsymbol{x}_1)\, d\boldsymbol{x}_1$, where the density function $\rho_1(\boldsymbol{x}_1)$ is defined by

$$\rho_1(\boldsymbol{x}_1) = N \int \Psi(\boldsymbol{x}_1, \boldsymbol{x}_2, \ldots, \boldsymbol{x}_N)\Psi^*(\boldsymbol{x}_1, \boldsymbol{x}_2, \ldots, \boldsymbol{x}_N)\,\mathrm{d}\boldsymbol{x}_2 \ldots \mathrm{d}\boldsymbol{x}_N. \quad (5.1.5)$$

It should be noted that $\boldsymbol{x}_1$ on the left refers to "point 1" at which the density is evaluated rather than to the coordinates of "electron 1", the indistinguishability of the electrons being properly accounted for by the factor N: we could of course use, say, $\boldsymbol{x}$ for an arbitrary point, but later we wish to refer to several different points in space and therefore retain the subscript, speaking of point 1, point 2, etc. The interpretation is always clear and no confusion should arise. Equation (5.1.5) is the many-electron generalization of (5.1.1).

The probability of finding an electron in $\mathrm{d}\boldsymbol{r}_1$ without reference to spin may again be obtained, as in establishing (5.1.3), by integrating over spin. Thus

$$P_1(\boldsymbol{r}_1) = \int \rho_1(\boldsymbol{x}_1)\,\mathrm{d}s_1 \quad (5.1.6)$$

is the probability per unit volume of finding an electron in $\mathrm{d}\boldsymbol{r}_1$ at point $\boldsymbol{r}_1$, regardless of spin. This is the ordinary *electron density function* measured by X-ray crystallographers.

We have put a subscript 1 on the density functions to indicate reference to *one* particle; but (exactly as in the theory of liquids) it is also possible to introduce probabilities for different configurations or "clusters" of any number of particles. Thus

$$\rho_2(\boldsymbol{x}_1, \boldsymbol{x}_2) = N(N-1) \int \Psi(\boldsymbol{x}_1, \boldsymbol{x}_2, \ldots, \boldsymbol{x}_N)\Psi^*(\boldsymbol{x}_1, \boldsymbol{x}_2, \ldots, \boldsymbol{x}_N)\,\mathrm{d}\boldsymbol{x}_3 \ldots \mathrm{d}\boldsymbol{x}_N \quad (5.1.7)$$

determines the probability of two electrons (*any* two) being found simultaneously at "points" $\boldsymbol{x}_1$, $\boldsymbol{x}_2$ (spin included), while

$$P_2(\boldsymbol{r}_1, \boldsymbol{r}_2) = \int \rho_2(\boldsymbol{x}_1, \boldsymbol{x}_2)\,\mathrm{d}s_1\,\mathrm{d}s_2 \quad (5.1.8)$$

is the probability of finding them at $\boldsymbol{r}_1$ and $\boldsymbol{r}_2$ (in ordinary space) with *any* combination of spins (one up, one down; both up, both down). The two-particle functions tell us how the motions of two different electrons are "correlated" as a result of their interactions, as will be apparent later.

Because electrons interact only in pairs (i.e. there are no specifically many-body effects), there is no need to consider distribution functions higher than the pair function P_2, or, if spin-dependent properties are considered, ρ_2. It is therefore useful to simplify the notation by using ρ and π (without subscripts) for the electron density and pair function (ρ_1 and ρ_2), with spin included, and corresponding capital letters, P and Π, for their spinless counterparts (P_1 and P_2). Thus

$$\rho_1, \rho_2 \rightarrow \rho, \pi,$$
$$P_1, P_2 \rightarrow P, \Pi,$$

and from now on we adopt this notation.

An example

To clarify the preceding ideas, we consider two states of the configuration 1s 2s of the helium atom. With A and B denoting the two (orthogonal) orbitals, the approximate wavefunctions of the singlet and triplet states (Section 1.2) are

$$^1\Psi(\mathbf{x}_1, \mathbf{x}_2) = \tfrac{1}{2}[A(\mathbf{r}_1)B(\mathbf{r}_2) + B(\mathbf{r}_1)A(\mathbf{r}_2)][\alpha(s_1)\beta(s_2) - \beta(s_1)\alpha(s_2)]$$

$$^3\Psi(\mathbf{x}_1, \mathbf{x}_2) = \tfrac{1}{2}\sqrt{2}\,[A(\mathbf{r}_1)B(\mathbf{r}_2) - B(\mathbf{r}_1)A(\mathbf{r}_2)] \times \begin{cases} \alpha(s_1)\alpha(s_2) & (M = +1), \\ \tfrac{1}{2}\sqrt{2}\,[\alpha(s_1)\beta(s_2) + \beta(s_1)\alpha(s_2)] & (M = 0), \\ \beta(s_1)\beta(s_2) & (M = -1). \end{cases}$$

Let us get ρ for the singlet function: from the definition (5.1.5),

$$\rho(\mathbf{x}_1) = 2 \times \tfrac{1}{4} \int |A(\mathbf{r}_1)B(\mathbf{r}_2) + B(\mathbf{r}_1)A(\mathbf{r}_2)|^2 |\alpha(s_1)\beta(s_2) - \beta(s_1)\alpha(s_2)|^2 \, \mathrm{d}\mathbf{r}_2 \, \mathrm{d}s_2$$

$$= \tfrac{1}{2}[|A(\mathbf{r}_1)|^2 + |B(\mathbf{r}_1)|^2][|\alpha(s_1)|^2 + |\beta(s_1)|^2].$$

A further spin integration gives the electron density (5.1.6):

$$P(\mathbf{r}_1) = \int \rho_1(\mathbf{x}_1) \, \mathrm{d}s_1 = |A(\mathbf{r}_1)|^2 + |B(\mathbf{r}_1)|^2,$$

which is simply the superposition of the densities due to one electron in orbital A and another in B, as might have been expected. Now the contribution to P that arises from the α-term in ρ_1, corresponding to the spin-($+\frac{1}{2}$) situation, is clearly half the total density. A similar result holds for the β-term, and the expression for ρ may be thus be written

$$\rho(\mathbf{x}_1) = P_\alpha(\mathbf{r}_1)\, |\alpha(s_1)|^2 + P_\beta(\mathbf{r}_1)\, |\beta(s_1)|^2,$$

where the first term vanishes unless $s_1 = +\frac{1}{2}$, the second unless $s_1 = -\frac{1}{2}$, and thus

$$P_\alpha(\mathbf{r}_1) = \tfrac{1}{2}P(\mathbf{r}_1) = \text{probability density for up-spin electrons},$$

$$P_\beta(\mathbf{r}_1) = \tfrac{1}{2}P(\mathbf{r}_1) = \text{probability density for down-spin electrons}.$$

These densities are always equal in a singlet state of any system; but in non-singlet states the α- and β-densities may differ, giving a resultant *spin density* (excess of α- over β-density). Thus for the first of the triplet states we get

$$P_\alpha(\boldsymbol{r}_1) = P(\boldsymbol{r}_1), \quad P_\beta(\boldsymbol{r}_1) = 0 \quad \text{(zero probability of spin } -\tfrac{1}{2} \text{ anywhere)},$$

while for the third we get

$$P_\beta(\boldsymbol{r}_1) = P(\boldsymbol{r}_1), \quad P_\alpha(\boldsymbol{r}_1) = 0 \quad \text{(zero probability of spin } +\tfrac{1}{2} \text{ anywhere)},$$

and for the second

$$P_\alpha(\boldsymbol{r}_1) = P_\beta(\boldsymbol{r}_1) = \tfrac{1}{2}P(\boldsymbol{r}_1) \quad \text{(equal probabilities of spin } \pm\tfrac{1}{2}),$$

as in the singlet state. The significance of these results will be discussed more fully in Section 5.9.

5.2 DENSITY MATRICES

To summarize, we have introduced density functions

$$\rho(\boldsymbol{x}_1) = N \int \Psi(\boldsymbol{x}_1, \boldsymbol{x}_2, \ldots, \boldsymbol{x}_N)\Psi^*(\boldsymbol{x}_1, \boldsymbol{x}_2, \ldots, \boldsymbol{x}_N)\, \mathrm{d}\boldsymbol{x}_2 \ldots \mathrm{d}\boldsymbol{x}_N, \quad (5.2.1a)$$

$$\pi(\boldsymbol{x}_1, \boldsymbol{x}_2) = N(N-1) \int \Psi(\boldsymbol{x}_1, \boldsymbol{x}_2, \ldots, \boldsymbol{x}_N)\Psi^*(\boldsymbol{x}_1, \boldsymbol{x}_2, \ldots, \boldsymbol{x}_N)\, \mathrm{d}\boldsymbol{x}_3 \ldots \mathrm{d}\boldsymbol{x}_N, \quad (5.2.1b)$$

and their spinless counterparts, obtained by integrating over spins,

$$P(\boldsymbol{r}_1) = \int \rho(\boldsymbol{x}_1)\, \mathrm{d}s_1, \quad \Pi(\boldsymbol{r}_1, \boldsymbol{r}_2) = \int \pi(\boldsymbol{x}_1, \boldsymbol{x}_2)\, \mathrm{d}s_1\, \mathrm{d}s_2. \quad (5.2.2)$$

The first of the spinless functions we have called the "electron density" (although strictly, of course, it is a probability density); the justification for the name lies in the fact that for purposes of calculation, the electrons may often be regarded as actually "smeared out", with a density P. The second function, Π, is the "pair function" and tells us how the motions of two electrons are correlated.

For some purposes it is necessary to generalize these functions slightly. To see why this is so, we again consider first a 1-electron system, an electron in spin-orbital ψ. The expectation value in this state of any quantity with operator F is given by

$$\langle \mathsf{F} \rangle = \int \psi^*(\boldsymbol{x})\mathsf{F}\psi(\boldsymbol{x})\, \mathrm{d}\boldsymbol{x}. \quad (5.2.3)$$

If F is just a multiplier (for example some function of coordinates) then we can write

$$\langle \mathsf{F} \rangle = \int \mathsf{F}\psi(\boldsymbol{x})\psi^*(\boldsymbol{x})\,\mathrm{d}\boldsymbol{x} = \int \mathsf{F}\rho(\boldsymbol{x})\,\mathrm{d}\boldsymbol{x}, \tag{5.2.4}$$

and the order of the factors does not matter: the expectation value is thus obtained simply by averaging $\mathsf{F} = F(\boldsymbol{x})$ over the electron density as defined in (5.2.1a), since $\rho(\boldsymbol{x})\,\mathrm{d}\boldsymbol{x}$ is the amount of charge in $\mathrm{d}\boldsymbol{x}$ and $F(\boldsymbol{x})$ is the corresponding value of F. It would be nice to do this when F is a "real" operator (e.g. involving differentation or integration), but this does not seem to be possible because in (5.2.3) the operator F works only on the factor that follows it and $\psi^*(\boldsymbol{x})$ cannot therefore be taken round to the right of F. In order to express everything in terms of the basic density function, we use a very simple device: we agree that F works on functions of $\boldsymbol{x}$ only, and we change the name of the variables in ψ^* from $\boldsymbol{x}$ to $\boldsymbol{x}'$ to make them immune from the effect of F. We can then rewrite (5.2.3) as

$$\langle \mathsf{F} \rangle = \int_{\boldsymbol{x}'=\boldsymbol{x}} \mathsf{F}\psi(\boldsymbol{x})\psi^*(\boldsymbol{x}')\,\mathrm{d}\boldsymbol{x},$$

where we put $\boldsymbol{x}' = \boldsymbol{x}$ *after* operating with F but before completing the integration. Initially we had $\psi(\boldsymbol{x})\psi^*(\boldsymbol{x}) = \rho(\boldsymbol{x})$; but there are now *two* sets of variables, primed and unprimed, and we denote $\psi(\boldsymbol{x})\psi^*(\boldsymbol{x}')$ by $\rho(\boldsymbol{x};\boldsymbol{x}')$, reserving $\rho(\boldsymbol{x})$ for the function obtained by identifying the two sets of variables, $\boldsymbol{x}$ and $\boldsymbol{x}'$:

$$\rho(\boldsymbol{x}) = \rho(\boldsymbol{x};\boldsymbol{x}).$$

With this definition, the expectation-value expression becomes

$$\langle \mathsf{F} \rangle = \int_{\boldsymbol{x}'=\boldsymbol{x}} \mathsf{F}\rho(\boldsymbol{x};\boldsymbol{x}')\,\mathrm{d}\boldsymbol{x}, \tag{5.2.5}$$

which is now exactly like (5.2.4) except for the generalization of the density function.

The same artifice may be used in the many-electron case. In particular (cf. (5.2.1a,b)), we define the generalized density functions

$$\rho(\boldsymbol{x}_1;\boldsymbol{x}_1') = N \int \Psi(\boldsymbol{x}_1, \boldsymbol{x}_2, \ldots, \boldsymbol{x}_N) \times \Psi^*(\boldsymbol{x}_1', \boldsymbol{x}_2, \ldots, \boldsymbol{x}_N)\,\mathrm{d}\boldsymbol{x}_2 \ldots \mathrm{d}\boldsymbol{x}_N, \tag{5.2.6}$$

$$\pi(\boldsymbol{x}_1, \boldsymbol{x}_2;\boldsymbol{x}_1', \boldsymbol{x}_2') = N(N-1) \int \Psi(\boldsymbol{x}_1, \boldsymbol{x}_2, \ldots, \boldsymbol{x}_N) \times \Psi^*(\boldsymbol{x}_1', \boldsymbol{x}_2', \ldots, \boldsymbol{x}_N)\,\mathrm{d}\boldsymbol{x}_3 \ldots \mathrm{d}\boldsymbol{x}_N, \tag{5.2.7}$$

in which we have put primes where necessary in Ψ^*. This protects Ψ^* from the action of an operator, as in the 1-electron case, and allows us to discuss *all* 1-electron properties in terms of the one density function $\rho(\mathbf{x}_1;\mathbf{x}_1')$ and *all* 2-electron properties in terms of $\pi(\mathbf{x}_1,\mathbf{x}_2;\mathbf{x}_1,\mathbf{x}_2')$. These functions reduce to the densities used previously as soon as the primes are removed, in which case we return to the original notation of Section 5.1, showing only the one unprimed set of variables

$$\rho(\mathbf{x}_1)=\rho(\mathbf{x}_1;\mathbf{x}_1), \qquad \pi(\mathbf{x}_1,\mathbf{x}_2)=\pi(\mathbf{x}_1,\mathbf{x}_2;\mathbf{x}_1,\mathbf{x}_2). \tag{5.2.8}$$

The generalized density functions therefore allow us to relate expectation values of operators directly to the electron distribution. It is also worth noting that successive density matrices are related; in particular,

$$(N-1)\rho(\mathbf{x}_1;\mathbf{x}_1')=\int \pi(\mathbf{x}_1,\mathbf{x}_2;\mathbf{x}_1',\mathbf{x}_2)\,\mathrm{d}\mathbf{x}_2, \tag{5.2.9}$$

$$(N-1)P(\mathbf{r}_1;\mathbf{r}_1')=\int \Pi(\mathbf{r}_1,\mathbf{r}_2;\mathbf{r}_1',\mathbf{r}_2)\,\mathrm{d}\mathbf{r}_2, \tag{5.2.10}$$

as follows easily from the definitions.

As an example, we consider the Hamiltonian operator (1.1.2), which (omitting external fields and spin effects) contains the one-electron term

$$\mathsf{h}(i)=-\tfrac{1}{2}\nabla^2(i)+\mathsf{V}(i)$$

for each electron, together with an electron-repulsion term

$$g(i,j)=1/r_{ij}$$

for each pair. The total Hamiltonian

$$\mathsf{H}=\sum_i \mathsf{h}(i)+\tfrac{1}{2}\sum_{i,j}{}' g(i,j) \tag{5.2.11}$$

thus contains 1- and 2-electron parts that are completely symmetrical in the electron labels. The expectation value of the 1-electron part $\langle\sum_i \mathsf{h}(i)\rangle$ is

$$\int \Psi^*(\mathbf{x}_1,\mathbf{x}_2,\ldots,\mathbf{x}_N)\left[\sum_i \mathsf{h}(i)\right]\Psi(\mathbf{x}_1,\mathbf{x}_2,\ldots,\mathbf{x}_N)\,\mathrm{d}\mathbf{x}_1\,\mathrm{d}\mathbf{x}_2\ldots\mathrm{d}\mathbf{x}_N,$$

and from the symmetry of $\Psi^*\Psi$ each value of i must give the same contribution. We therefore obtain N times the result for the first term in the sum:

$$N\int_{\mathbf{x}_1'=\mathbf{x}_1} \mathsf{h}(1)\Psi(\mathbf{x}_1,\mathbf{x}_2,\ldots,\mathbf{x}_N)\Psi^*(\mathbf{x}_1',\mathbf{x}_2,\ldots,\mathbf{x}_N)\,\mathrm{d}\mathbf{x}_1\,\mathrm{d}\mathbf{x}_2\ldots\mathrm{d}\mathbf{x}_N.$$

The right-hand side contains the density function $\rho(\mathbf{x}_1;\mathbf{x}_1')$, defined in

(5.2.6), and thus

$$\left\langle \sum_i \mathsf{h}(i) \right\rangle = \int_{\boldsymbol{x}_1'=\boldsymbol{x}_1} \mathsf{h}(1)\rho(\boldsymbol{x}_1;\boldsymbol{x}_1')\,\mathrm{d}\boldsymbol{x}_1. \tag{5.2.12}$$

A similar result is valid for any type of 1-electron operator, for example for kinetic-energy terms and potential-energy terms individually. If we were interested in the latter alone, we should have a slightly simpler result,

$$\left\langle \sum_i \mathsf{V}(i) \right\rangle = \int_{\boldsymbol{x}_1'=\boldsymbol{x}_1} \mathsf{V}(1)\rho(\boldsymbol{x}_1;\boldsymbol{x}_1')\,\mathrm{d}\boldsymbol{x}_1 = \int V(1)\rho(\boldsymbol{x}_1)\,\mathrm{d}\boldsymbol{x}_1, \tag{5.2.13}$$

because $\mathsf{V}(1) = V(\boldsymbol{r}_1)$ is then just a *factor* in the integrand and the prime may be removed, as in (5.2.8), to give the "ordinary" density (5.1.5).

In an exactly similar way, the electron-interaction energy, which involves *pairs* of electrons, may be written in terms of the pair function $\pi(\boldsymbol{x}_1, \boldsymbol{x}_2; \boldsymbol{x}_1', \boldsymbol{x}_2')$, which reduces to $\pi(\boldsymbol{x}_1, \boldsymbol{x}_2)$ when we put $\boldsymbol{x}_1' = \boldsymbol{x}_1$, $\boldsymbol{x}_2' = \boldsymbol{x}_2$. Since, in fact $g(1, 2)$ is just a factor in the integrand, like $\mathsf{V}(1)$, the primes may be dropped at once to give (in the notation of (5.2.8))

$$\left\langle \sum_{i,j}{}' g(i,j) \right\rangle = \int_{\substack{\boldsymbol{x}_1'=\boldsymbol{x}_1\\ \boldsymbol{x}_2'=\boldsymbol{x}_2}} g(1,2)\pi(\boldsymbol{x}_1, \boldsymbol{x}_2; \boldsymbol{x}_1', \boldsymbol{x}_2')\,\mathrm{d}\boldsymbol{x}_1\,\mathrm{d}\boldsymbol{x}_2$$

$$= \int g(1,2)\pi(\boldsymbol{x}_1, \boldsymbol{x}_2)\,\mathrm{d}\boldsymbol{x}_1\mathrm{d}\boldsymbol{x}_2. \tag{5.2.14}$$

As already noted, with only 1- and 2-body interactions, *we never need any density functions other than* $\rho(\boldsymbol{x}_1;\boldsymbol{x}_1')$ *and* $\pi(\boldsymbol{x}_1, \boldsymbol{x}_2; \boldsymbol{x}_1', \boldsymbol{x}_2')$. Moreover, in dealing with spin-independent effects, we can always go further by completing the spin integrations over the remaining one or two spin variables. The spinless densities analogous to those defined in (5.2.2) are

$$P(\boldsymbol{r}_1; \boldsymbol{r}_1') = \int_{s_1'=s_1} \rho(\boldsymbol{x}_1; \boldsymbol{x}_1')\,\mathrm{d}s_1, \tag{5.2.15}$$

$$\Pi(\boldsymbol{r}_1, \boldsymbol{r}_2; \boldsymbol{r}_1', \boldsymbol{r}_2') = \int_{\substack{s_1'=s_1\\ s_2'=s_2}} \pi(\boldsymbol{x}_1, \boldsymbol{x}_2; \boldsymbol{x}_1', \boldsymbol{x}_2')\,\mathrm{d}s_1\,\mathrm{d}s_2. \tag{5.2.16}$$

The notation employed in such equations is more cumbersome than strictly necessary: it may streamlined if we remember that the *primed* variables are only needed when an *operator* works on a density function, and that immediately before integration those primes are removed. With this understanding, we need no longer show the primed variables explicitly. Thus (5.2.12) and (5.2.14) will be written

$$\left\langle \sum_i \mathsf{h}(i) \right\rangle = \int \mathsf{h}\rho(\boldsymbol{x}_1)\,\mathrm{d}\boldsymbol{x}_1, \tag{5.2.17}$$

$$\left\langle \sum_{i,j}{}' \mathsf{g}(i,j) \right\rangle = \int \mathsf{g}\pi(\boldsymbol{x}_1, \boldsymbol{x}_2)\,\mathrm{d}\boldsymbol{x}_1\,\mathrm{d}\boldsymbol{x}_2 \tag{5.2.18}$$

for *any* 1- and 2-electron operators (h, g), while (5.2.15) and (5.2.16) may be abbreviated similarly, as in (5.2.2). In other words, we usually indicate only "diagonal elements" (5.2.8) of the density functions; if desired, the functions with $\boldsymbol{x}_1' \neq \boldsymbol{x}_1$ etc. may be written out explicitly simply by adding primes to variables associated with the Ψ^* factor.

With the above convention, the general expression for the energy of an N-electron system with the usual Hamiltonian (5.2.11) takes a very simple appearance:

$$E = -\tfrac{1}{2}\int \nabla^2\rho(\boldsymbol{x}_1)\,\mathrm{d}\boldsymbol{x}_1 + \int V\rho(\boldsymbol{x}_1)\,\mathrm{d}\boldsymbol{x}_1 + \tfrac{1}{2}\int g\pi(\boldsymbol{x}_1, \boldsymbol{x}_2)\,\mathrm{d}\boldsymbol{x}_1\,\mathrm{d}\boldsymbol{x}_2. \tag{5.2.19}$$

On completing the spin integrations, using (5.2.2), this gives

$$\boxed{\begin{aligned} E &= -\tfrac{1}{2}\int \nabla^2 P(\boldsymbol{r}_1)\,\mathrm{d}\boldsymbol{r}_1 + \int VP(\boldsymbol{r}_1)\,\mathrm{d}\boldsymbol{r}_1 + \tfrac{1}{2}\int g\Pi(\boldsymbol{r}_1, \boldsymbol{r}_2)\,\mathrm{d}\boldsymbol{r}_1\,\mathrm{d}\boldsymbol{r}_2 \\ &= T + V_{\mathrm{en}} + V_{\mathrm{ee}}, \end{aligned}} \tag{5.2.20}$$

where

V_{en} = potential energy of smeared-out charge of density P, in field of nuclei,

V_{ee} = average potential energy due to pairwise repulsions described by pair function Π.

The term T is the quantum-mechanical expectation value of the kinetic energy, and is the only term requiring knowledge of $P(\boldsymbol{r}_1; \boldsymbol{r}_1')$ for $\boldsymbol{r}_1' \neq \boldsymbol{r}_1$, while the others have a purely classical interpretation in terms of the distribution functions for a particle and for a pair of particles respectively. These results are valid for all kinds of wavefunctions, or approximate wavefunctions, for any state of any system; and because they involve the electron distribution directly it is often possible to get a useful interpretation of molecular properties in terms of the main features of the electron density, without detailed reference to the intricacies of the many-electron wavefunction. A chemical bond, for instance, arises from a concentration of electron density in the bond region, with a consequent lowering of potential energy—in full accord with the intuitive ideas of elementary valence theory.‡

We note in passing that the density functions such as $\rho_1(\boldsymbol{x}_1;\boldsymbol{x}_1')$ and $\rho_2(\boldsymbol{x}_1,\boldsymbol{x}_2;\boldsymbol{x}_1',\boldsymbol{x}_2')$ are special cases of the reduced density matrices defined by Husimi (1940). The density matrix§ was first introduced by von Neumann (1927) and by Dirac (1929) in the quantum-mechanical discussion of systems in incompletely specified states (e.g. in thermal equilibrium): thus, for a 1-particle system, with a probability w_i, say, of being found in state ψ_i, the "pure-state" density matrix $\rho(\boldsymbol{x},\boldsymbol{x}')= \psi(\boldsymbol{x})\psi^*(\boldsymbol{x}')$ for the definite state ψ is replaced by the "statistical" density matrix

$$\rho(\boldsymbol{x};\boldsymbol{x}')=\sum_i w_i\psi_i(\boldsymbol{x})\psi_i^*(\boldsymbol{x}'). \qquad (5.2.21)$$

The expectation-value expression (5.2.5) then retains its validity by including the further averaging due to indefiniteness of the state of the system. The definition is easily extended to systems of any kind, and the significance of the density matrix in statistical mechanics is discussed elsewhere (Tolman, 1938; ter Haar, 1954). In the applications that we shall make it is sufficient to consider the special case of a pure state. On the other hand, although we deal with many-electron systems, we are normally interested in probability functions referring to a few particles at a time; to distinguish the corresponding functions from the density matrix

$$\rho_{\text{sys}}(\boldsymbol{x}_1,\boldsymbol{x}_2,\ldots,\boldsymbol{x}_N;\boldsymbol{x}_1',\boldsymbol{x}_2',\ldots,\boldsymbol{x}_N') = \Psi(\boldsymbol{x}_1,\boldsymbol{x}_2,\ldots,\boldsymbol{x}_N)\Psi^*(\boldsymbol{x}_1',\boldsymbol{x}_2',\ldots,\boldsymbol{x}_N') \qquad (5.2.22)$$

for the *whole system*, it is customary in statistical mechanics to use the term "reduced" density matrix when referring to $\rho_1(\boldsymbol{x}_1;\boldsymbol{x}_1')$, $\rho_2(\boldsymbol{x}_1,\boldsymbol{x}_2;\boldsymbol{x}_1',\boldsymbol{x}_2')$ etc. The normalization we have used (McWeeny, 1954a) is convenient in dealing with systems of indistinguishable particles and is widely used in statistical mechanics (see e.g. Born and Green, 1947; Born, 1964, Yang, 1962); however, other conventions are also employed (Löwdin, 1955; ter Haar, 1961; Coleman, 1963).

‡ The physical interpretation of what happens in forming a chemical bond has been the subject of much controversy (see, in particular, Ruedenberg (1962); what is beyond dispute, however, is that the only *attractions* (i.e. negative terms in the energy) are contained in V_{en}.

§ The term "matrix" is in some ways unfortunate. It arises because $\rho(\boldsymbol{x};\boldsymbol{x}')$ is analogous to a matrix element ρ_{rs} in which the discrete indices r, s have been replaced by continuous variables $\boldsymbol{x}$, $\boldsymbol{x}'$.

5.3 DENSITY FUNCTIONS FOR 1-DETERMINANT WAVEFUNCTIONS

We recall that a single antisymmetrized product, or determinant, of 1-electron functions may sometimes provide a useful approximate wavefunction. This is the basis of MO theory, and it is therefore useful to obtain the corresponding orbital forms of the density matrices.

We consider the single spin-orbital configuration in which spin-orbitals $\psi_A, \psi_B, \ldots, \psi_X$ are occupied by the N electrons, each spin-orbital having an orbital factor that may be referred to in the present discussion as an MO, though its actual form is at present immaterial; and we write the corresponding 1-determinant wavefunction (3.3.3) as

$$\Psi(\boldsymbol{x}_1, \boldsymbol{x}_2, \ldots, \boldsymbol{x}_N) = \left(\frac{1}{N!}\right)^{1/2} \det |\psi_A(\boldsymbol{x}_1)\ \psi_B(\boldsymbol{x}_2)\ \ldots\ \psi_X(\boldsymbol{x}_N)|. \quad (5.3.1)$$

Instead of calculating the density matrices directly, from (5.2.6) and (5.2.7), it is sufficient to consider the energy expression already obtained in (3.3.7a):

$$E = \langle \Psi | \mathsf{H} | \Psi \rangle$$
$$= \sum_R \langle \psi_R | \mathsf{h} | \psi_R \rangle + \tfrac{1}{2} \sum_{R,S} (\langle \psi_R \psi_S | g | \psi_R \psi_S \rangle - \langle \psi_R \psi_S | g | \psi_S \psi_R \rangle), \quad (5.3.2)$$

where R, S run over all occupied spin-orbitals, and 1- and 2-electron integrals are defined in general by

$$\langle \psi_R | \mathsf{h} | \psi_S \rangle = \int \psi_R^*(\boldsymbol{x}_1) \mathsf{h}(1) \psi_S(\boldsymbol{x}_1)\, d\boldsymbol{x}_1, \quad (5.3.3)$$

$$\langle \psi_R \psi_S | g | \psi_T \psi_U \rangle = \int \psi_R^*(\boldsymbol{x}_1) \psi_S^*(\boldsymbol{x}_2) g(1,2) \psi_T(\boldsymbol{x}_1) \psi_U(\boldsymbol{x}_2)\, d\boldsymbol{x}_1\, d\boldsymbol{x}_2. \quad (5.3.4)$$

Expression (5.3.2) affords a simple method of writing down the density functions, for the two sums arise as the expectation values of the 1- and 2-electron parts of H respectively, and may be compared directly with the general expressions (5.2.12) and (5.2.14). For example, keeping primed variables for clarity, we may suppose $\rho(\boldsymbol{x}_1; \boldsymbol{x}_1')$ to be expanded in spin-orbitals in the form

$$\rho(\boldsymbol{x}_1; \boldsymbol{x}_1') = \sum_{R,S} \rho_{RS} \psi_R(\boldsymbol{x}_1) \psi_S^*(\boldsymbol{x}_1'), \quad (5.3.5)$$

where the ρ_{RS} are numerical coefficients and the factors ψ_R and ψ_S^* arise from Ψ and Ψ^* respectively, after completing the integrations in (5.2.6): the *general* 1-electron energy term therefore has the orbital form

$$\int \mathsf{h}(1)\rho(\boldsymbol{x}_1;\boldsymbol{x}_1')\,\mathrm{d}\boldsymbol{x}_1 = \sum_{R,S} \int \rho_{RS}\psi_S^*(\boldsymbol{x}_1)\mathsf{h}(1)\psi_R(\mathbf{x}_1)\,\mathrm{d}\boldsymbol{x}_1,$$

this step being permitted since h works only on ψ_R. Any expectation value such as (5.2.12) can thus be written as the *trace* of a matrix product:

$$\left\langle \sum_i \mathsf{h}(i) \right\rangle = \sum_{R,S} \rho_{RS}\langle \psi_S|\,\mathsf{h}\,|\psi_R\rangle = \mathrm{Tr}\,\boldsymbol{\rho}\mathbf{h}. \qquad (5.3.6)$$

The required coefficient ρ_{RS} in (5.3.5) is thus simply the coefficient of $\langle \psi_S|\,\mathsf{h}\,|\psi_R\rangle$ in the usual orbital expression for the 1-electron energy; and by comparison with (5.3.2) we see that in the 1-determinant approximation $\rho_{RS} = \delta_{RS}$. Trace expressions such as (5.3.6) are, however, very general and will be met frequently. The 1-determinant approximation to $\rho(\boldsymbol{x}_1;\boldsymbol{x}_1')$ may now be written

$$\boxed{\rho(\boldsymbol{x}_1;\boldsymbol{x}_1') = \sum_{R\,(\mathrm{occ})} \psi_R(\boldsymbol{x}_1)\psi_R^*(\boldsymbol{x}_1'),} \qquad (5.3.7)$$

which is said to be "diagonal", corresponding to zero coefficients for $R \neq S$. The matrix of coefficients $\boldsymbol{\rho}$ is not *in general* diagonal and is in fact a true matrix defining the function $\rho(\boldsymbol{x}_1;\boldsymbol{x}_1')$; moreover, both the matrix and the function are representations of an abstract density *operator*, as will become evident presently (Section 6.4).

The expression for π follows in a similar way. We find

$$\pi(\boldsymbol{x}_1,\boldsymbol{x}_2;\boldsymbol{x}_1',\boldsymbol{x}_2') = \sum_{R,S} [\psi_R(\boldsymbol{x}_1)\psi_S(\boldsymbol{x}_2)\psi_R^*(\boldsymbol{x}_1')\psi_S^*(\boldsymbol{x}_2') - \psi_R(\boldsymbol{x}_2)\psi_S(\boldsymbol{x}_1)\psi_R^*(\boldsymbol{x}_1')\psi_S^*(\boldsymbol{x}_2')], \qquad (5.3.8)$$

or, in terms of ρ,

$$\pi(\boldsymbol{x}_1,\boldsymbol{x}_2;\boldsymbol{x}_1',\boldsymbol{x}_2') = \rho(\boldsymbol{x}_1;\boldsymbol{x}_1')\rho(\boldsymbol{x}_2;\boldsymbol{x}_2') - \rho(\boldsymbol{x}_2;\boldsymbol{x}_1')\rho(\boldsymbol{x}_1;\boldsymbol{x}_2'). \qquad (5.3.9)$$

This "factorization" of the 2-electron density in terms of the 1-electron ρ is peculiar to the 1-determinant approximation: it means that in this approximation everything is determined by the function $\rho(\boldsymbol{x}_1;\boldsymbol{x}_1')$, which is often called the *Fock–Dirac density matrix* (Fock, 1930; Dirac, 1930; Lennard-Jones, 1931). It is in fact clear that the "reduced" density matrix for all N electrons may be written as‡

$$\rho_N(\boldsymbol{x}_1, \ldots, \boldsymbol{x}_N; \boldsymbol{x}'_1, \ldots, \boldsymbol{x}'_N) = N!\Psi(\boldsymbol{x}_1, \ldots, \boldsymbol{x}_N)\Psi^*(\boldsymbol{x}'_1, \ldots, \boldsymbol{x}'_N)$$

$$= \begin{vmatrix} \rho_1(\boldsymbol{x}_1; \boldsymbol{x}'_1) & \rho_1(\boldsymbol{x}_1; \boldsymbol{x}'_2) & \cdots & \\ \rho_1(\boldsymbol{x}_2; \boldsymbol{x}'_1) & \rho_1(\boldsymbol{x}_2; \boldsymbol{x}'_2) & \cdots & \\ \vdots & \vdots & \ddots & \vdots \\ & & \cdots & \rho_1(\boldsymbol{x}_N; \boldsymbol{x}'_N) \end{vmatrix}, \qquad (5.3.10)$$

and it may be shown (for example, by induction) that in general

$$\rho_n(\boldsymbol{x}_1, \ldots, \boldsymbol{x}_n; \boldsymbol{x}'_1, \ldots, \boldsymbol{x}'_n)$$

$$= \begin{vmatrix} \rho_1(\boldsymbol{x}_1; \boldsymbol{x}'_1) & \rho_1(\boldsymbol{x}_1, \boldsymbol{x}'_2) & \cdots & \\ \rho_1(\boldsymbol{x}_2; \boldsymbol{x}'_1) & \rho_1(\boldsymbol{x}_2; \boldsymbol{x}'_2) & \cdots & \\ \vdots & \vdots & \ddots & \vdots \\ & & \cdots & \rho_1(\boldsymbol{x}_n; \boldsymbol{x}'_n) \end{vmatrix} \qquad (5.3.11)$$

of which (5.3.9) is the special case $n = 2$. It should be noted that, even when the primes are removed, the diagonal elements of the 2- and many-electron density matrices always contain *off*-diagonal elements $\rho(\boldsymbol{x}_1; \boldsymbol{x}_2)$ etc. of the fundamental 1-electron density matrix.

In all the above results spin has been included implicitly through the use of spin-orbitals. Let us now admit in (5.3.1) two types of spin-orbital, $\psi_R = \phi_R\alpha$ and $\bar{\psi}_R = \phi_R\beta$, and derive the explicit form of the sum (5.3.7). Evidently (cf. the example on p. 118) the result may be written, for the case $\boldsymbol{x}'_1 = \boldsymbol{x}_1$,

$$\rho(\boldsymbol{x}_1) = P_\alpha(\boldsymbol{r}_1)\alpha(s_1)\alpha^*(s_1) + P_\beta(\boldsymbol{r}_1)\beta(s_1)\beta^*(s_1), \qquad (5.3.12)$$

where the densities of up- and down-spin electrons are

$$P_\alpha(\boldsymbol{r}_1) = \sum_{R(\alpha)} \phi_R(\boldsymbol{r}_1)\phi^*_R(\boldsymbol{r}_1), \qquad P_\beta(\boldsymbol{r}_1) = \sum_{R(\beta)} \phi_R(\boldsymbol{r}_1)\phi^*_R(\boldsymbol{r}_1). \quad (5.3.13)$$

Here the summations are over orbitals with α and β spin factors respectively.

On integrating (5.3.12) over spin, we obtain the electron density in the form

$$P(\boldsymbol{r}_1) = P_\alpha(\boldsymbol{r}_1) + P_\beta(\boldsymbol{r}_1) \qquad (5.3.14)$$

—a sum of up- and down-spin densities. It is also useful to define

$$Q_z(\boldsymbol{r}_1) = \tfrac{1}{2}[P_\alpha(\boldsymbol{r}_1) - P_\beta(\boldsymbol{r}_1)], \qquad (5.3.15)$$

‡ Transpose and star to get Ψ^* and multiply the determinants.

which is the excess of up-spin over down-spin density, with a weight factor $\frac{1}{2}$ (for spin-$\frac{1}{2}$ particles); this quantity, fully discussed later, is in fact the density of spin angular momentum around the z axis (in units of $\hbar$). The functions (5.3.14) and (5.3.15), off-diagonal elements included, determine all higher density functions and all properties of the electron distribution in a 1-determinant approximation.

For a *closed-shell* system, in which all orbitals are doubly occupied (once with α factor and once with β) $P_\alpha = P_\beta = \frac{1}{2}P$ and the spin density is everywhere zero. In this situation, typified by the majority of molecular ground states, we find immediately

$$P(\boldsymbol{r}_1) = 2 \sum_{R\,(\text{occ})} \phi_R(\boldsymbol{r}_1)\phi_R^*(\boldsymbol{r}_1), \tag{5.3.16}$$

and, putting (5.3.12) in (5.3.9) and integrating over spin,

$$\Pi(\boldsymbol{r}_1, \boldsymbol{r}_2) = P(\boldsymbol{r}_1)P(\boldsymbol{r}_2) - \tfrac{1}{2}P(\boldsymbol{r}_2; \boldsymbol{r}_1)P(\boldsymbol{r}_1; \boldsymbol{r}_2). \tag{5.3.17}$$

A closed-shell energy expression, often used in later sections, is obtained on substituting these results in (5.2.20):

$$\boxed{E = 2\sum_R \langle\phi_R|\,\mathsf{h}\,|\phi_R\rangle + \sum_{R,S} [2\langle\phi_R\phi_S|\,g\,|\phi_R\phi_S\rangle - \langle\phi_R\phi_S|\,g\,|\phi_S\phi_R\rangle],} \tag{5.3.18}$$

in agreement with the result obtained directly from (5.3.2) by including both types of spin-orbital (ψ_R, $\bar{\psi}_R$) and doing the spin integrations.

Before leaving the discussion of the 1-determinant approximation, we note that since P and Q_z are both of "sum-of-squares" form, being determined by (5.3.13), any 1-determinant wavefunction will have certain invariance properties. If we mix the α-orbitals (or the β-orbitals, or both sets) among themselves to get a new set of (orthonormal) orbitals $\bar{A}, \bar{B}, \ldots, \bar{R}, \ldots$ then the determinant based on the new orbitals will give exactly the same density functions as the original—in fact the two wavefunctions are, on expansion, identical. This invariance (Problem 3.5) is of considerable value in MO theory, linear combination of the MOs sometimes permitting the introduction of more localized orbitals (e.g. "bond orbitals") with a more immediate chemical interpretation.

5.4 TRANSITION DENSITIES. GENERALIZATIONS

So far we have been concerned with the properties of a single given state, and in particular with states represented by single determinants. Often, however, we need to discuss properties that depend jointly on two states, as in dealing with spectroscopic transitions and selection rules; and we also need to use many-determinant wavefunctions.

To this end, we introduce *transition* density matrices connecting pairs of states, Ψ_K and Ψ_L say, by equations exactly analogous to (5.2.6) and (5.2.7):

$$\rho(KL \mid \boldsymbol{x}_1; \boldsymbol{x}_1') = N \int \Psi_K(\boldsymbol{x}_1, \boldsymbol{x}_2, \ldots, \boldsymbol{x}_N)\Psi_L^*(\boldsymbol{x}_1', \boldsymbol{x}_2, \ldots, \boldsymbol{x}_N)\, d\boldsymbol{x}_2 \ldots d\boldsymbol{x}_N, \quad (5.4.1)$$

$$\pi(KL \mid \boldsymbol{x}_1, \boldsymbol{x}_2; \boldsymbol{x}_1', \boldsymbol{x}_2') = N(N-1) \int \Psi_K(\boldsymbol{x}_1, \boldsymbol{x}_2, \ldots, \boldsymbol{x}_N)\Psi_L^*(\boldsymbol{x}_1', \boldsymbol{x}_2', \ldots, \boldsymbol{x}_N)\, d\boldsymbol{x}_3 \ldots d\boldsymbol{x}_N. \quad (5.4.2)$$

Clearly, for $L = K$ we obtain the density functions for a single state, as used previously without any label: $\rho(KK \mid \boldsymbol{x}_1; \boldsymbol{x}_1')$ is the density function $\rho(\boldsymbol{x}_1; \boldsymbol{x}_1')$ for state $\Psi = \Psi_K$. The transition-density functions determine off-diagonal matrix elements‡ of ail operators just as the density functions used so far determine the diagonal elements (i.e. expectation values). Thus, with the usual convention (p. 123), we obtain in place of (5.2.17) and (5.2.18) respectively

$$\boxed{\left\langle \Psi_L \middle| \sum_i \mathsf{h}(i) \middle| \Psi_K \right\rangle = \int \mathsf{h}(1)\rho(KL \mid \boldsymbol{x}_1)\, d\boldsymbol{x}_1,} \quad (5.4.3)$$

$$\boxed{\left\langle \Psi_L \middle| \sum_{i,j}{}' \mathsf{g}(i,j) \middle| \Psi_K \right\rangle = \int \mathsf{g}(1,2)\pi(KL \mid \boldsymbol{x}_1, \boldsymbol{x}_2)\, d\boldsymbol{x}_1\, d\boldsymbol{x}_2.} \quad (5.4.4)$$

Again we may integrate over spins (when dealing with spinless operators) to obtain transition densities in ordinary space:

$$P(KL \mid \boldsymbol{r}_1) = \int \rho(KL \mid \boldsymbol{x}_1)\, ds_1 \quad (5.4.5)$$

$$\Pi(KL \mid \boldsymbol{r}_1, \boldsymbol{r}_2) = \int \pi(KL \mid \boldsymbol{x}_1, \boldsymbol{x}_2)\, ds_1\, ds_2 \quad (5.4.6)$$

It should be noted that, whereas the function $P(KK \mid \boldsymbol{r}_1)$ integrates to give the total number of electrons, $P(KL \mid \boldsymbol{r}_1)$ $(K \neq L)$ integrates to *zero*

‡ It should be noted that the KL transition densities give the LK matrix elements. This is the "natural" order corresponding to the starred function (Ψ_L^*) on the right or on the left respectively.

since Ψ_K and Ψ_L are orthogonal. Thus there is no net transition charge, the transition density being positive in some regions of space, but negative in others.

As an obvious and important example of the use of transition densities, we observe that the function $P(KL \mid \boldsymbol{r}_1)$ determines a transition probability between the states Ψ_K and Ψ_L: with radiation polarized in, say, the z direction, the probability of a transition $K \to L$ depends upon the transition moment

$$\left\langle \Psi_L \middle| \sum_i z_i \middle| \Psi_K \right\rangle = \int z_1 P(KL \mid \boldsymbol{r}_1)\, \mathrm{d}\boldsymbol{r}_1, \tag{5.4.7}$$

which is simply the moment (z component) of a static distribution of charge density $P(KL \mid \boldsymbol{r}_1)$. We note also that transition *spin* densities may be defined along similar lines (cf. Section 5.9) and that these will determine the absorption of radiation by spin systems (Chapter 11).

For any wavefunction built up from spin-orbital products, the single-state density matrices of Section 5.3 take the forms (cf. (5.3.5) *et seq.*)

$$\rho(\boldsymbol{x}_1) = \sum_{R,S} \rho_{RS} \psi_R(\boldsymbol{x}_1) \psi_S^*(\boldsymbol{x}_1), \tag{5.4.8a}$$

$$\pi(\boldsymbol{x}_1, \boldsymbol{x}_2) = \sum_{R,S,T,U} \pi_{RS,TU} \psi_R(\boldsymbol{x}_1) \psi_S(\boldsymbol{x}_2) \psi_T^*(\boldsymbol{x}_1) \psi_U^*(\boldsymbol{x}_2). \tag{5.4.8b}$$

The *transition* densities will assume similar forms

$$\rho(KL \mid \boldsymbol{x}_1) = \sum_{R,S} \rho_{RS}^{KL} \psi_R(\boldsymbol{x}_1) \psi_S^*(\boldsymbol{x}_1), \tag{5.4.9a}$$

$$\pi(KL \mid \boldsymbol{x}_1, \boldsymbol{x}_2) = \sum_{R,S,T,U} \pi_{RS,TU}^{KL} \psi_R(\boldsymbol{x}_1) \psi_S(\boldsymbol{x}_2) \psi_T^*(\boldsymbol{x}_1) \psi_U^*(\boldsymbol{x}_2), \tag{5.4.9b}$$

and it follows from (5.4.3) and (5.4.4) that typical off-diagonal matrix elements are given by

$$\begin{aligned} \left\langle \Psi_L \middle| \sum_i \mathsf{h}(i) \middle| \Psi_K \right\rangle &= \sum_{R,S} \rho_{RS}^{KL} \langle \psi_S | \, \mathsf{h} \, | \psi_R \rangle, \\ \left\langle \Psi_L \middle| {\sum_{i,j}}' g(i,j) \middle| \Psi_K \right\rangle &= \sum_{R,S,T,U} \pi_{RS,TU}^{KL} \langle \psi_T \psi_U | \, g \, | \psi_R \psi_S \rangle. \end{aligned} \tag{5.4.10}$$

It is not essential of course that Ψ_K and Ψ_L refer to actual *states*; they may, for example, be CFs (Φ_κ, Φ_λ) or even single determinants, and it is evident that (using $\sum \mathsf{h}$ and $\sum g$ for the operator sums)

$$\begin{aligned} \rho_{RS}^{\kappa\lambda} &= \text{coefficient of } \langle \psi_S | \, \mathsf{h} \, | \psi_R \rangle \text{ in } \langle \Phi_\lambda | \textstyle\sum \mathsf{h} \, | \Phi_\kappa |, \\ \pi_{RS,TU}^{\kappa\lambda} &= \text{coefficient of } \langle \psi_T \psi_U | \, g \, | \psi_R \psi_S \rangle \text{ in } \langle \Phi_\lambda | \textstyle\sum g \, | \Phi_\kappa \rangle. \end{aligned} \tag{5.4.11}$$

When the wavefunctions are expressed in terms of CFs, according to

$$\Psi_K = \sum_{\kappa} C_\kappa^K \Phi_\kappa, \qquad \Psi_L = \sum_{\lambda} C_\lambda^L \Phi_\lambda, \tag{5.4.12}$$

the matrix elements (5.4.10) become quadratic forms in the expansion coefficients; and exactly the same expressions hold good, with

$$\begin{aligned} \rho_{RS}^{KL} &= \sum_{\kappa,\lambda} C_\kappa^K C_\lambda^{L*} \rho_{RS}^{\kappa\lambda}, \\ \pi_{RS,TU}^{KL} &= \sum_{\kappa,\lambda} C_\kappa^K C_\lambda^{L*} \pi_{RS,TU}^{\kappa\lambda}, \end{aligned} \tag{5.4.13}$$

and the density-matrix elements on the right defined as in (5.4.11).

As a simple example of the use of (5.4.11), we may consider the transition densities for two determinants Φ_κ, Φ_λ describing spin-orbital configurations

$$\kappa = (A, B, \ldots, R, \ldots), \qquad \lambda = (A', B', \ldots, R', \ldots).$$

According to Slater's rules (3.3.7), the densities vanish (with orthogonal orbitals) unless there are 0, 1 or 2 spin-orbital differences between κ and λ: the corresponding densities (diagonal elements) are

(i) $\kappa = \lambda = (A, B, \ldots, R, \ldots)$

$$\rho(\kappa\kappa \mid \boldsymbol{x}_1) = \sum_R \psi_R(\boldsymbol{x}_1)\psi_R^*(\boldsymbol{x}_1),$$

$$\pi(\kappa\kappa \mid \boldsymbol{x}_1, \boldsymbol{x}_2) = \tfrac{1}{2}\mathsf{A}_{12}\mathsf{A}'_{12} \sum_{R,S} \psi_R(\boldsymbol{x}_1)\psi_S(\boldsymbol{x}_2)\psi_R^*(\boldsymbol{x}_1)\psi_S^*(\boldsymbol{x}_2);$$

(ii) $\kappa = (A, B, \ldots, R, \ldots)$, $\lambda = (A, B, \ldots, R', \ldots)$

$$\rho(\kappa\lambda \mid \boldsymbol{x}_1) = \psi_R(\boldsymbol{x}_1)\psi_R'^*(\boldsymbol{x}_1),$$

$$\pi(\kappa\lambda \mid \boldsymbol{x}_1, \boldsymbol{x}_1) = \mathsf{A}_{12}\mathsf{A}'_{12} \sum_{S(\neq R)} \psi_R(\boldsymbol{x}_1)\psi_S(\boldsymbol{x}_2)\psi_R'^*(\boldsymbol{x}_1)\psi_S^*(\boldsymbol{x}_2);$$

(iii) $\kappa = (A, B, \ldots, R, \ldots, S, \ldots)$, $\lambda = (A, B, \ldots, R', \ldots, S', \ldots)$

$$\pi(\kappa\lambda \mid \boldsymbol{x}_1, \boldsymbol{x}_2) = \mathsf{A}_{12}\mathsf{A}'_{12}\psi_R(\boldsymbol{x}_1)\psi_S(\boldsymbol{x}_2)\psi_R'^*(\boldsymbol{x}_1)\psi_S'^*(\boldsymbol{x}_2);$$

(5.4.14)

where the results have been written in a compact form by introducing $\mathsf{A}_{12} = 1 - \mathsf{P}_{12}$, which antisymmetrizes with respect to variables in the unstarred spin-orbitals, and A'_{12}, which are likewise for the starred spin-orbitals. As usual (p. 123), off-diagonal elements follow on adding primes to the variables in the starred spin-orbitals.

These results give some insight into the effects of mixing determinants in a CI expansion. For example, interaction of determinants that differ in only *one* spin-orbital will modify the electron density, through ρ, but with two or more differences the density will not be affected. For non-orthogonal orbitals, the corresponding results follow from (3.3.17) and (3.3.18).

The transition densities for CFs play a conspicuous part in all CI calculations. Thus when

$$\Psi = \sum_{\kappa} C_{\kappa} \Phi_{\kappa} \tag{5.4.15}$$

is the variational wavefunction for a single state of interest (usually the ground state) the energy functional $E[\Psi]$ takes the general form, using (5.4.10), (5.4.13) and dropping the redundant state labels (K, L),

$$\boxed{E = \sum_{R,S} \rho_{RS} \langle \psi_S | \mathsf{h} | \psi_R \rangle + \tfrac{1}{2} \sum_{R,S,T,U} \pi_{RS,TU} \langle \psi_T \psi_U | g | \psi_R \psi_S \rangle,} \tag{5.4.16}$$

$$\rho_{RS} = \sum_{\kappa,\lambda} C_{\kappa} C_{\lambda}^{*} \rho_{RS}^{\kappa\lambda}, \qquad \pi_{RS,TU} = \sum_{\kappa,\lambda} C_{\kappa} C_{\lambda}^{*} \pi_{RS,TU}^{\kappa\lambda}. \tag{5.4.17}$$

To pass to the spinless forms, appropriate at the usual non-relativistic level of approximation, we need only make the replacements

$$\left.\begin{aligned} \rho \rightarrow P, \qquad & \langle \psi_S | \mathsf{h} | \psi_R \rangle \rightarrow \langle \phi_S | \mathsf{h} | \phi_R \rangle, \\ \pi \rightarrow \Pi, \qquad & \langle \psi_T \psi_U | g | \psi_R \psi_S \rangle \rightarrow \langle \phi_T \phi_U | g | \phi_R \phi_S \rangle, \end{aligned}\right\} \tag{5.4.18}$$

where $\phi_R, \phi_S, \ldots$ are molecular orbitals (without the spin factors). With these changes, all previous equations are easily transcribed to spinless form. Thus, for example (cf. (5.4.17)),

$$P_{RS} = \sum_{\kappa,\lambda} C_{\kappa} C_{\lambda}^{*} P_{RS}^{\kappa\lambda}, \tag{5.4.19}$$

where $P_{RS}^{\kappa\lambda}$ is the coefficient of $\langle \phi_S | \mathsf{h} | \phi_R \rangle$ in $\langle \Phi_\lambda | \mathsf{H} | \Phi_\kappa \rangle$.

Finally, now that we know how to obtain density and transition-density matrices for wavefunctions expressible in terms of Slater determinants, it is easy to make contact with the second-quantization approach developed in Section 3.6. Thus, by comparison of (5.4.16) with (3.6.10), it is clear that

$$\rho_{RS} = \langle \Psi | \mathsf{a}_S^{\dagger} \mathsf{a}_R | \Psi \rangle, \qquad \pi_{RS,TU} = \langle \Psi | \mathsf{a}_T^{\dagger} \mathsf{a}_U^{\dagger} \mathsf{a}_S \mathsf{a}_R | \Psi \rangle, \tag{5.4.20}$$

where Ψ is now regarded as a vector of Fock space, while the creation and annihilation operators refer to the spin-orbitals ψ_R employed in the expansion of the wavefunction. More generally, the transition density matrices for Ψ_K, Ψ_L will have elements

$$\rho_{RS}^{KL} = \langle \Psi_L | \mathrm{a}_S^\dagger \mathrm{a}_R | \Psi_K \rangle, \qquad \pi_{RS,TU}^{KL} = \langle \Psi_L | \mathrm{a}_T^\dagger \mathrm{a}_U^\dagger \mathrm{a}_S \mathrm{a}_R | \Psi_K \rangle \quad (5.4.21)$$

and the CF transition densities in (5.4.11a,b) will have elements defined similarly but with CFs Φ_κ, Φ_λ in place of the states Ψ_K, Ψ_L. For two Slater determinants Φ_κ, Φ_λ, the rules given in (5.4.14) follow readily from the anticommutation properties of the creation and annihilation operators. Examples of such reductions occur in later chapters. At this point, we only remark that the *order* of the subscripts and superscripts, in all equations involving density matrices, is not "random"; it follows a logical pattern and should be carefully respected.

5.5 LOCALIZED ORBITALS. POPULATION ANALYSIS

The energy expression (5.4.16) is often used as a starting point in qualitative discussions of chemical bonding: the second term contains only interelectronic repulsion energy, and is therefore positive, while the first term contains the energy of attraction (p. 123) between the nuclei and the electronic charge cloud—which is usually regarded as the source of the bonds. After spin integrations, the charge density appears in the form

$$P(\boldsymbol{r}) = \sum_{R,S} P_{RS} \phi_R(\boldsymbol{r}) \phi_S^*(\boldsymbol{r}), \quad (5.5.1)$$

with P_{RS} obtained from (5.4.19). Since, however, the orbitals ϕ_R are in general delocalized (usually being the *molecular* orbitals used in a CI expansion), the coefficients in (5.5.1) give no information on the spatial distribution of charge among the atoms and bonds. To get such information, we must pass to the localized basis functions $\{\chi_i\}$ (e.g. AOs) from which the MOs are constructed.

In Chapter 6, where we develop MO theory systematically, we express the n MOs used in an approximate wavefunction Ψ in terms of m basis functions (frequently AOs) according to

$$\phi_R = \sum_i \chi_i T_{iR}. \quad (5.5.2)$$

The spinless counterpart of (5.4.16) then becomes

$$E = \sum_{i,j} P_{ij}^{\chi} \langle \chi_j | \mathrm{h} | \chi_i \rangle + \tfrac{1}{2} \sum_{i,j,k,l} \Pi_{ij,kl}^{\chi} \langle \chi_k \chi_l | g | \chi_i \chi_j \rangle, \quad (5.5.3)$$

in which a superscript has been added to indicate that the density matrices refer to the χ-basis, and

$$P^{\chi}_{ij} = \sum_{R,S} T_{iR} P_{RS} T^{\dagger}_{Sj}, \tag{5.5.4}$$

$$\Pi^{\chi}_{ij,kl} = \sum_{R,S,T,U} T_{iR} T_{jS} \Pi_{RS,TU} T^{\dagger}_{Tk} T^{\dagger}_{Ul}. \tag{5.5.5}$$

The 1- and 2-electron integrals are now all defined directly over basis functions, and (5.5.3) represents the final reduction of the energy functional, valid for any wavefunction constructed using a finite basis.

The density function (5.5.1) may now be written as

$$P(\boldsymbol{r}) = \sum_{i,j} P_{ij} \chi_i(\boldsymbol{r}) \chi^*_j(\boldsymbol{r}), \tag{5.5.6}$$

where (cf. (5.4.19) *et seq.*) P_{ij} is the coefficient of $\langle \chi_j | \mathsf{h} | \chi_i \rangle$ in (5.5.3) and is given by (5.5.4), the superscript χ being dropped when the basis is clear from the context. The density (5.5.6) is of course the most general that can arise from any orbital-type wavefunction constructed from basis functions $\{\chi_i\}$, no matter how many determinants are included. It is therefore possible to define the electron density for a wavefunction containing many thousands of determinants by specifying the single matrix **P**. When the basis consists of AOs (or similar functions with a well-defined localization in space) the elements of **P** provide a means of visualizing and quantitatively describing the form of the electron distribution.

Let us define normalized orbital and overlap densities (assuming for convenience real orbitals)

$$d_i(\boldsymbol{r}) = \chi_i(\boldsymbol{r})^2, \qquad d_{ij}(\boldsymbol{r}) = \frac{\chi_i(\boldsymbol{r})\chi_j(\boldsymbol{r})}{S_{ij}}, \tag{5.5.7}$$

where S_{ij} is the overlap integral for orbitals χ_i and χ_j, and write (5.5.6) in the form

$$P(\boldsymbol{r}) = \sum_i q_i d_i(\boldsymbol{r}) + \sum_{i<j} q_{ij} d_{ij}(\boldsymbol{r}), \tag{5.5.8}$$

where the numerical coefficients are

$$q_i = P_{ii}, \qquad q_{ij} = 2S_{ij}P_{ij}. \tag{5.5.9}$$

Since integration of $P(\boldsymbol{r})$ over all space must yield N (the number of electrons), it follows from (5.5.8) that

$$\int P(\boldsymbol{r})\,\mathrm{d}\boldsymbol{r} = \sum_i q_i + \sum_{i<j} q_{ij} = N, \tag{5.5.10}$$

and hence that the total charge is divided out in such a way that an amount q_i arises from each orbital density d_i and q_{ij} from each overlap density d_{ij}. These quantities (McWeeny, 1951a,b, 1952, 1954a) are usually referred to as orbital and overlap "populations". The idea of "population analysis" was extensively developed by Mulliken (1955a,b), and is often a useful tool in discussing and comparing the results of electronic structure calculations.

We note that the populations assigned to the various regions of space are in no way unique, depending entirely on the choice of basis orbitals. Given two bases, connected by $\bar{\boldsymbol{\chi}} = \boldsymbol{\chi}\mathbf{V}$, the density may be expressed in the alternative forms

$$P(\boldsymbol{r}) = \sum_{i,j} P_{ij}\chi_i(\boldsymbol{r})\chi_j^*(\boldsymbol{r}) = \sum_{i,j} \bar{P}_{ij}\bar{\chi}_i(\boldsymbol{r})\bar{\chi}_j^*(\boldsymbol{r}), \tag{5.5.11}$$

where the density matrices are related by

$$\mathbf{P} = \mathbf{V}\bar{\mathbf{P}}\mathbf{V}^\dagger. \tag{5.5.12}$$

This equivalence has interesting consequences. If, for example, the orbitals $\{\chi_i\}$ are assumed to be the (orthonormal) MOs of a 1-determinant wavefunction, and $\mathbf{V}$ is taken to be a unitary matrix that separately mixes the singly and doubly occupied orbitals, it follows that $\mathbf{P}$ and $\bar{\mathbf{P}}$ will have identical diagonal form; in each case the density will be represented as a superposition of orbital contributions (d_i terms), with coefficients 1 or 2, with no "cross-terms" of d_{ij} type. Even the choice of *molecular* orbitals, then, is arbitrary to within a unitary transformation—a point to which we shall return in Chapter 6.

There is another important consequence of the lack of uniqueness indicated by (5.5.12). If the basis functions, although localized essentially on the atoms, have been orthogonalized (for example, by using the construction (2.2.21)) then the overlap densities will not be normalizable and the amount of charge associated with any "bond" *i—j* will be zero. The total charge N will then be divided out formally among the atoms alone. The overlap terms in (5.5.6) may nevertheless have a profound effect on the electron density in each overlap region: it is therefore clear that conclusions about the nature of chemical bonds cannot be drawn simply from an inspection of bond populations, or even of the corresponding indices P_{ij}, the actual forms of the orbitals being equally important.

5.6 NATURAL EXPANSIONS

The density matrices have many interesting and fundamental properties, some of which we refer to later (Section 6.4). At this point we note that they allow us to formulate conditions for the optimum convergence of expansions such as (3.1.4).

As in (5.3.5), the 1-electron density matrix is in general expressible in the form

$$\rho(\boldsymbol{x}_1;\boldsymbol{x}_1') = \sum_{R,S} \rho_{RS}\psi_R(\boldsymbol{x}_1)\psi_S^*(\boldsymbol{x}_1'), \tag{5.6.1}$$

where the ψ_R are spin-orbitals of some complete orthonormal set. Since the matrix with elements ρ_{RS} must be Hermitian, a basis change leads to a unitary transformation of $\boldsymbol{\rho}$, exactly parallel to that described in (5.5.11) and (5.5.12). If, moreover, we choose $\bar{\boldsymbol{\psi}} = \boldsymbol{\psi}\mathbf{T}$ where the columns $\mathbf{t}_R$ of the transformation matrix are eigenvectors of $\boldsymbol{\rho}$,

$$\boldsymbol{\rho}\mathbf{t}_R = n_R\mathbf{t}_R, \tag{5.6.2}$$

then $\bar{\boldsymbol{\rho}} = \mathbf{T}^\dagger\boldsymbol{\rho}\mathbf{T}$ will assume *diagonal* form and we shall obtain

$$\rho(\boldsymbol{x}_1;\boldsymbol{x}_1') = \sum_{R} n_R\bar{\psi}_R(\boldsymbol{x}_1)\bar{\psi}_R^*(\boldsymbol{x}_1'). \tag{5.6.3}$$

The (orthonormal) spin-orbitals thus defined (Löwdin, 1955) are called the *natural spin-orbitals* (NSOs), and their populations n_R clearly satisfy $\sum_R n_R = N$ (as follows by putting $\boldsymbol{x}_1' = \boldsymbol{x}_1$ and integrating).

The importance of the NSOs rests on the fact that any spin-orbital whose population is negligible may be omitted from a CI expansion without appreciably affecting the accuracy of the expansion; there is therefore a "natural" criterion for selecting a finite number of spin-orbitals, such that a severely truncated expansion may approach the complete set expansion as closely as possible. In other words, the convergence of a CI expansion may be improved by going over from an arbitrary spin-orbital set to the NSOs obtained as eigenfunctions of the 1-electron density matrix.

The precise formulation of these ideas depends on a theorem due to Schmidt (1907) and rediscovered by Coleman (1963). We wish to approximate an arbitrary function‡ in terms of a finite set of products in the form

$$\Psi(\boldsymbol{x}_1, \boldsymbol{x}_2, \ldots, \boldsymbol{x}_N) \simeq \sum_{i=1}^{n}\sum_{j=1}^{n'} a_{ij}f_i(\boldsymbol{x}_1)g_j(\boldsymbol{x}_2, \boldsymbol{x}_3, \ldots, \boldsymbol{x}_N). \tag{5.6.4}$$

The "least-squares" error is defined by

$$\Delta = \int \left|\Psi(\boldsymbol{x}_1, \boldsymbol{x}_2, \ldots, \boldsymbol{x}_N) - \sum_{i,j} a_{ij}f_i(\boldsymbol{x}_1)g_j(\boldsymbol{x}_2, \boldsymbol{x}_3, \ldots, \boldsymbol{x}_N)\right|^2 \mathrm{d}\boldsymbol{x}_1 \ldots \mathrm{d}\boldsymbol{x}_N, \tag{5.6.5}$$

‡ We omit symmetry considerations; antisymmetry may be imposed subsequently in the usual way (Section 3.1).

and the theorem states that this takes its least value, for given n and n', when the f_i are the "natural" functions $\bar{\psi}_R$ defined above, with the highest eigenvalues, and that the corresponding natural expansion may be written in the form

$$\Psi(\boldsymbol{x}_1, \boldsymbol{x}_2, \ldots, \boldsymbol{x}_N) \simeq \sum_{i=1}^{n} c_i f_i(\boldsymbol{x}_1) g_i(\boldsymbol{x}_2, \boldsymbol{x}_3, \ldots, \boldsymbol{x}_N), \qquad (5.6.6)$$

where n is the lesser of n, n'. In its quantum-mechanical context, this theorem implies that with any given number of spin-orbitals the correspondingly truncated expansion will achieve maximum accuracy when they are chosen to be the *natural* spin-orbitals with the highest populations.

The existence of rapidly convergent expansions does not, of course, imply that they can be found easily; for the NSOs are defined in principle through an *exact* density matrix, which, with existing techniques, is usually approached by a complicated energy optimization. Nevertheless, it is often profitable to diagonalize the *approximate* density matrix at any stage in a CI calculation; selection of configurations involving the most heavily populated NSO may then lead to a dramatic shortening of the expansion (see e.g. Löwdin and Shull, 1956; Davidson, 1962), which may in turn simplify subsequent optimization procedures. Examples of such applications may be found in the work of Ruedenberg *et al.* (1979) and Roos *et al.* (1980), to which we refer in Chapter 8.

It is possible to define natural spin "geminals" (NSG) as eigenvectors of the matrix associated with $\boldsymbol{\pi}$; but these have proved to be of less immediate value than the NSOs. On the other hand, it is often advantageous to eliminate spin as soon as possible and to work in terms of *natural orbitals* (NOs) obtained as eigenfunctions of the spinless density matrix P (i.e. by diagonalizing the matrix $\mathbf{P}$, which represents P in an orbital basis). These give near-optimum convergence and are in many ways more convenient than the NSOs (McWeeny and Kutzelnigg, 1968).

5.7 MOLECULAR PROPERTIES: AN INTRODUCTION

So far we have considered only the "internal" properties of an isolated system—how the electrons and their spins may be described, and how their distribution determines the total electronic energy. Many important properties, however, relate to the *response* of a system when we do something to it; to the way it changes when we change its Hamiltonian from $\mathsf{H} = \mathsf{H}_0$ to $\mathsf{H} = \mathsf{H}_0 + \mathsf{H}'$ by interacting with it in some way. For example, we may switch on an electric field and observe the consequent change in energy due to interaction with the field. The theory of various electronic properties, and of response in general, is taken up in detail in later chapters: here we make only a preliminary approach in order to gain familiarity with the methods developed so far and to reveal the special importance of the electron density.

Let us suppose that the density functions ρ and π for the isolated molecule with Hamiltonian H_0 have been determined variationally so that the energy

$$E = \int \mathsf{h}\rho(x_1)\,\mathrm{d}\boldsymbol{x}_1 + \tfrac{1}{2}\int g\pi(\boldsymbol{x}_1, \boldsymbol{x}_2)\,\mathrm{d}x_1\,\mathrm{d}\boldsymbol{x}_2 \tag{5.7.1}$$

is stationary against any small changes $\delta\rho$ and $\delta\pi$ consistent with any essential constraints (for example that the wavefunction remains normalized, with a particular orbital form): thus

$$\int \mathsf{h}\delta\rho(\boldsymbol{x}_1)\,\mathrm{d}\boldsymbol{x}_1 + \tfrac{1}{2}\int g\delta\pi(\boldsymbol{x}_1, \boldsymbol{x}_2)\,\mathrm{d}\boldsymbol{x}_1\,\mathrm{d}\boldsymbol{x}_2 = 0. \tag{5.7.2}$$

This is precisely equivalent to saying that the wavefunction Ψ has been optimized by adjustment of all the parameters it contains. Then we suppose the interaction switched on: this is a change that affects every electron in the same way, $\mathsf{h}(i) \to \mathsf{h}(i) + \delta\mathsf{h}(i)$, but does not affect the mutual repulsion between electrons, $g(i, j)$. The densities ρ and π will change, and the energy change will be, to first order

$$\delta E = \int \mathsf{h}\delta\rho(\boldsymbol{x}_1)\,\mathrm{d}\boldsymbol{x}_1 + \tfrac{1}{2}\int g\delta\pi(\boldsymbol{x}_1, \boldsymbol{x}_2)\,\mathrm{d}\boldsymbol{x}_1\,\mathrm{d}\boldsymbol{x}_2 + \int \delta\mathsf{h}\rho(\boldsymbol{x}_1)\,\mathrm{d}\boldsymbol{x}_1.$$

But the original densities, by assumption, possessed the stationary property expressed by (5.7.2), and therefore only one term remains:

$$\delta E = \int \delta\mathsf{h}\rho(\boldsymbol{x}_1)\,\mathrm{d}\boldsymbol{x}_1. \tag{5.7.3}$$

This is just the expectation or average value of the interaction operator, $\delta\mathsf{h}$, *taken over the electron density function of the isolated molecule,* as would have been anticipated from perturbation theory. This is a simple, powerful and extremely general result, which provides a basis for discussion of all atomic and molecular properties that involve a *first-order* response, and is applicable even when we do not possess exact wavefunctions; it is usually referred to as a generalized Hellmann–Feynman theorem (see e.g. Epstein, 1974), which states that the first-order change in energy when $\mathsf{H} \rightarrow \mathsf{H} + \delta\mathsf{H}$ is given by

$$\delta E = \langle \Psi | \, \delta\mathsf{H} \, | \Psi \rangle, \tag{5.7.4}$$

provided that Ψ has been *fully optimized,* before application of the perturbation $\delta\mathsf{H}$, with respect to all the parameters it may contain. Higher-order effects depend on *change* of the density functions and are very much more difficult to discuss generally: here we shall be content to examine the immediate implications of (5.7.3). We consider the case where $\delta\mathsf{h}$ is spin-independent and arises merely from a change in the potential-energy function, in order to illustrate the great conceptual value of the density functions—in particular of the electron density $P(\boldsymbol{r}_1)$.

Change in potential field: Examples

We suppose that $\delta\mathsf{h}(1) = \delta V(\boldsymbol{r}_1)$, the change in the expression for the potential energy $V(\boldsymbol{r}_1)$ of a single electron of the system at point $\boldsymbol{r}_1$, is simply a function of position. This may arise in many ways, for example by applying an external electric‡ field or by changing the positions of the nuclei. The spin integration in (5.7.3) is immediate, and the result is

$$\boxed{\delta E = \int \delta V(\boldsymbol{r}_1) P(\boldsymbol{r}_1) \, \mathrm{d}\boldsymbol{r}_1.} \tag{5.7.5}$$

The integrand may be regarded as the modification in the potential energy of an amount of charge $P(\boldsymbol{r}_1)\,\mathrm{d}\boldsymbol{r}_1$ (since $\delta V(\boldsymbol{r}_1)$ is the change in energy of unit charge, one electron, at point $\boldsymbol{r}_1$); and δE is therefore mathematically identical with the change in potential energy of a "charge cloud" of density $P(\boldsymbol{r}_1)$ (in electrons per unit volume). In other words: *in calculating the first-order response of the system to any modification in the potential field, the electron distribution may be treated simply as a smeared-out electric charge of density P.* The importance of this result can hardly be overestimated. It provides the basis for almost the whole of qualitative valence theory, in which the properties of a molecule are discussed in terms of the electronic "charge cloud". We develop two aspects of this association.

First, let us suppose that δV arises from displacement of nucleus n through a distance δX_n in the x direction. Dividing (5.7.5) throughout by δX_n and letting $\delta X_n \rightarrow 0$, we note that $\delta V(\boldsymbol{r}_1)/\delta X_n$, which appears in the

‡ Magnetic fields introduce velocity-dependent perturbations and are considered in Chapter 11.

integrand, becomes the rate of change of the potential energy between nucleus n and an electron at point $\boldsymbol{r}_1$. Thus

$$-\frac{\partial V}{\partial X_n} = F_{nx}(\boldsymbol{r}_1),$$

where $F_{nx}(\boldsymbol{r}_1)$ is the x component of force on nucleus n due to one electron at point $\boldsymbol{r}_1$. We note also that in the actual molecular situation, the force exerted on nucleus n by the electrons may be defined in terms of derivatives of the *total* electronic energy E: thus

$$-\frac{\partial E}{\partial X_n} = F_{nx},$$

where F_{nx} is the x component of total force on nucleus n due to electron–nuclear terms in the Hamiltonian. The differential form of (5.7.5), obtained by letting $\delta X_n \to 0$ and inserting the above limiting values, therefore reads

$$F_{nx} = \int F_{nx}(\boldsymbol{r}_1) P(\boldsymbol{r}_1)\, \mathrm{d}\boldsymbol{r}_1 \qquad (5.7.6)$$

—the total x component of force exerted on nucleus n may be computed by adding the contributions due to each element of the charge cloud. This remarkable result, due to Hellmann (1937) and Feynman (1939), shows that the forces holding the nuclei together in a molecule may be given an entirely classical *interpretation* once the electron density has been computed by quantum mechanics. These forces are exactly balanced, in equilibrium, by the repulsions between different nuclei—which we have excluded from the electronic Hamiltonian, but whose additive contribution must of course be included when we require the *total* energy of the system.

One or two remarks are necessary. This result is apparently valid even for an approximate wavefunction if we regard F_{nx} as the force calculated as a gradient of the *variationally computed energy* E (i.e. as an *approximation* to the actual force); but for this to be true the computed energy E must be stationary (see (5.7.1) *et seq.*) against variation of *all* the parameters contained in the approximate wavefunction Ψ from which P is obtained. If this is the case then the energy of a molecule may be calculated either (i) directly from (5.2.19) as an expectation value of the Hamiltonian, or (ii) by calculating the forces on the nuclei, according to (5.7.6) etc., and integrating these as the atoms are brought together from infinity—and the results should agree. It might appear that (ii) is preferable because the forces are easier to calculate than the electron interactions (described by the pair function Π), which appear only in (i). But unfortunately the results only agree if P is obtained from a variationally determined wavefunction; and the direct variational calculation of E cannot therefore be avoided. It must also be realized that Ψ normally contains parameters (positions of orbital centres) that "anchor" the orbitals used in constructing Ψ to the nuclei: if these are varied to get the best Ψ, the orbitals are found to float away from the nuclei (very little, but enough to completely vitiate any force calculations). The resultant functions are said to be "stable" under the perturbation (Hall, 1961, 1964); in general it is desirable that the variation function be flexible enough to ensure stability. The direct use of the Hellmann–Feynman theorem is therefore accompanied by serious difficulties, and its value is largely conceptual. Other forms of the theorem may be derived, and full discussions are available in the literature.‡

As the second illustration of the value of the "charge-cloud" concept, we consider the definition of dipole and multipole moments, field gradients at points in the molecule, etc. These can all be discussed by evaluating the response of the system to an infinitesimal "test charge". If a test charge q is brought to any point 0 then the change in the Hamiltonian is (noting that in atomic units the electron charge is -1)

$$\mathsf{H}' = \sum_i \delta V(i) = \sum_i \left(\frac{-q}{r_{0i}}\right),$$

and the change in energy of the system (molecule + test charge) is

$$\delta E = -q \int \frac{1}{r_{01}} P(\boldsymbol{r}_1)\, \mathrm{d}\boldsymbol{r}_1 = q\phi_0, \tag{5.7.7}$$

where, by definition, ϕ_0 is the electric potential at point 0. The electric potential at any point in space may therefore be calculated as if, once again, the electrons were smeared out into a static distribution of charge of density P. This result contains many others. If $1/r_{01}$ is expanded in terms of coordinates of the point $\mathbf{r}_1$ relative to the centroid of P then we find that ϕ_0 is the same as that due to a point charge of N electrons, a point dipole, point quadrupole, etc. (all located at the centroid), where the dipole, quadrupole and multipole moments are just those computed classically for a charge distribution of density P. Thus, for example, the

‡ Floating functions were first used by Hurley (1954). The Hellmann–Feynman theorem has a long history and has been reviewed extensively by Epstein *et al.* (1967). See also Deb (1981) and Epstein (1974).

electronic part of the dipole moment in the z direction is

$$\left\langle \sum_i x_i \right\rangle = -\int z_1 P(\boldsymbol{r}_1)\, \mathrm{d}\boldsymbol{r}_1. \tag{5.7.8}$$

Other quantities, such as the field gradient at a nucleus (which determines the nuclear quadrupole coupling observed in quadrupole resonance experiments), may be computed similarly in a purely classical way.

5.8 THE PAIR FUNCTION. ELECTRON CORRELATION

We now turn to 2-electron properties, which are completely determined by the function $\pi(\boldsymbol{x}_1, \boldsymbol{x}_2)$. It may be shown (McWeeny and Mizuno, 1961) that for a state with definite spin (quantum numbers S, M) π is a six-component quantity. Let us write explicitly the off-diagonal components of π (p. 122), to avoid any ambiguity, putting

$$\begin{aligned}\pi(\boldsymbol{x}_1, \boldsymbol{x}_2; \boldsymbol{x}'_1, \boldsymbol{x}'_2) = {} & \Pi_{\alpha\alpha,\alpha\alpha}(\boldsymbol{r}_1, \boldsymbol{r}_2; \boldsymbol{r}'_1, \boldsymbol{r}'_2)\alpha(s_1)\alpha(s_2)\alpha^*(s'_1)\alpha^*(s'_2) \\ & + \Pi_{\alpha\beta,\alpha\beta}(\boldsymbol{r}_1, \boldsymbol{r}_2; \boldsymbol{r}'_1, \boldsymbol{r}'_2)\alpha(s_1)\beta(s_2)\alpha^*(s'_1)\beta^*(s'_2) \\ & + \text{etc.}\end{aligned} \tag{5.8.1}$$

This notation resembles that used in (5.3.12), but the subscripts after the commas now indicate the spin factors for the primed variables. We refer to the spinless factor in the first term as the $\alpha\alpha$, $\alpha\alpha$ component, that in the second term as the $\alpha\beta$, $\alpha\beta$ component etc., and it follows at once, on integrating over spins, that Π is the sum of four components only. With the usual brief notation for diagonal elements, namely

$$\Pi_{\alpha\alpha}(\boldsymbol{r}_1, \boldsymbol{r}_2) = \Pi_{\alpha\alpha,\alpha\alpha}(\boldsymbol{r}_1, \boldsymbol{r}_2, \boldsymbol{r}_1, \boldsymbol{r}_2) \quad \text{etc.,}$$

the pair function may then be written as

$$\Pi(\boldsymbol{r}_1, \boldsymbol{r}_2), = \Pi_{\alpha\alpha}(\boldsymbol{r}_1, \boldsymbol{r}_2) + \Pi_{\alpha\beta}(\boldsymbol{r}_1, \boldsymbol{r}_2) + \Pi_{\beta\alpha}(\boldsymbol{r}_1, \boldsymbol{r}_2) + \Pi_{\beta\beta}(\boldsymbol{r}_1, \boldsymbol{r}_2), \tag{5.8.2}$$

in which all components have an immediate physical interpretation: $\Pi_{\alpha\alpha}(\boldsymbol{r}_1, \boldsymbol{r}_2)$ is the contribution to the pair function Π that arises from electrons at $\boldsymbol{r}_1$ and $\boldsymbol{r}_2$ *both with spin* $+\frac{1}{2}$; similarly $\Pi_{\alpha\beta}(\boldsymbol{r}_1, \boldsymbol{r}_2)$ arises from electrons at $\boldsymbol{r}_1$ and $\boldsymbol{r}_2$ with spins $+\frac{1}{2}$ and $-\frac{1}{2}$ respectively. Each component therefore determines a probability of finding electrons simultaneously at two points in space with a given configuration of spins ($\alpha\alpha$, $\alpha\beta$, $\beta\alpha$ or $\beta\beta$). The other two of the six non-zero terms in (5.8.1) (namely the $\alpha\beta$, $\beta\alpha$ and $\beta\alpha$, $\alpha\beta$ components) do not contribute to the pair function, disappearing in the spin integration, but are important in determining expectation values of spin-dependent operators, which can reverse the spins (see Problem 4.3).

Let us write down the pair-function components for the simple case of a system described by one determinant of spin-orbitals, with orbital factors $\phi_A, \phi_B, \ldots, \phi_R, \ldots$. The 1-electron density matrix has the form (5.3.12), namely (adding primes since off-diagonal elements will be needed)

$$\rho(\boldsymbol{x}_1; \boldsymbol{x}_1') = P_\alpha(\boldsymbol{r}_1; \boldsymbol{r}_1')\alpha(s_1)\alpha^*(s_1') + P_\beta(\boldsymbol{r}_1; \boldsymbol{r}_1')\beta(s_1)\beta^*(s_1'), \quad (5.8.4)$$

where the two components‡ are given in (5.3.13) as sums over the orbitals with α and β spin factors respectively. Substitution in (5.3.9) yields an expression for π from which the required components may be picked out:

$$\Pi_{\alpha\alpha}(\boldsymbol{r}_1, \boldsymbol{r}_2) = P_\alpha(\boldsymbol{r}_1)P_\alpha(\boldsymbol{r}_2) - P_\alpha(\boldsymbol{r}_2; \boldsymbol{r}_1)P_\alpha(\boldsymbol{r}_1'; \boldsymbol{r}_2), \quad (5.8.5)$$

$$\Pi_{\alpha\beta}(\boldsymbol{r}_1, \boldsymbol{r}_2) = P_\alpha(\boldsymbol{r}_1)P_\beta(\boldsymbol{r}_2), \quad (5.8.6)$$

together with similar terms obtained by interchanging αs and βs.

From these results we can obtain a clear picture of the kind of *electron correlation* recognized in the 1-determinant approximation. Thus, for electrons of different spin, the form of $\Pi_{\alpha\beta}$ shows that the probability of two volume elements being occupied simultaneously by electrons, the first spin up the second down, is just the product of the probabilities of each of the two events occurring independently, i.e. without reference to the other. We say there is *no correlation* between the positions of electrons of opposite spin. This absence of correlation in the 1-determinant approximation is clearly a defect, since electrons repel each other and we should expect the probability of finding two of them close together to be reduced below the value for independent particles. Electrons of like spin ($+\frac{1}{2}$) are, however, described by a *correlated* pair function, namely $\Pi_{\alpha\alpha}$; and this clearly vanishes for $\boldsymbol{r}_2 \to \boldsymbol{r}_1$, since then the two terms become equal and cancel exactly. This special type of correlation prevents two electrons of like spin being found at the same point in space, and applies whenever the particles are "fermions" with antisymmetric wavefunctions; it is described as *Fermi correlation.*

The proper description of correlation has been one of the main obstacles to progress in quantum chemistry, and it is therefore important to be familiar with its general features. If we suppose that Ψ is an exact many-electron wavefunction then the following results hold (McWeeny, 1960).

‡ For a definite spin state there are only two components (α, α and β, β), and the single-subscript notation in (5.8.4) is thus unambiguous.

(1) $\Pi_{\alpha\alpha}(\boldsymbol{r}_1, \boldsymbol{r}_2)$ and $\Pi_{\beta\beta}(\boldsymbol{r}_1, \boldsymbol{r}_2)$ both vanish like r_{12}^2 for $\boldsymbol{r}_1 \rightarrow \boldsymbol{r}_2$, giving zero probability of finding two electrons of like spin at the same point in space. This follows from the antisymmetry of Ψ and the resulting antisymmetry of

$$\Pi_{\alpha\alpha,\alpha\alpha}(\boldsymbol{r}_1, \boldsymbol{r}_2; \boldsymbol{r}_1', \boldsymbol{r}_2') \quad \text{and} \quad \Pi_{\beta\beta,\beta\beta}(\boldsymbol{r}_1, \boldsymbol{r}_2; \boldsymbol{r}_1', \boldsymbol{r}_2')$$

in both pairs of variables, and may be regarded as the most general statement of the exclusion principle.

(2) If we write

$$\Pi_{\alpha\alpha}(\boldsymbol{r}_1, \boldsymbol{r}_2) = P_\alpha(\boldsymbol{r}_1)P_\alpha(\boldsymbol{r}_2)[1 + f^{\alpha\alpha}(\boldsymbol{r}_1, \boldsymbol{r}_2)]$$

and define the "correlation hole"

$$\frac{\Pi_{\alpha\alpha}(\boldsymbol{r}_1, \boldsymbol{r}_2)}{P_\alpha(\boldsymbol{r}_2)} - P_\alpha(\boldsymbol{r}_1) = P_\alpha(\boldsymbol{r}_1)f^{\alpha\alpha}(\boldsymbol{r}_1, \boldsymbol{r}_2)$$

(which is the difference between the probability of finding an electron at $\boldsymbol{r}_1$, with spin up, *when one is known to be at* $\boldsymbol{r}_2$, *also with spin up*, minus the same probability in the absence of the second electron) then the "hole" integrates to -1:

$$\int P_\alpha(\boldsymbol{r}_1)f^{\alpha\alpha}(\boldsymbol{r}_1, \boldsymbol{r}_2)\,\mathrm{d}\boldsymbol{r}_1 = -1 \quad (\text{any } \boldsymbol{r}_2). \tag{5.8.7}$$

The probability of finding an electron at point $\boldsymbol{r}_1$, in the vicinity of a "reference electron" at point $\boldsymbol{r}_2$, with the same spin, is therefore strongly reduced. Clearly, the exclusion principle requires that the correlation approaches -100% as $\boldsymbol{r}_1 \rightarrow \boldsymbol{r}_2$, i.e. that $f^{\alpha\alpha}(\boldsymbol{r}_1, \boldsymbol{r}_2) \rightarrow -1$ for $\boldsymbol{r}_1 \rightarrow \boldsymbol{r}_2$. A similar result holds, of course, for down-spin electrons. The form of the correlation factor is known for a dilute electron gas (see e.g. Slater, 1951), but much less is known about the form of the hole in regions of strongly varying potential.

(3) If we write, similarly,

$$\Pi_{\alpha\beta}(\boldsymbol{r}_1, \boldsymbol{r}_2) = P_\alpha(\boldsymbol{r}_1)P_\beta(\boldsymbol{r}_2)[1 + f^{\alpha\beta}(\boldsymbol{r}_1, \boldsymbol{r}_2)]$$

then the correlation hole for an electron of up spin, in the vicinity of one of down spin, integrates to *zero*, again for all positions of the down-spin reference electron:

$$\int P_\alpha(\boldsymbol{r}_1)f^{\alpha\beta}(\boldsymbol{r}_1, \boldsymbol{r}_2)\,\mathrm{d}\boldsymbol{r}_1 = 0 \quad (\text{any } \boldsymbol{r}_2). \tag{5.8.8}$$

The probability is expected to be decreased (hole function negative) near the reference electron, owing to the coulomb repulsion, but enhanced further away so that the integral will vanish. The hole function is also known to have a discontinuity of slope, or "cusp" for $\boldsymbol{r}_1 \rightarrow \boldsymbol{r}_2$, as may be shown from work by Kato (1957) and others.

There was little detailed work on correlation factors before the nineteen-seventies, when many-electron wavefunctions of high precision were rarely available. But such factors may now be calculated fairly easily for CI-type wavefunctions, and inspection of their forms (see e.g. Cooper and Pounder, 1980) clearly indicates what kind of correlation a given function introduces. Even at a lower level of accuracy, however, analysis of the pair functions is worthwhile. A trivial but interesting example is provided by the Heitler–London–Wang calculation on the hydrogen molecule. The hole in the up-spin distribution due to a reference electron of down spin on the left-hand nucleus decreases the chance of finding the electron on the same centre, enhancing that of finding it on the right-hand centre. This kind of correlation may be termed "longitudinal" (i.e. it operates along the bond direction). In the same way, it is possible to speak of "angular" and "radial" correlation, but with orbital wavefunctions it is exceedingly difficult to get a correct description of the cusp for $\boldsymbol{r}_1 \rightarrow \boldsymbol{r}_2$, and the computed functions give only the broadest features of the correlation.

The mathematical description of correlation (see also McWeeny and Kutzelnigg, 1968; McWeeny, 1976) sometimes leads to results that, at first sight, are unexpected. Thus, for example, $\Pi_{\alpha\alpha}(\boldsymbol{r}_1, \boldsymbol{r}_2)$ in the example just considered is everywhere zero; this indicates a strong correlation keeping like-spin electrons apart, with $f^{\alpha\alpha}(\boldsymbol{r}_1, \boldsymbol{r}_2) = -1$ everywhere and a Fermi hole exactly like the up-spin electron density $P_\alpha(\boldsymbol{r}_1)$ but with the sign reversed. For a 2-electron system this correlation is built into the wavefunction by imposing the spin-eigenfunction condition $S = M = 0$; intuitively there is no chance of both electrons having the same spin because this would correspond to a triplet function. The correlation factor is required in the "bookkeeping", however, because $P_\alpha(\boldsymbol{r}_1)P_\alpha(\boldsymbol{r}_2)$ is clearly non-zero; as far as either electron is aware it has a probability of having its spin either way up, but the correlation factor ensures that when one spin is known the other must be different. This point of view, in which the spins are classically interpreted as particle labels, has been developed by Kutzelnigg *et al.* (1968). Much detailed work evidently remains to be done before it will be possible to give a completely coherent physical analysis of correlation effects in molecules.

In the meantime, it is usual in discussing correlation effects to fall back on a more pragmatic definition of "correlation energy" due to Löwdin (1955), in which $E_{\text{corr}} = E_{\text{exact}} - E_{\text{HF}}$, where E_{HF} is a best one-determinant total energy while E_{exact} is the "exact" energy from a high-precision (e.g. extended CI) calculation based on the usual non-relativistic Hamiltonian. This definition is untenable in many situations; in the dissociation of the hydrogen molecule, for example, there is physically no correlation for two independent 1-electron atoms—but the MO approximation leads to a large and spurious value of E_{corr} (McWeeny, 1967, 1976). For molecular ground states and equilibrium geometries, however, the Löwdin definition has a certain operational value.

The development of "density-functional theory" (see e.g. Parr, 1983) has given an added impetus to calculations of the correlation hole, and it is already apparent (Colle and Salvetti, 1975, 1979; McWeeny, 1976; Cohen and Frishberg, 1976a,b; Rajagopal, 1980; do Amaral and McWeeny, 1983; Gritsenko *et al.*, 1986) that use of an appropriate functional form often permits a rapid and accurate estimation of correlation energies E_{corr}, even for many-electron molecules. The status of the density-functional approach has not yet been fully established, however, and certain semi-empirical elements still remain; for further references in this promising field see Dahl and Avery (1982).

5.9 SPIN DENSITY

It was noted in Section 5.3 that when the 1-electron density matrix is written in the form (5.3.12) the difference between the α and β components allows us to define a resultant spin density, essentially as the excess density of up-spin electrons compared with down-spin. In recent years, with the development of magnetic resonance techniques, this quantity has acquired great importance. To indicate its origin we note that the expectation value of the z component of spin angular momentum may be written, using for clarity the explicit form (5.2.12),

$$\langle \mathsf{S}_z \rangle = \int_{\boldsymbol{x}_1' = \boldsymbol{x}_1} \mathsf{S}_z(1)\rho(\boldsymbol{x}_1, \boldsymbol{x}_1')\, d\boldsymbol{x}_1.$$

Now for any spin eigenstate, ρ may be expressed in terms of its components according to (5.3.12), and the spin integration may then be performed at once to give

$$\langle \mathsf{S}_z \rangle = \int_{r_1'=r_1} \tfrac{1}{2}[P_\alpha(\boldsymbol{r}_1; \boldsymbol{r}_1') - P_\beta(\boldsymbol{r}_1; \boldsymbol{r}_1')]\,\mathrm{d}\boldsymbol{r}_1.$$

For $\boldsymbol{r}_1' = \boldsymbol{r}_1$, the integrand represents the contribution to the expectation value arising from the volume element $\mathrm{d}\boldsymbol{r}_1$ in ordinary 3-dimensional space; we therefore define (cf. (5.3.15))

$$Q_z(\boldsymbol{r}_1; \boldsymbol{r}_1') = \tfrac{1}{2}[P_\alpha(\boldsymbol{r}_1; \boldsymbol{r}_1') - P_\beta(\boldsymbol{r}_1; \boldsymbol{r}_1')] \qquad (5.9.1)$$

as the *spin-density matrix*. We have already noted that the diagonal element ($\boldsymbol{r}_1' = \boldsymbol{r}_1$) is simply the *density of spin angular momentum* about the z axis. It is also clear that Q_z is extracted from ρ simply by applying the spin operator S_z before integrating over spin: returning to the briefer notation (p. 123),

$$Q_z(\boldsymbol{r}_1) = \int \mathsf{S}_z \rho(\boldsymbol{x}_1)\,\mathrm{d}s_1. \qquad (5.9.2)$$

Since, for a spin eigenstate, the expectation-value expression yields

$$\langle \mathsf{S}_z \rangle = \int Q_z(\boldsymbol{r}_1)\,\mathrm{d}\boldsymbol{r}_1 = M, \qquad (5.9.3)$$

Q_z may be normalized. Thus

$$Q_z(\boldsymbol{r}_1) = M D_S(\boldsymbol{r}_1), \qquad (5.9.4)$$

where D_S is the *normalized* spin density first defined in a rather different way by McConnell (1958) (see also Weissman, 1956).

The spin density defined above in (5.9.1) is in fact just one component of a *vector* density $\boldsymbol{Q} = (Q_x, Q_y, Q_z)$; the other two components may be calculated by generalizing (5.9.2) and writing

$$Q_\alpha(\boldsymbol{r}_1) = \int \mathsf{S}_\alpha \rho(\boldsymbol{x}_1)\,\mathrm{d}s_1 \quad (\alpha = x, y, z). \qquad (5.9.5)$$

It is then found that, for any spin eigenstate with spin quantized in the z direction, Q_x and Q_y are identically zero at all points in space; only Q_z is non-zero. The spin distribution is therefore polarized everywhere along the axis of spin quantization (normally fixed by an applied magnetic field—as in magnetic resonance experiments) and there is zero probability of finding a spin-angular-momentum component transverse to the axis. This is no longer true for a Hamiltonian that contains spin–orbit coupling terms (Chapter 11), and in such cases the spin distribution may be "warped" in an interesting way (McWeeny, 1973). In the present section attention will be confined to spin eigenstates.

For $S \neq 0$ there is of course a degenerate set of $2S+1$ states characterized by M-values ranging from $-S$ to $+S$, and the density of spin angular momentum will depend on which state we consider. In fact, however, the spin densities are all the same except for a proportionality constant, in accordance with (5.9.4). Often it is convenient to express the spin density in terms of that for the "top" state of the multiplet, with $M = S$, and in this case, introducing the state label κ explicitly and using $\bar{\kappa}$ to denote the state with $M = S$ $(S \neq 0)$, we obtain

$$Q_z(\kappa\kappa \mid \boldsymbol{r}_1; \boldsymbol{r}_1') = \frac{M}{S} Q_z(\bar{\kappa}\bar{\kappa} \mid \boldsymbol{r}_1; \boldsymbol{r}_1') = \frac{M}{S} Q_S(\boldsymbol{r}_1; \boldsymbol{r}_1'), \qquad (5.9.6)$$

where Q_S will always refer to the *standard* state with $M = S$. The reason for the proportionality is essentially group-theoretical. In a matrix element such as $\langle \Phi_\kappa |\mathsf{S}_z| \Phi_\kappa \rangle$, each of the three parts transforms, under rotation of the axis of spin quantization, according to a particular irreducible representation of the 3-dimensional rotation group; for Φ_κ, S determines the representation while M labels its basis vectors; on the other hand, S_z behaves like the $M = 0$ component of a basis corresponding to $S = 1$. If we denote the species of the operator by (s, m) then, according to (A3.27), the matrix elements have the property

$$\langle \Phi_\kappa |\mathsf{S}_z| \Phi_\kappa \rangle = \begin{pmatrix} S & s & | & S \\ M & m & | & M \end{pmatrix} \times \text{constant}$$

where the constant is a quantity independent of M and m. The coefficient is the Clebsch–Gordan coefficient that occurs in coupling angular momenta (S, M) and (s, m) to a resultant (S, M) as in (A3.28); and for $s = 1$, $m = 0$ reduces to M. Expectation values of S_z, in different states, are therefore proportional to M, and the density functions defined in terms of these matrix elements exhibit the same proportionality.

This last result, although trivial in itself (since Φ_κ is an eigenfunction of S_z), points the way to a further generalization, in which we introduce *transition* spin densities analogous to the transition (charge) densities $P(\kappa\kappa' \mid \boldsymbol{r}_1)$ defined in (5.4.5). The spin operators S_x, S_y and S_z may have non-zero matrix elements between functions Φ_κ and $\Phi_{\kappa'}$, with *different* spin quantum numbers, (S, M) and (S', M'), and these are most easily discussed in terms of the "spherical" components

$$\mathsf{S}_{+1} = -\frac{\mathsf{S}_x + i\mathsf{S}_y}{\sqrt{2}}, \qquad \mathsf{S}_0 = \mathsf{S}_z, \qquad \mathsf{S}_{-1} = \frac{\mathsf{S}_x - i\mathsf{S}_y}{\sqrt{2}}, \qquad (5.9.7)$$

which are chosen (with due regard to "phase conventions") so as to behave exactly like spin-eigenfunctions, with $s = 1$ and $m = +1, 0, -1$. It is now possible to reduce any 1-electron matrix element containing spin operators. We simply write

$$\left\langle \Phi_{\kappa'} \left| \sum_i \mathsf{f}(i)\mathsf{S}_m(i) \right| \Phi_\kappa \right\rangle = \int \mathsf{f}\mathsf{S}_m \rho(\kappa\kappa' \mid \boldsymbol{x}_1)\, \mathrm{d}\boldsymbol{x}_1,$$

and perform the spin integration first to obtain

$$\boxed{\left\langle \Phi_{\kappa'} \left| \sum_i \mathsf{f}(i)\mathsf{S}_m(i) \right| \Phi_\kappa \right\rangle = \int \mathsf{f} Q_m(\kappa\kappa' \mid \boldsymbol{r}_1)\, \mathrm{d}\boldsymbol{r}_1,} \qquad (5.9.8)$$

where

$$\boxed{Q_m(\kappa\kappa' \mid \boldsymbol{r}_1) = \int \mathsf{S}_m \rho(\kappa\kappa' \mid \boldsymbol{x}_1)\, \mathrm{d}s_1.} \qquad (5.9.9)$$

Equation (5.9.9) defines the transition spin-density function connecting Φ_κ and $\Phi_{\kappa'}$, and is the immediate generalization of (5.9.2). The generalization of (5.9.6) is readily found to be

$$Q_m(\kappa\kappa' \mid \boldsymbol{r}_1) = \begin{bmatrix} S & 1 & S' \\ M & m & M' \end{bmatrix} Q_S(\bar{\kappa}\bar{\kappa}' \mid \boldsymbol{r}_1), \qquad (5.9.10)$$

where the proportionality factor is simply a ratio of CG coefficients:

$$\begin{bmatrix} S & 1 & S' \\ M & m & M' \end{bmatrix} = \begin{pmatrix} S & 1 & S' \\ M & m & M' \end{pmatrix} \begin{pmatrix} S & 1 & S' \\ S & \bar{m} & S' \end{pmatrix}^{-1} \quad (\bar{m} = S' - S). \qquad (5.9.11)$$

Since angular momenta with quantum numbers S and 1 may be coupled only to S or $S \pm 1$, the transition spin density vanishes unless $S' = S$, $S \pm 1$. The subscript m on the left in (5.9.10) is really superfluous, since the coefficients and the densities vanish unless $m = M' - M$, $\bar{m} = S' - S$: the standard density in (5.9.10) connects states $\Phi_{\bar{\kappa}}$ and $\Phi_{\bar{\kappa}'}$ in which $M = S$ and $M' = S'$ respectively, and is thus

$$Q_S(\bar{\kappa}\bar{\kappa}' \mid \boldsymbol{r}_1) = \int_{x_1 = s_2} \mathsf{S}_{\bar{m}} \rho(\bar{\kappa}\bar{\kappa}' \mid \boldsymbol{x}_1)\, \mathrm{d}s_1 \quad (\bar{m} = S' - S). \qquad (5.9.12)$$

The other transition spin densities each give the one non-zero matrix element involving the spin component ($m = M' - M$) that connects the two states concerned.

Since, using (5.9.7), the cartesian components S_x, S_y are linear combinations of S_{+1}, S_{-1} (which yield non-zero spin-density contributions $Q_{\pm 1}$ between functions Φ_κ, Φ_λ with different spin eigenvalues), it is evident that the transverse components Q_x, Q_y will generally be non-zero for any state represented by a mixture of different multiplets S or multiplet components M. Such mixing occurs in a CI calculation when the Hamiltonian contains, typically, strong spin–orbit coupling terms.

Any matrix element involving spin-dependent one-electron operators may now be written as a one-electron integral (5.9.8) involving a *spatial* operator working on a *spatial* density function—the transition spin density (5.9.10); and this density may be calculated by using just one state from each multiplet ($M = S$, $M' = S'$) and then multiplying by the numerical factor defined in (5.9.11). The Clebsch–Gordan coefficients are tabulated in many places (e.g. Condon and Shortley, 1935), and the ratio (5.9.11) frequently takes a very simple form, as in the case $S' = S$, $M' = M$ where it reduces to M/S as used in (5.9.6). It now appears that spin-dependent properties of molecules may also be handled along the lines of Section 5.7, if we replace charge densities by spin densities. We can also understand the assertion (pp. 68, 90) that, in evaluating matrix elements, only the "top" states with $M = S$ need be considered. These results will find a natural application in Chapter 11.

It is clear, as in the case of the electron density, that the spin angular momentum may be divided out among orbitals and overlap regions by writing

$$Q_z(\boldsymbol{r}) = \sum_{R,S} Q_{z,RS}\phi_R(\boldsymbol{r})\phi_S^*(\boldsymbol{r}) = \sum_{i,j} Q_{z,ij}^{\chi}\chi_i(\boldsymbol{r})\chi_j^*(\boldsymbol{r}) \qquad (5.9.13)$$

(where the second form contains *basis* functions) and similarly for Q_x, Q_y. The spin populations are defined in terms of the matrix elements $Q_{z,ij}^{\chi}$ just as the electron populations were defined (Section 5.5) in terms of the P_{ij}^{χ}. Thus, in a pure spin state with $\langle S_z \rangle = M$, and with orthonormal basis functions, $Q_{z,ii}^{\chi}$ is the amount of spin angular momentum associated with the region $d_i = |\chi_i|$ and

$$\mathrm{tr}\,\mathbf{Q}_z^{\chi} = \sum_i Q_{z,ii}^{\chi} = M, \qquad (5.9.14)$$

which is a "conservation equation" for the spin density.

It might be anticipated that, just as the correlation between the spatial motions of electrons was recognized in the pair function, there ought to be a 2-electron function that would recognize the correlation between *spins* of electrons located in different volume elements. Such functions have indeed been defined and used in discussion of spin-coupling effects that arise when relativistic terms are included in the Hamiltonian (McWeeny, 1965). Here we need only note that *two* functions are required, a spin–orbit coupling function Q^{SL}, which fully determines spin–orbit interactions, and a spin–spin coupling function Q^{SS}, which describes the effects of magnetic dipole–dipole coupling between the electrons. These functions are defined (cf. (5.9.8)) by

$$\left\langle \Phi_{\kappa'} \left| \sum_{i,j}{}' \mathsf{F}(i,j)\mathsf{S}_m(j) \right| \Phi_\kappa \right\rangle = \int \mathsf{F} Q_m^{SL}(\kappa\kappa' \mid \boldsymbol{r}_1, \boldsymbol{r}_2)\, \mathrm{d}\boldsymbol{r}_1\, \mathrm{d}\boldsymbol{r}_2, \quad (5.9.15)$$

$$\left\langle \Phi_{\kappa'} \left| \sum_{i,j}{}' \mathsf{G}(i,j)\mathsf{S}_m^{(2)}(i,j) \right| \Phi_\kappa \right\rangle = \int \mathsf{G} Q_m^{SS}(\kappa\kappa' \mid \boldsymbol{r}_1, \boldsymbol{r}_2)\, \mathrm{d}\boldsymbol{r}_1\, \mathrm{d}\boldsymbol{r}_2, \quad (5.9.16)$$

where $\mathsf{F}(i, j)$, $\mathsf{G}(i, j)$ are arbitrary spatial operators. The $\mathsf{S}_m^{(2)}(i, j)$ ($m = 0, \pm 1, \pm 2$) are "rank-2 tensor operators" and are linear combinations of spin products $\mathsf{S}_{m_1}(i)\mathsf{S}_{m_2}(j)$ that transform like spin eigenfunctions with $S = 2$. They occur (Chapter 11, p. 390) in the dipole–dipole interaction operator. Again, the functions corresponding to all non-zero matrix elements are simply numerical multiples of those that connect the top states ($M = S$) of each multiplet system:

$$Q_m^{SL}(\kappa\kappa' \mid \boldsymbol{r}_1, \boldsymbol{r}_2) = \begin{bmatrix} S & 1 & S' \\ M & m & M' \end{bmatrix} Q_{SL}(\bar{\kappa}\bar{\kappa}' \mid \boldsymbol{r}_1, \boldsymbol{r}_2), \quad (5.9.17)$$

$$Q_m^{SS}(\kappa\kappa' \mid \boldsymbol{r}_1, \boldsymbol{r}_2) = \begin{bmatrix} S & 2 & S' \\ M & m & M' \end{bmatrix} Q_{SS}(\bar{\kappa}\bar{\kappa}' \mid \boldsymbol{r}_1, \boldsymbol{r}_2), \quad (5.9.18)$$

where the first coefficient is defined in (5.9.11) while

$$\begin{bmatrix} S & 2 & S' \\ M & m & M' \end{bmatrix} = \begin{pmatrix} S & 2 & S' \\ M & m & M' \end{pmatrix} \begin{pmatrix} S & 2 & S' \\ S & \bar{m} & S' \end{pmatrix}^{-1} \quad (\bar{m} = S' - S). \quad (5.9.19)$$

Again the densities are non-zero for $m = M' - M$ and $\bar{m} = S' - S$, and the standard densities are defined by equations analogous to (5.9.12).

It is interesting to note that, for a given state ($\Phi_{\kappa'} = \Phi_\kappa$), the diagonal element $Q_z^{SL}(\boldsymbol{r}_1, \boldsymbol{r}_2)$ is a "conditional" spin density, giving the excess probability of finding an electron at $\boldsymbol{r}_1$ with spin up rather than down, *given a second electron at point* $\boldsymbol{r}_2$: similarly, $Q_m^{SS}(\boldsymbol{r}_1, \boldsymbol{r}_2)$ measures the anisotropy of the coupling between the spins of electrons at points $\boldsymbol{r}_1$ and $\boldsymbol{r}_2$. We shall not pursue these interpretations here, but discussions are available in the literature (McWeeny and Mizuno, 1961; McLachlan, 1963; McWeeny, 1965). It is also worth noting that, since integration of $Q_z^{SL}(\boldsymbol{r}_1, \boldsymbol{r}_2)$ over $\boldsymbol{r}_2$ leads to a spin density at point $\boldsymbol{r}_1$, properties that depend on the spin density at *one point* in space (and are very difficult to compute accurately using orbital approximations) may be re-expressed in terms of integrals involving Q_z^{SL}; point properties calculated in this way then depend less critically on point imperfections in the wavefunction and accuracy may be greatly enhanced (see e.g. Harriman, 1980; Ishida 1985).

5.10 OTHER DENSITY FUNCTIONS‡

The charge density $P(\boldsymbol{r})$ and spin density $Q_z(\boldsymbol{r})$ are examples of "point properties" or "sub-observables" (Hirschfelder, 1977). Their values are inferred (never directly measured) by reference to an integral such as V_{en} in (5.2.20) that defines an expectation value. Thus $VP(\boldsymbol{r})$ in (5.2.20) may be interpreted as a "potential-energy density", since it is the contribution per unit volume to V_{en} evaluated at point $\boldsymbol{r}$; and similarly, in (5.9.3), $Q_z(\boldsymbol{r})$ is interpreted as a density of spin angular momentum.

It might have been expected that other property densities could be defined in exactly the same way. Thus, referring again to (5.2.20),

$$T(\boldsymbol{r}) = \left[\frac{\mathrm{p}^2}{2m} P(\boldsymbol{r};\boldsymbol{r}')\right]_{\boldsymbol{r}'=\boldsymbol{r}} \qquad (\mathrm{p}^2 = \mathrm{p}_x^2 + \mathrm{p}_y^2 + \mathrm{p}_z^2)$$

would appear to be a kinetic-energy density, since $T(\boldsymbol{r})\,\mathrm{d}\boldsymbol{r}$ is the contribution to $\langle T\rangle$ associated with volume element $\mathrm{d}\boldsymbol{r}$ at position $\boldsymbol{r}$. There are, however, difficulties with such definitions in so far as the point contributions to real observables are not necessarily real. This is because the reality of expectation values of Hermitian operators (and even the definition of Hermitian character) depends on integration over the whole domain of the variables. Alternative definitions of a property density that give the same expectation value when integrated over all space may give very different (and non-real) values *locally*.

The simplest acceptable definition of a kinetic-energy density would appear to be

$$T(\boldsymbol{r}) = \mathrm{Re}\left[\frac{\mathrm{p}^2}{2m} P(\boldsymbol{r};\boldsymbol{r}')\right]_{\boldsymbol{r}'=\boldsymbol{r}}. \tag{5.10.1a}$$

This gives the usual expectation value $\langle \mathrm{T}\rangle$ when integrated over all space, but has the conceptual advantage of being both real and positive at all points.

For some purposes (5.10.1a) is not completely satisfactory: it contains a term (Problem 5.19) that depends on the Laplacian of the electron density (making no reference to electronic *motion*), and on integrating over any finite region this term gives a contribution to $\langle \mathrm{T}\rangle$ that depends

‡ In this section, where necessary, the off-diagonal elements of the density matrices are shown explicitly.

on the *boundary* of the region. The "non-physical" term may be eliminated most neatly by using an alternative definition

$$T(\boldsymbol{r}) = \left[\frac{\mathsf{p}\cdot\mathsf{p}^\dagger}{2m}P(\boldsymbol{r};\boldsymbol{r}')\right]_{\boldsymbol{r}'=\boldsymbol{r}}, \tag{5.10.1b}$$

where the adjoint operator (obtained by reversing the sign of i in $(\hbar/i)\boldsymbol{\nabla}$) is understood to work on the *primed* variable only.

Another density function that we shall meet is the (probability) *current density*,

$$\begin{aligned} J_\alpha(\boldsymbol{r}) &= \frac{1}{2m}[(\mathsf{p}_\alpha + \mathsf{p}_\alpha^\dagger)P(\boldsymbol{r};\boldsymbol{r}')]_{\boldsymbol{r}'=\boldsymbol{r}} \\ &= \frac{1}{m}\operatorname{Re}[\mathsf{p}_\alpha P(\boldsymbol{r};\boldsymbol{r}')]_{\boldsymbol{r}'=\boldsymbol{r}} \quad (\alpha = x, y, z). \end{aligned} \tag{5.10.2}$$

Like $\boldsymbol{Q}$, this three-component quantity defines a vector density $\boldsymbol{J} = (J_x, J_y, J_z)$. For a system in a stationary state described by a real wavefunction $\boldsymbol{J} = \boldsymbol{0}$, everywhere, but for a system in the presence of a magnetic field (Chapter 11) circulating currents, responsible for diamagnetism, are always present.

The more general form of (5.10.2) is obtained on introducing field terms in the Hamiltonian via the "gauge-invariant momentum operator", whose components (see Section 11.1) are

$$\pi_\alpha = \mathsf{p}_\alpha + eA_\alpha \quad (\alpha = x, y, z). \tag{5.10.3}$$

Here A_α is the α-component of the *magnetic vector potential* from which the applied field is derived. The corresponding form of the current density then becomes, using the usual more compact notation (p. 123)

$$\boxed{J_\alpha(\boldsymbol{r}) = m^{-1}\operatorname{Re}[\pi_\alpha P(\boldsymbol{r})].} \tag{5.10.4}$$

For a one-electron system (wavefunction ϕ) this reduces to

$$J_\alpha(\boldsymbol{r}) = \frac{\hbar}{2mi}\left(\phi^*(\boldsymbol{r})\frac{\partial\phi(\boldsymbol{r})}{\partial r_\alpha} - \phi(\boldsymbol{r})\frac{\partial\phi^*(\boldsymbol{r})}{\partial r_\alpha}\right) + \frac{e}{m}A_\alpha\phi^*(\boldsymbol{r})\phi(\boldsymbol{r}),$$

which is the usual expression given in elementary textbooks.

The kinetic-energy density is also modified by the presence of the field; thus, for example, (5.10.1b) becomes

$$T(\boldsymbol{r}) = \left[\frac{\boldsymbol{\pi}\cdot\boldsymbol{\pi}^\dagger}{2m}P(\boldsymbol{r};\boldsymbol{r}')\right]_{\boldsymbol{r}'=\boldsymbol{r}} \tag{5.10.5}$$

—a form that is used in Section 11.6.

Finally, it may be shown (Problem 5.20) that for a wavefunction that develops in time according to (1.1.12) the probability density $P(\boldsymbol{r})$ and the current density $\boldsymbol{J}(\boldsymbol{r})$ satisfy the conservation equation

$$\operatorname{div} \boldsymbol{J}(\boldsymbol{r}) = -\frac{\partial P(\boldsymbol{r})}{\partial t}, \tag{5.10.6}$$

which is well known in classical hydrodynamics.

It must be stressed that P and $\boldsymbol{J}$ are *one*-electron density functions, obtained from general many-electron wavefunctions by integration, and that (5.10.6) refers to ordinary 3-dimensional space: it implies that the flux of electron density out of any region is equal to the rate of decrease of charge (number of electrons) within the region. When Ψ is an exact stationary-state function, P is time-independent and the net flux of density out of any region will be zero; but with $\mathbf{A} \neq \mathbf{0}$ there will in general be a steady-state distribution of non-zero *currents,* satisfying the usual continuity equation $\operatorname{div} \boldsymbol{J} = 0$.

We note, in closing, that current densities may be defined for the individual "up-spin" and "down-spin" components of P, and consequently for the *spin density*; the spin current, obtained by replacing $P(\boldsymbol{r}; \boldsymbol{r}')$ in (5.10.4) by the various components of spin density, will clearly be a 9-component tensor density. Such densities occur in the discussion of properties that depend jointly on spin and electron velocity.

PROBLEMS 5

5.1 Use the singlet and triplet wavefunctions (Ψ_1, Ψ_2) for H_2, in the VB approximation given on p. 211, to obtain (see (5.1.5) and (5.1.6)) corresponding electron density functions. Write the results in the form (5.5.6), χ_1 and χ_2 being the hydrogen 1s AOs; identify the orbital and overlap populations, and verify the charge conservation property (5.5.10).

5.2 Obtain the same density functions as in Problem 5.1 by starting from the energy expectation value, in terms of spin-orbitals $\{(\psi_r)\}$ (i.e. $\chi_1\alpha$, $\chi_1\beta$, etc.), noting that ρ_{rs} is the coefficient of $\langle \psi_s | \mathsf{h} | \psi_r \rangle$; the spinless functions follow using the first equation of (5.2.2). Identify in each case the up-spin and down-spin components of the density and hence obtain corresponding spin densities. [*Hint: Keep* the spin factors, even for spinless operators, until you are ready to do the spin integrations.]

5.3 Make calculations parallel to those in Problems 5.1 and 5.2, using MO functions. Compare the populations with those already found, for a few values of the overlap integral S. [*Hint:* Obtain the densities first in terms of MOs ($\phi_1\alpha$, $\phi_1\beta$, . . .); then insert the LCAO forms.]

5.4 Add "ionic" terms, $|\chi_1\alpha\chi_1\beta|$ and $|\chi_2\alpha\chi_2\beta|$, to the VB singlet function in Problem 5.1 to obtain (taking account of symmetry) a 1-parameter (λ, say) wavefunction. Show how the orbital and overlap populations depend on λ.

5.5 Refine the MO singlet function in Problem 5.3 by admitting CI. Show that only one configuration contributes (as a result of symmetry) and hence obtain an alternative 1-parameter wavefunction (cf. Problem 5.4). Show that the MO and VB functions (with "full CI") give exactly the same electron density, provided that the parameters are correctly chosen.

5.6 In the singlet ground state the H_2 molecule is stable, while in in the first excited triplet state it dissociates spontaneously into two hydrogen atoms. Interpret this fact in the light of the preceeding results and the Hellmann–Feynman theorem (p. 138). Use the approach of Problem 5.3 to discuss the ground states of the molecular species He_2^{2+}, He_2^{+}, He_2. Observe the trend in overlap populations and interpret the fact that the neutral molecule is not found.

5.7 Consider a 3-electron system AB, constructing a pair of MOs $\phi_1 = C_1(\chi_1 + \lambda\chi_2)$, $\phi_2 = C_2(\chi_2 + \mu\chi_1)$ from AOs χ_1, χ_2 on A and B respectively. Obtain 1-parameter expressions for the orbital and overlap populations in the species $AB[\phi_1^2\phi_2]$, $AB^+[\phi_1^2]$, and $AB^-[\phi_1^2\phi_2^2]$. How might you measure a flow of charge in the direction $1 \rightarrow 2$? Show that the wavefunction is equivalent in each case to a combination of polar and non-polar VB structures (an "unpaired" electron being shown as a dot) and indicate how it could be improved. [*Hint:* Keep only the parameter λ, using orthogonality and normalization conditions to eliminate the others; λ will indicate the polarity of the bond.]

5.8 Show that the general form of (5.2.9) is

$$(N-n)\rho_n(\boldsymbol{x}_1, \ldots, \boldsymbol{x}_n; \boldsymbol{x}'_1, \ldots, \boldsymbol{x}'_n) = \int \rho_{n+1}(\boldsymbol{x}_1, \ldots, \boldsymbol{x}_{n+1}; \boldsymbol{x}'_1, \ldots, \boldsymbol{x}_{n+1})\, d\boldsymbol{x}_{n+1}.$$

5.9 Show that the Fock–Dirac density matrix (5.3.7), for one determinant of orthonormal spin-orbitals, has the fundamental property

$$\int \rho(\boldsymbol{x}_1; \boldsymbol{x}_2)\rho(\boldsymbol{x}_2; \boldsymbol{x}_3)\, d\boldsymbol{x}_2 = \rho(\boldsymbol{x}_1; \boldsymbol{x}_3).$$

Show also that when (using an orthonormal basis) $\rho(\boldsymbol{x}; \boldsymbol{x}')$ is written in the form $\sum_{i,j} \rho_{ij}\chi_i(\boldsymbol{x})\chi_j^*(\boldsymbol{x}')$ then the coefficients ρ_{ij} must have a similar property, $\sum_j \rho_{ij}\rho_{jk} = \rho_{ik}$. The *matrix* of coefficients is thus "idempotent", $\boldsymbol{\rho\rho} = \boldsymbol{\rho}$.

5.10 Complete the inductive proof of (5.3.11) by using the results obtained in Problems 5.8 and 5.9. [*Hint:* Assume (5.3.11) to be true. Write the determinant as $\sum_{\mathsf{P}} \epsilon_{\mathsf{P}}\mathsf{P}$ working on the unprimed variables in the product $\rho(\boldsymbol{x}_1; \boldsymbol{x}'_1) \ldots \rho(\boldsymbol{x}_n; \boldsymbol{x}'_n)$, noting that the sum can be replaced by $\sum_{\mathsf{P}'} \epsilon_{\mathsf{P}'}\mathsf{P}'(1 - \sum_{j=1}^{n-1} \mathsf{P}_{jn})$, where P' permutes only the first $n-1$ variables while P_{jn} exchanges $\boldsymbol{x}_j$, $\boldsymbol{x}_n$. Perform the integration (for $\boldsymbol{x}'_n = \boldsymbol{x}$) to get ρ_{n-1} and find $N - n + 1$ times the determinant in (5.3.11) but with n replaced by $n - 1$.]

5.11 Return to Problem 5.3 and obtain the MO function for the first excited *singlet* state (Ψ_3, say). Obtain the transition densities $\rho(KL \mid \mathbf{x}_1; \mathbf{x}_1')$ for transitions in which (i) K represents the ground state (Ψ_1) and L the first excited singlet (Ψ_3), and (ii) K is Ψ_1 and L is one of the three states ($\Psi_{2,M}$, $M = 0, \pm 1$) of the first excited triplet. By making appropriate spin integrations, obtain the transition "charge" densities, $P(KL \mid \mathbf{r}_1)$, and the transition spin densities $Q_\alpha(KL \mid \mathbf{r}_1)$. What kinds of perturbation would produce transitions between the various pairs of states considered? [*Hint:* Use (5.4.5) and (5.9.5) to obtain charge and spin densities and remember that transitions occur when off-diagonal matrix elements such as (5.4.7) have non-zero values.]

5.12 When spin–orbit interaction is admitted, the orbital and spin angular momenta in an atom may be coupled to a resultant $\boldsymbol{J} = \boldsymbol{L} + \boldsymbol{S}$. What wavefunctions would you use to describe the p electron in the configuration Na[(Ne)3p] of the sodium atom (i) in the state with $J = M_J = \frac{3}{2}$, and (ii) in the state with $J = M_J = \frac{1}{2}$? Show that the spin density in the first state is everywhere along the axis of quantization (conventionally the z axis), but that in the second state the x and y components may also be non-zero. Draw arrows to indicate the direction of spin polarization at points on a spherical surface at a given distance from the nucleus. [*Hint:* Use standard methods (Appendix 2) to obtain $\psi_1 = p_{+1}\alpha$, $\psi_2 = (p_0\alpha - \sqrt{2}\, p_{+1}\beta)/\sqrt{3}$ and derive the spin-density components from (5.9.5). The distribution will be symmetrical around the z axis, and the ratio of the contributions to $\langle \mathsf{S}_x \rangle$ and $\langle \mathsf{S}_z \rangle$, at any point, will indicate a local spin direction (being the cosine of the inclination to the z axis).]

5.13 When the distance R between two hydrogen atoms is substantially greater than the equilibrium bond length, a wavefunction of the form det $|\phi^{(\alpha)}\alpha \;\; \phi^{(\beta)}\beta|$, with $\phi^{(\beta)} \neq \phi^{(\alpha)}$, gives a lower variational energy than the standard MO function with $\phi^{(\alpha)} = \phi^{(\beta)} = \phi$: the determinant represents an "unrestricted Hartree–Fock" function in which the doubly occupied MO has been "split". Show that the UHF function can give the correct energy in the limit $R \to \infty$ but gives a spin density that behaves wrongly. [*Hint:* Use, noting symmetry, the AO combinations $\phi^{(\alpha)} = C(\chi_1 + \lambda\chi_2)$, $\phi^{(\beta)} = C(\chi_2 + \lambda\chi_1)$ to obtain the electron and spin densities.]

5.14 Repeat the analysis of Problem 5.13 after coupling the two spins to obtain a singlet state, showing that the electron density is unchanged but that the spin density is then (correctly) zero at all points in space. (Remember that the orbitals are not orthogonal!)

5.15 Show that the density matrices (5.4.8a,b) can be written, in second-quantization language, as

$$\rho(\mathbf{x}_1; \mathbf{x}_1') = \langle \Psi \mid \psi^\dagger(\mathbf{x}_1')\psi(\mathbf{x}_1) \mid \Psi \rangle,$$

$$\pi(\mathbf{x}_1, \mathbf{x}_2; \mathbf{x}_1', \mathbf{x}_2') = \langle \Psi \mid \psi^\dagger(\mathbf{x}_1')\psi^\dagger(\mathbf{x}_2')\psi(\mathbf{x}_2)\psi(\mathbf{x}_1) \mid \Psi \rangle,$$

where $\psi(\mathbf{x}) = \sum_R \psi_R(\mathbf{x})\mathsf{a}_R$ and $\psi^\dagger(\mathbf{x}') = \sum_S \psi_S^*(\mathbf{x}')\mathsf{a}_S^\dagger$ are so-called "field operators" and $\{\psi_R\}$ is any complete orthonormal set.

5.16 Introduce spin explicitly in the field operators (Problem 5.15), putting for example $\psi_\sigma(\mathbf{x}) = \sum_R \psi_{R\sigma}(\mathbf{x})\mathsf{a}_{R\sigma}$ ($\sigma = \alpha, \beta$) and verify that, in any eigenstate of S_z, ρ and π are respectively 2-component and 6-component functions (pp. 127,

142). [*Hint:* First show that $2S_z = N_\alpha - N_\beta$, where, for example, $N_\alpha = \sum_R a^\dagger_{R\alpha} a_{R\alpha}$ is a *number operator* whose eigenvalue gives the number of up-spin electrons. Then use the fact that the operators whose expectation values appear in Problem 5.15 must commute with S_z, since the results of operating with S_z on the bra or the ket must be identical.]

5.17 Obtain the function $\Pi_{\alpha\beta}(\mathbf{r}_1, \mathbf{r}_2)$ for the two electrons in a 2-centre bonding MO ϕ_σ of σ type (i.e. symmetric around the bond axis). Then admit CI with configurations (i) $[\phi'^2_\sigma]$, where ϕ'_σ is an antibonding partner of ϕ_σ, and (ii) $[\phi^2_\pi]$, where $\phi_{\pi x}$, $\phi_{\pi y}$ are π-type MOs pointing perpendicular to the bond axis (z). What sort of correlation is described in the two cases? [*Hint:* Put one electron (i) near a nucleus, or (ii) at a point in the (x, z) plane, and look at the probability density for finding another electron simultaneously at a second point. Note that the appropriate CF of configuration $[\phi^2_\pi]$ must contain the spatial factor $\phi'_x(\mathbf{r}_1)\phi'_x(\mathbf{r}_2) + \phi'_y(\mathbf{r}_1)\phi'_y(\mathbf{r}_2)$. Why?]

5.18 A σ bond, linking atoms A and B, is represented by putting two electrons in the bonding MO of a bonding/antibonding pair (ϕ_1, ϕ_2). An "odd" electron (up-spin, say) is present in a 2p AO ϕ on atom A, pointing perpendicular to the bond. Show that a wavefunction det $|\phi_1\alpha\ \phi_1\beta\ \phi\alpha|$ can give no spin density at nucleus B. What kind of CI should be admitted to yield an improved doublet-state wavefunction with *non-zero* spin density at nucleus B? [*Hint:* An electron can be promoted to ϕ_2 and the three spins can be coupled in two ways. The resultant CFs will mix with the 1-configuration approximation (with small coefficients, λ_1, λ_2, say). Use Slater's rules and the formalism of Section 5.4 to calculate contributions to the spin density.]

5.19 By considering the contribution made to $\langle T \rangle$ from a finite region of space, show that the form (5.10.1a) for T yields a boundary-dependent term while the form (5.10.1b) does not. [*Hint:* Integrate by parts.]

5.20 Calculate the current-density components for an electron in a p-type AO $(p_{\pm 1} = (p_x \pm ip_y)/\sqrt{2})$, showing that the current flows in circles around the z axis. Then consider the MOs (Problem 2.13) for a ring of hydrogen atoms, showing that the complex solutions correspond to "ring currents" while the real combinations do not. [*Hint:* Start from (5.10.2). In the case of the ring, take p_α as a local momentum pointing from one atom to the next.]

5.21 Derive the conservation equation (5.10.6) for the probability and probability-current densities. [*Hint:* Start from $H\Psi = ih\, \partial\Psi/\partial t$ and its complex conjugate, multiplying by Ψ^* and Ψ respectively and integrating over all variables except $\mathbf{r}_1$. The difference, $\langle \Psi | H\Psi \rangle_1 - \langle H\Psi | \Psi \rangle_1$, say, will yield the time derivative of ρ. Potential-energy terms cancel, and the difference of the kinetic-energy terms yields, using the definition (5.10.2) and various vector identities, the required result. A complete discussion (McWeeny, 1986) is non-trivial and also involves the definition of the kinetic-energy density.]

6 Self-Consistent Field Theory

6.1 HARTREE–FOCK THEORY. THE INDEPENDENT-PARTICLE MODEL

In Chapters 1 and 3 we introduced the idea of a molecular orbital (MO), as a 1-electron wavefunction, generally delocalized, describing an electron free to move over the whole molecule in some appropriate "effective field"; and in Section 1.3 a simple example was used to show how such orbitals could be constructed. In this chapter we develop various forms of the self-consistent field (SCF) method, whose aim is to *optimize* the MOs in order to obtain a "best" many-electron function of given form. This given form will usually be a single determinant; in other words we shall consider the representation of the wavefunction by a *single term* of the complete set expansion (3.1.4), choosing the MOs so as to obtain the best 1-term approximation to the exact molecular wavefunction.

The most commonly encountered molecular ground state (with an even number of electrons) is a singlet, possessing the full spatial symmetry of the system; such a state is well represented by a single determinant of doubly occupied MOs and is usually referred to as a "closed-shell ground state". Instead of turning immediately to this special case, however, we shall consider generally the problem of optimizing the spin-orbitals (without separating space and spin factors) in a one-determinant wavefunction; this will expose all the main features of the theory developed by Hartree (1928), Slater (1930) and Fock (1930).

We assume a wavefunction of the form (Section 3.3)

$$\Psi(\mathbf{x}_1, \mathbf{x}_2, \ldots, \mathbf{x}_N) = (N!)^{-1/2} \det |\psi_1(\mathbf{x}_1) \;\; \psi_2(\mathbf{x}_2) \;\; \ldots \;\; \psi_N(\mathbf{x}_N)|, \quad (6.1.1)$$

in which ψ_i is the ith occupied spin-orbital, and consider the variational optimization of the individual factors. The aim will be to show that the best orbitals to use are then eigenfunctions of a *one-electron* eigenvalue equation

$$\mathsf{F}\psi = \epsilon\psi, \quad (6.1.2)$$

in which the operator F (first given in complete form by Fock, 1930) is the Hamiltonian for a single electron in some kind of "effective field" due to the nuclei *and* the remaining electrons; in this way we obtain a specific form of the independent-particle model (IPM) referred to in Section 1.2, in which electron interaction is allowed for to a considerable degree without disturbing the product form of the wavefunction. The condition (6.1.2) is usually referred to as a Hartree–Fock (HF) equation.

To derive (6.1.2), we start from the variational energy approximation (3.3.7a) associated with the function (6.1.1), namely,

$$E = \sum_i \langle \psi_i | \mathsf{h} | \psi_i \rangle + \tfrac{1}{2} \sum_{i,j} \langle \psi_i \psi_j \| \psi_i \psi_j \rangle, \tag{6.1.3}$$

where the summations run over all the occupied spin-orbitals and we have introduced the "antisymmetrized" 2-electron integral

$$\langle \psi_i \psi_j \| \psi_i \psi_j \rangle = \langle \psi_i \psi_j | g | \psi_i \psi_j \rangle - \langle \psi_i \psi_j | g | \psi_j \psi_i \rangle \tag{6.1.4}$$

To obtain a stationary value, we let $\psi_k \rightarrow \psi_k + \delta\psi_k$ and set $\delta E = 0$ (to first order) for all $\delta\psi_k$, remembering that (6.1.3) remains valid only when the spin-orbitals remain normalized and orthogonal.

With the change $\delta\psi_k$, we obtain a corresponding first-order variation of the energy

$$\delta E^{(k)} = \langle \delta\psi_k | \mathsf{h} | \psi_k \rangle + \text{c.c.}$$
$$+ \frac{1}{2}\Big(\sum_j \langle \delta\psi_k \psi_j \| \psi_k \psi_j \rangle + \sum_i \langle \psi_i \delta\psi_k \| \psi_i \psi_k \rangle \Big) + \text{c.c.},$$

where each c.c. is the complex conjugate of the preceding term. The two sums in the parentheses, however, are seen to be identical because the integrals are invariant against exchange of the indices on *both* sides of the ||, simultaneously, while the name of the summation index (i or j) is immaterial. Thus

$$\delta E^{(k)} = \Big(\langle \delta\psi_k | \mathsf{h} | \psi_k \rangle + \sum_i \langle \delta\psi_k \psi_i \| \psi_k \psi_i \rangle \Big) + \text{c.c.}, \tag{6.1.5}$$

and the first-order change δE arising from changes in all the spin-orbitals will be simply the sum of all such terms.

To proceed, it is convenient to introduce two new operators. The first of these is a "coulomb operator" J_i associated with an electron in ψ_i: working on an *arbitrary* function ψ of the variable $\boldsymbol{x}_1$, it has the effect

$$\mathsf{J}_i(1)\psi(\boldsymbol{x}_1) = \left[\int g(1, 2)\psi_i(\boldsymbol{x}_2)\psi_i^*(\boldsymbol{x}_2)\, d\boldsymbol{x}_2 \right] \psi(\boldsymbol{x}_1). \tag{6.1.6}$$

In other words, $\psi(\boldsymbol{x}_1)$ is simply multiplied by the potential energy of an electron at $\boldsymbol{x}_1$ in the field due to a charge distribution $|\psi_i(\boldsymbol{x}_2)|^2$ (in electrons per unit volume). The second operator, the "exchange operator" K_i, is more complicated and is often said to describe a "non-local" potential: its effect on $\psi(\boldsymbol{x}_1)$ is to exchange the variables $\boldsymbol{x}_1$ and $\boldsymbol{x}_2$ in the ψ and ψ_i factors of the last equation and then to complete the $\boldsymbol{x}_2$ integration with inclusion of the factor $\psi(\boldsymbol{x}_2)$ outside the square brackets. Thus

$$\mathsf{K}_i(1)\psi(\boldsymbol{x}_1) = \int g(1, 2)\psi_i(\boldsymbol{x}_1)\psi_i^*(\boldsymbol{x}_2)\psi(\boldsymbol{x}_2)\,\mathrm{d}\boldsymbol{x}_2. \tag{6.1.7}$$

When this result is written in the form

$$\mathsf{K}_i(1)\psi(\boldsymbol{x}_1) = \int K_i(\boldsymbol{x}_i;\boldsymbol{x}_2)\psi(\boldsymbol{x}_2)\,\mathrm{d}\boldsymbol{x}_2 \tag{6.1.8}$$

the function $K_i(\boldsymbol{x}_1;\boldsymbol{x}_2)$ is called the *kernel* of the operator (which is an "integral operator"); the associateion is expressed in the form

$$\mathsf{K}_i \rightarrow K_i(\boldsymbol{x}_1;\boldsymbol{x}_2) = g(1, 2)\psi_i(\boldsymbol{x}_1)\psi_i^*(\boldsymbol{x}_2). \tag{6.1.9}$$

It is also convenient to introduce *total* coulomb and exchange operators

$$\mathsf{J} = \sum_i \mathsf{J}_i, \qquad \mathsf{K} = \sum_i \mathsf{K}_i, \tag{6.1.10}$$

where the summation is over all occupied spin-orbitals. We note that both operators are Hermitian. It is also clear that these operators can be expressed in terms of the 1-electron density matrix (5.3.7). Thus, operating on an arbitrary function

$$\boxed{\begin{aligned} \mathsf{J}(1)\psi(\boldsymbol{x}_1) &= \left(\int g(1, 2)\rho(\boldsymbol{x}_2;\boldsymbol{x}_2)\,\mathrm{d}\boldsymbol{x}_2\right)\psi(\boldsymbol{x}_1), \\ \mathsf{K}(1)\psi(\boldsymbol{x}_1) &= \int g(1, 2)\rho(\boldsymbol{x}_1;\boldsymbol{x}_2)\psi(\boldsymbol{x}_2)\,\mathrm{d}\boldsymbol{x}_2 \end{aligned}} \qquad \begin{aligned}(6.1.11\text{a})\\(6.1.11\text{b})\end{aligned}$$

(cf. (6.1.6), (6.1.7)), the coulomb operator multiplies $\psi(\boldsymbol{x}_1)$ by the potential energy of an electron at $\boldsymbol{x}_1$ due to the whole electron distribution, and the exchange operator likewise depends on the density matrix for the whole system (rather than on the individual orbitals).

From the definition of the operators J and K, it readily follows that the 2-electron terms in the energy expectation value (6.1.3) can be expressed in terms of expectation values of a *one*-electron operator: thus

$$\begin{aligned} \sum_j \langle \psi_i\psi_j \| \psi_i\psi_j\rangle &= \sum_j \langle \psi_i|\, \mathsf{J}_j\, |\psi_i\rangle - \sum_j \langle \psi_i|\, \mathsf{K}_j\, |\psi_i\rangle \\ &= \langle \psi_i|\, (\mathsf{J} - \mathsf{K})\, |\psi_i\rangle, \end{aligned}$$

and consequently

$$E = \sum_i \langle \psi_i | (\mathsf{h} + \tfrac{1}{2}\mathsf{G}) | \psi_i \rangle, \qquad \mathsf{G} = \mathsf{J} - \mathsf{K}, \tag{6.1.12}$$

which is a sum of *orbital* expectation values of the operator $\mathsf{h} + \frac{1}{2}(\mathsf{J} - \mathsf{K})$, which evidently contains the electron interactions implicitly.

In the same way, the first-order variation resulting from $\psi_k \rightarrow \psi_k + \delta\psi_k$, given in (6.1.5), becomes

$$\delta E^{(k)} = \langle \delta\psi_k | \mathsf{F} | \psi_k \rangle + \langle \psi_k | \mathsf{F} | \delta\psi_k \rangle, \tag{6.1.13}$$

where the Fock Hamiltonian is

$$\boxed{\mathsf{F} = \mathsf{h} + \mathsf{J} - \mathsf{K} = \mathsf{h} + \mathsf{G}.} \tag{6.1.14}$$

Since (6.1.13) evidently gives the first-order variation of a 1-electron energy expectation value

$$\epsilon_k = \langle \psi_k | \mathsf{F} | \psi_k \rangle, \tag{6.1.15}$$

the Fock operator may be regarded as the *effective* Hamiltonian for a single electron moving in the field of the nuclei (contained in h) together with an effective "coulomb-exchange" field representing the presence of the other electrons. The requirement that this quantity be stationary strongly suggests that the *orbital energies* ϵ_k might be eigenvalues of F, in accordance with (6.1.2); but to verify this conjecture we must take proper account of the orthonormality constraints.

The spin-orbital ψ_k is normalized and orthogonal to all other ψ_j provided that

$$\langle \psi_k \mid \psi_k \rangle = 1, \qquad \langle \psi_k \mid \psi_j \rangle = \langle \psi_j \mid \psi_k \rangle = 0 \quad (\text{all } j \neq k),$$

and, applying these conditions to $\psi_k + \delta\psi_k$, we obtain, to first order,

$$\left.\begin{aligned} &\langle \delta\psi_k \mid \psi_k \rangle + \langle \psi_k \mid \delta\psi_k \rangle = 0, \\ &\left.\begin{aligned} &\langle \delta\psi_k \mid \psi_j \rangle = 0 \\ &\langle \psi_j \mid \delta\psi_k \rangle = 0 \end{aligned}\right\} \quad (\text{all } j \neq k). \end{aligned}\right\} \tag{6.1.16}$$

These constraints may be combined with $\delta E^{(k)} = 0$ by using the method of Lagrangian multipliers: we multiply the equations (6.1.16) by arbitrary constants ϵ_{kk}, ϵ_{jk} and ϵ_{kj} respectively and subtract the results from (6.1.13). On collecting terms with $\delta\psi_k$ on the left, we obtain

$$\langle \delta\psi_k | \mathsf{F} | \psi_k \rangle - \sum_j \varepsilon_{jk} \langle \delta\psi_k | \psi_j \rangle = 0, \tag{6.1.17a}$$

and from terms with $\delta\psi_k$ on the right

$$\langle \psi_k | \mathsf{F} | \delta\psi_k \rangle - \sum_j \epsilon_{kj} \langle \psi_j \mid \delta\psi_k \rangle = 0. \tag{6.1.17b}$$

The conditional stationary values arise when both equations are satisfied, for all choices of k. Since F is Hermitian we may take the complex conjugate of (6.1.17b), noting that $\langle \psi_k | \mathsf{F} | \delta\psi_k \rangle^* = \langle \delta\psi_k | \mathsf{F} | \psi_k \rangle$, and subtract the result from (6.1.17a) to obtain

$$(\epsilon_{jk} - \epsilon_{kj}^*) \langle \delta\psi_k \mid \psi_j \rangle = 0 \quad \text{(all } j, k).$$

In other words, the Lagrangian multipliers must form a Hermitian matrix, $\boldsymbol{\epsilon}^\dagger = \boldsymbol{\epsilon}$.

To establish the Hartree–Fock (HF) equation (6.1.2), it is sufficient to note that (6.1.17a) is the scalar product of $\delta\psi_k$ and a function that (for $\delta\psi_k$ arbitrary) must evidently vanish:

$$\mathsf{F}\psi_k - \sum_j \epsilon_{jk} \psi_j = 0 \quad \text{(all } k). \tag{6.1.18}$$

This equation is not, as it stands, an eigenvalue equation of the desired form (6.1.2); but it *would* present the ψ_k as eigenfunctions of F if the matrix $\boldsymbol{\epsilon}$ were diagonal. To see whether such a form can be found, we express (6.1.18) in the matrix notation used in Section 2.2: on collecting the spin-orbitals into a row matrix $\boldsymbol{\psi} = (\psi_1 \ \ \psi_2 \ \ \ldots \ \ \psi_N)$, we may write

$$\mathsf{F}\boldsymbol{\psi} = \boldsymbol{\psi}\boldsymbol{\epsilon}. \tag{6.1.19}$$

It is then possible to exploit the fact that a 1-determinant wavefunction is invariant under unitary mixing of the spin-orbitals (Problem 3.4) by introducing a new set $\bar{\boldsymbol{\psi}} = \boldsymbol{\psi}\mathbf{U}$ and writing (6.1.19) as

$$\mathsf{F}\bar{\boldsymbol{\psi}} = \bar{\boldsymbol{\psi}}(\mathbf{U}^\dagger \boldsymbol{\epsilon} \mathbf{U}) = \bar{\boldsymbol{\psi}}\bar{\boldsymbol{\epsilon}}.$$

For we know (Section 2.3) that any Hermitian matrix $\boldsymbol{\epsilon}$ can be brought to a *diagonal* form $\bar{\boldsymbol{\epsilon}}$ by a unitary transformation. Thus, without changing the many-electron wavefunction itself, we may always change the spin-orbitals so that (dropping the bars) (6.1.19) assumes the form

$$\boxed{\mathsf{F}\boldsymbol{\psi} = \boldsymbol{\psi}\boldsymbol{\epsilon} \quad (\boldsymbol{\epsilon} \text{ diagonal}).} \tag{6.1.20}$$

In other words, every spin-orbital satisfies

$$\mathsf{F}\psi_i = \varepsilon_i \psi_i, \tag{6.1.21}$$

and is thus an eigenfunction of (6.1.2) with eigenvalue ϵ_i. It is usual to refer to (6.1.20) as the "canonical form" of the HF equations.

The meaning of the term "independent-particle model" may now be clarified. If we consider a system in which every electron is described by the same 1-electron HF Hamiltonian (6.1.14), and there are (formally) no electron interactions terms (i.e. no 2-body operators), then the corresponding IPM Hamiltonian will be

$$\mathsf{H}_{\mathrm{M}} = \sum_i \mathsf{F}(i), \tag{6.1.22}$$

and it is immediately verified that (6.1.1) will be an exact *many*-electron eigenfunction of this model Hamiltonian, provided that the spin-orbitals satisfy (6.1.2).

It should be noted that the energy of the system in the HF approximation is not simply the expectation value of the IPM Hamiltonian (6.1.22), since

$$\langle \Psi | \mathsf{H}_{\mathrm{M}} | \Psi \rangle = \sum_i \langle \psi_i | \mathsf{F} | \psi_i \rangle = \sum_i \epsilon_i = E_{\mathrm{orb}}, \tag{6.1.23}$$

which is the sum of the (occupied) orbital energies. The variational energy, however, is given by a formally similar expression (6.1.12) in which F is replaced by an operator

$$\mathsf{F}_{\mathrm{ad}} = \mathsf{h} + \tfrac{1}{2}(\mathsf{J} - \mathsf{K}), \tag{6.1.24}$$

where the factor $\frac{1}{2}$ avoids a "double-counting" of electron interaction terms. Thus

$$E = \sum_i \langle \psi_i | \mathsf{F}_{\mathrm{ad}} | \psi_i \rangle = \tfrac{1}{2} \sum_i (\langle \psi_i | \mathsf{F} | \psi_i \rangle + \langle \psi_i | \mathsf{h} | \psi_i \rangle), \tag{6.1.25}$$

the first form showing that an "additive partitioning" of the total energy into orbital contributions can be obtained, but only by introducing a modified Fock operator (6.1.24).

The immense value of the IPM description of atomic and molecular structure, with orbitals determined from the canonical HF equations rests mainly upon the following results.

(i) The HF eigenvalues ϵ_k have an immediate physical significance: $-\epsilon_k$ is a first approximation to the ionization energy I_k needed to produce a positive ion by removing an electron from ψ_k. This result, due to Koopmans (1933), follows from (6.1.3) on separating the terms corresponding to $i, j = k$ from the summations; it follows that

$$E = E_k^+ + \epsilon_k,$$

where E_k^+ corresponds to the positive ion, and hence that

$$I_k = E_k^+ - E = -\epsilon_k. \tag{6.1.26}$$

This is a legitimate approximation to I_k only when the spin-orbitals satisfy the canonical HF equations; for then E, ϵ_k, and consequently the *difference* E_k^+, are all stationary against spin-orbital variations, and the *same* spin-orbitals therefore serve to describe both the neutral system and its ions.

(ii) The eigenvalues ϵ_m and eigenfunctions ψ_m that correspond to solutions of (6.1.2) *not used* in the ground-state function Ψ, normally being of higher energy than the occupied orbitals ψ_k, have a significance of their own; they represent "virtual" orbitals for an extra "ghost" electron moving in the HF field of the molecule. The difference between two orbital energies, ϵ_m (virtual) minus ϵ_k (occupied) gives a rough estimate of the excitation energy $\Delta E(k \to m)$ for an electronic transition from the ground state to a state in which the spin-orbital ψ_k is replaced by ψ_m; differences in *total* electronic energies may thus be interpreted pictorially in terms of "jumps" between *one*-electron energy levels. More accurately,

$$\Delta E(i \to m) = \epsilon_m - \epsilon_i + \langle mi \,\|\, mi \rangle, \tag{6.1.27}$$

as may be confirmed (Problem 3.9) by expressing the difference of the two 1-determinant energies in terms of orbital energies.

(iii) The electron density function in (5.2.8) takes the form (5.3.7); namely

$$\rho(\boldsymbol{x}) = \sum_i \psi_i(\boldsymbol{x})\psi_i^*(\boldsymbol{x}), \tag{6.1.28}$$

each electron making its own contribution $|\psi_i|^2$ to the total density. Knowledge of the spin-orbitals then leads, as will become clear in Chapter 11, to simple interpretations of many electronic properties. The relatively high accuracy of this result rests upon a theorem attributed to Brillouin. If we use $\Psi(i \to m)$ to indicate the determinant in which the occupied MO ψ_i has been replaced by the virtual MO ψ_m then it is easy to show that

$$\langle \Psi(i \to m) | \mathsf{H} | \Psi \rangle = \langle \psi_m | \mathsf{F} | \psi_i \rangle, \tag{6.1.29}$$

where F is the Fock operator (6.1.14). But when the MOs are eigenfunctions of F the off-diagonal elements on the right vanish. This means that on trying to correct Ψ by admitting CI with all possible "single-excitation" functions $\Psi(i \to m)$, which according to (5.4.14) *et seq.* are the only determinants that can lead to a modification of the electron density, there will be no effect: the vanishing of the matrix elements (6.1.29) means that, to the first order of perturbation theory (see Section 2.5), the 1-determinant result (6.1.28) is the best that can be obtained.

In view of the above results, molecular orbital theory, with orbitals and orbital energies determined from the canonical HF equations, provides the natural tool for the interpretation of electronic spectroscopy; it was indeed developed and used for this purpose, long before accurate solution of the HF equations became feasible, in the classic papers of R.S. Mulliken during the nineteen twenties and thirties.

6.2 FINITE-BASIS APPROXIMATIONS. CLOSED-SHELL SYSTEMS

In the approximations with which we are almost exclusively concerned, each MO is constructed from a *finite basis* of AOs centred (usually) on the various nuclei; such orbitals are not necessarily the orbitals used in a single-atom calculation (e.g. hydrogen-like orbitals or Slater-type orbitals (STOs)), but they are of single-centre form and are, broadly speaking, "atomic" in character. The construction of MOs in the "linear combination of atomic orbitals" (LCAO) approximation, in this conventional terminology, is characteristic of most of the methods of calculation at present in use. We note, however, that the procedure of using a finite set of basis functions is applicable quite generally, not just for the construction of MOs, and applies equally well to the determination of *atomic orbitals* for a free atom using any suitable set of basis functions (see e.g. Clementi and Roetti, 1974). In fact, the LCAO MO approach is but one typical example of a finite-basis method.

First we develop the finite-basis approach in a purely formal way, assuming that the n occupied spin-orbitals $\{\psi_K\}$‡ are expressed in terms of a spin-orbital *basis* $\{\chi_r\}$ containing m functions, in order to obtain a matrix form of the HF equations. Thus we write

$$\boldsymbol{\psi} = \boldsymbol{\chi}\mathbf{T}, \qquad \mathbf{T} = (\mathbf{c}_1 \,|\, \mathbf{c}_2 \,|\, \ldots \,|\, \mathbf{c}_K \,|\, \ldots), \tag{6.2.1}$$

where $\mathbf{T}$ is an $m \times n$ matrix whose Kth column $\mathbf{c}_K$ contains the coefficients in the expansion of ψ_K. The density matrix (5.3.7) then takes the form

$$\rho(x_1; x_1') = \sum_K \sum_{rs} \chi_r(x_1) T_{rK} T_{sK}^* \chi_s^*(x_1')$$

‡ It is often useful to use capital letters to label the MOs in order to distinguish subscripts referring to MOs and AOs.

or

$$\rho(\boldsymbol{x}_1;\boldsymbol{x}_1') = \sum_{r,s} \rho_{rs}\chi_r(\boldsymbol{x}_1)\chi_s^*(\boldsymbol{x}_1'), \tag{6.2.2}$$

where ρ_{rs} is an element of the $m \times m$ matrix

$$\boldsymbol{\rho} = \mathbf{T}\mathbf{T}^\dagger = \sum_K \boldsymbol{c}_K\boldsymbol{c}_K^\dagger. \tag{6.2.3}$$

This matrix is a matrix representative of the 1-electron density operator ρ, as mentioned in Section 5.3; in the basis $\{\psi_K\}$ the operator is represented (according to 5.3.7) by an n-dimensional unit matrix, but after the basis change (6.2.1) this matrix assumes the more general form (6.2.3).

The energy expression (6.1.3) may also be expressed in terms of basis-function integrals: thus

$$\begin{aligned} E = {} & \sum_K \sum_{r,s} T_{rK}^* \langle \chi_r | \, \mathsf{h} \, | \chi_s \rangle T_{sK} \\ & + \tfrac{1}{2} \sum_{K,L} \sum_{r,s,t,u} T_{rK}^* T_{sL}^* \langle \chi_r\chi_s \, \| \, \chi_t\chi_u \rangle T_{tK} T_{uL}, \end{aligned} \tag{6.2.4}$$

which may be written, introducing the matrix trace (p. 126), as

$$E = \operatorname{tr} \boldsymbol{\rho}\mathbf{h} + \tfrac{1}{2} \operatorname{tr} \boldsymbol{\rho}\mathbf{G}(\boldsymbol{\rho}), \tag{6.2.5}$$

where $\mathbf{G}(\boldsymbol{\rho})$ indicates an electron-interaction matrix that depends on the density matrix $\boldsymbol{\rho}$. In full,

$$\begin{aligned} G(\boldsymbol{\rho})_{rt} &= \sum_{u,s} \rho_{us}(\langle \chi_r\chi_s | \, g \, | \chi_t\chi_u \rangle - \langle \chi_r\chi_s | \, g \, | \chi_u\chi_t \rangle) \\ &= J(\boldsymbol{\rho})_{rt} - K(\boldsymbol{\rho})_{rt}. \end{aligned} \tag{6.2.6}$$

In other words, $\mathbf{G}$ is the difference of coulomb and exchange *matrices*:

$$\mathbf{G}(\boldsymbol{\rho}) = \mathbf{J}(\boldsymbol{\rho}) - \mathbf{K}(\boldsymbol{\rho}). \tag{6.2.7}$$

Again, it is easily verified that $\mathbf{J}(\boldsymbol{\rho})$ and $\mathbf{K}(\boldsymbol{\rho})$ are matrix representatives of the *operators* introduced in (6.1.10) whose effect on a one-electron function is defined through (6.1.11).

In a first-order variation of the orbitals, i.e. of the matrix $\mathbf{T}$ in (6.2.1), the density matrix changes by

$$\delta\boldsymbol{\rho} = \delta\mathbf{T}\mathbf{T}^\dagger + \mathbf{T}\delta\mathbf{T}^\dagger, \tag{6.2.8}$$

and the corresponding change in electronic energy is

$$\delta E = \operatorname{tr} \delta\boldsymbol{\rho}\mathbf{h} + \tfrac{1}{2} \operatorname{tr} \delta\boldsymbol{\rho}\mathbf{G}(\boldsymbol{\rho}) + \tfrac{1}{2} \operatorname{tr} \boldsymbol{\rho}\mathbf{G}(\delta\boldsymbol{\rho}), \tag{6.2.9}$$

where the last term arises because $\mathbf{G}$ in (6.2.5) depends on the matrix $\boldsymbol{\rho}$. This result simplifies at once, owing to the following interchange property, which follows from the definition of $\mathbf{G}$ in (6.2.6):

$$\text{tr}\,[\mathbf{AG}(\mathbf{B})] = \text{tr}\,[\mathbf{BG}(\mathbf{A})]. \tag{6.2.10}$$

Thus, exchanging $\boldsymbol{\rho}$ and $\delta\boldsymbol{\rho}$ in the last term of (6.2.9),

$$\delta E = \text{tr}\;\delta\boldsymbol{\rho}\mathbf{h} + \text{tr}\;\delta\boldsymbol{\rho}\mathbf{G} = \text{tr}\;\delta\boldsymbol{\rho}\mathbf{F}, \tag{6.2.11}$$

where

$$\boxed{\mathbf{F} = \mathbf{h} + \mathbf{G}} \tag{6.2.12}$$

is the finite-basis analogue of the Fock operator (6.1.14) and is an $m \times m$ matrix.

To maintain the parallel with the operator equations, we may consider a variation of one orbital ψ_K i.e. of the column $\mathbf{c}_K$ of the matrix $\mathbf{T}$. From (6.2.3), the corresponding special form of (6.2.8) is

$$\delta\boldsymbol{\rho}^{(K)} = \delta\mathbf{c}_K\mathbf{c}_K^\dagger + \mathbf{c}_K\delta\mathbf{c}_K^\dagger, \tag{6.2.13}$$

and substitution in (6.2.11) gives

$$\delta E^{(K)} = \delta\mathbf{c}_K^\dagger\mathbf{F}\mathbf{c}_K + \mathbf{c}_K^\dagger\mathbf{F}\delta\mathbf{c}_K, \tag{6.2.14}$$

which is the analogue of (6.1.13).‡ Again this change of *total* electronic energy coincides with the first-order change of a one-electron *orbital* energy

$$\epsilon_K = \mathbf{c}_K^\dagger\mathbf{F}\mathbf{c}_K, \tag{6.2.15}$$

corresponding to the expectation value (6.1.15).

The rest of the argument is exactly like that in the last section. The preceding equations do not depend on either the dimension or the orthonormality of the *basis* $\{\chi_r\}$; but orthonormality of the occupied orbitals $\{\psi_K\}$ requires (cf. Section 2.3)

$$\mathbf{c}_K^\dagger\mathbf{S}\mathbf{c}_L = \delta_{KL}, \tag{6.2.16}$$

where $\mathbf{S}$ is the overlap matrix with elements $S_{rs} = \langle\chi_r \mid \chi_s\rangle$. Instead of (6.1.16), we then find

$$\left.\begin{array}{l} \delta\mathbf{c}_K^\dagger\mathbf{S}\mathbf{c}_K + \mathbf{c}_K^\dagger\mathbf{S}\delta\mathbf{c}_K = 0, \\ \left.\begin{array}{l}\delta\mathbf{c}_K^\dagger\mathbf{S}\mathbf{c}_L = 0 \\ \mathbf{c}_L^\dagger\mathbf{S}\delta\mathbf{c}_K = 0\end{array}\right\} \quad (\text{all } L \neq K), \end{array}\right\} \tag{6.2.17}$$

‡ The trace of a matrix product is invariant under cyclic permutation of the factors; the products in (6.2.14) yield 1×1 matrices.

and combination of these auxiliary conditions, with Lagrangian multipliers collected in a matrix $\boldsymbol{\epsilon}$, leads to

$$\mathbf{F}\mathbf{c}_K - \sum_L \mathbf{S}\mathbf{c}_L \varepsilon_{LK} = \mathbf{0}, \tag{6.2.18}$$

which corresponds to the HF equation (6.1.18). On collecting these conditions, for all the occupied orbitals (i.e. columns $\mathbf{c}_K$ of $\mathbf{T}$), we obtain the matrix analogue of (6.1.20), namely

$$\mathbf{F}\mathbf{T} = \mathbf{S}\mathbf{T}\boldsymbol{\epsilon}, \tag{6.2.19}$$

in which the presence of $\mathbf{S}$ allows for a non-orthonormal basis.

Again, a "canonical form" may be obtained by the transformation $\mathbf{T} \to \mathbf{T}\mathbf{U}$ ($\mathbf{U}$ being an $n \times n$ unitary matrix that mixes the columns of $\mathbf{T}$, i.e. the occupied orbitals), and simultaneously $\boldsymbol{\epsilon} \to \mathbf{U}\boldsymbol{\epsilon}\mathbf{U}^\dagger$ (diagonal): in this case (6.2.19) holds with $\boldsymbol{\epsilon}$ diagonal, and the individual columns of $\mathbf{T}$ satisfy

$$\boxed{\mathbf{F}\mathbf{c} = \epsilon\mathbf{S}\mathbf{c},} \tag{6.2.20}$$

which is the usual matrix eigenvalue equation (see Section 2.3) for vectors referred to a non-orthonormal basis. The Hartree–Fock equations in this matrix form were first derived by Hall (1951) and Roothaan (1951); they cannot, of course, reproduce the results of solving the *operator* HF equations unless the basis is complete (infinite), but in practice they can yield approximations of high accuracy even for quite large molecules.

Before turning to solution methods, we note that other equivalent statements of the stationary-value conditions, not involving eigenvalue equations, may be formulated. In particular, on choosing (without loss of generality) an orthonormal basis ($\mathbf{S} = \mathbf{1}$), multiplying (6.2.19) from the right by $\mathbf{T}^\dagger$ and its Hermitian conjugate from the left by $\mathbf{T}$, and subtracting, we obtain

$$\mathbf{F}\boldsymbol{\rho} - \boldsymbol{\rho}\mathbf{F} = \mathbf{0}. \tag{6.2.21}$$

In other words, the Fock Hamiltonian must commute with the density matrix. This equation is necessary but not sufficient; it must be combined with the auxiliary condition that the occupied orbitals $\{\psi_K\}$ remain orthonormal, which may be written (with $\mathbf{S} = \mathbf{1}$)

$$\boldsymbol{\rho}\boldsymbol{\rho} = (\mathbf{T}\mathbf{T}^\dagger)(\mathbf{T}\mathbf{T}^\dagger) = \mathbf{T}\mathbf{1}\mathbf{T}^\dagger = \boldsymbol{\rho}$$

or

$$\boldsymbol{\rho}^2 = \boldsymbol{\rho} \tag{6.2.22}$$

It may be shown (McWeeny, 1956, 1960) that (6.2.2) and (6.2.22) provide necessary and sufficient conditions for a stationary value of the energy, and that these *density matrix* equations may be solved without reference to the individual occupied orbitals. The *idempotency* condition (6.2.22) is characteristic of a projection operator, an interpretation that is discussed in later sections.

Closed shells. The SCF method

The HF equations of Section 6.1 were first solved, for atoms, by a "self-consistent field" (SCF) method devised by Hartree. The same approach and terminology is used in dealing with the finite-basis equations of the present section, (6.2.19) and (6.2.20) often being referred to as the Roothan form of the "SCF equations". To clarify the concept of the self-consistent field, we turn to the closed-shell case (Roothan, 1951), in which spin (implicit in the development so far) is eliminated and the problem is to determine the *spatial* factors in the occupied spin-orbitals.

All the basic equations are analogous to those already developed; but now the occupied spin-orbitals occur in pairs $\psi_K = \phi_K\alpha$, $\bar{\psi}_K = \phi_K\beta$ with a common orbital factor ϕ_K. The density matrix $\rho(\boldsymbol{x};\boldsymbol{x}')$ takes the form (5.3.12), namely

$$\rho(\boldsymbol{x};\boldsymbol{x}') = \tfrac{1}{2}P(\boldsymbol{r};\boldsymbol{r}')[\alpha(s)\alpha^*(s') + \beta(s)\beta^*(s')], \qquad (6.2.23a)$$

with

$$P(\boldsymbol{r};\boldsymbol{r}') = 2\sum_K \phi_K(\boldsymbol{r})\phi_K^*(\boldsymbol{r}'), \qquad (6.2.23b)$$

the factor 2 arising from the addition of (equal) up-spin and down-spin densities. On performing the spin integrations in the equations used so far, we easily find an expression equivalent to (5.3.18), namely

$$E = 2\sum_K \langle\phi_K|\,(\mathsf{h} + \tfrac{1}{2}\mathsf{G})\,|\phi_K\rangle, \qquad \mathsf{G} = \mathsf{J} - \tfrac{1}{2}\mathsf{K} \qquad (6.2.24)$$

for the total energy, and

$$\mathsf{F}\phi = \epsilon\phi, \qquad \mathsf{F} = \mathsf{h} + \mathsf{G} \qquad (6.2.25)$$

for the orbital eigenvalue equation. The operators J and K are now spin-independent; thus J(1), K(1) operate on $\phi(\boldsymbol{r}_1)$ according to (6.1.11) but with ρ replaced by P. The fact that $\tfrac{1}{2}$K appears in (6.2.24) instead of K follows from the spin-integrations, which lead to a factor P in (6.1.11a) but $\tfrac{1}{2}P$ in (6.1.11b) (the α or β term, according to the spin factor in

$\psi(\boldsymbol{x})$). The orbital energy (6.1.15) yields a corresponding expression

$$\epsilon_K = \langle \phi_K | \mathsf{F} | \phi_K \rangle, \tag{6.2.26}$$

which is the energy of *one* electron in the doubly occupied orbital ϕ_K, in the effective field represented by the Fock Hamiltonian of (6.2.25).

Evidently to pass from the spin-orbital HF equations to their spinless analogues, for a closed-shell system, it is only necessary to

(i) replace spin-orbitals (ψ_K) by their orbital factors (ϕ_K), adding a factor 2 to sums over occupied orbitals (e.g. in (6.2.24)); and
(ii) replace ρ by P in defining coulomb and exchange operators, adding a factor $\frac{1}{2}$ to the latter ($\mathsf{K} \rightarrow \frac{1}{2}\mathsf{K}$).

Similar rules apply to the finite-basis equations; the basis now becomes a set of *spatial* functions, all one- and two-electron integrals being interpreted accordingly; the matrix $\mathbf{T}$ in (6.2.1) expresses the *orbitals* $\{\phi_K\}$ in terms of these basis functions; and $\boldsymbol{\rho}$ in (6.2.3) is replaced by

$$\mathbf{P} = 2\mathbf{T}\mathbf{T}^\dagger = 2 \sum_K \mathbf{c}_K \mathbf{c}_K^\dagger, \tag{6.2.27}$$

being an occupied-orbital sum. Thus, instead of (6.2.5), we obtain

$$E = \operatorname{tr} \mathbf{Ph} + \tfrac{1}{2} \operatorname{tr} \mathbf{PG}, \tag{6.2.28}$$

where

$$\mathbf{G} = \mathbf{G}(\mathbf{P}) = \mathbf{J}(\mathbf{P}) - \tfrac{1}{2}\mathbf{K}(\mathbf{P}), \tag{6.2.29}$$

and the definitions are similar to those in (6.2.6), namely

$$\left.\begin{aligned} J(\mathbf{P})_{rs} &= \sum_{t,u} P_{ut} \langle \chi_r \chi_t | g | \chi_s \chi_u \rangle, \\ K(\mathbf{P})_{rs} &= \sum_{t,u} P_{ut} \langle \chi_r \chi_t | g | \chi_u \chi_s \rangle. \end{aligned}\right\} \tag{6.2.30}$$

The expression for the Fock matrix $\mathbf{F}$ and the orbital energy ϵ_K are formally identical with those in (6.2.12) and (6.2.15) respectively, and the (spinless) orbital eigenvalue equation takes the matrix form (6.2.20).

Since F in (6.2.25) depends on the electron density P through the electron-interaction terms, the eigenvalue equation must be solved *iteratively*; this is the essence of the SCF method. The procedure first proposed and adopted by Hartree and coworkers, for atomic ground states, was to initiate the interaction with a first "guess" of the occupied orbitals, computing the density P and the effective Hamiltonian F; the eigenvalue equation (6.2.25) was then solved numerically to obtain new approximations to the occupied orbitals; starting from the new orbitals, the "cycle" was then recommenced, recalculating the density and resultant effective field (contained in the G term of F), and then once again solving the eigenvalue equation; the iteration was continued until the density and effective field computed from the orbitals at the end of a cycle agreed with those used in constructing the effective Hamiltonian (i.e. those from the previous cycle); the effective field was then said to be self-consistent and the iteration was terminated. This procedure, whether used with the operator equations or the matrix equations, is the characteristic feature of all SCF methods of solving the HF equations, although many refinements have been added in order to speed convergence.

In fact, the matrix (i.e. the finite-basis) equations provide the most widely used method of obtaining approximate solutions of the HF equations. This approach owes its popularity mainly to the relative ease with which the eigenvalue equation (6.2.20) may be solved and the fact that algebraic and iterative procedures are ideally suited for electronic computers. There is also an important by-product of this approach: with m basis functions, the eigenvalue equation possesses $m - n$ solutions in addition to those that define the n occupied orbitals. The extra solutions correspond to (unoccupied) orbitals with higher orbital energies, which are usually called *virtual orbitals*. The virtual orbitals in effect describe the possible states of a "test" electron moving in the field of the whole neutral molecule (effective Hamiltonian F), and are often used in CI studies, particularly of the excited states. They are not ideally suited for this purpose, since an "excited" electron does not feel the field of a *neutral* molecule, but more nearly that of a residual positive ion. The virtual orbitals are consequently rather too diffuse and tend to yield slowly convergent CI expansions; nevertheless, they provide a useful orthogonal complement to the set of occupied orbitals, and may be employed in many ways.

6.3 UNRESTRICTED HARTREE–FOCK THEORY

In passing from the spin-orbital equations to the closed-shell orbital equations in Section 6.2, a constraint was introduced: it was assumed that all spin-orbitals were "pure" space–spin products of the form $\phi_K\alpha$ or $\phi_K\beta$ and that these were occupied in *pairs* with a common orbital factor ϕ_K. In other words, we developed a *restricted* Hartree–Fock (RHF) theory. It is possible, however, that an optimized single-determinant wavefunction in which such restrictions were absent would yield a lower variational energy, the spin-orbitals being neither pure nor paired. In that case, without constraints, the variational wavefunction would be *unstable,* the RHF solution corresponding to a saddle-point on the energy surface rather than a true minimum. A well-known example of this phenomenon occurs during molecular dissociation: the RHF function usually gives a true minimum at the equilibrium geometry, but under a bond-breaking deformation a lower energy is often achieved by allowing each doubly occupied MO to "split" into two, giving spin-orbitals $\phi_K\alpha$, $\bar{\phi}_K\beta$ with different spatial factors. For a system containing an odd number of electrons such a splitting, when permitted, is bound to occur. The determination of spin-orbitals of this type is the aim of *unrestricted* Hartree–Fock (UHF) theory. This designation, although established by tradition, is not strictly correct since all spin-orbitals are still restricted to be of simple product form. In this section we start by allowing each ψ_K to take the form

$$\psi_K = \phi_K^\alpha \alpha + \phi_K^\beta \beta, \tag{6.3.1}$$

and to avoid confusion of terminology use the acronym GUHF to denote the generalized UHF theory to which this assumption leads. Clearly, the equations of UHF theory will arise as a limiting case in which one term or other in each ψ_K is dominant, the other component being negligible or absent.

Before formulating the stationary-value conditions, we note that a single determinant of 2-component spin-orbitals (6.3.1) will not in general be an eigenfunction of the operator S^2, nor even of S_z. Such functions are therefore, strictly speaking, unsuitable for representing real spectroscopic states; typically, a GUHF or UHF function may yield an expectation value $\langle \mathsf{S}^2 \rangle$ very close to $S(S+1)$, with S integral or half-integral, but there will still be a small "contamination" by eigenfunctions of other spin-multiplicity. The use of such functions in discussing physical processes and properties (particularly spin properties) is therefore always open to criticism. These defects may always be removed by applying spin-projection operators, as in Section 4.2, to generate a pure spin state with the required eigenvalues of S^2 and S_z; but in practice this procedure is very cumbersome and leads to *many*-determinant wavefunctions—thus destroying the most valuable features of the HF approximation. When solution of the extended equations indicates a serious breakdown of the RHF approximation, it is more satisfactory to turn to *multi*configuration methods (Chapter 8) in which spin requirements are introduced at the outset and the spin-orbitals are of the usual type, with double or single occupation. In less extreme situations the simplicity of the UHF method gives it a certain appeal, and it is therefore worth sketching the basis of the method in its most general form.

The derivations, which we present in finite-basis language, run parallel to those of the last section. We use a spin-orbital basis

$$\boldsymbol{\chi} = (\chi_1\alpha\ \chi_2\alpha\ \ldots\ \chi_m\alpha \mid \chi_1\beta\ \chi_2\beta\ \ldots\ \chi_m\beta) = (\boldsymbol{\chi}^\alpha \mid \boldsymbol{\chi}^\beta), \quad (6.3.2)$$

and then, assuming N electrons, express the occupied 2-component spin-orbitals (6.3.1) as

$$\boldsymbol{\psi} = \boldsymbol{\chi}\mathbf{T} = (\boldsymbol{\chi}^\alpha\ \boldsymbol{\chi}^\beta)\begin{pmatrix}\mathbf{T}_\alpha \\ \mathbf{T}_\beta\end{pmatrix}, \quad (6.3.3)$$

where the $2m \times N$ matrix $\mathbf{T}$ is partitioned into $m \times N$ blocks $\mathbf{T}_\alpha$, $\mathbf{T}_\beta$ whose Kth columns contain the expansion coefficients for ϕ_K^α and ϕ_K^β respectively, in terms of the basis functions $\chi_1, \chi_2, \ldots, \chi_m$. On defining the $m \times m$ matrices

$$\mathbf{R}_{\sigma\sigma'} = \mathbf{T}_{\sigma'}\mathbf{T}_{\sigma'}^\dagger \quad (\sigma, \sigma' = \alpha, \beta), \quad (6.3.4)$$

the energy expression corresponding to (6.2.5) may be presented, after performing spin integrations throughout, in the form

$$E = \operatorname{tr}[\mathbf{R}(\mathbf{h} + \tfrac{1}{2}\mathbf{G})], \quad (6.3.5)$$

where each $2m \times 2m$ matrix consists of four $m \times m$ blocks. For example,

$$\mathbf{R} = \begin{pmatrix}\mathbf{R}_{\alpha\alpha} & \mathbf{R}_{\alpha\beta} \\ \mathbf{R}_{\beta\alpha} & \mathbf{R}_{\beta\beta}\end{pmatrix}. \quad (6.3.6)$$

The blocks of the $\mathbf{G}$ matrix are

$$\left.\begin{aligned}
\mathbf{G}_{\alpha\alpha} &= \mathbf{J}(\mathbf{R}_{\alpha\alpha} + \mathbf{R}_{\beta\beta}) - \mathbf{K}(\mathbf{R}_{\alpha\alpha}),\\
\mathbf{G}_{\beta\beta} &= \mathbf{J}(\mathbf{R}_{\alpha\alpha} + \mathbf{R}_{\beta\beta}) - \mathbf{K}(\mathbf{R}_{\beta\beta}),\\
\mathbf{G}_{\alpha\beta} &= -\mathbf{K}(\mathbf{R}_{\alpha\beta}),\\
\mathbf{G}_{\beta\alpha} &= -\mathbf{K}(\mathbf{R}_{\beta\alpha}),
\end{aligned}\right\} \quad (6.3.7)$$

where the coulomb and exchange matrices are defined exactly as in (6.2.30) in terms of the 2-electron integrals $\langle\chi_r\chi_t|\,g\,|\chi_s\chi_u\rangle$. The requirement that E be stationary, subject to orthonormality of the occupied spin-orbitals then leads to a matrix GUHF equation

$$\begin{pmatrix}\mathbf{F}_{\alpha\alpha} & \mathbf{F}_{\alpha\beta} \\ \mathbf{F}_{\beta\alpha} & \mathbf{F}_{\beta\beta}\end{pmatrix}\begin{pmatrix}\mathbf{c}^\alpha \\ \mathbf{c}^\beta\end{pmatrix} = \epsilon\begin{pmatrix}\mathbf{S} & \mathbf{0} \\ \mathbf{0} & \mathbf{S}\end{pmatrix}\begin{pmatrix}\mathbf{c}^\alpha \\ \mathbf{c}^\beta\end{pmatrix}, \quad (6.3.8)$$

where we have assumed the canonical form corresponding to a diagonal matrix of Lagrangian multipliers. The matrix $\mathbf{S}$ contains the overlap integrals $\langle\chi_r \mid \chi_s\rangle$, while the blocks of the Fock matrix are

$$\mathbf{F}_{\sigma\sigma'} = \delta_{\sigma\sigma'}\mathbf{h} + \mathbf{G}_{\sigma\sigma'} \quad (\sigma, \sigma' = \alpha, \beta). \quad (6.3.9)$$

The matrix $\mathbf{h}$, which occurs only in the diagonal blocks, is the usual matrix of 1-electron integrals $\langle\chi_r|\,\mathrm{h}\,|\chi_s\rangle$. Equation (6.3.8) may be solved by the usual SCF procedure. The Kth eigenvector $\mathbf{c}_K$, comprising $\mathbf{c}_K^\alpha$ and $\mathbf{c}_K^\beta$, then contains the $2m$ expansion coefficients of ψ_K, the 2-component spin-orbital (6.3.1).

Another approach to GUHF theory has been given by Seeger and Pople (1977). In practice, however, with a spinless Hamiltonian, it is seldom necessary to admit the 2-component functions since the UHF forms are usually stable (see e.g. Cook, 1981, 1984). The question of stability in the general case, where the wavefunction is not even an eigenfunction of S_z, has been discussed by Fukutome and others (e.g. Fukutome, 1981; Calais, 1985); but the value and physical significance of the wavefunction then remains open to question (cf. Problem 6.12).

The UHF equations, which are still widely used, follow from (6.3.6–6.3.8) on striking out all the off-diagonal ($\alpha\beta$, $\beta\alpha$) blocks. There are then two sets of equations,

$$\mathbf{F}_\alpha\mathbf{c}^\alpha = \epsilon\mathbf{S}\mathbf{c}^\alpha, \qquad \mathbf{F}_\beta\mathbf{c}^\beta = \epsilon\mathbf{S}\mathbf{c}^\beta, \tag{6.3.10}$$

which determine the orbitals

$$\phi_K = \phi_K^\alpha = \boldsymbol{\chi}\mathbf{c}_K^\alpha, \qquad \bar{\phi}_K = \phi_K^\beta = \boldsymbol{\chi}\mathbf{c}_K^\beta, \tag{6.3.11}$$

which occur in the 1-determinant wavefunction with spin factors α and β respectively. The only coupling between the α- and β-solutions arises indirectly from the electron-interaction matrices $\mathbf{G}_{\alpha\alpha}$, $\mathbf{G}_{\beta\beta}$ in (6.3.7), which involve both α and β contributions to the electron density.

One of the main uses of the UHF approach is the determination of simple first approximations to the spin density for molecular ions and radicals in which non-singlet ground states commonly occur. UHF orbitals are also sometimes used (e.g. Handy *et al.*, 1985a,b) as a basis CI.

6.4 SOME FURTHER PROPERTIES OF DENSITY MATRICES

In Chapter 5 we discussed various properties of density matrices such as $\rho(\boldsymbol{x}_1;\boldsymbol{x}_1')$. We must now turn to the *density operators,* of which these matrices—or more correctly *kernels*—provide representations. Let us first consider the general form (5.3.5), namely

$$\rho(\boldsymbol{x}_1;\boldsymbol{x}_1') = \sum_{r,s} \rho_{rs}\psi_r(\boldsymbol{x}_1)\psi_s^*(\boldsymbol{x}_1'), \tag{6.4.1}$$

where $\{\psi_r\}$ is any spin-orbital basis, and regard $\rho(\boldsymbol{x}_1;\boldsymbol{x}_1')$ as the kernel of an integral operator ρ. This simply means that ρ operating on $\psi(\boldsymbol{x}_1)$ gives a new function $\psi'(\boldsymbol{x}_1)$, which we choose to define by

$$\psi'(\boldsymbol{x}_1) = \rho\psi(\boldsymbol{x}_1) = \int \rho(\boldsymbol{x}_1;\boldsymbol{x}_1')\psi(\boldsymbol{x}_1')\,\mathrm{d}\boldsymbol{x}_1'. \tag{6.4.2}$$

In other words, we change the variable from $\boldsymbol{x}_1$ to $\boldsymbol{x}_1'$, multiply by the kernel, and integrate to get rid of $\boldsymbol{x}_1'$—the result being a new function of $\boldsymbol{x}_1$. Such operators are linear operators, just like the various differential operators we have used, and play an important part in the general formulation of quantum mechanics. We call ρ the 1-electron *density operator*.

It has been noted (footnote § on p.124) that $\rho(\boldsymbol{x}_1;\boldsymbol{x}_1')$ formally resembles a matrix element, in which $\boldsymbol{x}_1$ and $\boldsymbol{x}_1'$ play the part of (continuous) row and column indices; in this sense it provides a particular *representation* of ρ. We now note that, on introducing any orthonormal set $\{\psi_r(\boldsymbol{x}_1)\}$, the array of coefficients appearing in (6.4.1) simply provides a true matrix representation of the operator ρ, in which $\boldsymbol{x}_1$, $\boldsymbol{x}_1'$ are replaced by the *discrete* indices r, s. This follows easily from the definition of the matrix elements of an operator; since, using the orthonormality property,

$$\begin{aligned}\langle\psi_r|\,\rho\,|\psi_s\rangle &= \int \psi_r^*(\boldsymbol{x}_1)\int \sum_{t,u} \rho_{tu}\psi_t(\boldsymbol{x}_1)\psi_u^*(\boldsymbol{x}_1')\psi_s(\boldsymbol{x}_1')\,\mathrm{d}\boldsymbol{x}_1'\,\mathrm{d}\boldsymbol{x}_1.\\ &= \sum_{t,u} \delta_{rt}\delta_{us}\rho_{tu}\\ &= \rho_{rs}.\end{aligned}$$

Thus, the full array $\boldsymbol{\rho}$ is the matrix representing the density operator ρ, with kernel (6.4.1), referred to the basis $\{\psi_r\}$. An exactly similar interpretation may be placed upon the 2-electron density kernel $\pi(\boldsymbol{x}_1, \boldsymbol{x}_2;\boldsymbol{x}_1', \boldsymbol{x}_2')$, the *operator* π being defined by an equation analogous to (6.4.2) and the *matrix* representing π simply being the array of coefficients $\pi_{rs,tu}$ when we write

$$\pi(\boldsymbol{x}_1, \boldsymbol{x}_2;\boldsymbol{x}_1', \boldsymbol{x}_2') = \sum_{r,s,t,u} \psi_r(\boldsymbol{x}_1)\psi_s(\boldsymbol{x}_2)\pi_{rs,tu}\psi_t^*(\boldsymbol{x}_1')\psi_u^*(\boldsymbol{x}_2') \tag{6.4.3}$$

in terms of the products $\{\psi_r(\boldsymbol{x}_1)\psi_s(\boldsymbol{x}_2)\}$ that define a basis in 2-electron space. We are most concerned, however, with ρ, and note that the definition of a matrix representation implies that (6.4.2) has a matrix counterpart $\mathbf{c}' = \boldsymbol{\rho}\mathbf{c}$, where $\mathbf{c}$ and $\mathbf{c}'$ are the columns of expansion coefficients defining ψ and ψ' in the given basis. The other properties of matrix representations, established in Section 2.2, may be taken over in a similar way. In particular, we define the product of two integral operators, with kernels $A(\boldsymbol{x};\boldsymbol{x}')$ and $B(\boldsymbol{x};\boldsymbol{x}')$, by‡

$$C(\boldsymbol{x};\boldsymbol{x}') = \int A(\boldsymbol{x};\boldsymbol{x}'')B(\boldsymbol{x}'';\boldsymbol{x}')\,\mathrm{d}\boldsymbol{x}'', \tag{6.4.4}$$

and it is easily verified that the matrices associated with the operators then satisfy $\mathbf{C} = \mathbf{AB}$, thus establishing the basic representation property.

The Fock–Dirac density matrix (5.3.7) corresponding to a 1-determinant wavefunction has a very special property. We regard the N occupied spin-orbitals as the first N members of a complete orthonormal 1-electron basis and consider

$$\rho(\boldsymbol{x}_1;\boldsymbol{x}_1') = \sum_{R\ \mathrm{occ}} \psi_R(\boldsymbol{x}_1)\psi_R^*(\boldsymbol{x}_1') \tag{6.4.5}$$

as the kernel of the corresponding density operator. Let us take an arbitrary (space–spin) function and expand it in the form

$$\psi(\boldsymbol{x}_1) = \sum_S c_S\psi_S(\boldsymbol{x}_1),$$

where S runs over the complete set. Operation with ρ then yields, using the orthonormality property,

$$\rho\psi(\boldsymbol{x}_1) = \sum_{S\ \mathrm{occ}} c_S\psi_S(\boldsymbol{x}_1). \tag{6.4.6}$$

Consequently all components of an arbitrary function are annihilated except those that involve the occupied spin-orbitals defining ρ. We refer to this process (remembering the geometrical interpretation) as "projection onto the subspace spanned by the occupied spin-orbitals". The simplest projector arises from a single spin-orbital:

$$\rho_R(\boldsymbol{x}_1;\boldsymbol{x}_1') = \psi_R(\boldsymbol{x}_1)\psi_R^*(\boldsymbol{x}_1')$$

will eliminate all components except one, producing from any spin-orbital a multiple of ψ_R (or zero), as may be seen from the equation

$$\rho_R\psi(\boldsymbol{x}_1) = \int \psi_R(\boldsymbol{x}_1)\psi_R^*(\boldsymbol{x}_1')\psi(\boldsymbol{x}_1')\,\mathrm{d}\boldsymbol{x}_1' = \psi_R(\boldsymbol{x}_1)\langle\psi_R\,|\,\psi\rangle. \tag{6.4.7}$$

It is often convenient to employ a symbolic ("dyadic") notation, writing

$$\rho_R = \psi_R\psi_R^* \tag{6.4.8}$$

for the *operator* (suppressing the variables) and combining ψ_R^* with any

‡ Note the order of the variables, which corresponds to the "chain rule" for matrix subscripts: $C_{rs} = \sum_t A_{rt}B_{ts}$.

function immediately to the right of it to form the scalar product. Thus

$$\rho_R \psi = (\psi_R \psi_R^*)\psi = \psi_R \langle \psi_R \mid \psi \rangle \tag{6.4.9}$$

is an abbreviated statement of (6.4.7).

The characteristic property of a projection operator, such as ρ defined through (6.4.5), is that a repeated projection is exactly equivalent to a single projection. In symbols,

$$\rho^2 = \rho, \tag{6.4.10}$$

which is easily verified by taking the product of ρ with itself, according to (6.4.4). The matrix associated with ρ according to (6.4.1) is *diagonal* when we use the Fock–Dirac form (6.4.5), with N unit elements and the rest zero. It is thus obvious that the density *matrix* in this particular approximation shares the property (6.4.10) of *idempotency*,

$$\boldsymbol{\rho}^2 = \boldsymbol{\rho} \tag{6.4.11}$$

More generally, with the usual basis-set expansion

$$\boldsymbol{\psi} = \boldsymbol{\chi}\mathbf{T}, \tag{6.4.12}$$

we obtain as in (6.2.3)

$$\boldsymbol{\rho} = \mathbf{T}\mathbf{T}^\dagger, \tag{6.4.13}$$

and again, provided that the basis is orthonormal, the matrices themselves will exhibit the characteristic property (6.4.11). It is also seen that for a non-orthonormal basis (6.4.10) leads to the more general result

$$\boldsymbol{\rho}\mathbf{S}\boldsymbol{\rho} = \boldsymbol{\rho}. \tag{6.4.14}$$

Evidently there are considerable attractions in using an orthonormal basis with $\mathbf{S} = \mathbf{1}$. When such a basis is not immediately available, however, it may always be constructed by the Löwdin transformation (2.2.21). Thus, using a bar to distinguish quantities in the orthonormal basis $\bar{\boldsymbol{\chi}} = \boldsymbol{\chi}\mathbf{S}^{-1/2}$, the MOs become

$$\boldsymbol{\psi} = \boldsymbol{\chi}\mathbf{T} = \boldsymbol{\chi}\mathbf{S}^{-1/2}\mathbf{S}^{1/2}\mathbf{T} = \bar{\boldsymbol{\chi}}\bar{\mathbf{T}},$$

and the corresponding density matrix is expressed as

$$\bar{\boldsymbol{\rho}} = \bar{\mathbf{T}}\bar{\mathbf{T}}^\dagger = \mathbf{S}^{1/2}\mathbf{T}\mathbf{T}^\dagger\mathbf{S}^{1/2} = \mathbf{S}^{1/2}\boldsymbol{\rho}\mathbf{S}^{1/2}. \tag{6.4.15}$$

This matrix then reflects the idempotency property (6.4.11):

$$\bar{\boldsymbol{\rho}}\bar{\boldsymbol{\rho}} = \mathbf{S}^{1/2}\boldsymbol{\rho}\mathbf{S}^{1/2}\mathbf{S}^{1/2}\boldsymbol{\rho}\mathbf{S}^{1/2} = \mathbf{S}^{1/2}(\boldsymbol{\rho}\mathbf{S}\boldsymbol{\rho})\mathbf{S}^{1/2} = \bar{\boldsymbol{\rho}}.$$

No representation of the density operator is unique, of course, and it is clear that in a *non*-orthonormal basis the matrix $\boldsymbol{\rho}\mathbf{S}$ is also idempotent.‡

$$(\boldsymbol{\rho}\mathbf{S})(\boldsymbol{\rho}\mathbf{S}) = (\boldsymbol{\rho}\mathbf{S}\boldsymbol{\rho})\mathbf{S} = \boldsymbol{\rho}\mathbf{S}, \tag{6.4.16}$$

as follows from (6.4.14). The main attraction of the form (6.4.15) is the Hermitian symmetry of the matrix.

Sometimes it is useful to divide the space spanned by any number of functions into subspaces by means of projection operators. We form the projection operator ρ from the functions of a given subspace according to (6.4.5), in which the functions were designated as "occupied", and recall the property (6.4.6). It then follows readily that the operator $\rho' = 1 - \rho$ has the complementary property

$$\rho'\psi = (1-\rho)\psi = \psi - \rho\psi = \sum_{S\ \text{unocc}} c_S\psi_S.$$

The division into two sets of orthonormal spin-orbitals (here designated "occ" and "unocc") is of course quite arbitrary. We say that $\mathcal{V} = \{\psi_S\ (S\ \text{occ})\}$ and $\mathcal{V}' = \{\psi_U (U\ \text{unocc})\}$ are *complementary subspaces,* of the space spanned by a complete set of spin-orbitals, and that ρ and ρ' project onto $\mathcal{V}$ and $\mathcal{V}'$ respectively. It follows that ρ and ρ' have the properties:

$$\left.\begin{aligned} \rho^2 = \rho, \qquad \rho'^2 &= \rho', \\ \rho\rho' = \rho'\rho &= 0, \\ \rho + \rho' &= 1, \end{aligned}\right\} \tag{6.4.17}$$

where 0 and 1 are used as the zero and unit *operators* (i.e. multiply by 0 or 1, to annihilate any function or leave it unchanged respectively). The first line states the projection-operator property (idempotency); the second that the projection of any Ψ onto one subspace has zero component in the other (the subspaces are non-overlapping or orthogonal); and the third states that by projecting onto $\mathcal{V}$ and $\mathcal{V}'$ and adding the parts together we retrieve the original ψ (i.e. multiply by 1). The latter result is usually described as "resolution of the identity". All the properties of the projection operators must of course be shared by the matrices that represent them, whatever choice of basis we care to make. Moreover, these observations are valid for all the linear spaces we have occasion to use, whether they refer to one particle or any number, to spatial or spin variables or both.

‡ At first sight, the representation of the density operator by the matrix $\boldsymbol{\rho}\mathbf{S}$ appears to be inconsistent with the discussion in Section 2.2. In fact, however, the elements of $\boldsymbol{\rho}$ express the density *operator* (with integral kernel $\rho(\boldsymbol{x};\boldsymbol{x}')$) in the dyadic form (6.4.1); the metrically defined matrix associated with the operator, with elements $\rho^{\mathrm{S}}_{pq} = \langle\chi_p|\,\rho\,|\chi_p\rangle$, is thus $\boldsymbol{\rho}^{\mathrm{S}} = \mathbf{S}\boldsymbol{\rho}\mathbf{S}$; and from (2.2.15) it follows that the matrix that *represents* ρ in the basis $\boldsymbol{\chi}$ is $\mathbf{S}^{-1}\boldsymbol{\rho}^{\mathrm{S}} = \boldsymbol{\rho}\mathbf{S}$—in agreement with (6.4.16).

In practice, we are usually concerned with the *orbital* space spanned by a set of m AOs, or other basis functions, spin integrations having been performed already. The *spinless* density matrix in the closed-shell form (6.2.27) is not itself idempotent, but on writing

$$\mathbf{P} = 2\mathbf{R} \quad (\mathbf{R} = \mathbf{T}\mathbf{T}^{\dagger}), \tag{6.4.18}$$

it follows that $\mathbf{R}$ is the proper analogue of $\boldsymbol{\rho}$ and, for an orthonormal basis, has the usual property

$$\mathbf{R}^2 = \mathbf{R}. \tag{6.4.19}$$

This $m \times m$ matrix represents a projection operator onto the space spanned by the (doubly) occupied MOs. The *unit* operator in the full m-dimensional space is of course the $m \times m$ unit matrix $\mathbf{1}$, and we may therefore define a projection operator

$$\mathbf{R}' = \mathbf{1} - \mathbf{R} \tag{6.4.20}$$

for the complementary subspace spanned by any $m - n$ linear combinations of the AOs that are linearly independent of, and orthogonal to, the n MOs of the occupied subspace. Clearly, since $\mathbf{R}^2 = \mathbf{R}$,

$$\mathbf{R}'^2 = (\mathbf{1} - \mathbf{R})(\mathbf{1} - \mathbf{R}) = \mathbf{1} - 2\mathbf{R} + \mathbf{R}^2 = \mathbf{1} - \mathbf{R} = \mathbf{R}',$$

$$\mathbf{R}'\mathbf{R} = (\mathbf{1} - \mathbf{R})\mathbf{R} = \mathbf{R} - \mathbf{R}^2 = \mathbf{0},$$

and the matrices thus possess the definitive properties listed in (6.4.17).

In a matrix representation $\mathbf{c}' = \mathbf{R}\mathbf{c}$ should be a column of coefficients representing a function of the occupied subspace (i.e. one that can be expressed as a linear combination of the occupied MOs). If $\mathbf{c}$ represents an arbitrary function in the space of the m AOs, it may certainly be written as a linear combination of the columns $\mathbf{c}_A$, $\mathbf{c}_B$, . . . , representing a *full set* of MOs—including the n occupied MOs and any further $m - n$ independent orthonormal functions, which we may refer to as the unoccupied MOs:

$$\mathbf{c} = \sum_S \mu_S \mathbf{c}_S. \tag{6.4.21}$$

We then note that $\mathbf{R}$ is a sum of terms $\mathbf{c}_A\mathbf{c}_A^{\dagger} + \mathbf{c}_B\mathbf{c}_B^{\dagger} + \ldots$, *each* of which represents a projection operator onto a one-dimensional subspace: thus (from the orthonormality property)

$$(\mathbf{c}_A\mathbf{c}_A^{\dagger})\mathbf{c} = \sum_S \mu_S \mathbf{c}_A(\mathbf{c}_A^{\dagger}\mathbf{c}_S) = \mu_A \mathbf{c}_A,$$

and represents a multiple of orbital ϕ_A, in other words the "A component" of the arbitrary function. Each term in $\mathbf{R}$ projects a component for one of the occupied MOs, and consequently

$$\mathbf{c}' = \mathbf{R}\mathbf{c} = \sum_{A \text{ occ}} \mu_A \mathbf{c}_A \tag{6.4.22}$$

describes an orbital that is a linear combination of the occupied MOs, all other components being annihilated.

Finally, we note that any matrix **A**, associated with some operator in the full space, may be resolved into various "projected parts" or components. Thus, since $\mathbf{R} + \mathbf{R}' = \mathbf{1}$ (resolution of the identity), multiplication from left and right by unity yields

$$\mathbf{A} = \mathbf{RAR} + \mathbf{RAR}' + \mathbf{R}'\mathbf{AR} + \mathbf{R}'\mathbf{AR}'. \tag{6.4.23}$$

This resolution is a unique and **A** vanishes if and only if each of its projected parts vanishes: thus equality of two matrices implies equality of all corresponding projected parts, just as equality of two vectors implies equality of all corresponding components.

There is of course no reason why a space should not be divided into several subspaces. We may, for instance, wish to distinguish doubly occupied, singly occupied and empty MOs, and could introduce for this purpose matrices $\mathbf{R}_1$, $\mathbf{R}_2$, $\mathbf{R}_3$ with the properties

$$\mathbf{R}_i^2 = \mathbf{R}_i, \qquad \mathbf{R}_i\mathbf{R}_j = \mathbf{0} \quad (i \neq j), \qquad \sum_i \mathbf{R}_i = \mathbf{1} \quad (i, j = 1, 2, 3). \tag{6.4.24}$$

This possibility will prove useful in developing open-shell SCF theory in the following sections.

Some of the preceding observations may seem somewhat abstruse. However, they provide a concise, powerful and geometrically transparent calculus for performing operations (both algebraic and numerical) that would otherwise appear extremely cumbersome.

6.5 RESTRICTED HF THEORY. AN OPEN-SHELL SYSTEM

We now discuss the determination of optimum MOs for a state constrained to be of the form (6.1.1) in which n_1 orbitals are doubly occupied and the remaining n_2 are singly occupied with spins parallel. In this context, the sets of doubly and singly occupied orbitals are usually referred to respectively as the "closed shell" and the "open shell". Other states, in which the closed shell is fixed but there are several determinants corresponding to various orbital occupations and coupling schemes in the open shell, may sometimes be dealt with in the same way but first we consider the simple 1-determinant case. We also assume the wavefunction to be non-degenerate, although later we consider many-determinant functions and remove this restriction.

The MOs of the closed shell and the open shell may be collected in matrices $\mathbf{T}_1$ and $\mathbf{T}_2$, with n_1 and n_2 columns respectively, and we define density matrices for each shell,

$$\mathbf{R}_1 = \mathbf{T}_1\mathbf{T}_1^\dagger, \qquad \mathbf{R}_2 = \mathbf{T}_2\mathbf{T}_2^\dagger. \tag{6.5.1}$$

The energy expression that replaces (5.3.18) is then found to be (using $A, B, \ldots$ and $U, V, \ldots$ for closed- and open-shell orbitals respectively)

$$\begin{aligned} E = \nu_1\Big[\sum_A \langle\phi_A|\,\mathsf{h}\,|\phi_A\rangle + \sum_{A,B} (\langle\phi_A\phi_B|\,g\,|\phi_A\phi_B\rangle - \tfrac{1}{2}\langle\phi_A\phi_B|\,g\,|\phi_B\phi_A\rangle)\Big]. \\ + \nu_2\Big[\sum_U \langle\phi_U|\,\mathsf{h}\,|\phi_U\rangle + \sum_{U,V} (\phi_U\phi_V|\,g\,|\phi_U\phi_V\rangle - \langle\phi_U\phi_V|\,g\,|\phi_V\phi_U\rangle)\Big] \\ + \nu_1\nu_2\Big[\sum_{A,U} (\langle\phi_A\phi_U|\,g\,|\phi_A\phi_U\rangle - \tfrac{1}{2}\langle\phi_A\phi_U|\,g\,|\phi_U\phi_A\rangle)\Big], \end{aligned} \tag{6.5.2}$$

where we have introduced closed- and open-shell occupation numbers ($\nu_1 = 2,\ \nu_2 = 1$) to reveal the symmetry of the expression. From this we obtain, proceeding as in the derivation of (6.2.28),

$$E = \nu_1 \operatorname{tr} \mathbf{R}_1(\mathbf{h} + \tfrac{1}{2}\mathbf{G}_1) + \nu_2 \operatorname{tr} \mathbf{R}_2(\mathbf{h} + \tfrac{1}{2}\mathbf{G}_2), \tag{6.5.3}$$

where the two shells have slightly different electron interaction matrices,

$$\left.\begin{aligned} \mathbf{G}_1 &= \mathbf{G}(\nu_1\mathbf{R}_1) + \mathbf{G}(\nu_2\mathbf{R}_2) \quad \text{(closed shell)}, \\ \mathbf{G}_2 &= \mathbf{G}(\nu_1\mathbf{R}_1) + \mathbf{G}'(\nu_2\mathbf{R}_2) \quad \text{(open shell)}, \end{aligned}\right\} \tag{6.5.4}$$

expressed in terms of coulomb and exchange matrices (6.2.30) through

$$\left.\begin{aligned} \mathbf{G}(\mathbf{R}) &= \mathbf{J}(\mathbf{R}) - \tfrac{1}{2}\mathbf{K}(\mathbf{R}), \\ \mathbf{G}'(\mathbf{R}) &= \mathbf{J}(\mathbf{R}) - \mathbf{K}(\mathbf{R}). \end{aligned}\right\} \tag{6.5.5}$$

In $\mathbf{G}_2$, for instance, the $\mathbf{G}$ term represents the coulomb-exchange effect of the closed-shell electrons (density described by $\nu_1\mathbf{R}_1$), while the $\mathbf{G}'$, term refers to the other *open*-shell electrons.

The orthonormality conditions may be replaced by equivalent conditions on the density matrices: these represent projections onto the subspaces spanned by doubly and singly occupied MOs and therefore satisfy the conditions (6.4.24), namely

$$\mathbf{R}_1^2 = \mathbf{R}_1, \qquad \mathbf{R}_2^2 = \mathbf{R}_2, \qquad \mathbf{R}_1\mathbf{R}_2 = \mathbf{0}, \tag{6.5.6}$$

where the third condition expresses the orthogonality of the closed- and open-shell MOs. For optimum orbitals (6.5.3) must be stationary subject to the constraints (6.5.6). In close analogy to (6.2.11), we find a first-order variation

$$\delta E^{(1)} = \nu_1 \operatorname{tr} \mathbf{F}_1 \delta \mathbf{R}_1 + \nu_2 \operatorname{tr} \mathbf{F}_2 \delta \mathbf{R}_2, \tag{6.5.7}$$

where

$$\mathbf{F}_1 = \mathbf{h} + \mathbf{G}_1, \qquad \mathbf{F}_2 = \mathbf{h} + \mathbf{G}_2. \tag{6.5.8}$$

There are consequently two Hartree–Fock matrices, one for each shell, instead of the single $\mathbf{F}$ of closed-shell theory. We see also that if the MOs of the two shells happen to be orthogonal by symmetry, so that the orthogonality condition for the two subspaces ($\mathbf{R}_1\mathbf{R}_2 = \mathbf{0}$) is satisfied identically, the equations to be solved at any stage will be formally the same as those for two separate closed-shell systems—one for $\mathbf{R}_1$ and one for $\mathbf{R}_2$ (though of course $\mathbf{F}_1$ and $\mathbf{F}_2$ involve both matrices implicitly). The orthogonality constraint generally leads, however, to a considerable complication, introducing a strong coupling between the equations for the two shells. In this case it is not immediately clear that the stationary conditions can be expressed by means of an eigenvalue equation: to explore this possibility we shall partition the m-dimensional space spanned by the basis into three subspaces, $\mathcal{V}_1$, $\mathcal{V}_2$, $\mathcal{V}_3$, using the projection operators defined in (6.4.24).

Let us consider a variation in which $\mathbf{R}_i \to \mathbf{R}_i + \delta\mathbf{R}_i$ ($i = 1, 2, 3$) and require that the conditions (6.4.24) remain valid to first order. Attention may in fact be confined to $i, j = 1, 2$ since the last condition defines $\mathbf{R}_3$ for the virtual-orbital subspace as $\mathbf{R}_3 = \mathbf{1} - \mathbf{R}_1 - \mathbf{R}_2$; in other words, we may start from (6.5.6). Explicitly, first-order variation leads to the conditions

$$\mathbf{R}_1\delta\mathbf{R}_1 + \delta\mathbf{R}_1\mathbf{R}_1 = \delta\mathbf{R}_1, \tag{6.5.9a}$$

$$\mathbf{R}_2\delta\mathbf{R}_2 + \delta\mathbf{R}_2\mathbf{R}_2 = \delta\mathbf{R}_2, \tag{6.5.9b}$$

$$\mathbf{R}_1\delta\mathbf{R}_2 + \delta\mathbf{R}_1\mathbf{R}_2 = \mathbf{0}. \tag{6.5.9c}$$

We now consider, following (6.4.23), the equality of the "projected parts" of the two sides of each equation in turn. For example, the $\mathbf{R}_1(\ldots)\mathbf{R}_1$ projection of (6.5.9a) yields a 11-equality $2(\delta\mathbf{R}_1)_{11} = (\delta\mathbf{R}_1)_{11}$, showing that $(\delta\mathbf{R}_1)_{11} = \mathbf{0}$. On the other hand, the 12-projection arising from $\mathbf{R}_1(\ldots)\mathbf{R}_2$ gives only $(\delta\mathbf{R}_1)_{12} = (\delta\mathbf{R}_1)_{12}$, i.e. no constraint. In this way we find that (6.5.9a,b) imply

$$\left.\begin{aligned} \delta\mathbf{R}_1 &= (\delta\mathbf{R}_1)_{13} + (\delta\mathbf{R}_1)_{31} + (\delta\mathbf{R}_1)_{12} + (\delta\mathbf{R}_1)_{21}, \\ \delta\mathbf{R}_2 &= (\delta\mathbf{R}_2)_{23} + (\delta\mathbf{R}_2)_{32} + (\delta\mathbf{R}_2)_{21} + (\delta\mathbf{R}_2)_{12}. \end{aligned}\right\} \tag{6.5.10}$$

The essentially Hermitian character of the matrices effectively reduces the number of independent components, requiring that

$$(\delta\mathbf{R}_1)_{ji} = (\delta\mathbf{R}_1)^\dagger_{ij}, \qquad (\delta\mathbf{R}_2)_{ji} = (\delta\mathbf{R}_2)^\dagger_{ij}. \tag{6.5.11}$$

The last condition (6.5.9c) introduces the further constraint

$$(\delta\mathbf{R}_2)_{12} + (\delta\mathbf{R}_1)_{12} = \mathbf{0}. \tag{6.5.12}$$

On noting that the ij-projection of an arbitrary variation can be written $\mathbf{R}_i\mathbf{M}\mathbf{R}_j$, with an arbitrary $m \times m$ matrix $\mathbf{M}$, and introducing

$$\left.\begin{aligned} \mathbf{x}_1 &= (\delta\mathbf{R}_1)_{13} = \mathbf{R}_1\mathbf{X}_1\mathbf{R}_3, \\ \mathbf{x}_2 &= (\delta\mathbf{R}_2)_{23} = \mathbf{R}_2\mathbf{X}_2\mathbf{R}_3, \\ \mathbf{x} &= (\delta\mathbf{R}_1)_{12} = \mathbf{R}_1\mathbf{X}\mathbf{R}_2, \end{aligned}\right\} \tag{6.5.13}$$

the expressions (6.5.10) become (taking note of (6.5.11) and (6.5.12))

$$\left.\begin{aligned} \delta\mathbf{R}_1 &= (\mathbf{x}_1 + \mathbf{x}_1^\dagger) + (\mathbf{x} + \mathbf{x}^\dagger), \\ \delta\mathbf{R}_2 &= (\mathbf{x}_2 + \mathbf{x}_2^\dagger) - (\mathbf{x} + \mathbf{x}^\dagger). \end{aligned}\right\} \tag{6.5.14}$$

These equations embody, to first order, all the orthonormality constraints on the MOs of the closed-shell, the open-shell and the virtual space; they may be extended to any order (McWeeny, 1961a), but (6.5.14) will be sufficient for present purposes.

The correctly constrained energy variation follows on substituting (6.5.14) in (6.5.7):

$$\delta E^{(1)} = \mathrm{tr}\,[\nu_1\mathbf{F}_1(\mathbf{R}_3\mathbf{X}_1\mathbf{R}_1 + \mathbf{R}_2\mathbf{X}\mathbf{R}_1) + \nu_2\mathbf{F}_2(\mathbf{R}_3\mathbf{X}_2\mathbf{R}_2 - \mathbf{R}_2\mathbf{X}\mathbf{R}_1)] + \text{c.c.},$$

and, on remembering that the trace of a product is invariant against cyclic permutation of the factors, and that $\mathbf{X}_1, \mathbf{X}_2$ and $\mathbf{X}$ are arbitrary, the stationary condition $\delta E^{(1)} = 0$ yields at once

$$\mathbf{R}_1\mathbf{F}_1\mathbf{R}_3 = \mathbf{R}_2\mathbf{F}_2\mathbf{R}_3 = \mathbf{R}_1\mathbf{F}_{(12)}\mathbf{R}_2 = \mathbf{0}, \tag{6.5.15}$$

where we have introduced an intershell matrix

$$\mathbf{F}_{(12)} = \nu_1\mathbf{F}_1 - \nu_2\mathbf{F}_2. \tag{6.5.16}$$

Conditions (6.5.15) are necessary and sufficient to ensure that the energy E in (6.5.3) be stationary.

In generalizing the closed-shell equations, Roothaan (1960) was able to find a *single* Hamiltonian ($\mathbf{F}$, say) whose eigenvectors, obtained on solving an equation analogous to (6.2.20), determined the MOs of *both shells*; his Hamiltonian represents a special case of the one we are about to introduce. To this end (using V to label the virtual space $\mathscr{V}_3$), we consider the matrix

$$\mathbf{F} = \sum_K \mathbf{R}_K\mathbf{F}_K\mathbf{R}_K + \sum_K \alpha_K(\mathbf{R}_K\mathbf{F}_K\mathbf{R}_\mathrm{V} + \mathbf{R}_\mathrm{V}\mathbf{F}_K\mathbf{R}_K)$$

$$+ \sum_{K,L} \beta_{KL}\mathbf{R}_K\mathbf{F}_{(KL)}\mathbf{R}_L \quad (K, L = 1, 2), \tag{6.5.17}$$

where α_K and β_{KL} are arbitrary real parameters. From (6.5.16), $\mathbf{F}_{(LK)} = -\mathbf{F}_{(KL)}$ and the choice $\beta_{LK} = -\beta_{KL}$ thus ensures that $\mathbf{F}$ will be Hermitian. In this case there exists a unitary transformation that diagonalizes $\mathbf{F}$, i.e. we can find columns $\mathbf{c}_i$ such that $\mathbf{c}_i^\dagger\mathbf{F}\mathbf{c}_j = 0$ for $i \neq j$. Suppose that we find these vectors, in the usual way, and put them in ascending energy order so that the first n_1 correspond to the closed-shell MOs, the next n_2 to the open-shell MOs, and the remainder to the virtual orbitals. We then revise $\mathbf{F}$, recalculating the $\mathbf{F}_K$ with the updated $\mathbf{G}$ matrices and also updating the $\mathbf{R}$ matrices in (6.5.17). It is then clear, by construction, that when $\mathbf{c}_i$ and $\mathbf{c}_j$ belong, for example, to shells 1 and 3 (so that $\mathbf{R}_1\mathbf{c}_i = \mathbf{c}_i$, $\mathbf{R}_3\mathbf{c}_j = \mathbf{c}_j$, other projections vanishing)

$$\mathbf{c}_i^\dagger\mathbf{F}\mathbf{c}_j = \alpha_1\mathbf{c}_i^\dagger\mathbf{F}_1\mathbf{c}_j = 0,$$

and consequently, forming $\mathbf{c}_i(\mathbf{c}_i^\dagger\mathbf{F}_1\mathbf{c}_j)\mathbf{c}_j^\dagger$ and summing over all eigenvectors in $\mathcal{V}_1$ and $\mathcal{V}_3$,

$$\mathbf{R}_1\mathbf{F}_1\mathbf{R}_3 = \mathbf{0}.$$

Similarly, it follows term-by-term that *all* the projections in (6.5.15) will vanish; and the stationary-value conditions will thus be satisfied.

There is, of course, no more than in the closed-shell SCF procedure, any guarantee that the proposed iteration will converge; but the Hamiltonian in (6.5.17) possesses a remarkable flexibility, by virtue of the arbitrary parameters it contains, and this flexibility may be exploited in order to improve the convergence characteristics of the process. Before discussing such possibilities, it is convenient to write (6.5.17) in a simpler and more general form.

We first note that the first term in (6.5.17) plays no part in ensuring that the stationary conditions are satisfied; it merely ensures that the eigenvalues of $\mathbf{F}_K$ coincide with energy expectation values,

$$e_i = \langle \phi_i^K | \mathsf{F}_K | \phi_i^K \rangle = \mathbf{c}_i^{K\dagger}\mathbf{F}_K\mathbf{c}_i^K,$$

corresponding to the use of an appropriate effective Hamiltonian in each shell; without this term, all the eigenvalues would be zero and it would be difficult to put the MOs in any physically sensible "energy order".

However, a modification of $\mathbf{F}_K$ will have no effect on the solution—in particular on $\mathbf{R}_K$. If $\mathbf{F}_K$ is replaced by an *arbitrary* Hermitian matrix $\mathbf{d}_K$ then the eigenvectors $\mathbf{c}_i^K$ that span the $\mathbf{R}_K$ subspace will simply undergo a unitary transformation such that $\mathbf{d}_K$ will be brought to diagonal form instead of $\mathbf{F}_K$. We therefore replace the $\mathbf{F}_K$ in the first term of (6.5.17) by $\mathbf{d}_K$.

Secondly, the second sum in (6.5.17) may be included in the third by extending the summation indices to that $K, L = 1, 2, 3$ (thus including the virtual shell) and using a virtual-shell occupation number $\nu_V = 0$; since the extra terms in the third sum will then be

$$\begin{aligned}\sum_K (\beta_{KV}\mathbf{R}_K\mathbf{F}_{(KV)}\mathbf{R}_V + \beta_{VK}\mathbf{R}_V\mathbf{F}_{(VK)}\mathbf{R}_K) \\ = \sum_K (\beta_{KV}\nu_K\mathbf{R}_K\mathbf{F}_K\mathbf{R}_V + \beta_{VK}(-\nu_K)\mathbf{R}_V\mathbf{F}_K\mathbf{R}_K) \\ = \sum_K \nu_K\beta_{KV}(\mathbf{R}_K\mathbf{F}_K\mathbf{R}_V + \mathbf{R}_V\mathbf{F}_K\mathbf{R}_K),\end{aligned}$$

thus yielding the second sum. The most general effective Hamiltonian is thus

$$\boxed{\mathbf{F} = \sum_K \mathbf{R}_K\mathbf{d}_K\mathbf{R}_K + \sum_{K,L} \beta_{KL}\mathbf{R}_K\mathbf{F}_{(KL)}\mathbf{R}_L \quad (\beta_{KL} = -\beta_{LK}),} \tag{6.5.18}$$

where the intershell matrix is

$$\mathbf{F}_{(KL)} = \nu_K\mathbf{F}_K - \nu_L\mathbf{F}_L. \tag{6.5.19}$$

In this form, the method developed in this section is applicable (McWeeny, 1975) without change to a system described by *any number* of fully or partially occupied shells, the only requirement being that the energy expression can be cast in the form (6.5.3) with one term for each shell instead of only two.

In the practical implementation of the single-Hamiltonian method a useful choice of the arbitrary matrices $\mathbf{d}_K$ is

$$\mathbf{d}_K = \mathbf{F}_K + \alpha_K\mathbf{1}, \tag{6.5.20}$$

where the parameter α_K, first introduced in a more limited context by Guest and Saunders (1974), is called a "level shifter". With $\alpha_K = 0$, the K-shell eigenvalues have the interpretation noted above; they are orbital energies for an electron in the effective field contained in $\mathbf{F}_K$. But when $\alpha_K \neq 0$ the K-shell eigenvalues of $\mathbf{F}$ are all displaced upwards by an amount α_K. By increasing the separation of the eigenvalues of different shells in this way, it is possible not only to avoid confusion between shells, during the iteration, but often to greatly improve the convergence of the process.

The parameters β_{KL}, which have been called "damp factors" (Guest and Saunders, 1974), also have a bearing on convergence (Firsht and Pickup, 1977), and may be used to control the approach to various kinds of stationary point (for example, those representing excited states), but their application has not yet been fully explored.

Finally, it is important to note that although, for convenience, all equations have been derived using an orthonormal basis, the final results also apply to an arbitrary basis—with two small changes: the eigenvectors must be calculated using the general eigenvalue equation (6.2.20); and $\mathbf{F}$ in (6.5.18) must be replaced by $\mathbf{SFS}$. All quantities in (6.5.18), including the 1- and 2-electron integrals, are then defined over the *non*-orthonormal basis functions.

The simple open-shell system with parallel coupling of all electron spins in the singly occupied orbitals has been treated at some length because it is the prototype for the discussion of a number of other open-shell systems, to which the same solution method can be applied without change. We now consider a few important cases.

6.6 OTHER OPEN-SHELL SITUATIONS. ENSEMBLE AVERAGING

So far in this chapter we have been concerned with electronic states represented by 1-determinant wavefunctions. Such functions are not strictly appropriate, however, as became apparent in Section 6.3, except in single-configuration situations of high symmetry; the wavefunction should be *spatially* non-degenerate (belonging, in group-theoretical terms, to a unidimensional representation of the molecular point group), and, with an open shell, the *spin* symmetry should be that of the state of highest multiplicity (all spins parallel-coupled) as in the last section. In other cases the wavefunction must be represented by a linear combination of several—possibly many—determinants: to optimize the orbitals in such a function is more difficult, and the study of multiconfiguration (MC) SCF methods is deferred to a separate chapter. In some cases, however, the formalism and methodology used so far is still applicable; indeed, this will be so whenever the energy expression can be cast into a form similar to (6.5.3), possibly even with several shells instead of just two. In other words, from a practical standpoint, what matters is the availability of a particular type of energy expression—irrespective of the nature of the wavefunction. The following examples make this point clear.

Some special 2-shell states (Roothaan)

Roothaan (1960) first noted that certain states of atoms and linear molecules, even those requiring many-determinant wavefunctions, yield energy expressions that can be put in the form

$$E = \nu_1\Big[\sum_A \langle\phi_A|\,\mathsf{h}\,|\phi_A\rangle + \sum_{A,B}(\langle\phi_A\phi_B|\,g\,|\phi_A\phi_B\rangle - \tfrac{1}{2}\langle\phi_A\phi_B|\,g\,|\phi_B\phi_A\rangle)\Big]$$
$$+ \nu_2\Big[\sum_U \langle\phi_U|\,\mathsf{h}\,|\phi_U\rangle + \sum_{U,V}(a\langle\phi_U\phi_V|\,g\,|\phi_U\phi_V\rangle - b\langle\phi_U\phi_V|\,g\,|\phi_V\phi_U\rangle)\Big]$$
$$+ \nu_i\nu_2\Big[\sum_{A,U}(\langle\phi_A\phi_U|\,g\,|\phi_A\phi_U\rangle - \tfrac{1}{2}\langle\phi_A\phi_U|\,g\,|\phi_U\phi_A\rangle)\Big]. \tag{6.6.1}$$

The similarity between (6.6.1) and (6.5.2) is immediately apparent. In (6.6.1) ν_2 may be thought of as a "fractional occupation number" $0 \leqslant \nu_2 \leqslant 2$, and the constants a and b depend on the particular state considered.

Values of ν_2, a and b for various states may be obtained by setting up the appropriate combinations of determinants and writing down the expressions for the energy, and have been listed for many states by Roothaan (1960).‡ The SCF equations for all functions that give energy expressions of the form (6.6.1) can be obtained by the procedure developed in Section 6.5, with only trivial modifications. In fact, with the revised definition of ν_2, (6.5.3) and the analysis that follows may be adopted in its entirety, provided that $\mathbf{G}'$ in (6.5.4) is replaced by

$$\mathbf{G}'(\mathbf{R}) = a\mathbf{J}(\mathbf{R}) - \tfrac{1}{2}b\mathbf{K}(\mathbf{R}). \tag{6.6.2}$$

The iterative introduction of self-consistency, using the Hamiltonian (6.5.18), is unaffected by these minor changes.

The theoretical basis of this approach is not immediately obvious: what, for example, is the meaning of the "fractional occupation numbers"? To establish its status, let us take a simple example, the boron atom with electron configuration $1s^2\,2s^2\,2p^1$ and a ^2P ground state. If we use a 1-determinant wavefunction with $A = 1s$, $B = 2s$, $W = 2p_z$ then the energy expression will be

$$E = E_{\text{closed-shell}} + \langle W|\,\mathsf{h}\,|W\rangle + (\langle AW|\,g\,|AW\rangle - \langle AW|\,g\,|WA\rangle) \tag{6.6.3}$$

‡ Instead of ν_2, Roothaan uses an occupation number per *spin*-orbital, f, $0 \leqslant f \leqslant 1$. Thus $\nu_2 = 2f$ should be employed in the present formulation. See also Roothaan and Bagus (1964).

and the electron density will be

$$P = P_{\text{closed-shell}} + W(\boldsymbol{r})W^*(\boldsymbol{r}). \tag{6.6.4}$$

Since the open-shell term in P does not possess spherical symmetry, the effective Hamiltonian will contain a non-spherical potential; and as a result, even with initial orbitals of true central-field form (i.e. with spherical-harmonic angle dependence), the first cycle of an SCF iteration will destroy the symmetry properties of the orbitals—the solutions that give an improved energy will not be of pure s and p type but will be mixtures. This is a second example of a "symmetry-breaking" situation, akin to the spin polarization encountered in the UHF method. The resultant many-electron wavefunction will also lose the symmetry characteristic of a true spectroscopic state; there will be a *spatial* polarization of the $1s^2\,2s^2$ core and the predicted ground state will no longer be of pure P type, just as in the UHF calculation there will be a *spin* polarization and the exact spin multiplicity of the many-electron state will be lost. Of course, the *many*-electron Hamiltonian *does* possess spherical symmetry (i.e. invariance under rotations around the nucleus), and the reason for the symmetry breaking lies at the level of the one-electron (i.e. IPM-type) *model*—the effective field in the 1-electron Hamiltonian is a fiction rather than a reality.

In order to conserve symmetry in a "natural" way, without introducing complicated constraints on the individual orbitals, we may first observe that the three components of a P state, with open-shell orbitals U, V, $W = 2p_x, 2p_y, 2p_z$ respectively, must all possess exactly the same energy; and that a normalized combination $\Psi = a\Psi_x + b\Psi_y + c\Psi_z$ (subscripts indicating choice of the 2p orbital) would yield an identical energy

$$E = a^2\langle \Psi_x|\mathsf{H}|\Psi_x\rangle + b^2\langle \Psi_y|\mathsf{H}|\Psi_y\rangle + c^2\langle \Psi_z|\mathsf{H}|\Psi_z\rangle = \tfrac{1}{3}(E_U + E_V + E_W),$$

since off-diagonal terms are zero by symmetry, diagonal terms are equal, and $a^2 + b^2 + c^2 = 1$. On using this expression for the energy, however, we obtain, instead of (6.6.3),

$$E = E_{\text{closed-shell}} + \tfrac{1}{3}[(U\text{-term}) + (V\text{-term}) + (W\text{-term})], \tag{6.6.5}$$

where the W-term appears in (6.6.3) and the others are defined similarly. The energy expression is now completely symmetrical; and so will be the Hartree–Fock Hamiltonian—since this is determined by the corresponding density function P, which is now

$$P = 2A^2 + 2B^2 + \tfrac{1}{3}(U^2 + V^2 + W^2), \tag{6.6.6}$$

in which the three p orbitals, with *fractional occupancy* $\frac{1}{3}$, give a density that is spherically symmetrical.‡

What amounts to the same procedure, often referred to as "spherical averaging" when applied to atoms, was used in Hartree's early work on atomic structures. It is evidently easily generalized to molecules provided that the states involved are non-mixing (i.e. give zero off-diagonal elements of H), leading with suitable averaging to a density function P with the full symmetry of the molecule.

A further generalization of the above procedure is possible, even when the off-diagonal elements of H are *not* zero and a mixture of determinants would lead to a lowering of the energy (and to an energy functional of less tractable form). In this case it is expedient, with slight loss of accuracy, to consider an "ensemble average" in which the alternative 1-determinant energy expressions are combined with appropriate weight factors $w_1, w_2, \ldots$ (cf. the derivation of (6.6.5)) to yield a new expression of the desired form. Physically, this expression will correspond to the average energy of an ensemble of independent systems, with fractional numbers $w_1, w_2, \ldots$ in the various states considered. The next example illustrates the use of this approach.

Configurational averaging

First let us consider a 2-shell system with n_1 doubly occupied orbitals and n_2 orbitals available in the open-shell part of the configuration; n_2 will be the number of orbitals when *all members* of any partially occupied degenerate sets are included. With only n ($<2n_2$) electrons in the open shell, certain linear combinations could be set up to describe the possible "states of the configuration". Instead, however, we now consider an ensemble average of the energies of these states. To see why it is useful to do so, we consider another example—the carbon atom. From the discussion in Section 3.4, it is clear that in the first approximation (i.e. neglect of electron interaction) the carbon atom can be described by the orbital configuration $1s^2\,2s^2\,2p^2$. It also appears that a single determinant constructed from these orbitals would not in general have a definite symmetry. In fact, the argument in Problem 3.10 shows that it is possible to construct functions representing ^1S, ^3P and ^1D states. These "spectroscopic states" of the configuration p^2 are generally represented as linear combinations of determinants. The optimum orbitals for each of these states could in this case be determined by the procedure just discussed, but if this were done each state would have its own set of optimum

‡ This is a consequence of Unsöld's theorem (see e.g. Pauling and Wilson, 1935, p. 150).

orbitals, and different sets would not be mutually orthogonal. Thus when calculating quantities like transition probabilities we should have to use the complicated formulae appropriate to matrix elements between non-orthogonal functions (Section 3.3). In many cases the optimum orbitals and the state functions constructed from them would also lose their pure symmetry properties (as in the UHF case) and would no longer be completely satisfactory for describing spectroscopic states. It would clearly be more convenient for many purposes if we could treat the various states in some average way and so find a single set of orbitals appropriate to all the states, even though they may give a slightly inferior energy value for any single state. Even when the incompletely filled set of orbitals is extended to include sets that are not strictly degenerate, the ensemble-averaging approach can often lead to a good "compromise set" of near-optimal orbitals for describing a number of close-lying spectroscopic states. We therefore formulate the corresponding SCF equations for a single open shell defined by an arbitrary set of orbitals, and then extend the approach to include any number of open and closed shells.

Let us use n_2 for the number of open-shell orbitals, now including *all* those that belong to the multidimensional representation; and let us use n for the number of electrons in the open shell. Thus, for an atomic p^2 configuration, $n_2 = 3$ since there are three p orbitals and $n = 2$. We now use the fact that the *sum* of the energies obtained by solving a secular problem for the functions Φ_κ of an orbital configuration is equal to the sum of the diagonal matrix elements $H_{\kappa\kappa}$. If n' $(=2n_2)$ is the number of open-shell *spin*-orbitals then there will be $n'!/n!(n'-n)!$ determinantal functions (which we take as the Φ_κ) for the configuration, and it follows easily (Slater, 1960) that for the open shell by itself the mean energy would be, after performing the spin integrations,

$$\frac{n}{n'}\sum_U 2\langle U|\,\mathrm{h}\,|U\rangle + \frac{n(n-1)}{2n'(n'-1)}\sum_{U,V}(4\langle UV|\,g\,|UV\rangle - 2\langle UV|\,g\,|VU\rangle).$$

When the closed shell is taken into account we obtain

$$\begin{aligned}E_{\text{av}} = {} & 2\sum_A \langle A|\,\mathrm{h}\,|A\rangle + \sum_{A,B}(2\langle AB|\,g\,|AB\rangle - \langle AB|\,g\,|BA\rangle) \\ & + \frac{n}{n'}\left[2\sum_U \langle U|\,\mathrm{h}\,|U\rangle + \frac{n-1}{n'-1}\sum_{U,V}(2\langle UV|\,g\,|UV\rangle - \langle UV|\,g\,|VU\rangle)\right] \\ & + \frac{2n}{n'}\left[\sum_{A,V}(2\langle AU|\,g\,|AU\rangle - \langle AU|\,g\,|UA\rangle)\right]. \qquad (6.6.7)\end{aligned}$$

This may be written in a form resembling (6.5.2) on introducing $\nu_2 = n/n'$ as the *fractional* occupation number of the open-shell orbitals;

if there were just sufficient electrons to fill each orbital singly we should have $\nu_2 = 1$, but more generally ν_2 may take a range of values between 0 and 2. The final result is then formally identical with (6.5.3):

$$E_{\text{av}} = \nu_1 \operatorname{tr} \mathbf{R}_1(\mathbf{h} + \tfrac{1}{2}\mathbf{G}_1^{\text{av}}) + \nu_2 \operatorname{tr} \mathbf{R}_2(\mathbf{h} + \tfrac{1}{2}\mathbf{G}_2^{\text{av}}). \tag{6.6.8}$$

The only difference is that

$$\left.\begin{aligned} \mathbf{G}_1^{\text{av}} &= \mathbf{G}(\nu_1\mathbf{R}_1) + \mathbf{G}(\nu_2\mathbf{R}_2), \\ \mathbf{G}_2^{\text{av}} &= \mathbf{G}(\nu_1\mathbf{R}_1) + \mathbf{G}(\nu_2'\mathbf{R}_2), \end{aligned}\right\} \tag{6.6.9}$$

in which the open-shell coulomb-exchange energy is described by a $\mathbf{G}$ term, instead of the $\mathbf{G}'$ appropriate to parallel coupling of the spins, and by an "effective" occupation number

$$\nu_2' = \frac{2(n-1)}{2n_2 - 1}.$$

This ensures that there is no contribution when there is only one electron in the open shell, while $\nu_2' \to \nu_2$ for sufficiently large numbers. The minimization of E_{av}, subject to the orthonormality constraints embodied in (6.5.6), may be performed exactly as in Section 6.5, using a correspondingly modified Hamiltonian (6.5.18). This Hamiltonian will be a total symmetric operator, whose eigenfunctions will be symmetry orbitals at all stages of the iteration; functions of definite symmetry species, representing the spectroscopic states of the configuration, may then be set up without difficulty.

There is nothing in the above derivation that requires the open-shell orbitals to be degenerate. For example, the whole group of states arising from promotion of up to four electrons from the top two filled MOs of a closed-shell system into the lowest pair of empty MOs may be dealt with by this method on putting $n = 4$, $n_2 = 4$. In this way it is possible to obtain MOs that are better adapted to the description of excited states than are the virtual MOs of a closed-shell calculation, whose defects for this purpose have been noted already (p. 172).

The generalization to many-shell systems (McWeeny, 1974) is straightforward. We suppose that there are m_K available orbitals ($2m_K$ spin-orbitals) and n_K electrons in shell K, and obtain the average-energy expression (cf. (6.6.7))

$$E_{\text{av}} = \sum_K \left(\frac{n_K}{m_K} H_K + \frac{n_K(n_K - 1)}{m_K(2m_K - 1)} G_{KK} \right) + \tfrac{1}{2} \sum_{K,L}{}' \frac{n_k}{m_k} \frac{n_L}{m_L} G_{KL}, \tag{6.6.10}$$

where, with a notation like that used in (6.1.4),

$$\left.\begin{aligned} H_K = \sum_i \langle \phi_i^K | \mathsf{h} | \phi_i^K \rangle, \quad G_{KL} = \sum_{i,j} \langle \phi_i^K \phi_j^L \| \phi_i^K \phi_j^L \rangle, \\ \langle \phi_i^K \phi_j^L \| \phi_i^K \phi_j^L \rangle = \langle \phi_i^K \phi_j^L | g | \phi_i^K \phi_j^L \rangle - \tfrac{1}{2} \langle \phi_i^K \phi_j^L | g | \phi_j^L \phi_i^K \rangle. \end{aligned}\right\} \tag{6.6.11}$$

The first summation in (6.6.10) represents the energy of the shell-K electrons alone in the field of the nuclei; the double sum contains the intershell coulomb-exchange interactions with other shells. The averaging used gives equal weight to all the states that can be formed by linear combination of determinants in which a fixed number n_K of electrons is assigned to each set of $2m_K$ spin-orbitals.

When all orbitals are expressed in terms of a common set of basis functions, with

$$\boldsymbol{\phi}^K = \boldsymbol{\chi} \mathbf{T}_K \quad (\text{all } K), \tag{6.6.12}$$

the analogue of (6.5.3) becomes

$$E_{\text{av}} = \sum_K \nu_K \operatorname{tr} \mathbf{R}_K (\mathbf{h} + \tfrac{1}{2} \mathbf{G}_K), \tag{6.6.13}$$

where

$$\mathbf{G}_K = \mathbf{G}(\nu'_K \mathbf{R}_K) + \sum_{L(\neq K)} \mathbf{G}(\nu_L \mathbf{R}_L). \tag{6.6.14}$$

Here the **R** and **G** matrices are defined as usual, while the fractional occupation numbers are

$$\nu_K = \frac{n_K}{m_K}, \qquad \nu'_K = \frac{2(n_K - 1)}{2m_K - 1}, \tag{6.6.15}$$

the latter having a "reduced" value that vanishes for $n_K = 1$, thus ensuring automatically that a shell with only one electron contains no self-interaction term.

The first-order variation of the energy takes the usual form (cf. (6.5.7) and (6.5.8))

$$\delta E_{\text{av}} = \sum_K \nu_K \operatorname{tr} \mathbf{F}_K \mathbf{R}_K \quad (\mathbf{F}_K = \mathbf{h} + \mathbf{G}_K), \tag{6.6.16}$$

and the stationary-value conditions, with intra- and intershell orthonormality constraints included, follow exactly as in Section 6.5. These conditions are satisfied when the orbitals of *all* shells are determined from the eigenvectors (used in forming the $\mathbf{T}_K$ in (6.6.12)) of the single Hamiltonian matrix defined in (6.5.18) and (6.5.19). As usual (cf.

p. 187), the choice of basis (orthonormal or non-orthonormal) is immaterial provided that the eigenvalue equation includes the overlap matrix and **F** is replaced by **SFS**.

The many-shell generalization has a considerable field of application. In discussing the excited states encountered in photoelectron spectroscopy, for example, electrons may be missing from one or more inner shells as a result of ionization or transfer into an incompletely filled valence shell. Even though upper-bound theorems do not apply to the resultant highly excited states, the calculated energies (Firsht and McWeeny, 1976) appear to be physically satisfactory and to take full account of the large orbital "relaxation" that occurs. Thus, by calculating the stationary values of an appropriate energy functional, it is possible to study with considerable precision electron ionization, attachment and transfer processes, which are extremely difficult to handle by CI and related methods.

6.7 STATES OF GIVEN SPIN

In Section 6.5 the closed shell contained g doubly occupied orbitals while the open shell consisted of $N-2g$ different orbitals, all with spin factor α, leading to a state of maximum multiplicity with $S=M=\frac{1}{2}(N-2g)=S_{\text{max}}$. We now consider the more general problem in which the n open-shell spins are coupled to give any allowed resultant spin ($S_{\text{max}}, S_{\text{max}}-1, \ldots, 0$ (N even) or $\frac{1}{2}$ (N odd)), and we wish to optimize the orbital forms for any chosen S.

This problem may be approached using tableaux (Section 4.4). For any given shape, the set of standard tableaux will indicate the coupling schemes that lead to all f_S^n linearly independent spin functions $\Theta_i^{(S)}$ of given S (and any chosen M). On adding g 2-cell columns to the left of each possible shape, we may also include the closed shell, thus obtaining a very much larger number of standard tableaux (f_S^N with $N=2g+n$). This number will be effectively reduced when we take account of the symmetry of the spatial wavefunction, resulting from the presence of the doubly occupied orbitals; but for the moment we ignore any peculiarities of the spatial wavefunction, whose form will be assumed to be arbitrary.

To construct an antisymmetric wavefunction for any given spatial factor Φ, it is sufficient to attach a spin function $\Theta_i^{(S)}$ and antisymmetrize, obtaining

$$\begin{aligned}\Psi_i^{(S)} &= \sum_{\mathsf{P}} \epsilon_{\mathsf{P}}(\mathsf{P}\Phi)(\mathsf{P}\Theta_i^{(S)}) \\ &= \sum_{\mathsf{P}} \epsilon_{\mathsf{P}} \sum_j D_S(\mathsf{P})_{ji}(\mathsf{P}\Phi)\Theta_j^{(S)} = \sum_j \Phi_{j(i)}^{(\tilde{S})}\Theta_j^{(S)}. \qquad (6.7.1)\end{aligned}$$

This is not the exact wavefunction in (4.3.5) because the spatial factor is not an exact solution of the spinless Schrödinger equation; $\Phi^{(\tilde{S})}_{j(i)}$ is in fact one of a set of functions that carry the dual representation $D_{\tilde{S}}$, and there are d_S such sets, one for each of the d_S possible values of the index i. The d_S independent functions $\Psi^{(S)}_i$ obtained in (6.7.1) should therefore be allowed to mix, in setting up a variational approximation, and the mixing coefficients will not be determined by a symmetry but will follow as usual from the secular equations

$$\mathbf{HC} = E\mathbf{MC}. \tag{6.7.2}$$

The matrix elements reduce as follows (remembering that the antisymmetrizer A is idempotent):

$$\begin{aligned} H_{ij} &= \langle \Psi^{(\tilde{S})}_i | \mathsf{H} | \Psi^{(\tilde{S})}_j \rangle = \sum_{\mathsf{P}} \epsilon_{\mathsf{P}} \langle \Phi\Theta^{(S)}_i | \mathsf{H} | (\mathsf{P}\Phi)(\mathsf{P}\Theta^{(S)}_j \rangle \\ &= \sum_{\mathsf{P}} \epsilon_{\mathsf{P}} \langle \Phi | \mathsf{H} | \mathsf{P}\Phi \rangle \langle \Theta^{(S)}_i | \mathsf{P} | \Theta^{(S)}_j \rangle, \end{aligned} \tag{6.7.3}$$

where in the last step space and spin factors have been separated. The spin matrix element, multiplied by ϵ_{P}, is an element of $\mathbf{D}_{\tilde{S}}(\mathsf{P})$, and the full Hamiltonian matrix is thus

$$\mathbf{H} = \sum_{P} H_{\mathsf{P}} \mathbf{V}(\mathsf{P}), \tag{6.7.4}$$

where

$$H_{\mathsf{P}} = \langle \Phi | \mathsf{H} | \mathsf{P}\Phi \rangle, \qquad \mathbf{V}(\mathsf{P}) = \mathbf{D}_{\tilde{S}}(\mathsf{P}) = \epsilon_{\mathsf{P}} \mathbf{D}_S(\mathsf{P}). \tag{6.7.5}$$

Since H contains only 1- and 2-body operators, only *transpositions* are needed in evaluating the quantities H_{P}. The elements of the matrix **M** follow on omitting the operator H in (6.7.5); **M** is thus diagonal: $M_{ik} = \delta_{ik} M_{ii}$.

In evaluating the elements of **H** and **M**, the special form of the spatial function Φ may now be recognized. By assuming that electrons 1 and 2 occupy the first orbital, 3 and 4 the second, and so on, we impose a symmetry on the spatial function Φ. If Φ is symmetric under transposition (12), it will be necessary to ensure that the spin factor is *anti*symmetric under (12); this must be so for each doubly occupied orbital, and the first g columns of any Young tableau describing an associated spin eigenfunction will thus read

1	3	5	...	$2g-1$
2	4	6	...	$2g$

Tableaux containing this common "fixed part", followed by an arbitrary 2-row pattern for the open-shell electrons, will be called "acceptable": in terms of the branching diagram (Fig. 4.3), for the example $g=2$, $n=5$, they correspond to coupling schemes such as

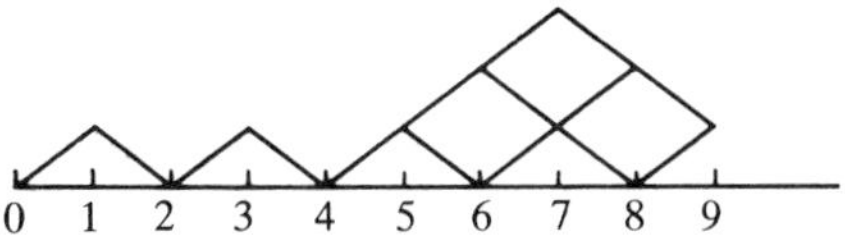

The number of *acceptable* standard tableaux is thus determined by n rather than N, and the dimension of the matrices $\mathbf{V}(\mathsf{P})$ in (6.7.4) will be relatively small. The remaining standard tableaux may be listed after the acceptable tableaux but in fact are not required,‡ since all that we need is the effect of the permutations on the acceptable tableaux functions, which carry the representation D_S whose dual appears in (6.7.5). The dimension of these irreps is f_S^n instead of f_S^N, and as long as we consider only permutations of the n variables labelled $2g+1, 2g+2, \ldots, N$, the matrices $\mathbf{V}(\mathsf{P})$ will be those of the standard irrep D_S, which may be assumed known. The only difficulty is that P in (6.7.4) runs over *all* $N!$ permutations of the group S_N; so transpositions that mix open- and closed-shell indices must also be considered.

To proceed, we recognize that the presence of the g doubly occupied orbitals in Φ means that Φ, and consequently H_P, is invariant under a subgroup (G, say) of S_N that consists of all 2^g permutations $(12)^{\nu_1}$ $(34)^{\nu_2}\ldots(2g-1, 2g)^{\nu_g}$ with $\nu_i = 0, 1$. Let us therefore write the general permutation P in the form

$$\mathsf{P} = \mathsf{P}_g \mathsf{Q}, \tag{6.7.6}$$

where Q is one of the 2^g permutations of the subgroup G, while P_g is one of the $N!/2^g$ generators that, combined with each Q in turn, lead to the full group of $N!$ permutations. Since Q multiplies any of the spin functions by $(-1)^q$, q being the number of transpositions in Q, $\mathbf{D}_S(\mathsf{Q})$ must be $(-1)^q$ times the unit matrix; consequently (multiplying again by $(-1)^q$) $\mathbf{V}(\mathsf{Q})$ will be the unit matrix itself. Thus H_{ij} in (6.7.3) will reduce as follows:

$$H_{ij} = \sum_{\mathsf{P}_g,\mathsf{Q}} \langle \Phi | \mathsf{H} | \mathsf{P}_g \mathsf{Q} \Phi \rangle V(\mathsf{P}_g\mathsf{Q})_{ij} = \sum_{\mathsf{P}_g} 2^g \langle \Phi | \mathsf{H} | \mathsf{P}_g \Phi \rangle \sum_k V(\mathsf{P}_g)_{ik} V(\mathsf{Q})_{kj},$$

‡ They are "unacceptable" in the sense that they lead to space–spin functions that vanish identically (Pauncz, 1979).

and hence

$$H_{ij} = 2^g \sum_{\mathsf{P}_g} H_{\mathsf{P}_g} V(\mathsf{P}_g)_{ij}, \tag{6.7.7}$$

which (cf. (6.7.3)) is now a sum over only $N!/2^g$ permutations. The matrix **M** reduces in the same way to give

$$M_{ij} = 2^g \sum_{\mathsf{P}_g} \langle \Phi | \, \mathsf{P}_g \, | \Phi \rangle V(\mathsf{P}_g)_{ij} = 2^g \delta_{ij}, \tag{6.7.8}$$

where orthonormal orbitals have been assumed and the only contribution is thus from the identity element. In what follows we may therefore renormalize the functions $\Psi_i^{(S)}$ in (6.7.1) so that the factors 2^g in the last two equations can be discarded, **M** becoming the unit matrix.

For a Hamiltonian of the usual form the main term in (6.7.7) arises when P_g is the identity. In this case (on using r, s as general indices, i,j for those of the closed shell, and p, q for those of the open shell) we find at once

$$\begin{aligned}\left\langle \Phi \middle| \left(\sum_r \mathsf{h}(r) + \sum_{r<s} g(r,s) \right) \middle| \Phi \right\rangle &= 2 \sum_i \langle \phi_i | \, \mathsf{h} \, | \phi_i \rangle + \sum_p \langle \phi_p | \, \mathsf{h} \, | \phi_p \rangle \\ &\quad + 4 \sum_{i<j} \langle \phi_i \phi_j | \, g \, | \phi_i \phi_j \rangle + \sum_i \langle \phi_i \phi_i | \, g \, | \phi_i \phi_i \rangle \\ &\quad + 2 \sum_{i,p} \langle \phi_i \phi_p | \, g \, | \phi_i \phi_p \rangle \\ &\quad + \sum_{p<q} \langle \phi_p \phi_q | \, g \, | \phi_p \phi_q \rangle. \end{aligned} \tag{6.7.9}$$

The other non-zero terms arise when P_g is one of three types of transposition: (ij), (ip) or (pq). The contributions to the sum in (6.7.7) arise as follows.

(i) (pq) *type*:

$$\langle \Phi | \, \mathsf{H} \, | \mathsf{P}_{pq} \Phi \rangle V(\mathsf{P}_{pq})_{kl} = \langle \phi_p \phi_q | \, g \, | \phi_q \phi_p \rangle (-1) D_S(\mathsf{P}_{pq})_{kl},$$

where the two-electron integral comes from the $g(p, q)$ term in H, while S corresponds to the Young shape with the "fixed part" omitted. The representation matrices have been tabulated (Kotani *et al.*, 1955) for up to 10 open-shell electrons.

(ii) (ip) *type*: The contributions occur in pairs: using i and i' for electrons in the doubly occupied orbital ϕ_i, transpositions (ip) and $(i'p)$ lead to terms with the same 2-electron integral. The final result (see Problem 6.18) is

$$\langle \Phi | \, \mathsf{H} \, | \mathsf{P}_{ip} \Phi \rangle [V(\mathsf{P}_{ip})_{kl} + V(\mathsf{P}_{i'p})_{kl}] = \langle \phi_i \phi_p | \, g \, | \phi_p \phi_i \rangle \times (-1)$$

—which is to be summed in (6.7.7) over all i and all p.

(iii) (ij) *type*: As in (ii), the same integral appears more than once and terms may be collected. It is then found (Problem 6.18) that

$$\langle \Phi | \mathsf{H} | \mathsf{P}_g \Phi \rangle V(\Sigma \mathsf{P}_g)_{kl} = -2 \langle \phi_i \phi_j | g | \phi_j \phi_i \rangle \delta_{kl},$$

the sum being over four transpositions (ij), $(i'j)$, (ij') and $(i'j')$. These contributions are to be summed for all $i < j$.

On collecting the above results and adding (6.7.9), we obtain

$$\begin{aligned} H_{kl} = \delta_{ij} \Big\{ & 2 \sum_i \langle \phi_i | \mathsf{h} | \phi_i \rangle + \sum_i \langle \phi_i \phi_i | g | \phi_i \phi_i \rangle + 4 \sum_{i<j} \langle \phi_i \phi_j | g | \phi_i \phi_j \rangle \\ & + \sum_p \langle \phi_p | \mathsf{h} | \phi_p \rangle + 2 \sum_{i,p} \langle \phi_i \phi_p | g | \phi_i \phi_p \rangle + \sum_{p<q} \langle q_p \phi_q | g | \phi_p \phi_q \rangle \\ & - \sum_p \langle \phi_i \phi_p | g | \phi_p \phi_i \rangle - 2 \sum_{i<j} \langle \phi_i \phi_j | g | \phi_j \phi_i \rangle \Big\} \\ & - \sum_{p<q} D_S(\mathsf{P}_{pq})_{kl} \langle \phi_p \phi_q | g | \phi_q \phi_p \rangle, \end{aligned} \qquad (6.7.10)$$

where only the last term is dependent on spin coupling.

The orbital optimization now starts from a more general energy functional than in earlier examples. In fact, the normalized solutions of (6.7.2) give

$$E(\Psi) = \sum_{k,l} C_l C_k^* H_{kl}, \qquad (6.7.11)$$

where different sets of expansion coefficients (i.e. eigenvectors of $\mathbf{H}$) represent alternative states of given spin. Wavefunctions determined in this way may be called "spin-optimized" since they involve the mixing of all linearly independent spin functions of given S and M. For *fixed* orbitals, the optimized mixing coefficients are contained in the eigenvectors of $\mathbf{H}$, the lowest-energy eigenvector giving the lowest spin state of given S. On inserting (6.7.10) in (6.7.11), the energy expression becomes (adding a superscipt S to indicate spin value)

$$E^S = E_0 + \sum_{p<q} A_{pq}^S \langle \phi_p \phi_q | g | \phi_q \phi_p \rangle, \qquad (6.7.12)$$

where E_0 stands for the curly-bracket term in (6.7.10), which is independent of spin coupling, while

$$A_{pq}^S = -\sum_{k,l} C_k^* D_S(\mathsf{P}_{pq})_{kl} C_l. \qquad (6.7.13)$$

In order to optimize the *orbitals,* we now wish to put (6.7.12) in a matrix form analogous to (6.5.3).

By suitably grouping closed- and open-shell terms, and expressing all integrals in terms of integrals over basis functions, we obtain an expression of the desired form; but now every open-shell orbital must be regarded as an independent "shell" *in itself.* Thus

$$E^S = 2\,\mathrm{tr}\,\mathbf{R}_c(\mathbf{h} + \tfrac{1}{2}\mathbf{G}_c) + \sum_p \mathrm{tr}\,\mathbf{R}_p(\mathbf{h} + \tfrac{1}{2}\mathbf{G}_p^S), \qquad (6.7.14)$$

where $\mathbf{R}_c$ is the usual closed-shell R-matrix, while $\mathbf{R}_p = \mathbf{c}_p\mathbf{c}_p^\dagger$ is the single-orbital density matrix for ϕ_p. The electron-interaction matrices are

$$\left.\begin{aligned} \mathbf{G}_c &= \mathbf{G}(2\mathbf{R}_c) + \sum_p \mathbf{G}(\mathbf{R}_p), \\ \mathbf{G}_p^S &= \mathbf{G}(2\mathbf{R}_c) + \sum_{q(\neq p)} [\mathbf{J}(\mathbf{R}_q) + A_{pq}^S\mathbf{K}(\mathbf{R}_q)], \end{aligned}\right\} \qquad (6.7.15)$$

and the effective Hamiltonian for every open shell thus depends, through the weight factors determined in (6.7.13), on the mixing coefficients (C_i) of the linearly independent spin-coupled functions. Since (6.7.14) is of the general form (6.5.3), all orbitals may be optimized by solving a single eigenvalue equation, with a Hamiltonian of type (6.5.18). After such an optimization, it will normally be necessary to revise the mixing coefficients (by obtaining eigenvectors of the recalculated matrix **H**) and to continue the process until self-consistency is achieved. This type of "two-step" optimization (of expansion coefficients and orbitals, in turn) is a common procedure in the multiconfiguration SCF methods discussed in Chapter 8.

It is possible to avoid the need for repeated orbital optimization, and also the need for explicit knowledge of the representation matrices, by various types of ensemble averaging. Thus by averaging over all states of given S we obtain a mean value

$$\overline{A_{pq}^S} = -\chi_S(\mathsf{P}_{pq})/d_S, \qquad (6.7.16)$$

where $\chi_S(\mathsf{P}_{pq})$ is the *character* of the single transpositions (such as (12)), and this single value may then be used in (6.7.15). With 5 electrons in the open shell, as in the example used earlier, the corresponding weight of the open-shell exchange terms is

$$\overline{A_{pq}^S} = \begin{cases} -1 & (S = \tfrac{5}{2}), \\ -\tfrac{1}{2} & (S = \tfrac{3}{2}), \\ -\tfrac{1}{5} & (S = \tfrac{1}{2}), \end{cases}$$

while averaging over all 32 spin states gives the value $\overline{A^S_{pq}} = -\frac{1}{2}$. More generally, the characters and the dimensions of the irreps may be obtained from standard formulae (see e.g. Coleman, 1968), and the mean value (6.7.16) may thus be evaluated easily for any number of open-shell orbitals.

6.8 PHYSICAL SIGNIFICANCE OF MOs. LOCALIZATION

When the SCF methods of the preceding sections are applied to many-electron polyatomic molecules, the MOs that arise by solution of the Hartree–Fock eigenvalue problem are in general completely delocalized, spreading over the whole molecule. This happens even in cases where the chemist would regard the molecule as held together by *localized* bonds. Indeed, classical chemistry is built on the belief, supported by a vast amount of experimental evidence, that many compounds can be described in terms of localized bonds that possess highly specific properties largely independent of environment; the C–H bond, for example, appears to be much the same whether it occurs in methane or ethane. A description in terms of delocalized molecular orbitals would appear to offer no explanation of such observations.

On the other hand, the MOs that arise from solution of an SCF eigenvalue equation (i.e. the canonical MOs) are unique only in a *mathematical* sense. Energy expressions such as (6.5.3) or (6.6.13) contain only *density matrices* for one or more shells, and such matrices are invariant under mixing of the orbitals *within* a shell. The orbitals $\{\boldsymbol{\phi}^K\}$ of shell K define a certain n_K-dimensional subspace of the space spanned by the m basis functions; $\mathbf{R}_K$ represents a projection operator onto the subspace; and any unitary transformation

$$\boldsymbol{\phi}^K \rightarrow \bar{\boldsymbol{\phi}}^K = \boldsymbol{\phi}^K \mathbf{U}_K \tag{6.8.1}$$

will leave $\mathbf{R}_K$, and hence the total electronic energy, unchanged. The same is true, obviously, of the total charge density $\mathbf{P}$, and the choice of "which" orbitals are occupied is thus in a sense arbitrary; different choices correspond only to different ways of resolving the shell densities into sums of orbital contributions. In this section we shall be concerned mainly with the possible advantages of using *localized* combinations of the MOs, more obviously related to chemical notions such as "electron-pair bonds" or "lone pairs".

Before proceeding, we note that the most valuable features of the MOs themselves only become clear when we consider not a single electronic state but rather a *transition* in which there is change of orbital occupation numbers. Thus, Koopmans' theorem (p. 164) attributes a physical significance to an orbital energy ϵ_i by showing that the canonical choice of ψ_i ensures that *both* E (for the N-electron determinant) and E_i^+ (for the $(N-1)$-electron determinant with ψ_i missing) are stationary—so that their difference is a realistic estimate of an ionization energy. In the same way, (6.1.28) shows that differences of MO energies (with correction terms) can be related to spectroscopic transition energies at the IPM level. It has also been shown (Section 6.2) that similar results apply in a finite-basis approximation, at least in the closed-shell case.

It is not immediately clear whether orbital energies obtained by solving canonical *open-shell* equations, based on an effective Hamiltonian such as (6.5.18), can be given any physical interpretation at all; since the orbital energies can be moved at will by means of level shifters etc. But here again it is sometimes possible to exploit this freedom to obtain orbital energies that relate to electron transfer or ionization processes (Dodds and McWeeny, 1972; Hirao, 1974a,b; McWeeny, 1975). Thus for the 2-shell case treated in Section 6.5 the eigenvalue ϵ_X associated with any eigenvector $\mathbf{c}_X$ of the Hamiltonian (6.5.18) will coincide with a negative ionization energy $-I_X$ provided that the $\mathbf{d}$ matrices are chosen according to

$$\mathbf{d}_1 = 2\mathbf{F}_1 - \mathbf{F}_2, \qquad \mathbf{d}_2 = \mathbf{F}_2. \tag{6.8.2}$$

In this case the same closed- and open-shell orbitals yield stationary energies for both the neutral system and its positive ion—whether the electron be taken from the closed shell or the open shell. In this case the proper choice of $\mathbf{d}$ matrices has *forced* the eigenvalues to satisfy Koopmans' theorem.

For more general situations, where the energy formula may involve ensemble averaging as in Section 6.6, the eigenvalues of the effective Hamiltonian cannot be given such a precise significance; but it is still possible to use "ensemble-averaged" orbital energies (McWeeny, 1974; Firsht and McWeeny, 1976) in discussing the main features of intershell electron transfer processes. Beyond this point, IPM concepts begin to lose their validity, and the more detailed study of electron ionization and attachment processes then requires more sophisticated techniques (Chapter 13).

Localized orbitals

It is possible to devise various criteria for "maximum localization" of the linear combinations $\bar{\boldsymbol{\phi}}^K$ occurring in a transformation equation such as (6.8.1). In cases of high symmetry, for example, it is frequently possible to find "equivalent orbitals" (Lennard-Jones, 1949a,b) that are well localized around bonds that are equivalent in the crystallographic sense (i.e. being permuted by symmetry operations). Transformation to equivalent orbitals appears to have been employed first by Coulson (1937), who showed that MOs on the methane molecule could be replaced by four equivalent "localized MOs", identical in form and each describing one of the four C–H bonds. In other cases, where there is no symmetry, some kind of "energy criterion" is commonly adopted. For example, the coulomb repulsion energy of electron pairs occupying two different orbitals ϕ_K, ϕ_L will be reduced to a minimum when ϕ_K and ϕ_L are localized in essentially non-overlapping regions of space, as far apart as possible (consistent with normalization and orthogonality). This suggests the criterion, for a closed-shell system,

$$\sum_{K<L} \langle \phi_K\phi_L| g |\phi_K\phi_L\rangle = \text{minimum}, \tag{6.8.3}$$

the sum being over all occupied orbitals. An alternative criterion for localization of a charge distribution $|\phi_K|^2$ is that the self-energy $\langle \phi_K\phi_K| g |\phi_K\phi_K\rangle$ should be a *maximum*; and this suggests we choose the orbitals so that

$$\sum_{K} \langle \phi_K\phi_K| g |\phi_K\phi_K\rangle = \text{maximum}. \tag{6.8.4}$$

It is easy to show that the last two conditions are precisely equivalent. Thus, if we express the total coulomb energy as

$$E_c = \int \frac{P(\boldsymbol{r}_1)P(\boldsymbol{r}_2)}{r_{12}}\, d\boldsymbol{r}_1\, d\boldsymbol{r}_2 = 2\,\text{tr}\,\mathbf{RJ}(\mathbf{R})$$

—which is invariant (like $\mathbf{R}$ itself) under a unitary transformation of orbitals—and then write $\mathbf{R}$ as a sum of orbital contributions ($\mathbf{R}_K = \mathbf{c}_K\mathbf{c}_K^\dagger$), we find

$$E_c = 2\sum_{K} \mathbf{c}_K^\dagger \mathbf{J}(\mathbf{R}_K)\mathbf{c}_K + 4\sum_{K<L} \mathbf{c}_K^\dagger \mathbf{J}(\mathbf{R}_L)\mathbf{c}_K. \tag{6.8.5}$$

The separate sums, which are simply the finite-basis forms of those in (6.8.3) and (6.8.4), are not *individually* invariant; and consequently when the first is a maximum the second is a minimum.

Let us now use the condition (6.8.4) in the finite-basis form

$$\sum_{K} \text{tr}\,\mathbf{R}_K\mathbf{J}(\mathbf{R}_K) = \text{maximum}, \tag{6.8.6}$$

and consider an orbital variation in which

$$\mathbf{c}_K \rightarrow \sum_L \mathbf{c}_L U_{LK} \quad (\mathbf{U} \text{ unitary}). \tag{6.8.7}$$

On equating to zero the first-order variation of the sum in (6.8.6), we obtain (Problem 6.20) the stationary conditions

$$D_{KL} = |\mathbf{c}_K^\dagger[\mathbf{J}(\mathbf{R}_K) - \mathbf{J}(\mathbf{R}_L)]\mathbf{c}_L| = 0 \quad (\text{all } K, L), \tag{6.8.8}$$

where the modulus is taken so as to obtain a corresponding matrix $\mathbf{D}$ that is symmetric. To satisfy the conditions (6.8.8), we regard the occupied MOs (corresponding to $\mathbf{c}_K$, $\mathbf{c}_L$) as a basis, defining the occupied orbital subspace: the *canonical* MOs will *not* make $D_{KL} = 0$ ($K \neq L$), since they diagonalize instead the Fock matrix $\mathbf{F}$. But if we solve

$$\mathbf{Du} = \alpha\mathbf{u}, \tag{6.8.9}$$

we shall obtain eigenvectors $\mathbf{u}_1, \mathbf{u}_2, \ldots$ that *do* bring $\mathbf{D}$ to diagonal form. These eigenvectors define the appropriate linear combinations of the $\mathbf{c}_K$, and hence the columns of the unitary matrix $\mathbf{U}$ in (6.8.7), which determine varied MOs with a higher degree of localization. This procedure will be repeated, after revising the matrix $\mathbf{D}$, until convergence is obtained, and is clearly a variant of the effective-Hamiltonian method of Section 6.5, in which (taking the MOs as a basis) the arbitrary nature of the matrix $\mathbf{d}_K$ in (6.5.18) is exploited in order to obtain K-shell orbitals with specific properties.

The method described above is essentially that due to Edmiston and Ruedenberg (1963). Another commonly used localization criterion (Boys and Foster, 1960) is that of maximum separation of the centroids of the transformed orbitals, and other methods have also been successfully applied (see e.g. Gilbert, 1964; Magnasco and Perico, 1967; McWeeny and Del Re, 1968).

Usually, when applied to molecular closed-shell ground states, the various localization methods lead to orbitals that are concentrated either around individual nuclei (for example inner-shell orbitals not very different from those in the free atoms) or in the "valence regions" (for example lone-pair orbitals, mainly on *one* centre, and bond-pair orbitals, confined mainly to two adjacent centres). Naturally, there are exceptions, in which such a high degree of localization cannot be attained (notably in electron-deficient molecules like the boron hydrides, and in conjugated systems), but in such cases the remaining delocalization is associated with very particular molecular properties.

Whenever a transformation to strongly localized orbitals is possible, it has interesting physical implications. Let us consider a closed-shell situation and assume that localized orbitals have been found, with the property $\bar{\phi}(\boldsymbol{r}) \to 0$, when $\boldsymbol{r}$ goes outside the region of localization of orbital $\bar{\phi}$ and that the different regions are exclusive in the sense that no point lies inside more than one such region (i.e. overlap is negligible). We now look for a classically acceptable interpretation of the energy expression (5.2.20) compatible with a localized-bond description. The electron–nuclear potential-energy term raises no difficulty; P is a sum of contributions, and it is immaterial whether it is resolved into localized or non-localized parts. We are therefore at liberty to picture the bonds as localized charge distributions, each amounting to two electrons and contributing a corresponding amount of potential energy, if a suitable transformation can be found. The pair function is less simple because it contains dynamical effects due to correlation and therefore appears to be incompatible with a model based on localized and static charge distributions. Let us take, for example, the part of Π that refers to electrons both with spin $+\frac{1}{2}$. This is, by (5.8.5),

$$\Pi_{\alpha\alpha}(\boldsymbol{r}_1, \boldsymbol{r}_2) = P_\alpha(\boldsymbol{r}_1)P_\alpha(\boldsymbol{r}_2) - P_\alpha(\boldsymbol{r}_2;\boldsymbol{r}_1)P_\alpha(\boldsymbol{r}_1;\boldsymbol{r}_2), \tag{6.8.10}$$

where the last term ensures that two up-spin electrons, even though assigned to MOs that extend over the same region of space, do not simultaneously come close together: this is the term that does not allow us to discuss energy relationships in terms of the interaction between static charge distributions. Suppose, however, that the orbitals $\bar{\boldsymbol{\phi}}$ are strongly *localized* combinations of the MOs. By definition, (5.3.16) then gives, for $\boldsymbol{r}_1$ and $\boldsymbol{r}_2$ both in $\bar{\phi}_X$,

$$P_\alpha(\boldsymbol{r}_1;\boldsymbol{r}_2) \simeq \bar{\phi}_X(\boldsymbol{r}_1)\bar{\phi}_X^*(\boldsymbol{r}_2),$$

and consequently

$$\begin{aligned}\Pi_{\alpha\alpha}(\boldsymbol{r}_1, \boldsymbol{r}_2) &\\ &\simeq \bar{\phi}_X(\boldsymbol{r}_1)\bar{\phi}_X^*(\boldsymbol{r}_1)\bar{\phi}_X(\boldsymbol{r}_2)\bar{\phi}_X^*(\boldsymbol{r}_2) - \bar{\phi}_X(\boldsymbol{r}_2)\bar{\phi}_X^*(\boldsymbol{r}_1)\bar{\phi}_X(\boldsymbol{r}_1)\bar{\phi}_X^*(\boldsymbol{r}_2)\\ &= 0.\end{aligned}$$

This result has a purely classical interpretation. If we cannot assign two electrons with the same spin to the same orbital (Pauli principle) then there should be no electron-interaction term associated with such a situation: the above cancellation simply ensures that this will be so. The only interaction term expected, for this orbital, is that between electrons of *opposite* spin, and this is exactly what arises from the $\alpha\beta$ and $\beta\alpha$ components of Π: thus, from (5.8.6),

$$\Pi_{\alpha\beta}(\boldsymbol{r}_1, \boldsymbol{r}_2) = P_\alpha(\boldsymbol{r}_1)P_\beta(\boldsymbol{r}_2) = \bar{\phi}_X(\boldsymbol{r}_1)\bar{\phi}_X^*(\boldsymbol{r}_1)\bar{\phi}_X(\boldsymbol{r}_2)\bar{\phi}_X^*(\boldsymbol{r}_2),$$

with a similar result for $\Pi_{\beta\alpha}$. On the other hand, when r_1 and r_2 are in different localized orbitals $(\bar{\phi}_X, \bar{\phi}_Y)$, the second term in $\Pi_{\alpha\alpha}$ vanishes identically and no cancellation occurs; thus

$$\Pi_{\alpha\alpha}(r_1, r_2) \simeq \bar{\phi}_X(r_1)\bar{\phi}_X^*(r_1)\bar{\phi}_Y(r_2)\bar{\phi}_Y^*(r_2),$$

and there is a corresponding energy contribution (coulomb integral) representing the electrostatic interaction between electrons of the same spin, *provided they occupy different orbitals* $(\bar{\phi}_X, \bar{\phi}_Y)$.

The actual terms that remain, accepting the above approximation, lead to a very simple expression for the energy of a molecule, as pointed out by Lennard-Jones (1949a,b). In particular, the electron-interaction energy reduces to

$$V_{ee} = \sum_X \langle \bar{\phi}_X\bar{\phi}_X | g | \bar{\phi}_X\bar{\phi}_X \rangle + 2 \sum_{X,Y}{}' \langle \bar{\phi}_X\bar{\phi}_Y | g | \bar{\phi}_X\bar{\phi}_Y \rangle, \quad (6.8.11)$$

which contains the coulomb interactions between the electrons in every orbital $\bar{\phi}_X$ plus the coulomb interactions between the pairs of electrons in *different* orbitals $(\bar{\phi}_X, \bar{\phi}_Y)$. The exchange integrals over the localized orbitals have apparently disappeared as a result of the localization assumption, but this is not quite so, for the "exchange" integrals with $\bar{\phi}_Y = \bar{\phi}_X$ have been retained in cancelling the spurious Coulomb integrals representing two electrons of like spin in the same orbital (which would appear if we simply discarded the exchange term in (6.8.10)). The total "exchange energy" has already been written down as the last term in (5.3.18) and is of course invariant, like P itself, against change of orbitals: but the individual integrals are not, and by choosing sufficiently well localized orbitals we are reducing the genuine exchange integrals $(\bar{\phi}_Y \neq \bar{\phi}_X)$ to negligible values and maximizing the "self-energy" integrals that occur when $\bar{\phi}_Y = \bar{\phi}_X$. How closely this idealized picture of localization can be approached will of course depend on the particular electronic system considered; it is a fairly accurate picture in many saturated molecules, but in other situations (e.g. conjugated molecules) there may be an essential delocalization.

There are evidently good reasons for seeking combinations of MOs with a maximum localization, particularly if the localized orbitals correspond to chemically identifiable entities such as bonds, lone pairs, etc.; for then the energy relationships that determine the shape and stability of a molecule are exposed with maximum clarity. We have noted already that the forces responsible for bonding arise from the electron density $P(r_1)$, whose association with the bond regions is particularly transparent in a localized orbital description. But it is now apparent also that the electron-interaction energy V_{ee} may be estimated by putting one up-spin and one down-spin electron into each localized orbital and supposing them to interact classically, like smeared-out static charge distributions. It follows that, while the coarse features of the bonds are determined by $P(\boldsymbol{r}_1)$, their equilibrium configuration must be largely dependent on V_{ee} (and hence on $\Pi(\boldsymbol{r}_1, \boldsymbol{r}_2)$) and that this energy term may be visualized classically, in a localized orbital description, in terms of electrostatic interactions among bond pairs and lone pairs. This analysis provides the theoretical basis of the very successful semiclassical discussions of molecular shapes proposed by Sidgwick and Powell (1940) and developed at length by Nyholm and coworkers (e.g. Nyholm and Gillespie, 1957; Gillespie, 1973).

PROBLEMS 6

6.1 Use the argument indicated in the text (p. 164) to establish the Koopmans theorem (6.1.26).

6.2 Obtain the Brillouin theorem (6.1.29). [*Hint*: Use Slater's rules together with the definition of the Fock operator and the properties of the operators J and K.]

6.3 Demonstrate the **G**-matrix property defined in (6.2.10).

6.4 Derive equations analogous to those that lead up to (6.2.20), for a closed-shell state, starting from the spinless energy expression (5.3.18). Hence justify the conclusions stated on p. 171. [*Hint*: Introduce *spinless* J and K operators and verify the alternative form (6.2.24). Insert finite basis approximations in the 1- and 2-electron integrals and pass to the equivalent matrix forms (p. 171).]

6.5 Make the following rough approximations (Pople, 1953) in formulating the matrix SCF equations for the π electrons of a conjugated molecule: (i) the π electrons move in an "effective field", due to a σ-bonded "core" or "framework" of positive ions, which may be recognized by using a core Hamiltonian h_c in place of h; (ii) h_c has diagonal elements $\alpha_i = \langle \chi_i | \mathsf{h}_c | \chi_i \rangle \sim \omega_i - \sum_{j(\neq i)} Z_i / R_{ji}$, where ω_i is characteristic of framework atom i, Z_j is the number of π electrons provided by atom j, R_{ji} is the ij distance,‡ while off-diagonal elements (β_{ij}, say) are characteristic of the type of link i—j (C—C, C—N, etc.); (iii) the $2p_\pi$ AOs $\{\chi_i\}$ are orthonormal, and 2-electron integrals can be neglected unless of type $\langle \chi_i \chi_j | g | \chi_i \chi_j \rangle$ $(= \gamma_{ij})$. Derive the closed-shell SCF equations from (6.2.25) *et seq.* and use them in a calculation on the *trans*-butadiene molecule, using the following parameter values:

$$\omega_C = -11.20\ \text{eV}, \quad \beta_{CC} = -4.75\ \text{eV}, \quad \gamma_{11} = 10.60\ \text{eV},$$
$$\gamma_{12} = 7.30\ \text{eV}, \quad \gamma_{13} = 5.46\ \text{eV}, \quad \gamma_{14} = 4.90\ \text{eV}.$$

‡ A better approximation is to replace $1/R_{ji}$ by γ_{ji} (see (iii)).

Perform two or three cycles of the SCF calculation. [*Hint*: First neglect all 2-electron integrals, obtaining an IPM Hamiltonian (cf. Problem 2.10) and a first approximation to the MOs. Then calculate **P**, revise the elements of **F**, and start a new cycle.]

6.6 Use the results of Problem 6.5 to discuss the $\pi \rightarrow \pi^*$ transitions of the butadiene molecule that arise on promoting an electron from a bonding (π) MO to an antibonding (π^*) MO. (For molecules of this kind there are normally four main bands in the electronic spectrum, designated (Clar) by α, p, β, β'.) [*Hint*: It will be necessary to couple spins to singlet or triplet (cf. Problem 3.8), and possibly to allow for degeneracy among the "excited" CFs.]

6.7 Use the IPM MOs for a 6-atom ring (Problem 2.13) as a starting approximation for solving the SCF equations, obtained in Problem 6.5, for the benzene π electrons. Show that (i) the MOs are already self-consistent, (ii) the diagonal elements of **P** all have the value unity, and (iii) the nearest-neighbour off-diagonal elements all have the same value $\frac{2}{3}$ (compared with 1 for the ethene π bond). (The diagonal and off-diagonal elements of **P** are the (π) "charges" and "bond orders" and have played a central role in theoretical organic chemistry.)

6.8 Show that if the atoms in a conjugated system are assumed to be (approximately) electrically neutral, and the bond orders are assigned a common average value, then the SCF equations reduce to those of Hückel (1931). (Hückel MOs are easy to obtain, requiring trivial arithmetic and no iteration, and still retain a certain value in qualitative MO theory. See also Problem 2.10.)

6.9 How would the equations in Problem 6.5 be modified in dealing with a system with one *singly* occupied MO, (i) in the UHF approximation (p. 175), and (ii) in the RHF approximation (Section 6.5)? Examine the detailed forms of the equations for a 3-electron π system such as $C_4H_6^+$, performing an SCF cycle (parameter values from Problem 6.5) to see how the MOs of the neutral molecule would change.

6.10 The approximations in Problem 6.5 may be generalized to include *all* valence electrons, the prototype of many approximation schemes being that of Pople, Santry and Segal (1965). Derive a corresponding SCF Hamiltonian as follows: (i) use $i = a, b, \ldots$ to label AOs on atoms A, B, . . . , respectively, and neglect any 2-electron integral that contains an overlap density of type $\chi_a\chi_{a'}$ (intraatomic) or $\chi_a\chi_b$ (interatomic)—this being a "complete neglect of differential overlap" (CNDO) approximation; (ii) use a common average value γ_{AA}, or γ_{AB}, for integrals such as γ_{aa}, or γ_{ab}, for *any* valence AOs on each atom; (iii) introduce Z_A, P_A for the number of valence electrons provided by A and the valence population ($P_A) = \sum_a P_{aa}$) of A *in the molecule*; (iv) assume $\beta_{ab} \sim \beta^0_{AB}S_{ab}$ (β_{AB} characteristic of pair A, B, and S_{ab} putting in an empirical overlap dependence). The remaining parameters (U_a, corresponding to ω_a in Problem 6.5) are estimated by a variety of methods. Whole books (e.g. Pople and Beveridge, 1970; Murrell and Harget, 1972) have been devoted to semi-empirical procedures of this kind.

6.11 The GUHF determinant, in which the N spin-orbitals have the general form (6.3.1), leads to a 1-electron density matrix of the form

$$\rho = R_{\alpha\alpha}\alpha\alpha^* + R_{\alpha\beta}\alpha\beta^* + R_{\beta\alpha}\beta\alpha^* + R_{\beta\beta}\beta\beta^*,$$

where, more fully, the terms are $R_{\alpha\alpha}(\boldsymbol{r}_1;\boldsymbol{r}_1')\alpha(s_1)\alpha^*(s_1')$ etc., and are determined in finite basis form by the matrices (6.3.4). Show that the spin expectation values are given by

$$\langle \mathsf{S}_x \rangle = \tfrac{1}{2}\int [R_{\alpha\beta}(\boldsymbol{r}) + R_{\beta\alpha}(\boldsymbol{r})]\, \mathrm{d}\boldsymbol{r},$$

$$\langle \mathsf{S}_y \rangle = \tfrac{1}{2}i\int [R_{\alpha\beta}(\boldsymbol{r}) - R_{\beta\alpha}(\boldsymbol{r})]\, \mathrm{d}\boldsymbol{r},$$

$$\langle \mathsf{S}_z \rangle = \tfrac{1}{2}\int [R_{\alpha\alpha}(\boldsymbol{r}) - R_{\beta\beta}(\boldsymbol{r})]\, \mathrm{d}\boldsymbol{r},$$

and that *all* the spin-density components may be non-zero at any point in space.

6.12 In a rotation of axes of spin quantization, $(\alpha\ \beta) \rightarrow (\alpha\ \beta)\mathbf{U}(\mathsf{R})$, where $\mathbf{U}(\mathsf{R})$ is a unitary matrix with $U_{11} = a$, $U_{22} = a^*$, $U_{21} = b$, $U_{12} = -b^*$ and $aa^* + bb^* = 1$. Obtain the components of ρ (Problem 6.11) after the spin rotation and hence show that (i) the electron density remains unchanged, (ii) the spin-density component Q_z is unchanged for any transformation that corresponds to rotation around the z axis in real space, and (iii) the components Q_x and Q_y behave, under the same transformation, like the components of a vector of fixed length, lying in a plane containing the z axis. Hence show that the spin distribution, while polarized predominantly along the axis of quantization (z axis), may contain a non-physical transverse component that is rotated merely by changing the phase of the wavefunction: this component would be eliminated on projecting the GUHF function to obtain an eigenstate of S_z, the $\alpha\beta$- and $\beta\alpha$-components of ρ then vanishing (p. 143) and is thus an artefact. Note that on passing to a UHF approximation (S_z definite) the transverse component disappears, and on projecting a singlet state ($S^2 = 0$) *all* components vanish! [*Hint*: On rotating the unit vectors along the cartesian axes according to $\mathbf{e} \rightarrow \mathbf{eR}$, the components of a vector change to $\mathbf{V}' = \mathbf{R}^{-1}\mathbf{V}$, where (orthogonal transformation) the inverse is obtained by transposition of $\mathbf{R}$; expectation values $\langle \mathsf{S}_\alpha \rangle$ and hence the operators S_α change in this way, as a result of change of axes. When the vectors of spin space are rotated by a unitary operator U ($= \mathsf{U}(\mathsf{R})$, with matrix $\mathbf{U}(\mathsf{R})$ as indicated), the operators change to $\mathsf{S}_\alpha' = \mathsf{U}\mathsf{S}_\alpha\mathsf{U}^\dagger$ and the Pauli matrices to $\mathbf{U}\mathbf{S}_\alpha\mathbf{U}^\dagger$. By choosing $a = e^{i\theta}$, $b = 0$, show that $\mathsf{S}_x' = \cos 2\theta\, \mathsf{S}_x - \sin 2\theta\, \mathsf{S}_y$ etc. and conclude that the new x and y axes are rotated about the z axis through an angle $\phi = -2\theta$. Write the new density function $\rho(\boldsymbol{x};\boldsymbol{x}') = R_{\alpha\alpha}(\boldsymbol{r};\boldsymbol{r}')\alpha'(s)\alpha'(s')^* + \ldots$ in terms of the original α and βs and look at the new components.]

6.13 Establish the open-shell energy expression (6.5.2) directly from Slater's rules and, by putting in finite-basis forms of the orbitals (cf. (6.2.1)), derive the matrix forms (6.5.3) *et seq*.

6.14 Follow through the derivation of the orthogonality-constrained variations of the matrices $\mathbf{R}_1$, $\mathbf{R}_2$ in (6.5.14), and generalize the result to include terms quadratic in $\boldsymbol{x}_1$, $\boldsymbol{x}_2$, $\boldsymbol{x}$.

6.15 Write out the effective Hamiltonian (6.5.18) for an RHF approximation in which there are two shells, with Fock matrices (6.5.8), and show that by choosing $\mathbf{d}_1 = 2\mathbf{h}_1 - \mathbf{h}_2$, $\mathbf{d}_2 = \mathbf{h}_2$ the Koopmans theorem remains valid for removal of an electron from either the closed or the open shell. [*Hint*: Take one electron from the orbital ϕ_r, represented by the eigenvector $\mathbf{c}_r$, to obtain $\mathbf{R}_1$, $\mathbf{R}_2$ for the positive ion (noting that if ϕ_r is doubly occupied the electron left behind will then belong to the open shell of the ion). Choose $\mathbf{d}_K$ so that the K-shell eigenvalues give the correct energy difference, whichever shell the electron is taken from.]

6.16 Derive the effective Hamiltonian corresponding to (6.5.18) in the case where the basis functions $\{\chi_i\}$ have an overlap matrix $\mathbf{S}$. [*Hint*: The easiest way to do this is to introduce an orthonormal basis $\bar{\boldsymbol{\chi}} = \boldsymbol{\chi}\mathbf{S}^{-1/2}$, for which (6.5.18) (with bars added to all quantities) is appropriate, and then to transform everything to the χ basis. Note that the eigenvalue equation to be solved will then be $\mathbf{Fc} = \epsilon\mathbf{Sc}$.]

6.17 Set up a 1-configuration MO function for the oxygen molecule in the $^3\Sigma$ ground state and show that the energy can be written in the form (6.6.1) by suitably choosing the a, b coefficients. How would the coefficients change on averaging over all states of the configuration? Write down matrix-eigenvalue equations to determine the MOs in the two cases. How would your equations be modified, in the second case, if an electron were removed from the K shell of one of the oxygen atoms (for example in an ESCA experiment), thus creating a second open shell? [*Hint*: It is convenient to use MOs in real form, using π_x, π_y for the open shell and $\{\phi_i\}$ $(i = 1, \ldots, 7)$ for the closed shell.]

6.18 Complete the derivations (p. 197) leading to (6.7.10), and then obtain, by introducing a finite-basis approximation, the result (6.7.14), which may be handled by the standard method. [*Hint*: Terms of type (ii) (p. 197) involve matrix elements of permutations P_{ip}, $\mathsf{P}_{i'p}$, taken between basis functions $(\Phi_k \Phi_l)$ belonging to the irrep D_S. These differ only in sign from those between branching-diagram spin functions (Θ_k, Θ_l) of the form $[\alpha(i)\beta(i') - \beta(i)\alpha(i')][f\alpha(p) + g\beta(p)]$, where the f and g factors do not involve i, i', p; and for such functions (verify it!) the sum $\mathsf{P}_{ip} + \mathsf{P}_{i'p}$ is equivalent to the identity operator. For type (iii) the permutations act only on the paired-spin factors, and the matrix elements (zero unless $\Theta_k = \Theta_l$) may be obtained by simple spin algebra or by using the algorithm (7.3.1).]

6.19 Consider a CI wavefunction including all "pair excitations" from a closed-shell ground state, namely $\Psi = c_0\Phi_0 + \sum_{i,m} c_i^m \Phi_{ii}^{mm}$, where in Φ_{ii}^{mm} *both* electrons have been moved from ϕ_i (occupied) to ϕ_m (virtual). Show that the energy functional can again be cast in the form (6.6.13), with each pair constituting a "shell", and the occupation numbers are $\nu_i = 2[1 - \sum_m (a_i^m)^2]$ and $\nu_m = 2\sum_i (a_i^m)^2$. Identify the forms of the $\mathbf{G}$ matrices. (This type of wavefunction was first used by Das and Wahl (1967) and by Clementi and Veillard (1967); it may be optimized by various methods, including (see e.g. Cook, 1975) the effective-Hamiltonian method of Section 6.5.)

6.20 Using (6.8.6) with (6.8.7), obtain the stationary conditions (6.8.8).

7 Valence Bond Theory

7.1 THE HEITLER–LONDON CALCULATION

The valence bond (VB) theory is a direct development from the hydrogen-molecule calculation of Heitler and London, already referred to in Sections 3.4 and 3.5, and played an extremely important part in the early history of molecular quantum mechanics. The physical basis of VB theory is the notion that a chemical bond is associated with the pairing of spins of the electrons in the (singly occupied) valence orbitals of the atoms concerned, and its aim is to construct wavefunctions in which all possible bonds are described in terms of spin pairing. Mathematically, this means that we must deal with *many*-determinant wavefunctions constructed directly from *atomic* orbitals by admitting all allocations of spin factors and coupling the spins in pairs to a resultant $S = 0$. This approach has already been contrasted (Section 3.5) with that of MO theory, in which the aim is to achieve a good *one*-determinant wavefunction by allowing its orbitals (MOs) to be arbitrary delocalized mixtures of the basic AOs.

Whereas, however, the MOs can always be chosen orthonormal, without loss of generality, the VB approach is plagued by an intrinsic non-orthogonality among the AOs on different atoms; and the development of the VB method as a means of constructing *ab initio* molecular wavefunctions has been severely handicapped by this so-called "non-orthogonality problem". In later sections various ways of dealing with this problem will be discussed; but first we trace the historical development of VB theory. First we review the Heitler–London results for the hydrogen molecule, taking over the notation used in Section 3.5. With neglect of CI, the lowest-energy singlet and triplet states are represented by the wavefunctions $\Phi_1^{\rm g}$ and $\Phi_4^{\rm u}$ in (3.5.3). On expanding the determinants, these may be written as products of space and spin factors thus:

$$\left.\begin{aligned}\Psi_{\rm g}(\mathbf{x}_1,\mathbf{x}_2) &= N_{\rm g}[a(\mathbf{r}_1)b(\mathbf{r}_2)+b(\mathbf{r}_1)a(\mathbf{r}_2)]\\ &\quad\times[\alpha(s_1)\beta(s_2)-\beta(s_1)\alpha(s_2)] \qquad \text{(singlet)},\\ \Psi_{\rm u}(\mathbf{x}_1,\mathbf{x}_2) &= N_{\rm u}[a(\mathbf{r}_1)b(\mathbf{r}_2)-b(\mathbf{r}_1)a(\mathbf{r}_2)]\\ &\quad\times[\alpha(s_1)\beta(s_2)+\beta(s_1)\alpha(s_2)] \qquad \text{(triplet)},\end{aligned}\right\} \tag{7.1.1}$$

and it is a simple matter to obtain the corresponding approximate energies. The results are usually written as

$$E = \frac{Q \pm K}{1 \pm \Delta}, \tag{7.1.2}$$

where

$$\left.\begin{aligned} Q &= \langle ab|\,\mathsf{H}\,|ab\rangle, \\ K &= \langle ab|\,\mathsf{H}\,|ba\rangle, \\ \Delta &= \langle ab \mid ba\rangle = S_{ab}^2. \end{aligned}\right\} \tag{7.1.3}$$

These matrix elements are between simple orbital products (not determinants) and S_{ab} is the overlap integral between the AOs. The upper and lower signs in (7.1.2) refer to singlet and triplet respectively. The quantity Q is usually referred to as a "coulomb integral" because expansion gives

$$\begin{aligned} Q &= \int a^*(\boldsymbol{r}_1)b^*(\boldsymbol{r}_2)[\mathsf{h}(1) + \mathsf{h}(2) + g(1, 2)]a(\boldsymbol{r}_1)b(\boldsymbol{r}_2)\,\mathrm{d}\boldsymbol{r}_1\,\mathrm{d}\boldsymbol{r}_2 \\ &= \langle a|\,\mathsf{h}\,|a\rangle + \langle b|\,\mathsf{h}\,|b\rangle + \langle ab|\,g\,|ab\rangle, \end{aligned}$$

where the first term is the expectation energy of an electron in a (in the field of both nuclei), the second is that of an electron in b, while the third is their coulomb repulsion energy. The "exchange integral" K is often said to describe "non-classical" effects; but this is dangerous terminology for we know that the Hamiltonian contains only *coulomb* interactions and (Section 5.2) that in this case even the exact energy expression can always be given a classical electrostatic interpretation.

It is customary to neglect Δ ($= S_{ab}^2$) in the denominator of (7.1.2) and to write the electronic energy in the more approximate form

$$E = Q \pm K. \tag{7.1.4}$$

The total energy of the system, nuclear repulsion energy (V_{nn}) included, is thus

$$E_{\mathrm{tot}} = Q + V_{\mathrm{nn}} \pm K. \tag{7.1.5}$$

At large internuclear distances, K is negligible while $Q + V_{\mathrm{nn}}$ represents the energy of the separate atoms plus their coulombic interaction. At shorter distances, $Q + V_{\mathrm{nn}}$ varies only slowly, showing a very shallow minimum at the normal internuclear distance; but K becomes large and negative and accounts for over 90% of the binding energy of the molecule in its singlet state (upper sign). In the triplet state, K appears with the opposite sign and leads to repulsion at all distances, and hence to spontaneous dissociation of the molecule into two hydrogen atoms. The "exchange energy" in this way acquired, during the early days of molecular quantum mechanics, an apparently crucial importance. On the other hand, K is defined only for a particular *approximate* wavefunction, in contrast with the approach in Section 5.2, where it was possible to attribute the chemical binding quite generally to attraction between the nuclei and a "charge cloud". These apparently conflicting views of the origin of the bond will be resolved presently. Historically, however, the "exchange energy" interpretation gained a further impetus from results due to Dirac and van Vleck, to which we now turn.

The coupling of the spins to give a singlet or triplet state may be described by writing the operator for the square of the total spin in the form

$$\mathsf{S}^2 = \mathsf{S}^2(1) + \mathsf{S}^2(2) + 2\mathbf{S}(1)\cdot\mathbf{S}(2), \tag{7.1.6}$$

where $\mathbf{S}(1)\cdot\mathbf{S}(2)$ is the spin scalar-product operator:

$$\mathbf{S}(1)\cdot\mathbf{S}(2) = \mathsf{S}_x(1)\mathsf{S}_x(2) + \mathsf{S}_y(1)\mathsf{S}_y(2) + \mathsf{S}_z(1)\mathsf{S}_z(2). \tag{7.1.7}$$

Since the effect of $\mathsf{S}^2(i)$ on any spin-orbital product (and so on any wavefunction Ψ) is merely to multiply by $\frac{3}{4}$, it follows that in a state with spin quantum number S

$$\langle \Psi | \mathsf{S}^2 | \Psi \rangle = 2(\tfrac{3}{4}) + 2\langle \Psi | \mathbf{S}(1)\cdot\mathbf{S}(2) | \Psi \rangle = S(S+1) \tag{7.1.8}$$

and that in singlet and triplet states ($S = 0, 1$) the expectation values of the spin scalar product are consequently $-\frac{3}{4}$ and $+\frac{1}{4}$ respectively.

The energy expression (7.1.4) may thus be rewritten as the expectation value of a "spin Hamiltonian" H_s:

$$E = \langle \Psi | \mathsf{H}_s | \Psi \rangle, \tag{7.1.9}$$

where

$$\mathsf{H}_s = (Q - \tfrac{1}{2}K) - 2K\mathbf{S}(1)\cdot\mathbf{S}(2). \tag{7.1.10}$$

This device is nowadays used frequently in the interpretation of spin-resonance experiments: in essence the actual Hamiltonian is replaced by an artificial Hamiltonian, containing only *spin* operators and numerical parameters, so chosen as to give the correct approximate energies as its eigenvalues. Thus (7.1.9) yields the singlet and triplet values $E = Q \pm K$, given by (7.1.4), on putting in the appropriate expectation value ($-\frac{3}{4}$ or $+\frac{1}{4}$) of the spin scalar product. The complexities of a detailed energy calculation are in this way "absorbed" into the numerical parameters Q and K, and the orbital nature of the wavefunction is immaterial in using

(7.1.9). It must be stressed, however, that this is a *purely formal* device and that (disregarding the small relativistic effects, which are taken up in Chapter 11) *there is no physical coupling between the electron spins.* There are obvious attractions in a theory that so neatly introduces a formal "model", complete with empirical parameters, and apparently reduces the intricacies of chemical binding to the coupling of pairs of electron spins; but there are many pitfalls in the interpretation of this result.

To appreciate the significance of the "exchange energy", in terms of the electron distribution, it is only necessary to recall the solution of Problem 5.1, in which the amount of charge associated with the overlap density ab is found to be $q_{ab} = 2S_{ab}^2/(1 + S_{ab}^2)$ for the Heitler–London singlet function. It is at this point that inconsistencies in the approximations begin to appear: in dropping Δ, we assumed $S_{ab}^2 \simeq 0$, but now it appears that the enhancement of charge in the bond region is roughly proportional to S_{ab}^2, so that non-zero overlap is essential for bonding. Moreover, if the orbitals are assumed orthogonal then the exchange integral K must be positive‡, while the value actually calculated (to which the bond is attributed) is large and negative. The resolution of these apparent contradictions depends on the fact that in the non-empirical calculation of Heitler and London the orthogonality assumption was not made: the expression for K then reduces to (for real orbitals)

$$K = \langle ab|\, g\, |ba\rangle + 2S_{ab}\langle a|\, \mathsf{h}\, |b\rangle,$$

and only the first term is a *true* exchange integral in the sense that it involves *two* electrons and an interchange of variables. The main term, which is proportional to overlap, contains only 1-electron integrals and includes the energy of attraction between the nuclei and the overlap charge in the bond; it is large and negative and completely outweighs the real exchange term, giving K the negative value normally assumed in the semi-empirical development of Heitler–London theory. It is now clear that, although the Heitler–London wavefunction may be rather satisfactory, neglect of overlap can lead to serious inconsistencies, and also that the introduction of an "exchange integral" effectively disguises the actual factors involved in chemical bonding by indiscriminately mixing together terms representing kinetic energy, electron–nuclear attraction energy and electron–electron repulsion energy. For the present we shall disregard these difficulties and indicate the formal generalizations leading to the VB method.

‡ See Slater (1960, Vol. 1, App. 19). K represents the "self-energy" of a charge distribution of density $\rho = ab$.

7.2 GENERALIZATIONS

To generalize the method of Heitler and London, we consider any configuration containing N singly occupied‡ orbitals $\phi_1, \phi_2, \ldots, \phi_N$. These are normally valence AOs, analogous to the 1s orbitals in the H_2 calculation, but for the moment are assumed orthogonal. Spin factors may then be allocated in 2^N possible ways. We now set up antisymmetric space–spin functions as in Section 4.2 (p. 88), starting from an orbital function

$$\Omega(\boldsymbol{r}_1, \boldsymbol{r}_2, \ldots, \boldsymbol{r}_N) = \phi_1(\boldsymbol{r}_1)\phi_2(\boldsymbol{r}_2)\ldots\phi_N(\boldsymbol{r}_N), \tag{7.2.1}$$

adding the various linearly independent spin functions $\Theta_1, \Theta_2, \ldots, \Theta_\kappa, \ldots$ and antisymmetrizing. The resultant functions

$$\Phi_\kappa(\boldsymbol{x}_1, \boldsymbol{x}_2, \ldots, \boldsymbol{x}_N) = (N!)^{-1/2} \sum_{\mathsf{P}} \epsilon_{\mathsf{P}} \mathsf{P}\Omega(\boldsymbol{r}_1, \boldsymbol{r}_2, \ldots, \boldsymbol{r}_N)\Theta_\kappa(s_1, s_2, \ldots, s_N), \tag{7.2.2}$$

for all choices of Θ_κ, are then allowed to mix in constructing a wavefunction of the form

$$\Psi = \sum c_\kappa \Phi_\kappa. \tag{7.2.3}$$

The mixing coefficients and approximate energies are obtained by solving the 2^N-dimensional secular problem

$$\mathbf{Hc} = E\mathbf{Mc}, \tag{7.2.4}$$

where the elements of the matrices $\mathbf{H}$ and $\mathbf{M}$ are, as usual, $K_{\kappa\lambda} = \langle \Phi_\kappa | \mathsf{H} | \Phi_\lambda \rangle$ and $M_{\kappa\lambda} = \langle \Phi_\kappa | \Phi_\lambda \rangle$. It should be noted that CI in the sense of Section 3.4 has not been admitted so far: we are working within a single orbital configuration, allowing changes of *spin* function but not of orbital occupation. The resolution of the corresponding spin degeneracy is the essence of the VB method in its simple form; generalizations are considered later.

Wavefunctions of the form (7.2.2) have been used already in Section 6.7, where it was shown that the matrices in (7.2.4) could be expressed in terms of permutation matrices:

$$\mathbf{H} = \sum_{\mathsf{P}} \langle \Omega | \mathsf{H} | \mathsf{P}\Omega \rangle \mathbf{V}(\mathsf{P}), \tag{7.2.5}$$

$$\mathbf{M} = \sum_{\mathsf{P}} \langle \Omega | \mathsf{P}\Omega \rangle \mathbf{V}(\mathsf{P}), \tag{7.2.6}$$

‡ The presence of doubly occupied orbitals is admitted later.

where

$$V(\mathsf{P})_{\kappa\lambda} = \epsilon_{\mathsf{P}} D(\mathsf{P})_{\kappa\lambda} = \epsilon_{\mathsf{P}} \langle \Theta_\kappa | \mathsf{P} | \Theta_\lambda \rangle. \tag{7.2.7}$$

If the 2^N possible spin products are used as the spin functions then the matrices $\mathbf{D}(\mathsf{P})$ provide a 2^N-dimensional representation of the symmetric group (Section 4.3), but if we use spin-eigenfunctions (such as the spin-paired functions in Section 4.2) then the representation will be *reduced* and the matrices (7.2.5), (7.2.6) and secular equations (7.2.4) will assume diagonal block form, each block being labelled by corresponding spin eigenvalues (S, M). For example, with $N = 6$ there are 64 $(= 2^6)$ independent spin products, but from these (according to (4.2.2)) we can construct only 5 spin-eigenfunctions with $S = M = 0$: if we are looking for singlet states and choose the appropriate Θs, it is therefore only necessary to handle a 5×5 secular equation (instead of 64×64) and this could be written down directly from a knowledge of the corresponding irreducible representation of the symmetric group.

The early development of the group-theoretical approach was due almost entirely to Serber (1934), while the necessary permutation matrices were first systematically tabulated, for the branching-diagram functions of Section 4.2, by Kotani *et al.* (1955). An alternative to the group-theoretical approach is the direct construction of spin-paired functions, possibly with expansion of (7.2.2) as a linear combination of Slater determinants: but in all cases the procedure to be followed from this point on will depend on whether or not the orbitals are assumed to be orthogonal. In the historical development, overlap was formally neglected (in spite of the inconsistencies already referred to), and in the next section we shall proceed as if the AO basis were indeed orthonormal, retaining also the approximation of a single orbital configuration. The removal of such restrictions will be the subject of later sections.

7.3 PERFECT PAIRING AND RESONANCE

Even with a single configuration of orthonormal orbitals, there are two important cases to consider. These may be distinguished by reference to the type of spin coupling to be adopted, remembering the guiding principle that paired spins will be associated with chemical bonds. For molecules in singlet states the implication of using an orbital product (7.2.1) with a spin eigenfunction of the form (4.2.5), namely

$$\Theta_\kappa = \theta(s_i, s_j)\theta(s_k, s_l) \ldots,$$

where

$$\theta(s_i, s_j) = [\alpha(s_i)\beta(s_j) - \beta(s_i)\alpha(s_j)]/\sqrt{2},$$

is that the spins associated with ϕ_i and ϕ_j are paired, as are those of ϕ_k, ϕ_l, etc., leading to a total spin $S = 0$. The coupling scheme is in this way associated with the *orbitals* and is unaffected by antisymmetrizing to obtain the corresponding Φ_κ in (7.2.2). The paired spins consequently label regions in the molecule where, by analogy with the hydrogen-molecule calculation, we might expect chemical bonds to occur. This association is purely formal unless it can be justified from the energy- and charge-density expressions to which it leads, but nevertheless a function constructed in this way is usually referred to as a "structure", and is associated pictorially with a possible way of drawing classical chemical bonds in order to accommodate all the electrons from a set of singly occupied "valence" orbitals. In this way, each Rumer diagram (p. 93) acquires a certain "chemical" significance.

It is now plausible to assume that in favourable cases a *single* spin-paired structure may give a reasonably good description of the electronic state. This might be so, for instance, in the water molecule, where the structure in which the electrons of the singly occupied AOs on the oxygen and hydrogen atoms may have their spins paired to give two O—H bonds would seem intuitively satisfactory. The structure in which the two hydrogen atoms have their spins paired would represent a less satisfactory scheme, with "formal" bonding between AOs that have little overlap. When *one structure* is used as an adequate description of a given electronic state we speak of the "perfect-pairing approximation". More generally, of course, the wavefunction will be represented as a linear combination of a number of structures, as in (7.2.3), whose weights are determined by solution of a secular problem (7.2.4). Since each structure is associated with a particular set of bonds ("real" or formal), it has become customary in chemistry to regard the electronic state as a mixture or "hybrid" of classically feasible situations in which the bonds have been drawn in all possible ways and, for example, to speak of "resonance" of the molecule among the alternative structures. This is graphic but inaccurate terminology and does not of course imply any kind of "oscillation". In fact, any reference to "resonance" merely implies that the wavefunction cannot be well represented by a single VB structure.

The whole of VB theory rests upon one simple formula, which gives the value of the scalar product $\langle \Theta_\kappa \mid \Theta_\lambda \rangle$ for any two spin-paired functions.‡ We first present this formula and then apply it, first in the

‡ We continue to use spin-paired functions (rather than the branching-diagram functions used in Section 6.7) in order to derive the standard matrix-element rules employed in simple VB theory.

perfect-pairing approximation and then in the general case. To this end we must introduce the idea of the *superposition pattern* associated with Θ_κ and Θ_λ: this is obtained simply by superimposing their Rumer diagrams to obtain a pattern consisting of "islands" and "chains". For example, two possible 6-electron triplet spin functions and their superposition pattern would be

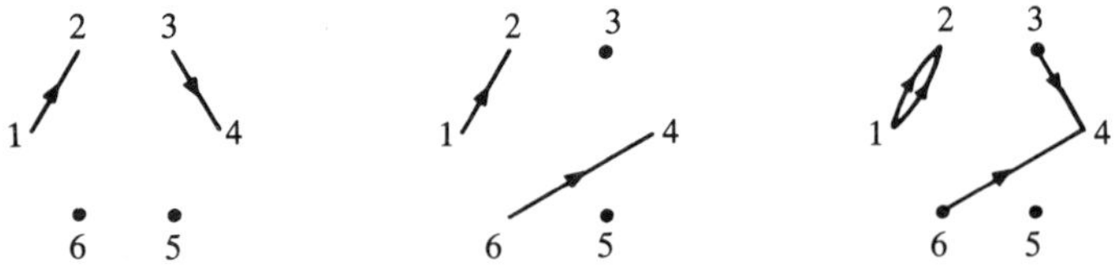

Θ_κ Θ_λ superposition pattern

Such patterns may not always occur in the "matching form", in which all arrows come head-to-head and tail-to-tail'; but, since any arrow $i \rightarrow\!\!- j$ indicates a spin factor $\alpha(s_i)\beta(s_j) - \beta(s_i)\alpha(s_j)$, its reversal corresponds merely to multiplying the spin function by -1 to obtain a factor $\alpha(s_j)\beta(s_i) - \beta(s_j)\alpha(s_i)$ (with arrow $j \rightarrow\!\!- i$).

In the pattern we distinguish *islands*, each formed by a closed sequence of arrows and *chains*, each formed by an open sequence: the latter are of two types: O-chains containing an odd number of centres, and E-chains containing an even number. The required result is then (Cooper and McWeeny, 1966a,b) for normalized spin eigenfunctions

$$\boxed{\langle \Theta_\kappa \mid \Theta_\lambda \rangle = \delta_{\mathrm{E}}(-1)^{\nu_{\kappa\lambda}} 2^{n_{\kappa\lambda} - g} = \langle \Theta_\lambda \mid \Theta_\kappa \rangle,} \tag{7.3.1}$$

where

$$\nu_{\kappa\lambda} = \text{number of arrow reversals to achieve matching},$$
$$n_{\kappa\lambda} = \text{number of islands in superposition pattern},$$
$$g = \text{number of spin pairs in each function},$$
$$\delta_{\mathrm{E}} = 1 \quad (\text{no E-chains}), \text{ or } 0 \quad (\text{otherwise}).$$

This key formula is easily established by considering the expansion of each function in terms of spin products and evaluating the number of matching terms. The appearance of an E-chain (for example an arrow in one diagram falling between two dots on the other) implies that matching terms cannot occur, giving a zero matrix element. The scalar product between the two triplet spin-eigenfunctions considered above is evidently $\langle \Theta_\kappa \mid \Theta_\lambda \rangle = 2^{-1}$ (no reversals, 1 island, no E-chains, 2 spin pairs). We now use (7.3.1) to obtain the perfect-pairing energy formula.

The perfect-pairing approximation

We take the wavefunction Ψ to be a single structure Φ_κ in which, for generality, the spins associated with some orbitals are paired and the remainder are parallel-coupled to give a non-zero total spin S. As a concrete example, we may consider a 6-electron case in which the paired spins are associated with orbitals (ϕ_1, ϕ_2) and (ϕ_3, ϕ_4). This coupling scheme corresponds to the Rumer diagram shown above for the function Θ_κ, the unconnected dots referring to electrons in ϕ_5 and ϕ_6, both with α spin factors, giving a triplet state with $S = M = 1$. We expect to find a generalization of Heitler and London's result (7.1.4), with an exchange integral K for *each* pair and possibly with interaction terms. It is clear that this will be so, for on putting $\lambda = \kappa$ in (7.2.5) the energy expectation value associated with Φ_κ may be written (with orthogonal orbitals and hence only single-interchange‡ terms)

$$E = H_{\kappa\kappa} = Q - \sum_{ij} \mu_{ij} K_{ij}, \tag{7.3.2}$$

where

$$Q = \langle \Omega | \mathsf{H} | \Omega \rangle = \int \phi_1^*(\boldsymbol{r}_1)\phi_2^*(\boldsymbol{r}_2) \ldots \phi_N^*(\boldsymbol{r}_N) \times \mathsf{H}\phi_1(\boldsymbol{r}_1)\phi_2(\boldsymbol{r}_2) \ldots \phi_N(\boldsymbol{r}_N)\, \mathrm{d}\boldsymbol{r}_1\, \mathrm{d}\boldsymbol{r}_2 \ldots \mathrm{d}\boldsymbol{r}_N \tag{7.3.3}$$

and, for each interchange $i \leftrightarrow j$,

$$K_{ij} = \langle \Omega | \mathsf{H} | \mathsf{P}_{ij}\Omega \rangle = \int \phi_1^*(\boldsymbol{r}_1) \ldots \phi_i^*(\boldsymbol{r}_i) \ldots \phi_j^*(\boldsymbol{r}_j) \ldots \phi_N^*(\boldsymbol{r}_N) \times \mathsf{H}\phi_1(\boldsymbol{r}_1) \ldots \phi_i(\boldsymbol{r}_j) \ldots \phi_j(\boldsymbol{r}_i) \ldots \phi_N(\boldsymbol{r}_N)\, \mathrm{d}\boldsymbol{r}_1\, \mathrm{d}\boldsymbol{r}_2 \ldots \mathrm{d}\boldsymbol{r}_N. \tag{7.3.4}$$

These are the analogues of the Q and K in (7.1.3), while

$$\mu_{ij} = \langle \Theta_\kappa \,|\, \mathsf{P}_{ij}\Theta_\kappa \rangle \tag{7.3.5}$$

and depends only on the spin-coupling scheme. In (7.3.2) it is assumed that $\langle \Phi_\kappa \,|\, \Phi_\kappa \rangle = 1$, which is a consequence of orbital orthonormality. The "exchange integral" K_{ij} reduces, for the same reason, to

$$K_{ij} = \langle \phi_i\phi_j | g | \phi_j\phi_i \rangle, \tag{7.3.6}$$

only the $1/r_{ij}$ term in H giving a non-zero contribution to the integral. It remains only to determine the coefficients μ_{ij} with which the various exchange integrals appear in the energy formula (7.3.2).

‡ Note that the summation in (7.3.2) is over distinct single interchanges, i.e. each pair ij is counted only once.

We note that $P_{ij}\Theta_\kappa = \Theta_\lambda$ is simply another structure‡ in which the ends of any arrows attached to points i and j have been interchanged, and we may therefore use (7.3.1) to obtain $\mu_{ij} = \langle \Theta_\kappa \mid P_{ij}\Theta_\kappa \rangle = \langle \Theta_\kappa \mid \Theta_\lambda \rangle$ from the superposition pattern of Θ_κ and Θ_λ. For interchanges $1 \leftrightarrow 2$ and $3 \leftrightarrow 4$ the result is trivial, for $P_{ij}\Theta_\kappa$ will differ from Θ_κ only by a factor -1 due to one arrow reversal: hence $\mu_{12} = \mu_{34} = -1$. A more interesting case is the interchange $2 \leftrightarrow 3$, for which we obtain

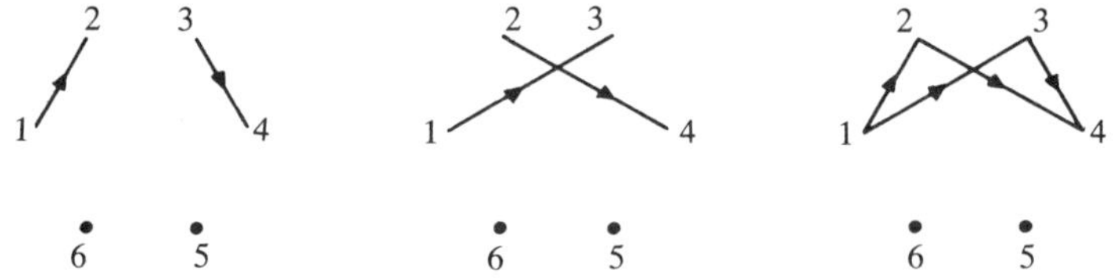

Θ_κ $\Theta_\lambda = \mathbf{P}_{23}\Theta_\kappa$ superposition pattern

Here $\nu_{\kappa\lambda} = 2$ (two reversals to match the arrows) and there is only one island: thus $\mu_{23} = \frac{1}{2}$. similar arguments apply to all interchanges, and the results are independent of the number of electrons, depending only on the forms of the Rumer diagrams. On inserting the values of the coefficients, (7.3.2) takes the final form

$$E = Q + \sum_{\substack{(ij)\\ \text{paired}}} K_{ij} - \frac{1}{2}\sum_{\substack{(ij)\\ \text{uncoupled}}} K_{ij} - \sum_{\substack{(ij)\\ \text{parallel}}} K_{ij}, \tag{7.3.7}$$

where (ij) refers to any pair of positions (each counted once only) and "uncoupled" means i, j belong to *different* coupled groups (parallel-coupled or paired). The conventional interpretation of this result, by comparison with (7.1.4), is that there is a binding energy due to exchange (K_{ij} being supposed negative, as in the Heitler–London calculation) for every "bond" i—j, while the remaining positive-energy contributions represent "exchange repulsions between non-bonded atoms".

The perfect-pairing formula (7.3.7) has been widely employed in qualitative discussions of the interactions determining the shape and stability of polyatomic molecules and in the interpretation of empirical "additivity rules" etc., which apply in many instances and appear to support the validity of a wavefunction representing a single well-defined set of localized electron pair bonds. It must be remembered, however, that the derivation rests upon an orthogonality assumption that intro-

‡ It is useful to apply the term indiscriminately for the wavefunction Φ_κ, the spin-paired function Θ_κ from which it is derived, and the Rumer diagram depicting the coupling scheme.

duces serious inconsistencies into the theory even in the case of only two electrons (Section 7.1). Thus, if the orbitals are truly orthogonal then all the Ks become positive and every spin pair becomes *anti*bonding; while if non-orthogonal orbitals are admitted (consistent with negative Ks) then multiple permutations must be included in (7.2.5) and no simple energy formula emerges. Thus, although the *wavefunction* of a perfect-pairing approximation might give a good description of the system (when non-orthogonal orbitals are employed) the energy formula derived above will not stand the test of non-empirical evaluation.

Before leaving this topic, it is important to note that the validity of a single structure as a good molecular wavefunction rests essentially upon the choice of orbital forms. It is often possible to make an intelligent choice of the set $\phi_1, \phi_2, \ldots, \phi_N$ on intuitive grounds, starting from the premise that localized chemical bonds are associated with regions of high electron density between adjacent atoms and that these may best be described in terms of *strongly overlapping* pairs of orbitals on the individual atoms. In the context of VB theory this suggests that the optimum ϕs should be directed towards each other in pairs, so permitting an unambiguous choice of spin-pairing schemes to describe bonds in the regions thus defined. Directed orbitals of this kind can, of course, be set up by mixing the AOs on any given atom, and it is standard practice in qualitative valence theory to introduce *hybrid* AOs (for example four tetrahedrally directed hybrids on a carbon atom, instead of the 2s and three 2p orbitals) to interpret stereochemical situations that cannot be readily discussed in terms of the overlapping of unmixed AOs (for example the carbon atom in the tetrahedral molecule methane). The concept of hybridization is most useful at a *qualitative* level and will not be developed in this book; historically, it arose from efforts to retain the perfect-pairing picture in situations where, without hybridization, an unambiguous allocation of electron-pair bonds would be impossible; but its impact on chemistry has been enormous.

The general case

In many situations, there are no strong grounds for preferring one structure to another, the approximation of perfect pairing breaks down, and the wavefunction must be represented as a mixture of many VB structures. In full generality, we should consider not only alternative spin-coupling schemes but also new orbital configurations arising from transfer of electrons between orbitals. The inclusion of the resultant "ionic" or "polar" structures has been anticipated in Section 3.5; but for the present we admit only *non*-polar structures.

To obtain the basic matrix element expressions, assuming orthonormal orbitals, we again start from (7.2.5), and systematically evaluate the matrix elements (7.2.7) of the spin permutations. For the moment, we restrict the discussion to the singlet structures of one orbital configuration. We write (7.2.5) as

$$H_{\kappa\lambda} = \langle \Theta_\kappa \mid \Theta_\lambda \rangle Q - \sum_{(ij)} \langle \Theta_\kappa \mid \mathsf{P}_{ij} \Theta_\lambda \rangle K_{ij} \tag{7.3.8}$$

and assume that the structures Θ_κ and Θ_λ have been defined and their superposition pattern drawn. In general the pattern will contain one or more islands, and the values of $\langle \Theta_\kappa \mid \mathsf{P}_{ij} \Theta_\lambda \rangle$ will depend on the position of i and j. As a simple example, we consider two structures for a 6-electron system, and their corresponding superposition pattern:

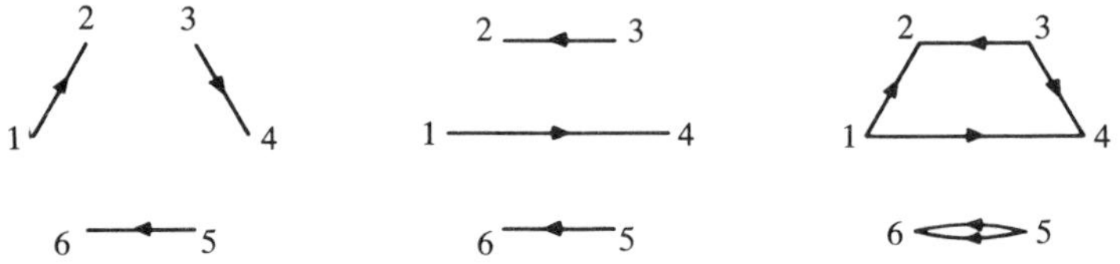

Θ_κ Θ_λ superposition pattern

The pattern contains two islands, and no arrow reversals are required; thus $n_{\kappa\lambda} = 2$ and $\nu_{\kappa\lambda} = 0$, $\langle \Theta_\kappa \mid \Theta_\lambda \rangle = 2^{n_{\kappa\lambda} - g} = \frac{1}{2}$. To obtain the remaining coefficients in (7.3.8), we may compare the number of islands in the pattern for $\langle \Theta_\kappa \mid \Theta'_\lambda \rangle$, where $\Theta'_\lambda = \mathsf{P}_{ij}\,\Theta_\lambda$, with that in the pattern for $\langle \Theta_\kappa \mid \Theta_\lambda \rangle$. If this number is increased by one, the matrix element is multiplied by 2^1, if decreased by one, the factor is 2^{-1}. The sign may also be changed if rematching of arrow heads and tails requires an odd number of arrow reversals. Since P_{ij} simply crosses the arrows attached to positions i and j in structure Θ_λ, the effect on the superposition pattern is always apparent. In the present example there are only three distinct *types* of interchange, leading to the following results:

Matrix element	$\langle \Theta_\kappa \mid \mathsf{P}_{12} \Theta_\lambda \rangle$	$\langle \Theta_\kappa \mid \mathsf{P}_{13} \Theta_\lambda \rangle$	$\langle \Theta_\kappa \mid \mathsf{P}_{15}\, \Theta_\lambda \rangle$
Pattern			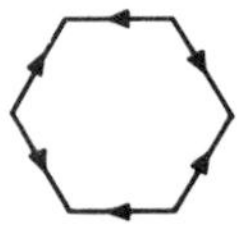
Comment	Number of islands unchanged, one arrow reversal necessary	Number of islands increased by 1, no reversal necessary	Number of islands decreased by 1, no reversal necessary
Value	$-1 \times \langle \Theta_\kappa \mid \Theta_\lambda \rangle$	$2 \times \langle \Theta_\kappa \mid \Theta_\lambda \rangle$	$2^{-1} \times \langle \Theta_\kappa \mid \Theta_\lambda \rangle$

These observations are completely general, for non-polar singlet structures, depending only on the topological relationship of each pair (ij) in the superposition pattern for $\langle \Theta_\kappa \mid \Theta_\lambda \rangle$. From them, noting from (7.2.6) that

$$M_{\kappa\lambda} = \langle \Theta_\kappa \mid \Theta_\lambda \rangle = (-1)^{v_{\kappa\lambda}} 2^{n_{\kappa\lambda}-g}, \tag{7.3.9}$$

we obtain

$$\begin{aligned} H_{\kappa\lambda} = M_{\kappa\lambda}\{QQ + \textstyle\sum K_{ij} \quad & (i, j) \text{ same island, odd number of links apart)} \\ - \tfrac{1}{2} \textstyle\sum K_{ij} \quad & (i, j \text{ different islands)} \\ -2 \textstyle\sum K_{ij} \quad & (i, j \text{ same island, even number of links apart)}\}, \end{aligned} \tag{7.3.10}$$

which is the matrix element rule in the form due to Pauling (1933). Most semi-empirical applications of VB theory have been based on the use of (7.3.10), Q and the Ks being regarded as numerical parameters to be identified by fitting experimental data. From a theoretical standpoint this is not a very satisfactory procedure; apart from the fundamental inconsistencies concerned with orbital non-orthogonality, there is no reason to assume that Q is constant from molecule to molecule or that K_{ij} is characteristic of the orbital pair (ϕ_i, ϕ_j) and independent of molecular environment. The most important applications have been to conjugated hydrocarbons and to a discussion of the delocalized bonding provided by the π electrons, but even here (where closely similar C—C links appear in a whole series of molecules) the assumption of constancy of parameter values is dubious. In order to make *non*-empirical calculations, comparable to those based on *ab initio* SCF theory, it is absolutely essential to remove the assumption of orthonormal orbitals.

7.4 ONE-CONFIGURATION VB THEORY. NON-ORTHOGONAL ORBITALS

So far we have considered VB approximations of the form

$$\left.\begin{aligned} \Psi &= \sum_\kappa C_\kappa \Phi_\kappa, \\ \Phi_\kappa &= (N!)^{1/2} \mathsf{A}[\Omega(\boldsymbol{r}_1, \boldsymbol{r}_2, \ldots, \boldsymbol{r}_N)\Theta_\kappa(s_1, s_2, \ldots, s_N)], \end{aligned}\right\} \tag{7.4.1}$$

where the spatial function represents a single configuration of *orthonormal* orbitals,

$$\Omega(\boldsymbol{r}_1, \boldsymbol{r}_2, \ldots, \boldsymbol{r}_N) = \phi_1(\boldsymbol{r}_1)\phi_2\boldsymbol{r}_2) \ldots \phi_N(\boldsymbol{r}_N), \tag{7.4.2}$$

and *in this case* (7.4.1) is a normalized function. It is convenient to retain the same "normalization" in what follows, even when the orbitals are no longer orthonormal, and to develop the theory in a number of well-defined steps.

Elimination of closed shells

When orbital orthonormality is relaxed (7.2.4)–(7.2.6) remain valid, but the matrix element expressions contain non-zero contributions from *all* $N!$ permutations. It is therefore of great importance to keep N, the number of electrons to be dealt with *explicitly*, as small as possible; this may be done by noting (McWeeny, 1954b, 1959; Gallup *et al.*, 1982) that any configuration of *doubly occupied* orbitals (which may indeed be common to a large number of different orbital products of the type (7.4.2)) may be "eliminated" at the outset simply by redefining the one-electron Hamiltonian h. The number of electrons to be considered (N) then includes only the electrons outside the ""core" of doubly occupied orbitals, moving in an "effective field" embodied in a new Hamiltonian h_{eff}. This "core–valence" separation is derived most easily and generally as a special case of a result established in Chapter 14: the matrix elements of h_{eff} are thus found to be

$$\langle \phi_r | \mathsf{h}_{\text{eff}} | \phi_s \rangle = \langle \phi_r | \mathsf{h} | \phi_s \rangle + \langle \phi_r | (\mathsf{J}_c - \tfrac{1}{2}\mathsf{K}_c) | \phi_s \rangle, \tag{7.4.3}$$

where the core coulomb and exchange terms are

$$\left.\begin{aligned} \langle \phi_r | \mathsf{J}_c | \phi_s \rangle &= \sum_{t,u} P^c_{tu} \langle \phi_r \phi_u | g | \phi_s \phi_t \rangle, \\ \langle \phi_r | \mathsf{K}_c | \phi_s \rangle &= \sum_{t,u} P^c_{tu} \langle \phi_r \phi_u | g | \phi_t \phi_s \rangle, \end{aligned}\right\} \tag{7.4.4}$$

and $\mathbf{P}^c$ is the 1-electron density matrix describing the core in terms of a basis $\{\phi_t\}$. When the ϕ_t are the core orbitals themselves (orthonormal and doubly occupied) $P^c_{ut} = 2\delta_{ut}$, and the above results simplify accordingly. The total electronic energy of the system is obtained by adding to the calculated "valence-electron" energy a "core energy" E_c, calculated as if the electrons of the doubly occupied core orbitals were *alone* in the field of the nuclei.

The above results apply in the present context because any VB-type function (7.4.1) is equivalent to a sum of Slater determinants formed by attaching various spin products to the orbital factor Ω: every term corresponds to an antisymmetrized product of two determinants (one for the core and one for the valence electrons), and on extracting the common core factor the whole VB function assumes the "generalized-product" form considered in Chapter 14. The core and valence orbitals may be assumed‡ to be mutually orthogonal *with no loss of generality,* as has been stressed by Gallup *et al.* (1982), their separation being simply a convenient mathematical device. The enormous reduction achieved in this way can hardly be overemphasized.

The open-shell wavefunction

After formally removing the core electrons, by introducing the effective Hamiltonian, we return to the approximation (7.4.1) and now suppose all orbitals to be non-orthogonal and different. The variational problem will no longer be the same as that considered in Section 6.7, in spite of a general similarity, because we shall wish to optimize the orbitals *without* any orthonormality constraint. The general problem has been considered in great detail by Goddard (1967a,b, 1968a,b), Gerratt (1971) and Kaplan (1975).

To connect the approach with that used in Chapter 4 (pp. 98–99) we remember that, instead of using a space–spin antisymmetrizer as in (7.4.1), we may apply a Wigner-type operator (cf. (A3.23) *et seq.*)

$$\rho_{\lambda\kappa} = \left(\frac{f_S^N}{N!}\right)^{1/2} \sum_{\mathsf{P}} V(\mathsf{P})_{\lambda\kappa} \mathsf{P}, \tag{7.4.5}$$

where the normalization is such that when working on a product of *orthonormal* orbitals the resultant symmetry function will be normalized in the usual way: the matrices $\mathbf{V}(\mathsf{P})$ $(=\epsilon_{\mathsf{P}}\mathbf{D}_S(\mathsf{P}))$ provide the irrep dual to that carried by the spin functions. On applying $\rho_{\lambda\kappa}$ to the function (7.4.2), we obtain

$$F_\lambda^{(\kappa)} = \rho_{\lambda\kappa}\Omega = \left(\frac{f_S^N}{N!}\right)^{1/2} \sum_{\mathsf{P}} V(\mathsf{P})_{\lambda\kappa} \mathsf{P}\Omega, \tag{7.4.6}$$

which transforms (for any fixed choice of κ) like the λth basis vector of the irrep dual to D_S. The sum (p. 99)

$$\Phi_\kappa = (f_S^N)^{-1/2} \sum_{\lambda} F_\lambda^{(\kappa)} \Theta_\lambda \tag{7.4.7}$$

(with the same conventional normalization) will then be a fully antisymmetric function associated with a particular choice of κ. It is easy to see that this coincides with (7.4.1). Thus, from the last two equations

‡ In practice, the valence orbitals are conveniently orthogonalized against the core by a projection method (cf. p. 180).

(noting the representation property (4.3.1)),

$$\Phi_\kappa = (f_S^N)^{-1/2}\left(\frac{f_S^N}{N!}\right)^{1/2} \sum_{\mathsf{P}} \epsilon_{\mathsf{P}} \sum_\lambda \Theta_\lambda D_S(\mathsf{P})_{\lambda\kappa} \mathsf{P}\Omega$$

$$= (N!)^{-1/2} \sum_{\mathsf{P}} \epsilon_{\mathsf{P}} (\mathsf{P}\Omega)(\mathsf{P}\Theta_\kappa) = (N!)^{1/2} A[\Omega\Theta_\kappa] \qquad (7.4.8)$$

—in agreement with (7.4.1).

The above equivalence confirms that, with a spinless Hamiltonian, we may obtain the same energy expectation value either (i) by expanding (7.4.1) in terms of determinants and then using the rules in Section 3.3; or (ii) by using a linear combination of purely *spatial* functions (7.4.6) of appropriate symmetry. The second approach is essentially that of "spin-free quantum chemistry" (Matsen, 1964), which is considered in more detail in later sections. In the present case a first-principles argument will lead to the required matrix-element expressions.

It is at once clear that, although we have chosen only a single product in (7.4.2), there are in fact f_S^N independent functions Φ_κ (or $F_\lambda^{(\kappa)}$, for any desired λ) that may be included with arbitrary coefficients C_κ in (7.4.1). Goddard (1967a,b) was concerned mainly with *one* such function, corresponding to the "diagonal" Wigner operator with $\lambda = \kappa = 1$. Gerratt (1971) instead considered a general linear combination, optimizing both orbitals and the mixing coefficients C_κ, which must satisfy the secular equation (7.2.4). There are other differences of detail between the two approaches: here we give a simplified derivation that includes both.

Matrix-element evaluation

To evaluate the general matrix element $\langle \Phi_\lambda | \mathsf{H} | \Phi_\kappa \rangle$ and overlap integral $\langle \Phi_\lambda | \Phi_\kappa \rangle$, it will be sufficient to start from the latter, writing

$$M_{\lambda\kappa} = \langle \Phi_\lambda | \Phi_\kappa \rangle = \sum_{\mathsf{P}} \langle \Omega | \mathsf{P}\Omega \rangle V(\mathsf{P})_{\lambda\kappa}. \qquad (7.4.9)$$

It is convenient to introduce at this point permutation operators that act on the *orbital* indices rather than the electronic variables, using for example $\bar{\mathsf{P}}_{rs}$ to denote the operator that interchanges ϕ_r and ϕ_s. The two types of operator $(\mathsf{P}, \bar{\mathsf{Q}})$ will then commute because they work on different sets of indices. Moreover, two operators $\mathsf{P}, \bar{\mathsf{P}}$ that correspond to the same permutation may be applied simultaneously to Ω without changing it:

$$\mathsf{P}\bar{\mathsf{P}} \equiv 1, \qquad \bar{\mathsf{P}} \equiv \mathsf{P}^{-1} \qquad (7.4.10)$$

—an orbital permutation is equivalent to the inverse permutation applied to corresponding particles. It is also useful to note that transpositions P_{rs}, which play a special role in the theory of the symmetric group (Section 4.5), are self-inverse and so $\bar{\mathsf{P}}_{rs} \equiv \mathsf{P}_{rs}$.

Let us now apply a particular transposition P_{iN} to the variables of Ω and $\mathsf{P}\Omega$ in (7.4.9): nothing is changed, but we may then write

$$\langle \Omega \mid \mathsf{P}\Omega \rangle = \langle \mathsf{P}_{iN}\Omega \mid \mathsf{P}_{iN}\mathsf{P}\Omega \rangle = \langle \bar{\mathsf{P}}_{iN}\Omega \mid \mathsf{P}\bar{\mathsf{P}}_{iN}\Omega \rangle. \tag{7.4.11}$$

The fact that

$$\bar{\mathsf{P}}_{iN}\Omega(\boldsymbol{r}_1, \boldsymbol{r}_2, \ldots, \boldsymbol{r}_N) = \Omega_i^{N-1}(\boldsymbol{r}_1, \boldsymbol{r}_2, \ldots, \boldsymbol{r}_{N-1})\phi_i(\boldsymbol{r}_N), \tag{7.4.12}$$

in which

$$\Omega_i^{N-1} = \phi_1(\boldsymbol{r}_1) \ldots \phi_N(\boldsymbol{r}_i) \ldots \phi_{N-1}(\boldsymbol{r}_{N-1}) \tag{7.4.13}$$

is an $(N-1)$-electron product with $\phi_i(\boldsymbol{r}_i)$ replaced by $\phi_N(\boldsymbol{r}_i)$, suggests the possibility of reducing (7.4.11) by "decoupling" one at a time the orbitals $\phi_N, \phi_{N-1}, \ldots$ from the original product Ω. To do this, we note that if N objects are divided into two sets, A and B, any permutation of the full group S_N may be expressed as

$$\mathsf{P} = \mathsf{Q}\mathsf{P}_A\mathsf{P}_B, \tag{7.4.14}$$

where P_A and P_B are permutations *within* the sets A and B respectively, while Q effects only transpositions (single, double, multiple) *between* the sets. On choosing for set A the first $N-1$ factors of Ω, and for B the last, it is clear that Q must be a single transposition of type P_{iN} $(i = 1, 2, \ldots, N-1)$. This allows us to write (7.4.9), making use of (7.4.11), as

$$\begin{aligned}
M_{\lambda\kappa} &= \sum_{\mathsf{P}} V(\mathsf{P})_{\lambda\kappa} \langle \Omega_i^{N-1}\phi_i(\boldsymbol{r}_N) \mid \mathsf{P}\mathsf{P}_{iN}\Omega \rangle \\
&= \sum_{\mathsf{P}'} V(\mathsf{P}'\mathsf{P}_{iN})_{\lambda\kappa} \langle \Omega_i^{N-1}\phi_i(\boldsymbol{r}_N) \mid \mathsf{P}'\Omega \rangle \quad (\mathsf{P}' = \mathsf{P}\mathsf{P}_{iN}) \\
&= \sum_{\mathsf{P}''}\sum_{j} V(\mathsf{P}''\mathsf{P}_{iN})_{\lambda\kappa} \langle \Omega_i^{N-1}\phi_i(\boldsymbol{r}_N) \mid \mathsf{P}_{jN}\mathsf{P}''\Omega \rangle \quad (\mathsf{P}' = \mathsf{P}_{jN}\mathsf{P}'', \mathsf{P}'' \in \mathrm{S}_{N-1}) \\
&= \sum_{\mathsf{P}}\sum_{j} V(\mathsf{P}_{jN}\mathsf{P}\mathsf{P}_{iN})_{\lambda\kappa} \langle \Omega_i^{N-1}\phi_i(\boldsymbol{r}_N) | \mathsf{P} | \Omega_j^{N-1}\phi_j(\boldsymbol{r}_N) \rangle \quad (\mathsf{P} \in \mathrm{S}_{N-1}) \\
&= \sum_{j} \left[\sum_{\mathsf{P} \in \mathrm{S}_{N-1}} V(\mathsf{P}_{jN}\mathsf{P}\mathsf{P}_{iN})_{\lambda\kappa} \langle \Omega_i^{N-1} | \mathsf{P} | \Omega_j^{N-1} \rangle \right] \langle \phi_i \mid \phi_j \rangle.
\end{aligned}$$

The final result is thus

$$M_{\lambda\kappa} = \langle \Phi_\lambda \mid \Phi_\kappa \rangle = \sum_j D_{ji}^{\kappa\lambda} \langle \phi_i \mid \phi_j \rangle \quad (\text{any } i), \tag{7.4.15a}$$

where the coefficients (which will presently be identified as density-matrix elements—apart from normalization) are

$$D_{ji}^{\kappa\lambda} = \sum_{\mathsf{P} \in \mathsf{S}_{n-1}} V(\mathsf{P}_{jN}\mathsf{P}\mathsf{P}_{iN})_{\lambda\kappa} \langle \Omega_i^{N-1} | \mathsf{P} | \Omega_j^{N-1} \rangle \tag{7.4.16}$$

and involve only $(N-1)$-electron products and a sum over permutations within S_{N-1}. The permutation-matrix elements are determined from the matrices $\mathbf{V}(\mathsf{P})$ of the *sub*group S_{N-1}, together with those for transpositions of type P_{iN}, which may be constructed by the methods of Section 4.5.

The fact that (7.4.15a) is independent of i follows from the invariance property (7.4.11), and this result may thus be written more symmetrically as

$$\langle \Phi_\lambda | \Phi_\kappa \rangle = N^{-1} \sum_{i,j} D_{ji}^{\kappa\lambda} \langle \phi_i | \phi_j \rangle. \tag{7.4.15b}$$

It is then clear that the coefficients are indeed elements of a 1-body transition density matrix, since, from the definitions (5.4.1) *et seq.*,

$$\langle \Phi_\lambda | \Phi_\kappa \rangle = N^{-1} \int \tilde{P}(\kappa\lambda | \boldsymbol{r}_1 ; \boldsymbol{r}_1)\, \mathrm{d}\boldsymbol{r}_1$$

$$= N^{-1} \sum_{i,j} \tilde{P}_{ji}^{\kappa\lambda} \int \phi_j(\boldsymbol{r}_1)\phi_i^*(\boldsymbol{r}_1)\, \mathrm{d}\boldsymbol{r}_1 = N^{-1} \sum_{i,j} \tilde{P}_{ji}^{\kappa\lambda} \langle \phi_i | \phi_j \rangle.$$

The tilde has been added to indicate that the functions Φ_λ and Φ_κ are not necessarily orthonormal; and, with this proviso, the result may be identified with (7.4.15b). The transition density matrices introduced in Chapter 5, for *normalized* wavefunctions, will thus have elements

$$P_{ji}^{\kappa\lambda} = (M_{\kappa\kappa}M_{\lambda\lambda})^{-1/2}\tilde{P}_{ji}^{\kappa\lambda} = (M_{\kappa\kappa}M_{\lambda\lambda})^{-1/2} D_{ji}^{\kappa\lambda}. \tag{7.4.17}$$

By evaluating the non-orthogonality integral (7.4.9), we have therefore been able to infer the form of the 1-electron density matrices, referred to the orbital basis $\{\phi_i\}$, and consequently the form of all 1-electron contributions to the matrix elements.

Clearly it is possible to obtain many-electron density matrices in the same way. Thus, by applying *two* transposition operators, we obtain instead of (7.4.11)

$$\langle \Omega | \mathsf{P}\Omega \rangle = \langle \mathsf{P}_{j,N-1}\mathsf{P}_{iN}\Omega | \mathsf{P}_{j,N-1}\mathsf{P}_{iN}P\Omega \rangle$$

$$= \langle \bar{\mathsf{P}}_{iN}\bar{\mathsf{P}}_{j,N-1}\Omega | \mathsf{P}\bar{\mathsf{P}}_{iN}\bar{\mathsf{P}}_{j,N-1}\Omega \rangle,$$

and, proceeding as in the 1-electron case, we easily find a second form of (7.4.15a):

$$\langle \Phi_\lambda \mid \Phi_\kappa \rangle = \sum_{k,l} D^{\kappa\lambda}_{kl,ij} \langle \phi_i \mid \phi_k \rangle \langle \phi_j \mid \phi_l \rangle \quad \text{(any } i, j\text{)}, \qquad (7.4.18a)$$

where

$$D^{\kappa\lambda}_{kl,ij} = \sum_{\mathsf{P}\in \mathsf{S}_{N-2}} V(\mathsf{P}_{kN}\mathsf{P}_{l,N-1}\mathsf{P}\mathsf{P}_{j,N-1}\mathsf{P}_{iN}) \langle \Omega^{N-2}_{ij} | \mathsf{P} | \Omega^{N-2}_{kl} \rangle. \qquad (7.4.19)$$

Again, there is a more symmetrical form of (7.4.18a), obtained by summing $N(N-1)$ identical terms:

$$\langle \Phi_\lambda \mid \Phi_\kappa \rangle = [N(N-1)]^{-1} \sum_{i,j,k,l} D^{\kappa\lambda}_{kl,ij} \langle \phi_i \mid \phi_k \rangle \langle \phi_j \mid \phi_l \rangle, \qquad (7.4.18b)$$

and the 2-electron density matrices for normalized functions will have elements

$$\Pi^{\kappa\lambda}_{kl,ij} = (M_{\kappa\kappa} M_{\lambda\lambda})^{-1/2} D^{\kappa\lambda}_{kl,ij}, \qquad (7.4.20)$$

which will provide us with the 2-electron contributions to the matrix elements.

The results of Section 5.4 may evidently be taken over at once. With a wavefunction of the general form (7.4.1) the 1- and 2-electron density matrices will have elements

$$P_{ji} = \sum_{k,\lambda} C_\kappa C^*_\lambda D^{\kappa\lambda}_{ji}, \qquad \Pi_{kl,ij} = \sum_{\kappa,\lambda} C_\kappa C^*_\lambda D^{\kappa\lambda}_{kl,ij}, \qquad (7.4.21)$$

in complete analogy with (5.4.7) *et seq.*

As we know from (5.2.10), it is in fact sufficient to calculate the *two*-electron density, obtaining the 1-electron quantity by integration; and indeed comparison of (7.4.15b) and (7.4.18b) shows that

$$\frac{1}{N} D^{\kappa\lambda}_{ki} = \sum_{j,l} \frac{1}{N(N-1)} D^{\kappa\lambda}_{kl,ij} \langle \phi_j \mid \phi_l \rangle, \qquad (7.4.22)$$

and consequently, adding coefficients and summing as in (7.4.21),

$$P_{ki} = \frac{1}{(N-1)} \sum_{j,l} \Pi_{kl,ij} \langle \phi_j \mid \phi_l \rangle, \qquad (7.4.23)$$

which is the finite-basis analogue of (5.2.10). Alternatively, it is possible to avoid one summation by using the less symmetrical forms (7.4.15a) and (7.4.18a), thus removing the j summation in (7.4.23) and eliminating the factor $(N-1)^{-1}$.

The existence of relations such as (7.4.22) between successive density matrices has been used by Gerratt (1971) and collaborators to set up a recurrence scheme for numerical calculation. This depends on starting from a p-electron density matrix with a high value of p (e.g. $p = N - 2$), which accordingly contains only permutations in the small subgroup S_{N-p} and may be evaluated directly, and then proceeding downwards. The higher density matrices needed in this approach all have forms similar to those in (7.4.16) and (7.4.19). Thus the D-coefficients needed in forming the *triplet* distribution function are

$$D^{\kappa\lambda}_{kln,ijm} = \sum_{\mathsf{P}\in \mathsf{S}_{N-3}} V(\mathsf{P}_{kN}\mathsf{P}_{l,N-1}\mathsf{P}_{n,N-2}\mathsf{P}\mathsf{P}_{m,N-2}\mathsf{P}_{j,N-1}\mathsf{P}_{iN})\langle \Omega^{N-3}_{ijm}|\,\mathsf{P}\,|\Omega^{N-3}_{kln}\rangle, \tag{7.4.24}$$

and the recurrence relations among the 1-, 2- and 3-electron densities are found to be

$$\left.\begin{aligned} D^{\kappa\lambda}_{ki} &= \sum_{l} D^{\kappa\lambda}_{kl,ij}\langle \phi_j \mid \phi_l\rangle, \\ D^{\kappa\lambda}_{kl,ij} &= \sum_{n} D^{\kappa\lambda}_{kln,ijm}\langle \phi_m \mid \phi_n\rangle \quad (\text{any } j,\, m). \end{aligned}\right\} \tag{7.4.25}$$

With efficient programming (**V**-matrix elements being computed once only and filed sequentially), there are no insuperable difficulties in dealing with up to about 15 electrons outside a closed shell.

Optimization of the wavefunction

The last step in any calculation is the optimization of the wavefunction, in particular of the orbitals, by minimization of the energy. We start from the energy expression

$$E = \sum_{i,j} P_{ji}\langle \phi_i|\,\mathsf{h}\,|\phi_j\rangle + \tfrac{1}{2}\sum_{i,j,k,l} \Pi_{kl,ij}\langle \phi_i\phi_j|\,g\,|\phi_k\phi_l\rangle, \tag{7.4.26}$$

which has a deceptive simplicity. It has the standard form (5.4.16) corresponding to any many-determinant orbital approximation; but the density-matrix elements (defined through (7.4.16), (7.4.19) and (7.4.21)) now contain all the overlap integrals that occur in the construction of the D-coefficients. Consequently, if we vary the orbitals, we must vary the *coefficients* in (7.4.26) as well as the 1- and 2-electron integrals!

Let us introduce explicitly the D-coefficients, using (7.4.21), and write the energy functional (7.4.26) in the form

$$E = \sum_{\kappa,\lambda} C_\kappa\left[\sum_{i,j} D^{\kappa\lambda}_{ji}\langle \phi_i|\,\mathsf{h}\,|\phi_j\rangle + \tfrac{1}{2}\sum_{i,j,k,l} D^{\kappa\lambda}_{kl,ij}\langle \phi_i\phi_j|\,g\,|\phi_k\phi_l\rangle\right]C^*_\lambda, \tag{7.4.27}$$

remembering that any variation must also preserve normalization:

$$\Delta = \langle \Psi \mid \Psi \rangle = \sum_{\kappa,\lambda} C_\kappa C_\lambda^* \langle \Phi_\lambda \mid \Phi_\kappa \rangle = 1. \qquad (7.4.28)$$

On subtracting the variation of (7.4.28), with a Lagrange multiplier E, from the variation of (7.4.27), the stationary-value condition becomes

$$\delta L = \left\{ \sum_\kappa C_\kappa \delta C_\lambda^* \langle \Phi_\lambda | (\mathsf{H} - E) | \Phi_\kappa \rangle \right.$$
$$\left. + \sum_{\kappa,\lambda} C_\kappa C_\lambda^* \langle \delta\Phi_\lambda | (\mathsf{H} - E) | \Phi_\kappa \rangle \right\} + \{\text{c.c.}\} = 0, \quad (7.4.29)$$

where {c.c.} contains similar (complex–conjugate) terms with the variations in the ket factors. Since the variations and their complex conjugates are linearly independent, the last equation yields

$$\sum_\kappa \langle \Phi_\lambda | (\mathsf{H} - E) | \Phi_\kappa \rangle C_\kappa = 0, \qquad (7.4.30)$$

$$\sum_{\kappa,\lambda} C_\kappa C_\lambda^* \langle \delta\Phi_\lambda | (\mathsf{H} - E) | \Phi_\kappa \rangle = 0, \qquad (7.4.31)$$

together with two similar equations resulting from the term {c.c.}. The first condition is immediately recognized as the set of secular equations for the expansion coefficients C_κ. It is in the second equation (7.4.31) that difficulties arise from the complicated dependence of the D-coefficients upon the orbital variations.

The simplest term to consider in (7.4.31) is $-E\langle \delta\Phi_\lambda \mid \Phi_\kappa \rangle$, and this will be a sum of contributions resulting from orbital variations $\phi_r^* \to \phi_r^* + \delta\phi_r^*$, one at a time, in the product Ω from which Φ_λ is generated. The typical result will be (using (7.4.15a))

$$\langle \delta\Phi_\lambda \mid \Phi_\kappa \rangle = \sum_j D_{jr}^{\kappa\lambda} \langle \delta\phi_r \mid \delta_j \rangle \quad (\phi_r^* \to \phi_r^* + \delta\phi_r^*),$$

where we have put $i = r$, noting that, according to (7.4.16), $D_{jr}^{\kappa\lambda}$ is independent of the varied orbital ϕ_r^*. The complete E-term in (7.4.31) will thus be

$$-E \sum_{k,\lambda} C_\kappa C_\lambda^* \langle \delta\Phi_\lambda \mid \Phi_\kappa \rangle = -E \sum_{k,\lambda} \sum_{j,r} C_\kappa C_\lambda^* D_{jr}^{\kappa\lambda} \langle \phi_r \mid \phi_j \rangle$$
$$= -E \sum_{j,r} P_{jr} \langle \delta\phi_r \mid \phi_j \rangle. \qquad (7.4.32)$$

The variation of the normalization integral (the "zero-electron" density matrix) thus involves variations of overlap integrals multiplied by elements of the *one*-electron density matrix; and this provides a clue to the evaluation of the other terms in (7.4.31).

For the 1-electron terms, again varying ϕ_r^* alone, we obtain

$$\left\langle \delta\Phi_\lambda \middle| \sum \mathsf{h} \middle| \Phi_\kappa \right\rangle = \sum_j D_{jr}^{\kappa\lambda} \langle \delta\phi_t | \mathsf{h} | \phi_j \rangle + \sum_{i,j(\neq r)} \delta_r D_{ji}^{\kappa\lambda} \langle \phi_i | \mathsf{h} | \phi_j \rangle,$$

where $\delta_r D_{ji}^{\kappa\lambda}$ is the first-order variation in the *coefficient of* $\langle \phi_i | \mathsf{h} | \phi_j \rangle$ $(i, j \neq r)$: but this will be expressible in terms of *two*-body density-matrix elements through the first of the recurrence relations in (7.4.25). Thus

$$\delta_r D_{ji}^{\kappa\lambda} = \sum_l D_{jl,ir}^{\kappa\lambda} \langle \delta\phi_r \mid \phi_l \rangle \quad (i \neq r, j \neq l),$$

and on substituting in the previous expression and using the result in (7.4.31) we find, after summation,

$$\sum C_\kappa C_\lambda^* \left\langle \delta\Phi_\lambda \middle| \sum \mathsf{h} \middle| \Phi_\kappa \right\rangle = \sum_{r,j} P_{jr}^{\kappa\lambda} \langle \delta\phi_r | \mathsf{h} | \phi_j \rangle$$

$$+ \sum_{i,r} \sum_{j,l} \Pi_{jl,ir}^{\kappa\lambda} \langle \delta\phi_r \mid \phi_l \rangle \langle \phi_i | \mathsf{h} | \phi_j \rangle. \qquad (7.4.33)$$

A similar procedure may be used for the 2-electron terms, and on collecting the results the variation (7.4.31) becomes

$$\sum_{k,\lambda} C_\kappa C_\lambda^* \langle \delta\Phi_\lambda | (\mathsf{H} - E) | \Phi_\kappa \rangle$$

$$= \sum_{r,s} P_{sr} \langle \delta\phi_r | \mathsf{h} | \phi_s \rangle + \sum_{r,t}{}' \sum_{s,u}{}' \Pi_{su,rt} \langle \delta\phi_r | V_{tu} | \phi_s \rangle$$

$$- E \sum_{r,s} P_{sr} \langle \delta\phi_r \mid \phi_s \rangle + \sum_{r,t}{}' \sum_{s,u}{}' \Pi_{su,rt} \langle \phi_t | \mathsf{h} | \phi_u \rangle \langle \delta\phi_r \mid \phi_s \rangle$$

$$+ \tfrac{1}{2} \sum_{s,u,w}{}' \sum_{r,t,v}{}' T_{uws,tvr} \langle \phi_t \phi_v | g | \phi_u \phi_w \rangle \langle \delta\phi_r \mid \phi_s \rangle, \qquad (7.4.34)$$

where no two indices in a primed summation are equal and T denotes a "triplet" density. The quantity V_{tu} is a 1-electron potential defined by

$$V_{tu}(1) = \int \frac{1}{r_{12}} \phi_t^*(\boldsymbol{r}_2) \phi_u(r_2) \, \mathrm{d}\boldsymbol{r}_2. \qquad (7.4.35)$$

The stationarity condition requires that (7.4.34) be zero for every $\delta\phi_r^*$.

At this point it is convenient, with no loss of generality, to assume normalized orbitals and to introduce the constraint $\langle \delta\phi_r \mid \phi_r \rangle = \langle \phi_r \mid \delta\phi_r \rangle = 0$, which conserves normalization to first order. The terms with $r = s$ in the last three summations in (7.4.34) may then be omitted. It is also expedient to define the operator F_r such that

$$\mathsf{F}_r | \phi_r \rangle = \sum_s P_{sr} (\mathsf{h} - E \eta_{rs}) | \phi_s \rangle + \sum_{s,t,u} \Pi_{su,rt} (\langle \phi_t | \mathsf{h} | \phi_u \rangle \eta_{rs} + V_{tu}) | \phi_s \rangle$$

$$+ \tfrac{1}{2} \sum_{s,t,u,v,w} (\eta_{rs} T_{uws,tvr} \langle \phi_t \phi_v | g | \phi_u \phi_w \rangle) | \phi_s \rangle, \qquad (7.4.36)$$

where the symbol η_{rs} ($=0$ for $r=s$, $=1$ for $r \neq s$) serves to exclude the unwanted terms with $r=s$. This definition of an effective 1-electron Hamiltonian, reminiscent of the Fock operator in Chapter 6, is the one adopted by Gerratt (1971); it is to be noted that, as in Section 6.7, there is a distinct operator for every orbital ϕ_r.

It is now evident that the condition for the vanishing of (7.4.34) may be written alternatively as

$$\langle \delta\phi_r | \mathsf{F}_r | \phi_r \rangle = 0 \quad (\text{all } r), \tag{7.4.37}$$

the only constraint being that $\delta\phi_r$ preserves normalization. But this is a *one-body* equation, and, as we know from Section 2.4, it simply expresses the stationary-value property of an "orbital energy"

$$\epsilon_r = \langle \phi_r | \mathbf{F}_r | \phi_r \rangle$$

for an electron in some kind of effective field (defined by F_r), under variations of the normalized function ϕ_r. The normalization constraint may be eliminated, as usual, by considering instead the functional

$$\epsilon_r = \frac{\langle \phi_r | \mathsf{F}_r | \phi_r \rangle}{\langle \phi_r | \phi_r \rangle}, \tag{7.4.38}$$

whose stationary values are, in turn, eigenvalues of the equation

$$\mathsf{F}_r \phi = \epsilon \phi. \tag{7.4.39}$$

These equations, one for every ϕ_r in the orbital configuration, serve to optimize the whole orbital set, while simultaneous solution of (7.4.30) optimizes the mixing of the "structures" Φ_κ corresponding to the various spin-coupling schemes.

With only a few electrons outside a closed shell, the above equations may be solved iteratively in essentially the same manner as the Hartree–Fock equations. Usually an algebraic approximation is used in which every ϕ_r is expressed as a linear combination of basis functions as in Section 6.2; and this leads to matrix equations that are more suitable for computational purposes. The actual techniques of optimization are similar to those used in multiconfiguration versions of MO SCF theory (Chapter 8): the more rapidly convergent procedures require also *second* derivatives of the energy expression, but these may be obtained in the same way as the first derivatives.

It should be noted that in the 1-configuration approximation used in this section only *one* solution of (7.4.39) is needed for each operator F_r—the ϕ_r that appears in the ground-state wavefunction. As in Hartree–Fock theory, however, "virtual" orbitals (ϕ_r', ϕ_r'', . . .) will also be obtained, and it appears (Gerratt and Raimondi, 1980) that such orbitals provide an excellent basis for improvement of the 1-configuration function by admitting CI. Multiconfiguration VB functions are considered in the next two sections.

Finally, we mention in passing that the "orbital energy" introduced in (7.4.38) is simply related neither to the total energy (7.4.26) nor to an ionization energy. After some rearrangement, making use of (7.4.25), it turns out that

$$\epsilon_r = P_{rr}E - \sum_{t,u} \Pi_{ru,rt}\langle \phi_r | \mathsf{h} | \phi_u \rangle - \tfrac{1}{2} \sum_{t,u,v,w} T_{uwr,tvr}\langle \phi_t\phi_v | g | \phi_u\phi_w \rangle, \tag{7.4.40}$$

which involves both the pair and triplet distribution functions. To obtain ionization potentials from a VB calculation, it is necessary either to perform a *second* calculation, for the ion, or to use quite different techniques (Chapter 13) to extract the desired information from the ground-state VB function.

7.5 MULTICONFIGURATION VB THEORY. ORTHOGONAL ORBITALS

In earlier sections of this chapter we have considered only *single*-configuration approximations, the many interacting CFs (Φ_κ) all containing the same occupied orbitals and differing only through the great variety of spin-coupling schemes for electrons in the singly occupied set. In the language of traditional VB theory, these CFs are "non-polar structures": when the orbitals are spatially localized on different atoms (for example the carbon 2p AOs in a conjugated hydrocarbon) the structures involve no electron transfer from one atom to another. On going beyond this simple approximation, however, we shall need to admit changes of orbital occupation that do correspond to electron transfer and can be pictured as "polar" or "ionic" structures. Indeed, such structures have been used already (p. 74) in the hydrogen-molecule calculation. Generally, an electron configuration $[\phi_1^1\phi_2^1 \ldots \phi_i^2 \ldots \phi_j^0 \ldots \phi_N^1]$ gives rise to spin-paired CFs in which link i—j of a non-polar structure becomes a "dummy" $i^- \ldots j^+$; and it will be necessary to consider all types of singly, doubly and multiply polar structures. Moreover, when the orbital set $\phi_1, \phi_2, \ldots, \phi_N$ is extended, by admitting a complementary set $\{\phi_1', \phi_2', \ldots\}$, we shall also have to consider configurations such as $[\phi_1^1\phi_2^1 \ldots \phi_i'^1 \ldots \phi_N^1]$ that describe a "local excitation" $\phi_i \to \phi_i'$ (and which may be polar or non-polar according to the spatial location of ϕ_i' relative to ϕ_i).

To implement a CI calculation using VB techniques, we shall need to evaluate matrix elements of the Hamiltonian between functions of the type (7.4.1) but with all possible selections of N orbitals (including repetitions) in the basic product Ω. It must be stressed at the outset that the *nature* of the orbitals is irrelevant at this level of generality. The connotations "MO" and "VB" refer essentially to the 1-configuration *reference function,* which in MO theory usually contains only doubly occupied orbitals but in VB theory only singly occupied orbitals: it is when the orbitals of this reference function are *optimized* that they become delocalized in the MO case but localized in the VB case. The techniques we develop here are therefore immediately applicable in the general CI approach, which is the subject of Chapter 10. Again, the calculation of matrix elements is immensely more difficult when the orbitals are allowed to be non-orthogonal. We therefore begin by assuming an *orthonormal* set as in Sections 7.2 and 7.3. In this case it is possible to derive a compact algorithm, based on the superposition patterns already introduced, for obtaining matrix elements of the Hamiltonian between spin-paired CFs belonging to arbitrary configurations. This result, in one form or another (Boys, 1955; Reeves, 1957, 1966; McWeeny, 1954b, 1955b; Cooper and McWeeny, 1966a,b; Sutcliffe, 1966), has been widely used in large-scale CI calculations (see e.g. Diercksen and Sutcliffe, 1974) and has been re-derived in various ways (e.g. Wormer, 1981).

To formulate the algorithm, we first recognize that, whereas a single label κ was sufficient to characterize the CFs in (7.2.2) and (7.4.1), a second label will now be required to indicate the orbital configuration. The general CF will be written

$$\Phi_{\kappa}^{(K)} = (N!\,2^{d_K})^{-1/2} \sum_{\mathsf{P}} \epsilon_{\mathsf{P}}(\mathsf{P}\Omega_K)(\mathsf{P}\Theta_\kappa), \qquad (7.5.1)$$

where κ still labels the spin eigenfunctions (normalized) but now the corresponding latin letter K specifies the orbital factor to which they are attached. More fully,

$$\Omega_K(\mathbf{r}_1, \mathbf{r}_2, \ldots, \mathbf{r}_N) = k_1(\mathbf{r}_1)k_2(\mathbf{r}_2) \ldots k_N(\mathbf{r}_N). \qquad (7.5.2)$$

and the label K therefore stands for a particular sequence of orbitals, denoted by $k_1k_2 \ldots k_N$, chosen from an arbitrary orbital basis $\{\phi_1, \phi_2, \ldots, \phi_m\}$, no orbital appearing more than twice. The normalizing factor in (7.5.1) is appropriate when d_K is the number of "doubles" (i.e. doubly occupied orbitals) in Ω_K.

The general matrix element for two functions, $\Phi_{\kappa}^{(K)}$ and $\Phi_{\lambda}^{(L)}$ (Ω_L

containing orbitals $l_1, l_2, \ldots, l_N$) may now be written

$$\langle \Phi_\lambda^{(L)} | \mathsf{H} | \Phi_\kappa^{(K)} \rangle = (2^{d_L+d_K})^{-1/2} \sum_{\mathsf{P}} \epsilon_{\mathsf{P}} \langle \Omega_L | \mathsf{H} | \mathsf{P}\Omega_K \rangle \langle \Theta_\lambda | \mathsf{P}\Theta_\kappa \rangle, \quad (7.5.3)$$

where the properties of the antisymmetrizer have been used (eliminating the $N!$) as in the derivation of (7.2.5). It will be assumed that orbitals *common* to Ω_K and Ω_L appear in coincident positions (i.e. they are in fully "matched" order); if this is not so, they may be matched by applying a "line-up" permutation to orbitals and corresponding spin factors in one of the functions, adding a phase factor ± 1 to the matrix element according to parity. With orthonormal orbitals the only non-trivial permutations needed in (7.5.3) are single transpositions involving at least one singly occupied orbital; and the corresponding spin-matrix elements follow easily from (7.3.1).

With the usual spinless Hamiltonian, the results (Cooper and McWeeny, 1966a,b) may be written in a simple and general form reminiscent of Slater's rules (p. 63). The overlap integral vanishes except within a configuration: it has the value

$$\langle \Phi_\lambda^{(L)} | \Phi_\kappa^{(K)} \rangle = \delta_E \Delta_{\lambda\kappa} \delta_{LK}, \quad (7.5.4)$$

where

$$\Delta_{\lambda\kappa} = (-1)^{\nu_{\lambda\kappa}} 2^{n_{\lambda\kappa} - g}, \quad (7.5.5)$$

and the quantities involved are defined in (7.3.1) *et seq.*; this factor depends only on the spin-coupling schemes.

The matrix element (7.5.3) is non-zero only if there are no more than two E-chains, and then reduces to

$$\langle \Phi_\lambda^{(L)} | \mathsf{H} | \Phi_\kappa^{(K)} \rangle = \delta_E \omega_{LK} \Delta_{\lambda\kappa} \Big[\sum_i Q_i^{LK} \langle l_i | \mathsf{h} | k_i \rangle + \tfrac{1}{2} \sum_{i,j}{}' Q_{ij}^{LK} \langle l_i l_j | g | k_i k_j \rangle \Big]$$

$$- \tfrac{1}{2} \omega_{LK} \Delta_{\lambda\kappa} \Big[\sum_{i,j}{}' Q_{ij}^{LK} x_{ij}^{\lambda\kappa} \eta_{ij}^{LK} \langle l_i l_j | g | k_j k_i \rangle \Big]. \quad (7.5.6)$$

The Q-coefficients have the value unity within a configuration ($l_p = k_p$, all p), while more generally

$$\left. \begin{aligned} Q_i^{LK} &= \begin{cases} 1 & (l_p = k_p, \text{ all } p \neq i), \\ 0 & \text{otherwise}; \end{cases} \\ Q_{ij}^{LK} &= \begin{cases} 1 & (l_p = k_p, \text{ all } p \neq i,j) \\ 0 & \text{otherwise}. \end{cases} \end{aligned} \right\} \quad (7.5.7)$$

The sums in (7.5.6) consequently include only terms that contain 0, 1 or 2

Table 7.1 Coefficients $x_{ij}^{\lambda\kappa}$.

	i	j	$x_{ij}^{\lambda\kappa}$
No E-chains	Island I	Island I	$\frac{1}{2}(3p_{ij}+1)$‡
	O-chain J	O-chain J	$\frac{1}{2}(3p_{ij}+1)$
	Island I	Island I′ (≠I) or O-chain J	$\frac{1}{2}$
	O-chain J	O-chain J′ (≠J)	$\frac{1}{2}(p_{ij}+1)$
Two E-chains	E-chain K	E-chain K′ (≠K)	p_{ij}

‡ The relative positions of points i, j in the superposition pattern are indicated in the two central columns; and $p_{ij}=p_ip_j$, where the "parity factor" $p_i=\pm1$ is assigned by giving $+1$ to an arbitrary position in an island, or to an *end* point in a chain, and then proceeding along the sequence giving ±1 to alternate positions. For E-chains, the endpoint chosen is the one where the arrow points *into* the chain.

mismatching orbitals in Ω_K and Ω_L (cf. Slater's rules, p. 63). The quantity ω_{LK}, which is a *configurational* factor depending on Ω_L and Ω_K, has the value

$$\omega_{LK}=2^{(d_L+d_K-2d_{LK})/2}, \tag{7.5.8}$$

where d_L and d_K are, as previously, the numbers of doubles in Ω_L and Ω_K, while d_{LK} is the number of *coincident* doubles§ when (as assumed) the two products have been matched by the line-up permutation. The other configurational factor needed is

$$\eta_{ij}^{LK}=\begin{cases}1 & (l_i=k_i \text{ and } l_j=k_j),\\ 0 & (\text{otherwise})\end{cases} \tag{7.5.9}$$

The remaining coefficients in (7.5.6), like $\Delta_{\lambda\kappa}$, depend only on the spin-coupling schemes in Θ_λ and Θ_κ: they are

$$x_{ij}^{\lambda\kappa}=\langle\Theta_\lambda|\,\mathsf{P}_{ij}\,|\Theta_\kappa\rangle, \tag{7.5.10}$$

and arguments similar to those used in establishing the rule on p. 223 lead to the results summarized in Table 7.1.

In spite of recent developments in matrix-element evaluation (Chapter 10), (7.5.6) remains probably the simplest and most compact algorithm available: it may be used to obtain the formula for any desired matrix

§ For example, using a temporary notation with $\{\phi_i\}=\{a, b, c, \ldots\}$, Ω_L and Ω_K after matching might be *aa bb cc def* and *aa bb ce def*. In this case the *ee*-double in the second product has been "broken" by lining up the orbitals. Although $d_L=d_K=3$, the exclusion of the broken double leaves only $d_{LK}=2$ in coincident positions.

element (possibly involving, implicitly, many thousands of Slater determinants) in two or three minutes, using nothing more than a pencil and the back of an envelope!

More general rules, applicable to all the more common spin-dependent operators that occur as "small terms" in the Hamiltonian (see Chapter 11) are also available (Cooper and McWeeny, 1966a,b). The complete results, including those already presented, are conveniently expressed in terms of the transition densities $P(K\kappa, L\lambda \mid \mathbf{r}_1; \mathbf{r}_1')$ etc. connecting the various CFs, and from these it is easy to write down the matrix elements of any desired operator. The connection between the matrix elements and the density functions has been dealt with in Section 5.4. We shall not list all the results; but two are of special importance—the charge- and spin-density matrices. Thus, by inspection of the 1-electron terms in (7.5.6), it is evident (cf. (5.4.11)) that

$$P(K\kappa, L\lambda \mid \mathbf{r}_1; \mathbf{r}_1') = \delta_E \omega_{LK} \Delta_{\lambda\kappa} \sum_i Q_i^{LK} k_i(\mathbf{r}_1) l_i^*(\mathbf{r}_1'), \tag{7.5.11}$$

and there is a similar result for the spin density:

$$Q_S(K\kappa, L\lambda \mid \mathbf{r}_1; \mathbf{r}_1') = \delta_E \omega_{LK} \Delta_{\lambda\kappa} \sum_i x_i^{\lambda\kappa} Q_i^{LK} k_i(\mathbf{r}_1) l_i^*(\mathbf{r}_1'), \tag{7.5.12}$$

where, with the parity factor defined in Table 7.1,

$$x_i^{\lambda\kappa} = \begin{cases} \frac{1}{2}p_i & \text{(position } i \text{ in an O-chain),} \\ 0 & \text{otherwise.} \end{cases} \tag{7.5.13}$$

These results applying to spin states with $M = S$) are invariant against change of M, except for (7.5.12), which for other states of a multiplet is multiplied by a suitable factor as in (5.9.10). The other density functions defined in Chapter 5 all follow from the more general rules cited above.

In order to apply the algorithm (7.5.6) in the original context of VB theory, where the orbitals employed are essentially the AOs on the various atoms, it will clearly be necessary to start by *orthonormalizing* the orbitals, since orthonormality has been a cornerstone in all the derivations. It is certainly possible to do this, using for example the Löwdin transformation (2.2.21), which leads to orbitals (OAOs) that retain an essentially "atomic" localization and still behave under symmetry operations in the same way as the original AOs. But a disappointing feature of this approach (McWeeny, 1955a; Campion and Karplus, 1973; Raimondi and Simonetta, 1976) is the very slow convergence of the CI calculation. In particular, the use of "covalent" structures alone fails to describe bonding (leading to negative overlap populations and a negative binding energy relative to the free atoms). This discouraging situation is only remedied when large numbers of *polar* structures are admitted. For this reason, coupled with the fact that a transformation of all 1- and 2-electron integrals to an OAO basis is just as time-consuming as a transformation to an MO basis (which has alternative advantages), VB methods based on orthogonalized AOs have not been widely adopted. Whether these objections remain valid in the light of improved techniques for large-scale CI calculations (using 10^4–10^6 CFs) remains to be seen; since it is certainly true that VB structures possess a pictorial significance (not shared by MO CFs) that permits the use of a certain degree of "physical intuition" in choosing an appropriate set. At this point, however, we return to the more traditional VB approach and to more direct methods of meeting the difficulties arising from non-orthogonality.

7.6 MULTICONFIGURATION VB THEORY. NON-ORTHOGONAL ORBITALS

The difficulties of dealing with non-orthogonal orbitals are once again of practice rather than principle. With the same notation as in the last section, we now need overlap and matrix-element formulae for two functions of type (7.5.1). As in (7.5.3), we start from

$$\langle \Phi_\lambda^{(L)} | \mathsf{H} | \Phi_\kappa^{(K)} \rangle = \sum_{\mathsf{P}} \epsilon_{\mathsf{P}} \langle \Omega_L | \mathsf{H} | \mathsf{P}\Omega_K \rangle \langle \Theta_\lambda | \mathsf{P}\Theta_\kappa \rangle, \qquad (7.6.1\text{a})$$

$$\langle \Phi_\lambda^{(L)} | \Phi_\kappa^{(K)} \rangle = \sum_{\mathsf{P}} \epsilon_{\mathsf{P}} \langle \Omega_L | \mathsf{P}\Omega_K \rangle \langle \Theta_\lambda | \mathsf{P}\Theta_\kappa \rangle, \qquad (7.6.1\text{b})$$

and consider the problem of including all permutations. The spin functions may be of the type used in the last section, this being the traditional Heitler–London–Weyl–Rumer approach, or may be the branching-diagram functions, which lend themselves more readily to the use of group-theoretical methods. The latter have been used most widely; but so far no explicit formula, even for the overlap integral (7.6.1b), has been obtained by any method.

In this somewhat unsatisfactory situation, there are two main ways of proceeding. In the first, every function $\Phi_\kappa^{(K)}$ is expressed as a linear combination of Slater determinants (by expanding each Θ_κ in terms of elementary spin products, attaching the spatial factor Ω_K, and antisymmetrizing): the matrix elements then become sums of contributions from all pairs of determinants, and these may be evaluated, even with non-orthogonal spin-orbitals, by using (3.3.11), (3.3.17) and (3.3.18). This method has been developed into a useful tool by Simonetta *et al.*

(1968), Balint-Kurti and Karplus (1968, 1974) and by Gerratt and Raimondi and coworkers (see e.g. Pyper and Gerratt, 1977; Raimondi *et al.*, 1975, 1985; Gerratt and Raimondi, 1980; Cooper *et al.*, 1987), but although its efficient implementation poses severe and interesting problems (for example the calculation of the determinantal cofactors for spin-orbital sets whose overlap matrix may be singular), the basic theory itself requires no further discussion.

The second method, which we now develop, is due mainly to Gallup and coworkers (for a review see Gallup *et al.*, 1982), and may be regarded as another application (cf. p. 226) of the "spin-free" approach (Matsen, 1964): it leads to rather efficient procedures, and also involves ideas needed in Chapter 10 that are conveniently introduced at this point.

The CFs that appear in (7.6.1) may be written, as in Section 7.4, in two forms:

$$\Phi_\kappa^{(K)} = (N!)^{1/2} A[\Omega_K \Theta_\kappa], \qquad (7.6.2a)$$

which is comparable to (7.4.1), or

$$\Phi_\kappa^{(K)} = (f_S^N)^{-1/2} \sum_\mu F_\mu^{(K\kappa)} \Theta_\mu, \qquad (7.6.2b)$$

which corresponds to (7.4.7). The spatial factor in (7.6.2b) is a projected function, obtained from the product Ω_K by using the operator (7.4.5):

$$F_\mu^{(K,\kappa)} = \rho_{\mu\kappa}\Omega_K = \left(\frac{f_S^N}{N!}\right)^{1/2} \sum_{\mathsf{P}} V(\mathsf{P})_{\mu\kappa} \mathsf{P}\Omega_K, \qquad (7.6.3)$$

with the usual definition of the matrices $\mathbf{V}(\mathsf{P})$. The exact space–spin wavefunction may therefore be expanded in the form

$$\Psi = \sum_{K,\kappa} C_{K\kappa} \Phi_\kappa^{(K)} = (f_S^N)^{-1/2} \sum_{K,\kappa,\mu} C_{K\kappa} F_\mu^{(K,\kappa)} \Theta_\mu = (f_S^N)^{-1/2} \sum_\mu F_\mu \Theta_\mu \quad (7.6.4)$$

(cf. (4.3.3)), in which the spin-free wavefunction F_μ (in principle an eigenfunction of H) is

$$F_\mu = \sum_{K,\kappa} C_{K\kappa} F_\mu^{(K,\kappa)}. \qquad (7.6.5)$$

Since the matrix elements of H between different terms in (7.6.4) vanish, through spin orthogonality, while the f_S^N diagonal terms must be identical by symmetry (Appendix 3), the energy becomes (for a normalized F_μ)

$$E_\mu = \langle F_\mu | \mathsf{H} | F_\mu \rangle,$$

and is the same for all μ. It will thus be sufficient to give μ any convenient value in the expansion (7.6.5), determining the coefficients from secular equations that contain matrix elements between the spinless functions $F_{\mu}^{(K,\kappa)}$.

At this point a simplification is possible. The f_S^N functions $F_{\mu}^{(K,\kappa)}$ (variable κ) are produced, according to (7.6.3), from a *single* configurational product Ω_K in which, by convention, the index set K may be ordered, with $k_1 < k_2 < k_3 \ldots$; but they apparently require knowledge of the Wigner operators $\rho_{\mu\kappa}$ for all κ. Instead, however, it is possible (Problem 7.18) to use a *single Wigner operator* $\rho_{\mu\mu}$ working in turn on f_S^N different products Ω_K in which K will contain the same indices but in non-standard order. We shall see later how the non-standard products may be chosen: but for the present it is sufficient to note that (7.6.5) may now be replaced, on identifying μ and κ and dropping the redundant label, by

$$F = \sum_K C_K F_K, \qquad (7.6.6)$$

where, making the choice $\mu = \kappa = 1$ in (7.6.3),

$$F_K = F_1^{(K,1)} = \rho_{11}\Omega_K. \qquad (7.6.7a)$$

The spatial function (7.6.6) transforms under permutations like the *first* basis vector of the irrep with matrices $\mathbf{V}(\mathsf{P})$; but the choice $\mu = \kappa = 1$ is purely conventional.

In Gallup's development the choice $\mu = 1$ corresponds to the first (i.e. "fundamental") tableau for the related spin function Θ_μ: and the corresponding "dual" tableau for F_μ in (7.6.4) is obtained by an interchange of rows and columns. For the example $N = 5$, $S = \frac{1}{2}$ the tableaux (Section 4.4) associated with $F_1^{(K,\kappa)}$ and Θ_1 in (7.6.4) will thus be as in Fig. 7.1(a,b). The orbital product Ω_K may then be indicated by putting the indices k_1, k_2, k_N in the boxes that hold the spatial variables $1, 2, \ldots, N$ to obtain the "Weyl tableau" shown in Fig. 7.1(c). Other products for use in (7.6.7) will then be obtained by taking the same

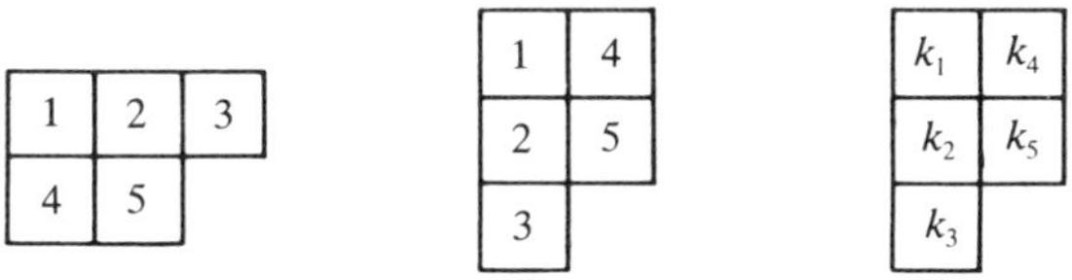

(a) (b) (c)

Fig. 7.1 Tableaux for $N = 5$, $S = \frac{1}{2}$: (a) fundamental tableau for D_S; (b) dual tableau; (c) Weyl tableau.

selection of indices (i.e. configuration K) but putting them in a different order; thus if k_2 and k_4 are interchanged in Fig. 1(c) the function $F_{K'}$ associated with the new tableau will be generated as in (7.6.7) but from the product $\Omega_{K'} = k_1(1)k_4(2)k_3(3)k_2(4)k_5(5)$. Here we use the Weyl tableaux only as a means of identifying the CFs F_K; their deeper significance is considered in Chapter 10.‡

Instead of using the Wigner operator, Gallup (1973) found it expedient to employ a particular kind of Young idempotent (Section 4.4), namely the operator NPN—rather than the usual NP—in which the row and column permutations move the *indices* in the *Weyl tableau* indicated in Fig. 7.1(c). The NPN operator is a true idempotent when multiplied by the factor $f_S^N/N!$, and the projected function (not normalized) will be written

$$F_K = \theta \mathsf{NPN} \Omega_K \quad (\theta = f_S^N/N!). \tag{7.6.7b}$$

It remains to discuss (i) the form of the F_K, (ii) the evaluation of matrix elements $\langle F_L | \mathsf{H} | F_K \rangle$ and (iii) the choice of linearly independent set for the CI calculation. With extensive CI, the optimization of the orbitals becomes less important (cf. Section 7.4, where only *one* configuration was admitted) and will not be considered.

The projected function

Intuitively, since an N operator *anti*symmetrizes with respect to numbers in the two columns (which are normally longer than the rows in the present case), an NPN operator is expected to produce an orbital function in some sense "closest" to a single determinant. And, in order to exploit the properties of determinants, it is advantageous to replace Ω_K in (7.6.7) by an orbital determinant D_K. Thus, with a partitioning suggested by the 2-column tableau

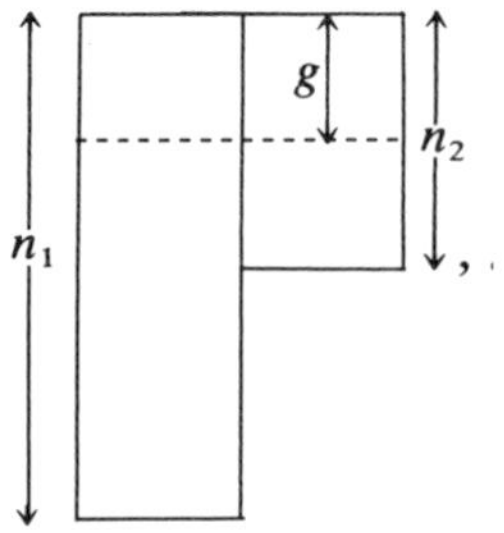

‡ It is also noteworthy that Gallup's choice of tableau ($\mu = 1$) may be replaced by others; the *final* tableau (Problem 7.18) offers other advantages.

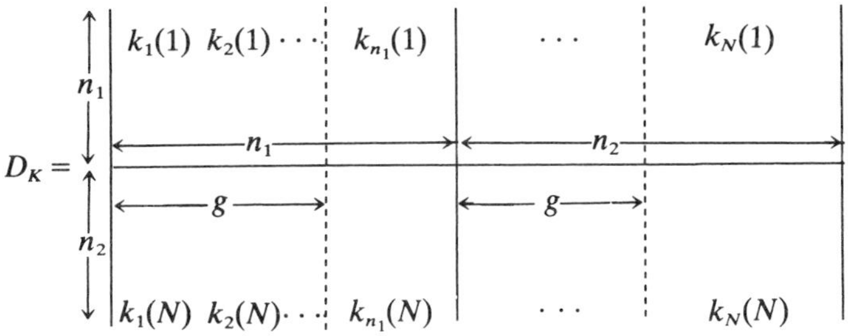

. (7.6.8)

The blocks of dimension n_1 and n_2 are partitioned further (broken lines) in the presence of any doubly occupied orbitals (g pairs, say), which may always be written in the first g rows of the tableau. Clearly,

$$n_1 + n_2 + N, \qquad n_1 - n_2 = 2S. \tag{7.6.9}$$

It may then be shown (Gallup, 1973) that the effect of $\theta(\mathsf{NPN})$ on D_K is to produce a linear combination of modified or "masked" determinants. When $g = 0$ the expansion is

$$F_K = \theta(\mathsf{NPN})D_K = \theta \sum_{\alpha=1}^{\alpha_{\max}} AD_K^{\mathrm{M}}(q_\alpha), \tag{7.6.10}$$

where the masked determinant, which contains a real variable q, is defined by

$$D_K^{\mathrm{M}}(q) = \mathbf{w}(q):D_K = \begin{array}{|c|c|} \hline D_K^{11} & D_K^{12} \\ \hline D_K^{21} & qD_K^{22} \\ \hline \end{array} \tag{7.6.11}$$

The effect of the "masking matrix" $\mathbf{w}(q)$ is simply to multiply the 22-block of D_K by the variable q. The values of $d_\alpha(N, S)$, $q_\alpha(N, S)$ and $\alpha_{\max}(N, S)$ have been tabulated (Gallup *et al.*, 1982) for up to 10 electrons, with total spin up to $S = \frac{5}{2}$. In such cases (7.6.10) includes no more than three terms and seems to be the most economical representation possible in terms of determinants. When $g = 0$ there is a further reduction: the first g columns of the 21- and 12-blocks of D_k in (7.6.11) are replaced by zeros, the parameters are calculated using $N - 2g$ instead of N, and the constant θ is multiplied by $(\frac{1}{2}N + S + 1)/(\frac{1}{2}N + S + 1 - g)$.

Matrix elements

To set up the secular equations, using the expansion (7.6.6), we need

$$H_{LK} = \langle F_L | \mathsf{H} | F_K \rangle = \langle \Omega_L \mid F_K \rangle, \tag{7.6.12a}$$

$$M_{LK} = \langle F_L \mid F_K \rangle = \langle \Omega_L \mid F_K \rangle, \tag{7.6.12b}$$

where the projected functions (7.6.7) are obtained from the products $\Omega_L = l_1 l_2 \ldots l_N$ and $\Omega_K = k_1 k_2 \ldots k_N$, and the idempotency of the operator has been used in passing to the final forms. Evidently H_{LK} and M_{LK} depend on the expansion of F_K (not F_L), and maximum reduction will occur when the product with the greater number of doubly occupied orbitals is put on the right (as we shall assume); the results will then involve the masking matrix $\mathbf{w}_K(q)$ for the *ket* function F_K.

As in Section 3.3, a key role is played by the overlap determinant

$$D_{LK} = \det |\langle l_1 | k_1 \rangle \ \langle l_2 | k_2 \rangle \ \ldots \ \langle l_N | k_N \rangle|. \tag{7.6.13}$$

Now, however, the determinant will be masked as in (7.6.11), becoming

$$D^M_{LK}(q) = \det |w_{rs}(q) \langle l_r | k_s \rangle|, \tag{7.6.14}$$

where w_{rs} ($=1$ or q as appropriate) is attached to the rs-element of D_{LK}. The non-orthogonality integral is then given by

$$M_{LK} = \theta \sum_\alpha d_\alpha D^{\mathrm{M}}_{LK}(q_\alpha). \tag{7.6.15}$$

In the presence of doubly occupied orbitals the parameters are modified as indicated above and the masked determinants acquire two zero-blocks.

Matrix elements of the Hamiltonian may be written in the usual form (cf. p. 132)

$$H_{LK} = \sum_{r,s} A_{sr} \langle l_r | \mathrm{h} | k_s \rangle + \tfrac{1}{2} \sum_{r,s,t,u} B_{tu,rs} \langle l_r l_s | g | k_t k_u \rangle, \tag{7.6.16}$$

where now the coefficients involve cofactors (signed minors) of the masked overlap determinants. In full,

$$\left.\begin{aligned} A_{sr} &= \theta \sum_\alpha d_\alpha w_{rs}(q_\alpha) [D^{\mathrm{M}}_{LK}(q_\alpha)]_{r,s}, \\ B_{tu,rs} &= \theta \sum_\alpha d_\alpha w_{rs}(q_\alpha) w_{tu}(q_\alpha) \epsilon_{rs} \epsilon_{tu} [D^{\mathrm{M}}_{LK}(q_\alpha)]_{tu,rs}, \end{aligned}\right\} \tag{7.6.17}$$

where, for example, $[D^{\mathrm{M}}_{lk}(q_\alpha)]_{r,s}$ is the cofactor obtained from the masked determinant by striking out the rth row and the sth column. As in Section 3.3, $\epsilon_{rs} = 1$ $(r < s)$, $= -1$ $(r > s)$.

It is now clear that, as in the direct method of expansion in Slater determinants (p. 239), the practical feasibility of calculations by this method will depend heavily on the availability of efficient algorithms for the computation of cofactors. This problem, which we do not consider further, has been studied in particular by King *et al.* (1967), Prosser and Hagstrom (1968), and again by Gallup *et al.* (1982).

Linearly independent sets

It was noted above that when using a single projector it is not sufficient to interpret the symbol K as a *single* sequence of orbitals (for example in the standard order $k_1 \leqslant k_2 \leqslant k_3 \ldots$); instead, K will denote a particular product in which the order of the factors is important. Thus, if Ω_K and $\Omega_{K'}$ are products containing the same selection of orbitals, but in a different order, the projected functions F_K and $F_{K'}$ may in general be linearly independent. For any given selection of orbitals then (i.e. for an orbital configuration as usually defined), we must find a set of permuted products that is just sufficient to yield the full set of linearly independent projected functions, with no repetitions or redundancies. In the case of 5 electrons, for example, there are 120 permutations of 5 distinct orbitals, but we can find only 5 CFs of given symmetry type.

This problem is solved in terms of the Weyl tableaux (Fig. 7.1c) derived from the first Young tableau by inserting *orbitals* (or their indices) in the boxes in place of the electron indices; thus, with 5 different orbitals, the products

$$\Omega_K = \phi_1\phi_2\phi_3\phi_4\phi_5, \qquad \Omega_{K'} = \phi_1\phi_2\phi_5\phi_3\phi_4$$

(electron indices always in the order $1, 2, \ldots, 5$) will yield two distinct Weyl tableaux

$$\begin{array}{|c|c|}\hline \phi_1 & \phi_4 \\ \hline \phi_2 & \phi_5 \\ \hline \phi_3 & \\ \cline{1-1} \end{array} \qquad \begin{array}{|c|c|}\hline \phi_1 & \phi_3 \\ \hline \phi_2 & \phi_4 \\ \hline \phi_5 & \\ \cline{1-1} \end{array}$$

and these may be used to indicate both the products Ω_K, $\Omega_{K'}$ and the functions F_K, $F_{K'}$ (of common symmetry) obtained from them by using the operator (7.6.3) with our conventional choice $\kappa = \mu = 1$ (or alternatively using (7.6.7b)).

Since, according to (7.4.10), permutations of electron labels are equivalent to inverse permutations of orbitals, it is not surprising that there is a one-to-one correspondence between the functions we need and the *standard* Young tableaux. If we adopt a similar convention, with orbital indices increasing along rows and down columns, we obtain standard Weyl tableaux

$$\begin{array}{|c|c|}\hline \phi_1 & \phi_4 \\ \hline \phi_2 & \phi_4 \\ \hline \phi_3 & \\ \cline{1-1} \end{array} \quad \begin{array}{|c|c|}\hline \phi_1 & \phi_3 \\ \hline \phi_2 & \phi_5 \\ \hline \phi_4 & \\ \cline{1-1} \end{array} \quad \begin{array}{|c|c|}\hline \phi_1 & \phi_2 \\ \hline \phi_3 & \phi_5 \\ \hline \phi_4 & \\ \cline{1-1} \end{array} \quad \begin{array}{|c|c|}\hline \phi_1 & \phi_3 \\ \hline \phi_2 & \phi_4 \\ \hline \phi_5 & \\ \cline{1-1} \end{array} \quad \begin{array}{|c|c|}\hline \phi_1 & \phi_2 \\ \hline \phi_3 & \phi_4 \\ \hline \phi_5 & \\ \cline{1-1} \end{array}$$

$$(\phi_1\phi_2\phi_3\phi_4\phi_5)\ (\phi_1\phi_2\phi_4\phi_3\phi_5)\ (\phi_1\phi_3\phi_4\phi_2\phi_5)\ (\phi_1\phi_2\phi_5\phi_3\phi_4)\ (\phi_1\phi_3\phi_5\phi_2\phi_4)$$

(the corresponding products being shown beneath each tableau), and the corresponding F_K, $F_{K'}$, . . . do in fact form a full independent set for the given orbital configuration.

When repeated orbitals occur the definition of the standard tableaux requires a very slight modification: standard Weyl tableaux are those in which the orbital indices *increase* down the columns and are *non-decreasing* along the rows. This definition excludes the possibility of identical orbitals occurring in the same column (such a function would be destroyed by the column antisymmetrizer), but permits identical orbitals in the same row (symmetry in spatial variables then corresponding to paired spins in the traditional formulation). As may be seen from the tableaux above, the number of Weyl tableaux needed is reduced when doubly occupied orbitals occur: thus on choosing $k_4 = k_2 = \phi_2$ (ϕ_2 doubly occupied), the second and last tableaux must be excluded, while the third differs from the first only in the non-standard order in the first column. There remains only the standard set

$$\begin{array}{|c|c|}\hline \phi_1 & \phi_2 \\ \hline \phi_2 & \phi_5 \\ \hline \phi_3 & \\ \cline{1-1} \end{array} \qquad \begin{array}{|c|c|}\hline \phi_1 & \phi_2 \\ \hline \phi_2 & \phi_3 \\ \hline \phi_5 & \\ \cline{1-1} \end{array}$$

as defined above. Evidently, there are only two linearly independent CFs for the configuration $[\phi_1^1\phi_2^2\phi_3^1\phi_5^1]$, and these may be generated from the products

$$\Omega_K = \phi_1\phi_2\phi_3\phi_2\phi_5, \qquad \Omega_{K'} = \phi_1\phi_2\phi_5\phi_2\phi_3.$$

The choice of linearly independent set is not of course unique, but the definition by means of standard Weyl tableaux has the merits of great simplicity and convenience.

Finally, it is possible to derive a simple formula (Weyl, 1956; Mulder, 1966) for the *total* number of linearly independent CFs of given S, M, with N electrons and m available orbitals: it is

$$\boxed{D(m, N, S) = \frac{2S+1}{m+1}\binom{m+1}{\frac{1}{2}N+S+1}\binom{m+1}{\frac{1}{2}N-S}} \qquad (7.6.18)$$

—a number that, owing to the factorials in the binomial coefficients, rapidly becomes astronomical even for small systems. We return in Chapter 10 to some of the topics considered in this section, and in particular to the solution of the very-high-order secular equations that determine the CI coefficients.

7.7 A NON-VARIATIONAL METHOD

It is by now clear that there is no simple solution to the non-orthogonality problem; the calculation of matrix elements in the VB secular equations, with its intrinsic dependence on $N!$ permutations, is bound to be long and laborious unless N is quite small. In this section we discuss a radically different approach in which non-orthogonality difficulties are completely removed—but only at the expense of sacrificing the variational principle and the upper-bound properties to which it leads. In this approach, based on the "method of moments" (for a brief review see Bangudu *et al.*, 1974; see also Dalton, 1980) the wavefunction expansion is unchanged but the expansion coefficients and energies are *not* obtained from the orthodox secular equations. Instead, a *second set* of expansion functions is utilized, these functions being in principle arbitrary but in practice designed to make life easier.

Let us start from the usual expansion‡ $\Psi = \sum_\kappa C_\kappa \Phi_\kappa$, which in the limit of a complete set may be assumed exact. In this case

$$(\mathsf{H} - E)\Psi = \sum_\kappa C_\kappa(\mathsf{H} - E)\Phi_\kappa = 0, \tag{7.7.1}$$

and on multiplying from the left by any member of an infinite set of *arbitrary* functions $\{\tilde{\Phi}_\lambda\}$ and integrating in the usual way,

$$\sum_\kappa \langle \tilde{\Phi}_\lambda | \mathsf{H} | \Phi_\kappa \rangle C_\kappa = E \sum_\kappa \langle \tilde{\Phi}_\lambda | \Phi_\kappa \rangle C_\kappa \quad (\text{all } \lambda). \tag{7.7.2}$$

This system of linear equations will determine the coefficients C_κ and corresponding eigenvalues E; but it is not the usual equation system; and when the set is truncated it is not generally possible to put bounds on the approximate eigenvalues i.e. the Hylleraas–MacDonald theorem (Section 2.3) no longer applies.

It is convenient to write (7.7.2) in the matrix form

$$\tilde{\mathbf{H}}\mathbf{C} = \mathbf{E}\tilde{\mathbf{M}}\mathbf{C}, \tag{7.7.3}$$

where the matrix elements contain the arbitrary function in the bra factor:

$$\tilde{H}_{\lambda\kappa} = \langle \tilde{\Phi}_\lambda | \mathsf{H} | \Phi_\kappa \rangle, \qquad \tilde{M}_{\lambda\kappa} = \langle \tilde{\Phi}_\lambda | \Phi_\kappa \rangle. \tag{7.7.4}$$

The choice of the bra functions is open so far. In some of the first applications of the approach in quantum chemistry (Boys, 1969; Boys and

‡ For ease of exposition, we resume the usual practice of labelling the CFs with a single subscript κ.

Handy, 1969), a *one*-term Ψ and its partner $\tilde{\Psi}$ were chosen as

$$\Psi = F\Phi, \qquad \tilde{\Psi} = F^{-1}\Phi,$$

where Φ was a single determinant of orthogonal orbitals and F a "correlation factor" (a symmetric function of interelectronic distances). Such functions are not of VB form, but serve to illustrate the approach. The matrices in (7.7.3) are then 1-dimensional, and the approximate energy is

$$E = \frac{\langle \tilde{\Psi} | \mathsf{H} | \Psi \rangle}{\langle \tilde{\Psi} | \Psi \rangle}. \tag{7.7.5}$$

The normalization integral $\langle F^{-1}\Phi \mid F\Phi \rangle$ clearly reduces to $\langle \Phi \mid \Phi \rangle$, while the single element of $\tilde{\mathbf{H}}$ becomes $\langle F^{-1}\Phi | \mathsf{H} | F\Phi \rangle$, and, although not trivial, is considerably easier to evaluate than the normal expectation value $\langle F\Phi | \mathsf{H} | F\Phi \rangle$. If Ψ $(=F\Phi)$ were *exact,* (7.7.5) would of course give the exact energy, no matter what the choice of $\tilde{\Psi}$. In other cases E is much closer (Boys, 1969) to the rather accurate energy obtainable from the correlated function Ψ than to the poor expectation energy associated with the "anticorrelated" function $\tilde{\Psi}$. Rigorous error bounds, however, have not been established, and, owing also to various computational problems, this particular method does not seem to have fulfilled its initial promise.

With a finite set of functions $\{\Phi_\kappa\}$, the truncated from of (7.7.3) may yield upper bounds to energy eigenvalues in one important case. If the bra and ket functions span the same many-electron space (i.e. form alternative linearly independent bases) then we may assume

$$\tilde{\mathbf{\Phi}} = \mathbf{\Phi}\mathbf{V}, \tag{7.7.6}$$

where $\mathbf{V}$ is a transformation matrix (in general non-unitary). From (7.7.4) it then follows that

$$\tilde{\mathbf{H}} = \mathbf{V}^\dagger\mathbf{H}, \qquad \tilde{\mathbf{M}} = \mathbf{V}^\dagger\mathbf{M}, \tag{7.7.7}$$

where $\mathbf{H}$ and $\mathbf{M}$ are the matrices as usually defined for the single basis $\{\Phi_\kappa\}$. In this case (7.7.3) yields

$$\mathbf{V}^\dagger(\mathbf{HC} - E\mathbf{MC}) = \mathbf{0}, \tag{7.7.8}$$

which is satisfied when and only when E and $\mathbf{C}$ are eigenvalues and eigenvectors of $\mathbf{H}$ (with metric $\mathbf{M}$) in the usual way: the solutions of (7.7.3) will then share the property of the usual eigenvalue equation in providing upper bounds to exact energy levels. This observation has been exploited in VB calculations by Norbeck and McWeeny (1975).

The functions $\{\tilde{\Phi}_\kappa\}$ and $\{\Phi_\kappa\}$ will span the same space when the individual orbitals ($\tilde{\phi}_i$, say) used in constructing the bra functions are linear combinations of the ϕ_i used in constructing the ket functions, and *when full CI is admitted.*‡ As a truncated set of CFs is expanded, the results will then converge towards those of a full-CI calculation: the convergence will not in general be monotonic, but, if sufficiently rapid, may be useful.

In the context of VB theory, the ket orbitals will normally be AOs, while it is most useful to construct the bra orbitals $\{\tilde{\phi}_i\}$ so that

$$\langle \tilde{\phi}_i \mid \phi_j \rangle = \delta_{ij}. \tag{7.7.9}$$

The two sets $\{\tilde{\phi}_i\}$ and $\{\phi_i\}$ are often said to be "bi-orthogonal", and the relationship between them is easily seen to be

$$\tilde{\boldsymbol{\phi}} = \boldsymbol{\phi}\mathbf{S}^{-1} = \boldsymbol{\phi}\mathbf{T}, \tag{7.7.10}$$

where $\mathbf{S}$ is the overlap matrix for the original set $\{\phi_i\}$. Bi-orthogonal sets have been used for various purposes (see e.g. Amos and Hall, 1961; Gouyet, 1973, 1981), and their possible use in VB theory (although along different lines) was first proposed by Moshinsky and Seligman (1971) and discussed further by Cantu *et al.* (1975) and Seligman (1981). The point of the transformation (7.7.10) is that, when every $\tilde{\Phi}_\kappa$ is defined exactly like Φ_κ except for the change of orbitals ($\phi_i \rightarrow \tilde{\phi}_i$), any matrix element $\langle \tilde{\Phi}_\lambda | \mathsf{H} | \Phi_\kappa \rangle$ may be evaluated using exactly the same formula as for $\langle \Phi_\lambda | \mathsf{H} | \Phi_\kappa \rangle$—derived with the assumption that both CFs are constructed from a common *orthonormal* set. This result follows because, in the reduction to a combination of 1- and 2-electron integrals, every orbital scalar product will be of the type (7.7.9), and no overlap terms will therefore appear. Consequently, the simple algorithm of Section 7.5 will be immediately applicable: non-orthogonality will be rigorously included merely by redefinition of the integrals. All the bra functions will carry a tilde, and we shall therefore need to use

$$\langle \tilde{\phi}_i | \mathsf{h} | \phi_j \rangle = \sum_m T^\dagger_{im} \langle \phi_m | \mathsf{h} | \phi_j \rangle, \tag{7.7.11a}$$

$$\langle \tilde{\phi}_i \tilde{\phi}_j | g | \phi_k \phi_l \rangle = \sum_{m,n} T^\dagger_{im} T^\dagger_{jn} \langle \phi_m \phi_n | g | \phi_k \phi_l \rangle. \tag{7.7.11b}$$

These "one-sided" transformations are computationally far less demanding than those usually encountered; for (7.7.11b) is only of "2-index"

‡ Any determinant of an arbitrary $\tilde{\Phi}_\kappa$ will then expand into a linear combination of *all possible* determinants based on the orbitals ϕ_i. See also Problem 3.6.

type (m^2 terms with m orbitals), compared with the normal 4-index transformation (m^4 terms per integral) of 2-electron integrals.

The main disadvantage of this approach is clearly the loss of the upper-bound property; but preliminary calculations (Norbeck and McWeeny, 1975) show that convergence, though oscillatory, may be rather rapid. With the development of improved computing strategies for large-scale CI calculations (Chapter 10), this line of attack is not without its attractions.

It has been pointed out (Seligman, 1981) that it is not *necessary* to sacrifice the variational principle, and that matrix elements in the usual secular equations (even with non-orthogonal orbitals) may also be evaluated using the properties of bi-orthogonal sets. The argument (given in second-quantization language) uses bi-orthogonality only in simplifying the anticommutation properties of the creation and annihilation operators for non-orthogonal orbitals, the CFs in the bra and the ket being left unchanged, by introducing destruction operators for the bi-orthogonal set $\{\tilde{\phi}_i\}$. The result is that the coefficients of the integrals in (7.7.11) in the expression for a *normal* matrix element $\langle \Phi_\lambda | \mathsf{H} | \Phi_\kappa \rangle$ may be written in terms of non-orthogonality integrals $\langle \Phi'_\lambda | \Phi_\kappa \rangle$, where Φ'_λ is obtained from Φ_λ using simple rules (in essence, Φ'_λ and Φ_λ are "structures" with different Rumer diagrams). This is a useful observation, but it does not solve the main problems: every integral coefficient in every matrix element requires an evaluation of a particular $\langle \Phi'_\lambda | \Phi_\kappa \rangle$—which still depends on a tedious and time-consuming evaluation of orbital-overlap determinants (Section 7.6). Since a full set of CFs is involved in every step, it would seem preferable to eliminate such complications completely by using the non-variational approach.

In conclusion, we note that in the full-CI limit it is easy to pass from the "bi-orthogonal" matrices $\tilde{\mathbf{H}}$ and $\tilde{\mathbf{M}}$ to the matrices $\mathbf{H}$ and $\mathbf{M}$ defined over the non-orthogonal basis. For this purpose, both the right and left eigenvectors of $\tilde{\mathbf{H}}$ (which is non-Hermitian) are required. In the full-CI limit there are two equivalent forms of Ψ,

$$\Psi = \sum_\kappa C_\kappa \Phi_\kappa = \sum_\kappa \tilde{C}_\kappa \tilde{\Phi}_\kappa, \tag{7.7.12}$$

and use of the second expansion in (7.7.1), followed by taking scalar products with the functions Φ_λ, leads to conditions that may be written (taking Hermitian conjugates) as

$$\tilde{\mathbf{C}}^\dagger \tilde{\mathbf{H}} = E \tilde{\mathbf{C}}^\dagger \tilde{\mathbf{M}}, \tag{7.7.13}$$

instead of (7.7.3). The *rows* of coefficients ($\tilde{\mathbf{C}}_1^* \ \tilde{\mathbf{C}}_2^* \ \ldots$) that satisfy

(7.7.13) are the *left* eigenvectors of $\tilde{\mathbf{H}}$ (with metric $\tilde{\mathbf{M}}$).‡ If we collect the right and left eigenvectors into two matrices

$$\mathbf{R} = (\mathbf{C}^{(1)} | \, \mathbf{C}^{(2)} \, | \ldots), \qquad \mathbf{L} = (\tilde{\mathbf{C}}^{(1)} | \, \tilde{\mathbf{C}}^{(2)} \, | \ldots) \tag{7.7.14}$$

then from (7.7.6) and (7.7.12) we find $\tilde{\mathbf{C}}^{(K)} = \mathbf{V}^{-1}\mathbf{C}^{(K)}$, for the Kth eigenvector, and thus

$$\mathbf{L} = \mathbf{V}^{-1}\mathbf{R}. \tag{7.7.15}$$

Since **L**, **R** and **V** define transformations between linearly independent bases in the full-CI space, their inverses exist and we obtain $\mathbf{V} = \mathbf{R}\mathbf{L}^{-1}$. Finally, then, from (7.7.7),

$$\mathbf{H} = \mathbf{W}\tilde{\mathbf{H}}, \qquad \mathbf{M} = \mathbf{W}\tilde{\mathbf{M}} \qquad (\mathbf{W} = \mathbf{L}\mathbf{R}^{-1}). \tag{7.7.16}$$

For systems containing only a few electrons outside a closed shell (the normal limit for complete VB calculations) this is a perfectly feasible and efficient way of proceeding. For example, with 6 orbitals and 6 electrons (the classic problem of the benzene π electrons, with full CI) the number of singlet CFs is only 175 and the computation involved is, by current standards, trivial.

PROBLEMS 7

7.1 Use the Dirac spin-exchange identity (Problem 4.4) to obtain an alternative proof of the energy formula (7.1.9). Extend your result to the case of N electrons, for any "perfect-pairing" wavefunction (p. 219), and express (7.3.7) as the expectation value of a spin Hamiltonian.

7.2 Use a Heitler–London function based on *orthogonalized* 1s AOs ($\bar{\chi}_1$, $\bar{\chi}_2$, say), to obtain singlet and triplet energies from (7.1.9), comparing the results with those obtained using the unmodified orbitals (χ_1, χ_2). Why is there agreement in the triplet case but disagreement for the singlet? Calculate also the electron density for the singlet function, in terms of χ_1, χ_2, and the populations of the orbital and overlap regions (p. 134); hence interpret the absence of the bonding. Do a CI calculation, still with the orthogonal orbitals, and verify that the bonding reappears when the polar structures are admitted. [*Hint:* Combinations $C(\chi_1 + \lambda\chi_2)$, $C(\chi_2 + \lambda\chi_1)$ will allow you to use symmetry most easily. You may use the following integral values (all for bond length $R = 1.4a_0$): $\langle\chi_1|\,\mathsf{h}\,|\chi_1\rangle = -1.1099$, $\langle\chi_1|\,\mathsf{h}\,|\chi_2\rangle = -0.9680$, $(\chi_1\chi_1, \chi_1\chi_1) = 0.6250$, $(\chi_1\chi_1, \chi_1\chi_2) = 0.4257$, $(\chi_1\chi_2, \chi_1\chi_2) = 0.5034$ (all in units of E_{h}).]

7.3 Use the perfect-pairing formula (7.3.7) to interpret the shape of the water molecule, assuming that only the singly occupied 2p AOs on the oxygen are

‡ Equivalently, if the stars are removed, they are right eigenvectors of $\tilde{\mathbf{H}}^{\dagger}$.

involved in the bonding. Write out the expressions for Q and for the relevant exchange integrals K_{ij} (in terms of 1- and 2-electron integrals), considering their probable signs and their contributions to the total energy. What would be the effect of orthogonalizing the hydrogen AOs, as in Problem 7.2?

7.4 Introduce a second VB structure in Problem 7.3, that in which there is an "internal" pairing between the oxygen 2p electrons. Set up secular equations, using (7.3.9) and (7.3.10) to obtain the matrix elements, and suggest reasons for the dominant participation of the structure used in Problem 7.3.

7.5 Look at Section 14.2, going only as far as (14.2.8), and use this result to justify the representation of a closed-shell "core" (p. 224) by means of an effective Hamiltonian. Obtain an explicit expression for E_c when the core orbitals are of LCAO form. Obtain matrix elements of the 1-electron operator h_{eff} relative to an AO basis for the electrons *outside* the core, in terms of the core density matrix $\mathbf{P}_c$. [*Hint:* do the spin integrations to pass from spin-orbital forms to orbital forms, and introduce finite-basis approximations in the usual way.]

7.6 For 4 orbitals and 4 electrons it is possible to construct 20 linearly independent singlet functions (CFs). Indicate a suitable set of 20 VB structures that could be used in discussing the singlet states of a square planar system of 4 hydrogen atoms (H_2 dimer), using $\pm$ signs to show which AOs are empty or doubly occupied in the polar structures. Write down linear combinations of Slater determinants to represent the non-polar structures and typical structures of the other types.

7.7 Whereas in MO theory the AOs of a molecule with symmetry are usually projected into symmetry *orbitals* (e.g. Problem 3.17, p. 85), it is the *structures* in VB theory that must be combined to form functions belonging to the various symmetry species. Use Wigner operators based on the irreps of C_{4v} (p. 85) to set up structure combinations of A_1, A_2, B_1, and B_2 symmetry for the 20 VB structures of Problem 7.6. [*Hint:* The orbitals usually employed in VB theory (AOs or more generally "hybrids", i.e. combinations of AOs on the same centre) are *permuted* under symmetry operations. Make a table to show how a symmetry operation sends each structure into a structure of the same type, multiplied by a phase factor $\pm$, noting that the operation produces a new arrow diagram (which may differ in arrow directions from one in the original set) and also permutes the orbitals in the Slater determinants.]

7.8 Consider the use of structures based on branching-diagram spin functions, instead of the "classical" VB structures (Rumer-type) used so far. The second non-polar function in Problem 7.6 would thus become (see Problem 4.7) a linear combination of the two classical structures. Show that the branching-diagram functions will not in general have simple spatial symmetry properties that can be inferred pictorially.

7.9 Reconsider the water molecule (Problem 7.4), setting up hybrid AOs (by mixing 2s and 2p AOs) that point along the OH bonds at the observed angle ($\sim 105°$). Take account of the two non-polar structures, using both Rumer and branching-diagram spin couplings, and set up combinations that belong to the irreps of the point group C_{2v} (p. 84). Extend your discussion to polar structures, using the method of Problem 7.7 for the Rumer functions. How do you deal with the branching-diagram functions?

7.10 Generalize the rule in Problem 7.7, showing (McWeeny, 1955a) that the 'image' of any structure under a symmetry operation is $(-1)^{\nu}(-1)^{p_e}$ times one of the original structures, where ν is the number of arrow reversals necessary to make the image coincide with the original and p_e is the number of *even* cycles in the orbital permutation produced by the operation. Verify that the rule applies to both polar and non-polar structures of any spin multiplicity.

7.11 In Problem 7.6 there are two non-polar structures Φ_1, Φ_2, with spin pairs $1\rightarrow 2$, $3\rightarrow 4$ and $2\rightarrow 3$, $4\rightarrow 1$ (with the AOs $\chi_1, \ldots, \chi_4$ reading anticlockwise, say, round the square). Show, by considering the generators P_{12}, P_{23}, P_{34} of the group S_4 of 24 permutations, how the 2×2 matrices $\mathbf{H}$ and $\mathbf{M}$ in (7.2.6) can be evaluated directly. Note that this (Rumer) basis is non-orthogonal: from the orthogonal functions of Problem 4.7 find (by transformation) the corresponding matrices ($\mathbf{V}(\mathsf{P}) = \epsilon_{\mathsf{P}}\mathbf{D}(\mathsf{P})$) of the generators in the standard branching-diagram basis. Hence obtain (by matrix multiplication) all 24 matrices of the irrep, not only for S_4 but also for the subgroups S_3 and S_2. Finally, calculate the 1-electron contribution to the matrix $\mathbf{H}$ from the summation in (7.2.5). Why is it not possible to proceed in the same way when using the Rumer basis?

7.12 Use the results of Problem 7.11 (i) to obtain the 1-electron density matrix for structure Φ_1, and (ii) to examine step by step the procedure described in Section 7.2 (pp. 215, 216). Obtain also an expression for the 1-electron part of the ground-state energy, using an appropriate linear combination of Φ_1 and Φ_2, and calculate the overlap populations (Section 5.5) of the H—H bonds. [*Hint:* Note that here there are only two distinct overlap integrals, $s_1 = S_{12}$, $s_2 = S_{13}$, and three 1-electron integrals, $\alpha = h_{11}$, $\beta_1 = h_{12}$, $\beta_2 = h_{13}$, owing to symmetry.]

7.13 Assume for simplicity that the overlap integrals in Problem 7.12 have values $s_1 = \frac{1}{2}$, $s_2 = \frac{1}{4}$ and set up an orthonormal basis by the Schmidt method (p. 33). Use structures ($\bar{\Phi}_1$, $\bar{\Phi}_2$, say) built from the orthonormal orbitals (formally similar to Φ_1, Φ_2) to obtain 1-electron terms in the matrix elements H_{ij}, by means of the algorithm (7.5.6) (or the special case (7.3.10)). Evaluate also the density-matrix elements for each structure, first in the orthonormal basis and then in the original AO basis. Note that the simple symmetry properties of the structures have been destroyed and show that the electron density will contain an overlap density such as $\chi_1\chi_2$ with a *negative* sign. To what extent could these matters be improved by using the Löwdin basis (2.2.21)?

7.14 Show that it is possible to express any Slater determinant, and hence any VB structure, in terms of similar quantities defined over an orthonormal basis and hence to calculate overlap and energy matrices by means of the algorithm (7.5.6). What advantage, if any, would the Schmidt basis (Problem 7.13) possess? [*Hint:* Invert the orbital transformation in Problem 7.13 and expand each determinant as in Problem 3.5, noting that the terms lead to (formally!) polar structures.]

7.15 Obtain the 1-electron contributions to $\mathbf{H}$ in Problem 7.11 by a direct method, using the non-orthogonal form of Slater's rules (p. 66). [*Hint:* Each structure contains four determinants. Consider them in pairs, forming the overlap determinants (D_{ab} on p. 64) and getting the cofactors. Set out the results in a 4×4 array and use the 4×2 matrix of expansion coefficients (one column for each structure) to derive the 2×2 matrix **H**. This is not as bad as it seems, each D_{ab} containing many zeros.]

7.16 Consider again the system in Problem 7.6. Write out the full set of 20 standard Weyl tableaux (p. 241) for use in Gallup's method of constructing VB functions of branching-diagram type and show that they may be put in one-to-one correspondence with the polar and non-polar structures. Apply the operator NPN to the orbital products associated with the first two tableaux and use this direct method to evaluate the quantities in (7.6.12) (1-electron terms only) with the parameters used in Problem 7.12. How do your results compare with those obtained in Problem 7.11? [*Hint:* NPN is idempotent, except for a numerical factor, and therefore need be used on only one of the two orbital factors, for example that in the ket. There are 20 products in each projection.]

7.17 How many triplet structures can you find for the system in Problem 7.6? And how many Weyl tableaux? Show that in each case the number agrees with that predicted by the Weyl formula (7.6.18). Show also that of four apparently distinct "short-bonded" non-polar structures only three are linearly independent. Construct the branching-diagram functions and show that the second and third, in *inverse* last-letter sequence, are linear combinations of the spin functions in the remaining short-bonded structures. Proceed as in Problem 7.7 to obtain combinations belonging to the irreps of C_{4v}.

7.18 Show that the 11-elements of the matrices $\mathbf{D}_S(\mathsf{P})$ in the standard (branching-diagram) irrep of S_4 (Θ_1 being the first function in *inverse* last-letter sequence, namely Θ_f of Section 4.6) may be evaluated very simply by using the spin scalar-product algorithm (7.3.1). Hence construct a 24-term Wigner operator (ρ, say) that will produce a spatial function "dual" to Θ_1 when acting on an arbitrary orbital product Ω. Verify that by taking $\Omega = \Omega_1 = \chi_1\chi_2\chi_3\chi_4$, the orbital product in Problem 7.11, the 1-electron part of the matrix element $\langle \rho\Omega | \mathsf{H} | \rho\Omega \rangle$ ($= H_{11}$) agrees with that already found. Show how, by using a second product $\Omega_2 = \chi_1\chi_3\chi_2\chi_4$ in the ket it will be possible to generate the whole matrix **H**. [*Hint:* Note that $D_S(\mathsf{P})_{11}$ in the standard irrep for spin S is the coefficient of Θ_1 in the permuted spin function $\Theta_1' = \mathsf{P}\Theta_1$, and Θ_1 is of spin-paired form. The corresponding 24 permuted *orbital* products give matrix element contributions that are merely sums of 1- and 2-electron integrals multiplied by chains of overlaps. This procedure is very easily implemented on a computer, and provides yet another viable method (McWeeny, 1988) of making *ab initio* VB calculations.]

8 Multiconfiguration SCF Theory

8.1 THE OPTIMIZATION PROBLEM

In Chapters 6 and 7 we have encountered two types of optimization problem: (i) that which arises when we vary the *orbitals* in a wavefunction of 1-determinant form; and (ii) that which results when we vary the linear expansion coefficients in a wavefunction of many-determinant or CI form such as (7.2.3). Optimization of the wavefunction with respect to linear parameters is a simple matter, depending only on solution of a large set of linear equations. But the optimization of even a relatively simple wavefunction with respect to *orbital* variations raises more difficult problems, typical of non-linear variation methods, as we have seen in both chapters.

In this chapter we turn to the problem of optimizing both CI coefficients and orbitals in a CI-type wavefunction, possibly of substantial length. We know from Section 3.4 that the choice of orbitals to use in setting up the wavefunction, assuming them to be constructed as linear combinations of a given set of basis functions, would be immaterial in the limit of "full CI": the electronic states obtained would represent the *basis-set limit.* In practice, however, full-CI calculations are usually out of the question, and in a *truncated* CI expansion it is of paramount importance to optimize both orbitals and CI coefficients. The optimization of orbitals in this context gives rise to various forms of multiconfiguration (MC) SCF method, the orbitals usually being adjusted iteratively until some kind of "self-consistency" is reached. The optimization of CI coefficients may be achieved simultaneously (in a "one-step" process) or in a separate step after each improvement of the orbitals (i.e. in a two-step process). The formalism for achieving these ends has been developed in earlier chapters; we now turn to its implementation, treating first the problem of orbital optimization.

8.2 ORBITAL VARIATION. STATIONARY-VALUE CONDITIONS

We start from the energy functional (5.4.16), corresponding to use of a CI-type variation function

$$\Psi = \sum_{\kappa} C_\kappa \Phi_\kappa, \tag{8.2.1}$$

and first sketch the conventional derivation of the MC SCF equations. Both (5.4.16) and its spinless counterpart have the form

$$\boxed{E = \sum_{r,s} \langle r|\, \mathsf{h}\, |s\rangle A_{sr} + \tfrac{1}{2} \sum_{r,s,t,u} \langle rt|\, g\, |su\rangle B_{su,rt}.} \tag{8.2.2}$$

When the integrals are defined over occupied *orbitals* $\{\phi_r\}$, the A and B coefficients are elements of the spinless density matrices P and Π; but if instead we define the integrals with respect to *spin*-orbitals $\{\psi_r\}$ then the appropriate coefficients will be elements of the density matrices ρ and π (*including* spin). Thus both variation problems may be discussed using the same formalism, merely by changing the interpretation of the integrals and their coefficients. In both cases, the A and B coefficients are defined in terms of similar (transition) quantities for all pairs of CFs $(\Phi_\kappa, \Phi_\lambda)$ in the expansion (8.2.1): for example,

$$A_{sr} = \sum_{\kappa,\lambda} C_\kappa C_\lambda^* A_{sr}^{\kappa\lambda}. \tag{8.2.3}$$

We know that for E to be stationary under variation of CI coefficients it is necessary to satisfy secular equations of the usual form (2.3.4); the problem that concerns us now is how to make E stationary under *orbital*‡ variations. We assume that the orbitals used in the expansion (8.2.1) are orthonormal.

First we note that when $\phi_r \to \phi_r + \delta\phi_r$ (all orbitals) the first-order variation is

$$\delta E = \Big(\sum_{r,s} \langle \delta r|\, \mathsf{h}\, |s\rangle A_{sr} + \sum_{r,s,t,u} \langle \delta rt|\, g\, |su\rangle B_{su,rt} \Big) + (\text{c.c.}), \tag{8.2.4}$$

where we have used the symmetry $\langle rt|\, g\, |su\rangle = \langle tr|\, g\, |us\rangle$ and noted that the labels are merely dummy indices. This result may be written in the form

$$\delta E = \sum_{r,s} \langle \delta r|\, \mathsf{F}_{rs}\, |s\rangle + \text{c.c.}, \tag{8.2.5}$$

‡ The term is now used, with the two alternative interpretations, to cover both orbitals and *spin*-orbitals.

where F_{rs} is a 1-electron operator defined by

$$\mathsf{F}_{rs}(1) = A_{sr}\mathsf{h}(1) + \sum_{t,u} B_{su,rt} V_{tu}(1), \tag{8.2.6}$$

in which $V_{tu}(1)$ is the potential energy of an electron at point 1 in the field of a charge distribution $\phi_t^*(\mathbf{r}_2)\phi_u(\mathbf{r}_2)$:

$$V_{tu}(1) = \int g(1, 2)\phi_t^*(\mathbf{r}_2)\phi_u(\mathbf{r}_2)\, \mathrm{d}\mathbf{r}_2. \tag{8.2.7}$$

Such operators are reminiscent of the Fock operator F used in Chapter 6; but there is one for *every pair* (r, s) of occupied orbitals.

The derivation depends on the assumed orthonormality of the orbitals (otherwise the A and B coefficients depend on overlap integrals, and hence on orbitals as in Section 7.4). To preserve orthonormality, we require

$$\langle \delta r \mid s \rangle = 0, \qquad \langle s \mid \delta r \rangle = 0 \quad (\text{all } s \neq r),$$
$$\langle \delta r \mid r \rangle + \langle r \mid \delta r \rangle = 0,$$

and on multiplying these variations by ϵ_{sr}, ϵ_{rs} and ϵ_{rr} respectively, and subtracting from (8.2.5), it follows that

$$\sum_s \langle \delta r | \mathsf{F}_{rs} | s \rangle - \sum_s \langle \delta r \mid s \rangle \epsilon_{sr} = 0 \quad (\text{all } r),$$

together with a complex-conjugate equation: this is the scalar product of an *arbitrary* vector $|\delta r\rangle$ with a vector that must consequently be the zero vector. Thus the conditional stationary-value conditions become (with the usual notation)

$$\boxed{\sum_s \mathsf{F}_{rs}\phi_s = \sum_s \phi_s \epsilon_{sr} \quad (\text{all } r).} \tag{8.2.8}$$

The complex-conjugate equation simply ensures (cf. Section 6.1) that the matrix of Lagrangian multipliers $\boldsymbol{\epsilon}$ is Hermitian.

Equation (8.2.8) is comparable to the Hartree–Fock equation

$$\mathsf{F}\phi_r = \sum_s \phi_s \epsilon_{sr} \quad (\text{all } r), \tag{8.2.9}$$

but the single Fock operator F is now replaced by a whole family of Fock-like operators (8.2.6). By choosing the orbitals so that $\mathbf{A}$ is diagonal (as in a natural-orbital expansion), it is sometimes possible to decouple the terms on the left in (8.2.8); but there is still one equation for every orbital ϕ_r, each with its own Fock-type operator F_{rr}, and the coupling terms $(r \neq s)$ on the right cannot be eliminated by bringing the matrix $\boldsymbol{\epsilon}$ simultaneously to diagonal form (since $\boldsymbol{\epsilon}$ and $\mathbf{A}$ will not in general commute). The Lagrange multipliers may in fact be identified by taking scalar products in (8.2.8), which leads to

$$\epsilon_{sr} = \sum_t \langle s| \mathsf{F}_{rt} |t\rangle = \sum_t \langle s| \mathsf{h} |t\rangle A_{tr} + \sum_{t,u,v} \langle sv| g |tu\rangle B_{tu,rv}. \quad (8.2.10)$$

It should be noted that all indices r, s, t, u, v in these equations refer to orbitals used in the expansion (8.2.1) since, by definition, the A and B coefficients must otherwise vanish: a complementary set of "virtual orbitals" does not therefore automatically result from solution of (8.2.8).

In fact, it is exceedingly difficult to solve the MC SCF equations in operator form. Instead, the n orbitals of the CI expansion are expanded in terms of m basis functions $\{\chi_\mu\}$, so that (cf. 6.2.1)

$$\boldsymbol{\phi} = \boldsymbol{\chi}\mathbf{T}, \quad (8.2.11)$$

where $\mathbf{T}$ is an $m \times n$ matrix, and the operator equations are replaced by a single matrix equation for $\mathbf{T}$. Instead of (8.2.8), we easily find, assuming for generality a non-orthogonal basis (overlap matrix $\mathbf{S}$).

$$\mathbf{X} = \mathbf{hTA} + \mathbf{Z} = \mathbf{ST}\boldsymbol{\varepsilon}, \quad (8.2.12)$$

where the electron-interaction matrix $\mathbf{Z}$ has elements

$$Z_{\mu r} = \sum_{\substack{s,t,u \\ \rho,\nu,\sigma}} T^*_{\rho t} \langle \mu\rho| g |\nu\sigma\rangle T_{\nu s} T_{\sigma u} B_{su,rt}, \quad (8.2.13)$$

and $\mathbf{h}$ is the usual matrix of the one-electron Hamiltonian. The matrices in (8.2.12) (including those that result from products) are $m \times n$, corresponding to the ranges of the Greek and Latin subscripts respectively.

The matrix of Lagrange multipliers may be eliminated by multiplying (8.2.12) from the left by $\mathbf{T}^\dagger$ and noting the orthonormality condition $\mathbf{T}^\dagger\mathbf{ST} = \mathbf{1}$. Consequently,

$$\boxed{\boldsymbol{\epsilon} = \mathbf{T}^\dagger\mathbf{X} = \mathbf{T}^\dagger(\mathbf{hTA} + \mathbf{Z}),} \quad (8.2.14)$$

while the total electronic energy (8.2.2) may also be expressed in finite-basis form in terms of $\mathbf{T}$:

$$E = \text{tr}\, \mathbf{T}^\dagger\mathbf{X}^{\text{ad}} = \text{tr}\, \mathbf{T}^\dagger(\mathbf{hTA} + \tfrac{1}{2}\mathbf{Z}). \quad (8.2.15)$$

These results are again reminiscent of Hartree–Fock theory: the Fock matrix $\mathbf{F} = \mathbf{h} + \mathbf{G}$ determines the orbital energies (ϵ), while $\mathbf{F}^{\text{ad}} = \mathbf{h} + \frac{1}{2}\mathbf{G}$ determines the *total* energy (the factor $\frac{1}{2}$ eliminating a "double-counting" of electron interaction terms).

All the preceding equations are quite old (McWeeny, 1955b); indeed they were implicit in the work of Frenkel (1934), and have been rediscovered many times (see e.g. Gilbert, 1965). Numerical solution of MC SCF equations became feasible much later, and the development of solution techniques has been stimulated by the use of Fock-space methods. Before turning to methods of solution, we therefore indicate the parallel derivations in Fock space.

The energy functional (8.2.2) is simply the expectation value (3.6.10) of the Hamiltonian (3.6.9), as follows from (5.4.20), provided that we now work in terms of spin-orbitals and interpret **A** and **B** as $\boldsymbol{\rho}$ and $\boldsymbol{\pi}$ respectively. When Ψ is interpreted as a general vector in Fock space, we must consider a general variation in which $|\Psi\rangle \rightarrow |\tilde{\Psi}\rangle = \mathsf{U}\,|\Psi\rangle$, where U must be a unitary operator to preserve normalization. The varied energy is then

$$\tilde{E} = \langle \tilde{\Psi}|\, \mathsf{H}\, |\tilde{\Psi}\rangle = \langle \Psi|\, \mathsf{U}^{\dagger}\mathsf{H}\mathsf{U}\, |\Psi\rangle, \qquad (8.2.16)$$

and is thus the Fock-space expectation value of a new *operator*, while the variation itself is

$$\delta E = \langle \Psi|\, (\mathsf{U}^{\dagger}\mathsf{H}\mathsf{U} - \mathsf{H})\, |\Psi\rangle = \langle \Psi|\, \delta\mathsf{H}\, |\Psi\rangle. \qquad (8.2.17)$$

We now ensure that U be unitary, to preserve orthonormality, by taking

$$\mathsf{U} = e^{\mathsf{R}}, \qquad \mathsf{R}^{\dagger} = -\mathsf{R}, \qquad (8.2.18)$$

which will automatically give $\mathsf{U}^{\dagger}\mathsf{U} = e^{-\mathsf{R}}e^{\mathsf{R}} = 1$. In terms of the anti-Hermitian operator R the stationarity condition $\delta E = 0$ becomes, from (8.2.17),

$$\delta E = \langle \Psi|\, [\mathsf{H}, \mathsf{R}]\, |\Psi\rangle = 0, \qquad (8.2.19)$$

which is sometimes called (see e.g. Kutzelnigg, 1979) a "generalized Brillouin condition", for reasons that will become clear presently.

Let us now take for R a general 1-body operator of the form

$$\mathsf{R} = \sum_{s,t} \Delta_{st}(\mathsf{a}_s^{\dagger}\mathsf{a}_t) = \mathbf{a}^{\dagger}\boldsymbol{\Delta}\mathbf{a}, \qquad (8.2.20)$$

where $\boldsymbol{\Delta}$ is an anti-Hermitian *matrix* and, as usual, $\mathsf{a}^{\dagger}$ and a are respectively row and column matrices of creation and annihilation operators. It is not at once obvious that the operator defined in (8.2.18) and (8.2.20) effects a rotation in Fock space equivalent to a unitary mixing of orbitals $\{\psi_r\}$,

$$\psi_r \rightarrow \tilde{\psi}_r = \sum_s \psi_s V_{sr}, \qquad (8.2.21)$$

in Schrödinger space. To see the connection, we first note that any N-electron $|\Psi\rangle$ is produced from the vacuum state by some operator A

built up from products of creation operators; so $|\Psi\rangle = \mathsf{A}\,|\mathrm{vac}\rangle$ and

$$\mathsf{U}\,|\Psi\rangle = \mathsf{UA}\,|\mathrm{vac}\rangle = \mathsf{UAU}^\dagger\,|\mathrm{vac}\rangle,$$

since $\mathsf{U} = 1 - \mathsf{R} + \ldots$, and all terms except the first contain destruction operators that destroy $|\mathrm{vac}\rangle$. The operator $\mathsf{A} = \mathsf{UAU}^\dagger$, which produces the rotated state vector, will contain new creation and annihilation operators, $\tilde{\mathsf{a}}_r^\dagger$, $\tilde{\mathsf{a}}_s$, in place of $\mathsf{a}_r^\dagger$, a_s, with‡

$$\tilde{\mathsf{a}}_r^\dagger = \mathsf{Ua}_r^\dagger\mathsf{U}^\dagger = e^{\mathsf{R}}\mathsf{a}_r^\dagger e^{-\mathsf{R}}. \tag{8.2.22}$$

Thus, using (8.2.20) and expanding the exponentials, we obtain

$$\tilde{\mathsf{a}}_r^\dagger = \mathsf{a}_r^\dagger + \sum_{s,t} \Delta_{st}\mathsf{a}_s^\dagger\mathsf{a}_t\mathsf{a}_r^\dagger - \sum_{s,t} \Delta_{st}\mathsf{a}_r^\dagger\mathsf{a}_s^\dagger\mathsf{a}_t + \ldots,$$

and hence, making use of the anticommutation relations,

$$\tilde{\mathsf{a}}_r^\dagger = \mathsf{a}_r^\dagger + \sum_s \mathsf{a}_s^\dagger\Delta_{sr} + \ldots = \sum_s \mathsf{a}_s^\dagger(\delta_{sr} + \Delta_{sr} + \ldots).$$

In matrix notation this may be written as

$$\tilde{\mathbf{a}}^\dagger = \mathbf{a}^\dagger\mathbf{V}, \tag{8.2.23}$$

with $\mathbf{V} = \mathbf{1} + \boldsymbol{\Delta} + \ldots$. The next term in the expansion is found to be $\frac{1}{2}\boldsymbol{\Delta}^2$, which suggests that

$$\mathbf{V} = \mathbf{1} + \boldsymbol{\Delta} + \tfrac{1}{2}\boldsymbol{\Delta}^2 + \ldots = e^{\boldsymbol{\Delta}} \quad (\boldsymbol{\Delta}^\dagger = -\boldsymbol{\Delta}). \tag{8.2.24}$$

This conjecture is confirmed most easily by noting that the first-order result must be valid for an *infinitesimal* rotation and that the finite rotation may be generated as a sequence of infinitesimal steps: on dividing $\boldsymbol{\Delta}$ by n and using n steps, the matrix describing the full rotation will thus be

$$\mathbf{V} = \lim_{n\to\infty}\left(\mathbf{1} + \frac{\boldsymbol{\Delta}}{n}\right)^n = e^{\boldsymbol{\Delta}},$$

as in (8.2.24).

Since $\mathsf{a}_r^\dagger\,|\mathrm{vac}\rangle = \psi_r$, the operator $\tilde{\mathsf{a}}_r^\dagger$ will produce the new spin-orbital $\tilde{\psi}_r = \sum_s \psi_s V_{sr}$. In other words, the action of U in (8.2.18) on any vector of Fock space is equivalent to a change of creation operators, and corresponding spin-orbitals, given by

$$\mathbf{a}^\dagger \to \tilde{\mathbf{a}}^\dagger = \mathbf{aV}, \qquad \boldsymbol{\psi} \to \tilde{\boldsymbol{\psi}} = \boldsymbol{\psi}\mathbf{V}, \tag{8.2.25}$$

with the matrix $\mathbf{V}$ defined in (8.2.24).

‡ Note that *all* operators must undergo the same transformation and that the new operators then have the same properties ("multiplication table") as the old. Thus if $\mathsf{A} = \mathsf{BC}$ then $\tilde{\mathsf{A}} = \mathsf{U(BC)U}^\dagger = (\mathsf{UBU}^\dagger)(\mathsf{UCU}^\dagger) = \tilde{\mathsf{B}}\tilde{\mathsf{C}}$. The relationship of the two sets of operators is an "isomorphism".

It is now a straightforward matter to obtain $\delta\mathsf{H}$ in (8.2.17), noting that here the variation is interpreted as a change of the *operator*: $\mathsf{H} \to \mathsf{U}^\dagger\mathsf{H}\mathsf{U}$. Thus, to first order,

$$\begin{aligned}\delta\mathsf{H} &= \mathsf{H}\sum_{s,t} \Delta_{st}(\mathsf{a}_s^\dagger \mathsf{a}_t) - \sum_{s,t} \Delta_{st}(\mathsf{a}_s^\dagger \mathsf{a}_t)\mathsf{H} \\ &= \sum_{s,t} \Delta_{st}(\mathsf{H}\mathsf{a}_s^\dagger \mathsf{a}_t - \mathsf{a}_s^\dagger \mathsf{H}\mathsf{a}_t + \mathsf{a}_s^\dagger \mathsf{H}\mathsf{a}_t - \mathsf{a}_s^\dagger \mathsf{a}_t \mathsf{H}) \\ &= \sum_{s,t} \Delta_{st}([\mathsf{H}, \mathsf{a}_s^\dagger]\mathsf{a}_t + \mathsf{a}_s^\dagger[\mathsf{H}, \mathsf{a}_t]),\end{aligned}$$

and this may be written as

$$\delta\mathsf{H} = \sum_{r,s} \Delta_{sr}(\mathsf{K}_{rs} - \mathsf{K}_{sr}^\dagger), \tag{8.2.26}$$

where

$$\mathsf{K}_{rs} = \mathsf{a}_s^\dagger[\mathsf{H}, \mathsf{a}_r] \tag{8.2.27}$$

is an important quantity known as the "Koopmans operator", which we encounter again in Chapter 13.

The connection with the Fock-like equations (8.2.8) becomes clear when we evaluate the expectation value of $\delta\mathsf{H}$. On inserting (3.6.9) in the Koopmans operator (8.2.27), and making use of the anticommutation relations, (3.6.7), we easily find first

$$[\mathsf{a}_r, \mathsf{H}] = \sum_s \langle r|\,\mathsf{h}\,|s\rangle \mathsf{a}_s + \sum_{s,t,u} \langle rs|\,g\,|tu\rangle \mathsf{a}_s^\dagger \mathsf{a}_u \mathsf{a}_t, \tag{8.2.28}$$

and then

$$-\langle \mathsf{K}_{rs}\rangle = \sum_t \langle r|\,\mathsf{h}\,|t\rangle \langle \mathsf{a}_s^\dagger \mathsf{a}_t\rangle + \sum_{t,u,v} \langle rv|\,g\,|tu\rangle \langle \mathsf{a}_s^\dagger \mathsf{a}_v^\dagger \mathsf{a}_u \mathsf{a}_t\rangle. \tag{8.2.29}$$

But since, by (5.4.20), the expectation values on the right-hand side are simply 1- and 2-electron density-matrix elements, we obtain at once (treating $\mathsf{K}_{sr}^\dagger$ in the same way)

$$\langle \mathsf{K}_{rs}\rangle = -\epsilon_{rs}, \qquad \langle \mathsf{K}_{sr}^\dagger\rangle = -\epsilon_{sr}^*, \tag{8.2.30}$$

which coincides with the definitions (8.2.10), (8.2.14). The stationary condition $\langle \delta\mathsf{H}\rangle = 0$ is then seen to be, from (8.2.26),

$$\boxed{\boldsymbol{\epsilon}^\dagger = \boldsymbol{\epsilon},} \tag{8.2.31}$$

a condition necessary and sufficient (as first noted by Hinze, 1973) to ensure that the operator equations in the Schrödinger form (8.2.8) are also satisfied. The second-quantization approach thus leads directly to the most succinct and economical statement of the conditional stationary-value conditions in the MC SCF theory.

The preceding analysis refers to *spin*-orbital variation; the spin may be eliminated, remembering the conventions in (8.2.2) *et seq.*, in the following way. The operators $a_r^\dagger$, a_s are replaced by $a_{r\sigma}^\dagger$, $a_{s\sigma}$, where the second subscripts indicate the spin state (α, β), and the variations considered are (cf. (8.2.22) and (8.2.23))

$$a_{r\sigma}^\dagger \rightarrow \tilde{a}_{r\sigma}^\dagger = U a_{r\sigma}^\dagger U^\dagger = \sum_s a_{s\sigma}^\dagger V_{sr}, \tag{8.2.32}$$

and similarly for $a_{s\tau} \rightarrow \tilde{a}_{s\tau}$. Here U works only on the orbital factors of the spin-orbitals, and the same transformation matrix **V** therefore applies for each spin factor. The Hamiltonian, in terms of *orbital* 1- and 2-electron integrals, is found from (3.6.9) to be

$$H = \sum_{r,s} \langle r|\,h\,|s\rangle \sum_\sigma a_{r\sigma}^\dagger a_{s\sigma} + \tfrac{1}{2} \sum_{r,s,t,u} \langle rs|\,g\,|tu\rangle \sum_{\sigma,\tau} a_{r\sigma}^\dagger a_{s\tau}^\dagger a_{u\tau} a_{t\sigma}, \tag{8.2.33}$$

and the anticommutation relations become

$$[a_{r\sigma}^\dagger, a_{s\tau}]_+ = \delta_{rs}\delta_{\sigma\tau} \tag{8.2.34}$$

etc. In an orbital variation described by the matrix **Δ** we then again find (8.2.30), but with

$$K_{rs} = \sum_\sigma a_{s\sigma}^\dagger [H, a_{r\sigma}] \tag{8.2.35}$$

instead of (8.2.27). The commutator that now appears is

$$[a_{r\sigma}, H] = \sum_s \langle r|\,h\,|s\rangle a_{s\sigma} + \sum_{s,t,u} \langle rs|\,g\,|tu\rangle \sum_\tau (a_{s\tau}^\dagger a_{u\tau}) a_{t\sigma}, \tag{8.2.36}$$

and the stationary-value condition (8.2.31), with ϵ_{rs} defined in (8.2.30), remains valid provided that we use the Koopmans operator (8.2.35). The matrix elements are then defined as in (8.2.29) with the substitutions

$$a_s^\dagger a_t \rightarrow \sum_\sigma a_{s\sigma}^\dagger a_{t\sigma}, \qquad a_s^\dagger a_v^\dagger a_u a_t \rightarrow \sum_{\sigma,\tau} a_{s\sigma}^\dagger a_{v\tau}^\dagger a_{u\tau} a_{t\sigma}, \tag{8.2.37}$$

as follows from (8.2.36).

Finally, we note that, in addition to (8.2.8) and (8.2.31), there is a third useful form of the stationary-value conditions. Thus, when (8.2.20) is inserted in (8.2.19) and the coefficients of the arbitrary Δ_{rs} are equated to zero, we obtain on putting $E_{sr} = a_s^\dagger a_r$

$$\langle [H, E_{sr}] \rangle = 0 \quad (\text{all } r, s). \tag{8.2.38}$$

This is equivalent to the Brillouin condition referred to in Section 6.1, for the case where the reference state $|\Psi\rangle$ is a single determinant; for then E_{sr} simply replaces an occupied spin-orbital ψ_r by a virtual orbital ψ_s (otherwise destroying $|\Psi\rangle$). In other words, H must have zero matrix elements between the ground-state determinant and all the single-excitation functions with $\psi_r \rightarrow \psi_s$. A spinless form of the condition follows on replacing E_{sr} in (8.2.38) by $\mathsf{E}_{sr} = \sum_\sigma \mathsf{a}^\dagger_{s\sigma}\mathsf{a}_{r\sigma}$, an operator‡ that replaces $\phi_r\alpha$ (or $\phi_r\beta$) by $\phi_s\alpha$ (or $\phi_s\beta$). We shall make considerable use of such "substitution" operators in later chapters.

It remains only to devise suitable methods of solving the MC SCF equations in one or other of the forms discussed above. We may distinguish three main families of methods: (a) those that aim to satisfy the operator equations (8.2.8) (or the equivalent condition (8.2.31)), normally in a finite-basis form such as (8.2.12); (b) those that minimize the energy directly by using steepest descent or more general gradient techniques; and (c) those that aim to satisfy the condition (8.2.38), which are usually described as Brillouin-condition methods. Many special techniques are available within each category; the examples in the next three sections illustrate these main approaches.

8.3 SOLUTION OF THE STATIONARY-VALUE EQUATIONS

Since the operator equations (8.2.8) are intractible, as they stand, we turn at once to finite-basis forms. There are two common choices of basis: the n occupied orbitals may be expressed directly in terms of a completely arbitrary set $\{\chi_r\}$ of m functions (e.g. AOs); or in terms of, say, "first approximations" $\{\phi_r^{(0)}\}$ to the ϕ_r (e.g. simple SCF MOs). Both possibilities may be handled using the same equations (8.2.11)–(8.2.15), but in the second case the intermediate set $\{\phi_r^{(0)}\}$ should contain *all* orbitals—occupied and, in some sense, virtual. This means that the rectangular matrix $\mathbf{T}$ must be filled up to give

$$\mathbf{T} = (\mathbf{T}_1 \mid \mathbf{T}_2) \tag{8.3.1}$$

an $m \times m$ matrix, in which the columns of $\mathbf{T}_1$ represent the occupied MOs, while the $m-n$ columns of $\mathbf{T}_2$ represent a complementary orthonormal set of virtual MOs. In this way the MOs may be related to the intermediate set according to

$$\boldsymbol{\phi} = \boldsymbol{\phi}^{(0)}\mathbf{V} = \boldsymbol{\chi}\mathbf{T}^{(0)}\mathbf{V} = \boldsymbol{\chi}\mathbf{T}, \tag{8.3.2}$$

‡ Which of the two definitions is to be used will be clear from the context, and a typographical distinction will not be made.

where the unitary matrix $\mathbf{V}$ describes a general rotation, conserving orthonormality, in the m-dimensional space spanned by either $\boldsymbol{\chi}$ or $\boldsymbol{\phi}^{(0)}$.

Very often $\boldsymbol{\phi}^{(0)}$ is revised during the course of an MC SCF calculation, being taken as the set of MOs at the beginning of each iterative cycle so that $\mathbf{V}$ is a "small" matrix determining a further correction; in this case all 1- and 2-electron integrals must be recalculated in every cycle. Alternatively, the fixed basis $\boldsymbol{\chi}$ may be retained explicitly, throughout the calculation, and the aim is then to find the improvement of the matrix $\mathbf{T}$. With a slight change of notation, we consider instead of (8.3.2)

$$\boldsymbol{\phi} \rightarrow \boldsymbol{\phi}' = \boldsymbol{\phi}\mathbf{V}, \qquad \mathbf{T} \rightarrow \mathbf{T}' = \mathbf{T}\mathbf{V}, \tag{8.3.3}$$

where the primes distinguish the "improved" quantities from those (unprimed) at the beginning of the cycle. Evidently, if $\mathbf{c}_r$ is the rth column of $\mathbf{T}$, this means $\mathbf{c}'_r = \sum_s \mathbf{c}_s V_{sr}$, which is the finite-basis analogue of $\phi'_r = \sum_s \phi_s V_{sr}$. In the finite-basis form the integrals are calculated once only, and no intermediate transformations are required; all quantities, in every cycle, refer to the "raw" basis functions $\{\chi_\mu\}$, which may be orthogonal or non-orthogonal. The two procedures lead to working equations that, although different in appearance, are fundamentally equivalent.

It is convenient to start from (8.2.31), with $\boldsymbol{\epsilon}$ defined in (8.2.10) or (8.2.14), according to the choice of basis. An obvious way to satisfy this equation would be to make a series of "2×2 rotations", as in the Jacobi method for matrix diagonalization, in order to reduce the elements of $\boldsymbol{\epsilon} - \boldsymbol{\epsilon}^\dagger$ to zero. This approach has been used by Hinze and coworkers (see e.g. Hinze, 1973; Hinze and Yurtsever, 1979), who choose the ϕ-basis and consider the reduction of

$$\epsilon_{rs} - \epsilon^*_{sr} = \sum_t [\langle r|\, \mathsf{F}_{st}\, |t\rangle - \langle t|\, \mathsf{F}^\dagger_{rt}\, |s\rangle]. \tag{8.3.4}$$

A typical 2×2 rotation is

$$\phi_r \rightarrow c\phi_r + s\phi_s, \qquad \phi_s \rightarrow -s\phi_r + c\phi_s,$$

where s and c stand for the sine and cosine of a rotation angle, which is then chosen so that $\epsilon_{rs} - \epsilon^*_{sr}$ becomes zero for this pair. In first approximations, the rotations for different pairs are calculated independently, and variation of the operators F_{rs} is neglected. The operators are recomputed after performing all rotations, and the cycle is recommenced using the updated quantities. Unfortunately, the convergence of such methods is very variable, largely owing to the neglect of operator variation. The computational effort involved is much heavier than in the Jacobi method, since a full recalculation of all matrix elements is needed in every cycle.

A more satisfactory procedure (Golebiewski *et al.*, 1979) is to adopt the full transformation (8.3.2), choosing the unitary matrix $\mathbf{V}$ so that in each cycle the transformed matrix $\boldsymbol{\epsilon}$ becomes more nearly Hermitian. Here we derive the required algorithm using the basis $\boldsymbol{\chi}$ and the transformation (8.3.3), with $\mathbf{T}$ in the partitioned form (8.3.1).

First we note that the matrices $\mathbf{T}$ and $\mathbf{X}$ in (8.2.14) are *rectangular*, their columns referring only to the n occupied orbitals. With the notation of (8.3.1) the $n \times n$ matrix $\boldsymbol{\epsilon}$ becomes $\mathbf{T}_1^\dagger\mathbf{X}_1$, $\mathbf{X}_1$ being the first block of an extended $\mathbf{X}$ matrix in which $\mathbf{X}_2$ is identically zero. In terms of the full matrices, the quantity of interest is thus simply the leading diagonal block of

$$\mathbf{T}^\dagger\mathbf{X} = \left(\begin{array}{c}\mathbf{T}_1^\dagger \\ \hline \mathbf{T}_2^\dagger\end{array}\right)\left(\begin{array}{c|c}\mathbf{X}_1 & \mathbf{0}\end{array}\right) = \left(\begin{array}{c|c}\mathbf{T}_1^\dagger\mathbf{X}_1 & \mathbf{0} \\ \hline \mathbf{T}_2^\dagger\mathbf{X}_1 & \mathbf{0}\end{array}\right) = \left(\begin{array}{c|c}\boldsymbol{\epsilon}_1 & \mathbf{0} \\ \hline \boldsymbol{\epsilon}_2 & \mathbf{0}\end{array}\right). \tag{8.3.5}$$

Let us confine attention to the non-zero part of this matrix, namely

$$\boldsymbol{\epsilon} = \begin{pmatrix}\boldsymbol{\epsilon}_1 \\ \boldsymbol{\epsilon}_2\end{pmatrix} = \mathbf{T}^\dagger\mathbf{X}_1, \tag{8.3.6}$$

and note that the transformation

$$\boldsymbol{\epsilon}_1 \rightarrow \boldsymbol{\epsilon}_1' = \mathbf{T}_1'^\dagger\mathbf{X}_1 \tag{8.3.7}$$

is unusual: in fact (8.3.3) leads to

$$\boldsymbol{\epsilon}' = \mathbf{T}'^\dagger\mathbf{X}_1 = \mathbf{V}^\dagger\mathbf{T}^\dagger\mathbf{X}_1 = \begin{pmatrix}\boldsymbol{\epsilon}_1' \\ \boldsymbol{\epsilon}_2'\end{pmatrix}. \tag{8.3.8}$$

The problem is how to choose $\mathbf{V}$ so that $\boldsymbol{\epsilon}_1'$ in which the transformation matrix $\mathbf{V}^\dagger$ appears only on the left) is Hermitian.

In the derivation given by Golebiewski *et al.* (1979) the choice of $\mathbf{V}$ was based on a "maximum-overlap" requirement. Here we simply note that any $m \times n$ matrix of rank n can be reduced to the above form by a so-called "singular-value decomposition" (see e.g. Amos and Hall, 1961; Stewart, 1973); the result is that $\boldsymbol{\epsilon}_1'$ is Hermitian while $\boldsymbol{\epsilon}_2'$ is a zero matrix. The required theorem states that any $m \times n$ rectangular matrix $\mathbf{M}$, of rank n, can be reduced as follows:

$$\mathbf{W}^\dagger\mathbf{M}\mathbf{U} = \begin{pmatrix}\mathbf{D} \\ \mathbf{0}\end{pmatrix}. \tag{8.3.9}$$

Here $\mathbf{W}$ is an $m \times m$ matrix whose columns are the eigenvectors of the Hermitian matrix $\mathbf{M}\mathbf{M}^\dagger$, while $\mathbf{U}$ is an $n \times n$ matrix constructed similarly using $\mathbf{M}^\dagger\mathbf{M}$; the $n \times n$ matrix $\mathbf{D}$ is diagonal, its elements being the square roots of the common eigenvalues of $\mathbf{M}\mathbf{M}^\dagger$ and $\mathbf{M}^\dagger\mathbf{M}$. The weaker form of the theorem, which is sufficient for our present purposes, is that

$$\mathbf{V}^\dagger\mathbf{M} = \begin{pmatrix} \boldsymbol{\lambda} \\ \mathbf{0} \end{pmatrix}, \tag{8.3.10}$$

where $\mathbf{V}$ is again an $m \times m$ unitary matrix but $\boldsymbol{\lambda}$ is Hermitian instead of diagonal. It is not difficult to show that a suitable $\mathbf{V}$ may be obtained in terms of the eigenvector matrices $\mathbf{W}$ and $\mathbf{U}$. If the first n columns ($\mathbf{W}_1$, say) of $\mathbf{W}$ correspond to the common eigenvalues, we find

$$\mathbf{V} = (\mathbf{W}_1\mathbf{U}^\dagger \mid \mathbf{W}_2), \tag{8.3.11}$$

although this choice is not, of course, unique.

The result of (8.3.10) applied to (8.3.8) shows the existence of a unitary matrix $\mathbf{V}$ such that

$$\mathbf{V}^\dagger\boldsymbol{\epsilon} = \begin{pmatrix} \boldsymbol{\epsilon}_1' \\ \mathbf{0} \end{pmatrix}. \tag{8.3.12}$$

Instead of calculating $\mathbf{V}$ according to (8.3.11), however, we formulate an alternative method, which leads directly to the algorithm of Golebiewski *et al.*

We multiply each side of (8.3.12) by its Hermitian transpose, and find, since $\mathbf{V}$ is unitary,

$$\boldsymbol{\epsilon}^\dagger\boldsymbol{\epsilon} = \boldsymbol{\epsilon}_1'^2. \tag{8.3.13}$$

Hence, multiplying (8.3.12) from the left by $\mathbf{V}$ and from the right by $\boldsymbol{\epsilon}_1'^{-1}$, we obtain

$$\boldsymbol{\epsilon}(\boldsymbol{\epsilon}^\dagger\boldsymbol{\epsilon})^{-1/2} = \begin{pmatrix} \mathbf{V}_1 & \big| & \mathbf{V}_2 \end{pmatrix}\begin{pmatrix} \mathbf{1}_n \\ \hline \mathbf{0} \end{pmatrix} = \mathbf{V}_1$$

where $\mathbf{V}_1$ comprises the first n columns of $\mathbf{V}$. Since $\mathbf{T}' = \mathbf{TV}$, and we need only the first n columns of $\mathbf{T}'$, this result is sufficient:

$$\mathbf{T}_1' = \mathbf{TV}_1 = \mathbf{T}\boldsymbol{\epsilon}(\boldsymbol{\epsilon}^\dagger\boldsymbol{\epsilon})^{-1/2}, \tag{8.3.14}$$

and this is the required algorithm for choosing $\mathbf{T}_1'$ in (8.3.7) to obtain a Hermitian $\boldsymbol{\epsilon}_1'$.

To obtain a more explicit form of the result, we substitute $\boldsymbol{\epsilon} = \mathbf{T}^\dagger\mathbf{X}_1$ from (8.3.6) and obtain from (8.3.7)

$$\mathbf{T}_1' = \mathbf{TT}^\dagger\mathbf{X}_1(\mathbf{X}_1\mathbf{TT}^\dagger\mathbf{X}_1)^{-1/2}. \tag{8.3.15}$$

But from the orthonormality conditions on the orbitals represented by the columns of $\mathbf{T}$, assuming a general basis $\boldsymbol{\chi}$ with overlap matrix $\mathbf{S}$, we have

$$\mathbf{T}^\dagger\mathbf{ST} = (\mathbf{S}^{1/2}\mathbf{T})^\dagger(\mathbf{S}^{1/2}\mathbf{T}) = \mathbf{1},$$

and hence $\mathbf{S}^{1/2}\mathbf{T}$ is a unitary matrix: consequently $\mathbf{S}^{1/2}\mathbf{T}\mathbf{T}^{\dagger}\mathbf{S}^{1/2}=\mathbf{1}$ and $\mathbf{T}\mathbf{T}^{\dagger}=\mathbf{S}^{-1}$. This allows us to eliminate the full matrix $\mathbf{T}$ from (8.3.15). The subscript 1 is then redundant and we may return to the original convention that $\mathbf{T}$ and $\mathbf{X}$ refer to the *rectangular* $m \times n$ matrices. The result is then

$$\boxed{\mathbf{T}'=\mathbf{S}^{-1}\mathbf{X}(\mathbf{X}^{\dagger}\mathbf{S}^{-1}\mathbf{X})^{-1/2},} \qquad (8.3.16)$$

which contains the overlap matrix $\mathbf{S}$, corresponding to use of a non-orthogonal basis, but otherwise coincides with the prescription given by Golebiewski *et al.* (1979).

The iterative procedure based on (8.3.16) is in fact the exact analogue, in MC SCF theory, of the Roothaan closed-shell SCF method: the singular-value decomposition corresponds to solution of the matrix eigenvalue problem, and the updating of the $\mathbf{X}$ matrix corresponds to the updating of the $\mathbf{G}$ matrix; in each case a single iteration would serve, were it not for the need to revise the electron interaction terms in every cycle.

It is worth noting that the algorithm (8.3.16) was first derived, in a more intuitive manner, by Mukherjee (1978) and that other variants (see e.g. Polezzo and Fantucci, 1975, 1979) have also been used.

Methods of the type discussed in this section have been employed widely and with considerable success, but (as in simple SCF calculations) their convergence is often slow and unreliable: this is basically because only *first*-order variations have been carried in determining each iterative step. What is needed is a higher-order procedure in which the variation of the $\mathbf{X}$ matrix is incorporated in each cycle of the iteration. The basis of such procedures is considered in the next section.

8.4 GRADIENT METHODS

Instead of attempting to solve the equations that define a stationary point, as in the last section, it is possible to find such a point by direct search on the energy surface; starting from an arbitrary point, a variation $E \rightarrow E+\delta E$ (δE negative) is visualized as a "descent", and the problem is to find an efficient strategy for proceeding to a minimum (for the ground state) in a convergent series of descents. Early applications in SCF theory (McWeeny, 1956) employed a *steepest*-descent procedure, every step being along a direction of maximum gradient; but this is not necessarily the best overall strategy, and even in small-scale MC SCF calculations it was soon found necessary to seek methods of accelerating the convergence (see e.g. Mukherjee and McWeeny, 1970). The steepest-descent procedure is in fact the precursor of a whole family of gradient methods (for a review see Garton and Sutcliffe, 1974). The most effective of these methods take account of the fact that knowledge of the gradient alone is not sufficient for the determination of an efficient strategy; it is also necessary to know the *curvature* of the energy surface, and this requires a knowledge of *second derivatives*. To this end, we develop the energy up to second order in the variation parameters around an arbitrary point corresponding to energy E: the condition that this be a *stationary* point (vanishing gradient) will lead back to the equations obtained in Section 8.2; but away from a stationary point we shall have a *quadratic* approximation to the energy surface, from which we can locate the minimum (in a quadratic approximation) in every step. In all such gradient procedures it is nowadays most common to incorporate orthonormality constraints from the outset by using the "exponential ansatz" (8.2.24) in describing the orbital variation.

In this section we indicate a simple quadratic procedure, starting from the finite-basis form (8.2.15) of the energy functional, in which the occupied orbitals are written explicitly as combinations (8.2.11) of the basis functions $\{\chi_\mu\}$, in order to maintain contact with the equations of the last section. Again we use the full $\mathbf{T}$ matrix (8.3.1), filling up the matrices $\mathbf{T}$ and $\mathbf{Z}$ in (8.2.15) with zeros as necessary, and consider a variation of the form (8.3.3) with $\mathbf{V}$ now in the explicit form (8.2.24).

Corresponding to $\mathbf{T} \rightarrow \mathbf{TV} = \mathbf{T} + \delta\mathbf{T}$, the variation of E is found to be

$$\delta \mathrm{E} = \mathrm{tr}\,(\delta\mathbf{T}^\dagger\mathbf{hTP} + \mathbf{T}^\dagger\mathbf{h}\delta\mathbf{TP}) + \tfrac{1}{2}\mathrm{tr}\,(\delta\mathbf{T}^\dagger\mathbf{Z} + \mathbf{T}^\dagger\delta\mathbf{Z}) + \mathrm{tr}\,(\delta\mathbf{T}^\dagger\delta\mathbf{Z}) + \mathrm{tr}\,(\delta\mathbf{T}^\dagger\mathbf{h}\delta\mathbf{TP}), \tag{8.4.1}$$

where $\delta\mathbf{Z}$ is the associated change in the $\mathbf{Z}$ matrix (8.2.13). Instead of passing immediately to the exponential form of $\delta\mathbf{T}$ ($\mathbf{T} + \delta\mathbf{T} = \mathbf{TV} = \mathbf{T}e^{\mathbf{\Delta}}$) and keeping terms up to second order in the rotation parameters Δ_{rs}, we note that δE in (8.4.1) is of second order in $\delta\mathbf{T}$ itself, and that if the elements of $\delta\mathbf{T}$ are regarded as the parameters then terms of third and higher order in $\mathbf{\Delta}$ will be implicitly included. This important fact, first stressed by Werner and Meyer (1980), provides a basis for the development of very rapidly convergent numerical techniques (Werner and Knowles, 1985; Knowles and Werner, 1985; Werner, 1987).

On using the definition of $\mathbf{Z}$ and noting that $\delta\mathbf{Z}$ contains four terms (each with one $\delta\mathbf{T}$ and three $\mathbf{T}$s), it is easy to rearrange (8.4.1) and to write the varied energy as

$$E = E_0 + \delta_1 E + \delta_2 E, \tag{8.4.2}$$

where E_0 refers to the initial point ($\delta\mathbf{T} = \mathbf{0}$) and, with $\mathbf{X} = \mathbf{hTP} + \mathbf{Z}$

(putting $\mathbf{A}=\mathbf{P}$ in (8.2.14) to pass to the orbital version),

$$\delta_1 E = \text{tr}\,(\delta\mathbf{T}^\dagger\mathbf{X}) + \text{c.c.}, \tag{8.4.3}$$

$$\delta_2 E = \tfrac{1}{2}[\text{tr}\,(\delta\mathbf{T}^\dagger\delta\mathbf{X}) + \text{c.c.}]. \tag{8.4.4}$$

The second-order variation includes all terms that contain two $\delta\mathbf{T}$ factors.

The results (8.4.3) and (8.4.4) are of central importance; they show that both first- and second-order energy variations, with orthonormality constraints included, may be expressed in compact form even for a general multiconfiguration wavefunction.

The first-order change (8.4.3) determines the gradient vector, whose components are the first derivatives of E with respect to parameters—namely the independent elements (Δ_{rs}, $r>s$ say) of the matrix $\mathbf{\Delta}$.

Close to a stationary point, we may use $\delta\mathbf{T}\simeq\mathbf{T}\mathbf{\Delta}$ (neglecting terms in $\mathbf{\Delta}^2$ etc.) and thus obtain, remembering that $\mathbf{\Delta}^\dagger=-\mathbf{\Delta}$,

$$\delta_1 E = \text{tr}\,\mathbf{\Delta}^\dagger(\mathbf{T}^\dagger\mathbf{X}-\mathbf{X}^\dagger\mathbf{T}). \tag{8.4.5}$$

The components $\partial E/\partial\Delta_{rs}$ are determined by the elements of the matrix in parentheses. In particular, the condition for a stationary point (vanishing gradient) becomes

$$\mathbf{T}^\dagger\mathbf{X}-\mathbf{X}^\dagger\mathbf{T}=\mathbf{0}. \tag{8.4.6a}$$

When the matrices are partitioned as in (8.3.1) this yields at once

$$\mathbf{T}_1^\dagger\mathbf{X}_1-\mathbf{X}_1^\dagger\mathbf{T}_1=\boldsymbol{\epsilon}-\boldsymbol{\epsilon}^\dagger=\mathbf{0}, \tag{8.4.6b}$$

where the "occupied rectangles" $\mathbf{T}_1$ and $\mathbf{X}_1$ are the matrices used (without subscripts) in (8.2.14) of the original derivation. We thus retrieve the stationary condition in the form (8.2.31).

Away from a stationary point, we may develop E up to second order in the rotation parameters Δ_{rs}, and then try to find the stationary point by using the quadratic approximation to the energy surface as in Section 2.4. This approach arises very naturally in the second-quantization formalism, as will be seen presently, and was pioneered by Jørgensen and collaborators (see e.g. Dalgaard and Jørgensen, 1978; Yeager and Jørgensen, 1979). It gives excellent convergence once a "quadratic basin" on the surface has been entered. If we use Δ_{rs} ($r>s$) as the independent parameters (d_μ, say, with $\mu=rs$), noting that $\Delta_{sr}=-d_\mu^*$ in consequence of (8.2.24), the first-order variation may be written in the form (2.4.16), namely

$$\delta_1 E = \sum_\mu [d_\mu^*(\nabla H)_\mu + (H\nabla)_\mu d_\mu], \tag{8.4.7a}$$

with

$$(H\nabla)_\mu = (\mathbf{T}^\dagger\mathbf{X} - \mathbf{X}^\dagger\mathbf{T})_\mu = (\nabla H)^*_\mu, \tag{8.4.7b}$$

and direct calculation of the variation thus allows us to identify the various derivatives defined in Section 2.4. A more cumbersome rearrangement of the terms in (8.4.4), following the introduction of $\mathbf{\Delta}$ from (8.4.2), allows us to identify the second derivatives $(\nabla H\nabla)_{\mu\nu}$, $(\nabla\nabla H)_{\mu\nu}$ etc., and thus to write the energy expression in the form (cf. (2.4.18))

$$E = E_0 + \begin{pmatrix}\mathbf{d}\\ \mathbf{d}^*\end{pmatrix}^\dagger \begin{pmatrix}\mathbf{a}\\ \mathbf{a}^*\end{pmatrix} + \frac{1}{2}\begin{pmatrix}\mathbf{d}\\ \mathbf{d}^*\end{pmatrix}^\dagger \begin{pmatrix}\mathbf{M} & \mathbf{Q}\\ \mathbf{Q}^* & \mathbf{M}^*\end{pmatrix}\begin{pmatrix}\mathbf{d}\\ \mathbf{d}^*\end{pmatrix} + \dots, \tag{8.4.8}$$

with $a_\mu = (H\nabla)_\mu$ a gradient component, and the square matrix in the second-order term as the Hessian matrix of second derivatives.

The analysis of Section 2.4 may now be taken over in its entirety. If the initial point is near a stationary point on the energy surface, and the surface is locally quadratic (at least in good approximation), then the stationary point may be reached by choosing the parameters so that (2.4.20) is satisfied. More commonly, when the parameters are *real,* it is sufficient to solve the reduced equation (2.4.21), which may be written

$$E^{(1)}_{rs} + \sum_{tu} E^{(2)}_{rs,tu}\,\Delta_{tu} = 0 \quad (\text{all } rs). \tag{8.4.9}$$

This system of simultaneous equations is in principle easily solved to obtain the approximate stationary point. Since the surface is not exactly quadratic, iteration will still, of course, be necessary; but this is a true second-order process and convergence will be rapid once a roughly quadratic basin has been reached.

Unfortunately the matrix $\mathbf{E}^{(2)}$ whose elements appear in (8.4.9) is usually rather large and solution by inversion is not convenient. There are various ways of avoiding direct solution of (8.4.9). The simplest is to assume that $\mathbf{E}^{(2)}$ is approximately diagonal, with elements $E^{(2)}_{rs,rs}$. The equations (8.4.9) are then uncoupled and $\Delta_{rs} = -E^{(1)}_{rs}/E^{(2)}_{rs,rs}$ (all rs). Since each element Δ_{rs} defines a 2×2 rotation, this approximation is equivalent to defining an optimum rotation for each pair rs and then performing all rotations simultaneously, as if they were independent. This is reminiscent of the methods introduced by Hinze and others (Section 8.3); but here the second-order terms are admitted in a more satisfactory way and the aim is to search systematically for the stationary point rather than to satisfy the stationary-value conditions.

The second way of avoiding direct solution of (8.4.9) is to construct recursively a sequence of approximations to the step vector $\mathbf{\Delta}$ without ever actually solving the equations themselves; this approach is characteristic of so-called "conjugate gradient" methods (of which there are many variants) in which the second-order terms are estimated numerically during iteration. Conjugate gradient techniques are well established in applied mathematics (see e.g. Hestenes, 1980); some applications in quantum chemistry have been discussed by Wormer *et al.* (1982) and Shepard (1987).

The methods described above often converge well near to a minimum. But in regions where the surface is non-quadratic, or the Hessian in (8.4.8) is not positive-definite, methods using expansion to second order are not entirely satisfactory, even when applied in their full form. The approach due to Werner and Meyer (1980) then has the advantage of more positive convergence.

Let us therefore return to (8.4.2)–(8.4.4) and consider the direct determination of $\delta\mathbf{T}$, starting from an arbitrary point $\mathbf{T}$ at which $\delta E = 0$. The variation of $\mathbf{T}$ will be

$$\delta\mathbf{T} = \mathbf{T}(e^{\boldsymbol{\Delta}} - \mathbf{1}) = \mathbf{TD}, \tag{8.4.10}$$

where $\mathbf{D} = \boldsymbol{\Delta} + \frac{1}{2}\boldsymbol{\Delta}^2 + \dots$. But, instead of using the expansion, we try to determine $\mathbf{D}$ directly by substituting (8.4.10) in (8.4.3), (8.4.4) and then requiring that, in any further variation $\mathrm{d}\mathbf{D}$, $\partial E/\partial D_{rs} = 0$ (all r, s). The varied parameters, $\mathbf{T} + \delta\mathbf{T} = \mathbf{TV} = \mathbf{T}(\mathbf{1} + \mathbf{D})$, will then correspond to the required stationary point in an approximation that includes important terms to infinite order in $\boldsymbol{\Delta}$.

The general reduction of (8.4.4) is awkward (Problem 8.8); but in the case most commonly encountered in practice, where all quantities are real, the high symmetry of the 2-electron integrals permits a simplification (Problem 8.10). The matrix $\delta\mathbf{Z}$ may then be expressed in terms of the familiar $\mathbf{J}$ and $\mathbf{K}$ matrices introduced in Chapter 6. On using $\mathbf{c}_r$, $\delta\mathbf{c}_r$ to denote typical columns of $\mathbf{T}$ and $\delta\mathbf{T}$, the second-order energy variation (8.4.2) takes the form

$$\boxed{E = E_0 + 2\sum_{r,s} \delta\mathbf{c}_r^\dagger \mathbf{F}^{(rs)} \mathbf{c}_s + \sum_{r,s} \delta\mathbf{c}_r^\dagger \mathbf{G}^{(rs)} \delta\mathbf{c}_s,} \tag{8.4.11}$$

where

$$\mathbf{F}^{(rs)} = P_{sr}\mathbf{h} + \sum_{t,u} \Pi_{su,rt} \mathbf{J}(\mathbf{R}_{ut}), \tag{8.4.12}$$

$$\mathbf{G}^{(rs)} = \mathbf{F}^{(rs)} + 2\sum_{t,u} \Pi_{su,tr} \mathbf{K}(\mathbf{R}_{ut}), \tag{8.4.13}$$

and the matrices $\mathbf{R}_{ut}$ are formally transition densities connecting orbitals ϕ_u and ϕ_t: $\mathbf{R}_{ut} = \mathbf{c}_u\mathbf{c}_t^\dagger$.

It will be recognized that $\mathbf{F}^{(rs)}$ is the matrix associated with the operator

F_{rs} defined in (8.2.6), with $A, B \rightarrow P, \Pi$. There is one such matrix for every pair of orbitals, and $\mathbf{h}$, $\mathbf{J}$, $\mathbf{K}$, $\mathbf{R}_{ut}$ are all $m \times m$ matrices defined over basis functions, although r, s, t, u run only over the "occupied" orbitals used in the CI expansion. Again it must be stressed that this final reduction is dependent on all orbitals and integrals being real and that in this context the dagger merely indicates transposition.

From (8.4.10) we may write $\delta\mathbf{c}_r = \sum_t \mathbf{c}_t D_{tr}$ etc., and substitution in (8.4.11) then gives (with m, n running over *all* orbitals)

$$E = E_0 + 2 \sum_{r,s,m} D_{mr}(\mathbf{c}_m^\dagger \mathbf{F}^{(rs)} \mathbf{c}_s) + \sum_{r,s,m,n} D_{mr}(\mathbf{c}_m^\dagger \mathbf{G}^{(rs)} \mathbf{c}_n) D_{ns}, \quad (8.4.14)$$

where each of the quantities in parentheses is a single number—a matrix element of a certain 1-electron operator between two occupied orbitals—and the summations are unrestricted. We may then differentiate with respect to a typical element D_{mr}, obtaining

$$\frac{\partial E}{\partial D_{mr}} = 2 \sum_s \mathbf{c}_m^\dagger \mathbf{F}^{(rs)} \mathbf{c}_s + \sum_{s,n} \mathbf{c}_m^\dagger \mathbf{G}^{(rs)} \mathbf{c}_n D_{ns} + \sum_{s,n} D_{ns} \mathbf{c}_n^\dagger \mathbf{G}^{(sr)} \mathbf{c}_m .$$

A little manipulation (Problem 8.11) shows that the last two terms are equal, and the condition that $\mathbf{D}$ give a stationary point becomes, denoting the above derivative by B_{mr},

$$\begin{aligned} B_{mr} &= 2 \sum_s \mathbf{c}_m^\dagger \mathbf{F}^{(rs)} \mathbf{c}_s + 2 \sum_{s,n} \mathbf{c}_m^\dagger \mathbf{G}^{(rs)} \mathbf{c}_n D_{ns} \\ &= 0 \quad (\text{all } m, n). \end{aligned} \quad (8.4.15)$$

As a first approximation we may of course take $D_{ns} = \Delta_{ns}$, noting that, since $\Delta_{sn} = -\Delta_{ns}$, half the parameters will be redundant; in that case the equation reduces to (8.4.9) with the coefficients explicitly identified.

Further away from a stationary point, $\mathbf{D}$ is not antisymmetric, and all elements must be determined. The orthonormality constraints (no longer embodied in an antisymmetric $\mathbf{D}$) must also be imposed explicitly. The procedure for doing so is well known from earlier chapters: the condition $\delta E = 0$ must be combined with the equations of constraint

$$\begin{aligned} \delta(\mathbf{V}^\dagger \mathbf{V})_{mn} &= (\delta \mathbf{V}^\dagger \mathbf{V} + \mathbf{V}^\dagger \delta \mathbf{V})_{mn} \\ &= (\delta \mathbf{D}^\dagger \mathbf{V} + \mathbf{V}^\dagger \delta \mathbf{D})_{mn} = 0 \quad (\text{all } r, s), \end{aligned} \quad (8.4.16)$$

and this may be done by attaching Lagrangian multipliers and subtracting to obtain the constrained condition

$$\delta E - \sum_{m,n} \delta(\mathbf{V}^\dagger \mathbf{V})_{mn} \varepsilon_{nm} = 0, \quad (8.4.17)$$

where the multipliers will be eliminated later, using $\mathbf{V}^\dagger \mathbf{V} = \mathbf{V}\mathbf{V}^\dagger = \mathbf{1}$.

Both terms in (8.4.17) may be written as traces; thus

$$\delta E = \sum_{m,n} B_{mn}\, \delta D_{mn} = \operatorname{tr} \mathbf{B}\, \delta \mathbf{D}^{\dagger},$$

and from (8.4.16)

$$\sum_{m,n} \delta(\mathbf{V}^{\dagger}\mathbf{V})_{mn} \varepsilon_{nm} = \operatorname{tr}(\delta \mathbf{D}^{\dagger}\mathbf{V}\varepsilon + \mathbf{V}^{\dagger}\mathbf{D}\varepsilon).$$

Since each trace in the last equation is a single real number, the second may be transposed to give $\operatorname{tr} \mathbf{V}^{\dagger}\, \delta\mathbf{D}\boldsymbol{\epsilon} = \operatorname{tr} \boldsymbol{\epsilon}^{\dagger}\, \delta\mathbf{D}^{\dagger}\mathbf{V} = \operatorname{tr} \delta\mathbf{D}^{\dagger}\mathbf{V}\boldsymbol{\epsilon}^{\dagger}$. Substitution in (8.4.17) then gives, noting that $\boldsymbol{\epsilon}$ may be chosen Hermitan since only the combination $\boldsymbol{\epsilon} + \boldsymbol{\epsilon}^{\dagger}$ appears in the equation,

$$\operatorname{tr} \delta\mathbf{D}^{\dagger}\mathbf{B} - 2 \operatorname{tr} \delta\mathbf{D}^{\dagger}\mathbf{V}\boldsymbol{\epsilon} = 0 \quad (\delta\mathbf{D} \text{ arbitrary, } \boldsymbol{\epsilon} \text{ Hermitian}).$$

The final equation to determine $\mathbf{V}$, yielding an improved matrix $\mathbf{T}' = \mathbf{TV}$, is thus

$$\mathbf{B} - 2\mathbf{V}\boldsymbol{\epsilon} = \mathbf{0}. \tag{8.4.18}$$

The multipliers may be eliminated on multiplying this equation from the left by $\mathbf{V}^{\dagger}$, and its Hermitian conjugate from the right by $\mathbf{V}$, and taking the difference. The result is

$$\boxed{\mathbf{V}^{\dagger}\mathbf{B} - \mathbf{B}^{\dagger}\mathbf{V} = \mathbf{0}.} \tag{8.4.19}$$

Iterative methods of solving this equation have been devised by Werner and Meyer (1980). An equivalent condition, however, is clearly

$$\mathbf{V}^{\dagger}\mathbf{B} = \begin{pmatrix} \mathbf{B}' \\ \mathbf{0} \end{pmatrix}, \quad \mathbf{B}' \text{ Hermitian} \tag{8.4.20}$$

where $\mathbf{B}$ now denotes the *rectangular* block whose columns are labelled by the *occupied* orbitals, the full matrix being partitioned as in Section 8.3 (see Problem 8.11). The problem is thus identical with one already solved. The matrix $\mathbf{V}$ that leads to a Hermitian $\mathbf{B}'$, and which therefore satisfies (8.4.19), is defined in (8.3.11) in terms of the eigenvectors of $\mathbf{BB}^{\dagger}$ and $\mathbf{B}^{\dagger}\mathbf{B}$. The solution, for given $\mathbf{B}$, thus involves nothing more than solving a matrix eigenvalue equation whose dimensions are those of the basis set. Of course, $\mathbf{B}$ depends on $\mathbf{D}$ through (8.4.15) and iteration is still required: when complete, $\mathbf{T}' = \mathbf{VT}$ defines a new starting point, closer to the minimum, and the process is then repeated.

The optimization of the orbitals by gradient methods of the kind discussed in this section forms the basis of highly efficient numerical techniques (see e.g. Werner and Knowles, 1985) in which simultaneous variation of CI coefficients may also be incorporated. The way in which the latter may be included is indicated in Section 8.6.

8.5 USE OF THE BRILLOUIN–LEVY–BERTHIER THEOREM

It was shown in Section 8.2 that the stationary condition used so far, namely (8.2.19), is equivalent to the requirement (8.2.38), namely

$$\langle [\mathsf{H}, \mathsf{E}_{sr}] \rangle = 0 \quad (\text{all } r, s). \tag{8.5.1}$$

We now show that this condition is essentially equivalent to the so-called Brillouin theorem in the form discussed by Levy and Berthier (1968); self-consistency may then be achieved using any procedure that ultimately gives a wavefunction satisfying this theorem.

First we recall that E_{sr} $(=\mathsf{a}_s^\dagger \mathsf{a}_r)$ replaces ψ_r by ψ_s, when ψ_r is present in an antisymmetric ket while ψ_s is not, but otherwise destroys the ket. On noting that $\mathsf{E}_{sr} = \mathsf{E}_{rs}^\dagger$, it follows from (8.5.1) that

$$\langle \Psi | \mathsf{HE}_{sr} | \Psi \rangle - \langle \mathsf{E}_{rs} \Psi | \mathsf{H} | \Psi \rangle = 0,$$

or, with an obvious notation,

$$\langle \Psi | \mathsf{H} | \Psi(r \to s) \rangle - \langle \Psi(s \to r) | \mathsf{H} | \Psi \rangle = 0. \tag{8.5.2}$$

In other words, the stationary condition on $\langle \Psi | H | \Psi \rangle$ is equivalent to a condition on the matrix elements connecting Ψ with "excited" Ψs formed by substituting an occupied spin-orbital by an unoccupied spin-orbital in all Slater determinants where it appears (destroying determinants in which it does not occur). This is reminiscent of the Brillouin theorem for a 1-determinant wavefunction, but is clearly a generalization; it is not, however, in the form given by Levy and Berthier (1968).

To obtain the Levy–Berthier form, we use (8.2.19) and (8.2.20), and the fact that $\Delta_{rs} = -\Delta_{sr}^*$, to write

$$\delta E = \sum_{r<s} \{ \Delta_{sr} \langle \Psi | (\mathsf{HE}_{sr} - \mathsf{E}_{rs}\mathsf{H}) | \Psi \rangle - \Delta_{sr}^* \langle \Psi | (\mathsf{HE}_{sr} - \mathsf{E}_{sr}\mathsf{H}) | \Psi \rangle \} = 0.$$

Since Δ_{rs} and Δ_{rs}^* are linearly independent, each coefficient must vanish separately, and we therefore retrieve (8.5.1). On the other hand, for Δ_{rs} arbitrary but *real,* we obtain

$$\delta E = \sum_{r<S} \Delta_{sr} \langle \Psi | \{ \mathsf{H}(\mathsf{E}_{sr} - E_{rs}) + (\mathsf{E}_{sr} - \mathsf{E}_{rs})\mathsf{H} \} | \Psi \rangle = 0$$

This condition may be written as

$$\delta E = \sum_{r<s} \Delta_{sr}\{\langle \Psi|\,\mathsf{H}\,[\Psi(r\to s) - \Psi(s\to r)]\rangle + \langle[\Psi(r\to s) - \Psi(s\to r)]|\,\mathsf{H}\,|\Psi\rangle\} = 0,$$

which is clearly the first-order energy variation corresponding to

$$\Psi \to \Psi + \sum_{r<s} \Delta_{sr}[\Psi(r\to s) - \Psi(s\to r)]. \qquad (8.5.3)$$

Since δE is real, the Brillouin condition in this form becomes

$$\langle \Psi|\,\mathsf{H}\,|[\Psi(r\to s) - \Psi(s\to r)]\rangle = 0 \quad (\text{all } r, s), \qquad (8.5.4)$$

which is the result derived by Berthier, Levy and others for real wavefunctions.

The Brillouin condition is commonly used in two ways as the basis of an iteration scheme for improving the orbitals in a limited CI function Ψ. Both involve the solution of a large set of secular equations to determine the optimum mixing betweeh $|\Psi\rangle$ and all the single-excitation functions

$$|\Psi^s_r\rangle = |[\Psi(r\to s) - \Psi(s\to r)]\rangle = |(\mathsf{E}_{sr} - \mathsf{E}_{rs})\Psi\rangle. \qquad (8.5.5)$$

This "super-CI problem", since it includes only *single* excitations from the occupied orbitals of the CI function Ψ, is not impossibly large; efficient methods of obtaining a solution have been discussed and implemented by Grein, Ruedenberg, Roos and others (see especially Grein and Banerjee, 1973; Ruedenberg *et al.* 1979; Roos *et al.*, 1980). Roos, in particular, has developed a "complete active space" (CAS) method in which Ψ admits full CI within a subspace of functions obtained using arbitrary occupations and coupling schemes for a chosen set of "active orbitals" (other orbitals being doubly occupied ("inactive") or virtual ("secondary")).

On solving the super-CI equations, by standard methods, we obtain a function

$$\Psi' = \Psi + \sum_{r<s} C^s_r \Psi^s_r. \qquad (8.5.6)$$

not very different from Ψ, which will have zero matrix elements connecting it with its partners (i.e. the other approximate solutions, which are predominantly mixtures of single-excitation terms). In other words, Ψ' will approximately satisfy the Brillouin condition. Since however, Ψ' is still written in terms of the original orbitals—but with a vastly increased number of configurations—we need to show that it may be rewritten, in good approximation, as a function $\tilde{\Psi}$ of *varied orbitals* but with the same structure as the original Ψ. The two procedures for obtaining such orbitals are due largely to Grein and coworkers, who use a transformation based on (8.5.6), and Ruedenberg *et al.* (1979), who use the properties of natural orbitals (Section 5.6).

To make use of (8.5.6), we perform a variation of $|\Psi\rangle$ using the unitary operator (8.2.18) with R of the form (8.2.20). The result is

$$\tilde{\Psi} = e^{\mathsf{R}}\Psi = \Psi + \sum_{s,r} \Delta_{sr}\mathsf{E}_{sr}\Psi = \Psi + \sum_{s>r} \Delta_{sr}\Psi_r^s,$$

with Ψ_r^s defined in (8.5.5). Comparison with (8.5.6) then shows that, provided the infinitesimal rotation R is chosen by putting

$$\Delta_{sr} = C_r^s, \tag{8.5.7}$$

the varied function $\tilde{\Psi}$ will coincide to first order with the function Ψ' obtained from the super-CI equations. Since the unitary operator $\mathsf{U} = e^{\mathsf{R}}$, working in one-electron space, yields the orbital transformation (8.2.21), it is clear that $\tilde{\Psi}$ is obtained on replacing ψ_r by

$$\tilde{\psi}_r = \sum_s \psi_s(\delta_{sr} + \Delta_{sr})$$

or

$$\tilde{\psi}_r = \psi_r + \sum_s C_r^s\psi_s. \tag{8.5.8}$$

Solution of the super-CI equations therefore determines to first order the orbital variation necessary to optimize Ψ; since, however, orbital variation changes the matrix elements in the super-CI equations, the whole cycle must be repeated until convergence is obtained. This method is practicable, even in its simplest form (see e.g. Grein and Chang, 1971), but, since it is essentially first-order, various refinements are necessary; in each cycle, for example, the orbitals must be re-orthogonalized as the transformation (8.5.8) is only approximately unitary.

The natural-orbital procedure developed by Ruedenberg and others is in some respects simpler and preserves orthogonality automatically. We again start from the super-CI function (8.5.6), but, instead of using the expansion coefficients to obtain the orbital transformation directly, according to (8.5.8), we work throughout in terms of the natural orbitals. The 1-body density matrix associated with any given Ψ may be brought to diagonal form by a certain unitary transformation of the occupied orbitals; it then appears in the form

$$\tilde{\rho}(\boldsymbol{x};\boldsymbol{x}') = \sum_k n_k\tilde{\psi}_k(\boldsymbol{x})\tilde{\psi}_k^*(\boldsymbol{x}'), \tag{8.5.9}$$

in which the n_k are positive occupation numbers and depend only on the structure of Ψ, and the CI coefficients, not on the orbital forms. The super-CI function (8.5.6), however, does not yield a diagonal form; instead we find (by standard methods)

$$\rho'(\boldsymbol{x};\boldsymbol{x}') = \sum_{r,s} \rho'_{rs}\psi_r(\boldsymbol{x})\psi_s^*(\boldsymbol{x}'), \tag{8.5.10}$$

where r, s run also over the virtual space. The varied function is then obtained by the criterion

$$\|\tilde{\rho} - \rho'\|^2 = \int |[\tilde{\rho}(\boldsymbol{x};\boldsymbol{x}') - \rho'(\boldsymbol{x};\boldsymbol{x}')]|^2 \, d\boldsymbol{x}\, d\boldsymbol{x}' = \text{minimum},$$

or, since ρ' and ρ are both individually normalized so that $\|\tilde{\rho}\|^2 = \|\rho'\|^2 = N$,

$$\int \tilde{\rho}(\boldsymbol{x};\boldsymbol{x}')\rho'(\boldsymbol{x}';\boldsymbol{x}) \, d\boldsymbol{x}\, d\boldsymbol{x}' = \text{maximum}. \tag{8.5.11}$$

When $\tilde{\rho}$ has the form (8.5.9), the summation extending only over the occupied orbitals, substitution yields the condition

$$\sum_{k,r,s} n_k \langle \tilde{\psi}_k \mid \psi_r \rangle \rho'_{rs} \langle \psi_s \mid \tilde{\psi}_k \rangle = \text{maximum}$$

The scalar products may be collected in an $m \times n$ matrix $\tilde{\mathbf{S}}$, and the condition then becomes

$$\sum_k n_k (\tilde{\mathbf{S}}^\dagger \boldsymbol{\rho}' \tilde{\mathbf{S}})_{kk} = \text{maximum}. \tag{8.5.12}$$

Since the n_k are independent of orbital forms and are non-zero only for the n natural orbitals appearing in Ψ, (8.5.12) may be satisfied in practice by finding the n highest stationary values of $\tilde{\mathbf{S}}^\dagger \boldsymbol{\rho}' \tilde{\mathbf{S}}$, bearing in mind the orthonormality constraint imposed on the orbitals ψ_k. This is a familiar problem: if the n varied orbitals are expressed in terms of the m orbitals used in the super-CI,

$$\tilde{\boldsymbol{\psi}} = \boldsymbol{\psi}\mathbf{T},$$

then $\tilde{\mathbf{S}} = \mathbf{ST}$ (with $\mathbf{S} = \mathbf{1}_m$), and we require

$$(\mathbf{T}^\dagger \boldsymbol{\rho}' \mathbf{T})_{kk} = \mathbf{c}_k^\dagger \boldsymbol{\rho}' \mathbf{c}_k = \text{maximum}, \tag{8.5.13}$$

where $\mathbf{c}_k$ is the kth column of $\mathbf{T}$. The stationary values of this quantity, subject to orthonormality requirements, are of course the eigenvalues of the matrix $\boldsymbol{\rho}'$; and therefore the optimum orbitals to use in Ψ are the n highest-eigenvalue natural orbitals of the super-CI problem.

The natural-orbital method must be used with care for states that are not totally symmetric in space and spin, since the natural orbitals‡ will not then be symmetry-adapted (see e.g. McWeeny and Kutzelnigg, 1968). In such cases symmetry-adapted orbitals may be obtained by diagonalizing the totally symmetric projection of the density matrix as discussed by Ruedenberg *et al.* (1979). In general, the procedure appears to be rather efficient.

8.6 VARIATION OF ALL PARAMETERS

So far we have discussed only optimization of a multiconfiguration reference function with respect to *orbital* variation, assuming that the CI coefficients (when not fully determined by symmetry) are optimized independently by repeated solution of the usual secular equations during the iterative improvement of the orbitals. This corresponds to the "two-step" process referred to in Section 8.1. The CI coefficients may, however, be optimized *simultaneously* within each iterative step, alongside the orbitals, in the "one-step" process that we now consider. To this end, it is convenient to employ the second-quantization formalism (following Dalgaard and Jørgensen, 1978; Dalgaard, 1979, 1980), which leads to elegant expressions for the required first and second derivatives. Further generalizations, in which non-linear parameters within the basis functions $\{\chi_\mu\}$ are also optimized, may be admitted in principle by using the tensor formalism of Section 2.4.

Let us start from the exponential-operator description of a variation of spin-orbitals, using (8.2.18) with the infinitesimal rotation R in the form (8.2.20). For a normalized reference $|\Psi\rangle$ the energy after variation may be written, using (8.2.16) and expanding the exponentials,

$$\tilde{E} = \langle \Psi | (\mathsf{H} - [\mathsf{R}, \mathsf{H}] + \tfrac{1}{2}[\mathsf{R}, [\mathsf{R}, \mathsf{H}]] - \ldots) | \Psi \rangle. \qquad (8.6.1)$$

This may be written in terms of the rotation parameters Δ_{rs}, using (8.2.20), which gives

$$\mathsf{R} = \sum_{r>s} (\Delta_{rs}\mathsf{E}_{rs} - \Delta_{rs}^{*}\mathsf{E}_{sr}).$$

On introducing a single-subscript convention

$$\mu = rs, \qquad \bar{\mu} = sr \quad (r > s), \qquad (8.6.2)$$

‡ Or natural *spin*-orbitals, depending on the details of the formulation.

and noting that $\mathsf{E}_{sr} = \mathsf{E}_{rs}^\dagger$, the expression may be written as

$$\mathsf{R} = \sum_\mu (d_\mu \mathsf{E}_\mu - d_\mu^* \mathsf{E}_\mu^\dagger), \tag{8.6.3}$$

where μ is the ordered pair rs and d_μ runs over the distinct rotation parameters Δ_{rs} $(r > s)$ in the lower triangle of the matrix $\mathbf{\Delta}$. It follows at once, on substituting (8.6.3) in (8.6.1) and comparing the result with (2.4.16), that the various energy derivatives are

$$(\nabla H)_\mu = \langle \Psi | [\mathsf{E}_\mu^\dagger, \mathsf{H}] | \Psi \rangle, \qquad (H\nabla)_\mu = -\langle \Psi | [\mathsf{E}_\mu, \mathsf{H}] | \Psi \rangle \tag{8.6.4}$$

and

$$\left.\begin{aligned}
(\nabla\nabla H)_{\mu\nu} &= \langle \Psi | [\mathsf{E}_\mu^\dagger, [\mathsf{E}_\nu^\dagger, \mathsf{H}]] | \Psi \rangle, \\
(H\nabla\nabla)_{\mu\nu} &= \langle \Psi | [\mathsf{E}_\mu, [\mathsf{E}_\nu, \mathsf{H}]] | \Psi \rangle, \\
(\nabla H\nabla)_{\mu\nu} &= -\tfrac{1}{2}\{\langle \Psi | [\mathsf{E}_\mu^\dagger, [\mathsf{E}_\nu, \mathsf{H}]] | \Psi \rangle + \langle \Psi | [\mathsf{E}_\nu, [\mathsf{E}_\mu^\dagger, \mathsf{H}]] | \Psi \rangle\}.
\end{aligned}\right\} \tag{8.6.5}$$

These commutator expressions for the derivatives are in widespread use and clearly provide a very convenient formalism, although considerable reduction is of course necessary to obtain more explicit expressions in terms of the density matrices $(\mathbf{A}, \mathbf{B})$ of the given reference function Ψ. On making such reductions, we naturally obtain the same expressions as in Section 8.4.

On turning to the variation of CI coefficients, it is clear that the linearly independent set of CFs $\{\Phi_\kappa\}$ (assumed orthonormal) in the expansion (8.2.1) may be subjected to unitary mixing in much the same way as the basis functions. With a unitary operator U_c in CI-space‡ we associate a matrix $\mathbf{U}_\mathrm{c}$ in the usual way:

$$\mathbf{\Phi}' = \mathsf{U}_\mathrm{c}\mathbf{\Phi} = \mathbf{\Phi}\mathbf{U}_\mathrm{c}, \qquad U^\mathrm{c}_{\kappa'\kappa} = \langle \Phi_{\kappa'} | \mathsf{U}_\mathrm{c} | \Phi_\kappa \rangle. \tag{8.6.6}$$

The corresponding rotation (variation) of the reference state $\Psi = \mathbf{\Phi}\mathbf{C}$ ($\mathbf{C}$ being the column of CI coefficients C_κ) will be

$$\Psi \rightarrow \Psi' = \mathsf{U}_\mathrm{c}\Psi = \mathsf{U}_\mathrm{c}\mathbf{\Phi}\mathbf{C} = \mathbf{\Phi}\mathbf{U}_\mathrm{c}\mathbf{C} = \mathbf{\Phi}\mathbf{C}'$$

and the variation of CI coefficients is thus

$$\mathbf{C} \rightarrow \mathbf{C}' = \mathbf{U}_\mathrm{c}\mathbf{C}. \tag{8.6.7}$$

Let us now introduce the infinitesimal operator corresponding to this rotation by writing

$$\mathsf{U}_\mathrm{c} = \exp(\mathsf{R}_\mathrm{c}) \quad (\mathsf{R}_\mathrm{c}^\dagger = -\mathsf{R}_\mathrm{c}), \tag{8.6.8}$$

‡ Henceforth we add labels "o" and "c" to distinguish similar quantities referring respectively to orbitals and coefficients.

where (cf. (8.6.3))

$$\mathsf{R}_c = \sum_\alpha (d^c_\alpha \mathsf{F}_\alpha - d^{c*}_\alpha \mathsf{F}^\dagger_\alpha). \tag{8.6.9}$$

Here $\alpha = \lambda\kappa$ $(\lambda > \kappa)$, d^c_α is a (lower-triangle) element of a rotation matrix (cf. $\mathsf{d}_\mu = \Delta_{rs}$) and F_α is a substitution operator analogous to E_μ; just as E_μ $(=\mathsf{E}_{rs})$ replaces a state containing an electron in ψ_s by one with an electron in ψ_r, F_α $(=\mathsf{F}_{\lambda\kappa})$ must replace $|\Phi_\lambda\rangle$ by $|\Phi_\lambda\rangle$ and may thus be written

$$\mathsf{F}_\alpha = \mathsf{F}_{\lambda\kappa} = |\Phi_\lambda\rangle\langle\Phi_\kappa|, \tag{8.6.10}$$

where $\mathsf{F}_{\kappa\lambda} = \mathsf{F}^\dagger_{\lambda\kappa}$ (cf. $\mathsf{E}_{sr} = \mathsf{E}^\dagger_{rs}$).

It is now clear that a simultaneous variation of both orbitals and expansion coefficients may still be described using (8.6.1) but with R replaced by $\mathsf{R}_o + \mathsf{R}_c$, with "orbital" and "coefficient" operators defined in (8.6.3) and (8.6.9) respectively. The set of distinct parameters will now be $\{d^o_\mu, d^c_\alpha\}$ and the tensor components in (8.6.4) and (8.6.5) will be supplemented by components associated with variation of the d^c_α; these will be given by analogous expressions, but with the F_α operators in place of the E_μ. For example, reserving $\mu, \nu, \ldots$ and $\alpha, \beta, \ldots$ for d^o and d^c components respectively, there will be new first derivatives

$$(\nabla H)_\alpha = \langle\Psi|\,[\mathsf{F}^\dagger_\alpha, \mathsf{H}]\,|\Psi\rangle, \qquad (H\nabla)_\alpha = (\nabla H)^*_\alpha. \tag{8.6.11}$$

and second derivatives such as

$$\left.\begin{aligned}(\nabla\nabla H)_{\mu\alpha} &= \langle\Psi|\,[\mathsf{E}^\dagger_\mu, [\mathsf{F}^\dagger_\alpha, \mathsf{H}]]\,|\Psi\rangle,\\ (\nabla\nabla H)_{\alpha\beta} &= \langle\Psi|\,[\mathsf{F}^\dagger_\alpha, [\mathsf{F}^\dagger_\beta, \mathsf{H}]]\,|\Psi\rangle.\end{aligned}\right\} \tag{8.6.12}$$

Again, the reduction of such expressions to the explicit forms required for computational purposes is straightforward, although the results may be somewhat cumbersome.

Once the tensor components have been evaluated, for some given point on the energy surface, the nearest stationary point (in quadratic approximation) can be found by solving the usual equations (cf. (2.4.20)). When the matrices are "blocked" according to "orbital" and "coefficient" indices, these equations may be written

$$\begin{pmatrix}\mathbf{a}_o\\ \mathbf{a}_c\\ \mathbf{a}^*_o\\ \mathbf{a}^*_c\end{pmatrix} + \left(\begin{array}{cc:cc}\mathbf{M}_{oo} & \mathbf{M}_{oc} & \mathbf{Q}_{oo} & \mathbf{Q}_{oc}\\ \mathbf{M}_{co} & \mathbf{M}_{cc} & \mathbf{Q}_{co} & \mathbf{Q}_{cc}\\ \hdashline \mathbf{Q}^*_{oo} & \mathbf{Q}^*_{oc} & \mathbf{M}^*_{oo} & \mathbf{M}^*_{oc}\\ \mathbf{Q}^*_{co} & \mathbf{Q}^*_{cc} & \mathbf{M}^*_{co} & \mathbf{M}^*_{cc}\end{array}\right)\begin{pmatrix}\mathbf{d}_o\\ \mathbf{d}_c\\ \mathbf{d}^*_o\\ \mathbf{d}^*_c\end{pmatrix} = \mathbf{0}, \tag{8.6.13}$$

in which $\mathbf{d}_o$ and $\mathbf{d}_c$ contain the orbital and coefficient rotation parameters and $\mathbf{a}_o$, $\mathbf{a}_c$ the corresponding gradient components $(\nabla H)_\mu$, $(\nabla H)_\alpha$; the elements of the $\mathbf{M}$ and $\mathbf{Q}$ matrices are the commutator matrix elements as defined in (8.6.5) and (8.6.12).

The dimensions of equation systems such as (8.6.13) (and even of the reduced form corresponding to (2.4.21)) are naturally very large even for a typical calculation on a small molecule. Other difficulties arise from possible redundancies among the variables and from near-singularities of the Hessian matrix, instabilities, etc.; techniques for dealing with such problems have been extensively discussed (see e.g. Dalgaard *et al.*, 1978; Jørgensen and Simons, 1981; Olsen *et al.*, 1983). When its implementation is computationally feasible the one-step procedure may have advantages over its predecessors; but the question remains open (Jørgen *et al.*, 1987).

In view of the large dimensions of (8.6.13), even when (with all quantities *real*) they have been halved as in (2.4.21), it is worth noting that a more economical method of optimizing the CI coefficients is possible. Thus, with M functions in the CI expansion, the rotation operator (8.6.9) contains $\frac{1}{2}M(M-1)$ distinct parameters; but in fact there are only $M-1$ independent CI coefficients C_κ.

To optimize the parameters C_κ directly, it is convenient to use the general procedure introduced in Section 2.4, taking d_κ as a variation in C_κ (rather than a rotation parameter). Thus derivatives of Ψ with respect to d_κ are

$$\Psi^\kappa = \frac{\partial \Psi}{\partial C_\kappa} = \Phi_\kappa, \tag{8.6.14}$$

while derivatives with respect to orbital rotation parameters d_μ ($\mu = rs$) and $d_{\bar{\mu}}$ are, noting (8.6.3),

$$\Psi^\mu = \frac{\partial \Psi}{\partial d_\mu} = \mathsf{E}_\mu \Psi, \qquad \Psi^{\bar{\mu}} = \frac{\partial \Psi}{\partial d_{\bar{\mu}}} = -\mathsf{E}_\mu^\dagger \Psi \tag{8.6.15}$$

etc. On using such derivatives in the defining equations (2.4.10) and (2.4.14), and taking account of the constraint $d_{\bar{\mu}} = -d_\mu^*$, it is not difficult to write the energy once again in the standard form (2.4.16) and to derive equations formally the same as (8.6.13). Instead of (8.6.11), however, the first derivatives for CI coefficient variation (contained in $\mathbf{a}_c$) are

$$(\nabla H)_\kappa = \langle \Phi_\kappa | \mathsf{H} | \Psi \rangle - EC_\kappa = a_\kappa^c \tag{8.6.16}$$

—a "residual" from the κ-row of the (unsatisfied) secular equations—while the additional second derivatives (contained in $\mathbf{M}_{oc}$, $\mathbf{M}_{cc}$, and $\mathbf{Q}_{oc}$, $\mathbf{Q}_{cc}$) are found to be respectively

$$\left.\begin{aligned} (\nabla H \nabla)_{\mu\kappa} &= \langle \Psi | [\mathsf{E}_\mu^\dagger, \mathsf{H}] | \Phi_\kappa \rangle - C_\kappa^* a_\mu^o, \\ (\nabla H \nabla)_{\lambda\kappa} &= \langle \Phi_\lambda | \mathsf{H} | \Phi_\kappa \rangle - a_\lambda^c C_\kappa^* - a_\kappa^{c*} C_\lambda, \end{aligned}\right\} \tag{8.6.17}$$

and

$$\left.\begin{aligned}(\nabla\nabla H)_{\mu\kappa} &= \langle \Phi_\kappa | [\mathsf{E}^\dagger_\mu, \mathsf{H}] | \Psi\rangle - C_\kappa a^{\circ}_{\mu},\\ (\nabla\nabla H)_{\lambda\kappa} &= 0.\end{aligned}\right\} \qquad (8.6.18)$$

These quantities are much simpler to calculate, and fewer in number, than those of the type (8.6.12). The terms that depend on the μ-parameters are calculated exactly as in (8.6.4), the only difference being that the ket (or the bra) contains Φ_κ instead of Ψ: when such matrix elements are reduced, in terms of transition density matrices, the density-matrix elements in (8.2.3) are simply replaced by *partial* sums such as

$$A^{(\kappa)}_{sr} = \sum_\lambda C^*_\lambda A^{\kappa\lambda}_{sr}. \qquad (8.6.19)$$

Thus (8.6.19) is the coefficient of $\langle r | \mathsf{h} | s\rangle$ in the matrix element $\langle \Psi | \mathsf{H} | \Phi_\kappa\rangle$.

It is possible, in principle, to generalize all of the preceding equations (without much change of formalism) in order to accommodate parameter variation in the basis functions themselves. Since, however, differentiating basis functions inevitably leads to orbitals with higher "quantum numbers", such a generalization would be equivalent to using an augmented AO basis (with corresponding "polarization functions"), and this would in turn involve a great increase in the number of basic 1- and 2-electron integrals to be computed. Other problems arise from the first-order violation of orthonormality constraints. Consequently, progress in this direction has been limited.

PROBLEMS 8

8.1 Show that the "all-pair-excitation" wavefunction (Problem 6.19) gives an energy expression of the form (8.2.2), first using a formulation in terms of spin-orbitals ($\psi_r = \phi_r\alpha$, $\psi_r = \phi_r\beta$) and then integrating over spin. Explicitly derive the generalized Fock operator in (8.2.6) and give a physical interpretation of the potential term.

8.2 Use the results of Problem 8.1 to set up a matrix form of the operator equations, by expanding the orbitals over a complete set of basis functions $\{\chi_\mu\}$ as in (8.2.11). Identify the elements of the matrix $\mathbf{Z}$. Is there are guarantee that on truncating the basis, and solving the *finite* matrix equations, the results will give a variational minimum of the total energy E?

8.3 Obtain a finite-basis approximation to the energy expression (8.2.2), using (8.2.11), and show that it takes the form (8.2.15). Then make a variation of the matrix **T** to derive the stationary-value condition (8.2.12). [*Hint:* Don't forget that the variation of **T** is constrained. The constraints may be introduced exactly as in Section 6.2 by varying the columns ($\mathbf{c}_r$, with μth element $T_{\mu r}$) and using the analogues of (6.2.17). Introduce Lagrange multipliers and collect the stationary-value conditions in the form (8.2.12), which is the analogue of (6.2.19).]

8.4 What combination of Fock-space operators will produce, when acting on a closed-shell state vector $|\Phi_0\rangle$, a resultant vector equivalent to the wavefunction used in Problem 8.1? Use the second-quantization form of the Hamiltonian (p. 82) and the anticommutation properties of the creation and annihilation operators (p. 81) to give an alternative derivation of the energy expression found in Problem 8.1. [*Hint:* Use operators $\mathsf{a}_{r\alpha}$, $\mathsf{a}_{r\beta}$, etc. to "destroy" or "create" up-spin or down-spin electrons in orbital ϕ_r. It is convenient to use indices $i, j, \ldots$ for orbitals in Φ_0 and $m, n, \ldots$ for the virtual set used in the CI.]

8.5 Derive stationary-value conditions for the total energy obtained in Problem 8.4 from first principles, using the exponential operator variation that leads to (8.2.31). [*Hint:* Obtain the varied energy to first order and equate to zero the coefficients of Δ_{mi} (m virtual, i occupied).]

8.6 Show that the result obtained in Problem 8.5 also follows from the general expression (8.2.29) *et seq.* when the expectation values on the right-hand side of the equation are replaced by appropriate density-matrix elements (see Problem 6.18). [*Hint:* Make use of the results in (5.4.20).]

8.7 Use the energy expression (8.2.15) to obtain the variation δE resulting from $\mathbf{T} \to \mathbf{T} + \delta\mathbf{T}$. Demonstrate the symmetry property $\operatorname{tr}(\delta\mathbf{T}^\dagger\mathbf{Z}) = \operatorname{tr}(\mathbf{T}^\dagger\delta\mathbf{Z})$ and hence obtain the first-order variation $\delta_1 E$ given in (8.4.3). [*Hint:* Use the explicit form of the matrix **Z** given in (8.2.13).]

8.8 Consider the second-order terms in δE of Problem 8.7 and obtain the expression (8.4.4) for the second-order change $\delta_2 E$. [*Hint:* Carefully read the paragraph containing (8.4.2)–(8.4.4).]

8.9 Parametrize the variations obtained in Problems 8.7 and 8.8 by using (8.3.3) with the exponential form of **V** given in (8.2.24). Hence find the gradient components and the elements of the Hessian matrix in explicit form. [*Hint:* Use $d_\mu = \Delta_{rs}$ $(r > s)$ as the distinct parameters, noting that $\Delta_{sr} = -d_\mu^*$, and pick out appropriate coefficients in $\delta_1 E$ and $\delta_2 E$ already obtained.]

8.10 Reconsider the derivation of $\delta_2 E$ in Problem 8.8, noting that when all quantities are real only two types of 2-electron terms remain—both of which may be expressed in terms of the matrices **J** and **K** defined in (6.2.30). Hence obtain (8.4.11). [*Hint:* For real matrices the dagger is equivalent to transposition, and a quantity $T_{rs}T_{tu}$ may thus be regarded as the rt-element of the matrix $\mathbf{c}_s\mathbf{c}_u^\dagger$, where $\mathbf{c}_s$ and $\mathbf{c}_u$ are the s and u columns of **T**. Remember also that for real orbitals $\langle rs|g|tu\rangle = (rt, su)$ has an 8-fold symmetry, as follows from the definition.]

8.11 Fill in the steps leading to (8.4.15) and show that **B** assumes the form $(\mathbf{B}_1 \mid \mathbf{0})$, where the submatrices are rectangular. [*Hint:* Carefully note which indices run over all orbitals and which are restricted to occupied orbitals; then consider B_{mn}.]

8.12 Show how the method of Werner *et al.* (p. 273) may be applied in simple closed-shell SCF theory in order to obtain a rapidly convergent procedure. How would you implement the method of Section 8.3 (singular-value decomposition) in an actual calculation? [*Hint:* Start from an energy expression (e.g. (5.3.18)); identify the density-matrix elements (e.g. $P_{rs} = 2\delta_{rs}$, r, s occupied, $=0$, otherwise); substitute in (8.4.12) and (8.4.13), and find the various elements of **B**.]

8.13 Extend the analysis in Problem 8.12 to an open-shell system, starting from any one of the energy expressions (6.5.2), (6.6.1) or (6.6.7).

9 Perturbation Theory and Diagram Techniques

9.1 PERTURBATION METHODS

In Section 3.1 it was noted that, in principle, an exact many-electron wavefunction could be expanded in terms of Slater determinants. So far, however, we have made no systematic attempt to exploit the completeness property of such expansions in constructing wavefunctions of very high precision; instead, we have been concerned mainly with finding good first approximations by careful optimization of the orbitals employed, and with the use of symmetry, especially the spin symmetry, in reducing the length of such expansions by combining sets of determinants into appropriate configurational functions (CFs). Thus, in the MO approach, the first approximation was a single determinant; while in the VB approach the wavefunction was constructed from spin-coupled CFs belonging to one orbital configuration (or, with polar structures included, to a small number of configurations). Even in the MC SCF methods of Chapter 8, the number of CFs admitted is usually modest by present-day computational standards.

It is now time to consider complete set expansions (with a fixed choice of basis, and hence of expansion functions) as a method of obtaining virtually exact solutions of the Schrödinger equation—at least for few-electron systems in a non-relativistic approximation. Although such expansions must of course be truncated, it is nowadays not uncommon to admit hundreds of thousands of CFs. There are basically two methods of handling these long expansions: (i) to determine the coefficients of the CFs directly, along with the first few energy levels, by direct solution of the secular equations, using the most powerful techniques available for evaluating matrix elements and solving the matrix-eigenvalue problem; and (ii) to start from a leading term Φ_0, which would be an exact eigenfunction of a "model" Hamiltonian H_0, regarding $\mathsf{H}-\mathsf{H}_0$ as a perturbation under whose influence the remaining coefficients C_κ $(\kappa > 0)$ will assume non-zero values—which can be estimated by *perturbation theory,* without actually solving secular equations. In the present chapter we develop the perturbation approach in its basic form, Φ_0 being taken as a non-degenerate ground-state function of Hartree–Fock type. The problems of "direct CI" are considered in Chapter 10.

Since the difference between the model Hamiltonian H_0, corresponding to a Hartree–Fock approximation, and the "true" Hamiltonian as defined in (1.1.2) is very substantial, it is unlikely that low-order Rayleigh–Schrödinger theory (Section 2.5) will be adequate for present purposes. In fact serious problems are encountered, as will be seen, and to obtain sufficient accuracy it is necessary to go to very high order. The traditional methods of obtaining high-order contributions require cumbersome analysis: to develop the perturbation approach, and to obtain a general insight into the nature of the expansion, it is preferable to introduce the "diagrammatic" methods used in many-body physics.

9.2 THE HARTREE–FOCK REFERENCE FUNCTION

We develop the perturbation theory by separating the N-electron Hamiltonian (1.1.2) into two parts:

$$\mathsf{H} = \mathsf{H}_0 + \mathsf{H}', \tag{9.2.1}$$

where we assume that H_0 corresponds to an independent-particle model (IPM) of Hartree–Fock type:

$$\mathsf{H}_0 = \sum_i \mathsf{F}(i). \tag{9.2.2}$$

It is possible to obtain a virtually exact ground-state wavefunction for this model, using the methods of Chapter 6. The resultant function Φ_0 is taken as the leading term or "reference function" in the CI expansion

$$\Psi = \Phi_0 + \sum_{\kappa>0} C_\kappa \Phi_\kappa, \tag{9.2.3}$$

and the remaining terms Φ_κ are constructed by promoting electrons from the occupied spin-orbitals of Φ_0, one at a time, two at a time, and so on, into the virtual orbitals (which are also available, at least in large numbers, from solution of the HF equations). It is assumed that the ground state is non-degenerate so that there is a *unique* single determinant Φ_0. More generally, it is possible to admit a linear combination of determinants (e.g. an MC SCF approximation) as the reference function; but the model is then less easily defined and much greater technical difficulties arise. We shall therefore consider only the case of a single-determinant reference function, this being the prototype for further developments. The low-order terms in the expansion were first given by Møller and Plessett (1934): but to obtain terms of higher order we use the formalism of Section 3.6, first using this formalism to rederive the properties of the Hartree–Fock reference function.

The existence of a convenient reference state Φ_0 means that when using second-quantization methods we do not need to start always from the vacuum state $|\text{vac}\rangle$; we can instead use

$$|0\rangle = \prod_{i=1}^{N} \mathrm{a}_i^\dagger \, |\text{vac}\rangle \tag{9.2.4}$$

as the reference state and produce the "excited states" by operating on $|0\rangle$ with substitution operators such as $\mathrm{a}_m^\dagger \mathrm{a}_i$, which, in Schrödinger language, changes the spin-orbital ψ_i of Φ_0 into ψ_m. In dealing with matrix elements among such states it is then useful to introduce the concept of a *normal product*.

To this end we shall adopt systematically the usual conventions: labels i, j, k, l will refer to spin-orbitals occupied in Φ_0; labels m, n, p, q to those unoccupied in Φ_0 (i.e. virtual); and r, s, t, u to those of either type. Briefly, ψ_i and ψ_m are respectively "occupied" and "virtual" (or "excited") spin-orbitals. We note that $\mathrm{a}_m^\dagger$ and a_i are both, in a sense, creation operators relative to the reference state: $\mathrm{a}_m^\dagger$ creates a particle in the "excited" spin-orbital ψ_m, while a_i creates a "hole" in the originally occupied spin-orbital ψ_i. Similarly a_m and $\mathrm{a}_i^\dagger$ are both annihilation operators in the sense that a_m destroys a particle in the excited spin-orbital ψ_m while $\mathrm{a}_i^\dagger$ "destroys a hole" by refilling the spin-orbital ψ_i of Φ_0. The reference state represented by $|0\rangle$ is thus a new "vacuum state", which contains neither holes nor particles. This point of view may be emphasized by a change of notation that is sometimes used:

$$\mathrm{a}_i,\ \mathrm{a}_m^\dagger \rightarrow \mathrm{b}_i^\dagger,\ \mathrm{b}_m^\dagger \quad \text{(hole and particle creation)},$$

$$\mathrm{a}_i^\dagger,\ \mathrm{a}_m \rightarrow \mathrm{b}_i,\ \mathrm{b}_m \quad \text{(hole and particle annihilation)}.$$

These new operators describe the creation and annihilation of "quasiparticles" (i.e. holes below a certain energy threshold, particles above), and their anticommutation properties are easily seen to be the same as for the original operators:

$$\left.\begin{aligned} &\mathrm{b}_r\mathrm{b}_s + \mathrm{b}_s\mathrm{b}_r = \mathrm{b}_r^\dagger\mathrm{b}_s^\dagger + \mathrm{b}_s^\dagger\mathrm{b}_r^\dagger = 0, \\ &\mathrm{b}_r^\dagger\mathrm{b}_s + \mathrm{b}_s\mathrm{b}_r^\dagger = \delta_{rs}. \end{aligned}\right\} \tag{9.2.5}$$

In terms of the quasiparticle operators, the definition of a *normal product* is immediate.

> The *normal product* of a sequence of creation and annihilation operators is a rearranged sequence with all annihilation operators standing to the right of the creation operators, prefixed by the sign $(-1)^p$, where p is the parity of the permutation required: it is denoted by N[. . .].

Thus, for example,

$$\mathrm{N}[\mathsf{b}_1\mathsf{b}_2^\dagger\mathsf{b}_3\mathsf{b}_4^\dagger\mathsf{b}_5] = (-1)^3\mathsf{b}_2^\dagger\mathsf{b}_4^\dagger\mathsf{b}_1\mathsf{b}_3\mathsf{b}_5,$$

as follows by application of (9.2.5). Any further permutation of factors gives of course an additional factor $(-1)^p$. The key property of any normal product is that it operates on $|0\rangle$ to give zero—since $|0\rangle$ contains no holes or particles to be destroyed.

It may be shown that *any* product can be expressed in terms of *normal* products. For this purpose two more definitions are needed. First we introduce the term "contraction".

> The *contraction* of a pair of operators is the expectation value of their product in the vacuum state $|0\rangle$.

Thus the pair $\mathsf{b}_r\mathsf{b}_s$ has a contraction that we denote by

$$\overparen{\mathsf{b}_r\mathsf{b}_s} = \langle 0|\,\mathsf{b}_r\mathsf{b}_s\,|0\rangle.$$

It is easily verified that only one type of contraction is non-zero, namely

$$\overparen{\mathsf{b}_r\mathsf{b}_s^\dagger} = \langle 0|\,\mathsf{b}_r\mathsf{b}_s^\dagger\,|0\rangle = \delta_{rs}, \tag{9.2.6}$$

which gives unity when $\mathsf{b}_r^\dagger$ creates a quasiparticle that is immediately destroyed by b_r.

Secondly, we must define the contraction of a normal product.

> A *contracted normal product* is one in which one or more operator pairs is replaced by the scalar factor obtained by contraction and is withdrawn from the normal product. With each contraction is associated a factor $(-1)^\nu$, where ν is the number of interchanges needed to bring the paired operators adjacent.

Thus, for example,

$$\begin{aligned}\mathrm{N}[\mathsf{b}_r\mathsf{b}_s\mathsf{b}_t^\dagger\mathsf{b}_u^\dagger\mathsf{b}_v^\dagger] \ (\text{contracted pairs } \mathsf{b}_r\mathsf{b}_t^\dagger,\ \mathsf{b}_s\mathsf{b}_u^\dagger) &= -\overparen{\mathsf{b}_s\mathsf{b}_u^\dagger}\,\mathrm{N}[\overparen{\mathsf{b}_r\mathsf{b}_t^\dagger}\mathsf{b}_v^\dagger]\\ &= -\overparen{\mathsf{b}_s\mathsf{b}_u^\dagger}\,\overparen{\mathsf{b}_r\mathsf{b}_t^\dagger}\,\mathsf{b}_v^\dagger \quad = -\delta_{su}\delta_{rt}\mathsf{b}_v^\dagger.\end{aligned}$$

A product is said to be "completely contracted" if no operators are left unpaired.

We are now in a position to state the central result, which is known as Wick's theorem. For just two operators, which we denote momentarily by x_1 and x_2 so as not to prejudge their characters (creation or annihilation), the result is simply

$$\mathsf{x}_1\mathsf{x}_2 = \mathrm{N}[\mathsf{x}_1\mathsf{x}_2] + \overparen{\mathsf{x}_1\mathsf{x}_2}. \tag{9.2.7}$$

This is demonstrated by using the anticommutation rules on all types of pair. In words, any 2-factor product can be written as the *normal* product plus its contraction. Wick's theorem is the generalization of this result to products of n factors, and may be proved by induction, starting from the case $n = 2$. The theorem is as follows.

> Any product of creation and annihilation operators can be written as the sum of the normal product plus all possible singly, multiply and fully contracted normal products.

In symbols,

$$\begin{aligned}
\mathsf{x}_1\mathsf{x}_2 \ldots \mathsf{x}_n = {} & \mathrm{N}[\mathsf{x}_1\mathsf{x}_2 \ldots \mathsf{x}_n] \\
& + \sum \mathrm{N}[\mathsf{x}_1 \ldots \overparen{\mathsf{x}_i \ldots \mathsf{x}_j} \ldots \mathsf{x}_n] && \text{(all single contractions)} \\
& + \sum \mathrm{N}[\mathsf{x}_1 \ldots \mathsf{x}_i \ldots \mathsf{x}_k \ldots \mathsf{x}_j \ldots \mathsf{x}_l \ldots \mathsf{x}_n] && \text{(all double contractions)} \\
& + \text{etc.}
\end{aligned} \tag{9.2.8}$$

Since (9.2.8) applies irrespective of the nature of the xs, it is immaterial whether we use the original operators (a_r, $\mathsf{a}_r^\dagger$) or the quasiparticle operators (b_r, $\mathsf{b}_r^\dagger$). In applying the theorem, however, it may be found easier to write the product in terms of the b operators, so that the only non-zero contractions are those containing pairs of type (9.2.6). Thus, for example,

$$\mathsf{b}_r^\dagger\mathsf{b}_s\mathsf{b}_t\mathsf{b}_u^\dagger = \mathrm{N}[\mathsf{b}_r^\dagger\mathsf{b}_s\mathsf{b}_t\mathsf{b}_u^\dagger] + \mathrm{N}[\mathsf{b}_r^\dagger\overparen{\mathsf{b}_s\mathsf{b}_t\mathsf{b}_u^\dagger}] + \mathrm{N}[\mathsf{b}_r^\dagger\mathsf{b}_s\overparen{\mathsf{b}_t\mathsf{b}_u^\dagger}]$$

(other contractions, not of type (9.2.6), giving zero), and this further simplifies to

$$\mathsf{b}_r^\dagger\mathsf{b}_s\mathsf{b}_t\mathsf{b}_u^\dagger = \mathsf{b}_r^\dagger\mathsf{b}_u^\dagger\mathsf{b}_s\mathsf{b}_t - \delta_{su}\mathsf{b}_r^\dagger\mathsf{b}_t + \delta_{tu}\mathsf{b}_r^\dagger\mathsf{b}_s.$$

In terms of the a operators, a product such as $\mathsf{a}_r^\dagger\mathsf{a}_s\mathsf{a}_u\mathsf{a}_t^\dagger$ can only be expanded when we know the ranges ($i, j, \ldots$ or $m, n, \ldots$) to which the indices r, s, t, u belong; there might be, for example, contractions of type $a_i^\dagger a_j$ (with the dagger on the *left*) as well as those of type $a_m a_n^\dagger$, because a_j is a hole *creation* operator for the reference state $|0\rangle$.

Clearly Wick's theorem achieves in a very general way results that could be obtained step by step, using the anticommutation rules. The great importance of (9.2.8) lies in the fact that, since normal products operating on $|0\rangle$ always give zero, the expectation value of any product arises only from *fully contracted products*. For example, the product considered above has zero expectation value because it contains no non-zero fully contracted terms. On the other hand, $b_r b_s b_t^\dagger b_u^\dagger$ would admit two potentially non-zero contractions, namely $\overbrace{b_r \overbrace{b_s b_t^\dagger} b_u^\dagger}$ and $\overbrace{b_r b_s b_t^\dagger b_u^\dagger}$, and hence

$$\langle 0|\, b_r b_s b_t^\dagger b_u^\dagger \,|0\rangle = -\delta_{rt}\delta_{su} + \delta_{st}\delta_{ru},$$

the first term having the parity factor -1.

There is an important corollary to Wick's theorem (Čížek, 1966).

> A product of operators $N[x_1x_2\ldots]N[y_1y_2\ldots]$, consisting of two factors already in normal-product form, can be expanded like the single product $x_1x_2\ldots y_1y_2\ldots$, according to (9.2.8), except that contractions *within* each of the sets $x_1x_2\ldots$ and $y_1y_2\ldots$ are excluded.

In symbols (cf. (9.2.8)),

$$\begin{aligned} N[x_1x_2\ldots]N[y_1y_2\ldots] = {} & N[x_1x_2\ldots y_1y_2\ldots] \\ & + \sum N[\overbrace{x_1x_2\ldots y_1}y_2\ldots] \quad \text{(all single x-y contractions)} \\ & + \sum N[\overbrace{x_1x_2\ldots y_1y_2}\ldots] \quad \text{(all double x-y contractions)} \\ & + \text{etc.,} \end{aligned} \qquad (9.2.9)$$

an expansion that will be useful in later sections.

As an example of the use of normal products, we rederive some of the basic properties of the Hartree–Fock approximation. To do this, we first rewrite the Hamiltonian (3.6.9) in normal-product form, noting that (9.2.8) applies whatever the nature of the xs. Thus

$$a_r^\dagger a_s = N[a_r^\dagger a_s] + a_r^\dagger a_s,$$

$$\begin{aligned} a_r^\dagger a_s^\dagger a_u a_t = {} & N[a_r^\dagger a_s^\dagger a_u a_t] + a_s^\dagger a_u N[a_r^\dagger a_t] + a_r^\dagger a_t N[a_s^\dagger a_u] \\ & - a_r^\dagger a_u N[a_s^\dagger a_t] - a_s^\dagger a_t N[a_r^\dagger a_u] + a_r^\dagger a_t a_s^\dagger a_u - a_r^\dagger a_u a_s^\dagger a_t. \end{aligned}$$

On inserting these expressions in (3.6.9), and remembering that $a_r^\dagger a_s = 0$ unless r, s both belong to the occupied set $(i, j, \ldots)$, we obtain

$$\begin{aligned}
\mathsf{H} = &\sum_{r,s} \langle r|\,\mathsf{h}\,|s\rangle \mathrm{N}[a_r^\dagger a_s] + \sum_i \langle i|\,\mathsf{h}\,|i\rangle \\
&+ \tfrac{1}{2} \sum_{r,s,t,u} \langle rs|\,g\,|tu\rangle \mathrm{N}[a_r^\dagger a_s^\dagger a_u a_t] + \tfrac{1}{2} \sum_{r,t,i} \langle ri|\,g\,|ti\rangle \mathrm{N}[a_r^\dagger a_t] \\
&+ \tfrac{1}{2} \sum_{s,u,i} \langle is|\,g\,|iu\rangle \mathrm{N}[a_s^\dagger a_u] - \tfrac{1}{2} \sum_{s,t,i} \langle is|\,g\,|ti\rangle \mathrm{N}[a_s^\dagger a_t] \\
&- \tfrac{1}{2} \sum_{r,u,i} \langle ri|\,g\,|iu\rangle \mathrm{N}[a_r^\dagger a_u] \\
&+ \tfrac{1}{2} \sum_{i,j} \langle ij|\,g\,|ij\rangle - \tfrac{1}{2} \sum_{i,j} \langle ij|\,g\,|ji\rangle.
\end{aligned}$$

Since the dummy indices may be renamed, and $\langle rs|\,g\,|tu\rangle = \langle sr|\,g\,|ut\rangle$, this reduces to

$$\boxed{\mathsf{H} = E_{\mathrm{HF}} + \sum_{r,s} \langle r|\,\mathsf{F}\,|s\rangle \mathrm{N}[a_r^\dagger a_s] + \tfrac{1}{2} \sum_{r,s,t,u} \langle rs|\,g\,|tu\rangle \mathrm{N}[a_r^\dagger a_s^\dagger a_u a_t],} \tag{9.2.10}$$

where

$$E_{\mathrm{HF}} = \sum_i \langle i|\,\mathsf{h}\,|i\rangle + \tfrac{1}{2} \sum_{i,j} (\langle ij|\,g\,|ij\rangle - \langle ij|\,g\,|ji\rangle) \tag{9.2.11}$$

is the usual Hartree–Fock energy expression while

$$\langle r|\,\mathsf{F}\,|s\rangle = \langle r|\,\mathsf{h}\,|s\rangle + \sum_i (\langle ri|\,g\,|si\rangle - \langle ri|\,g\,|is\rangle) \tag{9.2.12}$$

is a matrix element of a Fock-like Hamiltonian. Since normal products have zero expectation values in the reference state $|0\rangle$, it follows from (9.2.10) that

$$E_{\mathrm{HF}} = \langle 0|\,\mathsf{H}\,|0\rangle. \tag{9.2.13}$$

To complete the discussion, we need to know under what conditions E_{HF} has a stationary value.

It is known already, from Section 8.2, that the most general spin-orbital variation consistent with orthonormality leads to the new reference state‡

$$|\Phi\rangle = e^{\mathsf{R}}\,|\Phi_0\rangle \quad (\mathsf{R}^\dagger = -\mathsf{R}), \tag{9.2.14}$$

‡ Sometimes it is convenient to write the state (Φ or Φ_0) explicitly within the ket and the bra.

where R is the infinitesimal rotation operator (8.2.20). But now it is sufficient to choose the special form

$$\mathsf{R} = \sum_{i,m} \Delta_{im} \mathsf{a}_m^\dagger \mathsf{a}_i, \tag{9.2.15}$$

since terms with indices in the reverse order would clearly destroy the Hartree–Fock reference ket $|0\rangle$. The varied energy is then, by (8.2.19),

$$E_{\mathrm{HF}} + \delta E_{\mathrm{HF}} = \langle \Phi | \mathsf{H} | \Phi \rangle$$
$$= E_{\mathrm{HF}} + \sum_{i,m} \Delta_{im} [\langle \Phi_0 | \mathsf{H} \mathsf{a}_m^\dagger \mathsf{a}_i | \Phi_0 \rangle - \langle \Phi_0 | \mathsf{a}_m^\dagger \mathsf{a}_i \mathsf{H} | \Phi_0 \rangle],$$

and, on noting that $\langle \Phi_0 | \mathsf{a}_m^\dagger \mathsf{a}_i \mathsf{H} | \Phi_0 \rangle = \langle \mathsf{a}_i^\dagger \mathsf{a}_m \Phi_0 | \mathsf{H} | \Phi_0 \rangle = 0$ and writing $\mathsf{a}_m^\dagger \mathsf{a}_i | \Phi_0 \rangle = | \Phi_0 (i \to m) \rangle$, the Brillouin condition takes the form

$$\langle \Phi_0 | \mathsf{H} | \Phi_0 (i \to m) \rangle = 0, \tag{9.2.16}$$

The more usual form, in terms of the one-electron operator F, follows from (9.2.9) and (9.2.10):

$$\langle \Phi_0 | \mathsf{H} \mathsf{a}_m^\dagger \mathsf{a}_i | \Phi_0 \rangle = \sum_{r,s} \langle r | \mathsf{F} | s \rangle \langle \Phi_0 | \mathrm{N}[\mathsf{a}_r^\dagger \mathsf{a}_s] \mathsf{a}_m^\dagger \mathsf{a}_i | \Phi_0 \rangle$$
$$+ \sum_{r,s,t,u} \langle rs | g | tu \rangle \langle \Phi_0 | \mathrm{N}[\mathsf{a}_r^\dagger \mathsf{a}_s^\dagger \mathsf{a}_u \mathsf{a}_t] \mathsf{a}_m^\dagger \mathsf{a}_i | \Phi_0 \rangle.$$

But since $\mathsf{a}_m^\dagger \mathsf{a}_i = \mathrm{N}[\mathsf{a}_m^\dagger \mathsf{a}_i]$, there is only one fully contracted product, by the extended Wick theorem (9.2.9), that can give a non-zero result, namely that in the first term on the right:

$$\mathrm{N}[\mathsf{a}_s^\dagger \mathsf{a}_s] \mathrm{N}[\mathsf{a}_m^\dagger \mathsf{a}_i] = \delta_{ir} \delta_{ms}.$$

The Brillouin condition thus becomes, from (9.2.16),

$$\langle i | \mathsf{F} | m \rangle = 0 \tag{9.2.17}$$

—the Fock operator must have zero matrix elements between occupied and virtual spin-orbitals. Since unitary mixing of the occupied (or virtual) spin-orbitals among themselves makes no difference to $|\Phi_0\rangle$ and E_{HF}, it is possible to make the "canonical" choice, such that *all* off-diagonal elements vanish; and in this case, of course,

$$\mathsf{F} | r \rangle = \epsilon_r | r \rangle, \tag{9.2.18}$$

which is the usual Hartree–Fock eigenvalue equation.

We now define more precisely the "model" Hamiltonian to be used in the perturbation theory. In Schrödinger form this is simply $\sum_i \mathsf{F}(i)$ for a system of N independent electrons moving in the Hartree–Fock field, and its expectation value in the reference state is the sum of the orbital energies E_{orb} (not E_{HF}). Thus

$$E_{\text{orb}} = \langle \Phi_0 | \mathsf{H}_M | \Phi_0 \rangle = \sum_i \langle i | \mathsf{F} | i \rangle = \sum_i \epsilon_i. \tag{9.2.19}$$

In second-quantization form, the corresponding operator will be

$$\mathsf{H}_M = \sum_{r,s} \langle r | \mathsf{F} | s \rangle \mathsf{a}_r^\dagger \mathsf{a}_s = \sum_{r,s} \langle r | \mathsf{F} | s \rangle (\mathrm{N}[\mathsf{a}_r^\dagger \mathsf{a}_s] + \overbrace{\mathsf{a}_r^\dagger \mathsf{a}_s}),$$

or in other words, remembering (9.2.6),

$$\mathsf{H}_M = E_{\text{orb}} + \sum_{r,s} \langle r | \mathsf{F} | s \rangle \mathrm{N}[\mathsf{a}_r^\dagger \mathsf{a}_s], \tag{9.2.20}$$

and the full Hamiltonian (9.2.10) may therefore be written

$$\mathsf{H} = (E_{\text{HF}} - E_{\text{orb}}) + \mathsf{H}_{\text{M}} + \tfrac{1}{2} \sum_{r,s,t,u} \langle rs | g | tu \rangle \mathrm{N}[\mathsf{a}_r^\dagger \mathsf{a}_s^\dagger \mathsf{a}_u \mathsf{a}_t].$$

The natural partitioning of the Hamiltonian is now seen to be

$$\boxed{\mathsf{H} = \mathsf{H}_0 + \mathsf{H}',} \tag{9.2.21}$$

where

$$\boxed{\begin{aligned} \mathsf{H}_0 &= (E_{\text{HF}} - E_{\text{orb}}) + \mathsf{H}_M, \\ \mathsf{H}' &= \tfrac{1}{2} \sum_{r,s,t,u} \langle rs | g | tu \rangle \mathrm{N}[\mathsf{a}_r^\dagger \mathsf{a}_s^\dagger \mathsf{a}_u \mathsf{a}_t]. \end{aligned}} \tag{9.2.22}$$

The "unperturbed Hamiltonian" H_0 is thus, apart from the trivial energy-shift term, simply the model Hamiltonian H_{M} for any number of particles moving independently in the Hartree–Fock field. Since, by (9.2.20), the expectation value of H_{M} in the reference state is E_{orb}, the "unperturbed energy" is

$$E_0 = \langle \Phi_0 | \mathsf{H}_0 | \Phi_0 \rangle = E_{\text{HF}}, \tag{9.2.23}$$

and this coincides with the usual expression $\langle \Phi_0 | \mathsf{H} | \Phi_0 \rangle$ because the expectation value of H', which contains only a normal-product term, is identically zero.

Before considering the perturbation expansion, we list the following results, which are easily obtained from the definitions (9.2.20)–(9.2.22) and the properties already noted.

(i) The model energy eigenvalue is $E_{\rm orb}$, the sum of the orbital eigenvalues, but the Hartree–Fock energy $E_{\rm HF}$ is an eigenvalue of the unperturbed Hamiltonian defined in (9.2.22):

$$\mathsf{H}_{\rm M}\,|\Phi_0\rangle = E_{\rm orb}\,|\Phi_0\rangle, \qquad \mathsf{H}_0\,|\Phi_0\rangle = E_{\rm HF}\,|\Phi_0\rangle. \quad (9.2.24)$$

(ii) The expectation energy of the ion produced by operating on $|\Phi_0\rangle$ with a_i (i.e. destroying an electron in ψ_i) is obtained simply by removing its energy ϵ_i from $E_{\rm HF}$ for the neutral system:

$$E_i = \langle\Phi_0|\,\mathsf{a}_i^\dagger \mathsf{H} \mathsf{a}_i\,|\Phi_0\rangle = \langle\Phi_0|\,\mathsf{H}\,|\Phi_0\rangle - \epsilon_i = E_{\rm HF} - \epsilon_i. \quad (9.2.25)$$

Since Φ_0 is chosen to make both $E_{\rm HF}$ and ϵ_i stationary, E_i is an acceptable variational energy for the ion in the usual sense (Koopmans' theorem).

(iii) The energy of the state obtained by making the spin-orbital substitution $\psi_i \to \psi_m$ is, with the definition (6.1.4),

$$E(i\to m) = \langle\Phi_0|\,\mathsf{a}_i^\dagger \mathsf{a}_m \mathsf{H} \mathsf{a}_m^\dagger \mathsf{a}_i\,|\Phi_0\rangle = E_{\rm HF} + (\epsilon_m - \epsilon_i) + \langle mi\,\|\,mi\rangle. \quad (9.2.26)$$

The "excitation energy" is thus a difference of orbital energies plus a coulomb-exchange correction.

All the results established in Chapter 6 can thus be derived quite simply by using Fock-space methods.

9.3 THE CLUSTER DEVELOPMENT

First we consider some general properties of the CI expansion (9.1.3), in which Φ_0 is a single-determinant root function and the other functions Φ_κ are taken to be single determinants with single, double and multiple "excitations" from the occupied orbitals of Φ_0 (namely $i, j, \ldots$) to a complementary set of virtual orbitals ($m, n, \ldots$). This expansion is in principle exact (Section 3.1), and may evidently be written

$$\Psi = \Phi_0 + \sum_{i,m} C_i^m \Phi_i^m + \sum_{\substack{i<j\\ m<n}} C_{ij}^{mn} \Phi_{ij}^{mn} + \ldots, \quad (9.3.1)$$

where, for example, Φ_{ij}^{mn} is a normalized Slater determinant in which the pair ψ_i, ψ_j has been replaced by ψ_m, ψ_n. The mixing coefficients will in this chapter be determined by perturbation methods, but similar considerations will apply irrespective of the techniques of calculation.

The second-quantization equivalent of the expansion (9.3.1) may clearly be put in the form

$$|\Psi\rangle = \left(1 + \sum_{i_1} \mathsf{u}_{i_1} + \sum_{i_1<i_2} \mathsf{u}_{i_1 i_2} + \ldots\right)|0\rangle \tag{9.3.2}$$

in which, for example, u_{i_1} and $\mathsf{u}_{i_1 i_2}$ are operators that create singly and doubly excited functions, and in general

$$\mathsf{u}_{i_1 i_2 \ldots i_p} = \sum_{m_1<m_2\ldots<m_p} C_{i_1 i_2 \ldots i_p}^{m_1 m_2 \ldots m_p} \mathsf{a}_{m_1}^{\dagger} \mathsf{a}_{m_2}^{\dagger} \ldots \mathsf{a}_{i_2} \mathsf{a}_{i_1}. \tag{9.3.3}$$

The operator in parentheses in (9.3.2) thus creates an *exact* wavefunction out of the Hartree–Fock approximation (Φ_0) represented by $|0\rangle$: the individual terms u_{i_1}, $\mathsf{u}_{i_1 i_2}$, ..., $\mathsf{u}_{i_1 i_2 \ldots i_p}$ are said to generate 1-cluster, 2-cluster, ..., p-cluster corrections to Φ_0, describing the excitation of "clusters" of electrons from the Hartree–Fock "sea". The generality of the cluster functions (in Schrödinger language) allows them to recognize the *correlation* of electronic motions (Section 5.8), which is absent in an independent-particle model.

It is clear, however, that by simply *varying the orbitals* in Φ_0 we should also obtain a function Φ that could be expanded (in terms of the original orbitals) in the form (9.3.1); in such a case the multiple-cluster functions would be purely formal in the sense that they could be eliminated by redefining the orbitals, i.e. they would be *reducible* and would not describe any physically significant effects other than those already recognized in a suitably chosen Φ_0. The cluster functions for which we are looking must be *irreducible*, and the way to introduce them is suggested by the analysis of Section 9.2, in which the "1-cluster operator" ($\sum_{i_1,m_1} C_{i_1}^{m_1} \mathsf{a}_{m_1}^{\dagger} \mathsf{a}_{i_1}$) (which is evidently equivalent to (9.2.15)) is put into an *exponent*. By using the operator $\exp(\sum_{i_1} \mathsf{u}_{i_1})$ instead of $1 + \sum_{i_1} \mathsf{u}_{i_1}$, we automatically introduce all the multiple-cluster functions that appear as (antisymmetrized) products of 1-cluster functions (i.e. *orbital* corrections); by putting the *second* cluster operator in (9.3.2) into the exponent, we obtain all the multiple-cluster functions that appear simply as (antisymmetrized) products of 2-cluster functions (i.e. pair-function corrections); and so on. Clearly, to introduce real correlation effects, we must include at least the 2-cluster operators in (9.3.2), thus obtaining a development that cannot be reduced to 1-determinant form. Generally, recognizing that the numerical coefficients in the cluster operators will differ from those in (9.3.3) when we pass to the exponential representation, we shall write the alternative expansion

$$|\Psi\rangle = e^{\mathsf{T}}\,|0\rangle, \tag{9.3.4}$$

where

$$\mathsf{T} = \mathsf{T}_1 + \mathsf{T}_2 + \dots = \sum_{i_1} \mathsf{v}_{i_1} + \sum_{i_1<i_2} \mathsf{v}_{i_1 i_2} + \dots, \tag{9.3.5}$$

and the operators that generate the *irreducible* clusters are defined by (cf. (9.3.3))

$$\mathsf{v}_{i_1 i_2 \dots i_p} = \sum_{m_1<m_2\dots<m_p} D^{m_1 m_2 \dots m_p}_{i_1 i_2 \dots i_p} \mathsf{a}^{\dagger}_{m_1} \mathsf{a}^{\dagger}_{m_2} \dots \mathsf{a}_{i_2} \mathsf{a}_{i_1}. \tag{9.3.6}$$

By developing the exponential in (9.3.4) and analysing the results, we shall be able to identify the reducible and irreducible terms in the conventional CI expansion (9.3.1).

First we note that e^{T} may be written as $e^{\mathsf{T}_1} \times e^{\mathsf{T}_2} \times e^{\mathsf{T}_3} \dots$, since, as is easily verified, different Ts commute. We then also have

$$e^{\mathsf{T}_1} = \exp\left(\sum_{i_1} \mathsf{v}_{i_1}\right) = \prod_{i_1=1}^{N} (1 + \mathsf{v}_{i_1}), \tag{9.3.7}$$

since powers of any v_{i_1} (above the first) contain repeated operators and therefore destroy any operand. In a similar way,

$$e^{\mathsf{T}_2} = \exp\left(\sum_{i_1<i_2} \mathsf{v}_{i_1 i_2}\right) = \prod_{i_1<i_2=1}^{N} (1 + \mathsf{v}_{i_1 i_2}), \tag{9.3.8}$$

and so on. When such factors are multiplied out, and terms are collected for clusters of each order, we obtain

$$\begin{aligned} e^{\mathsf{T}} = 1 &+ \sum_{i_r} \mathsf{v}_{i_1} + \sum_{i_1>i_2} (\mathsf{v}_{i_1 i_2} + \mathsf{v}_{i_1}\mathsf{v}_{i_2}) \\ &+ \sum_{i_1>i_2>i_3} (\mathsf{v}_{i_1 i_2 i_3} + \mathsf{v}_{i_1}\mathsf{v}_{i_2 i_3} + \mathsf{v}_{i_2}\mathsf{v}_{i_1 i_3} + \mathsf{v}_{i_3}\mathsf{v}_{i_1 i_2} + \mathsf{v}_{i_1}\mathsf{v}_{i_2}\mathsf{v}_{i_3}) \\ &+ \dots \,. \end{aligned} \tag{9.3.9}$$

The explicit form of the cluster expansion follows on using (9.3.9) in (9.3.4).

It is instructive to look at the individual terms using Schrödinger language. To this end, we write

$$\Phi_0(\boldsymbol{x}_1, \boldsymbol{x}_2, \ldots, \boldsymbol{x}_N) = (N!)^{1/2}\mathsf{A}^{(N)}\Omega_0(\boldsymbol{x}_1, \boldsymbol{x}_2, \ldots, \boldsymbol{x}_N), \quad (9.3.10)$$

where Ω_0 is the basic spin-orbital product

$$\Omega_0(\boldsymbol{x}_1, \boldsymbol{x}_2, \ldots, \boldsymbol{x}_N) = \psi_1(\boldsymbol{x}_1)\psi_2(\boldsymbol{x}_2) \ldots \psi_N(\boldsymbol{x}_N), \quad (9.3.11)$$

and the antisymmetrizer (which is idempotent) is

$$\mathsf{A}^{(N)} = (N!)^{-1} \sum_{\mathsf{P}} \epsilon_{\mathsf{P}}\mathsf{P}. \quad (9.3.12)$$

When, as we assume, the spin-orbitals are orthonormal, the function (9.3.10) is a normalized single determinant.

The function corresponding to a single term $\mathsf{v}_{i_1 i_2 \ldots i_p}$ is then easily obtained from (9.3.6): it is a sum (with given coefficients) of normalized single determinants in which ψ_{i_1} has been replaced by ψ_{m_i}, ψ_{i_2} by ψ_{m_2}, etc. This sum may be written as

$$(N!)^{1/2}\mathsf{A}^{(N)} \sum_{m_1 > m_2 \ldots > m_p} D_{i_1 i_2 \ldots i_p}^{m_1 m_2 \ldots m_p}$$
$$\times [\psi_{m_1}(\boldsymbol{x}_{i_1}) \ldots \psi_{m_p}(\boldsymbol{x}_{i_p})]\Omega_{0(i_1 \ldots i_p)}(\boldsymbol{x}_1, \ldots, \boldsymbol{x}_N),$$

where $\Omega_{0(i_1 \ldots i_p)}$ denotes the product that remains when the factors $\psi_{i_1}, \ldots, \psi_{i_p}$ are struck out of Ω_0: it may be cast in more familiar form by introducing a p-electron antisymmetrizer $\mathsf{A}^{(p)}$ (noting that $\mathsf{A}^{(p)}\mathsf{A}^{(N)} = \mathsf{A}^{(N)}\mathsf{A}^{(p)} = \mathsf{A}^{(N)}$) and defining the antisymmetric p-cluster function

$$\phi_{i_1 i_2 \ldots i_p}(\boldsymbol{x}_{i_1}, \boldsymbol{x}_{i_2}, \ldots, \boldsymbol{x}_{i_p})$$
$$= \sum_{m_1 > m_2 \ldots > m_p} D_{i_1 i_2 \ldots i_p}^{m_1 m_2 \ldots m_p}(p!)^{1/2}\mathsf{A}^{(p)}[\psi_{m_1}(\boldsymbol{x}_{i_1}) \ldots \psi_{m_p}(\boldsymbol{x}_{i_p})]. \quad (9.3.13)$$

In terms of these antisymmetric irreducible cluster functions, the required correspondence is

$$\mathsf{v}_{i_1 i_2 \ldots i_p} |0\rangle \leftrightarrow (N!)^{1/2}\mathsf{A}^{(N)}\left\{\left(\frac{1}{p!}\right)^{1/2} \phi_{i_1 i_2 \ldots i_p}(\boldsymbol{x}_{i_1}, \boldsymbol{x}_{i_2}, \ldots, \boldsymbol{x}_{i_p})\right.$$
$$\left.\times\, \Omega_{0(i_1 \ldots i_p)}(\boldsymbol{x}_1, \ldots, \boldsymbol{x}_N)\right\}. \quad (9.3.14)$$

The *reducible* clusters correspond to the product terms in (9.3.9). For example, a 4-cluster function may arise as the product of two irreducible 2-cluster functions, the explicit correspondence being

$$\mathsf{v}_{i_1 i_2}\mathsf{v}_{i_3 i_4} |0\rangle \leftrightarrow (N!)^{1/2}\mathsf{A}^{(N)}\left\{\frac{1}{\sqrt{2}}\, \phi_{i_1 i_2}(\boldsymbol{x}_{i_1}, \boldsymbol{x}_{i_2}) \frac{1}{\sqrt{2}}\, \phi_{i_3 i_4}(\boldsymbol{x}_{i_3}, \boldsymbol{x}_{i_4})\right.$$
$$\left.\times\, \Omega_{0(i_1 \ldots i_4)}(\boldsymbol{x}_1, \ldots, \boldsymbol{x}_N)\right\}.$$

It is thus a simple matter to transcribe (9.3.4), using these results together with (9.3.9), into Schrödinger language. The resultant expansion was first used in molecular theory by Sinanoglou (1964).

It is worth recalling at this point (cf. p. 274; see also Primas, 1965) that, by making a particular choice of spin-orbitals in Φ_0, the 1-cluster corrections may be eliminated; the 3-cluster function then contains a single irreducible term, arising from $\mathsf{v}_{i_1i_2i_3}$ in (9.3.9), while the 4-cluster function reduces to

$$(N!)^{1/2}\mathsf{A}^{(N)}\Bigg\{\sum_{i_1>i_2>i_3>i_4}\Bigg[\phi_{i_1i_2i_3i_4}+\frac{1}{\sqrt{2}}\phi_{i_1i_2}\frac{1}{\sqrt{2}}\phi_{i_3i_4}+\frac{1}{\sqrt{2}}\phi_{i_2i_3}\frac{1}{\sqrt{2}}\phi_{i_3i_4}+\frac{1}{\sqrt{2}}\phi_{i_1i_4}\frac{1}{\sqrt{2}}\phi_{i_2i_3}\Bigg]\Omega_{0(i_1i_2i_3i_4)}\Bigg\}. \quad (9.3.16)$$

For electrons (unlike nucleons) the 3- and higher-body *irreducible* clusters are much less important than the 2-body. If the irreducible 3- and 4-cluster terms are neglected, the cluster expansion takes the rather simple form‡, using Schrödinger language,

$$\boxed{\Psi=(N!)^{1/2}A^{(N)}\Bigg\{\Omega_0+\sum_{(ij)}\frac{\phi_{ij}}{\sqrt{2}}\Omega_{ij}+\sum_{(ij)(kl)}\frac{\phi_{ij}}{\sqrt{2}}\frac{\phi_{kl}}{\sqrt{2}}\Omega_{ijkl}+\ldots\Bigg\},} \quad (9.3.17)$$

where the high-order cluster functions reduce simply to antisymmetrized products of the 2-cluster functions. If these "correlated pair" functions can be obtained accurately (for example by variational methods) then the higher-order corrections, and their contributions to energy can in principle be determined with great precision by use of (9.3.17).

We shall not consider in detail the many methods that have been devised for the actual computation of correlated pair functions: there has been a great proliferation of techniques and approximations (see e.g. Hurley (1976) for a full survey of the earlier work, and Meyer (1977), Kutzelnigg (1977) and Ahlrichs (1979, 1983) for reviews of further developments), but the underlying principles are quite simple. The most convenient starting point is (9.3.4), which, being formally exact, becomes

$$(\mathsf{H}-E)e^{\mathsf{T}}\,|\Phi_0\rangle=0. \quad (9.3.18)$$

The first two terms in the exponent, obtained from (9.3.5) and (9.3.6), may be written as

$$\mathsf{T}_1=\sum_{i,m}D_i^m\mathsf{a}_m^\dagger\mathsf{a}_i,\qquad \mathsf{T}_2=\sum_{\substack{i>j\\m>n}}D_{ij}^{mn}\mathsf{a}_m^\dagger\mathsf{a}_n^\dagger\mathsf{a}_j\mathsf{a}_i, \quad (9.3.19)$$

‡ The summation is over distinct combinations of pairs $i_1>i_2$, $i_3>i_4$, and i_1i_2 coming before i_3i_4 in "dictionary" order.

and the problem is to determine the expansion coefficients D_i^m, D_{ij}^{mn} (or only the latter if the choice of orbitals permits the neglect of the 1-cluster terms) and hence the correlated pair functions and the energy. To do this, we may take scalar products of (9.3.18) with functions that may in principle be arbitrary (this being essentially the "method of moments", p. 247) but in practice are chosen to provide the most tractable set of equations. If we put $\mathsf{T} = \mathsf{T}_1 + \mathsf{T}_2$ in the exponential and expand then the scalar products with zero-, single- and double-excitation functions give

$$\langle \Phi_0 | (\mathsf{H} - E) | (1 + \mathsf{T}_1 + \mathsf{T}_2 + \tfrac{1}{2}\mathsf{T}_1^2) | \Phi_0 \rangle = 0, \tag{9.3.20}$$

$$\langle \Phi_i^m | (\mathsf{H} - E) | (1 + \mathsf{T}_1 + \mathsf{T}_2 + \tfrac{1}{2}\mathsf{T}_1^2 + \mathsf{T}_1\mathsf{T}_2 + (3!)^{-1}\mathsf{T}_1^3) | \Phi_0 \rangle = 0, \tag{9.3.21}$$

$$\langle \Phi_{ij}^{mn} | (\mathsf{H} - E) | (1 + \mathsf{T}_1 + \ldots + (3!)^{-1}\mathsf{T}_1^3 + \tfrac{1}{2}\mathsf{T}_2^2 + \tfrac{1}{2}\mathsf{T}_1^2\mathsf{T}_2 + (4!)^{-1}\mathsf{T}_1^4) | \Phi_0 \rangle = 0 \tag{9.3.22}$$

—each equation adding terms to those in the previous equation. The last two equations (non-linear in the D-coefficients) serve to determine T_1 and T_2, and substitution in the first then gives the energy. The simplicity of the equations is deceptive, however; even with neglect of T_1 and use of a single-determinant (HF) Φ_0, the explicit form of (9.3.22) is

$$\begin{aligned}
(\epsilon_m + \epsilon_n - \epsilon_i - \epsilon_j) D_{ij}^{mn} = \langle mn \| ij \rangle &- \sum_{p>q} \langle mn \| pq \rangle D_{ij}^{pq} - \sum_{k>l} \langle kl \| ij \rangle D_{kl}^{mn} \\
&+ \sum_{k,p} (\langle kn \| jp \rangle D_{ik}^{mp} - \langle km \| jp \rangle D_{ik}^{np} \\
&- \langle kn \| ip \rangle D_{jk}^{mp} + \langle km \| ip \rangle D_{jk}^{np}) \\
&+ \sum_{\substack{k>l \\ p>q}} \langle kl \| pq \rangle [D_{ij}^{pq} D_{kl}^{mn} - 2(D_{ij}^{mp} D_{kl}^{nq} + D_{ij}^{nq} D_{kl}^{mp}) \\
&- 2(D_{ik}^{mn} D_{jl}^{pq} + D_{ik}^{pq} D_{jl}^{mn}) + 4(D_{ik}^{mp} D_{jl}^{nq} + D_{ik}^{nq} D_{jl}^{mp})],
\end{aligned} \tag{9.3.23}$$

in which the computational difficulties of solution are apparent.

Equations equivalent to (9.3.23) were first formulated explicitly by Čížek (1966), using diagram techniques to identify the many terms. These coupled-pair equations provide the basis for a whole hierarchy of approximations (for example an *independent*-pair approximation, in which only linear terms are kept and coupling between different pairs is thus neglected), some of which had been formulated and used earlier (see Sinanoglu, 1964; Nesbet, 1965a).

In another family of methods the antisymmetrized products in (9.3.17) are regarded as defining correlated pairs in the presence of a "Hartree–Fock sea" consisting of the remaining electrons, and the individual pair functions are optimized variationally (see e.g. McWeeny and Steiner, 1965). Coupling between the pairs is then taken into account to some extent by an iterative method that introduces "self-consistency" (Meyer, 1976; Dykstra, 1977a,b). It is noteworthy that greatly improved results are obtained on relaxing orthogonality requirements between different pair functions.

In all approaches to the calculation of correlated pair functions the computational machinery required is heavy, and applications in the forseeable future are likely to be limited to few-electron systems. A recent review has been given by Ahlrichs and Scharf (1987).

9.4 DIAGRAMMATIC PERTURBATION THEORY

The numerical coefficients in the cluster development (9.3.1) of the exact wavefunction may be determined in principle by perturbation theory, Φ_0 being an exact groundstate eigenfunction of the model Hamiltonian H_M. For this purpose, we write the full Hamiltonian

$$\mathsf{H} = \mathsf{H}_0 + \lambda \mathsf{H}', \tag{9.4.1}$$

where, as usual, we attach a "perturbation parameter" λ to H' ($\lambda \to 1$) to separate the orders, and, as in (9.2.22),

$$\mathsf{H}_0 = (E_0 - E_\mathrm{orb}) + \mathsf{H}_\mathrm{M}, \tag{9.4.2a}$$

$$\mathsf{H}' = \tfrac{1}{2} \sum_{r,s,t,u} \langle rs|\, g\, |tu\rangle \mathrm{N}[\mathsf{a}_r^\dagger \mathsf{a}_s^\dagger \mathsf{a}_u \mathsf{a}_t]. \tag{9.4.2b}$$

Thus H_0 describes the model (independent particles in the Hartree–Fock field) while H' introduces all correlation effects through the perturbation expansion. High-order terms in the perturbation series are nowadays most commonly represented by means of "graphs" or "diagrams". In this section and the next we derive the complete expansion (Goldstone, 1957) from the usual time-independent Rayleigh–Schrödinger perturbation theory.‡

It is convenient to employ the "canonical" orbitals, which satisfy (9.2.18). In this case

$$\mathsf{H}_\mathrm{M} = \sum_r \epsilon_r \mathsf{a}_r^\dagger \mathsf{a}_r, \tag{9.4.3}$$

and every ket produced from $|0\rangle$ by creation of holes and particles is an

‡ The derivation is essentially that of Moccia (1973a); a somewhat similar treatment has been given by Manne (1977).

eigenket of H_0 with an eigenvalue that is simply E_0 plus a sum of particle energies minus a sum of hole energies. Thus, for example,

$$\langle 0|\, \mathsf{a}_i^\dagger \mathsf{a}_m \mathsf{H}_0 \mathsf{a}_m^\dagger \mathsf{a}_i\, |0\rangle = E_0 + (\epsilon_m - \epsilon_i), \tag{9.4.4}$$

and similarly for multiple excitations.

We now apply normal Rayleigh–Schrödinger perturbation theory, using the partitioning of (9.4.2); the derivation is well known and will only be sketched. The exact ket $|\Psi\rangle$ should satisfy

$$(\mathsf{H}_0 + \lambda \mathsf{H}')\,|\Psi\rangle = E\,|\Psi\rangle, \tag{9.4.5}$$

and is derived from $|0\rangle$ by application of the so-called wave operator $\mathsf{W}(\lambda)$:

$$|\Psi\rangle = \mathsf{W}(\lambda)\,|0\rangle. \tag{9.4.6}$$

It is assumed that W and E are expressible as power series in λ:

$$\mathsf{W}(\lambda) = 1 + \lambda \mathsf{W}_1 + \lambda^2 \mathsf{W}_2 + \dots, \tag{9.4.7}$$

$$E(\lambda) = E_0 + \lambda E_1 + \lambda^2 E_2 + \dots, \tag{9.4.8}$$

where the choice $\mathsf{W}_0 = 1$ ensures that $|\Psi\rangle \to |0\rangle$ for $\lambda \to 0$; and, since the required eigenfunction of (9.4.1) can always be written $\Psi = \Phi_0 + \Phi_\perp$, where $\Phi_\perp$ is orthogonal to Φ_0, this choice is consistent with the normalization

$$\langle \Psi \,|\, 0\rangle = 1. \tag{9.4.9}$$

Substitution of (9.4.7) and (9.4.8) in (9.4.5), and separation of the orders, then leads easily to the general condition

$$(\mathsf{H}_0 - E_0)\mathsf{W}_n\,|0\rangle = -\mathsf{H}'\mathsf{W}_{n-1}\,|0\rangle + \sum_{j=1}^{n} E_j \mathsf{W}_{n-j}\,|0\rangle. \tag{9.4.10}$$

Since, with no loss of generality, the correction terms $\mathsf{W}_n\,|0\rangle$ (for $n > 0$) are taken orthogonal to $|0\rangle$, they are all destroyed by the operator $|0\rangle\langle 0|$ but unaffected by the complementary operator

$$\mathsf{P} = 1 - |0\rangle\langle 0|, \tag{9.4.11}$$

which has the usual projection-operator properties, $\mathsf{P}^2 = \mathsf{P}$ and $\mathsf{P}^\dagger = \mathsf{P}$. It is thus possible to define a "pseudo-inverse" of the operator $E_0 - \mathsf{H}_0$, that appears on the left of (9.4.10), in the form‡

$$\mathsf{Q} = \frac{\mathsf{P}}{E_0 - \mathsf{H}_0}. \tag{9.4.12}$$

‡ There is no ambiguity in the notation since P commutes with H_0, and any function of H_0, and hence $(z - \mathsf{H}_0)^{-1} = \mathsf{P}(z - \mathsf{H}_0)^{-1} = \mathsf{P}(z - \mathsf{H}_0)^{-1}\mathsf{P}$; this operator gives Q for $z \to E_0$. The *resolvent* (9.4.12) is a "pseudo-inverse" of $E_0 - \mathsf{H}_0$ in the sense that its effect is defined even for $z = E_0$, where $(z - \mathsf{H}_0)\,|0\rangle = 0$ and the inverse operator is therefore singular.

Operating on (9.4.10) with Q then yields

$$\boxed{\mathsf{W}_n\,|0\rangle = \mathsf{QH'W}_{n-1}\,|0\rangle - \sum_{j=1}^{n} E_j\mathsf{QW}_{n-j}\,|0\rangle,} \tag{9.4.13}$$

from which each correction term $\mathsf{W}_n\,|0\rangle$ may be derived from those preceding it. The energy corrections are obtained by forming the scalar product of (9.4.10) with $\langle 0|$ and noting that $\langle 0|\,\mathsf{W}_n\,|0\rangle = 0$ for $n \neq 0$:

$$\boxed{E_n = \langle 0|\,\mathsf{H'W}_{n-1}\,|0\rangle.} \tag{9.4.14}$$

The first-order quantities are

$$E_1 = \langle 0|\,\mathsf{H'}\,|0\rangle, \qquad \mathsf{W}_1\,|0\rangle = \mathsf{QH'}\,|0\rangle, \tag{9.4.15}$$

and higher-order terms follow by iteration.

To implement the perturbation theory, we use the partitioning indicated in (9.4.2), with the zeroth-order reference state (Φ_0)

$$|0\rangle = (\mathsf{a}_N^\dagger \ldots \mathsf{a}_2^\dagger\mathsf{a}_1^\dagger)\,|\text{vac}\rangle \tag{9.4.16}$$

corresponding to the one-determinant wavefunction Φ_0 in which the N lowest-energy spin-orbitals are occupied. Since the normal products in (9.4.2) have vanishing expectation values in $|0\rangle$, the first-order energy correction is

$$E_1 = \langle 0|\,\mathsf{H'}\,|0\rangle = 0.$$

The second-order correction requires the calculation of

$$|\Phi_1\rangle = (\mathsf{QH'})\,|0\rangle = \tfrac{1}{2}\sum_{r,s,t,u} \mathsf{Q}\langle rs|\,g\,|tu\rangle \mathrm{N}[\mathsf{a}_r^\dagger\mathsf{a}_s^\dagger\mathsf{a}_u\mathsf{a}_t]\,|0\rangle,$$

and since $|0\rangle$ is destroyed unless a_t, a_u produce holes (i.e. $t, u \rightarrow i, j$) and $\mathsf{a}_r^\dagger$, $\mathsf{a}_s^\dagger$ produce particles ($r, s \rightarrow m, n$), this means that

$$(\mathsf{QH'})\,|0\rangle = \tfrac{1}{2}\sum_{i,j,m,n} \mathsf{Q}\langle mn|\,g\,|ij\rangle \mathsf{a}_m^\dagger\mathsf{a}_n^\dagger\mathsf{a}_j\mathsf{a}_i\,|0\rangle. \tag{9.4.17}$$

Reading from right to left, the first four operators produce a "doubly-excited" ket $|\Phi_{ij}^{mn}\rangle$; the matrix element merely adds a numerical factor; and the Q operator (9.4.12), making use of (9.4.3) *et seq.*, multiplies the result by $(\epsilon_i + \epsilon_j - \epsilon_m - \epsilon_n)^{-1}$. Thus

$$|\Phi_1\rangle = \tfrac{1}{2}\sum_{i,j,m,n} \frac{\langle mn|\,g\,|ij\rangle}{\epsilon_i + \epsilon_j - \epsilon_m - \epsilon_n}\,|\Phi_{ij}^{mn}\rangle, \tag{9.4.18}$$

which is of course the familiar first-order correction of the wavefunction. The second-order energy correction then follows readily from (9.4.14). It is in proceeding to higher order that the general theory reveals its full power and that diagrammatic representations become useful.‡

To obtain $W_n |0\rangle$ we shall have to evaluate $(QH')^n |0\rangle$, a sequence of n QH′ operators acting on the reference ket. As an example, let us take $(QH')(QH') |0\rangle$.

With the expression (9.4.17) we associate a *diagram* as follows:

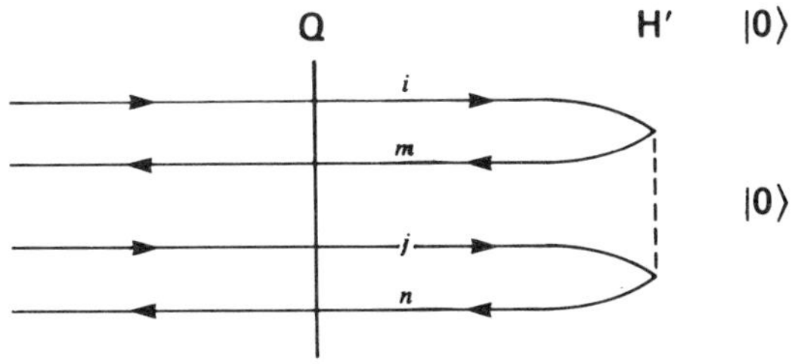

The broken line is a *vertex*, representing the matrix element $\langle mn| g |ij\rangle$ that appears in H′; the labels i, m are attached to the "entering" and "leaving" lines at one end of the vertex and j,n at the other; and these labelled lines indicate the excitations produced when this term in H′ operates on $|0\rangle$. The lines that *enter* from the left (going towards the operand) show that *holes* are produced in spin-orbitals i and j; while lines that *emerge* on the left show that *particles* appear in spin-orbitals m and n. The Q operator is indicated by a vertical stroke intersecting the hole and particle lines; and the numerical factor that it applies, namely $(\epsilon_i + \epsilon_j - \epsilon_m - \epsilon_n)^{-1}$, is obtained simply by reading the labels on the lines intersected. The effect of *all* the terms in QH′ is obtained simply by summation over all the labels associated with a single diagram of this particular "topological type". The result is (9.4.18).

Let us now look for a diagrammatic representation of the terms in $(QH')(QH') |0\rangle$ or, more explicitly, in

$$(QH')(QH')|0\rangle = (\tfrac{1}{2})^2 \sum_{r_2,s_2,t_2,u_2} \sum_{r_1,s_1,t_1,u_1} Q\langle r_2 s_2| g |t_2 u_2\rangle N[a^\dagger_{r_2} a^\dagger_{s_2} a_{u_2} a_{t_2}] \times Q\langle r_1 s_1| g |t_1 u_1\rangle N[a^\dagger_{r_1} a^\dagger_{s_1} a^\dagger_{u_1} a_{t_1}] |0\rangle. \quad (9.4.19)$$

We have noted already that the only choices of r_1, s_1, r_1, u_1 that do not destroy $|0\rangle$ are of the type $t_1 u_1 = i_1 j_1$ and $r_1 s_1 = m_1 n_1$, in which case the normal product produces a doubly excited ket $|\Phi^{m_1 n_1}_{i_1 j_1}\rangle$; the particle and hole lines, and their associated vertex, are shown in the diagram above. The second normal product in (9.4.19), with its Q operator, may be represented in a similar way by drawing a vertex and Q line "at level 2".

‡ An entertaining introduction has been given by Mattuck (1976).

But what arrangement of particle and hole lines will correspond to the non-zero contributions to $(\mathsf{QH}')(\mathsf{QH}')\,|0\rangle$? The result of the successive normal products in (9.4.19), operating on $|0\rangle$, may be obtained by expanding according to (9.2.9), and the effect of the various terms will be determined by the nature of the contractions. Thus if we take $r_2s_2t_2u_2 = m_2n_2i_2j_2$, we obtain

$$\mathrm{N}[\mathsf{a}^\dagger_{m_2}\mathsf{a}^\dagger_{n_2}\mathsf{a}_{j_2}\mathsf{a}_{i_2}]\mathrm{N}[\mathsf{a}^\dagger_{m_1}\mathsf{a}^\dagger_{n_1}\mathsf{a}_{j_1}\mathsf{a}_{i_1}] = \mathrm{N}[\mathsf{a}^\dagger_{m_2}\mathsf{a}^\dagger_{n_2}\mathsf{a}_{j_2}\mathsf{a}_{i_2}\mathsf{a}^\dagger_{m_1}\mathsf{a}^\dagger_{n_1}\mathsf{a}_{j_1}\mathsf{a}_{i_1}],$$

since none of the remaining contracted terms contain *non-zero* contractions of the type (9.2.6). The effect of this term is depicted below:

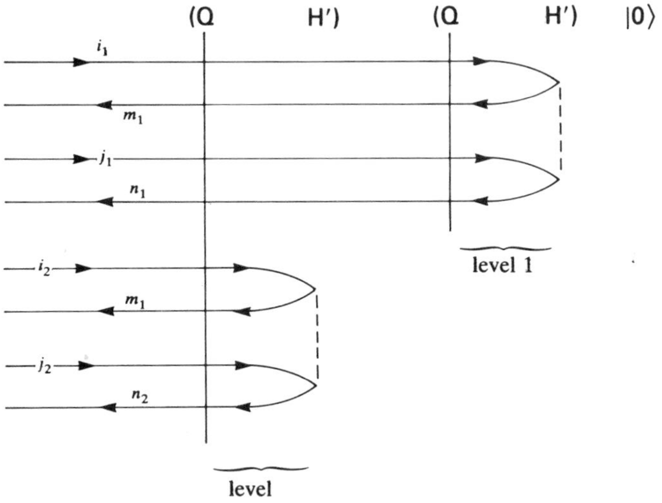

Thus each normal product creates its own double excitation to give a ket of the type $|\Phi^{m_1n_1m_2n_2}_{i_1j_1i_2j_2}\rangle$, as indicated by the hole and particle lines entering and leaving on the left, while the Q operators at levels 1 and 2 respectively supply numerical factors $(\epsilon_{i_1} + \epsilon_{j_1} - \epsilon_{m_1} - \epsilon_{n_1})^{-1}$ and $(\epsilon_{i_1} + \epsilon_{j_1} + \epsilon_{i_2} + \epsilon_{j_2} - \epsilon_{m_1} - \epsilon_{n_1} - \epsilon_{m_2} - \epsilon_{n_2})^{-1}$.

It is now clear how the argument proceeds. If at level 2 we take $r_2s_2t_2u_2 = m_2k_2j_2i_2$ then the uncontracted normal product will destroy $|0\rangle$ since a_{k2} tries to create a particle in an already occupied spin-orbital. On the other hand, a normal product with *one* contraction can give a non-zero result: thus

$$\begin{aligned}\mathrm{N}[\mathsf{a}^\dagger_{m_2}\mathsf{a}^\dagger_{k_2}\mathsf{a}_{j_2}\mathsf{a}_{i_2}]\mathrm{N}[\mathsf{a}^\dagger_{m_1}\mathsf{a}^\dagger_{n_1}\mathsf{a}_{j_1}\mathsf{a}_{i_1}] &= \mathrm{N}[\mathsf{a}^\dagger_{m_2}\mathsf{a}^\dagger_{k_2}\mathsf{a}_{j_2}\mathsf{a}_{i_2}\mathsf{a}^\dagger_{m_1}\mathsf{a}^\dagger_{n_1}\mathsf{a}_{j_1}\mathsf{a}_{i_1}] \\ &\quad + \mathrm{N}[\mathsf{a}^\dagger_{m_2}\overline{\mathsf{a}^\dagger_{k_2}\mathsf{a}_{j_2}\mathsf{a}_{i_2}\mathsf{a}^\dagger_{m_1}\mathsf{a}^\dagger_{n_1}\mathsf{a}_{j_1}}\mathsf{a}_{i_1}] \\ &\quad + \mathrm{N}[\mathsf{a}^\dagger_{m_2}\overline{\mathsf{a}^\dagger_{k_2}\mathsf{a}_{j_2}\mathsf{a}_{i_2}\mathsf{a}^\dagger_{m_1}\mathsf{a}^\dagger_{n_1}\mathsf{a}_{j_1}\mathsf{a}_{i_1}}],\end{aligned}$$

and, since $\widehat{a^{\dagger}_{k_2}a_{j_1}} = \delta_{k_2 j_1}$, $\widehat{a^{\dagger}_{k_2}a_{i_1}} = \delta_{k_2 i_1}$, the number of summation indices is effectively reduced in each non-zero term. The diagrammatic interpretation is indicated below for the first of these contracted products:

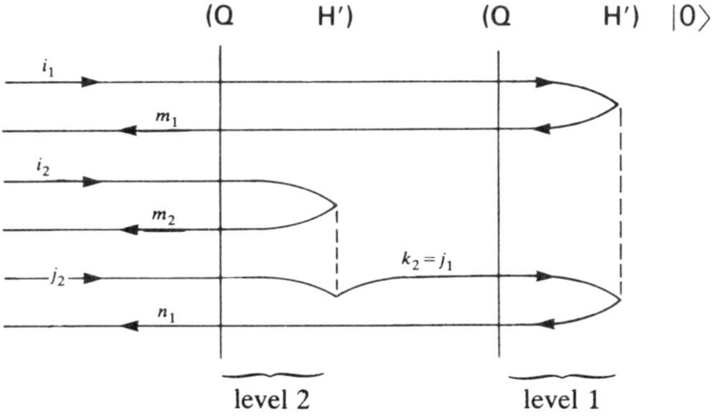

Clearly, *a non-zero contraction corresponds to joining together two hole (or particle) lines and identifying their labels.* The vertices at all levels after the first offer a variety of possible associations of hole and particle lines; for example,

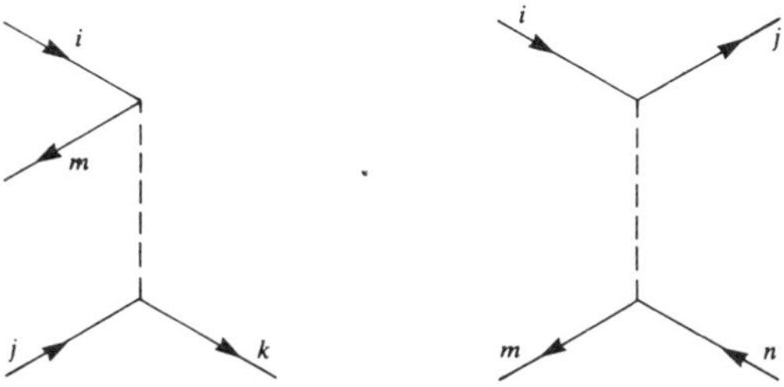

and when different vertices are joined in all allowable ways a considerable variety of diagrams will arise. The only non-zero terms in an expression such as (9.4.19) arise from diagrams in which the only "free" lines are those entering or leaving on the extreme left, all intermediate lines being connected and their labels identified. Evidently the diagram above corresponds to production of a triple excitation $|\Phi^{m_1m_2n_1}_{i_1i_2j_2}\rangle$; and the contribution from all labelled diagrams of this topological type is (adding the vertex and Q factors) given by

$$\sum_{\substack{i_1,i_2,j_1,j_2\\ m_1,m_2,m_1}} \frac{\langle m_2 j_1 | g | i_2 j_2\rangle}{(\epsilon_{i_1}+\epsilon_{i_2}+\epsilon_{j_2}-\epsilon_{m_1}-\epsilon_{m_2}-\epsilon_{n_1})} \frac{\langle m_1 n_1 | g | i_1 j_1\rangle}{(\epsilon_{i_1}+\epsilon_{j_1}-\epsilon_{m_1}-\epsilon_{n_1})} |\Phi^{m_1m_2n_1}_{i_1i_2j_2}\rangle.$$

Similarly, double-excitation terms will arise when there are two contractions (i.e. internal connections), and single-excitation terms when there are three contractions.

Before stating the general rules for working with diagrams, we note that many apparently different diagrams will yield algebraically identical results. For instance, in the evaluation of $(\mathsf{Q}\mathsf{H}')^3\,|0\rangle$ we encounter double-excitation diagrams such as

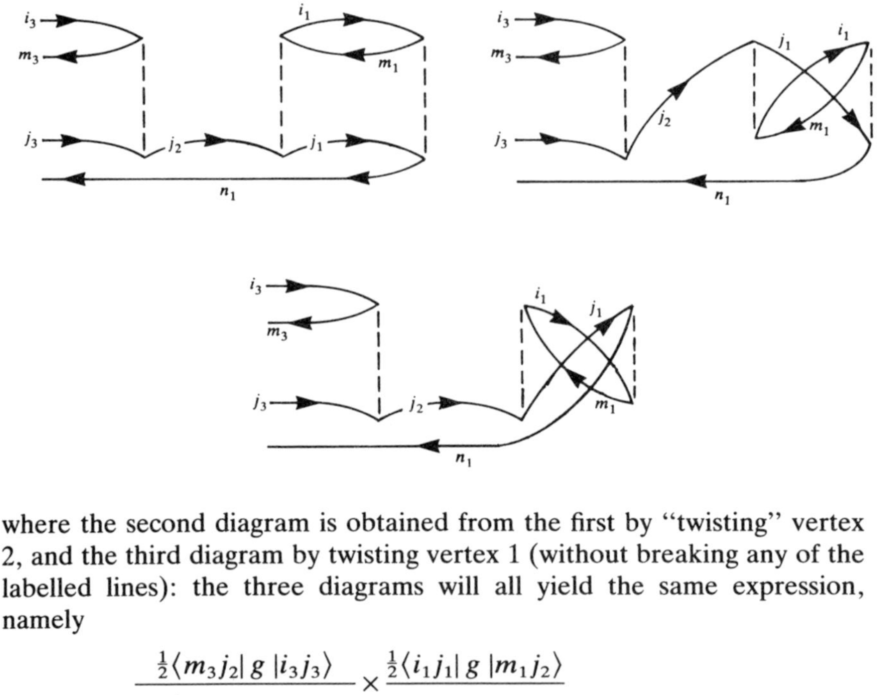

where the second diagram is obtained from the first by "twisting" vertex 2, and the third diagram by twisting vertex 1 (without breaking any of the labelled lines): the three diagrams will all yield the same expression, namely

$$\frac{\frac{1}{2}\langle m_3 j_2|\,g\,|i_3 j_3\rangle}{\epsilon_{i_3}+\epsilon_{j_3}-\epsilon_{m_3}-\epsilon_{n_1}}\times\frac{\frac{1}{2}\langle i_1 j_1|\,g\,|m_1 j_2\rangle}{\epsilon_{j_2}-\epsilon_{n_1}}\times\frac{\frac{1}{2}\langle m_1 n_1|\,g\,|i_1 j_1\rangle}{\epsilon_{i_1}+\epsilon_{j_1}-\epsilon_{m_1}-\epsilon_{n_1}}\,|\Phi_{i_3 j_3}^{m_3 n_1}\rangle.$$

If there are n vertices, each one may be twisted to give 2^n "topologically equivalent" diagrams. Thus it will be sufficient to count only topologically *distinct* diagrams and to multiply the result by 2^n; and this factor will exactly cancel the factor $(\frac{1}{2})^n$ coming from the matrix elements of H' in (9.4.2b).

Another simplification arises because diagrams that differ only through the *exchange* of two entering lines, or two leaving lines, attached to a given vertex, yield closely related results. Thus from the first of the above diagrams we could detach and interchange the lines with labels (m_1, n_1), (i_1, j_1), (j_2, m_3) respectively, to obtain

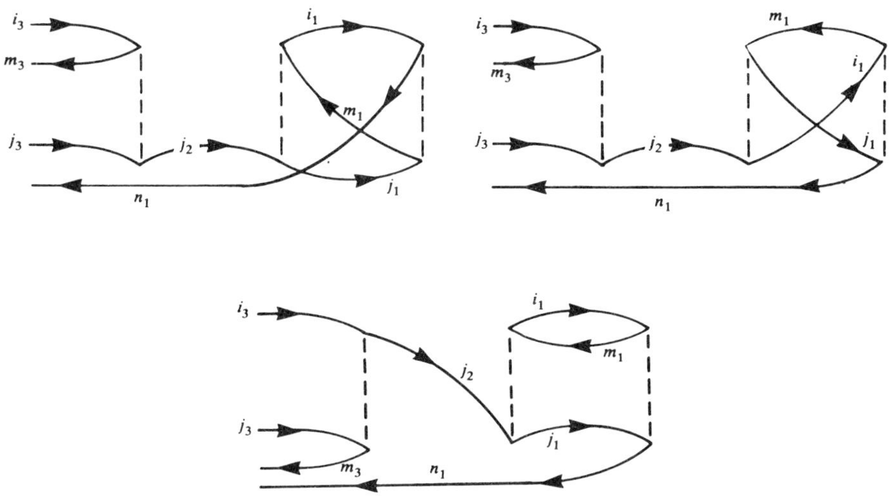

The associated algebraic terms are identical except that in each case a matrix element is replaced by its "exchange" partner (for example, in the first diagram, $\langle m_1 n_1 | g | i_1 j_1 \rangle$ is replaced by $\langle n_1 m_1 | g | i_1 j_1 \rangle$), and a factor -1 is attached. Consequently all such "exchange diagrams" may be ignored if we replace each vertex factor $\langle r_1 s_1 | g | t_1 u_1 \rangle$ by the antisymmetrized factor (6.1.4), namely $\langle r_1 s_1 \| t_1 u_1 \rangle = \langle r_1 s_1 | g | t_1 u_1 \rangle - \langle r_1 s_1 | g | u_1 t_1 \rangle$. Such diagrams, with "antisymmetrized vertices", are assumed in the rest of this chapter and referred to as Brandow-type diagrams (Brandow, 1967).

The reduction in the number of diagrams to be considered is often recognized by contracting the diagrams themselves, replacing each vertex by a bold dot that stands for a factor $\langle rs \| tu \rangle$. All the above diagrams are then absorbed into the single form

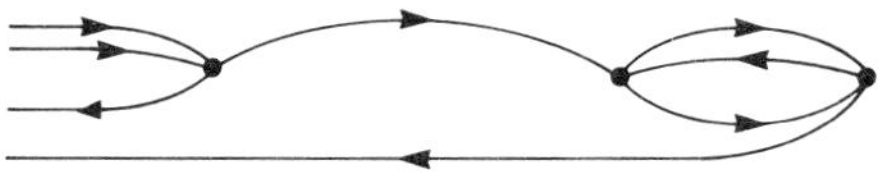

Such "Hugenholz diagrams" are fewer in number, and more easily distinguished in their topology, but the evaluation rules that we are about to give refer to the Brandow forms: to use them, each Hugenholz diagram must therefore be replaced by one of Brandow type, for example

the one that looks simplest. For example, the diagrams

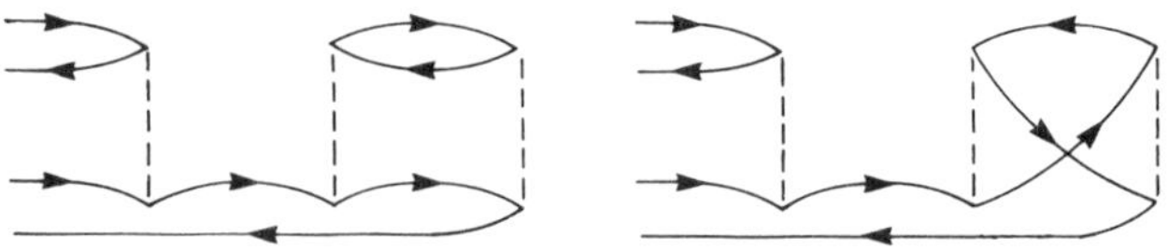

are distinct but have the same Hugenholz representation (given above); The first contains one *loop,* two *arcs* (i.e. continuous connected sequences of hole/particle lines) and two *internal hole lines* (i.e. hole lines connecting two vertices); the second contains no loops, two arcs and three internal hole lines. Both, however, lead to exactly the same algebraic result.

The rules for diagrammatic evaluation of $(\mathsf{QH}')^n\,|0\rangle$ may now be stated (in terms of Brandow-type diagrams) as follows.

(1) Draw n broken vertical lines (vertices), one for each H' operator.

(2) Attach two lines to the end of each vertex, travelling horizontally to left or right, except for the first vertex (extreme right), for which they travel to the left only.

(3) Add arrows so that one line points *into* each vertex end and the other points *out*: a line whose arrow points to the left is a "particle line" (and may be labelled $m, n, p, \ldots$), while one whose arrow points to the right is a "hole line" ($i, j, k, \ldots$).

(4) Join up all the arrowed lines to form internal arcs and loops, with arrows pointing in the same direction, except for one or more pairs of entering/leaving lines on the extreme left. Every way of doing this yields a possible diagram.

(5) If there are ν pairs of entering/leaving lines, (i_1, m_1), $(i_2, m_2), \ldots, (i_\nu, m_\nu)$, then the diagram corresponds to an excited ket

$$|\Phi^{m_1 m_2 \ldots m_\nu}_{i_1 i_2 \ldots i_\nu}\rangle = \mathsf{a}^\dagger_{m_\nu} \ldots \mathsf{a}^\dagger_{m_1} \mathsf{a}_{i_1} \ldots \mathsf{a}_{i_\nu}\,|0\rangle.$$

This is multiplied by a numerical factor, which is a product of factors $\langle r_1 s_1 \| t_1 u_1\rangle$ for the first vertex (with ingoing lines t_1, u_1 and outgoing lines r_1, s_1), $\langle r_2 s_2 \| t_2 u_2\rangle$ for the second, etc., and factors

$$\left[\sum (\text{hole energies}) - \sum (\text{particle energies})\right]^{-1}$$

from the hole and particle lines intercepted by each of the Q lines (not usually drawn in) between a pair of vertices.

(6) The result must be multiplied by a factor, which is usually $(-1)^{l+h}$, where l is the number of loops in the diagram and h is the number of internal hole lines. But if there are pairs of "equivalent lines" running in the same direction and joining the same two vertices then a further factor $\frac{1}{2}$ must be included for each such pair.

Any particular diagram indicates a particular type of contribution to $(\mathsf{QH}')^n\,|0\rangle$, consisting of a sum over all possible ways of labelling the particle and hole lines. The labels are usually omitted, summation being understood and the range of the indices being clear from the arrows (holes or particles). The basic rules are therefore simple, but the number and variety of diagrams increases rapidly with n; and it must not be forgotten that what we really need is the effect of the operator W_n in (9.4.13)—not that of $(\mathsf{QH}')^n$ itself. A great simplification arises from a theorem considered in the next section.

9.5 THE LINKED-CLUSTER (GOLDSTONE'S) THEOREM

Many of the possible diagrams that can be drawn using the rules in Section 9.4 are of no importance, since on summation the corresponding terms cancel exactly in all orders. In order to classify and eliminate them, and to proceed to the simplest expressions for the wavefunction and energy, we need the following definitions.‡

(i) A *subdiagram* is any set of vertices, connected only among themselves (i.e. to no other vertices) by particle/hole lines, that occurs as part of a larger diagram.

(ii) A (fully) *connected* diagram is one that contains no subdiagrams. For brevity, a diagram not of this type is called *unconnected.*

(iii) A *linked diagram* is one that, together with any subdiagrams it may possess, is "tethered" by external lines; no part of it may be removed without breaking such lines. A diagram is *unlinked* if it, or any of its subdiagrams, *can* be taken away without breaking external lines.

‡ Not all authors use the same terminology: that adopted here will be clear from the examples.

The following examples illustrate these definitions:

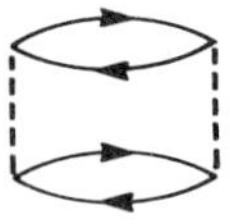
(a)

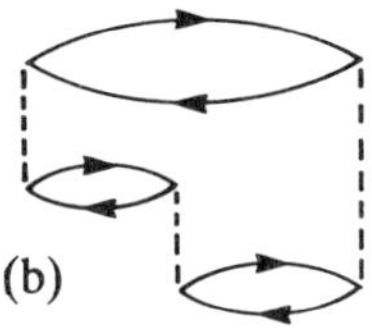
(b)

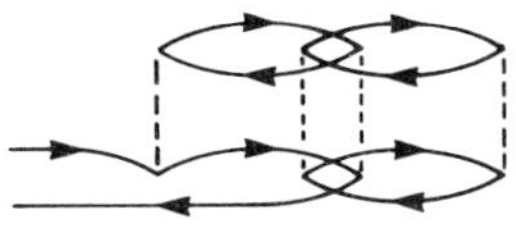
(c)

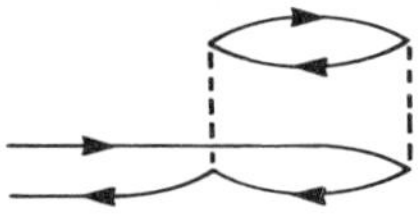
(d)

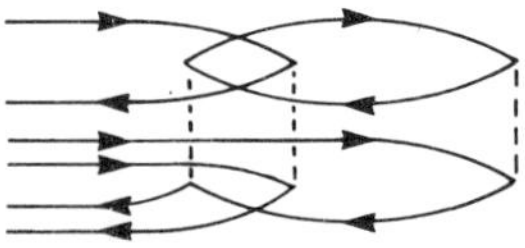
(e)

(a) and (b) are fully connected but unlinked, (c) is unconnected and unlinked, (d) is connected and linked, (e) is unconnected but linked.

It should be noted that, because it is only their topology‡ that is significant a diagram may be distorted to display its type more openly. Thus (b) and (c) are respectively equivalent to

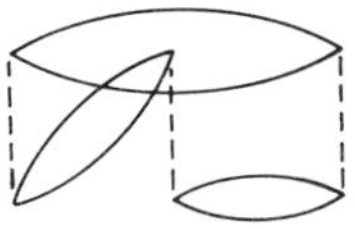

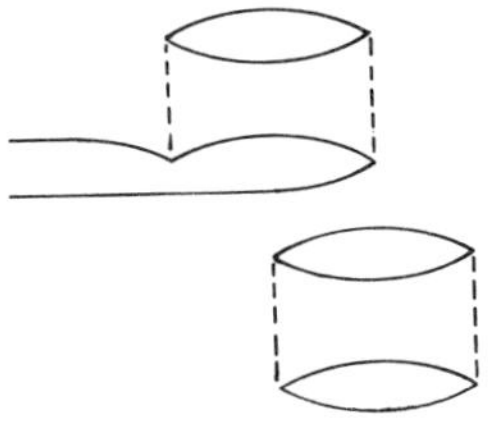

—the first looking more complicated, the second showing more clearly the presence of a subdiagram.

We now introduce a basic *factorization theorem* (Frantz and Mills, 1960), which refers to the sum of all contributions associated with a set of diagrams that differ among themselves only through the displacement (with its vertices) of a subdiagram. An example is the set

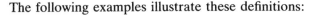

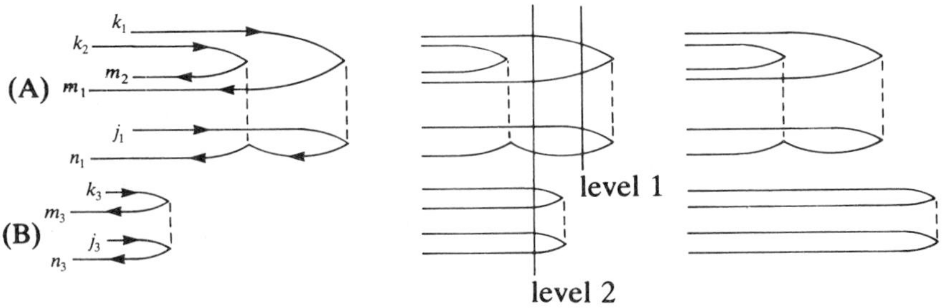

‡ I.e. the way the lines are connected to the given *ordered* set of vertices (changing the right-to-left order leads to new diagrams).

The upper and lower subdiagrams (A and B) show all possible orderings of the B vertex relative to the vertices in A: all three diagrams correspond to the same final state

$$\mathrm{a}^{\dagger}_{m_1}\mathrm{a}^{\dagger}_{m_2}\mathrm{a}^{\dagger}_{n_2}\mathrm{a}^{\dagger}_{m_3}\mathrm{a}^{\dagger}_{n_3}\mathrm{a}_{j_3}\mathrm{a}_{k_1}\mathrm{a}_{j_1}\mathrm{a}_{k_2}\mathrm{a}_{k_1}|0\rangle,$$

to the same sequence of matrix elements, and to the same sign, but they differ through their energy denominators. Let us denote by a_1, a_2 the denominators for subdiagram A at levels 1 and 2 (as indicated in the middle diagram), and similarly for subdiagram B. Then the denominator products for the three diagrams will evidently be (inserting Q lines between vertices from right to left)

$$a_1a_2(a_2+b_1), \qquad a_1(a_1+b_1)(a_2+b_1), \qquad b_1(a_1+b_1)(a_2+b_1).$$

On adding and simplifying, we find simply $1/a_1a_2b_1$, which is the product of the energy denominator factors associated with the *separate subdiagrams*. This is true whatever labels may be attached to the particle/hole lines, and the sum of the operators corresponding to the three diagrams may therefore be factorized in a similar way. The general result (Problem 9.16) may be stated as follows.

> The final state arising by summation of all the $(n_A+n_B)!/n_A!n_B!$ contributions whose diagrams differ only in the relative order of the vertices in A and B subdiagrams (containing n_A and n_B vertices, respectively, in given *internal* order), is obtained by letting the operators for the A and B subdiagrams work separately on $|0\rangle$. (9.5.1)

This result evidently arises from a purely algebraic identity, and the fact that some of the diagrams included in the summation may be "exclusion-principle violating diagrams" (EPVDs), when the same index appears more than once at some given level, is of no consequence.

The factorization theorem may now be used to establish the main result of this section (Goldstone, 1957).

> The nth-order correction $|\Phi_n\rangle$ may be represented using *linked diagrams only*; while E_{n+1} may be represented using only the *connected diagrams*, which are obtained by closing them on the left with a single vertex of type . Thus $E_{n+1}=\langle 0|\,\mathrm{H}'\mathrm{W}_n\,|0\rangle_{\mathrm{L}}$, where L means that the diagrams closed by $\langle 0|\,\mathrm{H}'$ must be *linked.* (9.5.2)

In other words, *all diagrams containing one or more unlinked subdiagrams can be ignored.*

This result (see e.g. March *et al.*, 1967), may be proved by induction, observing first that the diagram representing $\mathsf{W}_1 |0\rangle$ is linked and then showing that if $\mathsf{W}_{n-1} |0\rangle$ is represented by a sum of all possible linked diagrams, with $n-1$ vertices, then $\mathsf{W}_n |0\rangle$ must be similarly represented by diagrams with n vertices. Here we only indicate the proof.

First we note that in forming

$$E_n = \langle 0| \mathsf{H}'\mathsf{W}_{n-1} |0\rangle \tag{9.5.3}$$

the $\mathsf{W}_{n-1} |0\rangle$ diagram must be closed by a single vertex , since the last H' operator on the left must leave no external lines that would destroy the bra $\langle 0|$. Thus contributions can arise only from two types of (by hypothesis *linked*) diagrams in $\mathsf{W}_{n-1} |0\rangle$. These are

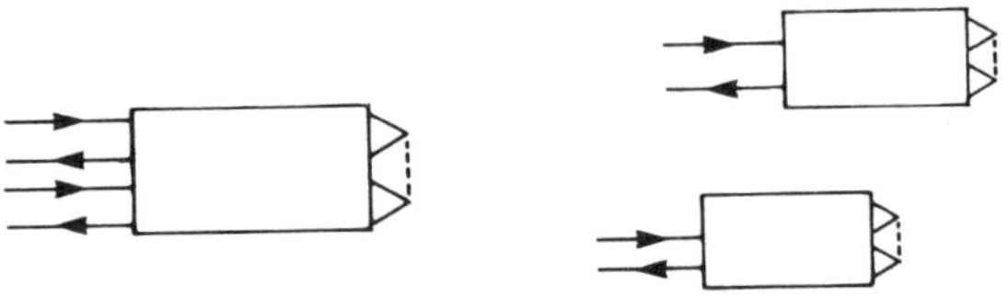

(a) linked – connected (b) linked – unconnected

since, with only number-conserving operators, the external lines attached to any subdiagram must appear in particle-hole pairs.

We now turn to $\mathsf{W}_n |0\rangle$ and note that any *un*linked terms must arise from the $\mathsf{H}'\mathsf{W}_{n-1}$ term in (9.4.13), namely

$$\mathsf{W}_n = \mathsf{Q}\mathsf{H}'\mathsf{W}_{n-1} - \sum_{k=0}^{n-1} E_{n-k}\mathsf{Q}\mathsf{W}_k, \tag{9.5.4}$$

and the extra vertex from H' must therefore close the $\mathsf{W}_n |0\rangle$ diagram thus:

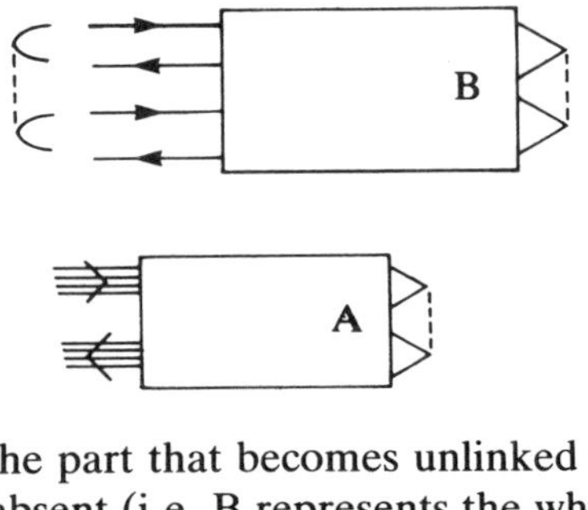

where B represents the part that becomes unlinked while A is the rest of the diagram. If A is absent (i.e. B represents the whole diagram) then the vertex that closes B leaves only a numerical multiple of $|0\rangle$, and the Q operator that follows then destroys the result: that is why we had to consider a *two*-part diagram.

It is now possible to show (Problem 9.13) that in summing over all orderings for the vertices of B relative to those of A (keeping the final H′ vertex always on the extreme left), the resultant contributions are exactly cancelled by the sum that appears in (9.5.4). The whole of $\mathsf{W}_n\,|0\rangle$ is consequently represented by part A alone, the unlinked B part disappearing, and the resultant diagram for $\mathsf{W}_n\,|0\rangle$ is *linked*.

The induction may now be completed. $\mathsf{W}_1\,|0\rangle$ contains no unlinked parts; if $\mathsf{W}_{n-1}\,|0\rangle$ contains no unlinked parts then none will be introduced in $\mathsf{W}_n\,|0\rangle$; and the theorem follows. To summarize, adding together the corrections $|\Phi_n\rangle = \mathsf{W}_n\,|0\rangle$ of all orders, we have the following.

The exact wavefunction, with particle interactions included to all orders of perturbation theory, is

$$|\Psi\rangle = \sum_{n=0}^{\infty} \sum_{\mathrm{L}} [(E_0 - \mathsf{H}_0)^{-1}\mathsf{PH}']^n\,|0\rangle, \tag{9.5.5}$$

where L means that in the diagrammatic representation of the operator only *linked* diagrams are to be included.

It is important to emphasize that the above result depends on the inclusion of diagrams that, when labelled, will apparently violate the exclusion principle; since without such diagrams the factorization theorem will not be valid. There is, however, nothing "unphysical" about this. The labels on the hole and particle lines occur as summation indices referring to contracted pairs in the normal products, and only the *emergent* lines correspond to actual states. It would be possible to restrict the summation indices so that no two hole or particle lines had the same index at any given level, but it is much more convenient to let the indices run freely over all available hole and particle states and to depend on cancellations to remove any redundant contributions. The antisymmetry requirement is of course built into the properties of the operators, and it is therefore impossible to create actual states that violate the exclusion principle. Equation (9.5.5) implies (9.5.2), the energy following from (9.4.14).

The implicit inclusion of EPVDs in the summations has implications in deciding which linked diagrams are distinct and therefore admissible. Let us consider two linked but unconnected diagrams that appear in fourth order and that differ only in the relative ordering of their vertices:

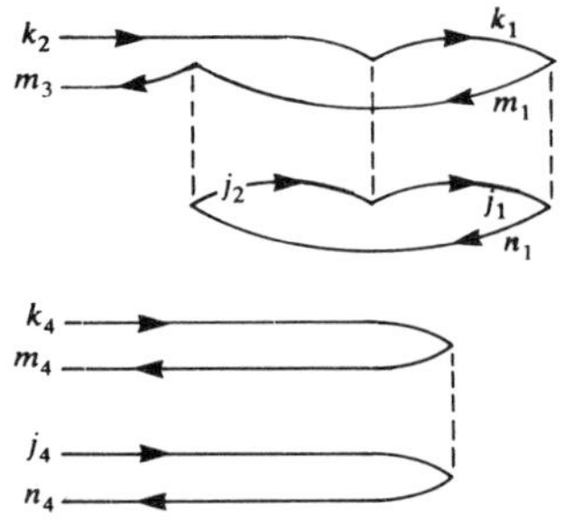

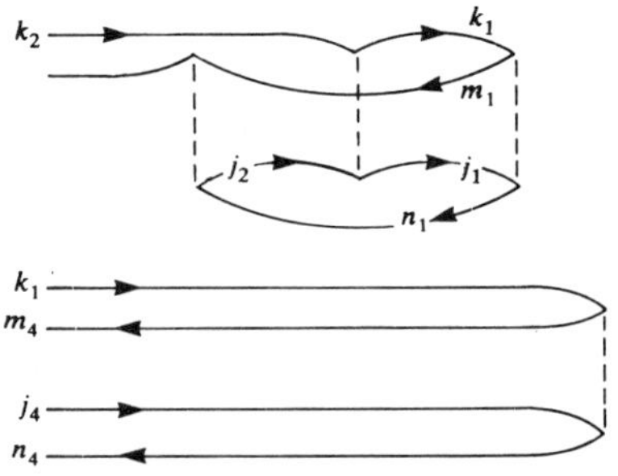

Both are EPVDs, and both must be admitted and counted as distinct: their contributions are in fact exactly cancelled, as is readily verified, by those that arise from the linked but *connected* diagrams

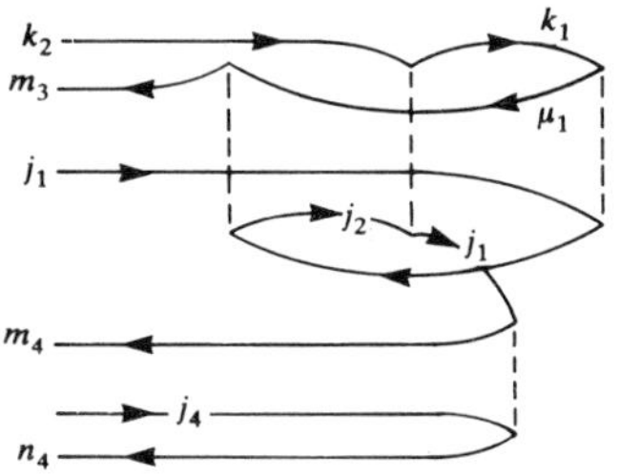

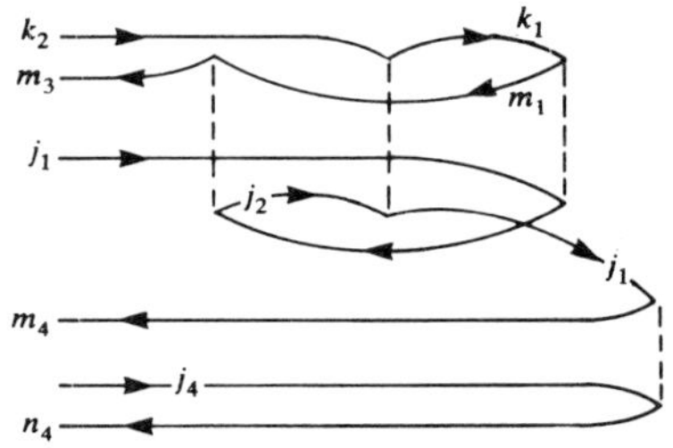

which are obtained from the two preceding diagrams respectively by cutting and rejoining the EPV lines labelled j_1.

In summary, the price to be paid for dropping all unlinked diagrams and freeing the summations over all hole and particle indices is not excessive: we must simply include all vertex orderings, even in unconnected subdiagrams (so long as they are linked), in order to ensure the cancellation of contributions not compatible with the exclusion principle.

9.6 EXTENSIONS OF THE DIAGRAMMATIC APPROACH

Diagrammatic methods were first developed for the study of "continuous" media, such as nuclear matter and solids, where the properties of interest are *extensive*. If, for example, the size of the system is doubled, keeping the particle density constant, then the total energy must also be doubled—remaining proportional to the number of particles N. Thus, for a free-electron gas, the total energy and the correlation energy contributions (in all orders of Rayleigh–Schrödinger perturbation theory) are rigorously N-proportional. There is then a correlation energy *per particle* that is independent of N. In such situations the energy contributions in all orders are often said to be "size-consistent", a term introduced by Pople *et al.* (1976). This is not necessarily the case for all approximation methods, or even for other forms of perturbation theory; it is a consequence of the grouping of R–S terms, in every order, into linked and unlinked clusters—the latter containing terms that are *not* N-proportional, but that cancel each other exactly and so may be ignored.

For atoms and molecules, the situation is not quite so clear-cut; the idea of increasing the size of a sodium atom, say, by adding electrons and keeping the density constant would obviously be nonsense, and the concept of extensivity does not apply. On the other hand, a system of n weakly interacting hydrogen molecules (with $N = 2n$ electrons) should clearly have a total energy proportional to N, so that the energy of the hydrogen gas would be an extensive property. This implies that if we calculate the energy of a system consisting of separate fragments, we should always use a size-consistent approach: it does *not* mean, however, that we *must* start from an independent-particle model (with an MO reference function) and then use perturbation theory to include all interactions—even those between the separate subsystems. Indeed, such a procedure could be a very slowly converging and inefficient method of dealing with a system containing weakly interacting fragments. As we know already, a more realistic reference function for such a system would recognize from the beginning the individuality of the subsystems (for example, by using a VB-type approach, with localized rather than delocalized orbitals); and this would also ensure size-consistency by dissociating correctly into a sum of fragment energies. The real issue in discussing dissociation is one of "separability" (rather than N-dependence) as was first pointed out by Primas (1965), and the whole problem of constructing separable wavefunctions for weakly interacting systems can be approached in a quite different way (Chapter 14). In summary, if we insist on using an IPM reference function in this context, with perturbation theory, then we must certainly use a linked-cluster form of the theory; but even then (as will be seen presently) the perturbation expansion may become *divergent* long before the dissociation limit is reached. In that case "size-consistency" in all orders will give only an illusion of "goodness", and the approach will still be invalid. The perturbation theory so far developed is in fact wholly satisfactory, for molecules, only in the vicinity of the equilibrium geometry; it does not provide a useful tool for the calculation of complete reaction surfaces.

The first accurate linked-cluster calculations of correlation energies were performed, for atoms, by Kelly (for a review see Kelly, 1969); later work is reviewed by Hibbert (1975), Kelly (1979), Lindgren and Morrison (1982) and Jankowski (1987). For molecules, many of the earlier calculations (e.g. Pople *et al.*, 1976) were essentially at the second-order (Møller–Plessett) level, and consequently did not require the use of diagrams. Some early applications of fully diagrammatic methods were made by Wilson (1977) and Bartlett *et al.* (1977) for the CO molecule, by Wilson (1978) for the HF molecule, by Silver and Wilson (1978) for N_2 and by Wilson and Silver (1979a,b) for H_2O. Usually, for computational reasons, such calculations are terminated at third order, but fourth- and higher-order calculations are nowadays not uncommon, at least for small molecules. Reviews of the more recent work on molecules can be found in the book by Wilson (1984) and in a rather comprehensive survey by Urban *et al.* (1987); detailed discussions of their computational aspects are also available (e.g. Wilson, 1983; Wilson *et al.*, 1983). The intention here, however, is not to enter into technical niceties, important as they may be for future work, and we return instead to the methodological aspects of the perturbation approach.

It is instructive to consider a very simple system for which, using a limited basis, a direct comparison of perturbation and full-CI calculations is possible. The CI approach is *finite* and leads at once to the basis-set limit, but the perturbation theory rests upon the expansions (9.4.7) and (9.4.8), which are *infinite* series in powers of a perturbation parameter and may or may not converge to the same limit.

An example: the hydrogen molecule

Let us consider the molecule H_2 and start from the IPM wavefunction in which the two electrons occupy the $1\sigma_g$ bonding MO, A say, which may be assumed to be an exact eigenfunction of the HF Hamiltonian. The virtual MOs $B, C, D, \ldots$ may then be added to provide an orbital basis for the CI/perturbation calculations; and to make matters more transparent we truncate this basis after the first two orbitals. There are then two occupied spin-orbitals $A\alpha$, $A\beta$ and two virtual $B\alpha$, $B\beta$, and the calculations run as follows.

(i) Full-CI approximation

From the Brillouin theorem, the CI involves only Φ_0 and one doubly excited function Φ_κ. These are respectively

$$\Phi_A = 2^{-1/2} \det |A\alpha \ A\beta|, \qquad \Phi_B = 2^{-1/2} \det |B\alpha \ B\beta|, \qquad (9.6.1)$$

and the secular equations are satisfied when

$$E^2 - E(H_{AA} + H_{BB}) + (H_{AA}H_{BB} - H_{AB}^2) = 0. \tag{9.6.2}$$

From Slater's rules, we find

$$H_{AA} = 2\epsilon_A - J_{AA}, \qquad H_{BB} = 2\epsilon_B + J_{BB} - 4J_{AB} + 2K_{AB}, \qquad H_{AB} = K_{AB}, \tag{9.6.3}$$

where ϵ_A, ϵ_B, J_{AB}, K_{AB}, etc. are HF orbital energies, coulomb, and exchange integrals defined in the usual way. On substituting these matrix elements in (9.6.2), solving, and (for comparison purposes) expanding the square root, we find the ground-state energy

$$E = 2\epsilon_A - \langle AA|\, g\, |AA\rangle + \Delta E, \tag{9.6.4}$$

where the first two terms give the model (HF) energy, while the level shift ΔE is

$$\Delta E = \frac{\frac{1}{2}\langle AA|g|BB\rangle^2}{(\epsilon_A - \epsilon_B - \frac{1}{2}\langle AA|g|AA\rangle - \frac{1}{2}\langle BB|g|BB\rangle + 2\langle AB|g|AB\rangle - \langle AB|g|BA\rangle)} \tag{9.6.5}$$

and represents the correlation energy. If we define

$$X = 2H_{AB}/(H_{BB} - H_{AA}) \tag{9.6.6}$$

then the approximation is good for $X^4 \ll 1$, which is certainly the case near the equilibrium geometry. In any case the accurate result is easily found by solving the quadratic equation (9.6.2).

(ii) Perturbation approximation

From (9.2.22) and (9.2.23), the unperturbed Hamiltonian H_0 gives a zero-order energy E_{HF}, namely,

$$E_0 = E_{\mathrm{HF}} = 2\epsilon_A - \langle AA|\, g\, |AA\rangle,$$

while the level shift in (9.6.5) must arise in second and higher order from the perturbation H' defined in (9.2.22), the first-order contribution vanishing. The second-order term E_2 is associated with diagrams that, when antisymmetrized vertices are used, become (using both Brandow

and Hugenholz forms)

The evaluation rules (p. 308) then yield

$$E_2 = \tfrac{1}{4} \sum_{\substack{i,j \\ m,n}} \frac{\langle ij \parallel mn \rangle \langle mn \parallel ij \rangle}{\epsilon_i + \epsilon_j - \epsilon_m - \epsilon_n}, \tag{9.6.7}$$

which is the so-called "second-order Møller–Plessett" expression for the correlation energy. In the present example, with only two occupied and two virtual spin-orbitals, $i, j \to A, \bar{A}$ and $m, n \to B, \bar{B}$. The summation becomes a single term, multiplied by 4 (two orderings for each pair); the spin integrations are immediate, and the final result is

$$E_2 = \frac{\langle AA|\, g\, |BB\rangle^2}{2(\epsilon_A - \epsilon_B)}, \tag{9.6.8}$$

which differs from (9.6.5) in having no 2-electron terms in the denominator.

The disagreement between the two results is of considerable interest. Both expressions are of second order in off-diagonal elements of the Hamiltonian: but all the machinery set up in the last twenty pages fails to give the results derived in a few lines from elementary principles. Why? Evidently it is necessary to go to higher orders of perturbation theory. It turns out that before we can resolve this mystery we must even consider selective summations *to infinite order*!

Infinite-order summations

Let us examine first the third-order perturbation energy for a general closed-shell system. There are three types of associated diagrams, of which the Goldstone (leading term only) and Hugenholz representations are

Brandow

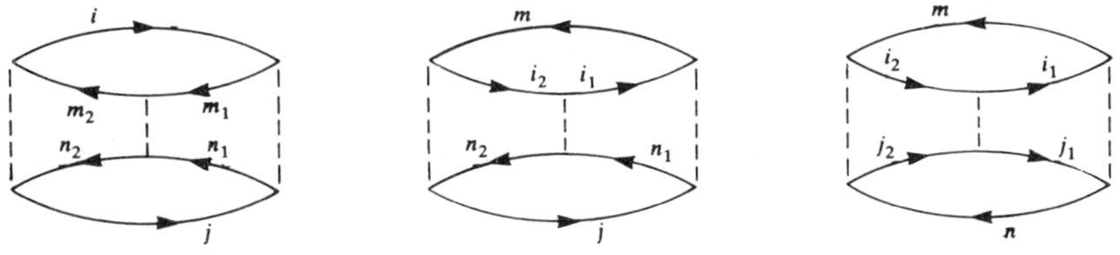

Hugenholz:

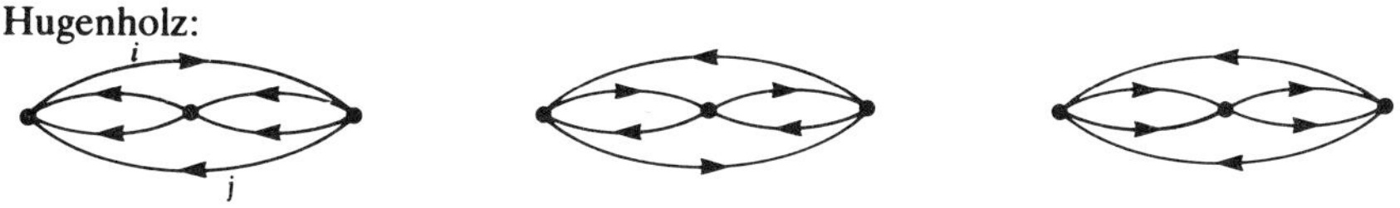

These are described respectively as "particle–particle" (pp), "particle–hole" (ph) and "hole–hole" (hh) diagrams, according to the nature of the central vertex. From the evaluation rules, the corresponding E_3 contributions are found to be, using D_{ij}^{mn} to denote a typical denominator,

$$E_3(\mathrm{pp}) = \tfrac{1}{8} \sum_{\substack{i,j \\ m_1,m_2,n_1,n_2}} \langle ij \| m_2n_2\rangle\langle m_2n_2 \| m_1n_1\rangle\langle m_1n_1 \| ij\rangle(D_{ij}^{m_2n_2}D_{ij}^{m_1n_1})^{-1},$$

$$E_3(\mathrm{ph}) = -\sum_{\substack{i_1,j,j_2 \\ n_1,m,n_2}} \langle i_2j \| mn_2\rangle\langle i_1n_2 \| i_2n_1\rangle\langle mn_1 \| i_1j\rangle(D_{i_2j}^{mn_2}D_{i_1j}^{mn_1})^{-1},$$

$$E_3(\mathrm{hh}) = \tfrac{1}{8} \sum_{\substack{i_1,i_2,j_1,j_2 \\ m,n}} \langle i_2j_2 \| mn\rangle\langle i_1j_1 \| i_2j_2\rangle\langle mn \| i_1j_1\rangle(D_{i_2j_2}^{mn}D_{i_1j_1}^{mn})^{-1}.$$

Instead of using the complete sums, let us pick out some of the main terms—those in which the central interaction is "diagonal" with $m_2n_2 = m_1n_1$ in the first expression, $i_1n_2 = i_2n_1$ in the second and $i_1j_1 = i_2j_2$ in the third. In this way we select diagrams such as

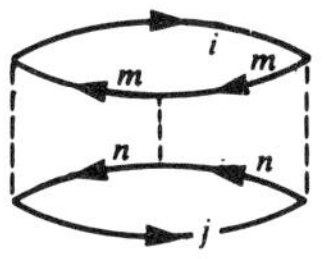

which correspond to "passive scattering" at the central interaction, the particle or hole states being unchanged. The resultant contributions are

$$\left.\begin{aligned} E_3(\mathrm{pp}) &= \tfrac{1}{4} \sum_{i,j,m,n} \langle ij \| mn\rangle\langle mn \| mn\rangle\langle mn \| ij\rangle(D_{ij}^{mn}D_{ij}^{mn})^{-1}, \\ E_2(\mathrm{ph}) &= -\sum_{i,j,m,n} \langle ij \| mn\rangle\langle in \| in\rangle\langle mn \| ij\rangle(D_{ij}^{mn}D_{ij}^{mn})^{-1}, \\ E_3(\mathrm{hh}) &= \tfrac{1}{4} \sum_{i,j,m,n} \langle ij \| mn\rangle\langle ij \| ij\rangle\langle mn \| ij\rangle(D_{ij}^{mn}D_{ij}^{mn})^{-1} \end{aligned}\right\} \quad (9.6.9)$$

On replacing $\langle in \| in\rangle$ in the last expression by $\tfrac{1}{4}(\langle in \| in\rangle + \langle jn \| jn\rangle + \langle im \| im\rangle + \langle jm \| jm\rangle)$, recognizing that the summation indices are dummies, the whole third-order energy from these selected diagrams is

$$E_3 = \tfrac{1}{4} \sum_{i,j,m,n} \langle ij \| mn\rangle d_{ij}^{mn}\langle mn \| ij\rangle(D_{ij}^{mn})^{-1}, \qquad (9.6.10)$$

where the numerator now contains a factor

$$d_{ij}^{mn} = \langle ij \parallel ij\rangle + \langle mn \parallel mn\rangle - \langle im \parallel im\rangle - \langle in \parallel in\rangle - \langle jm \parallel jm\rangle - \langle jn \parallel jn\rangle. \tag{9.6.11}$$

The result (9.6.7) could thus be modified so as to include (9.6.10) merely by inserting a factor $1 + d_{ij}^{mn}$ in the numerator.

The same argument may be repeated. In fourth order there is a second internal vertex, which can lead to any one of three types of factor in the numerator, together with a further D factor in the denominator. On summing in a similar way, we find that E_4 is obtained from E_3 simply by inserting another factor d_{ij}^{mn}/D_{ij}^{mn}. It is then clearly possible to sum the resultant terms to *infinite* order as a geometric progression: dropping the sub- and superscripts, for brevity,

$$E_2 + E_3 + \ldots = \tfrac{1}{4} \sum_{i,j,m,n} D^{-1}\langle ij \parallel mn\rangle\langle mn \parallel ij\rangle\left[1 + \frac{d}{D} + \left(\frac{d}{D}\right)^2 + \ldots\right]$$

$$= \tfrac{1}{4} \sum_{i,j,m,n} \frac{\langle ij \parallel mn\rangle\langle mn \parallel ij\rangle}{D(1 - d/D)}.$$

The final result may thus be written as

$$E_2 = \tfrac{1}{4} \sum_{i,j,m,n} \frac{\langle ij \parallel mn\rangle\langle mn \parallel ij\rangle}{\epsilon_i + \epsilon_j - \epsilon_m - \epsilon_n - d}, \tag{9.6.12}$$

and is usually referred to as the "shifted-demoninator" or Epstein–Nesbet form of the second-order energy (Epstein, 1926; Nesbet, 1955). The same summation technique may be applied to higher-order terms, every denominator being shifted by an appropriate d_{ij}^{mn}. The nth-order contributions that remain are then obtained from the general expressions, but with *omission* of the diagonal elements in the numerators. The Epstein–Nesbet form may indeed be obtained more directly from the partitioning approach of Section 2.5, in which diagonal elements of the perturbation were put into the denominators at an early stage.

On specializing to the hydrogen-molecule example, use of (9.6.12) now gives the revised denominator (after integrating over spin)

$$D \rightarrow D - d = 2(\epsilon_A - \epsilon_B) - \langle AA|g|AA\rangle - \langle BB|g|BB\rangle + 4\langle AB|g|AB\rangle - 2\langle AB|g|BA\rangle$$

—in exact agreement with the denominator in (9.6.5). Although, however, the results are now in agreement, the question of convergence has yet to be discussed—since both results break down in the dissociation limit!

Convergence of the perturbation series

The example considered above is illuminating in many ways. It is true that nobody would do a hydrogen-molecule calculation this way; but the results expose some of the basic difficulties of any perturbation approach. First of all, it is sometimes necessary to work quite hard to reproduce a very trivial result. Secondly, in spite of apparent size consistency in all orders, the perturbation method does *not* always have the property of separability. In fact, no amount of infinite summation can give the full-CI energy (i.e. the basis-set limit) once the distance between the hydrogen atoms increases beyond about twice the equilibrium bond length—simply because the expansion has a very limited radius of convergence.

To understand this failure, it is sufficient to note that (9.6.5), which has the characteristic form of second-order perturbation theory (a numerator of second degree in an off-diagonal element and a single energy denominator), arose from the expansion of a square root in solving the secular equation (9.6.2), and that a corresponding nth-order term (with an nth-degree numerator and $n-1$ energy differences in the denominator) follows in the same way. The difficulty is that the expansion converges only when the quantity X, defined in (9.6.6), is small compared with unity—while with increasing separation $H_{AA} - H_{BB} \to 0$ and this condition is violated. When the secular equations are solved *without* expansion of the square root, the difficulty does not arise and the calculated energy goes smoothly to the energy of two separate hydrogen atoms. But it is unavoidable in the perturbation calculation, which *starts* from the infinite-series representation of the wavefunction. This behaviour is typical *for molecules in bond-breaking geometries*‡, as may be verified by inspection of numerical results in the literature (see e.g. Wilson (1977) for the CO molecule, and Urban *et al.* (1987) (p. 237) for F_2), where computed energy curves begin to fail badly at separations of about $1\frac{1}{2}$ times the equilibrium bond length. The conclusion must be that, while the diagrammatic methods developed in this chapter can be carried to third and fourth order for small molecules, and can yield

‡ It should be noted that such difficulties are not encountered when the orbitals that become degenerate (for example A and B in the example above) *both* belong to the occupied set, and consequently (adding virtual orbitals $C, D, \ldots$) the dissociation of the He_2 into two closed-shell helium atoms could be studied with no extension of the formalism developed so far. This is because the energy denominators only contain differences of occupied/virtual type, and the coalescence of bonding and antibonding MOs therefore does no damage provided that *both* are doubly occupied. This is true generally whenever a closed-shell system dissociates into closed-shell fragments; the convergence difficulty arises only when covalent bonds are broken.

impressive results for near-equilibrium geometries, they can go seriously astray in bond-breaking situations (see especially Handy *et al.*, 1985).

9.7 GENERALIZATIONS

In general, convergence difficulties can only be overcome by making substantial generalizations of the theory. The simplest of these can be implemented without too much trouble; the others are much more fundamental.

Inclusion of 1-body vertices

First of all, it might be expected that convergence could be improved by making a new choice of *orbitals*; for we have noted elsewhere (e.g. p. 172) that the canonical HF orbitals are not well suited for use in CI calculations, owing to the diffuse nature of the virtual set. Kelly's early calculations on atoms did indeed verify this expectation; and in subsequent work it became common to use eigenfunctions of a modified Fock operator, containing a "V^{N-1} potential" obtained by removing one electron; the virtual orbitals then describe an electron moving in the field of a positive ion, are correspondingly "contracted", and often give improved convergence. The choice of orbitals is in fact quite open and can be changed at will by changing the partitioning of the Hamiltonian in (9.2.22).

Let us add an arbitrary 1-body potential u to the Fock operator to obtain $\mathsf{F} = \mathsf{F} + \mathsf{u}$ and a corresponding H_0 (defined in terms of the new IPM-type orbitals). The terms added to H_0 must then be subtracted from the H' term, and the whole perturbation theory will be unchanged except that the perturbation operator will now be

$$\mathsf{H}' = -\sum_{r,s} \langle r|\,\mathsf{u}\,|s\rangle \mathrm{N}[\mathsf{a}_r^\dagger \mathsf{a}_s] + \tfrac{1}{2} \sum_{r,s,t,u} \langle rs|\,g\,|tu\rangle \mathrm{N}[\mathsf{a}_r^\dagger \mathsf{a}_s^\dagger \mathsf{a}_u \mathsf{a}_t] = \mathsf{H}_1' + \mathsf{H}_2'.$$

With this new definition of H', the partitioning $\mathsf{H} = \mathsf{H}_0 + \mathsf{H}'$ is still valid, and the only change is that the operator products $(\mathsf{QH}')(\mathsf{QH}')\ldots$ now introduce *one*-body factors at every level of the perturbation expansion. For example, in second order,

$$(\mathsf{QH}')(\mathsf{QH}')\,|0\rangle = (\mathsf{QH}_1')(\mathsf{QH}_1')\,|0\rangle + (\mathsf{QH}_1')(\mathsf{QH}_2')\,|0\rangle + \ldots,$$

whereas, with canonical orbitals, only the first term was present.

The modification of the diagrams is straightforward. Possible H_1'

vertices are denoted by

and are incorporated in much the same way as the two-electron (H_2') vertices. Thus some of the new diagrams in the third-order energy term $\langle 0|\, H'(QH')(QH')\, |0\rangle$ will be

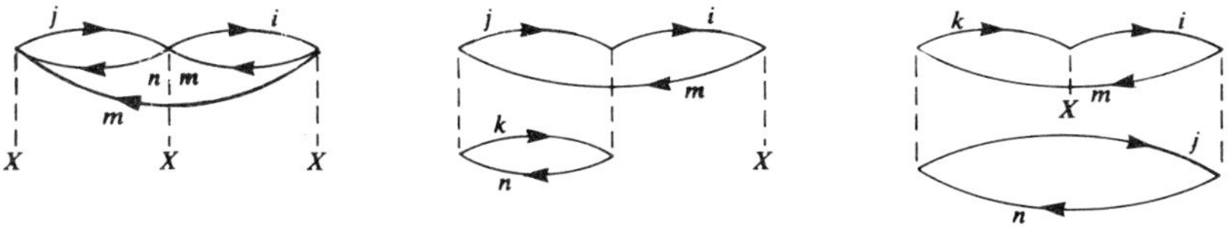

The rules for evaluating contributions from diagrams (p. 308) are unchanged except for the appearance of the two types of vertex factor; but clearly the number of possible diagrams in any order will be considerably increased.

Quasidegeneracy effects

The most severe convergence problems are undoubtedly those related to "quasidegeneracy" effects that occur when denominators become small. This is an indication, in turn, that the closed-shell reference function Φ_0, which we have assumed throughout, no longer provides a sound basis for the perturbation expansion: instead, a *multiconfiguration* reference function should be used. This is possible, and important progress has been made in this direction, notably by Brandow (1967, 1977), who developed a complete formalism—in many ways similar to that used so far—in terms of "folded" diagrams. The theory was first employed in molecular calculations by Kaldor (1975a,b). Details of such advances can be found in the books by Lindgren and Morrison (1982) and Wilson (1984), but, in view of the increasing number, variety and complexity of the diagrams involved, the future of this approach is still uncertain. The essence of the approach is the expansion of the "model space" from a single IPM determinant to an extended set of functions, which are degenerate eigenfunctions of a certain "model Hamiltonian". The eigenfunctions of the actual Hamiltonian are then related to those of the model by means of a wave operator, as in Section 9.4, and the wave operator is again expressed as a perturbation series. For the formulation, see Problems

10.14–10.17. For its diagrammatic implementation, the reader should consult the books cited above and the many references they contain.

In summary, diagrammatic methods provide a systematic and powerful approach to the calculation of high-accuracy wavefunctions and energies, and offer considerable flexibility in the choice of the model. They are particularly valuable for small closed-shell molecules in near-equilibrium geometries; but for larger systems, and for open-shell or quasidegenerate states, many problems are yet to be solved.

PROBLEMS 9

9.1 Work through all possible contractions of two operators $\mathrm{a}_r^\dagger$, a_s, showing that the results may be summarized in the form (9.2.6) by introducing the quasiparticle operators.

9.2 Verify (9.2.7) by using the anticommutation rules on all products of two operators $\mathrm{a}_r^\dagger$, a_s. Then show that if (9.2.8) is valid for a product of n operators, it will also be valid for $n+1$, thus confirming the inductive proof of Wick's theorem.

9.3 Read through the derivation of (9.2.10)–(9.2.12). Then close the book and repeat the derivation, filling in the details.

9.4 Why is it unnecessary to include ij-terms in (9.2.15) and what bearing does your answer have on (9.2.18)? Derive the results given in (i)–(iii) on p. 294. [*Hint*: Factorize the exponential operator and note the effect of e^{R} on Φ_0 when R contains only ij-terms.]

9.5 Obtain the 4-cluster terms in the exponential operator expansion (9.3.9) and show that if 1-cluster corrections to the reference function Φ_0 are neglected then considerable reduction occurs. Hence, neglecting irreducible cluster operators for more than 2 electrons, express the wavefunction in the correlated-pair form (9.3.17). What condition must the reference function satisfy in order to eliminate the 1-cluster terms? [*Hint*: The effect of the 1-cluster terms in the exponential operator is equivalent to an orbital variation in the operand. Look for a variational condition.]

9.6 Derive the conditions (9.3.20)–(9.3.22), noting why it is necessary to add extra terms in each successive equation. Show that, on assuming a 1-determinant reference function, (9.3.22) leads to algebraic equations to determine the coefficients D_{ij}^{mn} in the pair functions ϕ_{ij}. [*Hint*: Use the Hamiltonian (9.2.22), noting that Φ_0, Φ_i^m, Φ_{ij}^{mn} are eigenfunctions of H_0 with eigenvalues that are sums of orbital energies. You will also need to make extensive use of the anticommutation rules.]

9.7 Starting from the recursion formula (9.4.13), and the first-order results

(9.4.15), derive the wave operators W_2 and W_3. Then obtain general expressions for the second- and third-order terms in the energy.

9.8 Justify the statement following (9.4.18). [*Hint*: In terms of *determinants*, the corresponding expansion would include only terms with $m > n$, $i > j$ (ordered pairs), and the coefficient of Φ_{ij}^{mn} would be $\langle mn \| ij \rangle$.]

9.9 Obtain the explicit form of the second-order correction $|\Phi_2\rangle$ to the wavefunction for a system with Hartree–Fock reference function. [*Hint*: Proceed as in the derivation of (9.4.18).]

9.10 Use the expression (9.4.13) to calculate the second-order correction to the Hartree–Fock energy. Then use $|\Phi_2\rangle$ obtained in Problem 9.9 to extend your results to third order. The results obtained are usually called the second- and third-order Møller–Plesset corrections. [*Hint*: Form the scalar products (9.4.14), noting that any outgoing or incoming lines on the right of $\langle \Phi_0|$ will destroy it.]

9.11 Give a full diagrammatic interpretation of the second-order energy formula obtained in Problem 9.10, showing that two kinds of contribution occur and that the energy denominator is correctly given by the Q-line rule, (5) on p. 308. [*Hint*: When $|\Phi_\kappa\rangle\langle\Phi_\kappa|$ appears between two H' operators it acts like the unit operator; the first H' creates a "doubly excited" state $|\Phi_\kappa\rangle$, which is merely "carried" to the next H' by the projection operator $|\Phi_\kappa\rangle\langle\Phi_\kappa|$. The $H'H'$ then contains a product of two normal products, which can be reduced by using the extended Wick theorem.]

9.12 Suppose that a 1-electron perturbation is applied to the system discussed in earlier problems and that H in (9.2.10), and consequently H' in (9.4.2), is modified by the addition of a term $\sum_{r,s} \langle r|\, h\, |s\rangle N[a_r^\dagger a_s]$. How will the derivations made in Problems 9.9 and 9.10 be changed and what new terms will appear? Note that the new terms may be removed, to all orders, by starting from a new Hartree–Fock calculation in which the external perturbation is included. [*Hint*: Write $H' = H_1' + H_2'$ and evaluate terms that contain the 1-electron operator H' using the usual normal-product rules.]

9.13 In (9.5.4) show that any unlinked terms that arise from $QH'W_{n-1}$ are exactly cancelled by terms arising from the sum.

9.14 Give a diagrammatic interpretation of the analysis in Problem 9.12 by introducing 1-body vertices like those in Section 9.7.

9.15 Use the rules on p. 308 to give a purely diagrammatic derivation of the second-order correction $|\Phi_2\rangle = W_2|\Phi_0\rangle$. [*Hint*: Draw the possible diagrams that contain two (antisymmetrized) vertices and combine the associated terms with the first-order term, as in Problem 9.9]

9.16 Generalize the discussion of the factorization theorem (p. 311), first by taking *two* vertices in each of the subdiagrams, and then by taking n_A and n_B.

9.17 Use Goldstone's theorem (9.5.2) to obtain directly the third-order energy formula (Problem 9.10). [*Hint*: $\mathsf{W}_2|\Phi_0\rangle$ will contain 1-, 2-, 3- and 4-excitation terms. Look for appropriate linked diagrams and close them with a final vertex.]

9.18 Write the general Hamiltonian (3.6.9) in the form $\mathsf{H} = \mathsf{H}_0 + \mathsf{H}'$, where $\mathsf{H}_0 = \sum_{r,s} \langle r|\,(\mathsf{h}+\mathsf{u})\,|s\rangle \mathsf{a}_r^\dagger \mathsf{a}_s$, u being an arbitrary 1-electron operator, while $\mathsf{H}' = -\sum_{r,s} \langle r|\,\mathsf{u}\,|s\rangle \mathsf{a}_r^\dagger \mathsf{a}_s + \frac{1}{2}\sum_{r,s,t,u} \langle rs|\, g\, |tu\rangle \mathsf{a}_r^\dagger \mathsf{a}_s^\dagger \mathsf{a}_u \mathsf{a}_t$. Note that H_0 defines a new IPM. Obtain from first principles the contributions to $\mathsf{W}_1\,|0\rangle$, drawing a diagram for each one and using the convention on p. 323 for the 1-body terms. Show that when u is chosen so that $\mathsf{h}+\mathsf{u}=\mathsf{F}$ (the Fock operator) the 1-body vertex contributions are exactly cancelled by those that arise from 2-body vertices of type  in which the "bubble" is a hole line. [*Hint*: Consider all possible choices of the labels *r, s, t, u*; some will give repeated contributions (removing the factor $\frac{1}{2}$ in the 2-body term), others will give "exchange contributions"! Note the relationship to the discussion on p. 306 In particular, note how (using canonical HF orbitals) the introduction of normal products removes, in all orders, the diagrams whose contributions cancel—bubble diagrams (internal contractions) being excluded by the extended Wick theorem.]

9.19 Consider the third-order diagrams for a system described using orbitals that are *not* of canonical HF type but are instead eigenfunctions of $\mathsf{F}' = \mathsf{F} + \mathsf{u}$. There are two other diagrams, not shown on p. 323, which contain single 1-body vertices, and one other with three 1-body vertices. Draw them and evaluate their contributions to the energy.

10 Large-Scale CI and the Unitary-Group Approach

10.1 THE GENERAL APPROACH

The method of configuration interaction (CI) undoubtedly provides the simplest possible route, from a formal point of view, to the construction of high-accuracy electronic wavefunctions. It is sufficient to choose a basis set of orthonormal spin-orbitals $\{\psi_i\}$, take all the Slater determinants that can be formed from them, use Slater's rules to obtain matrix elements, and finally solve the secular equations to find energy levels and CI expansion coefficients. Since, with optimized orbitals, a single term in the expansion can give about 99% of the total energy, it might have been expected that at most a few hundred would give "chemical accuracy". As we know already, however, this is not the case; convergence is slow and it may be necessary to include *millions* of terms, even when the expansion has been shortened by exploiting symmetry and by optimizing in some sense (for example by MC SCF methods) the choice of orbitals.

The traditional approach, in which the whole matrix **H** is computed and eigenvalues and eigenvectors are found by standard diagonalization techniques, is obviously unable to deal with this situation (a matrix with 10^6 rows and columns would not even fit in the computer!). In the last chapter such difficulties were avoided by perturbation methods, but other difficulties remained; in particular, the expansion is infinite, convergence is uncertain, only selected terms can be summed, and for open-shell and excited states there are further problems.

In the present chapter we return to the "simple" CI approach, but now turn to the newer strategies that began to evolve during the nineteen seventies. The distinguishing feature of such strategies is that they are "global": they consider *all* the CFs (of any desired symmetry) that can be constructed from a given set of orbitals; the rapid construction, as and when they are needed, of *all* types of matrix-element contribution (without ever setting up the full matrix); and the efficient iterative refinement of all expansion coefficients. In other words, the full-CI space is looked at *as a whole*—even if truncation is considered at some later stage.

There are essentially three problems in handling expansions of, say, 10^6 terms: (i) how to label uniquely and economically (for example by sequential numbering) every one of the CFs, for all orbital configurations and all spin couplings; (ii) how to associate 1- and 2-electron integrals with pairs of CFs, and hence with matrix elements $H_{\lambda\kappa}$, so that all (or at least many) matrix-element contributions can be generated simultaneously; and (iii) how to construct from any given estimate of an eigenvector C_κ an improved estimate C'_κ. To attack these problems, a large amount of mathematical machinery, much of it quite abstruse and using the theory of Lie groups (see e.g. Louck, 1970), has been developed.

Here we adopt a simpler approach, which starts from the ideas introduced in earlier chapters. Briefly, the CFs employed in a full-CI calculation are constructed from an orbital basis $\{\phi_i\}$ of m functions by setting up *orbital products*, adding spin factors, antisymmetrizing, and combining the resultant Slater determinants according to symmetry requirements. But we know that, *provided* the *full* CI space is used, any new linearly independent mixtures $\{\bar{\phi}_i\}$ of the original orbitals will serve just as well; the fact that MO and VB approaches converge to the same full-CI limit is but one example of this equivalence. A transformation of orbitals $\phi_i \rightarrow \bar{\phi}_i$ (1-electron basis functions) induces a transformation of CFs $\Phi_\kappa \rightarrow \bar{\Phi}_\kappa$ (N-electron basis functions), and an arbitrary function in the full-CI space is invariant; with suitably transformed expansion coefficients,

$$\Psi = \sum_\kappa C_\kappa \Phi_\kappa = \sum_\kappa \bar{C}_\kappa \Phi_\kappa. \tag{10.1.1}$$

The approach that we follow begins with a general discussion of such transformations, and then turns to their relationship to two groups: (i) the symmetric group (S_N) of permutations of electronic variables; and (ii) the unitary group ($U(m)$) of transformations that lead from one orthonormal basis to another. It is the intimate relationship between these two groups that has been exploited in recent years both in classifying the CFs and in finding new procedures for matrix-element evaluation.

10.2 TRANSFORMATIONS AND TENSORS

Let us start from the m-dimensional space $\{\phi_i\}$ and consider the basis change (2.2.16), namely

$$\boldsymbol{\phi} \rightarrow \bar{\boldsymbol{\phi}} = \boldsymbol{\phi}\mathbf{T}.$$

The components $\mathbf{c}$ of any given vector (i.e. a general element of the vector space) must then be changed according to (2.2.18), becoming

$$\bar{\mathbf{c}} = \mathbf{T}^{-1}\mathbf{c}.$$

In the systematic study of such transformations it is usual to adopt the language of tensor algebra, expressing both transformation rules in a common form for ease of comparison. If we put $\mathbf{R} = \mathbf{T}^{-1}$ then we may write

$$\bar{\phi}_i = \sum_j \breve{R}_{ij}\phi_j, \tag{10.2.2a}$$

$$\bar{c}_i = \sum_j R_{ij}c_j, \tag{10.2.2b}$$

where, with the tilde denoting transposition,

$$\breve{\mathbf{R}} = \widetilde{\mathbf{R}^{-1}} = \tilde{\mathbf{R}}^{-1}. \tag{10.2.3}$$

The two transformations in (10.2.2) are said to be *contragredient*, the first (typical of basis vectors) being *covariant* and the second (typical of vector components) being *contravariant*. The relationship is clearly reflexive in the sense that if we put $\breve{\mathbf{R}} = \mathbf{U}$ then $\mathbf{R} = \breve{\mathbf{U}}$; and we can just as well write (10.2.2) as

$$\bar{\phi}_i = \sum_j U_{ij}\phi_j, \tag{10.2.4a}$$

$$\bar{c}_i = \sum_j \breve{U}_{ij}c_j \tag{10.2.4b}$$

—the terminology co- and contravariant is purely conventional. In the present context (see e.g. Hamermesh, 1962, Chap. 11) it is most common to adopt (10.2.4a) as the standard (formally contravariant) transformation rule.

When $\mathbf{U}$ in (10.2.4) is a general non-singular $m \times m$ matrix the infinite set $\{\mathbf{U}\}$ forms a matrix group, the *full linear group in m dimensions*, denoted by GL(m). If the matrices are chosen to be *unitary* (thus leaving invariant any Hermitian scalar product, as we know from Section 2.2) then we obtain the *unitary group* U(m); and in this case the matrices of the covariant transformation in (10.2.4b) are

$$\breve{\mathbf{U}} = \tilde{\mathbf{U}}^{-1} = \tilde{\mathbf{U}}^{\dagger} = \mathbf{U}^{*}. \tag{10.2.5}$$

A further reduction occurs when we use only *real unimodular* matrices; we then obtain the *real orthogonal group* R(m), in which $\breve{\mathbf{U}} = \mathbf{U}$, and the distinction between co- and contravariant transformations disappears.

The group R(3), for example, describes all possible rotations in 3-dimensional space. To include reflections, the unimodular condition must be relaxed and the group becomes $R^{\pm}(3)$, which includes the "improper" rotations with $\det \mathbf{U} = -1$. It is because all these groups are *infinite* and contain matrices with elements that are *continuously variable* that the results obtained for finite groups (Appendix 3) are not always immediately applicable.

Sets of quantities $\{A_i\}$ and $\{B_i\}$ that behave according to (10.2.4a) or (10.2.4b) respectively under the given group of transformations are said to form the components of a *rank-1 contravariant tensor* in case (a) or a *rank-1 covariant tensor* in case (b). This anticipates the definition of "rank-n" tensors—which follow more general transformation rules but are very easily found. Thus for a 2-electron system we may construct an approximate (spatial) wavefunction using all m^2 product functions $\{\phi_i(\boldsymbol{r}_1)\phi_j(\boldsymbol{r}_2)\}$; and if new orbitals are introduced according to (10.2.4a) then the new products will be

$$(\bar{\phi}_i\bar{\phi}_j) = \sum_{k,l} U_{ik}U_{jl}(\phi_k\phi_l). \tag{10.2.6}$$

This transformation law characterizes *rank-2* tensor quantities, with respect to the group U(m) of transformations in the m-dimensional vector space $\{\phi_i\}$; the product functions are said to span an m^2-dimensional *tensor space.* Any set of double-index quantities A_{ij} that transform in the same way, i.e. according to

$$\bar{A}_{ij} = \sum_{k,l} U_{ik}U_{jl}A_{kl}, \tag{10.2.7}$$

defines a *rank-2 contravariant tensor.*‡ If this equation is written in matrix form $\bar{\mathbf{A}} = \mathbf{MA}$ then the matrix $\mathbf{M}$ (rows and columns labelled by double indices) will have the general element $M_{ij,kl} = U_{ik}U_{jl}$. This matrix is called the "Krönecker square" of $\mathbf{U}$, and is written $\mathbf{M} = \mathbf{U} \times \mathbf{U}$.

Let us consider next a wavefunction expansion in which

$$F(\boldsymbol{r}_1, \boldsymbol{r}_2) = \sum_{m,n} C_{mn}\phi_m(\boldsymbol{r}_1)\phi_n(\boldsymbol{r}_2). \tag{10.2.8}$$

The same function may be expressed equally well in terms of the new orbital set $\{\bar{\phi}_i\}$ in (10.2.4a), *provided* that we can find coefficients, $\bar{C}_{ij}$ say, such that

$$F(\boldsymbol{r}_1, \boldsymbol{r}_2) = \sum_{i,j} \bar{C}_{ij}\bar{\phi}_i(\boldsymbol{r}_1)\bar{\phi}_j(\boldsymbol{r}_2). \tag{10.2.9}$$

‡ The transformation law refers to tensor *components* in some reference system (cf. vectors and vector components), but it is common to refer simply to "tensor quantities" as one refers to "vector quantities".

That this is possible follows from (10.2.6); since when each basis set is linearly independent the matrix **U** is non-singular and has an inverse; and we can multiply each side by $(U^{-1})_{mi}(U^{-1})_{nj}$ and sum over i and j to obtain

$$\sum_{i,j} (U^{-1})_{mi}(U^{-1})_{nj}(\bar{\phi}_i\bar{\phi}_j) = (\phi_m\phi_n).$$

On using this result in (10.2.8), we obtain

$$F = \sum_{m,n} C_{mn} \sum_{i,j} (U^{-1})_{mi}(U^{-1})_{nj}(\bar{\phi}_i\bar{\phi}_j) = \sum_{i,j} \Big(\sum_{m,n} \breve{U}_{im}\breve{U}_{jn}C_{mn}\Big)(\bar{\phi}_i\bar{\phi}_j),$$

and comparison with (10.2.8) shows that the required coefficients *can* be obtained, in the form

$$\bar{C}_{ij} = \sum_{m,n} \breve{U}_{im}\breve{U}_{jn}C_{mn}. \tag{10.2.10}$$

This is the standard transformation rule for components of a rank-2 *covariant* tensor, differing from (10.2.7) only through use of the matrix $\breve{\mathbf{U}}$ in place of **U**. The process may obviously be continued, leading to the definition of higher-rank tensors of each type and of "mixed" tensors with, for example, one degree of covariance and one of contravariance.

The fact that the same function may be represented equally well in two different ways, as in (10.2.8) and (10.2.9), is an example of *invariance* under a group of transformations. An invariant is a tensor of *rank zero*. Thus, using $\bar{F}$ to denote the function constructed from the barred quantities in (10.2.6) and (10.2.10),

$$\begin{aligned}\bar{F} = \sum_{i,j} \bar{C}_{ij}(\bar{\phi}_i\bar{\phi}_j) &= \sum_{i,j} \sum_{m,n,p,q} \breve{U}_{im}\breve{U}_{jn}U_{ip}U_{jq}C_{mn}(\phi_p\phi_q) \\ &= \sum_{m,n} C_{mn}(\phi_m\phi_n) = F,\end{aligned} \tag{10.2.11}$$

where it has been noted that

$$\sum_{i} \breve{U}_{im}U_{ip} = \sum_{i} (U^{-1})_{mi}U_{ip} = \delta_{mp} \tag{10.2.12}$$

etc. The array of all possible products $C_{mn}(\phi_p\phi_q)$ $(=D_{mn,pq}$, say) provides an example of a rank-4 tensor with 2 degrees of covariance and 2 of contravariance. But in (10.2.11) there occurs the transformed tensor $\bar{\mathbf{D}}$ with two pairs of indices (one co- and one contravariant in each pair) identified and summed. The process of identifying covariant and contravariant indices and then summing is known as *contraction* and leads to a tensor of lower rank; in this case, performed twice, it leads to the rank-zero tensor $\bar{F} = F$. The opposite process, in which we construct a new tensor (e.g. $D_{mn,pq}$) from the totality of products of lower-rank tensors, is called "forming a Krönecker product".

These basic notions will be sufficient for present purposes. In the context of full CI, for an N-electron system, a wavefunction (without spin) will be constructed from the m^N-orbital products

$$\{\Omega_{ij\ldots p}\} = \{\phi_i(\boldsymbol{r}_1)\phi_j(\boldsymbol{r}_2)\ldots\phi_p(\boldsymbol{r}_N)\}, \tag{10.2.13}$$

in which the orbital indices are chosen in all possible ways, N at a time, from the set of 1-electron basis functions. The orbitals span a vector (i.e. rank-1 tensor) space while the set (10.2.13) spans a rank-N tensor space, which is the Nth-degree Krönecker product of the orbital spaces for electron 1, electron 2, . . . , electron N. According to (10.2.7) *et seq.*, a common transformation $\mathbf{U}$ of all orbitals will induce a transformation of the N-index products (10.2.13) described by a matrix with elements

$$(\mathbf{U}\times\mathbf{U}\times\ldots\times\mathbf{U})_{ij\ldots p,\,i'j'\ldots p'} = U_{ii'}U_{jj'}\ldots U_{pp'}, \tag{10.2.14}$$

which is the Krönecker Nth power of $\mathbf{U}$. Any N-electron function in the full-CI space will be invariant under a change of orbitals (10.2.4a), provided that the coefficients of the $\Omega_{i,j\ldots p}$ are changed in a contragredient manner, as in (10.2.10) but with N $\bar{\mathbf{U}}$-factors.

It is a simple matter to include spin, if we wish, by noting that the transformations $(\alpha\ \beta)\rightarrow(\bar{\alpha}\ \bar{\beta})$ are described by matrices of the special‡ unitary group SU(2) and that transformations of the *spin*-orbital set $\phi_1\alpha, \phi_1\beta, \phi_2\alpha, \ldots, \phi_m\beta$ will then form a group U(2m) that is a Krönecker product U(m) × SU(2). We have seen, however, that the orbital and spin factors in a wavefunction can always be separated provided we insist that the spatial functions have the appropriate *symmetry* (under electron permutations), so that they may be combined with spin eigenfunctions (as in Section 4.3) to yield space–spin wavefunctions that satisfy the Pauli principle. In the present chapter we shall not need the spin factors: we consider the space spanned by the orbital products (10.2.13), and as in Section 7.6 seek linear combinations, from all possible configurations, that possess appropriate symmetry under electron permutations.

A simple example clarifies the procedure. If we wish to construct a 2-electron singlet using the m^2 products Ω_{ij}, we shall need only symmetric combinations (matching the spin *anti*symmetry)

$$\left.\begin{aligned}\Omega_{ii}^{(\mathrm{S})} &= \phi_i(\boldsymbol{r}_1)\phi_1(\boldsymbol{r}_2),\\ \Omega_{ij}^{(\mathrm{S})} &= \phi_i(\boldsymbol{r}_1)\phi_j(\boldsymbol{r}_2)+\phi_j(\boldsymbol{r}_1)\phi_i(\boldsymbol{r}_2)\quad (i\neq j).\end{aligned}\right\} \tag{10.2.15}$$

‡ The *special* unitary group contains only unimodular matrices, with det $\mathbf{U} = 1$.

For the triplet states, however, we should need the antisymmetric spatial functions

$$\Omega_{ij}^{(A)} = \phi_i(\boldsymbol{r}_1)\phi_j(\boldsymbol{r}_2) - \phi_j(\boldsymbol{r}_1)\phi_i(\boldsymbol{r}_2) \quad (i<j). \tag{10.2.16}$$

The sets in (10.2.15) and (10.2.16) are respectively components of rank-2 symmetric and antisymmetric tensors: there are $\frac{1}{2}m(m+1)$ components of the first type and $\frac{1}{2}m(m-1)$ of the second type, and the m^2-dimensional tensor space has in this way been reduced into symmetric and antisymmetric subspaces.

The general problem is now clear: the quantities $\Omega_{ij\ldots p}$ are tensor components, with respect to the group U(m), and we want to find linear combinations of these components that will display particular symmetries under electron permutations and hence under *index permutations*. Each set of symmetrized products, with a particular index symmetry, will provide a basis for constructing spin-free CFs (as in Section 7.6) for states of given spin multiplicity; and in this way the full-CI secular equations will be reduced into the desired block form, each block corresponding to an irreducible representation of U(m). It is therefore necessary to study both groups: U(m), which describes possible orbital transformations, and S_N, which provides a route (via the Young tableaux of Chapter 4) to the construction of rank-N tensors of particular symmetry type with respect to index permutations.

10.3 IRREDUCIBLE BASES IN CI SPACE

The construction of spin-free CFs that transform, under electron permutations, like the basis vectors of an irrep of S_N has been considered already in Section 7.6, and it remains only to fill in the details. Such functions are generated by "projection", using Wigner operators, and, because any permutation of variables in an orbital product is equivalent to the inverse permutation of *orbital indices*, it is possible to classify the CFs by means of *Weyl tableaux*, which contain indices (selected from $1, 2, \ldots, m$) instead of electron labels $(1, 2, \ldots, N)$. The functions to be used may thus be constructed, from the product $\Omega_K = k_1 k_2 \ldots k_N$, as follows:

$$F_\kappa^{(K,\kappa')} = \rho_{\kappa\kappa'}\Omega_K, \tag{10.3.1}$$

in which the subscript κ indicates a particular symmetry species ‡ (κth

‡ The label S, indicating the particular irrep, will be suppressed for clarity, being assumed fixed in all that follows.

basis vector of the irrep dual to D_S) while the double index (K, κ') labels the various CFs that can be projected (variable κ') from a given Ω_K.

When the orbital products are regarded as tensor components, with respect to the unitary group, the importance of the symmetry-adapted combinations (10.3.1) is that they retain their symmetry also under the unitary transformations, and therefore provide independent bases, classified according to spin eigenvalues, for the group U(m). These bases are in fact irreducible and therefore carry irreps of U(m). In any CI calculation (with a spinless Hamiltonian) there will be no mixing of CFs from different bases, and the CI secular equations will be reduced into block form with blocks of the lowest possible dimensions. There is an important extra bonus, considered in Section 10.4: the matrix elements of H between all CFs of each (permutation) symmetry species can be obtained from *group theory*, via the "generating elements" of the group U(m).

The transformation properties of the CFs defined above may be visualized schematically. For every species (κ) in (10.3.1) we shall have a set of CFs that may be written out in a row (K, κ' variable), and these rows, one for each value of κ, may be assembled into a rectangular array, thus:

$$\begin{array}{cccccccc} * & * & * & * & * & * & * & \cdots \\ * & * & * & * & * & * & * & \cdots \\ \vdots & \vdots & \vdots & \vdots & \vdots & \vdots & \vdots & \\ * & * & * & * & * & * & * & \cdots \end{array} \qquad (10.3.2)$$

The number of rows f_S^N is given by (4.2.2), while the number of columns $D(m, N, S)$ is given by (7.6.18). Under a unitary transformation (induced by an orbital basis change) the functions in each row are mixed among themselves and carry an irrep of U(m); while under a permutation of electrons (or orbital indices) the functions in each *column* are mixed among themselves and carry an irrep D_S of S_N. The number of functions in the array may be enormous; but the classification is simple.

The fact that, under the operations of the two groups, the functions are only mixed within the same row or column in (10.3.2) is a result of the commutation of permutation- and unitary-group elements (Problem 10.6). The behaviour of a row of functions, under unitary transformations, is thus unchanged if the functions are subjected to a common permutation. But the Wigner operators that we have used are simply linear combinations of permutations, and since $\rho_{\kappa\kappa'}$ leads from a function of species κ' to its partner of species κ, it follows that we may pass from the functions in row κ' of the array to those in row κ *without affecting* the way the functions behave under the unitary transformations. In other words, every row of functions will carry *the same irrep of* U(m), while every column (by construction) carries *the same irrep of* S_N.

The recipe for choosing a complete linearly independent set of CFs of the type (10.3.1) has been stated already (Section 7.6) and needs only slight amplification. For this purpose, a concrete example is helpful. We choose a system with 3 electrons and 4 orbitals ($N=3$, $m=4$), and consider the construction of doublet states ($S=\frac{1}{2}$). There are two Young tableaux (Section 4.4)

$$\mathrm{T}_1 = \begin{array}{|c|c|}\hline 1 & 2 \\ \hline 3 \\ \cline{1-1}\end{array} \qquad \mathrm{T}_2 = \begin{array}{|c|c|}\hline 1 & 3 \\ \hline 2 \\ \cline{1-1}\end{array}$$

and 4^3 possible products $\Omega_K = k_1(1)k_2(2)k_3(3)$, arising from the possible choices of k_1, k_2, k_3 (not all of them, of course, giving distinct non-zero results). Every product is characterized by its *index set* $K=(k_1, k_2, k_3)$; and with every index set we can associate a number of *index tableaux* (or *Weyl tableaux*) $\mathrm{W}_p^{(K)}$ by putting the ith index in the box containing electron index i, in Young tableau T_p, for all tableaux in turn, Thus, formally,

$$\mathrm{W}_p^{(K)} = (K, \mathrm{T}_p) \tag{10.3.3}$$

is the Weyl tableau associated with index set K and Young tableau T_p. In the present example,

$$\mathrm{W}_1^{(K)} = \begin{array}{|c|c|}\hline k_1 & k_2 \\ \hline k_3 \\ \cline{1-1}\end{array} \qquad \mathrm{W}_2^{(K)} = \begin{array}{|c|c|}\hline k_1 & k_3 \\ \hline k_2 \\ \cline{1-1}\end{array}$$

The importance of the Weyl tableaux is that they provide a unique labelling system for the *distinct* functions of a linearly independent set, which would be obtained (with many repetitions and zeros) by applying $\rho_{\kappa\kappa'}$ indiscriminately to all possible orbital products Ω_K. In the present example there would be 256 such products, but, of the resultant functions of given species κ, only 20 are linearly independent: these may be labelled by *standard* tableaux, in which the orbital indices appear in *increasing order* reading down the columns and *non-decreasing order* along the rows. A standard set of CFs is conveniently written (suppressing the label S, understood to be fixed)

$$F(p; \mathrm{W}_q^{(K)}) = \rho_{pq}\Omega_K \tag{10.3.4}$$

where K now includes *only ordered sets* of orbital indices ($k_1 \leqslant k_2 \leqslant k_3 \leqslant \ldots \leqslant k_N$) and therefore becomes a configuration symbol in the usual sense, while the tableaux $\mathrm{W}_q^{(K)}$ *must be standard*.

In the example that we are using there are 16 ordered index sets:

(112), (122), (113), (123), (223), (133), (233), (114),
(124), (224), (134), (234), (334), (144), (244), (344).

Let us take $p = q = 1$, so as to project from each Ω_K a function belonging to the first row of (10.3.2), i.e. of the first species ($p = 1$). The Weyl tableaux $W_1^{(K)} = (K, T_1)$ are automatically standard when K is an ordered set: they are‡

$$\begin{array}{|c|c|}\hline 1&1\\\hline 2\\\cline{1-1}\end{array}\quad \begin{array}{|c|c|}\hline 1&2\\\hline 2\\\cline{1-1}\end{array}\quad \begin{array}{|c|c|}\hline 1&1\\\hline 3\\\cline{1-1}\end{array}\quad \begin{array}{|c|c|}\hline 1&2\\\hline 3\\\cline{1-1}\end{array}\quad \begin{array}{|c|c|}\hline 2&2\\\hline 3\\\cline{1-1}\end{array}\quad \begin{array}{|c|c|}\hline 1&3\\\hline 3\\\cline{1-1}\end{array}\quad \begin{array}{|c|c|}\hline 2&3\\\hline 3\\\cline{1-1}\end{array}\quad \begin{array}{|c|c|}\hline 1&1\\\hline 4\\\cline{1-1}\end{array}$$

$$\begin{array}{|c|c|}\hline 1&2\\\hline 4\\\cline{1-1}\end{array}\quad \begin{array}{|c|c|}\hline 2&2\\\hline 4\\\cline{1-1}\end{array}\quad \begin{array}{|c|c|}\hline 1&3\\\hline 4\\\cline{1-1}\end{array}\quad \begin{array}{|c|c|}\hline 2&3\\\hline 4\\\cline{1-1}\end{array}\quad \begin{array}{|c|c|}\hline 3&3\\\hline 4\\\cline{1-1}\end{array}\quad \begin{array}{|c|c|}\hline 1&4\\\hline 4\\\cline{1-1}\end{array}\quad \begin{array}{|c|c|}\hline 2&4\\\hline 4\\\cline{1-1}\end{array}\quad \begin{array}{|c|c|}\hline 3&4\\\hline 4\\\cline{1-1}\end{array}$$

but, according to (7.6.18), $D(4, 3, \frac{1}{2}) = 20$—so 4 functions are missing. We can only get new tableaux by turning to T_2 and setting up $W_2^{(K)} = (K, T_2)$. Of the resultant tableaux, 6 are repetitions of those already found, 6 are non-standard and must be rejected, and the remaining 4 are

$$\begin{array}{|c|c|}\hline 1&3\\\hline 2\\\cline{1-1}\end{array}\quad \begin{array}{|c|c|}\hline 1&4\\\hline 2\\\cline{1-1}\end{array}\quad \begin{array}{|c|c|}\hline 1&4\\\hline 3\\\cline{1-1}\end{array}\quad \begin{array}{|c|c|}\hline 2&4\\\hline 3\\\cline{1-1}\end{array}$$

The functions of species $p = 1$ obtained from the first 16 Ω_K according to (10.3.4) are

$$F(1; W_1^{(K)}) = \rho_{11}\Omega_K, \tag{10.3.5}$$

while the remaining 4 functions are

$$F(1; W_2^{(k)}) = \rho_{12}\Omega_K. \tag{10.3.6}$$

At first sight, the present procedure seems to differ from that used by Gallup (p. 241), where only a single projector ρ_{11} was employed. In fact, however, the 4 functions in (10.3.6) can all be written in the form (10.3.5) by making an index permutation on Ω_K. Thus for $K = (1, 2, 3)$ the index permutation P_{23} gives a new set $K' = P_{23}K = (1, 3, 2)$: the associated Weyl tableau constructed using T_1 is $(K', T_1) = W_1^{(K')} =$ $\begin{array}{|c|c|}\hline 1&3\\\hline 2\\\cline{1-1}\end{array}$ $= W_2^{(K)}$; but there is a *unique* projected function of given species for each standard tableau, and therefore

$$F(1; W_2^{(K)}) = F(1; W_1^{(K')}) = \rho_{11}\Omega_{K'}. \tag{10.3.7}$$

It is therefore only necessary to permute the indices in Ω_K (using the

‡ Note that the Weyl tableaux based on T_1 display the factors of $\Omega_K = k_1k_2k_3\ldots$ in "book" order (i.e. reading left to right, row after row).

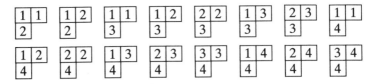

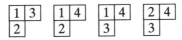

permutation that leads from one Young tableau to another), in order that all the independent CFs in any row of (10.3.2) may be obtained by using a *single* projector such as ρ_{11}. This was the procedure adopted in Section 7.6: the full set follows from (10.3.4) with $p = q = 1$ provided that K includes not only the ordered sequence $(k_1 \leqslant k_2 \leqslant k_3 \ldots)$ but also any new sequences obtained by a permissible permutation (i.e. one that leads to another *standard* Weyl tableau).

As in the case of the symmetric group, it is not usually necessary to actually construct the symmetry adapted functions (10.3.4) that are to be used as CFs in a CI calculation. Again, we use the basic theory only as a means of (i) classifying the vast numbers of CFs in a simple way, and (ii) finding the matrices in any desired irrep associated with the *generators* of the group (i.e. the elements of U(m) that play the part of the $(n-1, n)$ transpositions in the group S_N). There is indeed a very close parallel between the representation theory of U(m) and that of S_N. First we consider the possibility of classifying the basis vectors for the irreps of U(m) according to a subduction chain (cf. Section 4.5); then in the next section we turn to the generators themselves and their relationship to the CI method.

In Section 4.5 we observed that any basis vector of an irrep of S_N could be uniquely labelled by indicating the chain of irreps to which it belongs on specializing successively to the subgroups $S_{N-1}, S_{N-2}, \ldots, S_1$. The same process of subduction can be applied to U(m): and a basis vector of an irrep of U(m) can be specified by the chain of irreps to which it belongs on specializing to the subgroups U($m-1$), U($m-2$), ..., U(1), which involve the transformations of progressively fewer and fewer orbitals—only the first $m-1$, then the first $m-2$, and so on. For S_N the irreps were labelled by Young shapes and their basis vectors by the corresponding standard Young tableaux: for U(m) they are again labelled by Young shapes (which indicate the index symmetry of the tensor bases), while their basis vectors refer to standard *Weyl* tableaux.

The functions already defined, for the case $N = 3$, $m = 4$ and spin $S = \frac{1}{2}$, serve to illustrate the procedure. First we set out the Weyl tableaux in a standard order (cf. the last-letter sequence for S_N, p. 100), whose significance will be clear in a moment:

$$
\begin{array}{cc}1&1\\2&\end{array}\quad
\begin{array}{cc}1&2\\2&\end{array}\quad
\begin{array}{cc}1&1\\3&\end{array}\quad
\begin{array}{cc}1&2\\3&\end{array}\quad
\begin{array}{cc}2&2\\3&\end{array}\quad
\begin{array}{cc}1&3\\2&\end{array}\quad
\begin{array}{cc}1&3\\3&\end{array}\quad
\begin{array}{cc}2&3\\3&\end{array}\quad
\begin{array}{cc}1&1\\4&\end{array}\quad
\begin{array}{cc}1&2\\4&\end{array}
$$

$$
\begin{array}{cc}2&2\\4&\end{array}\quad
\begin{array}{cc}1&3\\4&\end{array}\quad
\begin{array}{cc}2&3\\4&\end{array}\quad
\begin{array}{cc}3&3\\4&\end{array}\quad
\begin{array}{cc}1&4\\2&\end{array}\quad
\begin{array}{cc}1&4\\3&\end{array}\quad
\begin{array}{cc}2&4\\3&\end{array}\quad
\begin{array}{cc}1&4\\4&\end{array}\quad
\begin{array}{cc}2&4\\4&\end{array}\quad
\begin{array}{cc}3&4\\4&\end{array}
\tag{10.3.8}
$$

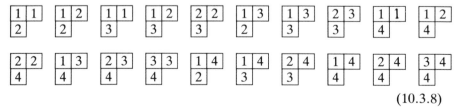

The functions associated with these tableaux according to (10.3.4) provide a 20-dimensional irrep of U(4), the group of transformations that mix the orbitals $\phi_1, \ldots, \phi_4$. To pass to the subgroup U(3) of transformations that mix only the first three orbitals ϕ_1, ϕ_2, ϕ_3, we simply "scratch out" the cells containing orbital index 4. The 20×20 irrep of U(4) then reduces to a block form in which the blocks provide irreps of U(3), whose basis vectors are labelled by indicating the "truncated" tableaux that remain. There is an 8×8 block, with index symmetry described by the Young shape $\begin{array}{|c|c|}\hline \ & \ \\ \hline \ \\ \cline{1-1}\end{array}$, whose basis vectors correspond to the truncated tableaux (still standard, and in standard order)

$$\begin{array}{|c|c|}\hline 1&1\\ \hline 2\\ \cline{1-1}\end{array}\quad \begin{array}{|c|c|}\hline 1&2\\ \hline 2\\ \cline{1-1}\end{array}\quad \begin{array}{|c|c|}\hline 1&1\\ \hline 3\\ \cline{1-1}\end{array}\quad \begin{array}{|c|c|}\hline 1&2\\ \hline 3\\ \cline{1-1}\end{array}\quad \begin{array}{|c|c|}\hline 2&2\\ \hline 3\\ \cline{1-1}\end{array}\quad \begin{array}{|c|c|}\hline 1&3\\ \hline 2\\ \cline{1-1}\end{array}\quad \begin{array}{|c|c|}\hline 1&3\\ \hline 3\\ \cline{1-1}\end{array}\quad \begin{array}{|c|c|}\hline 2&3\\ \hline 3\\ \cline{1-1}\end{array}$$

There is a 6×6 block, for shape $\begin{array}{|c|c|}\hline \ & \ \\ \hline\end{array}$, whose basis functions correspond to tableaux

$$\begin{array}{|c|c|}\hline 1&1\\ \hline\end{array}\quad \begin{array}{|c|c|}\hline 1&2\\ \hline\end{array}\quad \begin{array}{|c|c|}\hline 2&2\\ \hline\end{array}\quad \begin{array}{|c|c|}\hline 1&3\\ \hline\end{array}\quad \begin{array}{|c|c|}\hline 2&3\\ \hline\end{array}\quad \begin{array}{|c|c|}\hline 3&3\\ \hline\end{array}$$

then a 3×3 block, for shape $\begin{array}{|c|}\hline \ \\ \hline \ \\ \hline\end{array}$, with tableaux

$$\begin{array}{|c|}\hline 1\\ \hline 2\\ \hline\end{array}\quad \begin{array}{|c|}\hline 1\\ \hline 3\\ \hline\end{array}\quad \begin{array}{|c|}\hline 2\\ \hline 3\\ \hline\end{array}$$

and finally a 3×3 block, for shape $\begin{array}{|c|}\hline \ \\ \hline\end{array}$, with tableaux

$$\begin{array}{|c|}\hline 1\\ \hline\end{array}\quad \begin{array}{|c|}\hline 2\\ \hline\end{array}\quad \begin{array}{|c|}\hline 3\\ \hline\end{array}$$

Thus, on confining attention to the U(3) subgroup of transformations on ϕ_1, ϕ_2, ϕ_3, the irrep of U(4) is reduced to a direct sum of irreps of U(3); and each of the original functions may be classified according to which irrep of U(3) it falls into.

The process may evidently be continued until finally every basis function of the U(4) irrep is uniquely specified by giving a subgroup chain. Thus the first and the last functions in (10.3.8) have the respective genealogies

$$\begin{array}{llll}
\mathrm{U}(4) \rightarrow & \mathrm{U}(3) \rightarrow & \mathrm{U}(2) \rightarrow & \mathrm{U}(1) \\[4pt]
\begin{array}{|c|c|}\hline 1&1\\ \hline 2\\ \cline{1-1}\end{array} \rightarrow & \begin{array}{|c|c|}\hline 1&1\\ \hline 2\\ \cline{1-1}\end{array} \rightarrow & \begin{array}{|c|c|}\hline 1&1\\ \hline\end{array} \rightarrow & \begin{array}{|c|c|}\hline 1&1\\ \hline\end{array} \\[8pt]
\begin{array}{|c|c|}\hline 3&4\\ \hline 4\\ \cline{1-1}\end{array} \rightarrow & \begin{array}{|c|}\hline 3\\ \hline\end{array} & \quad - & \quad -
\end{array}$$

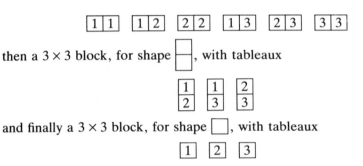

where it is sufficient to indicate the *shape* (i.e. the irrep) at each stage. This scheme for uniquely indexing (for example by a chain of integers) the vast number of CFs that arise from larger values of N and m has acquired great importance in large-scale CI calculations; but it is usually approached in a rather different manner, not through the symmetry adaptation of tensor spaces but rather from the theory of Lie groups—which is mathematically more demanding (see e.g. Louck, 1970). Here, therefore, we shall be content to indicate the alternative notation, which derives from the representation theory due to Gel'fand and Tsetelin (apparently first used in physics by Moshinsky—see Moshinsky, 1963, 1966, 1968).

In the approach due to Gel'fand and Tsetelin, the symmetry-adapted CFs of the standard irreps of U(m) are identical with those based on (10.3.4), but, instead of using a Young shape to indicate a particular irrep, a *Gel'fand tableau* is employed. Such a tableau is a triangular array of integers (there is one array for each basis function) of the form

$$\begin{array}{ccccccc}
\lambda_1 & \lambda_2 & \lambda_3 & \cdots\cdots\cdots\cdots & & & \lambda_m \\
& \lambda'_1 & \lambda'_2 & \lambda'_3 & \cdots\cdots\cdots & \lambda'_{m-1} & \\
& & \lambda''_1 & \lambda''_2 \; \lambda''_3 & \cdots\cdots & \lambda''_{m-2} & \\
& & & \cdots\cdots\cdots\cdots & & & \\
& & & \cdots\cdots\cdots & & &
\end{array}$$

in which each line of integers is called a "weight vector". The highest weight vector (first line), for any given CF, indicates the Young shape (λ_1 boxes in the first row, λ_2 in the second, etc., and $\lambda_n = 0$ if there is no nth row) for the irrep of U(m) to which it belongs; the next weight vector (second line) indicates the irrep of U($m-1$) to which it belongs for transformations that mix only the first $m-1$ orbitals; and so on. The Gel'fand tableau is thus simply another representation of the subduction chain described above in terms of Young shapes; going from one line to the next corresponds to scratching out the highest remaining integer in a set of Young tableaux.

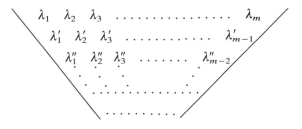

Again a simple example clarifies the procedure. The two basis functions corresponding to tableaux $\begin{array}{|c|c|}\hline 1 & 1 \\ \hline 2 \\ \cline{1-1}\end{array}$ and $\begin{array}{|c|c|}\hline 3 & 4 \\ \hline 4 \\ \cline{1-1}\end{array}$ in (10.3.8) will both have a highest weight vector 2 1 0 0, since they both belong to the irrep of U(4) with shape $\begin{array}{|c|c|}\hline \ & \ \\ \hline \ \\ \cline{1-1}\end{array}$ ($\lambda_1 = 2$, $\lambda_2 = 1$); but the first function will yield 2 1 0 (same Young shape for $m - 1 = 3$) as the *next* line of its Gel'fand tableau; while the second will yield instead 1 0 0 (shape $\square$ for $m-1=3$). The association of Gel'fand tableaux with the two standard Weyl tableaux is thus found to be

$$\begin{array}{|c|c|}\hline 1 & 1 \\ \hline 2 & \multicolumn{1}{c}{} \\ \cline{1-1}\end{array} \rightarrow \begin{array}{ccccccc} 2 & & 1 & & 0 & & 0 \\ & 2 & & 1 & & 0 & \\ & & 2 & & 1 & & \\ & & & 2 & & & \end{array}$$

$$\begin{array}{|c|c|}\hline 3 & 4 \\ \hline 4 & \multicolumn{1}{c}{} \\ \cline{1-1}\end{array} \rightarrow \begin{array}{ccccccc} 2 & & 1 & & 0 & & 0 \\ & 1 & & 0 & & 0 & \\ & & 0 & & 0 & & \\ & & & 0 & & & \end{array}$$

A less trivial example would be a 5-electron doublet function, constructed using 8 orbitals: a particular Weyl tableau (one out of 1008) and its Gel'fand equivalent might be

$$\begin{array}{|c|c|}\hline 1 & 3 \\ \hline 2 & 6 \\ \hline 8 & \multicolumn{1}{c}{} \\ \cline{1-1}\end{array} \rightarrow \begin{array}{ccccccccccccccc} 2 & & 2 & & 1 & & 0 & & 0 & & 0 & & 0 & & 0 \\ & 2 & & 2 & & 0 & & 0 & & 0 & & 0 & & 0 & \\ & & 2 & & 2 & & 0 & & 0 & & 0 & & 0 & & \\ & & & 2 & & 1 & & 0 & & 0 & & 0 & & & \\ & & & & 2 & & 1 & & 0 & & 0 & & & & \\ & & & & & 2 & & 1 & & 0 & & & & & \\ & & & & & & 1 & & 1 & & & & & & \\ & & & & & & & 1 & & & & & & & \end{array}$$

Integer arrays of this kind are particularly well suited for storing in a computer, and each array provides all the information required for constructing (if need be) a corresponding CF. The standard ordering of the CFs is also determined by a very simple arithmetic criterion. The weight vectors $\lambda_1, \lambda_2, \ldots$ (successive rows of any given Gel'fand tableau G, with corresponding Weyl tableau W) are read one after another as a single row with elements λ_i: G then comes before G′ (and hence W before W′) when the first difference $\lambda_i - \lambda_i'$ is positive. Thus the first of the 3-electron tableaux in (10.3.8) comes before the second because their λ-vectors

$$\lambda = (2\ 1\ 0\ 0\ 2\ 1\ 0\ 2\ 1\ 2),$$
$$\lambda' = (2\ 1\ 0\ 0\ 2\ 1\ 0\ 2\ 1\ 1)$$

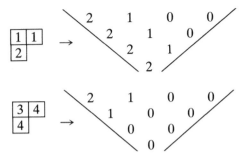

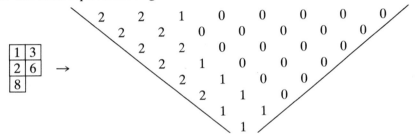

have a *positive* first difference (10th place). When this criterion is satisfied for every pair the tableaux are said to be in *lexical order*; and this lexical order is preserved at every stage when the "last index" is removed in the subduction process.

The standard tableaux of both types (Gel'fand and Weyl) are in fact of great generality, characterizing all possible irreps of U(m); but for electronic systems we can use only irreps with Weyl tableaux of 2-column form, and hence Gel'fand tableaux containing only 2s, 1s, and 0s, as we know from Section 4.3. In this case the Gel'fand tableaux contain mostly zeros and may be represented in a much simpler form due to Paldus (1975). A *Paldus tableau* contains only three columns, in which the integers (a_p, b_p, c_p) show the number of 2s, 1s and zeros in the pth row *from the bottom* of the corresponding Gel'fand tableau. The association between the Weyl tableau and the Paldus tableau for the 5-electron example given above would thus be

$$\begin{array}{|c|c|}\hline 1 & 3 \\ \hline 2 & 6 \\ \hline 8 & \multicolumn{1}{c}{} \\ \cline{1-1}\end{array} \quad \rightarrow \quad \left|\begin{array}{ccc} 2 & 1 & 5 \\ 2 & 0 & 5 \\ 2 & 0 & 4 \\ 1 & 1 & 3 \\ 1 & 1 & 2 \\ 1 & 1 & 1 \\ 0 & 2 & 0 \\ 0 & 1 & 0 \end{array}\right|$$

Clearly even the Paldus tableaux contains a redundant column since

$$a_p + b_p + c_p = p,$$

where p is the number of orbitals remaining (i.e. not scratched out in the subduction) at level p. For some purposes, however, it is useful to list all three integers.

The symmetry-adapted CFs introduced in this section, together with their classification according to the U(m) subduction chain, provide the foundations for the unitary-group approach (UGA) to CI calculations. A useful review is contained in a collection of papers edited by Hinze (1981), while the basic theory is fully presented (with many references) by Matsen and Pauncz (1986).

10.4 GENERATORS OF THE UNITARY GROUP. THE HAMILTONIAN

It is now time to consider the connection between the unitary group and the matrix elements of the Hamiltonian between CFs of type (10.3.4).

This connection rests upon the fact that the Hamiltonian may be expressed in terms of the *generators* of the transformations that comprise the group U(m); and the matrices of these generators, in an irrep spanned by the symmetry-adapted functions, can be determined using a simple algorithm closely analogous to Young's algorithm (p. 108).

Unitary-group generators have in fact already been encountered in Section 8.2, where they were represented as operators in Fock space describing the mappings of a spin-orbital basis, and, although we shall use a spin-free formalism, it is also useful to see how they occur in the second-quantization approach. In (8.2.18) *et seq.* it was established that the operator

$$\mathsf{U} = e^{\mathsf{R}}, \qquad \mathsf{R} = \sum_{i,j} \Delta_{ij} \mathsf{a}_i^\dagger \mathsf{a}_j \quad (\Delta_{ji} = -\Delta_{ij}^*) \tag{10.4.1}$$

produced a rotation of spin-orbitals in which $\boldsymbol{\psi} = (\psi_1, \psi_2, \ldots)$ is mapped into

$$\boldsymbol{\psi}' = \mathsf{U}\boldsymbol{\psi} = \boldsymbol{\psi}\mathbf{U}, \qquad \mathbf{U} = e^{\boldsymbol{\Delta}} = \mathbf{1} + \boldsymbol{\Delta} + \boldsymbol{\Delta}^2 + \ldots \quad . \tag{10.4.2}$$

The operator R in (10.4.1) and its related matrix $\boldsymbol{\Delta}$ in (10.4.2) thus define an *infinitesimal* rotation; and this operator itself is a linear combination, with coefficients Δ_{ij}, of the *generators*

$$\mathsf{E}_{ij} = \mathsf{a}_i^\dagger \mathsf{a}_j. \tag{10.4.3}$$

As noted in Section 8.2, E_{ij} is a "replacement" or "substitution" operator because, acting on any ket, it replaces an electron in ψ_j by one in ψ_i, where possible, otherwise acting as a destruction operator. The matrix representing E_{ij} in the spin-orbital basis has a single non-zero element, corresponding to $\Delta_{ij} = 1$ in (10.4.1). Thus

$$\mathsf{E}_{ij} \rightarrow \mathbf{E}_{ij}, \qquad (\mathbf{E}_{ij})_{kl} = \delta_{ik}\delta_{jl}, \tag{10.4.4}$$

and the matrix $\boldsymbol{\Delta}$, for any infinitesimal rotation, is obviously a linear combination, with coefficients Δ_{ij}, of these 1-element matrices. An arbitrary $\boldsymbol{\Delta}$ will contain m^2 (complex) parameters and thus $2m^2$ real numbers; but the requirement $\Delta_{ji} = -\Delta_{ij}^*$ imposes m^2 conditions. The general unitary matrix, obtained by exponentiation of $\boldsymbol{\Delta}$, is thus determined by m^2 real continuous variables: consequently U(m) is an "m^2-parameter continuous group".

The Hamiltonian, still in second-quantization form, is

$$\mathsf{H} = \sum_{r,s} \langle r|\, \mathsf{h}\, |s\rangle \mathsf{a}_r^\dagger \mathsf{a}_s + \tfrac{1}{2} \sum_{r,s,t,u} \langle rs|\, g\, |tu\rangle \mathsf{a}_r^\dagger \mathsf{a}_s^\dagger \mathsf{a}_u \mathsf{a}_t, \tag{10.4.5}$$

and the 1-electron integral is thus the coefficient of a generator E_{rs}. The

2-electron term may be expressed in a similar way by using the anticommutation relations (3.6.7): thus

$$\mathsf{a}_r^\dagger \mathsf{a}_s^\dagger \mathsf{a}_u \mathsf{a}_t = -\mathsf{a}_r^\dagger \mathsf{a}_s^\dagger \mathsf{a}_t \mathsf{a}_u = -\mathsf{a}_r^\dagger(\delta_{st} - \mathsf{a}_t \mathsf{a}_s^\dagger)\mathsf{a}_u = \mathsf{E}_{rt}\mathsf{E}_{su} - \delta_{st}\mathsf{E}_{ru},$$

and (10.4.5) becomes

$$\boxed{\mathsf{H} = \sum_{r,s} \langle r|\,\mathsf{h}\,|s\rangle \mathsf{E}_{rs} + \tfrac{1}{2} \sum_{r,s,t,u} \langle rs|\,g\,|tu\rangle(\mathsf{E}_{rt}\mathsf{E}_{su} - \delta_{st}\mathsf{E}_{ru}).} \quad (10.4.6)$$

The generators have the properties that define a "Lie algebra":

$$\boxed{[\mathsf{E}_{rs}, \mathsf{E}_{tu}] = \mathsf{E}_{ru}\delta_{st} - \mathsf{E}_{ts}\delta_{ru}.} \quad (10.4.7)$$

Since we wish to adopt a spin-free approach, (10.4.6) cannot be used as it stands; but to pass to the spin-free formalism only a few details need be changed. First, we are using orbitals not spin-orbitals, and so, writing $\mathsf{a}_{r\alpha}^\dagger$ as the creation operator for spin-orbital $\psi_r = \phi_r\alpha$, we ought to be using the analogue of (10.4.6) *after spin integrations*. This is easily found to be

$$\mathsf{H} = \sum_{r,s} \langle r|\,\mathsf{h}\,|s\rangle \mathsf{E}'_{rs} + \tfrac{1}{2} \sum_{r,s,t,u} \langle rs|\,g\,|tu\rangle(\mathsf{E}'_{rt}\mathsf{E}'_{su} - \delta_{st}\mathsf{E}'_{ru}), \quad (10.4.8)$$

where the integrals are now defined over *orbitals*, while E'_{rs} is a "spin-traced" quantity

$$\mathsf{E}'_{rs} = \mathsf{a}_{r\alpha}^\dagger \mathsf{a}_{s\alpha} + \mathsf{a}_{r\beta}^\dagger \mathsf{a}_{s\beta}, \quad (10.4.9)$$

which has the effect of replacing (where possible) an orbital ϕ_s by ϕ_r, whichever spin factor it may have. The operators E'_{rs} now effect only the spatial factors in a wavefunction, and are again completely characterized by their commutation property (exactly like (10.4.7))

$$[\mathsf{E}'_{rs}, \mathsf{E}'_{tu}] = \mathsf{E}'_{ru}\delta_{st} - \mathsf{E}'_{ts}\delta_{ru}. \quad (10.4.10)$$

No further reference to spin is necessary; and all primes will be dropped.

The connection with the Schrödinger formalism is now immediate, since we know (Section 3.10) that a Fock-space operator such as $\mathsf{a}_r^\dagger \mathsf{a}_s$ will have a Schrödinger equivalent $\sum_{i=1}^N \mathsf{E}_{rs}(i)$, where $\mathsf{E}_{rs}(i)$ works on functions of $\boldsymbol{x}_i$ (which can be expanded in terms of spin-orbitals $\psi_r(\boldsymbol{x}_i)$) and turns any component $c_s\psi_s$ into $c_s\mathsf{E}_{rs}\psi_s = c_s\psi_r$. Formally, E_{rs} may be represented as the ket–bra product $|\psi_r\rangle\langle\psi_s|$ or as an integral operator (cf. p. 177) with kernel $\psi_r(\boldsymbol{x}_i)\psi_s^*(\boldsymbol{x}'_i)$. Similarly, for a spatial function expanded in terms of orbitals $\phi_r(\boldsymbol{r}_i)$ we may introduce the operator $\mathsf{E}_{rs} = |\phi_r\rangle\langle\phi_s|$, the analogue of (10.4.9), such that

$$\mathsf{E}_{rs}(i)f(\boldsymbol{r}_i) = \int \phi_r(\boldsymbol{r}_i)\phi_s^*(\boldsymbol{r}'_i)f(\boldsymbol{r}'_i)\,\mathrm{d}\boldsymbol{r}'_i = \phi_r(\boldsymbol{r}_i)\langle\phi_s\,|f\rangle. \quad (10.4.11)$$

By using these operators, it is easy to transcribe (10.4.8) into Schrödinger language; the result is not in the usual differential-operator form—it is the *projection* of the usual Hamiltonian onto the full-CI *subspace* of the N-electron Hilbert space, and is thus a "model Hamiltonian" (prescribed by the choice of basis) whose eigenvalues and eigenfunctions will be full-CI limits.

It is clear that there is nothing in the introduction of the generators that actually requires an appeal to second-quantization ideas. It need only be noted that, in terms of the integral operators defined by (10.4.11),

$$\mathsf{I}(i) = \sum_{r=1}^{m} \mathsf{E}_{rr}(i) \qquad (10.4.12)$$

is the projection operator onto the subspace of functions $\{\phi_r(\boldsymbol{r}_i)\}$ $(r = 1, \ldots, m)$ for any function of $\boldsymbol{r}_i$; as long as we work *within* that subspace, it is the *unit* operator. Thus the 1- and 2-electron terms in the usual Hamiltonian are equivalent, within the full-CI space, to operators that may be defined as follows (Matsen, 1976): first

$$\sum_{i=1}^{N} \mathsf{h}(i) = \sum_{i=1}^{N} \sum_{r,s} |\phi_r(i)\rangle\langle\phi_r(i)|\, \mathsf{h}(i)\, |\phi_s(i)\rangle\langle\phi_s(i)|$$

$$= \sum_{r,s} \langle\phi_r|\, \mathsf{h}\, |\phi_s\rangle \mathsf{E}_{rs}, \qquad (10.4.13)$$

where E_{rs} is the usual symmetric sum of the 1-electron operators defined in (10.4.11), namely

$$\mathsf{E}_{rs} = \sum_{i=1}^{N} \mathsf{E}_{rs}(i). \qquad (10.4.14)$$

Similarly, the 2-electron term may be put in the form

$$\tfrac{1}{2}\sum_{i,j=1}^{N}{}' g(i, j) = \tfrac{1}{2} \sum_{r,s,t,u} \langle\phi_r\phi_s|\, g\, |\phi_t\phi_u\rangle \sum_{i,j=1}^{N}{}' \mathsf{E}_{rt}(i)\mathsf{E}_{su}(j).$$

Since the term $j = i$ is excluded, the sum can be rewritten in terms of the operators (10.4.14) as

$$\mathsf{E}_{rt}\mathsf{E}_{su} - \sum_{i=1}^{N} \mathsf{E}_{rt}(i)\mathsf{E}_{su}(i).$$

The sum that remains reduces at once to $\delta_{ts}\mathsf{E}_{ru}$, and hence

$$\tfrac{1}{2}\sum_{i,j=1}^{N}{}' g(i, j) = \tfrac{1}{2} \sum_{r,s,t,u} \langle\phi_r\phi_s|\, g\, |\phi_t\phi_u\rangle(\mathsf{E}_{rt}\mathsf{E}_{su} - \delta_{ts}\mathsf{E}_{ru}). \qquad (10.4.15)$$

On putting together (10.4.13) and (10.4.15), we find the full-CI space Hamiltonian (10.4.8), with the E-operators in their Schrödinger form (10.4.14), (10.4.11).

To actually evaluate the matrix elements of H in the full-CI space, remembering that these will be non-zero only within the diagonal blocks that refer to each possible spin multiplicity, it will be sufficient to know the matrices of the generators—since those of the generator *products* in (10.4.8) may then be obtained by multiplication of the (very sparse) matrices of the factors. As indicated at the beginning of this section, there is a simple algorithm for this purpose: it may be formulated either in terms of the Weyl tableaux of the CFs in the bra and the ket, or in terms of their Gel'fand tableaux. The analogy with Young's algorithm is most clearly revealed by using the Weyl tableaux. The matrix elements of the *elementary* generators $\mathsf{E}_{i-1,i}$, which correspond to the elementary permutations $\mathsf{P}_{n-1,n}$ in Young's algorithm, are determined from the rules given in Table 10.1. The integer d is the usual axial distance (p. 107) between two boxes (shaded and unshaded), while the orbital indices (not shown) in all other boxes must match exactly and must differ from i and $i-1$. All other matrix elements of $\mathsf{E}_{i-1,i}$ are zero.

The matrices of generators of the type $\mathsf{E}_{i,i-1}$ may be obtained using the identity

$$\langle \mathrm{T}'|\, \mathsf{E}_{ij}\, |\mathrm{T}\rangle = \langle \mathrm{T}|\, \mathsf{E}_{ji}\, |\mathrm{T}'\rangle, \tag{10.4.16}$$

while matrices of the *non*-elementary generators follow from repeated application of the commutation rule (10.4.10) in the form

$$[\mathsf{E}_{rs}, \mathsf{E}_{su}] = \mathsf{E}_{ru}, \tag{10.4.17}$$

using $[\mathsf{E}_{i,i-1}, \mathsf{E}_{i-1,i-2}] = \mathsf{E}_{i,i-2}$etc. Matrix elements of the generator products in (10.4.8) can then be obtained by matrix multiplication, special techniques being available when the matrices have so few non-zero elements. At the same time, it must be stressed that in this approach the *full* CI space is required (at least for the "intermediate states" over which the summation runs). When the space is truncated it is usually preferable to seek alternative algorithms for dealing with the generator products; such algorithms can be found in the literature cited below.

The problems involved in the efficient computation of CI matrix elements can now be appreciated; they are largely organizational in character and many brilliant strategies have been devised for their solution. The procedures based on the symmetric-group approach, used here, rest largely on the work of Wormer, Harter and Patterson, Sarma, Rettrup, Duch and Karwowski, and their collaborators; the use of alternative algorithms based on the Gel'fand tableaux has been de-

Table 10.1 Non-zero matrix elements of unitary-group generators between tableau functions.‡

(i) Identical tableaux, T′ = T

$\langle \mathrm{T} | \mathsf{E}_{ii} | \mathrm{T} \rangle$ = (number of *i*s in T)

(ii) Second-column mismatch, $i-1$ in T′ with i in T

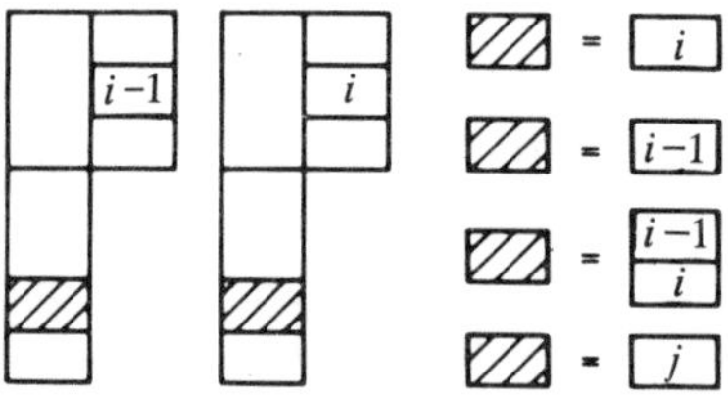

T′ T $(j \neq i, i-1)$

	$\langle \mathrm{T}' | \mathsf{E}_{i-1,i} | \mathrm{T} \rangle$
shaded = i	$[(d-1)/d]^{1/2}$
shaded = $i-1$	$[(d+1)/d]^{1/2}$
shaded = $i-1$ over i	1
shaded = j	1

(iii) First-column mismatch, $i-1$ in T′ with i in T

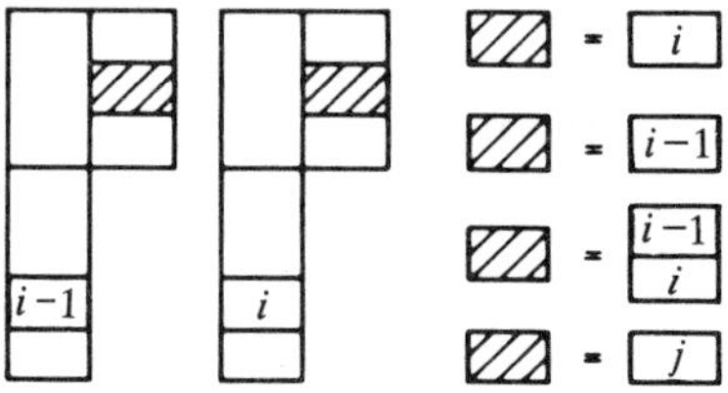

T′ T $(j \neq i, i-1)$

	$\langle \mathrm{T}' | \mathsf{E}_{i-1,i} | \mathrm{T} \rangle$
shaded = i	$[(d+1)/d]^{1/2}$
shaded = $i-1$	$[(d+1)/d]^{1/2}$
shaded = $i-1$ over i	1
shaded = j	1

(iv) Special case

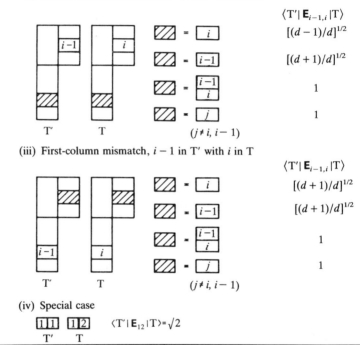

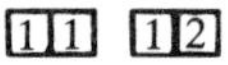

T′ T $\langle \mathrm{T}' | \mathsf{E}_{12} | \mathrm{T} \rangle = \sqrt{2}$

‡ Table contains matrix-element values for tableau functions (T′, T) with a single mismatching index ($i-1$, i); if either index is repeated the common shaded block must have the form shown. When d occurs it is the axial distance (shaded to unshaded) as in Table 4.1.

veloped mainly by Paldus, Shavitt, Siegbahn, Brooks *et al.*, and many others; a rather different approach, based on angular-momentum coupling techniques, was introduced by Drake and Schlessinger. Many of these developments depend heavily on graphical methods (pioneered by Shavitt), which lend themselves well to computer implementation, but whose discussion would take us too far afield. A good source of references to the work of all these authors is available in the conference proceedings edited by Hinze (1981). More recent reviews, with an emphasis on computing strategies, have been given by Robb and Niazi (1984) and by Duch and Karwowski (1985). A comprehensive account of the whole field is in preparation (Paldus and Wormer, 1989).

10.5 THE MATRIX-EIGENVALUE PROBLEM. DIRECT CI

At this point it may be assumed that the CFs (Φ_κ)‡ in a CI calculation can be efficiently indexed (for example by means of Paldus tableaux) and that any 1- or 2-electron integral in any matrix element $H_{\kappa\lambda}$ can be given its appropriate coefficient, namely a corresponding matrix element of a unitary-group generator or generator product. The only remaining problem is how to solve the very large matrix-eigenvalue equation

$$\mathbf{Hc} = E\mathbf{Mc}, \tag{10.5.1}$$

where $\mathbf{M}$ is most commonly the unit matrix but is included here for generality.

It is at once apparent that, with increasing size, methods of solving (10.5.1) are going to depend more and more on factors determined by computer architecture. Thus if all integrals were permanently on hand, in the fast memory, we could compute matrix elements one at a time and put them (in strings or blocks) into a backing store for future use. In fact, however, the integrals themselves are normally kept in a backing store, owing to their great number, and must be brought into the fast store "in buckets" (e.g. a 1000 at a time) for processing. Since the full integral list may need to be scanned in getting all contributions to a given matrix element, from an appropriate formula or algorithm, this type of "formula-driven" calculation rapidly becomes prohibitively time-consuming—most of the time being spent in fetching integrals from the backing store.

Roos (1972) was the first to propose a practical scheme for avoiding such difficulties, initially in a somewhat simpler context. He noted that, in CI with all single and double excitations from a closed-shell ground state, examination of any given integral $\langle \phi_r\phi_s | g | \phi_t\phi_u \rangle$ would serve, using appropriate rules, to determine (i) to which matrix element it made a contribution, and (ii) the coefficient with which it appeared. This observation opened up the possibility of an "integral-driven" approach in which, ideally, one reading of the integral list could be used to generate all matrix elements $H_{\kappa\lambda}$ (or at least a very large number of them) *simultaneously*. It also led to the increasing use of solution methods for the matrix-eigenvalue equation in which the whole matrix $\mathbf{H}$ (whose size leads to great storage problems) *is never generated.* Such methods are iterative and involve the repeated construction of single columns $\mathbf{b}' = \mathbf{Hb}$, where $\mathbf{b}$ is some kind of "trial" vector; and, instead of setting up the whole $\mathbf{H}$ matrix, all contributions to the sum $b'_\kappa = \sum_\lambda H_{\kappa\lambda} b_\lambda$, for all

‡ In this section it is sufficient to use a single-index notation for the CFs.

values of κ, are accumulated simultaneously during the reading of the integral list. The whole procedure for iterative improvement of an approximate eigenvector may thus be integral-driven from start to finish.

The integral-driven approach is sometimes referred to as "direct CI" to distinguish it from the more traditional approaches in which the whole of $\mathbf{H}$ is set up (as unnecessary intermediate step), by formula-driven methods, before beginning to calculate the CI coefficients. There are, however, many possibilities between these two extremes, as will be apparent on consulting the more specialist literature. Here we consider only the general nature of the iterative procedures.

Most of the methods currently favoured have certain features in common; they are well adapted for the calculation of a *small number* of eigenvalues (usually the lowest) and their eigenvectors; and they proceed by expanding the eigenvectors $\mathbf{c}_p$ in terms of a small number of trial vectors $\mathbf{b}_i$, chosen (and subsequently refined) to span approximately the subspace in which the desired eigenvectors lie. Thus at the nth iteration of the process the pth approximate eigenvector of (10.5.1) will be represented as a linear combination

$$\mathbf{c}_p^{(n)} = \sum_{i=1}^{k} \mathbf{b}_i^{(n)} a_{ip} = \mathbf{B}^{(n)} \mathbf{a}_p, \tag{10.5.2}$$

in which we use k columns $\mathbf{b}_i$, collected in the rectangular $(m \times k)$ matrix $\mathbf{B}^{(n)}$. The coefficients in the column $\mathbf{a}_p$ are then determined, in order to optimize, in some suitable sense, the approximation. This may be done by noting that (10.5.1) is equivalent to a stationary condition on the functional $\mathbf{c}^\dagger\mathbf{Hc}/\mathbf{c}^\dagger\mathbf{Mc}$ and writing $\mathbf{c}_p$, for the pth solution, in the form (10.5.2); the functional will then be stationary against variation of $\mathbf{a}_p$ when

$$\bar{\mathbf{H}}^{(n)}\mathbf{a}_p = E_p^{(n)}\bar{\mathbf{M}}^{(n)}, \tag{10.5.3}$$

where the matrices are now only k-dimensional, being defined as

$$\bar{\mathbf{H}}^{(n)} = \mathbf{B}^{(n)\dagger}\mathbf{H}\mathbf{B}^{(n)}, \qquad \bar{\mathbf{M}}^{(n)} = \mathbf{B}^{(n)\dagger}\mathbf{M}\mathbf{B}^{(n)}. \tag{10.5.4}$$

Equation (10.5.3) can be solved easily by conventional methods, while the matrices involved can be constructed without any need to hold the *whole* matrices $\mathbf{H}$ and $\mathbf{M}$ in the fast store. The results will be "optimized projections" of the eigenvectors (assumed to include $\mathbf{c}_p$) within the subspace spanned by the trial vectors. To proceed to the next iteration, it is necessary to improve the choice of trial vectors; and it is at this point that two main categories of method may be distinguished. We consider them in turn.

Relaxation methods

In this class of methods it is assumed that the full matrices **H** and **M** are available, at least in a backing store, so that one row at a time may be brought into use whenever it may be required. Consequently it is not well adapted to fully integral-driven procedures; nevertheless it is a powerful approach for matrices of not too large dimensions.

A two-term approximation is used in (10.5.2), the first column of **B** (i.e. $\mathbf{b}_1$) being the current approximation to the required eigenvector, while the second column is chosen to be a single element $(\mathbf{b}_2)_\kappa = \delta_{\kappa i}$ and therefore represents the ith basis vector in the full-CI space. The procedure is to perform a whole *cycle* of corrections to $\mathbf{b}_1$ by letting i run through the entire space; and then to use the fully updated $\mathbf{b}_1$ as the starting approximation for the next cycle. This is such a simple process that it can be performed very rapidly. The term "relaxation" is commonly used to describe such processes, in which one coordinate at a time (or a block of coordinates) is improved by relaxing constraints.

The subspace eigenvalue equation (10.5.3) now takes the form

$$\begin{pmatrix} E & a_i \\ a_i & H_{ii} \end{pmatrix}\begin{pmatrix} 1 \\ \alpha \end{pmatrix} = E'\begin{pmatrix} 1 & d_i \\ d_i & M_{ii} \end{pmatrix}\begin{pmatrix} 1 \\ \alpha \end{pmatrix}, \tag{10.5.5}$$

where E is the approximation to the desired eigenvalue associated with $\mathbf{b}_1$ (assumed to be normalized), while E' obtained by solving (10.5.5) will give a revised approximation. If we use **b** to denote the current approximation ($\mathbf{b}_1$) and $\mathbf{b}_i$ for the one-element column defined above then the eigenvalue approximation at the beginning of the cycle will be

$$E = \mathbf{b}^\dagger \mathbf{H}\mathbf{b}, \tag{10.5.6}$$

while a_i and d_i ($\mathbf{b}_i^\dagger\mathbf{H}\mathbf{b}$, $\mathbf{b}_i^\dagger\mathbf{M}\mathbf{b}$ respectively) will become

$$a_i = \sum_j H_{ij} b_j, \qquad d_i = \sum_j M_{ij} b_j. \tag{10.5.7}$$

It is convenient to put $E' = E + \Delta E$ and to solve (10.5.5) for ΔE, obtaining the lowest root in the form (Shavitt *et al.*, 1973; Davidson, 1983)

$$\Delta E = -\frac{2r_i^2}{s_i + (s_i^2 + t_i)^{1/2}}, \tag{10.5.8}$$

where r_i is a "residual",

$$r_j = \sum_j (H_{ij} - EM_{ij}) b_j = a_i - Ed_i, \tag{10.5.9}$$

while

$$s_i = H_{ii} - E - 2d_i r_i, \qquad t_i = 4(1 - d_i^2)r_i^2. \tag{10.5.10}$$

The correction to the approximate eigenvector **b** is then the single element

$$\Delta b_i = -\frac{r_i - \Delta E d_i}{H_{ii} - E}, \tag{10.5.11}$$

as follows on solving (10.5.5) for α.

It is a simple matter to update all elements in this way, to renormalize the revised vector **b**, and to recommence the cycle. Convergence is of course indicated by the vanishing of all residuals. When the residuals are sufficiently small the procedure reduces to a matrix perturbation method (cf. Section 2.5), in which, writing $a_i = H_{ib} = \mathbf{b}_i^\dagger \mathbf{H}\mathbf{b}$, etc. to make the connection more obvious,

$$\Delta E = -\sum_i \frac{(H_{ib} - E_b M_{ib})^2}{H_{ii} - E_b}, \qquad \Delta b_i = -\sum_i \frac{H_{ib} - E_b M_{ib}}{H_{ii} - E_b}. \tag{10.5.12}$$

This form of the method was first used in quantum chemistry by Nesbet (1965b) and is often referred to as the Nesbet method, although it was already known in applied mathematics (Cooper, 1948).

The method usually works well for the lowest root; and to obtain higher roots it is possible to proceed by a projection method. Thus if $\mathbf{c}_1$ is the lowest (normalized) eigenvector of **H** then the matrix

$$\mathbf{H}' = \mathbf{H} - E_1 \mathbf{M}\mathbf{c}_1\mathbf{c}_1^\dagger \mathbf{M} \tag{10.5.13}$$

will have a set of orthonormal eigenvectors, of which the first ($\mathbf{c}_1$) now gives $E_1 = \mathbf{c}_1^\dagger \mathbf{H}' \mathbf{c}_1 = 0$. By starting from an approximate eigenvector **b** that is orthogonal to $\mathbf{c}_1$ and using the relaxation process described above, the lowest eigenvector of $\mathbf{H}'$ ($\mathbf{c}_2$, say) will now be the next-lowest of **H**. Clearly, by repeating the process it will be possible to obtain successive eigenvalues, although with diminishing accuracy.

The disadvantage of relaxation methods is that they require access to the full matrices **H** and **M** in the construction, one at a time, of the quantities a_i, d_i, etc.; this is because the elements of the correction vector are calculated *sequentially*. In an integral-driven approach we should want to calculate all elements simultaneously, using random samples of integrals to calculate (via the elements of **H** and **M**) all the terms to which they contribute.

The Davidson method

In methods of the type introduced by Davidson (1975), the main step is the accurate solution of (10.5.3) within a subspace that may be somewhat larger than in relaxation methods (where it is usually 2-dimensional). The dimension of this subspace is in fact gradually increased, during the iteration, to some preset maximum (K, say), which may be 10–15 times larger than the number of eigenvectors actually required. If E_p and $\mathbf{c}_p$ are estimates of the pth eigenvalue and eigenvector, found by using the K-dimensional form of (10.5.3), the problem is to formulate an algorithm for generating revised estimates E'_p, $\mathbf{c}'_p$.

Davidson's basic procedure again involves the residuals used in the relaxation method; but now the *whole* vector is calculated. Thus, with $\mathbf{c}_p = \sum_i \mathbf{b}_i \alpha_{ip}$,

$$\mathbf{r} = (\mathbf{H} - E_p\mathbf{M})\mathbf{c}_p = \sum_i \alpha_{ip}(\mathbf{a}_i - E_p\mathbf{d}_i) = \sum_i \alpha_{ip}\mathbf{r}_i, \qquad (10.5.14)$$

where

$$\mathbf{a}_i = \mathbf{H}\mathbf{b}_i, \qquad \mathbf{d}_i = \mathbf{M}\mathbf{b}_i. \qquad (10.5.15)$$

The elements of $\mathbf{r}_i$ determine the gradient of the energy surface, for variation of the elements of $\mathbf{b}_i$, and the most effective way of improving $\mathbf{c}_p$ will therefore be to make a change of the form

$$\Delta\mathbf{c}_p = \sum_i \beta_i \bar{\mathbf{r}}_i,$$

where $\bar{\mathbf{r}}_i$ is the normalized gradient vector. The coefficients may be estimated by the partitioning method (p. 49), and the correction vector is then

$$\Delta\mathbf{c}_p = \sum_i \frac{\bar{\mathbf{r}}_i(H_{ip} - E_p M_{ip})}{E_p - H_{ii}}, \qquad (10.5.16)$$

where

$$H_{ii} = \bar{\mathbf{r}}_i^\dagger \mathbf{H} \bar{\mathbf{r}}_i, \qquad H_{ip} = \bar{\mathbf{r}}_i^\dagger \mathbf{H}\mathbf{c}_p, \qquad M_{ip} = \bar{\mathbf{r}}_i^\dagger \mathbf{M}\mathbf{c}_p. \qquad (10.5.17)$$

The correction vector (10.5.16) may now be used to extend the k-dimensional space; for this purpose it is usually normalized and then adopted as $\mathbf{b}_{k+1}$. This process of expansion of the basis is continued until the preset limit K is reached, and at this stage the cycle is complete. The final estimate of $\mathbf{c}_p$ is adopted as the starting vector $\mathbf{b}_1$ for a new cycle, and the process is continued until convergence is obtained.

The important advantage of the Davidson approach is that $\mathbf{H}$ only occurs in products such as $\mathbf{a}_i = \mathbf{H}\mathbf{b}_i$, where the elements of $\mathbf{b}_i$ are known; thus all contributions to the sum that determines any given element of $\mathbf{a}_i$ may be formed and accumulated *as they arise* (from the processing of integrals arriving in random order); all elements of $\mathbf{a}_i$ are generated simultaneously and the individual elements of $\mathbf{H}$ are never computed. Storage requirements are also modest, since no square matrices need be held in core—only, at most, $2K$ vectors.

We have described only the essential features of two important families of methods for obtaining the first few eigenvalues and eigenvectors of very large symmetric matrices. Many variants of these methods have evolved, however, in order to take advantage of special circumstances (for example the fact that the matrices may have a block structure that lends itself to a segmentation of the vectors and a reduction in storage requirements): details may be found elsewhere (see e.g. Liu, 1978; Raffinetti, 1979; Saxe *et al.*, 1982). The Davidson method has also been modified (Rettrup, 1982) for use with non-symmetric matrices, such as those that arose in Section 7.7. It is the development of such techniques, coupled with spectacular advances in computer technology, that have established large-scale CI as a realistic approach to the computation of highly accurate wavefunctions for small molecules.

Of course, even with the most efficient matrix-eigenvalue techniques, the tensor space discussed in Section 10.3 may be much too large for a full-CI calculation. The most promising method of proceeding in such cases is probably that of the *model space,* in which the complete calculation is performed only for a much smaller subspace built from a subset of the available orbitals, the presence of the complementary subspace being recognized by perturbation techniques akin to those used in Chapter 9. The subspace is referred to as the "model space" (defining a hypothetical system for which an "exact" calculation is possible), or (in the terminology of Roos *et al.* (1980) the "complete active space" (CAS); it comprises all products of the "active" orbitals—as distinct from the "virtual" orbitals (cf. the virtual orbitals in SCF theory), which form a complementary subset used only in the further refinement of the model-space reference function. The refinement of the reference function by perturbation methods then requires considerable development of the techniques used in Chapter 9 for the case of a *single-determinant* reference function. Such developments (indicated in Problems 10.14–10.17) are the object of much current research.

PROBLEMS 10

10.1 Consider the rotation of a coordinate frame, with real non-orthogonal basis vectors e_1, e_2, e_3, in which $\mathbf{e} \to \mathbf{eT}$ and the components of a fixed vector thus change according to $\mathbf{v} \to \mathbf{Rv}$ ($\mathbf{R} = \mathbf{T}^{-1}$). Classify the following sets of quantities according to their tensor character:

(i) the elements of the matrix $\mathbf{A}$ that relates two sets of vector components, $\mathbf{v}' = \mathbf{Av}$;
(ii) the 9 products of the components of two vectors, $\{u_i v_j\}$;
(iii) the elements of the metric matrix, $M_{ij} = \mathsf{e}_i \cdot \mathsf{e}_j$;

(iv) the scalar product of two vectors $\mathsf{u}\cdot\mathsf{v}=\tilde{\mathbf{u}}\mathbf{M}\mathbf{v}$;
(v) the quantities $u_i=\sum_j u_j M_{ji}$ (in tensor theory these are the "covariant components" of the vector whose "contravariant components" are u_i);
(vi) the components ∇_i of the gradient operator;
(vii) the ∇^2 operator (in its non-cartesian form).

10.2 Show that when an orthonormal set of functions $\{\phi_i\}$ undergoes a linear transformation that preserves orthonormality, the conjugate set $\{\phi_i^*\}$ transforms contragradiently. What is the tensor character of the set of matrix elements $\langle\chi_r|\,\mathsf{h}\,|\chi_s\rangle$ of an invariant operator h?

10.3 A set of three (cartesian) p-type AOs (χ_1, χ_2, χ_3) behaves, under the group of rotations in 3-dimensional space, like the set of orthonormal basis vectors $(\mathsf{e}_1, \mathsf{e}_2, \mathsf{e}_3)$. Identify the elements of the corresponding (real orthogonal) matrix $\mathbf{U}$ in (10.2.4), in terms of the direction cosines of the rotated vectors $(\bar{\mathsf{e}}_1, \bar{\mathsf{e}}_2, \bar{\mathsf{e}}_3)$ relative to the original set. If the function $\phi=\sum_i c_i\chi_i$ is to be basis-independent, how must the coefficients be changed?

10.4 What is the transformation law for the set of nine 2-electron product functions $\{\chi_i(\mathbf{r}_1)\chi_j(\mathbf{r}_2)\}$ when the factors change as in Problem 10.3? Show that the combination $\sum_i \chi_i(\mathbf{r}_1)\chi_i(\mathbf{r}_2)$ is invariant under all rotations (i.e. behaves like an S function), while the set of three functions $P_i=[\chi_j(\mathbf{r}_1)\chi_k(\mathbf{r}_2)-\chi_k(\mathbf{r}_1)\chi_j(\mathbf{r}_2)]$ (i, j, k cyclic) behaves in the same way as $\{\chi_i\}$ (i.e. like a set of P states); and that the six symmetric combinations $D_{jk}=[\chi_j(\mathbf{r}_1)\chi_k(\mathbf{r}_2)+\chi_k(\mathbf{r}_1)\chi_j(\mathbf{r}_2)]$ ($j\geqslant k$) behave like six *cartesian* d functions ($d_{x^2-y^2}, d_{y^2-x^2}, d_{z^2-x^2}, d_{yz}, d_{zz}, d_{xy}$).

10.5 Show that by suitably combining two functions of the set of six found in Problem 10.4 it is possible to obtain a function that behaves like d_{z^2}; and that it is possible to find another combination that is invariant under all rotations (i.e. is of S type). [*Note:* This problem and the previous one illustrate the classification of 2-electron wavefunctions, forming a "2-electron tensor space", with respect to their behaviour under transformations induced by *rotations*. In Chapter 10 the emphasis has been on transformations of the full linear group (or one of its subgroups); but in both cases symmetry with respect to *index permutations* is of importance in reducing the representation carried by the tensor space into its irreducible components.]

10.6 Prove that when the operators of the unitary group U(n) and the permutation group S_N act on the symmetry-adapted functions (10.3.1) the two types of operation commute; and hence that the CFs displayed schematically in (10.3.2) behave as indicated in the text. [*Hint:* The basic tensor products are Ω_K, with $K=k_1k_2\ldots k_N$, and the unitary operator U induces a transformation of the N-index tensor components in which

$$\Omega_{i_1i_2\ldots i_N}\rightarrow\bar{\Omega}_{i_1i_2\ldots i_N}=\mathsf{U}\Omega_{i_1i_2\ldots i_N}=\sum_{j_1j_2\ldots j_N}U_{i_1j_1}U_{i_2j_2}\ldots U_{i_Nj_N}\Omega_{j_1j_2\ldots j_N}.$$

Consider the corresponding quantities with an index permutation, $(\mathsf{P}\Omega)_{i_1i_2\ldots i_N}=\Omega_{\mathsf{P}(i_1i_2\ldots i_N)}$, and show that $\mathsf{PU}\Omega_{i_1i_2\ldots i_N}=\mathsf{UP}\Omega_{i_1i_2\ldots i_N}$. Note that dummy indices may be permuted in any way: and that a common permutation of the i- and j-indices merely puts the U-factors in a different order.]

10.7 Consider a 3-electron 4-orbital example (p. 335) and use the projection operator ρ_{11} to explicitly construct the 16 distinct functions whose tableaux are shown on p. 336. [*Hint:* Use the matrices derived in Problem 4.8 and put $p = q = 1$ in (10.3.3) and (10.3.4). Let K run through the 16 index sets at the top of p. 336.]

10.8 Obtain the 4 missing tableau functions, not constructed in Problem 10.7, by using the new operator ρ_{12}; and show that they are identical with those derived by using ρ_{11} in the manner indicated in the text. [*Hint:* Let ρ_{11} work on 4 new products (Ω_K), obtained by reading the index sets in "book order" from the 4 standard tableaux that were missing in Problem 10.7.]

10.9 Use the results from Problems 10.7 and 10.8 to verify explicitly, by considering orbital transformation that mix only ϕ_1, ϕ_2, ϕ_3, that the functions whose tableaux appear in (10.3.8) will fall into sets carrying representations of the subgroup U(3). Write out the six 3-electron functions corresponding to the

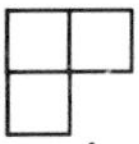

shape and use them to illustrate your conclusion. [*Hint:* The orbital transformation sends any Ω_K into a linear combination of products (Ω'_K) in which the indices k_1, k_2, k_3 (abbreviated to 1, 2, 3 in the tableaux) appear in a different order, while k_4 is unaffected. The tableau functions mixed in the transformation are therefore those that differ only in the positions of the indices 1, 2, 3.]

10.10 Set up the Gel'fand tableaux corresponding to the Weyl tableaux in (10.3.8) and show that they appear in standard order according to the weight-vector criterion (p. 340). Show also that this remains true within the subsets that remain (Problem 10.9) when index 4 is removed.

10.11 How many independent CFs will there be for the doublet states of a 5-electron system described using 6 orbitals? Set up the Weyl tableaux that can be derived using two choices of index set; $K = 12346$, 11346. Then form the equivalent Gel'fand and Paldus tableaux. [*Hint:* Use the Young tableaux in Fig. 4.2 and allocate orbital indices as on p. 335.]

10.12 Use the definition (10.4.3) and the anticommutation rules to derive the basic properties (10.4.7) of the operators E_{rs}. Show that the spin-traced operators in (10.4.9) have exactly the same commutation properties.

10.13 Calculate the matrix elements of the operators E_{rs} between the 5-electron tableau functions found in Problem 10.11; and hence obtain the 1-electron terms in the corresponding matrix elements of the Hamiltonian. Why is it so much more difficult to find the 2-electron terms? [*Hint:* Use Table 10.1 for the operators $\mathsf{E}_{i-1,i}$, and then (10.4.16) *et seq.*]

10.14 Formalize the concept of the "model space" (p. 352) by defining "active" and "virtual" subspaces in the full N-electron tensor space and introducing corresponding projection operators P and Q. Show, by defining $\mathsf{H}^{PP} = \mathsf{PHP}$, $\mathsf{H}^{PQ} = \mathsf{PHQ}$ etc., that a model system with the Hamiltonian

$$\mathsf{H}^{PP}_{\text{eff}} = \mathsf{H}^{PP} + \mathsf{H}^{PQ}(E - \mathsf{H}^{QQ})^{-1}\mathsf{H}^{QP}$$

(with non-zero elements only in the active space) will have energy eigenvalues coincident with those obtained in the full-CI limit. Show also that the corresponding full-CI wavefunction can be written in terms of the model-space projection ($\Psi^{\mathrm{F}} = \mathsf{P}\Psi$) as $\Psi = \mathsf{W}\Psi$, where the "wave operator" (p. 301) is defined by

$$\mathsf{W} = 1 + (E - \mathsf{H}^{\mathrm{QQ}})^{-1}\mathsf{H}^{\mathrm{QP}}.$$

How could you make use of these results and what difficulties would you foresee? [*Hint:* Review Section 2.5. Note that $\mathsf{P} = \sum_a |\Phi_a\rangle\langle\Phi_a|$ and $\mathsf{Q} = \sum_b |\Phi_b\rangle\langle\Phi_b|$. The Hamiltonian *matrix* **H** represents the operator $(\mathsf{P}+\mathsf{Q})\mathsf{H}(\mathsf{P}+\mathsf{Q})$ (i.e. the projection of the exact H on the *full*-CI space. Partition the operator and its eigenfunctions, writing the eigenvalue equation in 2-component form, and eliminate the component $\mathsf{Q}\Psi$ as in Section 2.5.]

10.15 Show that the Schrödinger equation in full-CI space (Problem 10.14), satisfied by the function Ψ_K say, can also be written as

$$(E_K - \mathsf{H}_0)\Phi_K = \mathsf{P}\mathsf{H}'\mathsf{W}\Phi_K,$$

where Φ_K is the projection ($\mathsf{P}\Psi_K$) of the full-CI solution, while $\mathsf{H}_0 = \mathsf{H}^{\mathrm{PP}}$ and $\mathsf{H}' = \mathsf{H}^{\mathrm{PQ}} + \mathsf{H}^{\mathrm{QP}} + \mathsf{H}^{\mathrm{QQ}}$. Then show that this equation is equivalent to

$$\mathsf{W}\mathsf{H}_0 - \mathsf{H}_0\mathsf{W} = (1 - \mathsf{W}\mathsf{P})\mathsf{H}'\mathsf{W},$$

an equation due to Kvasnicka (1974, 1977a,b) and Lindgren (1974, 1978). This equation for the wave operator W has come to be known as the "generalized Bloch equation". [*Hint:* Operate on the first equation from the left, with W and subtract from the original equation $E_K\Psi_K - \mathsf{H}\Psi_K = 0$ in order to eliminate E_K. The resultant equation is required to be valid for arbitrary functions Φ_K in the model space.]

10.16 Attach a perturbation parameter λ to the operator H′ in the equation for the wave operator W (Problem 10.15) and show how a solution may be obtained in the form

$$\mathsf{W}(\lambda) = 1 + \lambda\mathsf{W}_1 + \lambda^2\mathsf{W}_2 + \ldots$$

(where the normalization is such that $\mathsf{W}_0 = 1$) and derive the equations

$$[\mathsf{W}_1, \mathsf{H}_0]\mathsf{P} = \mathsf{Q}\mathsf{H}'\mathsf{P},$$
$$[\mathsf{W}_2, \mathsf{H}_0]\mathsf{P} = \mathsf{Q}\mathsf{H}'\mathsf{W}_1\mathsf{P} - \mathsf{W}_1\mathsf{P}\mathsf{H}'\mathsf{P}$$

for the first- and second-order terms. [*Hint:* Insert the expansion in the generalized Bloch equation and separate the terms of different order, requiring each one to vanish.]

10.17 Seek a matrix form of the equations obtained in Problem 10.16 by introducing an arbitrary orthonormal basis $\{\Phi_\kappa\}$ in the full tensor space and writing $\Psi_K = \sum_\mu C_\mu^K \Phi_\mu$. Show in particular that

$$\langle\Phi_\beta|\mathsf{W}_1|\Phi_\alpha\rangle = (E_\alpha - E_\beta)\langle\Phi_\beta|\mathsf{H}|\Phi_\alpha\rangle,$$

where $\{\Phi_\alpha\}$ and $\{\Phi_\beta\}$ define the model space and its complementary subspace respectively, and the energies (corresponding to $\lambda = 0$) are eigenvalues of the IPM operator H_0. Indicate how the matrix associated with W_2 could be obtained and discuss the connection between your results and those obtained in Problem 10.14. What advantages could the present approach have? [*Hint:* Start from the observation that $C_\kappa^K = \langle \Phi_\kappa \mid \Psi_K \rangle$ reduces to a matrix element of the wave operator. Note also that the partitioning is similar to that adopted in Section 2.5.]

11 Small Terms in the Hamiltonian: Static Properties

11.1 ELECTRIC AND MAGNETIC INTERACTIONS. THE CLASSICAL APPROACH

So far we have been concerned solely with the Schrödinger equation for the electrons in an *isolated* molecule with fixed nuclei, although we know (Section 1.1) that this equation is sometimes inadequate to describe the phenomena in which we are interested. In particular, we may need to consider the interaction of a molecule with an external electromagnetic field and also the internal electromagnetic interactions arising from spins and orbital motion.

To treat these effects properly, it is necessary to employ field theory and relativistic quantum mechanics. However, it is instructive at this stage to consider the incorporation in a purely classical way of an external field. Since some of the fields of interest are time-dependent, we shall start with the time-dependent Schrödinger equation for a collection of particles with charges q_i and masses m_i,‡ namely

$$\mathsf{H}\Psi = i\hbar\frac{\partial\Psi}{\partial t}, \qquad (11.1.1)$$

where

$$\mathsf{H} = \sum_i \frac{\mathsf{p}^2(i)}{2m_i} + \tfrac{1}{2}\sum_{i,j}{}' \frac{q_i q_j}{\kappa_0 r_{ij}}. \qquad (11.1.2)$$

From classical electromagnetic theory, it is known that the electric field strength $\boldsymbol{E}$ and the magnetic flux density $\boldsymbol{B}$ may be derived from a scalar

‡ It is convenient to consider particles of arbitrary charge and mass. For electrons $q_i = -e$, $m_i = m$.

potential ϕ and vector potential $\boldsymbol{A}$ according to ‡§

$$\left.\begin{aligned} \boldsymbol{E} &= -\text{grad}\,\phi - \frac{\partial \boldsymbol{A}}{\partial t}, \\ \boldsymbol{B} &= \text{curl}\,\boldsymbol{A}, \end{aligned}\right\} \tag{11.1.3}$$

or in components

$$\left.\begin{aligned} E_\lambda &= -\frac{\partial \phi}{\partial r_\lambda} - \frac{\partial A_\lambda}{\partial t}, \\ B_\lambda &= \frac{\partial A_\nu}{\partial r_\mu} - \frac{\partial A_\mu}{\partial r_\nu} \qquad (\lambda, \mu, \nu \text{ cyclic}), \end{aligned}\right\} \tag{11.1.4}$$

where Greek subscripts label the cartesian components ($r_\lambda = x, y, z$) and λ, μ, ν is any cyclic permutation of x, y, z.

For a uniform static field

$$B_x = \frac{\partial A_z}{\partial y} - \frac{\partial A_y}{\partial z} = \text{constant}, \tag{11.1.5}$$

with similar expressions for the y and z components, and by inspection we see that a satisfactory solution for $\boldsymbol{A}$ is given by

$$\boldsymbol{A} = \tfrac{1}{2}\boldsymbol{B} \times \boldsymbol{r}. \tag{11.1.6}$$

This solution however, is not unique, for two different choices of origin would give two alternative values of $\boldsymbol{A}$ at any given point in space, though the field itself must be independent of origin. More generally, replacement of the potentials by new potentials according to

$$\left.\begin{aligned} \boldsymbol{A} \to \boldsymbol{A}' &= \boldsymbol{A} + \text{grad}\, f, \\ \phi \to \phi' &= \phi - \frac{\partial f}{\partial t}, \end{aligned}\right\} \tag{11.1.7}$$

where f is an arbitrary function of position, leads to exactly the same fields. These equations define a *change of gauge*. It is clear that whenever the potentials occur in a Hamiltonian they must do so in such a way that all physical quantities, such as the energy and the electron density, are independent of choice of gauge—or *gauge-invariant*.

‡ Readers accustomed to "mixed Gaussian" (c.g.s.) units will note that (see page xiii *et seq.* for SI conventions) many familiar factors of $1/c$ are missing. See e.g. Coulson (1953) for a discussion.

§ For vector notation see e.g. Margenau and Murphy (1956, Chap. 4); ordinary 3-dimensional vectors (not considered as elements of a general vector space) are set in boldface italic.

To set up the Hamiltonian, we must return to the axiomatic foundations of quantum mechanics and find a set of canonically conjugate momenta and coordinates in the classical problem, in order to set up a quantum-mechanical Hamiltonian operator. To find the classical momenta and coordinates, we must find the Lagrangian L appropriate to charged particles moving in the field. The momentum conjugate to r_ν is then

$$p_\nu = \partial L/\partial \dot{r}_\nu, \tag{11.1.8}$$

and the classical Hamiltonian takes the form

$$H = \sum_\nu p_\nu \dot{r}_\nu - L, \tag{11.1.9}$$

where $\dot{r}_\nu = \mathrm{d}r_\nu/\mathrm{d}t = v_\nu$ is a cartesian velocity component.

The classical Lagrangian for a particle with charge q and mass m moving in a field specified by potentials ϕ and $\boldsymbol{A}$, with velocity $\boldsymbol{v}$, is (see e.g. Slater, 1960, Vol. 1, Appendix 4)

$$L = \frac{1}{2m}|m\boldsymbol{v}|^2 - q\phi + q\boldsymbol{v}\cdot\boldsymbol{A}, \tag{11.1.10}$$

so that

$$p_\nu = \frac{\partial L}{\partial v_\nu} = mv_\nu + qA_\nu, \tag{11.1.11}$$

$$H = \frac{1}{2m}(\boldsymbol{p} - q\boldsymbol{A})^2 + q\phi. \tag{11.1.12}$$

Converting this to operator form, summing over all particles, and including the potential energy of interaction, we get for the Hamiltonian operator

$$\mathsf{H} = \sum_i \frac{1}{2m_i}|\mathbf{p}(i) - q_i\boldsymbol{A}(i)|^2 + \sum_i q_i\phi(i) + \tfrac{1}{2}\sum_{i,j}{}' \frac{q_iq_j}{\kappa_0 r_{ij}}. \tag{11.1.13}$$

Here $\mathbf{p}(i)$ is the usual momentum operator, with components $(\hbar/i)\,\partial/\partial x_i$ etc., while the potentials $\boldsymbol{A}$ and ϕ are generally functions of position of the ith particle and are thus multiplicative factors, like the electron interaction terms.

Finally we must verify that the energies and densities obtained from H are indeed gauge-invariant. To do this, we first show that the *gauge-invariant momentum* operator (taking first a one-electron system)

$$\boldsymbol{\pi} = \mathbf{p} - q\boldsymbol{A} \tag{11.1.14}$$

has an invariant expectation value provided that the gauge change

(11.1.7) is accompanied by a phase change in which

$$\Psi \rightarrow \Psi' = e^{i\lambda}\Psi. \tag{11.1.15}$$

This result follows on noting that, after the gauge change,

$$\pi'_\nu \Psi' = \left(\pi_\nu - q\frac{\partial f}{\partial r_\nu}\right)e^{i\lambda}\Psi,$$

and by choosing

$$\lambda = \frac{q}{\hbar}f \tag{11.1.16}$$

we can then ensure

$$\pi'_\nu \Psi' = e^{iqf/\hbar}\pi_\nu \Psi,$$

and obtain the result

$$\langle \Psi' |\, \pi'_\nu \,| \Psi' \rangle = \langle \Psi |\, \pi_\nu \,| \Psi \rangle. \tag{11.1.17}$$

Thus, if the wavefunction is appropriately modified in the gauge change, the components of π (which define the cartesian velocity components) are properly invariant. To generalize the result to the N-particle case, we consider the gauge change

$$\boldsymbol{A}(i) \rightarrow \boldsymbol{A}'(i) = \boldsymbol{A}(i) + \operatorname{grad} f(i) \quad (i = 1, 2, \ldots, N),$$

where $f(i)$ refers to the ith particle only. The appropriate phase function in (11.1.15) is then

$$\lambda = \sum_{i=1}^{N} \frac{q}{\hbar} f(i), \tag{11.1.18}$$

and it follows immediately that

$$\boldsymbol{\pi} = \sum_i \boldsymbol{\pi}(i), \qquad \mathsf{T} = \sum_i \frac{1}{m}\pi^2(i)$$

are both gauge-invariant operators. Consequently, both the kinetic energy and the current density (see Problem 11.5) will display the invariance characteristics of physical observables. In particular, the time-independent‡ Schrödinger equation with Hamiltonian (11.1.13) is

‡ The time-dependent case requires further discussion. It is readily established that

$$\mathsf{H}'\Psi' - i\hbar\frac{\partial \Psi'}{\partial t} = \exp\left(\frac{iqf}{\hbar}\right)\left(\mathsf{H}\Psi - i\hbar\frac{\partial \Psi}{\partial t}\right),$$

and hence, *if Ψ is an exact solution of the time-dependent equation,* the right-hand side will vanish and the time development of Ψ' will be determined by an equation of the same form as the original time-dependent equation. When using *approximate* wavefunctions special care is needed to ensure gauge invariance.

gauge-invariant, i.e. in a gauge change, $\mathsf{H}\Psi = E\Psi$ becomes $\mathsf{H}'\Psi' = E\Psi'$ and all expectation values are left unchanged.

It is interesting to note that the transformation (11.1.15), with λ given in (11.1.18), corresponds to a very simple change in the density matrices when we specialize to the case of *electronic* systems (i.e. the usual fixed-nucleus approximation). For a gauge change in which $\boldsymbol{A} \rightarrow \boldsymbol{A} + \operatorname{grad} f$, the N-electron wavefunction must be multiplied by

$$\exp\left[-i\frac{e}{\hbar}\sum_i f(i)\right],$$

and, on deriving the density matrices by integration of $\Psi\Psi^*$, the phase factors will cancel for all variables over which integrations are performed. Thus, for the 1-electron density matrix, gauge invariance requires that

$$P'(\boldsymbol{r};\boldsymbol{r}') = \exp\left\{i\frac{e}{h}[f(\boldsymbol{r}) - f(\boldsymbol{r}')]\right\} P(\boldsymbol{r};\boldsymbol{r}'). \qquad (11.1.19)$$

One important example is provided by the displacement of a system through a uniform field (vector potential $\boldsymbol{A} = \frac{1}{2}\boldsymbol{B} \times \boldsymbol{r}$), in which $\boldsymbol{r} \rightarrow \boldsymbol{r} + \boldsymbol{R}$; this is also equivalent to a change of coordinates in which the origin is shifted to $-\boldsymbol{R}$. In this case the density matrices before and after the change should be related by

$$P'(\boldsymbol{r};\boldsymbol{r}') = \exp\left\{-i\frac{e}{2\hbar}\boldsymbol{B} \times \boldsymbol{R} \cdot (\boldsymbol{r} - \boldsymbol{r}')\right\} P(\boldsymbol{r};\boldsymbol{r}'). \qquad (11.1.20)$$

The phase factor vanishes, of course, whenever we calculate matrix elements of quantities such as coulomb interactions; but more generally it guarantees the invariance of velocity- (and hence field-) dependent terms.

It must be stressed that two *approximate* wavefunctions, calculated independently with different choices of origin, will not *automatically* yield density matrices that differ by the appropriate phase factor as in (11.1.20); and they will generally give different numerical estimates of field-dependent quantities. This is one aspect of the so-called "gauge problem" that we meet in later sections.

To proceed to the actual calculation of field-dependent effects, it is usual to expand the term $\pi^2(i)$ in the Hamiltonian, obtaining

$$\pi^2(i) = \frac{1}{2m_i}[\mathsf{p}^2(i) - q_i\mathbf{p}(i) \cdot \boldsymbol{A}(i) - q_i\boldsymbol{A}(i) \cdot \mathbf{p}(i) + q_i^2 A^2(i)]. \qquad (11.1.21)$$

Since $\mathsf{p}_v(i)$ differentiates all that follows it, any $\mathbf{p} \cdot \boldsymbol{A}$ operator must be interpreted so that

$$\mathbf{p} \cdot \boldsymbol{A}\Psi = \boldsymbol{A} \cdot \mathbf{p}\Psi + \Psi(\mathbf{p} \cdot \boldsymbol{A}).$$

It is usually convenient to restrict the choice of gauge so that the last term will vanish: this can always be done because any part of $\boldsymbol{A}$ that is expressible as the gradient of some scalar field f (see p. 358) makes zero contribution to $\boldsymbol{B}$. The scalar field may be chosen in such a way that $\boldsymbol{p}\cdot\boldsymbol{A}=(\hbar/i)\,\mathrm{div}\,\boldsymbol{A}=0$, and this choice defines the *coulomb gauge*. Thus

$$\mathbf{p}\cdot\boldsymbol{A}\Psi=\boldsymbol{A}\cdot\mathbf{p}\Psi \quad \text{(coulomb gauge)}. \tag{11.1.22}$$

We assume coulomb gauge throughout, so that (11.1.21) becomes

$$\pi^2(i)=\frac{1}{2m_i}[\mathsf{p}^2(i)-2q_i\mathbf{A}(i)\cdot\mathbf{p}(i)+q_i^2A^2(i)], \tag{11.1.23}$$

and the Hamiltonian including applied fields is thus

$$\mathsf{H}=\mathsf{H}_0+\sum_i q_i\phi(i)-\sum_i\frac{q_i}{m_i}\mathbf{A}(i)\cdot\mathbf{p}(i)+\sum_i\frac{q_i^2}{2m_i}A^2(i), \tag{11.1.24}$$

where H_0 is the field-free Hamiltonian (11.1.2).

So far we have treated all particles of the system on an equal footing, not requiring the fixed-nucleus approximation, and one or two comments are necessary at this point. The results of a fixed-nucleus calculation in the absence of a field can be used more or less directly in the solution of the complete problem, including nuclei, by the method of Born and Huang, referred to in Section 1.1. But the method depends for its utility on a factorization by means of which the internal motions are separated from the translational motion of the molecule as a whole, and, although this can be achieved by working in a centre-of-mass coordinate system (along the lines used in Problems 1.16 and 1.17) *in the absence of external fields,* it is not possible when fields are present. This difficulty is expected on physical grounds, because a molecule is a collection of charged particles that will experience forces due to their motion through the fields. In other words translation will affect the "internal" motions of the particles.

A consequence of inseparability is that the molecule may be described in a fixed frame only if we include extra terms in the Hamiltonian to allow for the overall molecular motion through the fields. Fortunately, at low fields and with neutral molecules moving at thermal velocities, these effects are estimated to be very small, and are nearly always neglected. The effects can of course be calculated (for example by perturbation theory) after solving the fixed-nucleus problem, and it is therefore clear once again that the fixed-nucleus problem must be solved first.

Similar remarks apply whether the molecule is subjected to static fields or to a radiation field, and in nearly every case we proceed from the solutions of the stationary state problem involving the field-free Hamiltonian H_0 given by (11.1.2) and used throughout the earlier chapters. Usually, we deal either with strong static fields or with weak oscillating fields (radiation). Some of the effects of static fields are discussed in later sections. Radiation, provided that it is weak and of fairly long wavelength (for example intensities of the order of $10^{-1}\,\mathrm{W\,m^{-2}}$ and less, and wavelengths of about 0.1 nm and longer) scarcely affects the energy levels, but simply induces *transitions* between them. Approximate solutions of the time-dependent equation (1.1.12) can then be obtained using time-dependent perturbation theory or by variational methods (Chapter 12): the corresponding (semiclassical) theory of transitions is discussed in most standard texts (e.g. Eyring *et al.*, 1944, Chap. 8; Slater, 1960, Vol. 1, Chap. 6) and will not be dealt with further here.

The classically inferred Hamiltonian (11.1.24), is still incomplete, though it is adequate for very many purposes. In particular, it omits internal electromagnetic interactions associated with the intrinsic (spin) magnetic moments of the particles: these may be "grafted on" to the Hamiltonian in a somewhat *ad hoc* way (Slater, 1960, Vol. 2, Chap. 24). Thus if a particle has magnetic moment $\boldsymbol{\mu}$ the Hamiltonian is expected to contain a term, $-\boldsymbol{\mu}\cdot\boldsymbol{B}$ representing its direct interaction with the external field (cf. p. 13), while the vector potential $\boldsymbol{A}$ that gives the field at all points in space will contain a corresponding dipole term. We shall not pursue this approach, since the terms in question are essentially non-classical in origin, and arise more naturally when the requirements of relativity theory are recognized.

11.2 RELATIVISTIC EFFECTS: ONE PARTICLE

We continue to assume that any fields are rather weak, so that it is meaningful in the relativistic context to distinguish between the field and the molecule, and also assume that the energies involved are very much less than mc^2, where m is the particle mass and c the velocity of light.

We start with the relativistic Lagrangian for a particle of rest mass m and charge q just as we started with the classical Lagrangian in Section 11.1. Instead of (11.1.10) we therefore use (Slater, 1960, Vol. 1, Appendix 4)

$$L = -mc^2\left(1-\frac{v^2}{c^2}\right)^{1/2} - q\phi + q\boldsymbol{v}\cdot\boldsymbol{A}, \tag{11.2.1}$$

which clearly reduces to (11.1.10) if $v \ll c$, since mc^2 is a constant and simply defines the energy zero. By precisely similar reasoning to that used in Section 11.1, the Hamiltonian is found to be

$$H = [m^2c^4 + |\boldsymbol{p} - q\mathbf{A}|^2 c^2]^{1/2} + q\phi. \tag{11.2.2a}$$

Because this is a square root, we cannot immediately obtain an operator form, but by squaring and rearranging, we obtain

$$c^{-2}(H - q\phi)^2 - m^2c^2 - |\boldsymbol{p} - q\mathbf{A}|^2 = 0, \tag{11.2.2b}$$

which is more suitable for conversion. The corresponding operator form is the Klein–Gordon equation, but, although this describes certain kinds of spinless particles, it does not describe an electron.

Dirac suggested that for electrons (11.2.2b) should be "factorized" in the form

$$\left(\pi_0 + \sum_{\mu=1}^{3} \alpha_\mu \pi_\mu + \alpha_4 mc\right)\left(\pi_0 - \sum_{\mu=1}^{3} \alpha_\mu \pi_\mu - \alpha_4 mc\right) = 0, \tag{11.2.3}$$

where

$$\pi_0 = (H - q\phi)/c, \tag{11.2.4a}$$

$$\pi_\mu = r_\mu - qA_\mu \quad (\mu = 1, 2, 3), \tag{11.2.4b}$$

and the αs are suitably chosen dimensionless operators not involving the time or the coordinates. In order that (11.2.3) reduce to (11.2.2b), they must satisfy the anticommutation relations

$$\alpha_\mu^2 = 1, \qquad \alpha_\mu\alpha_\nu + \alpha_\nu\alpha_\mu = 0 \qquad (\mu \neq \nu). \tag{11.2.5}$$

Dirac also proposed that the single factor nearest the operand

$$\pi_0 - \sum_{\mu=1}^{3} \alpha_\mu \pi_\mu - \alpha_4 mc \tag{11.2.6}$$

should be used in formulating the eigenvalue equation for the wavefunction ψ, the choice being conventional since use of either factor in (11.2.3) leads to exactly the same results. It is customary to introduce $\beta = \alpha_4$ and to use a vector notation, writing the eigenvalue equation in the form

$$(\pi_0 - \boldsymbol{\alpha} \cdot \boldsymbol{\pi} - \beta mc)\psi = 0, \tag{11.2.7}$$

where

$$\left.\begin{aligned} \pi_0 &= \frac{1}{c}\left(i\hbar\frac{\partial}{\partial t} - q\phi\right), \\ \pi_\mu &= \frac{\hbar}{i}\frac{\partial}{\partial r_\mu} - qA_\mu. \end{aligned}\right\} \tag{11.2.8}$$

The "scalar-product" operator $\boldsymbol{\alpha} \cdot \boldsymbol{\pi}$ simply stands for the sum

$\Sigma_{\mu=1}^{3} \alpha_\mu \pi_\mu$. Equation (11.2.7) may be written alternatively as

$$i\hbar \frac{\partial \psi}{\partial t} = (c\boldsymbol{\alpha} \cdot \boldsymbol{\pi} + q\phi + \beta mc^2)\psi, \tag{11.2.9}$$

which shows more clearly its similarity to the Schrödinger equation.

To satisfy the anticommutation rules (11.2.5), the α_μ may be regarded as operators working in a *spin space*, analogous to that used in the Pauli approach, but of four dimensions instead of two. However, it may be shown (Appendix 4) by the partitioning method of Section 2.5 that the four-dimensional equation may be replaced by an equivalent two-dimensional equation

$$i\hbar \frac{\partial \psi'}{\partial t} = \left(\frac{1}{2m}\boldsymbol{\pi} \cdot \mathsf{k}\boldsymbol{\pi} + \frac{i}{m}\mathbf{S} \cdot \boldsymbol{\pi} \times \mathsf{k}\boldsymbol{\pi} + q\phi\right)\psi', \tag{11.2.10}$$

where $\mathbf{S}$ stands for the Pauli spin operators with properties (1.3.20), and k is the operator defined by

$$\mathsf{k} = \left(1 + \frac{\pi_0}{2mc}\right)^{-1}. \tag{11.2.11}$$

The wavefunction ψ' is a two-component *spinor*

$$\psi' = \phi_\alpha(\boldsymbol{r})\alpha(s) + \phi_\beta(\boldsymbol{r})\beta(s) \tag{11.2.12}$$

of the form used in the Pauli theory. Thus $\phi_\alpha(\boldsymbol{r})$ and $\phi_\beta(\boldsymbol{r})$ are functions of space variables alone, and $\alpha(s)$ and $\beta(s)$ may be regarded as the spin functions introduced in Section 1.2. This form was used in (6.3.1).

A further reduction occurs when the potential ϕ is everywhere zero and $\boldsymbol{A}$ is independent of time. The system may then be in a stationary state, and the operator k becomes a numerical factor

$$k = \left(1 + \frac{E}{2mc^2}\right)^{-1}, \tag{11.2.13}$$

where E is the energy of the electron, excluding that due to its rest mass. Thus in the circumstances that interest us most, k will be exceedingly close to 1. This approximation yields, on putting $q = -e$ and using the operator identity

$$\boldsymbol{\pi} \times \boldsymbol{\pi} = iq\hbar \operatorname{curl} \boldsymbol{A} = iq\hbar \boldsymbol{B} \tag{11.2.14}$$

(which follows easily on considering individual components), the "Dirac–Pauli" equation for a free electron:

$$\left(\frac{1}{2m}\pi^2 + \frac{e\hbar}{m}\mathbf{S} \cdot \boldsymbol{B}\right)\psi' = E\psi'. \tag{11.2.15}$$

The second term in the operator may be compared with the classical expression $-\boldsymbol{\mu}\cdot\boldsymbol{B}$ for the energy of a dipole $\boldsymbol{\mu}$ in the field $\boldsymbol{B}$, and allows us to identify the dipole-moment operator of the electron:

$$\boldsymbol{\mu} = -\frac{e\hbar}{m}\mathbf{S} = -2\beta\mathbf{S}, \tag{11.2.16a}$$

where β is the Bohr magneton introduced in (1.2.21). Thus the electron spin arises naturally in a relativistic treatment, and we automatically obtain (very nearly, see p. 13) the right factor of proportionality between electron spin and dipole moment.

A more advanced quantum-electrodynamic calculation (see e.g. Akhiezer and Berestetskii, 1965) shows that the proportionality constant derived above is very slightly wrong. The corrected factor is approximately 2.0023, and it is therefore preferable to write (11.2.16a) as

$$\boldsymbol{\mu} = -g\beta\mathbf{S}, \tag{11.2.16b}$$

where g denotes the *free electron g factor*.

When the potential ϕ is non-zero, the reduction of (11.2.10) is less simple (Appendix 4). The corresponding Dirac–Pauli Hamiltonian is

$$\boxed{\begin{aligned}\mathsf{H} &= \frac{\pi^2}{2m} - e\phi + g\beta e\mathbf{S}\cdot\boldsymbol{B} - \frac{\pi^4}{8m^3c^2}\\ &\quad + \frac{g\beta e}{4mc}(\mathbf{S}\cdot\boldsymbol{\pi}\times\boldsymbol{E} - \mathbf{S}\cdot\boldsymbol{E}\times\boldsymbol{\pi} - \tfrac{1}{2}\hbar\,\mathrm{div}\,\boldsymbol{E}),\end{aligned}} \tag{11.2.17}$$

where $\boldsymbol{E}$ is the electric field vector derived from ϕ, at the position of the electron. The new terms allow for spin–orbit coupling effects and for relativistic mass variation at high velocities.

The above discussion does not apply to a nucleus; experimentally, many nuclei possess non-zero magnetic moments, indicating a resultant nuclear spin, but so far there is no satisfactory theory of the proportionality constants.

11.3 RELATIVISTIC TREATMENT OF MANY-PARTICLE SYSTEMS

We now turn to the many-electron generalization of the Dirac equation. It must be said at the outset that no fully satisfactory relativistic Hamiltonian has yet been derived, and that difficulties remain at a fundamental level; however, the somewhat *ad hoc* approach generally employed (including the reduction to Pauli form) leads to a large number of small terms that are consistent with a vast amount of experimental data. Detailed discussions are available elsewhere (e.g. Bethe and Salpeter, 1957; Itoh, 1965; Grant, 1970; Moss, 1973; Grant and Quinney, 1988).

In generalizing the classical one-particle Lagrangian to the many-particle case, in Section 11.1, the interaction between the electrons was taken to be independent of their velocities. We also ignored the fact that a moving charged particle generates a magnetic field and hence makes a contribution to the vector potential. In the relativistic approach, however, we must recognize that interactions between particles occur with a finite velocity, namely the velocity of light. In mathematical terms, the Lagrangian in Section 11.1 was set up on the assumption that the potentials at a point $\boldsymbol{r}_1$ due to a moving particle of charge q_2 at $\boldsymbol{r}_2$ are given by

$$\phi(\boldsymbol{r}_1) = \frac{q_2}{\kappa_0 r_{12}}, \quad \boldsymbol{A}(\boldsymbol{r}_1) = 0. \tag{11.3.1}$$

In fact, the relativistically correct potentials are the retarded or Liénard–Wiechert potentials (Darwin, 1920; Feynman *et al.*, 1964)

$$\phi(\boldsymbol{r}_1, t) = \frac{q_2}{\kappa_0(r_{12} + \boldsymbol{v}_2 \cdot \boldsymbol{r}_{12}/c)_{\text{ret}}}, \tag{11.3.2a}$$

$$\boldsymbol{A}(\boldsymbol{r}_1, t) = \frac{q_2 \boldsymbol{v}_2}{\kappa_0 c^2(r_{12} + \boldsymbol{v}_2 \cdot \boldsymbol{r}_{12}/c)_{\text{ret}}}, \tag{11.3.2b}$$

where $\boldsymbol{v}_2$ is the velocity of the particle at $\boldsymbol{r}_2$, and $\boldsymbol{r}_{12}$ is the interparticle distance vector pointing from 1 to 2. The distances and velocities on the right-hand side of these equations are evaluated at the "retarded" time t', earlier than t, such that $(t - t')c$ is the distance between the particles at the earlier time.

The retarded potentials are discussed in detail in the references cited. Here we need only note that the classical Lagrangian for a pair of particles interacting through these potentials is given by (Darwin, 1920)

$$L = L_1 + L_2 - \frac{q_1 q_2}{\kappa_0 r_{12}} + I, \tag{11.3.3}$$

where L_1 and L_2 are one-particle Lagrangians like (11.2.1) and I is given by

$$I = \frac{q_1 q_2}{2c^2 \kappa_0}\left(\frac{\boldsymbol{v}_1 \cdot \boldsymbol{v}_2}{r_{12}} + \frac{(\boldsymbol{v}_1 \cdot \boldsymbol{r}_{12})(\boldsymbol{v}_2 \cdot \boldsymbol{r}_{12})}{r_{12}^3}\right). \tag{11.3.4}$$

The interaction term I is an *approximation*, good up to terms of the order of v^2/c^2 only, and so may be regarded as a "perturbation". We therefore require a Hamiltonian that goes over in the limit into the sum of two isolated-particle Hamiltonians. Terms of the order v^2/c^2, arising from I, will be considered along with other perturbations of similar order.

The resultant Hamiltonian is found by straightforward but rather lengthy manipulation to be

$$H = h_1 + h_2 + \frac{q_1 q_2}{\kappa_0 r_{12}} - I, \tag{11.3.5}$$

where

$$h_i = m_i c^2 \left(1 - \frac{v_i^2}{c^2}\right)^{1/2} + \boldsymbol{v}_i \cdot \boldsymbol{\pi}_i + q_i \phi. \tag{11.3.6}$$

In the case where the two particles are electrons we can convert this Hamiltonian directly into quantum-mechanical form (Appendix 4), obtaining

$$\mathsf{H} = \mathsf{h}(1) + \mathsf{h}(2) + \frac{q_1 q_2}{\kappa_0 r_{12}} - \frac{q_1 q_2}{2\kappa_0}\left(\frac{\boldsymbol{\alpha}(1) \cdot \boldsymbol{\alpha}(2)}{r_{12}} + \frac{(\boldsymbol{\alpha}(1) \cdot \boldsymbol{r}_{12})(\boldsymbol{\alpha}(2) \cdot \boldsymbol{r}_{12})}{r_{12}^3}\right), \tag{11.3.7}$$

where

$$\mathsf{h}(i) = c\boldsymbol{\alpha}(i) \cdot \boldsymbol{\pi}(i) + q_i \phi(i) + m_i c^2 \beta(i). \tag{11.3.8}$$

The operators $\boldsymbol{\alpha}(i)$ and $\beta(i)$ are just like those used in (11.2.7), but they are taken to act on the spin variables of the ith particle only. The interaction term arising from I is called the Darwin–Breit term after its discoverers, and its form is confirmed by more advanced quantum-electrodynamic considerations.

The reduction of (11.3.7) to an approximate form involving the usual spin operators is both involved and uncertain (Appendix 4). Nevertheless, it leads to a Pauli-type Hamiltonian whose terms can be interpreted classically as field–dipole, dipole–dipole interactions and the like. This interpretation provides a basis, insecure though it may be, for a Hamiltonian when one or both particles are *nuclei,* provided that the nuclear spins and moments are regarded as phenomenological quantities with values to be inferred from experiment. It is then but a short step to the Hamiltonian for a many-particle system, so long as only pairwise interactions are present (as is currently believed to be the case).

The details of the complete Hamiltonian for electrons and nuclei are given in equations (A4.17), (A4.19) and (A4.21). The Hamiltonian may be written in the general form (1.1.13),

$$\mathsf{H} = \mathsf{H}_e + \mathsf{H}_n + \mathsf{H}_{en} = \mathsf{H}_0 + \mathsf{H}', \tag{11.3.9}$$

but now the contributions listed in (1.1.14), collected in H_0, are simply the leading terms in the more complete expressions, while the remaining "small terms" are embodied in H′ which is usually treated as a perturbation.

In the following sections we shall use a fixed-nucleus approximation in discussing some of the observable effects of the small terms. That we may do so, in spite of the fact that the presence of velocity-dependent terms precludes the rigorous separation of electronic and nuclear motion (even in the absence of *external fields*), is due to their dependence on higher powers of v/c—a ratio that is exceedingly small for nuclei. The terms that we shall in fact consider, some of which involve further reduction of the many contributions listed in Appendix 4, are conveniently collected at this point: those of most interest may be distinguished by writing H′ as

$$\boxed{\mathsf{H}' = \mathsf{H}_{\text{elec}} + \mathsf{H}_{\text{mag}} + \mathsf{H}_{SL} + \mathsf{H}_{SS} + \mathsf{H}_Z + \mathsf{H}_{\text{N}},} \tag{11.3.10}$$

where

- H_{elec} arises from external electric fields,
- H_{mag} from external magnetic fields interacting with electronic orbital motion,
- H_{SL} from interaction between electron spin and orbital motion,
- H_{SS} from electron spin–spin interactions,
- H_Z from (Zeeman) interaction between electron spin and magnetic field,
- H_{N} includes all "hyperfine" terms, arising from nuclear spins.

We assume a uniform magnetic field (**B** constant) with vector potential $\boldsymbol{A} = \frac{1}{2}\boldsymbol{B} \times \boldsymbol{r}$, and no external *electric* field (so that $\mathsf{H}_{\text{elec}} = 0$), and write down the corresponding forms of these terms.

H_{mag} arises from H_1^{c} as defined in (A4.17a): we assume coulomb gauge, write

$$\mathsf{H}_1^{\text{c}} = \frac{1}{2m}\sum_i [\mathsf{p}^2(i) + 2e\boldsymbol{A}(i)\cdot\mathbf{p}(i) + e^2A^2(i)] = \sum_i \frac{\mathsf{p}^2(i)}{2m} + \mathsf{H}_{\text{mag}},$$

and, on inserting $\boldsymbol{A}(i) = \frac{1}{2}\boldsymbol{B} \times \boldsymbol{r}_i$ and noting that $\hbar\mathbf{L}(i) = \boldsymbol{r}_i \times \mathbf{p}(i)$, where **L** is dimensionless (see p. 12), we find

$$\mathsf{H}_{\text{mag}} = \mathsf{H}'_{\text{mag}} + \mathsf{H}''_{\text{mag}}, \tag{11.3.11}$$

where

$$\mathsf{H}'_{\text{mag}} = \beta\sum_i \boldsymbol{B}\cdot\mathbf{L}(i), \qquad \mathsf{H}''_{\text{mag}} = \frac{e^2}{8m}\sum_i (\boldsymbol{B}\times\boldsymbol{r}_i)^2. \tag{11.3.12}$$

Here $\beta = e\hbar/2m$ and H'_{mag} gives the orbital paramagnetism of free atoms in states with non-zero angular momentum, but in second order also contributes to diamagnetism, reducing the main contribution which arises from $\mathsf{H}''_{\text{mag}}$: it also determines other effects, as will be evident shortly.

H_{SL} arises from the terms H_5^{e}, H_9^{e} and H_5^{en}, defined in (A4.17) and (A4.21). If there is no *external* electric field, as we are now assuming, then H_5^{e} vanishes, while the other terms may be combined to give

$$\mathsf{H}_{SL} = \frac{g\beta^2}{\kappa_0 c^2}\left[\sum_{n,i} \frac{Z_n \mathbf{S}(i)\cdot\mathbf{M}^n(i)}{r_{ni}^3} - \sum_{i,j}{}' \frac{2\mathbf{S}(i)\cdot\mathbf{M}^i(j) + \mathbf{S}(i)\cdot\mathbf{M}^j(i)}{r_{ij}^3}\right], \tag{11.3.13}$$

where we have introduced a gauge-invariant angular-momentum operator

$$\hbar\mathbf{M}^p(q) = \boldsymbol{r}_{pq} \times \boldsymbol{\pi}(q) \tag{11.3.14}$$

associated with the angular momentum of a particle at q about point p. It should be noted that $\boldsymbol{r}_{pq}$ has been used consistently throughout to denote the vector $\overrightarrow{pq} = \boldsymbol{r}_q - \boldsymbol{r}_p$, and that the opposite choice (resulting in sign changes) frequently occurs in the literature.

H_{Z} is simply the electron spin-field coupling given by H_3^{e} in (A4.17c),

$$\mathsf{H}_{\text{Z}} = g\beta \sum_i \boldsymbol{B}\cdot\mathbf{S}(i), \tag{11.3.15}$$

and gives the orbitally independent part of the Zeeman effect. The orbital contribution, sometimes included in a Zeeman term of the form $\beta\sum_i \boldsymbol{B}\cdot[\mathbf{L}(i) + g\mathbf{S}(i)]$, arises in fact from H'_{mag}, which is here dealt with separately.

H_{SS} arises from the magnetic dipole–dipole interaction between the electrons, $\mathsf{H}_{11}^{\text{e}}$ in (A4. 17k), and the associated contact correction $\mathsf{H}_{12}^{\text{e}}$. It is thus, introducing the 3-dimensional delta function $\delta(\boldsymbol{r}_{ij})$,

$$\mathsf{H}_{SS} = -\frac{g^2\beta^2}{2\kappa_0 c^2}\sum_{i,j}{}' \left[\frac{3(\mathbf{S}(i)\cdot\boldsymbol{r}_{ij})(\mathbf{S}(j)\cdot\boldsymbol{r}_{ij}) - r_{ij}^2\mathbf{S}(i)\cdot\mathbf{S}(j)}{r_{ij}^5} + \frac{8\pi}{3}\delta(\boldsymbol{r}_{ij})\mathbf{S}(i)\cdot\mathbf{S}(j)\right]. \tag{11.3.16}$$

It may be shown that the electron–electron contact term does not lead to any spin dependence of the energy; it is therefore of little interest in, say, the interpretation of ESR experiments (shifting different levels equally) and is frequently dropped.

H_{N}, describing (hyperfine) nuclear spin effects, arises from H_2^{n} (the nuclear Zeeman term), H_4^{n} (nuclear dipole–dipole term, with no contact correction), H_2^{en} and H_3^{en} (electron–nuclear dipole–dipole term, with contact correction) and $\mathsf{H}_4^{\mathrm{en}}$ (nuclear dipole interacting with electronic orbital motion). These terms are listed in (A4.15) and (A4.21). We shall use (with the direct-interaction $\mathsf{H}_4^{\mathrm{n}}$ written as $\mathsf{H}_{\mathrm{N}}^{\mathrm{direct}}$)

$$\mathsf{H}_{\mathrm{N}} = -\beta_{\mathrm{p}} \sum_n g_n \boldsymbol{B} \cdot \mathbf{I}(n) + \frac{2\beta\beta_{\mathrm{p}}}{\kappa_0 c^2} \sum_{n,i} g_n r_{ni}^{-3} \mathbf{I}(n) \cdot \mathbf{M}^{\mathrm{n}}(i)$$

$$+ \frac{g\beta\beta_{\mathrm{p}}}{\kappa_0 c^2} \sum_{n,i} g_n \Big\{ r_{ni}^{-5} [3(\mathbf{S}(i) \cdot \boldsymbol{r}_{ni})(\mathbf{I}(n) \cdot \boldsymbol{r}_{ni}) - r_{ni}^2 \mathbf{I}(n) \cdot \mathbf{S}(i)]$$

$$+ \tfrac{8}{3}\pi\delta(\boldsymbol{r}_{ni}) \mathbf{I}(n) \cdot \mathbf{S}(i) \Big\} + \mathsf{H}_{\mathrm{N}}^{\mathrm{direct}}. \qquad (11.3.17)$$

The electron–*nuclear* contact term, in contrast with that for the electrons alone, is of special importance in determining many observable effects.

The "small terms" in the Breit–Pauli Hamiltonian, collected in (11.3.10), clearly do not remain small for systems that include atoms of high atomic number Z_{n}; spin–orbit coupling and mass-variation terms then assume great importance, particularly for inner-shell electrons. In such cases *ab initio* methods of calculation encounter serious difficulties, perturbation techniques are not entirely appropriate, and efforts are often made to solve the relativistic equations more directly by using 2-component or even 4-component spin-orbitals in wavefunctions of Hartree–Fock type. For accounts of work in this area, reference may be made to the reviews by Grant (1970) and Pyykö (1978). Approximate wavefunctions constructed using the (non-relativistic) methods of earlier chapters must obviously be used with great caution in discussing relativistic corrections (regarded as a perturbation); and their limitations must be clearly recognized.

11.4 A VARIATION–PERTURBATION APPROACH

For the purposes of this section, we write the Hamiltonian in the form (11.3.9), namely

$$\mathsf{H} = \mathsf{H}_0 + \mathsf{H}', \qquad (11.4.1)$$

where H_0 is the usual fixed-nucleus electronic Hamiltonian while H' contains all the "small terms". The many terms in H' may be grouped into three categories: (i) those containing the external fields, in conjunction with electronic and nuclear variables; (ii) those involving small internal fields and depending only on electronic variables (leading to the "fine structure" of the energy levels); and (iii) those involving small internal fields associated with the *nuclei* (giving the "hyperfine structure").

There are many ways of dealing with the small terms, most of them perturbational in character and presupposing knowledge (at least in principle) of exact eigenfunctions of H_0. We develop only one approach, based on the variation–perturbation theory of Section 2.5, which has the merits of simplicity, flexibility and generality and does *not* require knowledge of exact unperturbed wavefunctions. This approach also lends itself well to the physical interpretation of the mode of interaction of the small terms.

Generally speaking, we shall be interested in some particular set of degenerate eigenstates of H_0, and the effect on these states of the perturbation H'. In the language of Section 2.5, these states form the "A set", containing, say, n_A approximate wavefunctions Φ_a, $\Phi_{a'}$, the "B set" $\{\Phi_b, \Phi_{b'}, \ldots\}$, and their union

$$\{\Phi_\kappa\} = \{\Phi_a, \Phi_{a'}, \ldots, \Phi_b, \Phi_{b'}, \ldots\},$$

which will be assumed to be complete and orthonormal for $n_B \rightarrow \infty$. The key results, which we restate here, are that, instead of solving the full set of $n_A + n_B$ secular equations (namely $\mathsf{Hc} = E\mathsf{c}$) to determine the energies E and expansion coefficients c_κ, we may write

$$\Psi = \sum_\kappa c_\kappa \Phi_\kappa = \sum_a c_a^A \Phi_a + \sum_b c_b^B \Phi_b \tag{11.4.2}$$

and solve the reduced set of secular equations

$$\mathbf{H}_{\text{eff}} \mathbf{c}^A = E\mathbf{c}^A, \tag{11.4.3}$$

in which $\mathbf{c}^A$ is the column of expansion coefficients c_a^A, while $\mathbf{H}_{\text{eff}}$ is an $n_A \times n_A$ matrix representing an "effective Hamiltonian":

$$\mathbf{H}_{\text{eff}} = \mathbf{H}^{AA} + \mathbf{H}^{AB}(E\mathbf{1} - \mathbf{H}^{BB})^{-1}\mathbf{H}^{BA}. \tag{11.4.4}$$

As a first approximation, E in (11.4.4) is given the value $E_a^{(0)}$ appropriate to the unperturbed A set, and the inverse matrix is expanded in order to obtain an explicit, order-by-order form of $\mathbf{H}_{\text{eff}}$.

It will now be assumed that the functions Φ_κ, although not exact eigenfunctions of H_0, are variationally determined approximations, in which case the matrix of H_0 will be diagonal:

$$\langle \Phi_\kappa | \mathsf{H}_0 | \Phi_{\kappa'} \rangle = \delta_{\kappa\kappa'} E_\kappa^{(0)}. \tag{11.4.5}$$

We shall also write the many-term perturbation in the form

$$\mathsf{H}' = \lambda \mathsf{H}^\lambda + \mu \mathsf{H}^\mu + \ldots, \tag{11.4.6}$$

where the perturbation parameters $\lambda, \mu, \ldots$ allow us conveniently to separate orders with respect to the various terms.

First let us consider a single term $\mathsf{H}' = \lambda \mathsf{H}^{\lambda}$. As in Section 2.5, the elements of $\mathbf{H}_{\text{eff}}$ are then found to be (using for simplicity a, a' in the matrix elements, to stand for Φ_a, $\Phi_{a'}$)

$$\langle a | \mathsf{H}_{\text{eff}} | a' \rangle = \delta_{aa'} E_a^{(0)} + \lambda \langle a | \mathsf{H}^{\lambda} | a' \rangle + \lambda^2 \sum_b \frac{\langle a | \mathsf{H}^{\lambda} | b \rangle \langle b | \mathsf{H}^{\lambda} | a' \rangle}{E_a^{(0)} - E_b^{(0)}} + \dots, \tag{11.4.7}$$

which is correct up to second order‡ in λ. If all energies are referred to $E_a^{(0)}$ as "zero", the first term in (11.4.7) may be discarded; the remaining terms determine, through solution of (11.4.3), the first- and second-order energy shifts due to the perturbation. The perturbed wavefunction, when required, is determined using (2.5.5), which takes the form

$$\mathbf{c}_a^B = (E_a^{(0)} \mathbf{1} - \mathbf{H}^{BB})^{-1} \mathbf{H}^{BA} \mathbf{c}_a^A, \tag{11.4.8}$$

where the subscript a indicates that the coefficients refer to the particular perturbed function $\tilde{\Phi}_a$ deriving from Φ_a. When, as we shall normally assume, the A-set functions are "correct zeroth-order combinations" in the usual sense, they will be non-mixing (off-diagonal elements of H' being zero), and $\mathbf{c}_a^A$ will contain only a single non-zero element (unity) corresponding to the particular function Φ_a. Consequently, on using (11.4.8) in (11.4.2) (with expansion of the inverse matrix), the perturbed functions take the form

$$\tilde{\Phi}_a^{(\lambda)} = \Phi_a + \lambda \sum_b \frac{\Phi_b \langle b | \mathsf{H}^{\lambda} | a \rangle}{E_a^{(0)} - E_b^{(0)}} = \Phi_a + \lambda \Phi_a^{\lambda}, \tag{11.4.9}$$

which is good to first order in λ.

Before considering the superposition of different perturbations in (11.4.6), we note that the elements of the effective Hamiltonian (11.4.7) may be expressed in an alternative form, which is often valuable in obtaining physical insight into the results. It is often stated that wavefunctions correct to first order yield energies correct to second order, and it might therefore have been expected that by taking matrix elements of H between the *first*-order perturbed functions given in (11.4.9) we should obtain the matrix $\mathbf{H}_{\text{eff}}$, with elements (11.4.7), whose eigenvalues are indeed second-order energies. This conjecture is easily

‡ Note that a term $\lambda \langle b | \mathsf{H}^{\lambda} | b \rangle$ in the denominator has now been dropped, the numerator being already of second order; expansion would lead to higher-order terms.

tested:

$$\begin{aligned}\langle \tilde{\Phi}_a^{(\lambda)} | \mathsf{H} | \tilde{\Phi}_{a'}^{(\lambda)} \rangle &= \langle (\Phi_a + \lambda \Phi_a^\lambda) | (\mathsf{H}_0 + \lambda \mathsf{H}^\lambda) | (\Phi_{a'} + \lambda \Phi_{a'}^\lambda) \rangle \\ &= E_a^{(0)} \delta_{aa'} + \lambda \langle \Phi_a | \mathsf{H}^\lambda | \Phi_{a'} \rangle \\ &\quad + \lambda^2 \langle \Phi_a^\lambda | \mathsf{H}^\lambda | \Phi_{a'} \rangle + \lambda^2 \langle \Phi_a | \mathsf{H}^\lambda | \Phi_{a'}^\lambda \rangle \\ &\quad + \lambda^2 \langle \Phi_a^\lambda | \mathsf{H}_0 | \Phi_{a'}^\lambda \rangle + O(\lambda^3).\end{aligned}$$

Other terms, involving H_0, vanish according to (11.4.5). On attaching a superscript $\lambda\lambda$ to denote the part of this expression quadratic in λ, we find

$$\begin{aligned}\langle \tilde{\Phi}_a^{(\lambda)} | \mathsf{H} | \tilde{\Phi}_{a'}^{(\lambda)} \rangle^{\lambda\lambda} &= 2\lambda^2 \sum_b \frac{\langle \Phi_q | \mathsf{H}^\lambda | \Phi_b \rangle \langle \Phi_b | \mathsf{H}^\lambda | \Phi_{a'} \rangle}{E_a^0 - E_a^{(0)}} \\ &\quad + \lambda^2 \sum_b \frac{\langle \Phi_a | \mathsf{H}^\lambda | \Phi_b \rangle E_b^{(0)} \langle \Phi_b | \mathsf{H}^\lambda | \Phi_{a'} \rangle}{(E_a^{(0)} - E_b^{(0)})^2},\end{aligned} \tag{11.4.10}$$

but this clearly does *not* coincide with the λ^2 term in (11.4.7).

The reason for this discrepancy is that the perturbed functions used are *not orthonormal* to second order; in determining energies to second order, the secular equation with matrix elements defined above would therefore be of the form (2.3.4), with overlap terms on the right. The matrix **M** will have elements

$$\langle \tilde{\Phi}_a^{(\lambda)} | \tilde{\Phi}_{a'}^{(\lambda)} \rangle = \langle \Phi_q | \Phi_{a'} \rangle + \lambda^2 \sum_b \frac{\langle \Phi_a | \mathsf{H}^\lambda | \Phi_b \rangle \langle \Phi_b | \mathsf{H}^\lambda | \Phi_{a'} \rangle}{E_a^{(0)} - E_b^{(0)}}, \tag{11.4.11}$$

and the secular equations will become

$$\sum_{a'} \left[\langle \tilde{\Phi}_a^{(\lambda)} | \mathsf{H} | \tilde{\Phi}_{a'}^{(\lambda)} \rangle - \lambda^2 E \sum_b \frac{\langle \Phi_a | \mathsf{H}^\lambda | \Phi_b \rangle \langle \Phi_b | \mathsf{H}^\lambda | \Phi_{a'} \rangle}{E_a^{(0)} - E_b^{(0)}} \right] c_{a'}^A = E c_a^A,$$

where the λ^2 term has been carried over to the left to make the result look like (11.4.3). But E on the left may be replaced by the zeroth-order approximation $E_a^{(0)}$, since it multiplies a term that already contains a λ^2 factor; and on combining the two terms in the square brackets, using (11.4.10), we immediately find the original expression (11.4.7) for $\langle a | \mathsf{H}_{\text{eff}} | a' \rangle$. Thus, on including overlap terms to order λ^2, the second sum on the right in (11.4.10) disappears, being absorbed in cancelling exactly half of the incorrect first sum.

In summary, with an abbreviated notation in which $\Phi_a \to a$, $\Phi_a^\lambda \to a_\lambda$, $\tilde{\Phi}_a^{(\lambda)} \to \tilde{a}_\lambda$,

$$\langle a | \mathsf{H}_{\text{eff}} | a' \rangle^{\lambda\lambda} = \tfrac{1}{2} \langle \tilde{a}_\lambda | \lambda \mathsf{H}^\lambda | \tilde{a}'_\lambda \rangle^{\lambda\lambda}. \tag{11.4.12a}$$

The zeroth-order *Hamiltonian* does not occur in this expression for the perturbation term. The use of first-order functions, with addition of the "correction factor" $\frac{1}{2}$, is in fact equivalent to performing a symmetric orthonormalization of the functions Φ_a (using (2.2.21)) and setting up the matrix of H in the new basis (working up to second order in λ). Thus we may also write

$$\langle a|\,\mathsf{H}_{\text{eff}}\,|a'\rangle^{\lambda\lambda} = \langle \bar{a}_\lambda|\,\lambda\mathsf{H}^\lambda\,|\bar{a}'_\lambda\rangle^{\lambda\lambda}, \tag{11.4.12b}$$

where the bar distinguishes the orthonormalized functions and the factor $\frac{1}{2}$ is absent. However, it is obviously more convenient to keep only first-order terms in the perturbed functions and to use the result (11.4.12a) as it stands. Whichever formulation is employed, the important fact is that the matrix elements of the *effective* Hamiltonian (which is a mathematical object, set up to determine energies to second order but with no intrinsic *physical* significance) are replaced by matrix elements of the *actual* Hamiltonian with respect to perturbed states.

Now let us consider a 2-term perturbation of the form (11.4.6): the first- and second-order parts of the effective Hamiltonian are evidently

$$\boxed{\langle a|\,\mathsf{H}_{\text{eff}}\,|a'\rangle^{(1)} = \lambda\langle a|\,\mathsf{H}^\lambda\,|a'\rangle + \mu\langle a|\,\mathsf{H}^\mu\,|a'\rangle,} \tag{11.4.13}$$

and

$$\begin{aligned}
\langle a|\,\mathsf{H}_{\text{eff}}\,|a'\rangle^{(2)} = {} & \lambda^2 \sum_b (E_a^{(0)} - E_b^{(0)})^{-1} \langle a|\,\mathsf{H}^\lambda\,|b\rangle\langle b|\,\mathsf{H}^\lambda\,|a'\rangle \\
& + \lambda\mu \sum_b (E_a^{(0)} - E_b^{(0)})^{-1} \langle a|\,\mathsf{H}^\lambda\,|b\rangle\langle b|\,\mathsf{H}^\mu\,|a'\rangle \\
& + \mu\lambda \sum_b (E_a^{(0)} - E_b^{(0)})^{-1} \langle a|\,\mathsf{H}^\mu\,|b\rangle\langle b|\,\mathsf{H}^\lambda\,|a'\rangle \\
& + \mu^2 \sum_b (E_a^{(0)} - E_b^{(0)})^{-1} \langle a|\,\mathsf{H}^\mu\,|b\rangle\langle b|\,\mathsf{H}^\mu\,|a'\rangle.
\end{aligned} \tag{11.4.14}$$

Evidently, in first order, the effects of different perturbations are simply additive; but in second order there are interesting interaction effects whose nature will become clear in later sections. Here we note only that second-order terms of both types (λ^2 and $\lambda\mu$) may again be expressed in terms of first-order perturbed functions. The typical λ^2 term has been given already in (11.4.12a), where it is sufficient to use the first-order functions $\tilde{\Phi}_a^{(\lambda)}$ calculated using only the $\lambda\mathsf{H}^\lambda$ perturbation. The $\lambda\mu$-terms (*both together*) may be determined using the *same* "λ-perturbed" functions but keeping only the $\mu\mathsf{H}^\mu$ term in H′:

$$\langle a|\,\mathsf{H}_{\text{eff}}\,|a'\rangle^{\lambda\mu} = \langle \tilde{a}_\lambda|\,\mu\mathsf{H}^\mu\,|\tilde{a}'_\lambda\rangle^{\lambda\mu}, \tag{11.4.15}$$

as may easily be verified by inserting (11.4.9) in the right-hand matrix element. It is also clear that there is an "interchange theorem": instead of using first-order functions $\bar{\Phi}_a^{(\lambda)}$, we could use "μ-perturbed" functions $\bar{\Phi}_a^{(\mu)}$, calculated as in (11.4.9) but with the perturbation $\mu \mathsf{H}^\mu$ alone, and then keep only the $\lambda \mathsf{H}^\lambda$ term in H'. The result (11.4.15) may thus be written alternatively as

$$\langle a| \mathsf{H}_{\text{eff}} |a'\rangle^{\lambda\mu} = \langle \tilde{a}_\mu| \lambda \mathsf{H}^\lambda |\tilde{a}'_\mu\rangle^{\lambda\mu}. \tag{11.4.16}$$

In either case it is clear that the "cross-terms" in (11.4.14) can be computed as if the two perturbations were applied *successively*, first-order wavefunctions being set up for the "first" perturbation and then used in calculating matrix elements for the "second".

The utility of the above results will become apparent in the applications to be considered presently. Evidently, in calculating the joint effect of two perturbations the last result offers the possibility of using whichever of the two perturbations is easier to handle in calculating the perturbed function. Moreover, the formulae (11.4.15) and (11.4.16) make no explicit reference to the full set of functions (including the B set) but only to the few states of interest (the A set), in the presence of *one* perturbation and calculated only to first order. The method of actually calculating the first-order perturbed states is thus left completely open (they are simply linear in the chosen perturbation parameter); and this again introduces a useful flexibility.

So far, the discussion has been based on the expansion (11.4.2) of the *electronic* wavefunction; but if we are to include the hyperfine structure of the energy levels, arising from the existence of nuclear spins, it will obviously be necessary to admit *nuclear* spin variables in the wavefunction and in the basis functions. The generalization of (11.4.2) is immediate: we expand the wavefunction in the usual way but in terms of electron–nuclear product functions

$$\Psi = \sum_{\kappa,\sigma} c_{\kappa\sigma} \Phi_\kappa \Theta_\sigma, \tag{11.4.17}$$

in which the Θ_σ are nuclear spin functions, which may in turn be products of spin functions for the individual nuclei. For a given number of nuclei there will be a finite number, p_n say, of such functions, and the number of terms in the original expansion (11.4.2) will therefore be multiplied by p_n.

The corresponding generalization of the matrix-eigenvalue equations is straightforward. If the A set contains p_e degenerate electronic functions Φ_a, admission of nuclear spin yields a manifold of $p_\text{e} \times p_\text{n}$ A-set products $\{\Phi_a \Theta_\sigma\}$, and this will be the dimension of the new AA-block in (11.4.4). The discussion then follows similar lines, except that the matrix elements in the various blocks are labelled by *double* indices, $a \to a\sigma$,

$b \rightarrow b\sigma$. Thus the effective Hamiltonian H_{eff} will have matrix elements (cf. (11.4.7))

$$\langle a\sigma| \mathsf{H}_{\text{eff}} |a'\sigma'\rangle = \delta_{aa'}\delta_{\sigma\sigma'}E_a^{(0)} + \langle a\sigma| \mathsf{H}' |a'\sigma'\rangle + \sum_{b,\sigma''} \frac{\langle a\sigma| \mathsf{H}' |b\sigma''\rangle\langle b\sigma''| \mathsf{H}' |a'\sigma'\rangle}{E_a^{(0)} - E_b^{(0)}}, \tag{11.4.18}$$

and all other equations will be modified in a similar way.

11.5 EFFECT OF A UNIFORM ELECTRIC FIELD

As a first illustration of the perturbation approach developed in Section 11.4, we consider the effect of a uniform electric field applied to a molecule in a non-degenerate ground state. The corresponding term in (11.3.10) will be $\mathsf{H}_{\text{elec}} = \sum_i[-e\phi(\boldsymbol{r}_i)]$, where $\phi(\boldsymbol{r}_i)$ is the electric potential at point $\boldsymbol{r}_i$ due to the applied field. In terms of field components, which conveniently serve as perturbation parameters, we may write (cf. (11.4.6))

$$\mathsf{H}_{\text{elec}} = F_x\mathsf{H}^x + F_y\mathsf{H}^y + F_z\mathsf{H}^z, \tag{11.5.1}$$

where $\mathsf{H}^x = -d_x$, etc., and

$$d_x = \sum_i d_x(i) = \sum_i (-ex_i), \tag{11.5.2}$$

is the x component of the electric-dipole-moment operator, referred to any convenient origin (at which $\phi = 0$) (similar expressions hold for d_y, d_z).

For a non-degnerate state the matrix $\mathbf{H}_{\text{eff}}$ in (11.4.3) has only one element, and it follows immediately that

$$E = \langle a| \mathsf{H}_{\text{eff}} |a\rangle = E^{(0)} + E^{(1)} + E^{(2)} + \ldots,$$

where, regarding (11.5.1) as a 3-term perturbation, the first- and second-order energies will follow from (11.4.13) and (11.4.14). We consider each order separately.

First order

The three terms in (11.5.1) give additive effects, and thus we find

$$E^{(1)} = -\boldsymbol{F} \cdot \boldsymbol{\mu}^0 = -(F_x\mu_x^0 + F_y\mu_y^0 + F_z\mu_z^0) \tag{11.5.3}$$

which corresponds to the classical expression for the interaction energy of the field and an electric dipole of moment $\boldsymbol{\mu}^0$ (the zero indicating a field-independent or "permanent" moment, appropriate to $\boldsymbol{F} \to \mathbf{0}$). A typical component of the electric moment is thus evidently

$$\mu_x^0 = -\langle a | \mathsf{H}^x | a \rangle = \langle a | d_x | a \rangle, \tag{11.5.4}$$

and since d_x is the sum of one-electron terms in (11.5.2) the usual reduction (5.2.17) allows us to write the expectation value as

$$\mu_x^0 = \int -exP(aa \mid \boldsymbol{r})\, \mathrm{d}\boldsymbol{r}, \tag{11.5.5}$$

where $P(aa \mid \boldsymbol{r})$ is the electron density associated with the unperturbed state a. The electric moment is evidently that of a classical charge distribution of density $-eP(aa \mid \boldsymbol{r})$, a result already anticipated in Section 5.7.

Second order

From (11.4.14), we obtain a second-order energy quadratic in the field components. Thus

$$E^{(2)} = \sum_{\lambda} E^{\lambda} + \sum_{\lambda<\nu} E^{\lambda\nu} \quad (\lambda, \nu = x, y, z), \tag{11.5.6}$$

where, for example,

$$E^x = -F_x^2 \sum_b \frac{\langle a | \mathsf{H}^x | b \rangle \langle b | \mathsf{H}^x | a \rangle}{\Delta E(a \to b)}, \tag{11.5.7}$$

$$E^{xy} = -F_x F_y \sum_b \frac{\langle a | \mathsf{H}^x | b \rangle \langle b | \mathsf{H}^y | a \rangle + \langle a | \mathsf{H}^y | b \rangle \langle b | \mathsf{H}^x | a \rangle}{\Delta E(a \to b)}. \tag{11.5.8}$$

Evidently E^{xy} is symmetric in the components, $E^{yx} = E^{xy}$, and (11.5.6) may thus be written as

$$E^{(2)} = \sum_{\lambda} E^{\lambda} + \tfrac{1}{2} \sum_{\lambda,\nu}{}' E^{\lambda\nu} = \tfrac{1}{2} \sum_{\lambda,\nu} E^{\lambda\nu}, \tag{11.5.9}$$

where we have noted that $E^x = \frac{1}{2}E^{xx}$ etc. and have taken such terms into the double sum.

It is customary to write the quadratic energy term in the form

$$E^{(2)} = -\tfrac{1}{2}\boldsymbol{F} \cdot \boldsymbol{\alpha} \cdot \boldsymbol{F} = -\tfrac{1}{2} \sum_{\lambda,\nu} F_{\lambda} \alpha_{\lambda\nu} F_{\nu}, \tag{11.5.10}$$

where $\alpha_{\lambda\nu}$ is one of the nine components of a symmetric *polarizability tensor.* By comparing (11.5.10) with (11.5.9) and using (11.5.8), with

$\mathsf{H}_x = -d_x$ etc., it follows that

$$\alpha_{\lambda\nu} = \sum_b \frac{\langle a|\, d_\lambda \,|b\rangle\langle b|\, d_\nu \,|a\rangle + \langle a|\, d_\nu \,|b\rangle\langle b|\, d_\lambda \,|a\rangle}{\Delta E(a \to b)}. \quad (11.5.11)$$

The matrix elements in (11.5.11) are off-diagonal, and their reduction thus required knowledge of the *transition* densities connecting the states a and b; thus, for example,

$$\langle b|\, d_x \,|a\rangle = \int -exP(ab \mid \boldsymbol{r})\, d\boldsymbol{r} \quad (11.5.12)$$

is the x component of the electric moment of a "transition charge cloud" of density $-eP(ab \mid \boldsymbol{r})$, associated with the transition $a \to b$.

It is because (11.5.11) involves transition energies and densities for all excited states b that the calculation of second-order quantities such as the polarizability is difficult, since usually the available Φ_b are only rough approximations to real excited states and their number is very limited. We note in passing, however, a connection with quantities encountered in spectroscopy: the intensity of a spectral line for the (electric-dipole-induced) transition $a \to b$, under the influence of isotropic radiation, is usually characterized by a dimensionless quantity, the "oscillator strength", and this is defined by

$$f_{ba} = \frac{2m}{\hbar^2 e^2} \frac{|\boldsymbol{d}_{ab}|^2}{3} \Delta E(a \to b), \quad (11.5.13)$$

where $\boldsymbol{d}_{ab}$ is the vector whose x component is given in (11.5.12). For rapidly tumbling molecules (fluid phase) only the averaged polarizability $\bar{\alpha} = \frac{1}{3}(\alpha_{xx} + \alpha_{yy} + \alpha_{zz})$ is observed, and, from (11.5.11),

$$\bar{\alpha} = 2 \sum_b \frac{|d_{ab}|^2}{3\, \Delta E(a \to b)}.$$

It is therefore possible to relate polarizability contributions to the strengths and energies of spectroscopic transitions; such relationships are sometimes of qualitative value.

For a classical system of electric charges there is an alternative interpretation of the components of the polarizability tensor—not as coefficients in the energy expression (11.5.10) but rather as proportionality constants describing an *induced* electric moment; thus, for example, α_{yx} measures the y component of an induced moment $\boldsymbol{\mu}^{\text{ind}}$ due to unit applied field in the x direction. The induced moment, linear in field components, is thus given by

$$\mu_x^{\text{ind}}(\boldsymbol{F}) = \alpha_{xx}F_x + \alpha_{xy}F_y + \alpha_{xz}F_z \quad (11.5.14)$$

etc. It is easily verified, by allowing the field to build up from zero to its final value and integrating the energy increase $\mathrm{d}E = -\mathrm{d}\boldsymbol{F} \cdot (\boldsymbol{\mu}^0 + \boldsymbol{\mu}^{\mathrm{ind}}(\boldsymbol{F}))$ (cf. (11.5.3)), that the quadratic part of the result is $-\frac{1}{2}\boldsymbol{F} \cdot \boldsymbol{\mu}^{\mathrm{ind}}$ ($\boldsymbol{F}$ and $\boldsymbol{\mu}^{\mathrm{ind}}$ having their *final* values), and is thus consistent with (11.5.10):

$$E^{(2)} = -\tfrac{1}{2}\boldsymbol{F} \cdot \boldsymbol{\mu}^{\mathrm{ind}}. \tag{11.5.15}$$

To show that the same alternative interpretation is valid in quantum mechanics, i.e. that α_{yx} measures the y polarization of the *electron density* due to unit field in the x direction, we must use the alternative expressions (11.4.12a) and (11.4.15) for the matrix elements. Thus, from (11.4.12a), the F_x-term in $E^{(2)}$ as given in (11.5.7) may also be written (using, for example, superscript xx for the term quadratic in F_x)

$$E^x = \tfrac{1}{2}\langle \tilde{a}_x | F_x \mathsf{H}^x | \tilde{a}_x \rangle^{xx}, \tag{11.5.16}$$

where $\tilde{a}_x$ corresponds to application of the x perturbation *alone* ($F_y = F_z = 0$). Similarly, from (11.4.15) and (11.4.16), the term E^{xy} in (11.5.8) may be obtained in *either* of the equivalent forms

$$E^{yx} = \langle \tilde{a}_x | \, F_y \mathsf{H}^y \, | \tilde{a}_x \rangle^{xy} = \langle \tilde{a}_y | \, F_x \mathsf{H}^x \, | \tilde{a}_y \rangle^{xy} = E^{xy}. \tag{11.5.17}$$

Thus, noting that $E^x = \frac{1}{2}E^{xx}$, the quadratic energy term may be written in terms of a symmetric tensor $\boldsymbol{\alpha}$:

$$E^{(2)} = \tfrac{1}{2}\sum_{\lambda,\nu} E^{\lambda\nu} = -\tfrac{1}{2}\sum_{\lambda,\nu} F_\lambda \alpha_{\lambda\nu} F_\nu. \tag{11.5.18}$$

The interpretation of the tensor components follows on writing the expectation values in (11.5.16) and (11.5.17) in terms of the perturbed electron density. Thus

$$P(\tilde{a}\tilde{a} \mid \boldsymbol{r}) = P(aa \mid \boldsymbol{r}) + F_x P_x(aa \mid \boldsymbol{r}) + \dots, \tag{11.5.19}$$

where the second term is the first-order *change* in the electron density due to the field component F_x. Hence

$$\alpha_{xx} = \int -exP_x(aa \mid \boldsymbol{r})\, \mathrm{d}\boldsymbol{r}. \tag{11.5.20}$$

In other words, α_{xx} is the x component of the electric moment of the *polarization* of the charge cloud produced by unit field in the x direction. A parallel reduction for the off-diagonal components gives

$$\alpha_{xy} = \int - exP_y(aa \mid \boldsymbol{r})\, \mathrm{d}\boldsymbol{r} = \int - exP_x(aa \mid \boldsymbol{r})\, \mathrm{d}\boldsymbol{r} = \alpha_{yx}, \tag{11.5.21}$$

where the first form corresponds to the x component of the density polarization P_y for unit field in the y direction, while the second has a similar interpretation but with the roles of x and y reversed.

The above results evidently confirm the validity of a classical interpretation of the effect of a uniform electric field; the polarization energy in (11.5.18) is actually associated with an induced electric moment, arising from a field-proportional distortion of the charge cloud, with components determined as in (11.5.21).

11.6 EFFECT OF A UNIFORM MAGNETIC FIELD

With a uniform magnetic field, the perturbation term linear in the field components is H'_{mag} defined in (11.3.12). The analogue of (11.5.1) is then

$$\mathsf{H}'_{\text{mag}} = -\boldsymbol{B}\cdot\mathbf{m} = -(B_x\mathsf{m}_x + B_y\mathsf{m}_y + B_z\mathsf{m}_z), \qquad (11.6.1)$$

where the components of the *magnetic*-moment operator are, for example,

$$\mathsf{m}_x = \sum_i \mathsf{m}_x(i) = \sum_i [-\beta\mathsf{L}_x(i)], \qquad (11.6.2)$$

where β is the Bohr magneton and the minus sign is associated with the negative charge of the electron.

The situation would appear to be precisely analogous to that discussed in the last section. Thus the analogues of (11.5.3) and (11.5.4) would be

$$E^{(1)} = -\boldsymbol{B}\cdot\boldsymbol{\mu}_m, \qquad (11.6.3)$$

where a typical component of the magnetic moment would be

$$\mu_{m,x} = \langle a|\,\mathsf{m}_x\,|a\rangle. \qquad (11.6.4)$$

For a free atom, for example, in which Φ_a might be an eigenstate of angular momentum around the nucleus (taken as origin in defining L), the system would behave like a small magnet of moment

$$\mu_{m,z} = -\beta M_L, \qquad (11.6.5)$$

β being the classically expected proportionality constant for circulating charged particles.

It is only for free atoms, however, that there is a unique "natural" origin, the nucleus, around which the angular momentum may be quantized. More generally, H'_{mag} in (11.3.12) contains $\boldsymbol{A} = \frac{1}{2}\boldsymbol{B}\times\boldsymbol{r}$, and is thus origin-dependent; the vector potential vanishes at the arbitrary origin of coordinates ($\boldsymbol{r} = \mathbf{0}$, $\boldsymbol{A} = \mathbf{0}$), but our results must be invariant, as indicated in Section 11.1, under change of origin and indeed under general gauge transformations of the vector potential.‡ Difficulties arise only in second and higher orders, however: in first order, the wavefunction used in (11.6.4) is field-independent, and if, under change of origin, $\mathbf{L} \to \mathbf{L} + (\boldsymbol{R} \times \mathbf{p})$ then (11.6.4) is unaffected as long as the linear-momentum components have zero expectation values (i.e. for bound states). In higher order the wavefunction acquires a field-dependence that depends on choice of origin, and there is also a second term in the perturbation, $\mathsf{H}''_{\text{mag}}$ in (11.3.12), which is already quadratic in field components and whose role is not immediately clear.

Before continuing, we note that for a molecule in a non-degenerate state, whose wavefunction may be assumed to be real, the components of the magnetic moment (11.6.4) are identically zero *whatever* the choice of origin; for L is i times a real operator, and Hermitian symmetry then requires $\langle \mathsf{L}_\lambda \rangle = 0$ ($\lambda = x, y, z$). In this case the angular momentum is said to be "quenched" and there is no corresponding "permanent" magnetism; molecules in non-degenerate ground states do not exhibit orbital *paramagnetism*. Molecules do, however, always exhibit *diamagnetism*; they acquire an *induced* magnetic moment, linearly dependent on the applied field components, and this may be described by a *magnetic*-polarizability tensor analogous to that used in describing the electric polarization. Evidently, diamagnetism is a second-order property and will depend on the first-order perturbation of the wavefunction. Since, in classical physics, diamagnetism is associated with induced electric currents, it is natural that we should look for a similar interpretation of the induced magnetic moment in quantum mechanics. We shall find that by introducing the probability current density (Section 5.10) the parallel with classical physics can be preserved and all results can be expressed in a gauge-invariant form.

The second-order contributions arising from H'_{mag} in (11.3.12) are§

$$E_1^{(2)} = \sum_{\lambda} E_1^{\lambda} + \sum_{\lambda<\nu} E_1^{\lambda\nu} \quad (\lambda, \nu = x, y, z), \tag{11.6.6}$$

where, for example,

$$E_1^x = -B_x^2 \sum_b \frac{\langle a|\, \mathsf{H}^x \,|b\rangle \langle b|\, \mathsf{H}^x \,|a\rangle}{\Delta E(a \to b)}, \tag{11.6.7}$$

‡ There is of course also a lack of uniqueness in defining the *electric* moment of the electron distribution in a molecule; but this causes no problems, for the centroid of charge offers a convenient reference point, and transformation to any other origin is trivial. This is not so for the magnetic moment; there are no magnetic "charges" (i.e. free poles), and the effect of gauge change requires further consideration.

§ Subscripts 1, 2 will indicate contributions from the first and second terms in $\mathsf{H}_{\text{mag}} = \mathsf{H}'_{\text{mag}} + \mathsf{H}''_{\text{mag}}$ of (11.3.12).

and similarly

$$E_1^{xy} = -B_x B_y \sum_b \frac{\langle a|\, \mathsf{H}^x\, |b\rangle\langle b|\, \mathsf{H}^y\, |a\rangle + \langle a|\, \mathsf{H}^y\, |b\rangle\langle b|\, \mathsf{H}^x\, |a\rangle}{\Delta E(a \to b)}. \quad (11.6.8)$$

So far, all equations resemble their counterparts (11.5.6)–(11.5.8) in the last section. However, we must now add a further contribution, also quadratic in field components, from $\mathsf{H}''_{\text{mag}}$ in (11.3.12). On writing this operator in the form

$$\mathsf{H}''_{\text{mag}} = \frac{e^2}{2m} \sum_i \boldsymbol{A}(i) \cdot \boldsymbol{A}(i) = \frac{1}{2}\left(\frac{e^2}{2m}\right) \sum_i \boldsymbol{B} \cdot \boldsymbol{r}_i \times \boldsymbol{A}(i) \quad (11.6.9)$$

(interchanging dot and cross in the vector triple product), the corresponding term in the energy will be

$$E_2^{(2)} = \langle a|\, \mathsf{H}''_{\text{mag}}\, |a\rangle = \frac{1}{2}\left(\frac{e^2}{2m}\right) \sum_\lambda B_\lambda \int [(\boldsymbol{r} \times \boldsymbol{A})_\lambda P(aa \mid \boldsymbol{r})]\, \mathrm{d}\boldsymbol{r}, \quad (11.6.10)$$

where $P(aa \mid \boldsymbol{r})$ is the electron density in the unperturbed state Φ_a.

Since induced currents will arise from the action of the applied field, the interpretation we seek requires that (11.6.7) and (11.6.8) be recast in terms of *field-perturbed* wavefunctions; and this may be done by using (11.4.12a) and (11.4.15) respectively. Thus we find

$$E_1^x = \tfrac{1}{2}\langle \tilde{a}_x | B_x \mathsf{H}^x | \tilde{a}_x \rangle^{xx}, \quad (11.6.11)$$

where $\tilde{a}_x$ denotes the perturbed function $\tilde{\Phi}_a^{(x)}$ resulting from the application of $B_x \mathsf{H}^x$ alone, exactly as in (11.5.16); and similarly

$$E_1^{yx} = \langle \tilde{a}_x|\, B_y \mathsf{H}^y\, |\tilde{a}_x\rangle^{xy} = \langle \tilde{a}_y|\, B_x \mathsf{H}^x\, |\tilde{a}_y\rangle^{xy} = E_1^{xy}, \quad (11.6.12)$$

as in (11.5.17). As usual, the superscripts xx and xy indicate that we must pick out the terms in B_x^2 and $B_x B_y$ respectively.

Let us now put, to first order in field components,

$$P(\tilde{a}\tilde{a} \mid \boldsymbol{r}) = P(aa \mid \boldsymbol{r}) + \sum_\nu B_\nu P_\nu(aa \mid \boldsymbol{r}) \quad (11.6.13)$$

for the perturbed density matrix when $\sum_\nu B_\nu H^\nu$ is applied, this being the analogue of (11.5.19). On using the explicit forms of H^x and H^y, for example

$$\mathsf{H}^x = -\mathsf{m}_x = \sum_i \beta \mathsf{L}_x(i) = \sum_i \frac{e}{2m} [\boldsymbol{r}_i \times \mathbf{p}(i)]_x, \quad (11.6.14)$$

we thus obtain, from (11.6.11) and (11.6.12), typical components

$$E_1^x = B_x^2 \frac{e}{2m} \int [(\boldsymbol{r} \times \mathbf{p})_x P_x(aa \mid \boldsymbol{r}; \boldsymbol{r}')]_{\boldsymbol{r}'=\boldsymbol{r}} \, d\boldsymbol{r}, \qquad (11.6.15)$$

$$E_1^{xy} = B_x B_y \frac{e}{2m} \int [(\boldsymbol{r} \times \mathbf{p})_x P_y(aa \mid \boldsymbol{r}; \boldsymbol{r}')]_{\boldsymbol{r}'=\boldsymbol{r}} \, d\boldsymbol{r} \qquad (11.6.16)$$

for use in (11.6.6). It then remains only to add the field-quadratic terms from $\mathsf{H}''_{\text{mag}}$, already obtained in (11.6.10), and to arrange the result so as to reveal the role of the current density defined in Section 5.10.

On noting that $E_1^\lambda = \frac{1}{2} E_1^{\lambda\lambda}$, we may write (11.6.6) in the form $E_1^{(2)} = \frac{1}{2} \sum_{\kappa,\nu} E_1^{\lambda\nu}$; and on adding (11.6.10) we then find

$$E^{(2)} = \frac{1}{2}\left(\frac{e}{2m}\right) \sum_\lambda B_\lambda \bigg[\int [(\boldsymbol{r} \times e\boldsymbol{A})_\lambda P(aa \mid \boldsymbol{r}; \boldsymbol{r}')]_{\boldsymbol{r}'=\boldsymbol{r}} \, d\boldsymbol{r}$$

$$+ \int [(\boldsymbol{r} \times \mathbf{p})_\lambda \sum_\nu B_\nu P_\nu(aa \mid \boldsymbol{r}; \boldsymbol{r}')]_{\boldsymbol{r}'=\boldsymbol{r}} \, d\boldsymbol{r} \bigg]. \qquad (11.6.17)$$

The current density, which will include both "permanent" and "induced" currents, of zero and first order respectively in field components, may now be introduced. The λ component of the current density is, from (5.10.4),

$$J_\lambda = \frac{1}{m} \operatorname{Re} \left[(\mathbf{p} + e\boldsymbol{A})_\lambda [P(aa \mid \boldsymbol{r}; \boldsymbol{r}') + \sum_\nu B_\nu P_\nu(aa \mid \boldsymbol{r}; \boldsymbol{r}')] \right]_{\boldsymbol{r}'=\boldsymbol{r}}, \qquad (11.6.18)$$

and, on separating field-independent and field-proportional terms, this breaks into

$$J_\lambda(\boldsymbol{r}) = J_\lambda^0(\boldsymbol{r}) + J_\lambda^{\text{ind}}(\boldsymbol{r}), \qquad (11.6.19)$$

where the permanent current (for $\boldsymbol{A}, \boldsymbol{B} \to \boldsymbol{0}$) is

$$J_\lambda^0(\boldsymbol{r}) = \frac{1}{m} \operatorname{Re} [\mathsf{p}_\lambda P(aa \mid \boldsymbol{r}; \boldsymbol{r}')]_{\boldsymbol{r}'=\boldsymbol{r}}, \qquad (11.6.20)$$

while the induced current is

$$J_\lambda^{\text{ind}}(\boldsymbol{r}) = \frac{1}{m} \operatorname{Re} [eA_\lambda P(aa \mid \boldsymbol{r}; \boldsymbol{r}') + \mathsf{p}_\lambda \sum_\nu B_\nu P_\nu(aa \mid \boldsymbol{r}; \boldsymbol{r}')]_{\boldsymbol{r}'=\boldsymbol{r}}. \qquad (11.6.21)$$

Evidently, as we expected, the quadratic energy term (11.6.17) depends on the *induced* current, which is field-proportional: in fact it may be written as

$$E^{(2)} = -\tfrac{1}{2} \int \tfrac{1}{2} \boldsymbol{B} \cdot \boldsymbol{r} \times (-e\boldsymbol{J}^{\text{ind}}) \, d\boldsymbol{r}, \qquad (11.6.22)$$

and this again has a classical form

$$E^{(2)} = -\tfrac{1}{2}\boldsymbol{B} \cdot \boldsymbol{\mu}_{\mathrm{m}}^{\mathrm{ind}} \tag{11.6.23}$$

(cf. (11.5.15), for an electric field) corresponding to the energy of interaction between the field $\boldsymbol{B}$ and an *induced* magnetic dipole of moment

$$\mu_{\mathrm{m},\lambda}^{\mathrm{ind}} = \tfrac{1}{2}\int [\boldsymbol{r} \times (-e\boldsymbol{J}^{\mathrm{ind}})]_\lambda \, \mathrm{d}\boldsymbol{r} \tag{11.6.24}$$

The *first*-order energy, on the other hand, has been obtained in (11.6.3) and (11.6.4), and may be written as

$$E^{(1)} = -\tfrac{1}{2}\boldsymbol{B} \cdot \boldsymbol{\mu}_{\mathrm{m}}^{0}, \tag{11.6.25}$$

which is the classical energy of interaction between the field and a *permanent* magnetic dipole, whose moment is seen to be (on reducing (11.6.4))

$$\mu_{\mathrm{m},\lambda}^{0} = -\frac{e\hbar}{2m}\int [L_\lambda P(aa \mid \boldsymbol{r};\boldsymbol{r}')]_{\boldsymbol{r}'=\boldsymbol{r}} \, \mathrm{d}\boldsymbol{r} = \tfrac{1}{2}\int [\boldsymbol{r} \times (-e\boldsymbol{J}^{0})]_\lambda \, \mathrm{d}\boldsymbol{r}. \tag{11.6.26}$$

The results are thus entirely analogous to those obtained for the electric field; the *scalar* moment of the electric *charge* density is simply replaced by the *vector* moment (torque) of the electric *current* density; the density in each case may be either permanent or induced—(11.6.24) being formally similar to (11.6.26).

The induced magnetic dipole may be described by means of a "magnetic-polarizability tensor" by putting

$$\mu_{\mathrm{m},\lambda}^{\mathrm{ind}} = \sum_{\nu} \alpha_{\lambda\nu}^{\mathrm{m}} B_\nu, \tag{11.6.27}$$

which is the analogue of (11.5.14). The typical tensor component is then (cf. (11.6.23) *et seq.*)

$$\alpha_{xy}^{\mathrm{m}} = \tfrac{1}{2}\int [\boldsymbol{r} \times (-e\boldsymbol{J}^{y})]_x \, \mathrm{d}\boldsymbol{r}, \tag{11.6.28}$$

as follows on putting (11.6.21) in (11.6.24), noting that $\boldsymbol{r} \times \boldsymbol{A} = \boldsymbol{0}$ when $\boldsymbol{A} = \tfrac{1}{2}\boldsymbol{B} \times \boldsymbol{r}$, and writing $\boldsymbol{J}^{y}$ for the part of the induced current density associated with the P_y term in (11.6.21). Comparison with (11.6.26) then shows that α_{xy}^{m} is the x component of the induced magnetic moment due to the currents arising from unit field along the y direction (P_y being the corresponding density change).

In summary, the energy up to terms quadratic in the applied field is

$$E = E_0 - \boldsymbol{\mu}_{\mathrm{m}}^{0} \cdot \boldsymbol{B} - \tfrac{1}{2}\boldsymbol{B} \cdot \boldsymbol{\alpha}_{\mathrm{m}} \cdot \boldsymbol{B}, \tag{11.6.29}$$

and the corresponding magnetic moment is

$$\boldsymbol{\mu}_{\mathrm{m}} = \boldsymbol{\mu}_{\mathrm{m}}^{0} + \boldsymbol{\mu}_{\mathrm{m}}^{\mathrm{ind}} = \boldsymbol{\mu}_{\mathrm{m}}^{0} + \boldsymbol{\alpha}_{\mathrm{m}} \cdot \boldsymbol{B}, \tag{11.6.30}$$

with the dipole components defined in (11.6.26) and (11.6.24) in terms of the quantum-mechanical current density. Expression (11.6.29) may be written in another way (interchanging dot and cross in (11.6.22) etc.),

$$E = E_0 - \int \boldsymbol{A} \cdot (-e\boldsymbol{J}^0)\, \mathrm{d}\boldsymbol{r} - \tfrac{1}{2} \int \boldsymbol{A} \cdot (-e\boldsymbol{J}^{\mathrm{ind}})\, \mathrm{d}\boldsymbol{r}, \tag{11.6.31}$$

and this is again of classical form: it represents the potential energy of a system of electric currents in a conducting medium in the presence of a magnetic field of vector potential $\boldsymbol{A}$. The result is in fact more general than the present derivation would suggest: the preceeding equations were derived for $\boldsymbol{A} = \frac{1}{2}\boldsymbol{B} \times \boldsymbol{r}$, but (11.6.31) does not depend on the uniformity of the applied field (McWeeny, 1986), being valid for any form of $\mathbf{A}$.

Finally, we note that all the results of this section are expressed in terms of current densities $(\boldsymbol{J}, \boldsymbol{J}^0, \boldsymbol{J}^{\mathrm{ind}})$, which are defined in a gauge-invariant manner. Under a change of origin, the density matrix P must be changed according to (11.1.20) and the current density (and its component parts $\boldsymbol{J}^0$, $\boldsymbol{J}^{\mathrm{ind}}$) will be unaffected as a result.

11.7 SOME SPIN-COUPLING EFFECTS (FIRST-ORDER)

We now turn to the spin-dependent terms listed in Section 11.3, and discuss some examples of their observable effects. Such effects are most commonly interpreted in terms of a "phenomenological" Hamiltonian, which usually contains only spin operators (for the various nuclei and for the *total* electron spin) and applied fields, together with numerical parameters that serve as "coupling constants". This *spin Hamiltonian* H_s describes a "model" spin system whose behaviour may be determined by solving

$$\mathsf{H}_s \Theta = E\Theta \tag{11.7.1}$$

in a basis of (electron-nuclear) spin functions. With a proper choice of coupling constants, the eigenvalues of (11.7.1) will fit the observed energy levels. Thus, for example, the spin Hamiltonian

$$\mathsf{H}_s = -\sum_{\lambda,\nu} \beta B_\lambda g_{\lambda\nu} \mathsf{S}_\nu + \sum_n \beta_{\mathrm{p}} g_n \boldsymbol{B} \cdot \mathsf{I}(n) + \sum_n \sum_{\lambda,\nu} a_{\lambda\nu}^{(n)} \mathsf{S}_\lambda \mathsf{I}_\nu(n) + \dots \tag{11.7.2}$$

contains an "electronic g tensor", whose components describe the (Zeeman) interaction between the applied field $\boldsymbol{B}$ and a *total* electron spin **S**; a *nuclear* Zeeman term for each nucleus n interacting with the field; and an electron–nuclear interaction with a coupling tensor whose components are $a_{\lambda\nu}^{(n)}$ for each nucleus.

The spin Hamiltonian, from which all the complexities associated with electronic motion have been eliminated, provides a convenient interface between theory and experiment; the task of the theoretician is to predict or interpret the values of the coupling constants, using the electronic wavefunctions of the real system. In this section we consider some of the effects that arise in first order in the effective-Hamiltonian formulation of Section 11.4.

Electron–nuclear contact coupling

The contact term in the expression (11.3.17) for H_{N} may be written, by simple rearrangement, as

$$\mathsf{H}_{\mathrm{N}}^{\mathrm{cont}} = \frac{8\pi g\beta\beta_{\mathrm{P}}}{3\kappa_0 c^2} \sum_n g_n \sum_{i,m} (-1)^m \delta(\boldsymbol{r}_{ni}) \mathsf{S}_m(i) \mathsf{I}_{-m}(n), \qquad (11.7.3)$$

where $m = 0$, ± 1 label the "spherical" components (p. 148) of the corresponding vector operators, spherical and cartesian components of a vector $\boldsymbol{V}$ being related by

$$V_{+1} = -\frac{V_x + iV_y}{\sqrt{2}}, \qquad V_0 = V_z, \qquad V_{-1} = \frac{V_x - iV_y}{\sqrt{2}}. \qquad (11.7.4)$$

The phase conventions adopted ensure that the V_m transform under rotation of axes (McWeeny, 1973) in exactly the same way as angular-momentum eigenfunctions (quantum numbers l, m) with $l = 1$, $m = 0$, ± 1; they therefore conform to the standard definition of "irreducible tensor operators". The contact term is of special importance in determining the nuclear hyperfine structure of ESR signals, since it produces isotropic effects that (unlike many others) do not average to zero for the rapidly tumbling molecules in gases and liquids. The reason for writing the scalar product in terms of tensor operators is that the matrix element expressions may then be reduced easily, using the results of Appendix 3, and of Section 5.9.

The contact term is just one part of the perturbation considered in (11.4.6), and makes its own contribution to $\mathsf{H}_{\mathrm{eff}}$ as indicated in (11.4.13). When nuclear spins are admitted, the effective Hamiltonian will contain corresponding matrix elements $\langle a\sigma | \mathsf{H}^{\mathrm{cont}} | a'\sigma' \rangle$, where a, a' stand for electronic wavefunctions Φ_a, $\Phi_{a'}$ and σ, σ' for nuclear spin functions Θ_σ, $\Theta_{\sigma'}$. If the electronic ground state comprises the p_{e} $(= 2S_a + 1)$

components of a spin multiplet, and there are p_n, say, possible nuclear spin factors Θ_σ, then the A-set will contain $p_e p_n$ degenerate functions; and the first-order splitting will be obtained from the eigenvalues of the $p_e p_n \times p_e p_n$ matrix $\mathbf{H}_{\text{eff}}$. On separating electronic and nuclear-spin factors, we obtain, for a given term in the sum over n and m in (11.7.3),

$$\left\langle a\sigma \left| \sum_i \delta(\boldsymbol{r}_{ni})\mathsf{S}_m(i)\mathsf{I}_{-m}(n) \right| a'\sigma' \right\rangle = \left\langle a \left| \sum_i \delta(\boldsymbol{r}_{ni})\mathsf{S}_m(i) \right| a' \right\rangle \left\langle \sigma \left| \mathsf{I}_{-m}(n) \right| \sigma' \right\rangle,$$

and the first factor, containing a rank-1 spin operator, may be reduced by (5.9.8) and (5.9.10) to

$$\left\langle a \left| \sum_i \delta(\boldsymbol{r}_{ni})\mathsf{S}_m(i) \right| a' \right\rangle = \begin{bmatrix} S'_a & 1 & S_a \\ M'_a & m & M_a \end{bmatrix} \int \delta(\boldsymbol{r}_{ni}) Q_s(\bar{a}\bar{a} \mid \boldsymbol{r}_1)\, d\boldsymbol{r}_1, \quad (11.7.5)$$

where $Q_S(\bar{a}\bar{a} \mid \boldsymbol{r}_1)$ is the spin density for the state $\bar{a}$ with $M_a = S_a$, and the numerical coefficient vanishes unless $M_a = M'_a + m$. The coefficient is a ratio of two Clebsch–Gordan coefficients (5.9.11), and may be eliminated in the present case by noting that the various matrix elements of pure total spin operators between pure spin eigenfunctions with quantum numbers (S, M) and (S, M') are related in a precisely similar way. Thus

$$\langle SM | \mathsf{S}_m | SM' \rangle = \begin{pmatrix} S & 1 & S \\ M' & m & M \end{pmatrix} \times \text{constant}$$

leads to

$$\begin{bmatrix} S & 1 & S \\ M' & m & M \end{bmatrix} = \begin{pmatrix} S & 1 & S \\ M' & m & M \end{pmatrix} \Big/ \begin{pmatrix} S & 1 & S \\ S & 0 & S \end{pmatrix}$$

$$= \frac{\langle SM | \mathsf{S}_m | SM' \rangle}{\langle SS | \mathsf{S}_z | SS \rangle} = \frac{\langle SM | \mathsf{S}_m | SM' \rangle}{S}. \quad (11.7.6)$$

This is clearly only possible in the case of matrix elements between functions with $S_a = S_b = S$, because otherwise the matrix elements of the pure spin operators all vanish and no proportionality can be established. The validity of the "replacement theorem" (A3.29), which is essentially another statement of the present results, is confined to similar situations.

The integral on the right, in (11.7.5), reduces to $Q_S(\bar{a}\bar{a} \mid \boldsymbol{R}_n)$, and hence, from (11.7.6),

$$\left\langle a \left| \sum_i \delta(\boldsymbol{r}_{ni})\mathsf{S}_m(i) \right| a' \right\rangle = \langle M_a | \mathsf{S}_m | M'_a \rangle D_S(aa \mid \boldsymbol{R}_n) \quad (11.7.7)$$

where $D_S(aa \mid \boldsymbol{R}_n) = Q_S(\bar{a}\bar{a} \mid \boldsymbol{R}_n)/S$ is the *normalized* spin density (common to all states of multiplet a) evaluated for $\boldsymbol{r}_1 = \boldsymbol{R}_n$, i.e. *at* nucleus n.

The contact contribution to $\mathbf{H}_{\text{eff}}$ then has matrix elements

$$\langle a\sigma|\, \mathsf{H}_{\mathrm{N}}^{\text{cont}}\, |a'\sigma'\rangle$$

$$= \frac{8\pi g\beta\beta_{\mathrm{P}}}{3\kappa_0 c^2} \sum_n g_n D_S(aa \mid \boldsymbol{R}_n) \sum_m (-1)^m \langle M_a|\, \mathsf{S}_m\, |M_a'\rangle \langle \sigma|\, \mathsf{I}_{-m}(n)\, |\sigma'\rangle$$

$$= \frac{8\pi g\beta\beta_{\mathrm{P}}}{3\kappa_0 c^2} \sum_n g_n D_S(aa \mid \boldsymbol{R}_n) \langle M_a\sigma|\, \mathbf{S}\cdot\mathbf{I}(n)\, |M_a'\sigma'\rangle.$$

In other words, the elements of $\mathbf{H}_{\text{eff}}$ arising from $\mathsf{H}_{\mathrm{N}}^{\text{cont}}$ are identical, in first order, with those of a spin-Hamiltonian term

$$\mathsf{H}_{\mathrm{s}}^{(1)}\,(\text{cont}) = \sum_n hA_n^{\text{cont}}\mathbf{I}(n)\cdot\mathbf{S}, \tag{11.7.8a}$$

taken in a basis of electron–nuclear spin states, the contact coupling constant for nucleus n (here expressed in Hz) being

$$A_n^{\text{cont}} = \frac{8\pi g\beta\beta_{\mathrm{P}}}{3\kappa_0 c^2 h} g_n D_S(aa \mid \boldsymbol{R}_n). \tag{11.7.8b}$$

The nuclear spins therefore provide "probes" by which the spin densities, at the various nuclei, can be experimentally inferred from the observed coupling constants.

Electron–nuclear dipole coupling

The dipole–dipole interaction between electrons and nuclei, given in (11.3.17), is

$$\mathsf{H}_{\mathrm{N}}^{\text{dip}} = \frac{g\beta\beta_{\mathrm{P}}}{\kappa_0 c^2} \sum_{n,i} g_n r_{ni}^{-5}[3(\mathbf{S}(i)\cdot\boldsymbol{r}_{ni})(\mathbf{I}(n)\cdot\boldsymbol{r}_{ni}) - r_{ni}^2\mathbf{S}(i)\cdot\mathbf{I}(n)], \tag{11.7.9}$$

and gives a first-order contribution to H_{eff} that is identical with that arising from a suitable term in a spin Hamiltonian, namely

$$\mathsf{H}_{\mathrm{s}}^{(1)}(\text{dip}) = \sum_n \sum_{\lambda,\mu} hA_n^{\text{dip}}{}_{\lambda\mu}\mathsf{S}_\lambda \mathsf{I}_\mu(n) \qquad (\lambda,\, \mu = x,\, y,\, z), \tag{11.7.10a}$$

where the coupling with each nucleus is described by a tensor with components (in Hz)

$$\left.\begin{aligned} A_{n,xx}^{\text{dip}} &= \frac{3g\beta g_n\beta_{\mathrm{P}}}{\kappa_0 c^2 h} \int r_{n1}^{-5}\,(x_{n1}^2 - \tfrac{1}{2}r_{n1}^2) D_S(aa \mid \boldsymbol{r}_1)\, \mathrm{d}\boldsymbol{r}_1, \\ A_{n,xy}^{\text{dip}} &= \frac{3g\beta g_n\beta_{\mathrm{P}}}{\kappa_0 c^2 h} \int r_{n1}^{-5}(x_{n1}y_{n1}) D_S(aa \mid \boldsymbol{r}_1)\, \mathrm{d}\boldsymbol{r}_1 \end{aligned}\right\} \tag{11.7.10b}$$

etc., and $\mathbf{r}_{n1}$ is the vector from nucleus n to the point $\mathbf{r}_1$ at which the spin density is evaluated. The coupling constants $A^{\text{dip}}_{n,\lambda\mu}$ therefore measure the moments and products of inertia (i.e. the second moments) of a distribution of spin of density $D_S(aa \mid \mathbf{r}_1)$, taken about nucleus n.

Electron spin–spin coupling (first order)

The dipolar coupling between the spins of different electrons, given in (11.3.16) may be written in terms of standard tensor operators.‡ We introduce the notation $[\mathbf{S}(i) \times \mathbf{S}(j)]^{(2)}_m$ to mean a linear combination of spin-operator products $\mathsf{S}_\mu(i)\mathsf{S}_\nu(j)$ transforming like a spin eigenfunction with quantum numbers (s, m) and $s = 2$ (i.e. a rank-2 tensor): such tensor operators are listed elsewhere (e.g. McWeeny, 1973). The dipolar term in (11.3.16) can then be put in the form (cf. (11.7.3); see also p. 551)

$$\mathsf{H}_{SS} = -\frac{3}{2\kappa_0 c^2} \sum_{i,j}{}' g^2\beta^2 r_{ij}^{-5} \sum_m (-1)^m [\mathbf{r}_{ij} \times \mathbf{r}_{ij}]^{(2)}_{-m} [\mathbf{S}(i) \times \mathbf{S}(j)]^{(2)}_m, \quad (11.7.11)$$

and a procedure exactly analogous to that used in establishing (11.7.7) then leads to

$$\left\langle a \left| \sum_{i,j}{}' r_{ij}^{-5} [\mathbf{r}_{ij} \times \mathbf{r}_{ij}]^{(2)}_{-m} [\mathbf{S}(i) \times \mathbf{S}(j)]^{(2)}_m \right| a' \right\rangle$$
$$= \frac{\langle M_a | \, [\mathbf{S} \times \mathbf{S}]^{(2)}_m \, | M'_a \rangle}{\langle S_a | \, [\mathbf{S} \times \mathbf{S}]^{(2)}_0 \, | S_a \rangle} \int r_{ij}^{-5} [\mathbf{r}_{ij} \times \mathbf{r}_{ij}]^{(2)}_{-m} Q_{SS}(\bar{a}\bar{a} \mid \mathbf{r}_1, \mathbf{r}_2) \, d\mathbf{r}_1 \, d\mathbf{r}_2, \quad (11.7.12)$$

where the $[\mathbf{S} \times \mathbf{S}]^{(2)}_m$ are formed from *total* spin components and $|M_a\rangle$ again denotes a spin eigenstate. Adopting unnormalized components,

$$[\mathbf{S} \times \mathbf{S}]^{(2)}_0 = 2\mathsf{S}_z^2 - \mathsf{S}_x^2 - \mathsf{S}_y^2 = 3\mathsf{S}_z^2 - \mathsf{S}^2,$$

and the lower matrix element in (11.7.12) reduces to $3S_a^2 - S_a(S_a + 1)$. The spin–spin coupling function Q_{SS} refers to the state with $M_a = S_a$ and has been defined in Section 5.9; again only the diagonal element ($\mathbf{r}'_1 = \mathbf{r}_1$, $\mathbf{r}'_2 = \mathbf{r}_2$) is required. On using (11.7.12) to reduce $\langle a\lambda | \, \mathsf{H}_{SS} \, | a'\lambda' \rangle$, the matrix element is apparently equivalent to that of a suitable spin Hamiltonian term $\mathsf{H}^{(1)}_s$ (spin–spin) taken between pure electron–nuclear spin states. For many purposes it is convenient to rewrite the result in terms of cartesian components, and when this is done we obtain

$$\mathsf{H}^{(1)}_s(\text{spin–spin}) = \sum_{\lambda,\mu} hD_{\lambda\mu} \mathsf{S}_\lambda \mathsf{S}_\mu \qquad (\lambda, \mu = x, y, z), \quad (11.7.13a)$$

‡ Discussions are available in books on angular-momentum theory. For a brief exposition see Talmi and de Shalit (1963) or McWeeny (1973, Section 5.8).

where

$$\left.\begin{aligned} D_{xx} &= \frac{2g^2\beta^2}{\kappa_0 c^2 h}\int r_{12}^{-5}(x_{12}^2 - \tfrac{1}{3}r_{12}^2)D_{SS}(aa \mid \boldsymbol{r}_1, \boldsymbol{r}_2)\,\mathrm{d}\boldsymbol{r}_1\,\mathrm{d}\boldsymbol{r}_2, \\ D_{xy} &= \frac{3g^2\beta^2}{\kappa_0 c^2 h}\int r_{12}^{-5}(x_{12}y_{12})D_{SS}(aa \mid \boldsymbol{r}_1, \boldsymbol{r}_2)\,\mathrm{d}\boldsymbol{r}_1\,\mathrm{d}\boldsymbol{r}_2, \end{aligned}\right\} \quad (11.7.13\text{b})$$

and $D_{SS}(aa \mid \boldsymbol{r}_1, \boldsymbol{r}_2)$ denotes the normalized function $[2S_a(S_a-1)]^{-1} \times Q_{SS}(\bar{a}\bar{a} \mid \boldsymbol{r}_1, \boldsymbol{r}_2)$ such that

$$\int D_{SS}(aa \mid \boldsymbol{r}_1, \boldsymbol{r}_2)\,\mathrm{d}\boldsymbol{r}_1\,\mathrm{d}\boldsymbol{r}_2 = 1.$$

Electron spin–spin coupling shifts the molecular Zeeman levels even in the limit $\boldsymbol{B}\rightarrow\boldsymbol{0}$, and therefore produces a "zero-field splitting"; this was first observed by Hutchison and Mangum (1958).

Spin–orbit coupling in diatomic molecules

In the preceeding examples the degeneracy of the electronic state arose only from spin. Sometimes, however, there may also be an *orbital* degeneracy associated with spatial symmetry. To illustrate this possibility, let us consider a diatomic molecule such as NO—in which the $^2\Pi$ ground state corresponds to $S=\frac{1}{2}$ and one unit of *orbital* angular momentum around the bond axis. Such states are characterized by quantum numbers Λ and Σ, which refer to the projections of orbital and spin angular momentum along the axis, and may be denoted by $|\Lambda, \Sigma\rangle$. Thus, for NO, with $\Lambda=0, \pm 1$ and $\Sigma=\frac{1}{2}$, there might appear to be six states in the A-set:

$$|1,\tfrac{1}{2}\rangle,\quad |1,-\tfrac{1}{2}\rangle,\quad |0,\tfrac{1}{2}\rangle,\quad |0,-\tfrac{1}{2}\rangle,\quad |-1,\tfrac{1}{2}\rangle,\quad |-1,-\tfrac{1}{2}\rangle.$$

Those with $\Lambda=0$, however, are *sigma* states; and, although all six would be degenerate for a free atom with $L=1$, the angular momentum is partially quenched by the molecular environment, and the states with $\Lambda=0$ no longer belong to the electron configuration of the ground state.‡ The A-set in this example therefore contains four states, which fall into two pairs:

$$|\pi^+\alpha\rangle,\ |\pi^-\beta\rangle, \qquad |\pi^+\beta\rangle,\ |\pi^-\alpha\rangle.$$

Each of these pairs, whose component states differ only through a

‡ In an orbital description they would involve promotion of the odd π electron into an antibonding σ-type MO.

common sign change of orbital and spin angular momentum, is called a "Kramers doublet" (Kramers, 1930), and remains degenerate in the absence of an applied magnetic field. The effect of spin–orbit coupling is to resolve the degeneracy between the *different* pairs.

We need the terms in the effective Hamiltonian $\mathbf{H}_{\text{eff}}$ that arise from H_{SL}. In the absence of the external field, this becomes

$$\mathsf{H}_{SL} = \frac{g\beta^2}{\kappa_0 c^2}\left[\sum_{n,i} \frac{Z_n \mathbf{S}_{(i)} \cdot \mathbf{L}^{(n)}(i)}{r_{ni}^2} - \sum_{i,j}{}' \frac{2\mathbf{S}(i) \cdot \mathbf{L}^{(i)}(j) + \mathbf{S}(i) \cdot \mathbf{L}^{(j)}(i)}{r_{ij}^3}\right], \tag{11.7.14}$$

and the corresponding elements of the effective Hamiltonian are then, to first order,

$$[H_{\text{eff}}(SL)]_{aM,a'M'} = \langle aM|\, \mathsf{H}_{SL}\, |a'M'\rangle, \tag{11.7.15}$$

where we use M, M' to label spin multiplet components ($\pm\frac{1}{2}$ for the doublets) and a, a' for the two orbitally degenerate states (π^+ and π^-). The matrix elements vanish, however, unless $a = a'$ and $M = M'$ (the bra and ket functions transforming, under rotations around the axis, like basis vectors of different irreps), while the remaining diagonal elements may be expressed in terms of the standard density functions of Section 5.9.

The first term in (11.7.14), written using spherical components, contains a factor of the form $F(i)S_m(i)$, which may be reduced, using (5.9.8), to an integral involving the spin density: the second depends instead upon the function Q_m^{SL}, according to (5.9.15). Since the density functions connecting states of different M are in general proportional (see (5.9.10) and (5.9.17)) to "standard" functions (defined for $M' = M = S$), with a proportionality factor that can be expressed as $\langle M|\, \mathsf{S}_z\, |M'\rangle/S$, we readily obtain the diagonal elements

$$\begin{aligned}[H_{\text{eff}}(SL)]_{aM,aM} &= \frac{g\beta^2}{\kappa_0 c^2}(Q_0 - Q_1 - 2Q_2)\langle M|\, \mathsf{S}_z\, |M\rangle \\ &= A\langle M|\, \mathsf{S}_z\, |M\rangle,\end{aligned} \tag{11.7.16}$$

where (with the usual convention, p. 123)

$$\left.\begin{aligned} Q_0 &= S^{-1}\sum_n Z_n \int r_{n1}^{-3}\mathsf{L}_z^{(n)}(1) Q_S(\bar{a}\bar{a}\mid \boldsymbol{r}_1)\, \mathrm{d}\boldsymbol{r}_1, \\ Q_1 &= S^{-1}\int r_{12}^{-3}\mathsf{L}_z^{(2)}(1) Q_{SL}(\bar{a}\bar{a}\mid \boldsymbol{r}_1, \boldsymbol{r}_2)\, \mathrm{d}\boldsymbol{r}_1\, \mathrm{d}\boldsymbol{r}_2, \\ Q_2 &= S^{-1}\int r_{12}^{-3}\mathsf{L}_z^{(1)}(2) Q_{SL}(\bar{a}\bar{a}\mid \boldsymbol{r}_1\, \boldsymbol{r}_2)\, \mathrm{d}\boldsymbol{r}_1\, \mathrm{d}\boldsymbol{r}_2. \end{aligned}\right\} \tag{11.7.17}$$

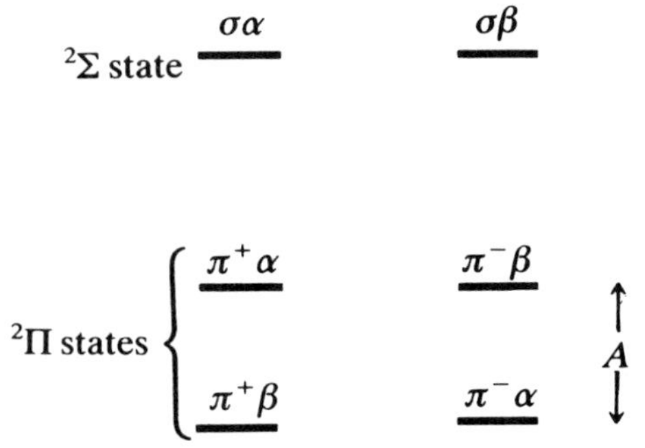

Fig. 11.1 Effect of spin–orbit coupling on $^2\Pi$ and $^2\Sigma$ states of a diatomic molecule (NO). The ground state comprises two Kramers doublets, whose separation measures the coupling parameter A. A sigma state is unaffected by spin–orbit coupling.

These integrals contain respectively operators for angular momentum (the axial component) of electron 1 around nucleus n, electron 1 around electron 2, and electron 2 around electron 1. They may all be evaluated readily (see Moores and McWeeny, 1973) for approximate wavefunctions of Gaussian form.

The first-order effect of spin-orbit coupling is now clear (Fig. 11.1). The energies of the π^+ states may be written as

$$\left.\begin{aligned} E(\pi^+\alpha) &= A\langle M|\,\mathsf{S}_z\,|M\rangle = \tfrac{1}{2}A \qquad (M=\tfrac{1}{2}), \\ E(\pi^+\beta) &= A\langle M|\,\mathsf{S}_z\,|M\rangle = -\tfrac{1}{2}A \quad (M=-\tfrac{1}{2}), \end{aligned}\right\} \tag{11.7.18}$$

and A thus measures the separation of the two Kramers doublets. The energies of the π^- partners are $E(\pi^-\alpha) = -\frac{1}{2}A$, $E(\pi^-\beta) = \frac{1}{2}A$, and the energies of all four states may thus be written, with the original notation in which Λ and Σ denote the eigenvalues of L_z and S_z,

$$\begin{aligned} E_{\Lambda\Sigma} &= \langle\Lambda\Sigma|\,\mathsf{H}_{\text{eff}}(\text{spin–orbit})\,|\Lambda\Sigma\rangle \\ &= \langle\Lambda\Sigma|\,A\mathsf{L}_z\mathsf{S}_z\,|\Lambda\Sigma\rangle. \end{aligned} \tag{11.7.19}$$

In this case, then, the *orbital* degeneracy can be described by introducing into the spin Hamiltonian the operator L_z for the unquenched part of the total orbital angular momentum. The term $A\mathsf{L}_z\mathsf{S}_z$ is the analogue of the commonly adopted free-atom form $\xi\mathbf{L}\cdot\mathbf{S}$, which is appropriate in a strictly central-field situation. Again, however, the effective Hamiltonian corresponds to a simple *model*, all the complexities of the real system being encountered only in the determination (via the density functions) of the numerical value of the spin–orbit coupling constant A.

Similar considerations apply in more complicated situations, for example in transition-metal complexes, where 3-fold orbital degeneracies often occur. Such degeneracy can sometimes be described by introducing into the spin Hamiltonian a *pseudo*-orbital angular-momentum operator L, whose sole function is to reproduce, using the model Hamiltonian in a basis of angular-momentum eigenstates, the matrix elements of the real Hamiltonian within the degenerate manifold (the A-set) of actual states. Further details are available in the standard texts (e.g. Griffith, 1961).

11.8 SOME SPIN-COUPLING EFFECTS (SECOND-ORDER)

In discussing second-order effects, which arise jointly from two perturbations, two forms of the perturbation theory of Section 11.4 are open to us. It will often be convenient, as in Sections 11.5 and 11.6, to use the form in which one perturbation is applied first and the second is used as a "probe" to study its effect. In this way, it is possible to gain valuable insight into the origin of various types of coupling. At the same time, the approach leads naturally towards the theory of linear response, taken up in the next chapter. Again, we consider one-by-one a few typical and important examples.

Magnetic shielding in NMR

The separation of levels that differ only in nuclear spins is due primarily to the nuclear Zeeman terms, given in (11.3.17), namely

$$\mathsf{H}_{\mathrm{NZ}} = -\beta_{\mathrm{P}} \sum_{n} g_n \boldsymbol{B} \cdot \mathsf{I}(n), \tag{11.8.1}$$

and depends linearly on applied field and nuclear spins. When the same nucleus occurs in different chemical environments, however, slightly different splittings occur and there is a "chemical shift" in the NMR signals; physically, these shifts may be associated with an environmental dependence of the actual field experienced by a nucleus, due to the presence of induced currents in the electron distribution. Each nucleus n is found to behave as if the field $\boldsymbol{B}$ were replaced by $\boldsymbol{B}_n = (1 - \sigma_n)\boldsymbol{B}$, and the "shielding constant" σ_n, although usually of the order of only a few parts in 10 000, may be measured accurately owing to the high precision of the experiments. More generally, the shielding may be anisotropic, and σ_n is then replaced by a shielding *tensor*. The term $-\boldsymbol{\sigma}_n \cdot \boldsymbol{B}$ evidently corresponds to a secondary field, usually opposing $\boldsymbol{B}$, generated by the induced currents already discussed in Section 11.6. We anticipate that this secondary field will follow from the classical Biot–Savart law, once the current distribution is known.

To verify this simple picture, we look for two perturbations among those listed in Section 11.3 that can lead to cross-terms in (11.4.14) bilinear in field components and nuclear-spin components: the obvious candidates are (using H_1, H_2 in place of $\lambda\mathsf{H}^\lambda$, $\mu\mathsf{H}^\mu$)

$$\mathsf{H}_1 = \sum_i \beta\boldsymbol{B}\cdot\mathbf{L}(i), \qquad \mathsf{H}_2 = \frac{\mu_0}{4\pi}\frac{e}{m}\sum_{i,n} g_n\beta_{\rm p} r_{ni}^{-3}\mathbf{I}(n)\cdot\hbar\mathbf{M}^n(i), \tag{11.8.2}$$

where we have regrouped the numerical constants in the second term, introducing the magnetic permeability of free space ($\mu_0/4\pi = 1/4\pi\epsilon_0 c^2$); $g_n\beta_{\rm p}\mathbf{I}(n)$ is the operator for the nuclear magnetic moment and, as usual, $\hbar\mathbf{M}^n(i) = \boldsymbol{r}_{ni}\times\boldsymbol{\pi}(i)$ is the gauge-invariant angular momentum of electron i around nucleus n. We know that H_1 (=$\mathsf{H}'_{\rm mag}$ considered fully in Section 11.6) will produce a current distribution, given by (11.6.18), as a result of the first-order change in the wavefunction, $\Phi_a \to \tilde{\Phi}_a$; and that the required second-order (H_1-H_2) terms in the effective Hamiltonian are then correctly given (cf. (11.4.15)) as

$$\langle a\sigma|\,\mathsf{H}_{\rm eff}\,|a\sigma'\rangle^{12} = \langle\tilde{a}_1\sigma|\,\mathsf{H}_2\,|\tilde{a}_1\sigma'\rangle^{12}, \tag{11.8.3}$$

where the ground state has been assumed degenerate only through the nuclear spin factor Θ_σ, and $\tilde{a}_1$ is the "H_1-perturbed" state ($\tilde{\Phi}_a$).

It will be sufficient to consider a typical term in H_2, referring to nucleus n, and we then have

$$\begin{aligned}
&\frac{\mu_0}{4\pi}\sum_m \left\langle \tilde{a}_1\sigma \left| \frac{e\hbar}{m}\sum_i r_{ni}^{-3}\mathsf{I}_{-m}(n)\mathsf{M}^n_m(i) \right| \tilde{a}_1\sigma'\right\rangle \\
&= \sum_m \langle\sigma|\,\mathsf{I}_{-m}(n)\,|\sigma'\rangle \left\langle \tilde{a}_1 \left| \frac{\mu_0}{4\pi}\frac{e\hbar}{m}\sum_i r_{ni}^{-3}\mathsf{M}^n_m(i) \right| \tilde{a}_1\right\rangle \\
&= \sum_m \langle\sigma|\,\mathsf{I}_{-m}(n)\,|\sigma'\rangle \left\langle \tilde{a}_1 \left| \frac{\mu_0}{4\pi}\frac{e}{m}\sum_i \frac{[\boldsymbol{r}_{ni}\times\boldsymbol{\pi}(i)]_m}{r_{ni}^3} \right| \tilde{a}_1\right\rangle \\
&= -\sum_m \langle\sigma|\,\mathsf{I}_{-m}(n)\,|\sigma'\rangle \frac{\mu_0}{4\pi}\int \frac{[\boldsymbol{r}_n\times(-e\boldsymbol{J}(\boldsymbol{r}))]_m}{r_n^3}\,d\boldsymbol{r} \\
&= -\langle\sigma|\,\mathbf{I}(n)\cdot\boldsymbol{B}(n)\,|\sigma'\rangle.
\end{aligned} \tag{11.8.4}$$

Here $\boldsymbol{J}(\boldsymbol{r})$ is the current density at point $\boldsymbol{r}$ (in the perturbed state $\tilde{a}_1$), with components‡ given in (11.6.18); $\boldsymbol{r}_n = \boldsymbol{r} - \boldsymbol{R}_n$ is the position vector of this

‡ It is immaterial at this point whether we use spherical or cartesian components.

point relative to nucleus n; and $\boldsymbol{B}(n)$ is the field at nucleus n due to the current distribution. Thus (11.8.3) reduces to the form (11.8.4), with

$$B_\lambda(n) = \frac{\mu_0}{4\pi} \int \frac{[\boldsymbol{r}_n \times (-e\boldsymbol{J}(\boldsymbol{r}))]_\lambda}{\mathrm{r}_n^3} \,\mathrm{d}r, \tag{11.8.5}$$

and this is the familiar Biot–Savart expression for the field due to circulating currents in a conducting medium.

It is possible to relate $\boldsymbol{B}(n)$ to $\boldsymbol{B}$ by introducing a tensor $\boldsymbol{\sigma}_n$ (cf. Section 11.6) and putting‡

$$B_\lambda(n) = -\sum_\nu \sigma_{\lambda\nu} B_\nu. \tag{11.8.7}$$

On using $\boldsymbol{J}^\nu(\boldsymbol{r})$, as in (11.6.28), to denote the current density produced by unit field along the ν axis, it follows readily that

$$\sigma_{\lambda\nu} = -\frac{\mu_0}{4\pi} \int \frac{[\boldsymbol{r} \times (-e\boldsymbol{J}^\nu(\boldsymbol{r})]_\lambda}{r^3} \,\mathrm{d}\boldsymbol{r}. \tag{11.8.8}$$

Finally, then, the spin-Hamiltonian term representing the interaction of $\mathsf{I}(n)$ with the field will follow on putting the preceding results in (11.8.3) and adding the *first*-order interaction $\langle a\sigma|\ -g_n\beta_\mathrm{p}\boldsymbol{B}\cdot\mathsf{I}(n)\ |a\sigma'\rangle$.

The final result, summing over all nuclei, is

$$\begin{aligned} \mathsf{H}_\mathrm{s}(\text{nuc. Zeeman}) &= -\sum_n g_n\beta_\mathrm{p}[\boldsymbol{B}\cdot\mathsf{I}(n) - \boldsymbol{B}(n)\cdot\mathsf{I}(n)] \\ &= -\sum_n g_n\beta_\mathrm{p}\mathsf{I}(n)\cdot(\mathbf{1}-\boldsymbol{\sigma}_n)\cdot\boldsymbol{B}. \end{aligned} \tag{11.8.9}$$

This is the experimentally observed form of the interaction: the shielding-tensor components can be calculated quantum-mechanically using (11.8.8) with the computed current densities.

Indirect nuclear spin–spin coupling

There are two main observable types of coupling between nuclear spins; the direct dipole–dipole coupling (11.3.17), which contains only nuclear spins and parameters and is therefore already in spin-Hamiltonian form, and a coupling that can be represented empirically by a scalar-product term $\mathsf{I}(n)\cdot\mathsf{I}(n')$. The former averages to zero for rapidly tumbling molecules, but the latter is isotropic and is responsible for the sharp NMR lines in solution spectra; it arises indirectly, through the electron distribution. Spin-Hamiltonian terms quadratic in the nuclear spins will clearly arise from the second-order sums in (11.4.14) when H_1 and H_2 are

‡ It is usual to include the minus sign since the induced field generally opposes the applied field.

each linear in nuclear spins; here we consider only the dominant isotropic contributions that arise from the contact interaction in (11.3.17), and, for simplicity, assume a singlet ground state.

Let us pick out the terms involving two nuclei, n and n', by taking

$$\mathsf{H}_n = c_n \sum_i \delta(r_{ni})\mathbf{S}(i) \cdot \mathbf{I}(n), \qquad \mathsf{H}_{n'} = c_{n'} \sum_i \delta(r_{n'i})\mathbf{S}(i) \cdot \mathbf{I}(n') \quad (11.8.10)$$

where, for example, $c_n = (8\pi g\beta\beta_{\rm p}/3\kappa_0 c^2)g_n$. If we use the direct form of the perturbation theory, a second-order term bilinear in the components of $\mathbf{I}(n)$ and $\mathbf{I}(n')$ will obviously follow from (11.4.14). Thus

$$\mathsf{H}_{\rm eff}(\text{spin–spin})_{a\sigma,a\sigma'} = -\sum_{b,\sigma''} \frac{\langle a\sigma|\, \mathsf{H}_n\, |b\sigma''\rangle\langle b\sigma''|\, \mathsf{H}_{n'}\, |a\sigma'\rangle}{\Delta E(a\to b)} - \sum (n \leftrightarrow n'),$$

where the second term is like the first except for interchange of H_n and $\mathsf{H}_{n'}$.

It will be sufficient to consider the first summation, noting that each matrix element separates into a product of electronic and nuclear factors; thus

$$\sum_{b,\sigma''} = c_n c_{n'} \sum_{b,\sigma''} \sum_{m,m'} (-1)^{m+m'} \frac{\left\langle a \middle| \sum_i \delta(r_{ni})\mathsf{S}_m(i) \middle| b \right\rangle \left\langle b \middle| \sum_i \delta(r_{n'i})\mathsf{S}_{m'}(i) \middle| a \right\rangle}{\Delta E(a\to b)}$$
$$\times \langle\sigma|\, \mathsf{I}_{-m}(n)\, |\sigma''\rangle\langle\sigma''|\, \mathsf{I}_{-m'}(n')\, |\sigma'\rangle. \quad (11.8.11)$$

The σ'' summation gives at once $\langle\sigma|\,\mathsf{I}_{-m}(n)\mathsf{I}_{-m'}(n')\,|\sigma'\rangle$, since the set of nuclear spin states is assumed complete. The sum (11.8.11) thus becomes

$$\sum_{b,\sigma''} = \sum_{m,m'} \langle\sigma|\, \mathsf{I}_{-m}(n)\mathsf{I}_{-m'}(n')\, |\sigma'\rangle C^{nn'}_{mm'},$$

where $C^{nn'}_{mm'}$ stands for the electronic factor, which is a sum over all excited states b. The presence of the electron-spin operators in the matrix elements means that only *triplet* excited states will enter, and each triplet level will comprise three states $|b, M\rangle$ with $M = 0, \pm 1$. On making use of (5.9.8) and (5.9.10), we then obtain

$$C^{nn'}_{mm'} = c_n c_{n'} \sum_b \frac{Q_S(\bar{b}\bar{a} \mid \boldsymbol{R}_n) Q_S(\bar{a}\bar{b} \mid \boldsymbol{R}_{n'})}{\Delta E(a\to b)}$$
$$\times \sum_M (-1)^{m+m'} \begin{bmatrix} 1 & 1 & | & 0 \\ M & m & | & 0 \end{bmatrix} \begin{bmatrix} 0 & 1 & | & 1 \\ 0 & m' & | & M \end{bmatrix}, \quad (11.8.12)$$

where the square-bracketed quantities are the usual ratios of Clebsch–Gordan coefficients (5.9.11). The coefficients are in this case trivial and lead to a factor $\sum_M (-1)^{m+m'}[-(-1)^m \delta_{M,m'}\delta_{M,-m}]$, namely $-(-1)^m$. On inserting this result in (11.8.11) and adding the second sum in (11.8.10), we obtain the $\sigma\sigma'$-element of the effective Hamiltonian; this contains only nuclear spin operators, and may consequently be written as the corresponding element of the *spin*-Hamiltonian term

$$\mathsf{H}_s(\text{nuc. spin–spin}) = hJ_{nn'}\mathbf{I}(n)\cdot\mathbf{I}(n'), \tag{11.8.13}$$

where the coupling constant is, with a standard density Q_S (cf. (5.9.6)),

$$J_{nn'} = -\frac{2g_n g_{n'}}{h}\left(\frac{8\pi g\beta\beta_{\mathrm{p}}}{3\kappa_0 c^2}\right)^2 \sum_{b(\text{triplet})} \frac{Q_S(\bar{a}\bar{b}\mid \boldsymbol{R}_{n'})Q_S(\bar{b}\bar{a}\mid \boldsymbol{R}_n)}{\Delta E(a\rightarrow b)} \tag{11.8.14}$$

There is, of course, one term of the form (11.8.13) for every pair of nuclei.

The alternative perturbation procedure reveals the physical origin of the indirect nuclear spin–spin coupling. The $a\sigma, a\sigma'$-element of the effective Hamiltonian may be regarded as that of $\mathsf{H}_{n'}$ relative to *perturbed* states produced by first applying H_n; or vice versa. In either case the ground state is contaminated by excited triplet functions with various possible nuclear spin factors; and the matrix element of the second perturbation therefore involves an electronic spin density, describing a *spin polarization* produced by the first perturbation. It is this electron spin polarization, transmitted through the electron distribution, that is picked up by the second nucleus via its Fermi contact term.

The g tensor and related effects

Let us now look for some second-order effects involving the spin-orbit term H_{SL}. In Section 11.7 we saw that this term could lead to a *first*-order splitting of an electronically degenerate ground state; but in this section we consider a *non*-degenerate ground state in which there is no first-order effect, the angular momentum being fully quenched. There are two important second-order effects: one arises jointly from H_{SL}, given in (11.3.13), and H'_{mag} in (11.3.12), and leads to a spin–field coupling that supplements the spin Zeeman interaction (11.3.15), turning the g factor into a *g tensor*: the other arises purely from H_{SL} and results in an electron spin–spin coupling that supplements the direct coupling studied in the last section (p. 390).

Nuclear spin factors may be omitted, and the perturbations of interest will be

$$\mathsf{H}_1 = \beta \boldsymbol{B} \cdot \mathbf{L}, \qquad \mathsf{H}_2 = \frac{g\beta^2}{\kappa_0 c^2} \sum_{n,i} \frac{Z_n \mathbf{S}(i) \cdot \mathbf{M}^n(i)}{r_{ni}^3}, \tag{11.8.15}$$

where at this stage we include only the main spin–orbit term (neglecting 2-electron effects). First we proceed directly, using the expression (11.4.14), with the two terms given above. The matrix elements we need are, typically, $\langle aM | \mathsf{H}_1 | bM'' \rangle$ and $\langle bM'' | \mathsf{H}_2 | aM' \rangle$, which we consider separately.

Since H_1 is spin-independent, the first matrix element vanishes unless $|aM\rangle$ and $|bM''\rangle$ have identical spin eigenvalues, and then takes the M-independent value

$$\begin{aligned}\langle aM | \mathsf{H}_1 | bM'' \rangle &= \delta_{MM''} \beta \sum_m (-1)^m B_{-m} \left\langle aS \middle| \sum_i \mathsf{L}_m(i) \middle| bS \right\rangle \\ &= \delta_{MM''} \beta \sum_m (-1)^m B_{-m} X_m^{ab},\end{aligned} \tag{11.8.16}$$

where, as usual, we have made the standard choice of states ($M = S$), using m to label spherical components, and

$$X_m^{ab} = \left\langle aS \middle| \sum_i \mathsf{L}_m(i) \middle| bS \right\rangle. \tag{11.8.17}$$

The second operator, on the other hand, is spin-dependent and

$$\langle bM'' | \mathsf{H}_2 | aM' \rangle = \frac{g\beta^2}{\kappa_0 c^2} \sum_m (-1)^m \left\langle bM'' \middle| \sum_{n,i} Z_n \frac{\mathsf{M}_{-m}^n(i) \mathsf{S}_m(i)}{r_{ni}^3} \middle| aM' \right\rangle. \tag{11.8.18}$$

The Wigner–Eckart theorem (Appendix 3) may now be applied, and gives, according to (A3.29),

$$\left\langle bM'' \middle| \sum_i f(i) \mathsf{S}_m(i) \middle| aM' \right\rangle = \text{constant} \times \langle M'' | \mathsf{S}_m | M' \rangle,$$

where S_m is a *total* spin component operator and the constant may be evaluated by making any single choice of M', M'', m. The easiest choice is $M' = M'' = S$, $m = 0$, which gives

$$\text{constant} = S^{-1} \left\langle bS \middle| \sum_i f(i) \mathsf{S}_z(i) \middle| aS \right\rangle.$$

On using the last two results in (11.8.18), we obtain easily

$$\langle bM'' | \mathsf{H}_2 | aM' \rangle = \frac{g\beta^2}{\kappa_0 c^2} \left\langle M'' \middle| \sum_m (-1)^m Y_{-m}^{ba} \mathsf{S}_m \middle| M' \right\rangle, \tag{11.8.19}$$

where

$$Y^{ba}_{-m} = S^{-1}\left\langle bS \left| \sum_{n,i} Z_n r_{ni}^{-3} M^n_{-m}(i) S_z(i) \right| aS \right\rangle. \qquad (11.8.20)$$

The required second-order sum (cf. (11.8.11)) thus becomes

$$\sum_b \frac{\langle aM|\, \mathsf{H}_1\, |bM''\rangle \langle bM''|\, \mathsf{H}_2\, |aM'\rangle}{\Delta E(a \rightarrow b)}$$

$$= \beta \frac{g\beta^2}{\kappa_0 c^2} \left\langle M \left| \sum_b \frac{\sum_m (-1)^m B_{-m} X^{ab}_m \sum_{m'} (-1)^{m'} Y^{ba}_{-m'} \mathsf{S}_{m'}}{\Delta E(a \rightarrow b)} \right| M' \right\rangle,$$

where X^{ab}_m and $Y^{ba}_{-m'}$ are sets of quantities analogous to vector components. The m, m' summations lead to scalar products in "spherical" form. The results are most commonly expressed in terms of cartesian components: on doing so, and remembering that (cf. (11.4.14)) we need *two* sums, with the operators interchanged, we find a second-order contribution to $\mathbf{H}_{\text{eff}}$ that may be represented as the matrix of a *spin*-Hamiltonian term

$$\mathsf{H}_s(\text{spin–field}) = \sum_{\kappa, \nu} \beta G_{\lambda\nu} B_\lambda \mathsf{S}_\nu \quad (\lambda, \nu = x, y, z). \qquad (11.8.21)$$

The g-tensor components, in second order, are

$$G_{\lambda\nu} = -\frac{g\beta^2}{\kappa_0 c^2} \sum_b \frac{X^{ab}_\lambda Y^{ab}_\nu + Y^{ab}_\nu X^{ba}_\lambda}{\Delta E(a \rightarrow b)}, \qquad (11.8.22)$$

where the X and Y quantities are evaluated using (11.8.17) and (11.8.20), but with cartesian rather than spherical components of the angular-momentum operator. On adding to (11.8.21) the *direct* spin–field interaction (i.e. the spin Zeeman term (11.3.15)), which appears in *first* order, we obtain the full Zeeman interaction

$$\mathsf{H}_s(\text{Zeeman}) = \beta \sum_{\lambda, \nu} g_{\lambda\nu} B_\lambda \mathsf{S}_\nu \quad (g_{\lambda\nu} = g\delta_{\lambda\nu} + G_{\lambda\nu}). \qquad (11.8.23)$$

The second-order interaction of the two terms in (11.8.15) thus leads to a spin–field term of *tensor* form, whose components will depend on the orientation of the molecule in the applied field. Although in (11.8.15) we only admitted the 1-electron terms of the full H_{SL}, the 2-electron terms are handled in a similar way and are easily added (Problem 11.15).

Both H_1 and H_2 produce, of course, their *individual* second-order effects: H_1 leads to field-quadratic terms, already dealt with in Section 11.6, while H_2 must lead to a *spin*-quadratic term and hence to an electron spin–spin interaction. We now deal with the latter, which will have the same general form as the *direct* spin–spin interaction that arises in first order (p. 390).

It is clearly necessary only to use the same operator H_2 in both matrix elements in the numerator of the second-order sum: evaluation then leads to a spin-Hamiltonian term, which is usually written in the form (frequency units)

$$\mathsf{H}_s(\text{spin–spin}) = \sum_{\lambda,\nu} hD_{\lambda\nu}\mathsf{S}_\lambda\mathsf{S}_\nu. \tag{11.8.24}$$

Here the coupling tensor has elements

$$D_{\lambda\nu} = -h^{-1}\left(\frac{g\beta^2}{\kappa_0 c^2}\right)^2 \sum_b \frac{Y_\lambda^{ab}Y_\nu^{ba}}{\Delta E(a\to b)} \quad (\lambda,\ \nu = x,\ y,\ z), \tag{11.8.25}$$

where the Y quantities are the cartesian analogues of those defined in (11.8.20). In the presence of heavy nuclei, with high-angular-momentum orbitals in the inner shells, this interaction may easily be much larger than that due to direct spin–spin interaction; observed spin–spin coupling in transition-metal complexes is thus usually *indirect.*

To conclude this discussion, let us turn to the physical interpretation of the expressions obtained above. We know that the effect of H_1 in (11.8.15) is to set up currents in the electron distribution: and the currents produced in each of the $2S+1$ multiplet states ($|aM\rangle$ to first order in field) can then be "detected" by the second operator H_2. In the present case, however, the second operator is spin-dependent and therefore discriminates between the up-spin and down-spin components of the current density. To proceed, we go back to the density matrix (for the perturbed state) including the spin.

Instead of (11.6.13), we have the first-order expression

$$\rho(\tilde{a}M,\tilde{a}M \mid \boldsymbol{x}) = \rho(aM, aM \mid \boldsymbol{x}) + \sum_\nu B_\nu\rho_\nu(aM, aM \mid \boldsymbol{x}) \tag{11.8.26}$$

for the perturbed density matrix in the presence of H_1, which would yield (11.6.13) on performing the usual spin integration. The matrix elements of H_2, with respect to these perturbed states, are then given by an expression analogus to (11.8.3), which reduces to

$$\langle\tilde{a}M|\,\mathsf{H}_2\,|\tilde{a}M'\rangle = \frac{g\beta^2}{\kappa_0 c^2}\left\langle M\left|\sum_m(-1)^m Y_{-m}^{\tilde{a}\tilde{a}}\mathsf{S}_m\right|M'\right\rangle, \tag{11.8.27}$$

where (cf. (11.8.20))

$$Y_m^{\tilde{a}\tilde{a}} = S^{-1}\left\langle\tilde{a}S\left|\sum_{n,i} Z_n r_{ni}^{-3}M_m^n(i)\mathsf{S}_z(i)\right|\tilde{a}S\right\rangle. \tag{11.8.28}$$

This is similar to a quantity encountered (p. 395) in discussing magnetic shielding, and it reduces in a similar way. The only difference is that it contains an extra operator S_z; and the analogous result will therefore contain a *spin*-current density instead of the usual (electron-) current density.

The standard spin density (for state $|\tilde{a}M\rangle$ with $M = S$) will be

$$
\begin{aligned}
Q_z(\tilde{a}S, \tilde{a}S \mid \boldsymbol{r}) &= \int \mathsf{S}_z \rho(\tilde{a}S, \tilde{a}S \mid \boldsymbol{x})\, \mathrm{d}s \\
&= \tfrac{1}{2}[P_\alpha(\tilde{a}S, \tilde{a}S \mid \boldsymbol{r}) - P_\beta(\tilde{a}S, \tilde{a}S \mid \boldsymbol{r})],
\end{aligned}
$$

and the corresponding spin-current density will be (cf. (5.10.4))

$$
\begin{aligned}
\boldsymbol{J}^{\mathrm{s}}(aS, aS \mid \boldsymbol{r}) &= m^{-1}\,\mathrm{Re}\,[\boldsymbol{\pi} Q_z(aS, aS \mid \boldsymbol{r})] \\
&= [\boldsymbol{J}_\alpha(aS, aS \mid \boldsymbol{r}) - \boldsymbol{J}_\beta(aS, aS \mid \boldsymbol{r})]. \qquad (11.8.29)
\end{aligned}
$$

Whereas the sum of the two currents $\boldsymbol{J}_\alpha$ and $\boldsymbol{J}_\beta$ determines the magnetic field at a *nuclear* dipole, any "probe" that contains *electron* spin variables can detect the *difference* ($\boldsymbol{J}^{\mathrm{s}}$) between $\boldsymbol{J}_\alpha$ and $\boldsymbol{J}_\beta$.

In terms of the spin current, the factor $Y_m^{\tilde{a}\tilde{a}}$ in (11.8.28) is easily reduced to give

$$
\begin{aligned}
\frac{g\beta^2}{\kappa_0 c^2} Y_m^{\tilde{a}\tilde{a}} &= \frac{\mu_0}{4\pi}(-g\beta)S^{-1} \sum_n Z_n \int \frac{[\boldsymbol{r}_x \times -e\boldsymbol{J}^{\mathrm{s}}(\boldsymbol{r})]_m}{r_n^3}\, \mathrm{d}\boldsymbol{r} \\
&= -g\beta \sum_n Z_n [\boldsymbol{B}_n(\boldsymbol{J}^{\mathrm{s}}/S)]_m, \qquad (11.8.30)
\end{aligned}
$$

where $\boldsymbol{r}_n = \boldsymbol{r} - \boldsymbol{R}_n$ and $\boldsymbol{B}_n(\boldsymbol{J}^{\mathrm{s}}/S)$ denotes the field at nucleus n derived from an electric-current distribution associated with the (normalized) spin density instead of with the electron density P.

The spin-Hamiltonian term that yields the matrix elements (11.8.28), and hence the required part of the effective Hamiltonian $\mathsf{H}_{\mathrm{eff}}$, is now seen to be

$$
\mathsf{H}_{\mathrm{s}}(\text{spin–field}) = \sum_n -g\beta \mathbf{S} \cdot (Z_n \boldsymbol{B}_n). \qquad (11.8.31)
$$

The spin magnetic dipole thus interacts principally‡ with *local dipoles* due to the induced currents that circulate around individual nuclei; contributions from different nuclei are additive and greatest for the heaviest

‡ We are still using only the main, Z-dependent, part of the full spin–orbit interaction (11.3.13).

atoms. By comparing (11.8.31) with (11.8.21), it is easily shown (Problem 11.16) that $G_{\lambda\nu}$ may be expressed alternatively in terms of the spin current.

The indirect electron spin–spin coupling may also be analysed in terms of spin currents (Problem 11.17). Briefly, one spin dipole produces, via the spin–orbit interaction, small circulating currents around the various nuclei, the currents being different for up-spin and down-spin components; the resulting spin polarization is picked up by a second spin and leads to the tensor interaction already derived.

Concluding remarks

From the examples considered above, it is clear that much information about the nature of the electron distribution, in particular about the distribution of spin and of induced currents, can be obtained from a careful study of spin-Hamiltonian parameters. At this point, then, it must be asked whether the parameters in the phenomenological spin Hamiltonian used by the experimentalist can *always* be accounted for in this way; or whether other types of coupling might appear.

At first sight it seems that the possibility of identifying the matrix elements of the effective Hamiltonian with those of a *spin* Hamiltonian rests entirely upon the "replacement theorem", which allowed us to express complicated matrix elements in terms of simple matrix elements of spin operators. But the example of nuclear spin–spin coupling (p. 396) shows that this is not so: in that case there occurred matrix elements between singlet and triplet electronic states—while equivalent spin operators $\mathbf{S}_m$ can only be used to connect states of the *same* spin. The important question is whether the Clebsch–Gordan coefficients, which appear on using the Wigner–Eckart theorem, can be eliminated in favour of suitable (possibly more complicated) spin operators. In the case of nuclear spin–spin coupling the coefficients were eliminated with no trouble. In the case of electron spin–spin coupling (above) the problem did not arise—simply because we only included intermediate states of the same spin as the ground state—and to include other multiplicities a further investigation is required. It is indeed possible to show (Griffith, 1961; McLachlan, 1962) that the effective-Hamiltonian terms arising from excited multiplets with $S_b' \neq S_a$ can still be matched by a spin-Hamiltonian term $\sum_{s,m} C_{sm}^{ab}[\mathbf{S}\times\mathbf{S}]_m^{(s)}$, where the various tensor operators arise by coupling two spin vectors in all possible ways; there are then 10 parameters to identify, corresponding to $s=0$, 1, 2 and $m=s$, $s-1$, $\ldots$, $-s$. Some useful equivalent operators have been listed by Atkins and Seymour (1973), but in discussing second-order effects each case should generally be considered on its own merits, in order to establish the theoretical possibility of setting up an equivalent spin Hamiltonian.

In conclusion, we note that in all the examples discussed in the present chapter the use of charge, current and spin densities has allowed us to obtain results that, besides exposing the physics, are of completely general form: they do not depend either on the use of simple orbital-type wavefunctions (see e.g. Slichter, 1964, Chap. 7) or on the availability (hypothetical) of complete sets of exact unperturbed functions. In other words, the treatment of a vast range of properties, using variational approximations of any form whatever, has been reduced in a unified way to the calculation of a small number of density functions.

11.9 VARIATIONAL METHODS. COUPLED HARTREE–FOCK THEORY

We conclude this chapter by considering the actual implementation of property calculations. Much progress has been made in this field, using a variety of apparently unrelated approaches, but almost invariably these employ orbital-type functions constructed from finite basis sets; and, although the property to be calculated is related to a perturbation term in the Hamiltonian, the calculation itself is variational in character, certain parameters in the wavefunction being optimized in order to represent the effect of the perturbation. There are consequently many alternatives to the straightforward CI-type approach adopted for expository purposes in Section 11.4, and to obtain a unified treatment we may adopt the general formalism of Section 2.4, showing how the most frequently used methods emerge as special cases.

In Section 2.4 it was shown that with a wavefunction Ψ, containing numerical parameters $\mathbf{p} = (p_1, p_2, \ldots)$, the variation of the usual energy functional around any point $\mathbf{p}_0$ in "parameter space" could be expressed as

$$E = E_0 + \begin{pmatrix} \mathbf{d} \\ \mathbf{d}^* \end{pmatrix}^\dagger \begin{pmatrix} \boldsymbol{\nabla} H \\ H\boldsymbol{\nabla} \end{pmatrix} + \tfrac{1}{2} \begin{pmatrix} \mathbf{d} \\ \mathbf{d}^* \end{pmatrix}^\dagger \begin{pmatrix} \mathbf{M} & \mathbf{Q} \\ \mathbf{Q}^* & \mathbf{M}^* \end{pmatrix} \begin{pmatrix} \mathbf{d} \\ \mathbf{d}^* \end{pmatrix} + \ldots, \quad (11.9.1)$$

where E_0 corresponds to the point $\mathbf{p}_0$. Here the $\mathbf{d}, \mathbf{d}^*$-columns contain parameter displacements such that $\mathbf{p} = \mathbf{p}_0 + \mathbf{d}$, while $\boldsymbol{\nabla} H$ and $H\boldsymbol{\nabla}$ contain gradient components and their complex conjugates; the square matrices $\mathbf{M}$ and $\mathbf{Q}$ contain second derivatives; and the expression (11.9.1) is valid up to second order in $\mathbf{d}$. The condition that E_0 be a stationary point is simply that the first derivatives, contained in $\boldsymbol{\nabla} H$ and $H\boldsymbol{\nabla}$, vanish at $\mathbf{p} = \mathbf{p}_0$. If, however, $\mathbf{p}_0$ is *not* a stationary point then such a point may be found (in quadratic approximation) by seeking a finite $\mathbf{d}$ that makes E stationary at the point $\mathbf{p} = \mathbf{p}_0 + \mathbf{d}$. The parameter changes needed are given by (2.4.20), namely

$$\left.\begin{aligned} \boldsymbol{\nabla} H + \mathbf{Md} + \mathbf{Qd}^* &= \mathbf{0}, \\ H\boldsymbol{\nabla} + \mathbf{Q}^*\mathbf{d} + \mathbf{M}^*\mathbf{d}^* &= \mathbf{0}, \end{aligned}\right\} \tag{11.9.2}$$

where it is understood that all derivatives are evaluated at $\mathbf{d} = \mathbf{0}$. This reduces to a single equation (2.4.21) when all quantities are real.

Let us now suppose that a variational approximation is available for the case $\mathsf{H} = \mathsf{H}_0$, and that the perturbation H' is added. The energy functional will become

$$H(p_1, p_2, \ldots) = H_0(p_1, p_2, \ldots) + H'(p_1, p_2, \ldots), \tag{11.9.3}$$

where

$$H' = \frac{\langle \Psi | \mathsf{H}' | \Psi \rangle}{\langle \Psi | \Psi \rangle} \tag{11.9.4}$$

is the expectation value of the perturbation. The derivative terms will likewise become sums of two parts, $\boldsymbol{\nabla} H \rightarrow \boldsymbol{\nabla} H_0 + \boldsymbol{\nabla} H'$ etc.; and we may write $\mathbf{d} = \mathbf{d}_0 + \mathbf{d}'$ ($\mathbf{d}_0$ applying when H' is absent). When, however, the gradient vanishes in the absence of H' (i.e. the unperturbed energy is a variational minimum, as will be assumed), $\boldsymbol{\nabla} H_0 = \mathbf{0}$ for $\mathbf{p} = \mathbf{p}_0$ ($\mathbf{d}_0 = \mathbf{0}$). Equation (11.9.2) will then become

$$\left.\begin{aligned} \boldsymbol{\nabla} H' + \mathbf{Md} + \mathbf{Qd}^* &= \mathbf{0}, \\ H'\boldsymbol{\nabla} + \mathbf{Q}^*\mathbf{d} + \mathbf{M}^*\mathbf{d}^* &= \mathbf{0}, \end{aligned}\right\} \tag{11.9.5}$$

where again all derivatives are evaluated at $\mathbf{d} = \mathbf{d}' = \mathbf{0}$ (i.e. in the absence of the perturbation) and we have dropped the primes (on the $\mathbf{d}$), which are now superfluous.

Example: Coupled Hartree–Fock theory

A simple example of the application of (11.9.5) is a derivation of the "coupled Hartree–Fock" (CHF) perturbation theory, first proposed by Peng (1941) and rediscovered, in various forms and with various generalizations, on many occasions. The essence of the approach is to start from a one-determinant wavefunction, optimized in the Hartree–Fock sense in the absence of the perturbation, and to seek the necessary first-order changes in the orbitals to maintain self-consistency when the perturbation is applied. The term "coupled" is used to indicate that, even if the perturbation contains only one-electron operators, the HF effective field must also change and will introduce a "coupling", through the electron interactions, between the perturbation and the electron density.

Let us start from a one-determinant function Ψ_0 of spin-orbitals $\{\psi_i\}$, assumed orthonormal, and introduce parameters by considering variations of the form $\boldsymbol{\psi} \to \boldsymbol{\psi} e^{\boldsymbol{\Delta}}$, as in Section 8.2, $\boldsymbol{\Delta}$ being an anti-Hermitian matrix that describes mixing of the occupied set $\{\psi_i\}$ with a complementary virtual set $\{\psi_m\}$. On expanding the exponential and taking the distinct elements Δ_{rs} $(r > s)$ as variational parameters, the varied wavefunction becomes

$$\Psi = \Psi_0 + \sum_{i,m} p_{mi} \Psi(i \to m) + \sum_{\substack{i \neq j \\ m \neq n}} p_{mi} p_{nj} \Psi(i \to m, j \to n) + \cdots . \tag{11.9.6}$$

Here the determinant of varied spin-orbitals has been expanded up to terms quadratic in the parameters, so that first and second derivatives can be evaluated around the point $\mathbf{p}_0 = \mathbf{0}$. We use the usual conventions (i, j occupied; m, n virtual; r, s general) and, for example, $\Psi(i \to m)$ differs from Ψ_0 through the replacement $\psi_i \to \psi_m$. Only the distinct parameters (p_{mi} but not also p_{im}) occur in (11.9.6) because when $\psi_i \to \sum_r \psi_r(\delta_{ri} + \Delta_{ri} + \ldots)$, r is restricted to the *virtual* $(r = m)$ set; since otherwise $\Psi(i \to m)$ etc. would contain repeated spin-orbitals and would vanish.

The derivatives needed in evaluating the tensor components of Section 2.4 are then, from (11.9.6),

$$\left.\begin{aligned} \Psi^{mi} &= \frac{\partial \Psi}{\partial p_{mi}} = \Psi(i \to m), \\ \Psi^{mi,nj} &= \frac{\partial^2 \Psi}{\partial p_{mi}\, \partial p_{nj}} = (1 - \delta_{mi,nj}) \Psi(i \to m, j \to n), \end{aligned}\right\} \tag{11.9.7}$$

and, using Slater's rules and remembering that the substituted functions will be orthogonal to Ψ_0, it follows easily that

$$\nabla H_{mi} = \langle \Psi(i \to m) |\, \mathsf{H} \,| \Psi_0 \rangle = (m |\, \mathsf{h} \,| i) + \sum_k \langle mk \,\|\, ik \rangle = (m |\, \mathsf{F} \,| i), \tag{11.9.8}$$

where F is the Fock operator. Similarly, using (2.4.13) and (2.4.14)‡, we find

$$\left.\begin{aligned} (\nabla\nabla H)_{mi,nj} &= \langle \Psi(i \to m, j \to n) |\, \mathsf{H} \,| \Psi_0 \rangle \\ &= \langle mn \,\|\, ij \rangle, \\ (\nabla H \nabla)_{mi,nj} &= \langle \Psi(i \to m) |\, \mathsf{H} \,| \Psi(j \to n) \rangle \\ &= \delta_{mi,nj}(\epsilon_m - \epsilon_i) + \langle mj \,\|\, in \rangle, \end{aligned}\right\} \tag{11.9.9}$$

‡ Note that exclusion of the case $mi = nj$, according to (11.9.7), is unnecessary in $(\nabla\nabla H)_{mi,nj}$ since the corresponding term is zero anyway.

where the orbital energy terms (e.g. $\epsilon_m = \langle m|\,\mathsf{F}\,|m\rangle$) arise from the energy expectation value associated with $\Psi(i \to m)$. The last three equations give the elements of $\boldsymbol{\nabla}H$, $\mathbf{M}$ and $\mathbf{Q}$ in (11.9.2).

In the absence of the perturbation, $\mathsf{H} = \mathsf{H}_0$ and the parameters p_{mi} will be chosen to make the unperturbed energy functional stationary: $\nabla H_{mi} = 0$, and consequently, from (11.9.8), $\langle m|\,\mathsf{F}\,|i\rangle = 0$, the usual Brillouin condition. This choice ensures that $\Psi = \Psi_0$ and $p^0_{mi} = 0$ (all m, i).

Suppose we now include a perturbation of one-electron form (this being the commonest case, though easily generalized):

$$\mathsf{H}' = \sum_i \mathsf{h}'(i). \tag{11.9.10}$$

The parameter changes d_{mi} $(=p_{mi} - p^0_{mi})$, determined according to (11.9.5), will then give spin-orbitals that make the energy functional stationary in the presence of the perturbation. These equations become, noting that $\nabla H'_{mi}$ is given by (11.9.8) but with h' instead of h and with no 2-electron terms,

$$\langle m|\,\mathsf{h}'\,|i\rangle + \sum_{j,n} [\delta_{mi,nj}(\epsilon_m - \epsilon_i) + \langle mj \,\|\, in\rangle]d_{nj} + \sum_{j,n} \langle mn \,\|\, ij\rangle d^*_{nj} = 0, \tag{11.9.11}$$

together with a complex-conjugate equation. When the parameters are real (11.9.11) alone serves to determine the perturbed HF function. Otherwise, the two equations together may be used to obtain the real and imaginary parts of the parameter variations. In the real case, (11.9.11) corresponds to (2.4.21), namely

$$\mathbf{a} + (\mathbf{M} + \mathbf{Q})\mathbf{d} = \mathbf{0}, \tag{11.9.12}$$

where $\mathbf{a}$ is the column of gradient components $\boldsymbol{\nabla}H'$ and $\mathbf{M} + \mathbf{Q}$ is the sum of the matrices with elements given in (11.9.9). Explicitly,

$$\left.\begin{aligned} a_{mi} &= \langle m|\,\mathsf{h}'\,|i\rangle, \\ (\mathbf{M} + \mathbf{Q})_{mi,nj} &= \delta_{mi,nj}(\varepsilon_m - \varepsilon_i) + \langle mj \,\|\, in\rangle + \langle mn \,\|\, ij\rangle. \end{aligned}\right\} \tag{11.9.13}$$

If m spin-orbitals are available, N being occupied, then the dimension of the matrix equation (11.9.12) will be $N(m - N)$: with basis sets of the size commonly employed in SCF calculations, this may easily reach 10^3–10^5. On the other hand, efficient numerical methods are available for the solution of such equations (see e.g. Pople *et al.*, 1979; Purvis and Bartlett, 1981; Wormer *et al.*, 1982) and have been used in high-quality property calculations (see e.g. Daborn and Handy, 1983).

Although the above derivation employs spin-orbitals, as in Section 6.1, closed-shell RHF theory leads to the same working equations (cf. p. 171), provided that the parameters d_{nj} are reinterpreted as elements of an *orbital* rotation matrix $\mathbf{\Delta}$ and the 1- and 2-electron integrals are redefined: the orbital energies then have their usual (closed-shell) significance.

In using the above equations, the perturbation h' may be any of the terms considered in earlier sections: it may, for example, represent an applied electric or magnetic field, and will then produce perturbed density functions such as (11.5.19) or (11.6.13), from which a variety of second-order properties can be calculated. It should be noted that the perturbation H'_{mag} is purely imaginary but that the equations remain real when $i\boldsymbol{B}$, instead of $\boldsymbol{B}$, is regarded as the perturbation parameter.

Example: SCF perturbation theory

Although based directly on the Hartree–Fock equations, the CHF method of dealing with a perturbation bears little resemblance to the SCF approach, whose key feature is *iteration* to self-consistency. This is because the *perturbed* HF equations have been *linearized* by working to first order in the parameter changes $\mathbf{d}$, and the resultant set of linear equations, although large, determines these changes directly, i.e. without the need for iteration. It is possible, however, to solve the perturbed HF equations in a manner that preserves the parallel with the usual SCF method, revising the orbitals and the $\mathbf{G}$ matrix iteratively until the results are self-consistent to any desired order in the perturbation. This approach has two advantages: (i) it does not require the 1- and 2-electron integrals to be transformed to the MO basis for the unperturbed molecule; and (ii) the dimensionality of the matrix equations is that of the *basis set,* and is consequently much smaller than in the CHF method. This approach was first formulated in density-matrix language (McWeeny, 1961a), and in that form has the additional advantage of leading directly to the perturbed density functions. Here we follow essentially the original derivation (see also Diercksen and McWeeny, 1966).

For comparison with the SCF equations in their usual form (Section 6.2), we consider a closed-shell ground state with MOs (ϕ_i) expressed in terms of basis functions $\{\chi_i\}$, at first assumed orthonormal. The key equations are then

$$\left.\begin{aligned} \boldsymbol{\phi} = \boldsymbol{\chi}\mathbf{T}, \qquad \mathbf{R} = \mathbf{T}\mathbf{T}^{\dagger}, \\ \mathbf{F} = \mathbf{h} + \mathbf{G}(\mathbf{R}) = \mathbf{h} + \mathbf{J}(\mathbf{R}) - \tfrac{1}{2}\mathbf{K}(\mathbf{R}), \end{aligned}\right\} \qquad (11.9.14)$$

where $\mathbf{R}$ is the idempotent density matrix and $\mathbf{F}$ represents the Fock Hamiltonian: the elements of the $\mathbf{G}$ matrix follow from (6.2.30). It is convenient to write the HF equations in the commutator form analogous to (6.2.21), namely

$$\mathbf{FR} - \mathbf{RF} = \mathbf{0}, \tag{11.9.15}$$

and the problem of SCF perturbation theory is then to restore self-consistency when (with a 1-electron perturbation) $\mathbf{h} \rightarrow \mathbf{h} + \mathbf{h}'$, other quantities changing accordingly. This problem is easily solved using the projection-operator properties developed in Section 6.4.

Let us assume that all quantities can be expanded in powers of a perturbation parameter λ ($\lambda \rightarrow 1$), so that

$$\left.\begin{aligned} \mathbf{h} &= \mathbf{h}_0 + \lambda \mathbf{h}^{(1)} + \lambda^2 \mathbf{h}^{(2)} + \ldots, \\ \mathbf{R} &= \mathbf{R}_0 + \lambda \mathbf{R}^{(1)} + \lambda^2 \mathbf{R}^{(2)} + \ldots, \\ \mathbf{F} &= \mathbf{F}_0 + \lambda \mathbf{F}^{(1)} + \lambda^2 \mathbf{F}^{(2)} + \ldots \\ &(\mathbf{F}^{(n)} = \mathbf{h}^{(n)} + \mathbf{G}(\mathbf{R}^{(n)})), \end{aligned}\right\} \tag{11.9.16}$$

where $\mathbf{h}_0$ is the unperturbed 1-electron Hamiltonian and $\mathbf{R}_0$ commutes with the HF matrix $\mathbf{F}_0 = \mathbf{h}_0 + \mathbf{G}(\mathbf{R}_0)$ (i.e. $\mathbf{R}_0$ is determined from the standard SCF equations for the unperturbed system). From the above equations, separating the orders, it follows readily that

$$\mathbf{F}_0\mathbf{R}^{(1)} - \mathbf{R}^{(1)}\mathbf{F}_0 + \mathbf{F}^{(1)}\mathbf{R}_0 - \mathbf{R}_0\mathbf{F}^{(1)} = \mathbf{0} \tag{11.9.17}$$

will determine the first-order change of the density matrix; but there will also be an auxiliary condition

$$\mathbf{R}_0\mathbf{R}^{(1)} + \mathbf{R}^{(1)}\mathbf{R}_0 = \mathbf{R}^{(1)}, \tag{11.9.18}$$

which arises from the requirement $\mathbf{R}^2 = \mathbf{R}$ and corresponds to an orthonormality constraint on the MOs. The constraint requires (Problem 11.21) that $\mathbf{R}^{(1)}$ be of the form ($\mathbf{M}$ arbitrary)

$$\mathbf{R}^{(1)} = \mathbf{x} + \mathbf{x}^{\dagger}, \qquad \mathbf{x} = (\mathbf{1} - \mathbf{R}_0)\mathbf{M}\mathbf{R}_0, \tag{11.9.19}$$

where $\mathbf{R}_0$ and $\mathbf{1} - \mathbf{R}_0$ represent projection operators onto the "occupied" and "virtual" subspaces (cf. Section 6.4). On substituting this result in (11.9.17) and multiplying from left and right by $\mathbf{1} - \mathbf{R}_0$ and $\mathbf{R}_0$ respectively, we obtain

$$\mathbf{F}_0\mathbf{x} - \mathbf{x}\mathbf{F}_0 + (\mathbf{1} - \mathbf{R}_0)\mathbf{F}^{(1)}\mathbf{R}_0 = \mathbf{0}, \tag{11.9.20}$$

where we have noted that $\mathbf{R}_0$ commutes with $\mathbf{F}_0$ and that $(\mathbf{1} - \mathbf{R}_0)\mathbf{R}_0 = \mathbf{0}$. This equation incorporates all constraints and completely determines the first-order change $\mathbf{R}^{(1)}$.

A solution is obtained most conveniently by writing $\mathbf{x}$ in terms of the eigenvectors of $\mathbf{F}_0$ (i.e. in "spectral" form). If $\mathbf{c}_r$ is the rth eigenvector of

$\mathbf{F}_0$, for eigenvalue ϵ_r then we may put (noting (11.9.19))

$$\mathbf{x} = \sum_{r,s} X_{rs}\mathbf{c}_r\mathbf{c}_s^\dagger = \sum_{m,i} X_{mi}\mathbf{c}_m\mathbf{c}_i^\dagger \tag{11.9.21}$$

and determine the coefficients by substituting in (11.9.20). Thus

$$\sum_{m,i} X_{mi}(\epsilon_m - \epsilon_i)\mathbf{c}_m\mathbf{c}_i^\dagger = -(\mathbf{1} - \mathbf{R}_0)\mathbf{F}^{(1)}\mathbf{R}_0,$$

and multiplication from left and right by particular eigenvectors, $\mathbf{c}_n$ and $\mathbf{c}_j$ say, gives at once $X_{jn} = -(\mathbf{c}_n^\dagger\mathbf{F}^{(1)}\mathbf{c}_j)/(\epsilon_n - \epsilon_j)$. In other words, from (11.9.21),

$$\mathbf{x} = \sum_{j,n} \frac{F_{nj}^{(1)}}{\epsilon_n - \epsilon_j}\mathbf{c}_n\mathbf{c}_j^\dagger, \tag{11.9.22}$$

where, from the definition in (11.9.16),

$$F_{nj}^{(1)} = \mathbf{c}_n^\dagger\mathbf{F}^{(1)}\mathbf{c}_j = \mathbf{c}_n^\dagger[\mathbf{h} + \mathbf{G}(\mathbf{R}^{(1)})]\mathbf{c}_j. \tag{11.9.23}$$

Although (11.9.22) is a formally correct first-order solution, $F_{nj}^{(1)}$ in (11.9.23) depends on $\mathbf{R}^{(1)}$ $(=\mathbf{x}+\mathbf{x}^\dagger)$; and the solution must therefore be obtained by iteration. With $\mathbf{x} = \mathbf{0}$ as a first approximation, convergence is usually rapid when there is a finite energy gap between occupied and virtual levels.

Once the self-consistent $\mathbf{x}$ matrix has been found, the perturbed energy can be obtained to *second* order in the form

$$E^{(1)} = 2\,\mathrm{tr}\,[\mathbf{h}^{(1)}\mathbf{R}_0], \qquad E^{(2)} = \mathrm{tr}\,[\mathbf{h}^{(1)}\mathbf{R}^{(1)}], \tag{11.9.24}$$

where we start from $E = 2\,\mathrm{tr}\,[(\mathbf{h} + \frac{1}{2}\mathbf{G})\mathbf{R}]$, substituting and making appropriate reductions (Problem 11.20)

It is readily verified (Problem 11.22) that the working equations (11.9.22) and (11.9.23) are unchanged on passing to a non-orthonormal basis, provided that the eigenvectors are interpreted as those appropriate to an equation of type (2.3.4) with the overlap matrix on the right.

11.10 GENERALIZATIONS OF CHF THEORY

The coupled Hartree–Fock method (in one or other of the forms discussed in the last section) may be generalized in various important ways, three of which we now consider in turn.

(i) Multiple perturbations

For the most common case of a 1-electron perturbation, the first expression in (11.9.16) will become‡

$$\mathbf{h} = \mathbf{h}_0 + \lambda\mathbf{h}^\lambda + \nu\mathbf{h}^\nu + \dots \qquad (11.10.1)$$

and the results may be derived directly from those given above by deleting λ, replacing $\mathbf{h}^{(1)}$ by $\lambda\mathbf{h}^\lambda + \nu\mathbf{h}^\nu + \dots$, and separating terms of the same degree in the various parameters. It readily follows, for example, that the first-order density change (cf. $\lambda\mathbf{R}^{(1)}$ in (8.6.16)) will now be a sum of terms $\lambda\mathbf{R}^\lambda + \nu\mathbf{R}^\nu + \dots$, each obtained in a similar way:

$$\mathbf{R}^\lambda = \mathbf{x}^\lambda + \mathbf{x}^{\lambda\dagger}, \qquad \mathbf{x}^\lambda = \sum_{j,n} \frac{F^\lambda_{nj}}{\epsilon_n - \epsilon_j}\mathbf{c}_n\mathbf{c}_j^\dagger, \qquad (11.10.2)$$

where, as in (11.9.23),

$$F^\lambda_{nj} = \mathbf{c}_n^\dagger\mathbf{F}^\lambda\mathbf{c}_j = \mathbf{c}_n^\dagger[\mathbf{h}^\lambda + \mathbf{G}(\mathbf{R}^\lambda)]\mathbf{c}_j, \qquad (11.10.3)$$

just as if only *one* perturbation $\mathbf{h}^\lambda$ were present. Knowledge of $\mathbf{x}^\lambda$, $\mathbf{x}^\nu, \dots$ is then sufficient to determine the required second-order parts of the density matrix, along with the energy up to *third* order: thus (cf. (11.9.19) *et seq.*) the 11- and 22-components of the second-order change in the density matrix are found to be

$$\mathbf{R}^{\lambda\nu}_{11} = -(\mathbf{x}^\lambda\mathbf{x}^{\nu\dagger} + \mathbf{x}^\nu\mathbf{x}^{\lambda\dagger}), \qquad \mathbf{R}^{\lambda\nu}_{22} = (\mathbf{x}^{\lambda\dagger}\mathbf{x}^\nu + \mathbf{x}^{\nu\dagger}\mathbf{x}^\lambda), \qquad (11.10.4)$$

while

$$E = E_0 + \sum_\lambda \lambda E^\lambda + \tfrac{1}{2}\sum_{\lambda\nu} \lambda\nu E^{\lambda\nu} + \frac{1}{3!}\sum_{\lambda,\nu,\rho} \lambda\nu\rho E^{\lambda\nu\rho} + \dots, \qquad (11.10.5)$$

where the general terms are

$$\left.\begin{aligned} E^\lambda &= 2\,\mathrm{tr}\,(\mathbf{h}^\lambda\mathbf{R}_0), \\ E^{\lambda\nu} &= \mathrm{tr}\,(\mathbf{h}^\lambda\mathbf{R}^\nu + \mathbf{h}^\nu\mathbf{R}^\lambda), \\ E^{\lambda\nu\rho} &= 2\,\mathrm{tr}\,[\mathbf{h}^\lambda(\mathbf{R}^{\nu\rho}_{11} + \mathbf{R}^{\nu\rho}_{22}) + \mathbf{h}^\nu(\mathbf{R}^{\lambda\rho}_{11} + \mathbf{R}^{\lambda\rho}_{22}) + \mathbf{h}^\rho(\mathbf{R}^{\lambda\nu}_{11} + \mathbf{R}^{\lambda\nu}_{22})]. \end{aligned}\right\} \qquad (11.10.6)$$

Equation (11.10.5) corresponds to the results obtained in Sect. 10.4, being a power series in the perturbation parameters, but is a *variational* approximation in which only the leading function Φ_0 of the CI expansion has been kept but the *orbitals* have been optimized—thus becoming perturbation-dependent.

‡ In some cases other terms such as $\lambda^2\mathbf{h}^{\lambda\lambda}$, $\lambda\nu\mathbf{h}^{\lambda\nu}, \dots$ may be present (for example a magnetic-field term in B^2), but their inclusion is straightforward (Dodds *et al.*, 1977a).

Clearly the whole approach is strictly analogous to the iterative solution of the Roothaan equations, and the approximation itself will have a similar status. The coefficients E^{λ}, $E^{\lambda\nu}$, ... provide *analytic derivatives* of the energy with respect to the applied perturbations; and the method of obtaining them is much superior to the use of numerical differentiation based on repeated solution of the Roothaan equations for a series of finite values of the parameters (the so-called "finite perturbation method" of Pople *et al.*, 1976). It is a straightforward matter to extend the treatment to higher orders though the results (Dodds *et al.* 1977a) become more cumbersome.

(ii) Basis-set variation

A second important generalization, first introduced by Gerratt (1967) and Gerratt and Mills (1968), allows the finite basis itself to depend on the perturbation: in the presence of an electric field, for example, each basis function may acquire a polarization; or if the nuclei are moved, the basis functions may follow them, or possibly float away from them.

Here we indicate the effect of basis-set variation at the closed-shell HF level, for a single 1-electron perturbation, in a form similar to that given above but with small changes in the definitions. Thus, the matrix elements of h will suffer a first-order change arising partly from the change in the *operator* and partly from the change of the *basis*. When we write $\mathsf{h} = \mathsf{h}_0 + \mathsf{h}^{(1)} + \ldots$ the first-order term will consequently be

$$h_{rs}^{(1)} = \langle \chi_r^{(0)} | \, \mathsf{h}^{(1)} \, | \chi_s^{(0)} \rangle + \langle \chi_r^{(1)} | \, \mathsf{h}_0 \, | \chi_s^{(0)} \rangle + \langle \chi_r^{(0)} | \, \mathsf{h}_0 \, | \chi_s^{(1)} \rangle, \qquad (11.10.7)$$

and similarly there will be changes in overlap and 2-electron integrals:

$$S_{rs}^{(1)} = \langle \chi_r^{(1)} \mid \chi_s^{(0)} \rangle + \langle \chi_r^{(0)} \mid \chi_s^{(1)} \rangle, \qquad (11.10.8)$$

$$g_{rs,tu}^{(1)} = \langle \chi_r^{(1)} \chi_s^{(0)} | \, g \, | \chi_t^{(0)} \chi_u^{(0)} \rangle + \ldots + \langle \chi_r^{(0)} \chi_s^{(0)} | \, g \, | \chi_t^{(0)} \chi_u^{(1)} \rangle. \qquad (11.10.9)$$

Since the **G**-matrix variation will now depend both on the change of **R** and the change of the 2-electron integrals, we shall have to use (cf. (11.9.16))

$$\mathbf{F}^{(1)} = \mathbf{h}^{(1)} + \mathbf{G}(\mathbf{R}^{(1)}, \mathbf{g}^{(0)}) + \mathbf{G}(\mathbf{R}^{(0)}, \mathbf{g}^{(1)}). \qquad (11.10.10)$$

Instead of (11.9.19), we then obtain (Dodds *et al.* 1977b)

$$\mathbf{R}^{(1)} = \mathbf{x} + \mathbf{x}^{\dagger} - \mathbf{R}_0 \mathbf{S}^{(1)} \mathbf{R}_0, \qquad (11.10.11)$$

where $\mathbf{x}$ is calculated iteratively using

$$\mathbf{x} = \sum_{n,j} \frac{F_{nj}^{(1)} - \epsilon_j S_{nj}^{(1)}}{\epsilon_n - \epsilon_j} \mathbf{c}_n \mathbf{c}_j^{\dagger}, \qquad (11.10.12)$$

which is not significantly more complicated than (11.9.22). The perturbation-dependence of the basis is explicitly included through the presence of $S_{nj}^{(1)} = \mathbf{c}_n^\dagger \mathbf{S}^{(1)} \mathbf{c}_j$ and $F_{nj}^{(1)} = \mathbf{c}_n^\dagger \mathbf{F}^{(1)} \mathbf{c}_j$, which include the first-order changes of overlap, 1-electron and 2-electron integrals.

First- and second-order changes of the total electronic energy are again obtained (Dodds *et al.*, 1977b) by substitution and reduction:

$$E^{(1)} = 2\,\mathrm{tr}\,[\mathbf{h}^{(1)}\mathbf{R}_0] + \mathrm{tr}\,[\mathbf{G}(\mathbf{R}_0, \mathbf{g}^{(1)})\mathbf{R}_0] - 2\,\mathrm{tr}\,[\mathbf{R}_0\mathbf{S}^{(1)}\mathbf{h}_0\mathbf{R}_0], \quad (11.10.13)$$

$$\begin{aligned} E^{(2)} = {} & \mathrm{tr}\,[\mathbf{h}^{(1)}\mathbf{R}^{(1)}] + \mathrm{tr}\,[\mathbf{G}(\mathbf{R}_0, \mathbf{g}^{(1)})\mathbf{R}_0] \\ & + \mathrm{tr}\,[\mathbf{G}(\mathbf{R}_0, \mathbf{g}^{(1)})\mathbf{R}^{(1)}] - 2\,\mathrm{tr}\,[\mathbf{h}_0\mathbf{R}_0\mathbf{S}^{(2)}\mathbf{R}_0] \\ & - \mathrm{tr}\,[\mathbf{h}^{(1)}\mathbf{R}_0\mathbf{S}^{(1)}\mathbf{R}_0 + \mathbf{S}^{(1)}\mathbf{R}_0\mathbf{h}_0\mathbf{R}^{(1)} + \mathbf{h}_0\mathbf{R}_0\mathbf{S}^{(1)}\mathbf{R}^{(1)}]. \quad (11.10.14) \end{aligned}$$

These results reduce to (11.9.24) in the absence of basis-set variation.

There are many important applications, Thus, by adding field-dependent factors to individual AOs, to obtain field-variant orbitals, it is possible to describe efficiently *local* polarization effects, in the case of electric fields (see e.g. Sadlej, 1977), and local currents in the case of magnetic fields (see e.g. Jaszunski and Sadlej, 1976, 1977; Jaszunski, 1976): and in the calculation of analytic derivatives of the energy with respect to nuclear displacements (see e.g. Gerratt and Mills, 1968; Pulay, 1977) it is possible to obtain accurate force constants and energy surfaces.

(iii) Multiconfiguration generalizations

So far, the possibility of optimizing the orbitals in the presence of a perturbation (i.e. of making "self-consistent" property calculations) has been considered only at the Hartree–Fock level. In many cases, however, it is necessary to use a many-determinant wavefunction, either because the IPM ground state is degenerate or because electron-correlation effects are too important to be ignored; and it is then desirable to optimize both CI coefficients and orbitals as in MC SCF theory (Section 8.6). To formulate the perturbation equations, both coefficients and orbitals will be expanded in terms of a perturbation parameter and the orders will be separated: the zeroth-order equations will be the MC SCF equations in the absence of the perturbation, while the first-order equations will determine the (optimized) response of the wavefunction, and will thus permit the calculation of second-order properties. Important progress had been made in this area (Jaszunski, 1978; Daborn and Handy, 1983), for particular types of perturbation and CI function. In fact, however, the equations in their most general form have been known for many years (Moccia, 1974), and are implicit in the stationary-value conditions formulated in Section 2.4. In conclusion, then, we sketch this generalization.

Let us start from (2.4.18), which expresses the energy expectation value, for a many-parameter wavefunction, in terms of parameter variations around a general point $\mathbf{p}_0$ in parameter space. The values $\mathbf{p}$ $(=\mathbf{p}_0+\mathbf{d})$ at a nearby *stationary* point will follow, in quadratic approximation, on solving (2.4.20), which may be written as

$$\nabla H + \mathbf{Md} + \mathbf{Qd}^* = \mathbf{0}, \tag{11.10.15}$$

together with a complex-conjugate equation. The first and second derivatives (contained in ∇H and in $\mathbf{M}$, $\mathbf{Q}$ respectively) are assumed to be evaluated at point $\mathbf{p}_0$. On taking, as usual,

$$\mathsf{H} = \mathsf{H}_0 + \lambda \mathsf{H}', \qquad \mathbf{d} = \mathbf{d}_0 + \lambda \mathbf{d}', \tag{11.10.16}$$

and noting that

$$\nabla H = \nabla H_0 + \lambda \nabla H', \qquad \mathbf{M} = \mathbf{M}_0 + \lambda \mathbf{M}', \qquad \mathbf{Q} = \mathbf{Q}_0 + \lambda \mathbf{Q}', \tag{11.10.17}$$

it is clear that, on substituting in (11.9.2) and separating the orders, there will be two conditions:

$$\nabla H_0 + \mathbf{M}_0\mathbf{d}_0 + \mathbf{Q}_0\mathbf{d}_0^* = \mathbf{0}, \tag{11.10.18}$$

$$\nabla H' + \mathbf{M}'\mathbf{d}_0 + \mathbf{M}_0\mathbf{d}' + \mathbf{Q}'\mathbf{d}_0^* + \mathbf{Q}_0\mathbf{d}'^* = \mathbf{0}, \tag{11.10.19}$$

along with their complex conjugates. The first condition determines the parameter values for a stationary point in the absence of the perturbation: we assume that $\mathbf{p}_0$ refers to this point, so that ∇H_0 (and $H_0\nabla$) vanish and $\mathbf{d}_0 = \mathbf{0}$. The first equation is then satisfied (i.e. we use an MC SCF reference function), while the second becomes

$$\nabla H' + \mathbf{M}_0\mathbf{d}' + \mathbf{Q}_0\mathbf{d}'^* = \mathbf{0}, \tag{11.10.20}$$

and (with its complex conjugate) determines the parameter changes needed to restore "self-consistency" (i.e. to reoptimize in the presence of the perturbation). The important thing about this equation is that the perturbation enters only in the gradient term: $\mathbf{M}_0$ and $\mathbf{Q}_0$ are second-derivative matrices for the *unperturbed* system, and are therefore known from the initial MC SCF calculation. As Daborn and Handy (1983) note, "this Hessian will already be available and thus all the hard work has been done!"; and although they used a CASSCF function (p. 275), with all orbitals and coefficients real, their observation is true quite generally.

When orbital variations are performed using the exponential operator (8.2.18), while CI coefficients are varied directly, the various blocks of the matrices $\mathbf{M}_0$ and $\mathbf{Q}_0$ follow from the expressions in Section 8.6. The only new quantities needed in (11.10.17) are thus the gradient components of H', the expectation value of the perturbation operator. With the notation of Section 8.6, the components with respect to orbital rotation parameters (Δ_μ, $\mu = rs$) are

$$(\nabla H')_\mu = \langle \Psi | [E^\dagger_\mu, H'] | \Psi \rangle, \tag{11.10.21}$$

and those with respect to variation of coefficients C_κ are

$$(\nabla H')_\kappa = \langle \Psi_\kappa | H' | \Psi \rangle - EC_\kappa, \tag{11.10.22}$$

where E ($=H_0$) is the unperturbed MC SCF energy. Of course, the actual evaluation of components, especially for the Hessian matrix, may be algebraically and computationally heavy; but the procedure itself is absolutely general. An important application, which has been formulated explicitly by Jørgensen and Simons (1983), is to the calculation of analytic derivatives of potential-energy surfaces.

PROBLEMS 11

11.1 Prove the Hellmann–Feynman theorem (p. 141) in its generalized form: The rate of change of the energy, with respect to some parameter in the Hamiltonian (describing a perturbation), may be calculated as the expectation value of the rate of change of the Hamiltonian relative to a variationally determined unperturbed wavefunction. [*Hint*: Let the Hamiltonian be $\mathsf{H} = \mathsf{H}_0 + \lambda \mathsf{H}'$ and suppose $\Psi = \Psi(\mathbf{p}; \boldsymbol{x})$, as in Section 2.4, where the variational parameters will depend on λ. Obtain a variational expression for E^λ ($=\mathrm{d}E/\mathrm{d}\lambda$) and assume E_0 to be stationary at $\lambda = 0$.]

11.2 Use an approach similar to that in Problem 11.1 to discuss the effect of "scaling" in which all the interparticle distances in the wavefunction are multiplied by a parameter p. Hence establish the virial theorem: For a system of particles with inverse distance interactions, $\langle V \rangle = -2\langle T \rangle$ and $\langle E \rangle = \frac{1}{2}\langle V \rangle$, either for a variationally optimized value of p or for an exact wavefunction. Then extend the theorem to admit external forces applied to the nuclei. [*Hint*: Show, by change of variables in the integrations, how the expectation values depend on p. Use a stationary-value condition and suppose that $p = 1$ for the exact wavefunction. Note that the nuclei may be held fixed by applying forces opposite to the (Hellmann–Feynman) forces exerted by the electrons. These forces must be included in forming the expectation value of the classical virial.]

11.3 Show that the energy of an electron in an atomic orbital ϕ, in the presence of a uniform magnetic field, is invariant when the atom is displaced from the origin to point $\boldsymbol{R}$ provided that ϕ is multiplied by $\exp[i(e/2h)\boldsymbol{B} \cdot \boldsymbol{R} \times \boldsymbol{r}]$. [*Hint*: After the move, an electron at point $\boldsymbol{r}$ relative to the nucleus will be at

$\boldsymbol{r}+\boldsymbol{R}$ relative to the origin; the vector potential at that point will be $\boldsymbol{A}' = \frac{1}{2}\boldsymbol{B}\times(\boldsymbol{r}+\boldsymbol{R})$, and the gauge function f can then be identified. Obtain the energy expectation value with the gauge-modified function.]

11.4 Consider a regular hexagon of 6 hydrogen atoms, with a magnetic field normal to the plane. Use a simple LCAO model (neglecting overlap and non-nearest-neighbour effects), with a mean value of the vector potential in the β_{rs} integrals, to discuss the field dependence of the total electronic energy of the ground state. Show that the secular equations are like those without the field, except that each β_{rs} is multiplied by a factor $\exp[i(e/h)BS_{rs}]$, where S_{rs} is the signed area (+ if $r\rightarrow s$ is right-handed around the field) of the triangle formed by nuclei r, s and the origin at which $\mathbf{A}=\mathbf{0}$. (Such approximations were introduced by London (1937) and are still widely used in simple discussions of magnetic properties.) [*Hint*: Use the gauge-modified orbitals of Problem 11.3, with the approximation indicated. Solve the secular equations and take E to be a sum of orbital energies.]

11.5 Show that the current density defined in (5.10.4) is invariant against a change of gauge.

11.6 A hydrogen atom, at the origin, experiences the field of a point charge Ze located at a distance R along the z axis. Make an expansion in terms of the unperturbed AOs and formulate the variation–perturbation procedure of Section 11.4 as a means of obtaining the polarization energy to order Z^2 and the perturbed wavefunction to order Z. Verify that the Z^2 term in the energy appears in the expectation value of the perturbation operator, with respect to the perturbed wavefunction, only if the latter is renormalized to order Z^2. [*Hint*: Set up the effective Hamiltonian (11.4.7) and use (11.4.3) for the energy and (11.4.9) for the wavefunction. Note (11.4.11) *et seq.*]

11.7 Proceed as in Problem 11.6, but put the point charge at (X, Y, Z) and consider the triply degenerate 2p state. Set up "correct zeroth-order combinations" of the functions $2p_x$, $2p_y$, $2p_z$ (what does this mean?) and obtain the 3×3 matrix $\mathbf{H}_{\text{eff}}$: verify that its elements are correctly given by (11.4.12a).

11.8 Derive the first-order change of electron density associated with the perturbed wavefunction obtained in Problem 11.6. Hence verify that the Z^2 term in the energy is the classical interaction energy between a point charge, building up from zero to its final value Ze, and the Z-proportional polarization density that it produces. [*Hint*: Obtain the analogue of (11.5.19) and express (11.4.12a) in terms of the polarization density. The classical interpretation of the factor $\frac{1}{2}$ is parallel to that following (11.5.14).]

11.9 Put charges Ze and $Z'e$ at points R and $-R'$ respectively on the z axis, with a hydrogen atom at the origin. Reformulate the perturbation problem (Problem 11.6) and show how to account for an energy term depending jointly on Z and Z'. Demonstrate the validity of a classical picture in which each charge interacts with a polarization density produced by the other. [*Hint*: With two perturbations, you will need the formulation that starts from (11.4.13). The interpretation will again be parallel to that for a uniform field (p. 380).]

11.10 Show how the spin–orbit coupling terms in the Dirac–Pauli Hamiltonian (11.2.17) reduce, for a particle with the central-field potential-energy function $V(r)$, to the form $\mathsf{H}_{SL} = f(r)\mathsf{S} \cdot \mathsf{L}$, where $f(r) = (g\beta^2/e^2c^2)r^{-1}\, dV/dr$. Verify that for an electron in the presence of a nucleus of charge $Z_n e$ this reduces further to give (in the absence of a magnetic field) the first term in (11.3.13).

11.11 Read carefully the discussion of the electron–nuclear contact coupling (pp 387–389) and make a similar analysis of the electron–nuclear *dipolar* coupling that arises from the term (11.7.9). Hence verify that the coupling with any nucleus is described by a coupling *tensor* whose components depend on the form of the spin density in the vicinity of the nucleus. [*Hint*: Express the scalar products in terms of spherical components as in (11.7.3) and (11.7.4), and reduce the matrix elements (within the degenerate manifold of electron–nuclear product functions) in a parallel fashion, focusing attention on the coefficient of $\mathsf{S}_m \mathsf{I}_{-m}$. The cartesian form can be obtained at the end.]

11.12 The quantization axes for electron and nuclear spin are fixed in the laboratory: the tensor components obtained in Problem 11.11 therefore depend on molecular orientation. Show, by admitting all orientations of the system, that for "rapidly tumbling" molecules the interaction (11.7.10) will average to zero. [*Hint*: Calculate the components for the rotating molecule by using coordinates $x'_{n1} = l_1 x_{n1} + m_1 y_{n1} + n_1 z_{n1}$ etc. in the expressions for $A^{\text{dip}}_{n,\lambda\mu}$. Infer the average values of the squares and products of the direction cosines by using the invariance of the length of a unit vector, and of its scalar product with a second (orthogonal) unit vector.]

11.13 For a many-electron atom, the main term in the spin–orbit coupling (the 1-electron term in (11.3.13)) is often represented in the form $\mathsf{H}_{SL} = \sum_i f(r_i)\mathsf{S}(i) \cdot \mathsf{L}(i)$, thus permitting the use of a non-coulombic central field. Show that within the set of $(2S+1)(2L+1)$ states of a multiplet (given S, L) the spin–orbit splitting of the levels may be correctly obtained using an equivalent operator $\xi\mathsf{S} \cdot \mathsf{L}$, which involves only *total* spin and orbital angular-momentum operators. Obtain the value of ξ in terms of $f(r)$ and the electron density function $\rho(\mathbf{r})$; and establish the Landé interval rule for the spacing of atomic-term values. [*Hint*: The states of the degenerate set may be denoted by $|L, M_L, S, M_S\rangle$. Use a result similar to (11.7.5) *twice*—once for spin and once for orbital angular momentum. Identify the Clebsch–Gordan coefficients in a similar way and express your results in terms of density functions for the unperturbed state with $M_L = L$, $M_S = S$.]

11.14 Fill in the details of the derivation of the spin–spin coupling formula (11.7.13). Obtain the coupling function $Q_{SS}(\bar{a}\bar{a} \mid \mathbf{r}_1, \mathbf{r}_2)$ for the standard state $\bar{a}$ with $M = S$, using a 1-determinant wavefunction with the spins of the open-shell electrons all parallel-coupled; and show that this function is then expressible in terms of the spin-density matrix $Q_S(\bar{a}\bar{a} \mid \mathbf{r}_1; \mathbf{r}'_1)$. What kind of integrals will you have to evaluate to obtain, in an LCAO approximation, the tensor components $D_{\lambda\mu}$ that determine the zero-field splitting of Zeeman levels? [*Hint*: Closed-shell electrons give no contribution to Q_S or Q_{SS}, while for electrons in orbitals with a common α factor the 2-electron density matrix π is related to ρ as in Section 5.3.

Spin integration leads to the required results. Simple integral approximations are sometimes possible (see e.g. McWeeny, 1961b).]

11.15 Extend the operator H_2 of (11.8.15) to include the 2-electron contributions shown in (11.3.13), and hence derive the full form of (11.8.23).

11.16 Express the g-tensor components in (11.8.22) in the alternative form (see (11.8.31) *et seq.*) that involves the field-induced spin current density near the nucleus.

11.17 Show how the coupling-tensor elements $D_{\lambda\nu}$ in (11.8.25) may be rewritten in terms of spin currents.

11.18 Reformulate the coupled Hartree–Fock equations (11.9.11) to admit non-canonical spin-orbitals, which do not diagonalize the zeroth-order Fock operator. What advantages might there be in using such orbitals? [*Hint*: With a large basis there may be a large number of very diffuse virtual orbitals.]

11.19 Carry through a derivation of the coupled SCF equations for a closed-shell reference function, eliminating spin at the beginning. Obtain also the modifications needed in dealing with a magnetic field, presenting all equations in real form. [*Hint*: Work to first order, using H'_{mag} in (11.3.12). With real unperturbed orbitals an angular-momentum operator is represented by i times a real antisymmetric matrix.]

11.20 Show that $E^{(1)}$ and $E^{(2)}$ are as given in (11.9.24) is the closed-shell SCF case.

11.21 Verify that (11.9.19) represents the most general first-order variation of $\mathbf{R}$ consistent with idempotency, and obtain a corresponding second-order term. Then confirm the first- and second-order results (11.9.24) and show that the *third-order* energy change can also be obtained from a first-order calculation of $\mathbf{x}$.

11.22 Show how the equations of coupled SCF perturbation theory must be modified when the basis functions are non-orthogonal. [*Hint*: The equations already derived would apply if the orbitals were *orthonormalized* (for example by the $\mathbf{S}^{-1/2}$ transformation, p. 35): reverse the procedure, transforming all equations to the *non*-orthogonal basis.]

11.23 Formulate all the necessary equations for calculating the tensor components of the electric polarizability and first hyperpolarizability of a molecule with a closed-shell Hartree–Fock ground state, using any of the methods discussed in Section 11.10.

12 Dynamic Properties and Response Theory

12.1 PRELIMINARIES. LINEAR RESPONSE

So far, attention has been concentrated on the calculation of (bound) stationary states and static electronic properties. In principle, a knowledge of the stationary-state wavefunctions provides a basis for the discussion of both static and dynamic properties; but in Chapter 11 some of the difficulties of the usual approach, in which the wavefunction is expanded in terms of the eigenfunctions of some suitable "unperturbed" Hamiltonian, were pointed out. In particlar, exact unperturbed functions are never available, and even good approximate wavefunctions are usually available only for a few of the lowest states, certainly not in a sufficient number to provide a reliable expansion of a perturbed wavefunction. Such difficulties, which were avoided in Chapter 11 by using finite-basis variational methods, prove to be even more serious in discussing dynamic properties and related quantities such as absorption frequencies and transition probabilities.

Instead of using the usual time-dependent perturbation theory, in which knowledge of a complete set of unperturbed functions is assumed, we shall therefore again base most of this chapter on a *variational* formulation of the time-dependent Schrödinger equation. This approach may seem eccentric, but it has a number of obvious advantages; not the least of these are, first, its ability to provide a unified formalism (providing "exact" results in an appropriate variational limit), and secondly the fact that all the results that we derive must pass continuously into those obtained for static perturbations (with similar approximations) on taking a suitable "zero-frequency" limit. We shall also be able to make contact in a very natural way with the Green's functions or "propagators", discussed in the following chapter, and with practical methods of calculating such quantities. First of all, however, we introduce some of the formal concepts of response theory, using simple time-dependent perturbation theory.

The dynamic properties of a system are associated with its response to a time-dependent perturbation. We therefore assume that

$$\mathsf{H} = \mathsf{H}_0 + \mathsf{H}'(t), \tag{12.1.1}$$

where H_0 is the Hamiltonian before applying the perturbation. The time evolution of the wavefunction is determined by solving

$$\mathsf{H}\Psi = i\hbar \frac{\partial \Psi}{\partial t}, \tag{12.1.2}$$

and it is assumed that Ψ may be expressed in terms of the complete set of time-independent wavefunctions $\{\Psi_n\}$. It is convenient to include explicitly in the time-dependent coefficients the usual oscillatory phase factors $\exp(-iE_n t/\hbar)$ and to write the perturbed ground state (to which we confine attention) as $\Psi_0' \exp(-iE_0 t/\hbar)$; this amounts to using the so-called "interaction picture" (see e.g. Davydov, 1976). Thus

$$\Psi_0' = \Psi_0 + \sum_{n(\neq 0)} c_n(t) e^{-i\omega_{0n}t} \Psi_n, \tag{12.1.3}$$

where the coefficients vary relatively slowly with time. $\mathsf{H}'(t)$ is usually turned on slowly, starting with the system in state Ψ_0 at $t = -\infty$. The quantities

$$\omega_{0n} = (E_n - E_0)/\hbar \tag{12.1.4}$$

are clearly the exact excitation frequencies of the unperturbed system; they are real and positive.

Substitution of (12.1.1) and (12.1.3) in (12.1.2), and separation of orders, yields the usual first-order equation for the coefficients, which, with the boundary conditions $c_0 = 1$, $c_m = 0$ $(n \neq 0)$ at $t = -\infty$, becomes

$$i\hbar \dot{c}_n = \langle n|\, \mathsf{H}'(t)\, |0\rangle \exp(i\omega_{0n} t), \tag{12.1.5}$$

where the matrix element is between time-independent functions Ψ_0, Ψ_n. This can be integrated at once when $\mathsf{H}'(t)$ has the simple form

$$\mathsf{H}'(t) = F(t)\mathsf{A}, \tag{12.1.6}$$

where the "fixed" Hermitian operator A determines the "shape" of the perturbation, while the time dependence is confined to the (real) "strength" factor $F(t)$. Thus, for a perturbation beginning at $-\infty$ and continuing up to time t, we obtain

$$c_n(t) = (i\hbar)^{-1} \int_{-\infty}^{t} \langle n|\, \mathsf{A}\, |0\rangle F(t') \exp(i\omega_{0n} t')\, \mathrm{d}t', \tag{12.1.7}$$

which determines, to first order, the perturbed wavefunction.

We shall be interested in the *response*, to the perturbation described by the operator A, of the expectation value of some quantity B with associated operator B; and the first-order result (12.1.7) will determine the *linear response* (linear in the strength parameter F). The time-dependent fluctuation $\langle \mathsf{B}\rangle - \langle \mathsf{B}\rangle_0$, at time t, is easily found to be

$$\delta\langle \mathsf{B}\rangle = \int_{-\infty}^{t} K(\mathsf{BA}\mid t-t')F(t')\,\mathrm{d}t' \qquad (12.1.8)$$

where

$$K(\mathsf{BA}\mid t-t') = (i\hbar)^{-1} \sum_{n(\neq 0)} [\langle 0|\,\mathsf{B}\,|n\rangle\langle n|\,\mathsf{A}\,|0\rangle \exp(-i\omega_{0n}(t-t')) - \langle 0|\,\mathsf{A}\,|n\rangle\langle n|\,\mathsf{B}\,|0\rangle \exp(i\omega_{0n}(t-t'))] \qquad (12.1.9)$$

and is a "time-correlation function", relating the fluctuation of $\langle \mathsf{B}\rangle$ at time t to the strength of the perturbation A at earlier time t'. It should be noted that $K(\mathsf{BA}\mid t-t')$ is defined only for $t'<t$, in accordance with the principle of causality (the response at time t depends only on the perturbation at *earlier* times), and is a function only of the *difference* $\tau = t-t'$.

It is often convenient to express $F(t)$ in (12.1.6) in terms of its Fourier transform $f(\omega)$, remembering the definitions (see e.g. Sneddon, 1951)

$$f(\omega) = \int_{-\infty}^{\infty} F(t)e^{i\omega t}\,\mathrm{d}t, \qquad F(t) = \frac{1}{2\pi}\int_{-\infty}^{\infty} f(\omega)e^{-i\omega t}\,\mathrm{d}\omega. \qquad (12.1.10)$$

In this case we use, instead of (12.1.6),

$$\mathsf{H}'(t) = \frac{1}{2\pi}\int_{-\infty}^{\infty} f(\omega)\tfrac{1}{2}(\mathsf{A}_\omega e^{-i\omega t} + \mathsf{A}_{-\omega}e^{i\omega t})\,\mathrm{d}\omega, \qquad (12.1.11)$$

where a label $\pm\omega$ has been added to the operators associated with the $e^{\mp i\omega t}$ components and it is necessary to take

$$\mathsf{A}_{-\omega} = \mathsf{A}_\omega^\dagger, \qquad f(-\omega) = f(\omega) \qquad (12.1.12)$$

in order to obtain a Hermitian H'. In this case (12.1.11), with $f(\omega)$ chosen according to (12.1.10), leads to (12.1.6) with

$$\mathsf{A} = \tfrac{1}{2}(\mathsf{A}_\omega + \mathsf{A}_{-\omega}), \qquad (12.1.13)$$

and $F(t)$ real. It is also clear that, instead of working in terms of time, we may consider the response to a single oscillatory perturbation

$$\mathsf{H}'(\omega) = \tfrac{1}{2}(\mathsf{A}_\omega e^{-i\omega t} + \mathsf{A}_{-\omega}e^{i\omega t}), \qquad (12.1.14)$$

seeking first a frequency-dependent response and then finally multiplying by $(2\pi)^{-1}f(\omega)$ and integrating over ω to obtain the time-dependent form (12.1.8). In other words, it is possible (and convenient) to work throughout in terms of Fourier transforms.

Let us now consider the effect of the perturbation (12.1.14). To ensure that $\mathsf{H}'(\omega)$ builds up gradually from zero at $t=-\infty$, we introduce a convergence factor $e^{\epsilon t}$ and, with the initial conditions $c_0 = 1$, $c_n = 0$ $(\epsilon > 0)$, easily obtain the first-order result for the coefficients in (12.1.3):

$$c_n(t) = \lim_{\epsilon\to 0}\left(-\frac{1}{2\hbar}\right)\left\{\frac{\langle n|\,\mathsf{A}_\omega\,|0\rangle}{\omega_{0n}-\omega-i\epsilon}\exp\left[i(\omega_{0n}-\omega-i\epsilon)\right]+(\omega\to-\omega)\right\}. \tag{12.1.15}$$

Again, the matrix elements are between time-independent Ψs, and $(\omega\to-\omega)$ indicates a similar term with ω replaced by $-\omega$. Instead of (12.1.8), we now find for the response to (12.1.14)

$$\delta\langle\mathsf{B}\rangle = \tfrac{1}{2}[\Pi(\mathsf{BA}_\omega\mid\omega)e^{-i\omega t}+\Pi(\mathsf{BA}_{-\omega}\mid-\omega)e^{i\omega t}], \tag{12.1.16}$$

where we have collected $\pm\omega$ terms to obtain a symmetrical result. The quantity $\Pi(\mathsf{BA}_\omega\mid\omega)$ is defined by

$$\boxed{\Pi(\mathsf{BA}_\omega\mid\omega) = \lim_{\epsilon\to 0}\frac{1}{\hbar}\sum_{n(\neq 0)}\left\{\frac{\langle 0|\,\mathsf{B}\,|n\rangle\langle n|\,\mathsf{A}_\omega\,|0\rangle}{\omega+i\epsilon-\omega_{0n}}-\frac{\langle 0|\,\mathsf{A}_\omega\,|n\rangle\langle n|\,\mathsf{B}\,|0\rangle}{\omega+i\epsilon+\omega_{0n}}\right\}} \tag{12.1.17}$$

and is sometimes called the frequency-dependent polarizability (FDP) of B with respect to A_ω at frequency ω. The second FDP in (12.1.16) is obtained simply by changing the sign of ω, remembering that $\mathsf{A}_{-\omega}=\mathsf{A}_\omega^\dagger$. When there is no risk of ambiguity the subscript may be dropped, A or $\mathsf{A}^\dagger$ being used as appropriate.

On integrating (12.1.16) over all frequencies, we obtain the alternative form of (12.1.8):

$$\delta\langle\mathsf{B}\rangle = (2\pi)^{-1}\int_{-\infty}\tfrac{1}{2}[\Pi(\mathsf{BA}_\omega\mid\omega)f(\omega)e^{-i\omega t} + \Pi(\mathsf{BA}_{-\omega}\mid-\omega)f(\omega)e^{i\omega t}]\,\mathrm{d}t. \tag{12.1.18}$$

To make the connection clearer, we extend the range of integration in (12.1.8) by writing

$$\delta\langle\mathsf{B}\rangle = \int_{-\infty}^{\infty}\theta(t-t')K(\mathsf{BA}\mid t-t')F(t')\,\mathrm{d}t', \tag{12.1.19}$$

where $\theta(t-t')$ is the Heaviside step function:

$$\theta(\tau)=\begin{cases}0 & (\tau<0),\\ 1 & (\tau>0).\end{cases} \tag{12.1.20}$$

On making use of (12.1.10), we can write‡

$$\begin{aligned}2\delta\langle \mathsf{B}\rangle &= \frac{1}{2\pi}\int_{-\infty}^{\infty}\int_{-\infty}^{\infty} \theta(t-t')K(\mathsf{BA}_\omega \mid t-t')f(\omega)e^{-i\omega t'}\,\mathrm{d}\omega\,\mathrm{d}t' \\ &\qquad + (\omega\to-\omega) \\ &= \frac{1}{\pi}\int_{-\infty}^{\infty}\left[\int_{-\infty}^{\infty} \theta(\tau)K(\mathsf{BA}_\omega \mid \tau)e^{i\omega\tau}\,\mathrm{d}\tau\right]f(\omega)e^{-i\omega t}\,\mathrm{d}\omega \\ &\qquad + (\omega\to-\omega),\end{aligned}$$

and comparison with (12.1.18) then yields

$$\Pi(BA_\omega \mid \omega)=\int_{-\infty}^{\infty}\theta(\tau)K(\mathsf{BA}_\omega \mid \tau)e^{i\omega\tau}\,\mathrm{d}\tau. \tag{12.1.21}$$

The FDP is thus essentially the Fourier transform of the time correlation function, and vice versa:

$$\theta(\tau)K(\mathsf{BA}_\omega \mid \tau)=(2\pi)^{-1}\int_{-\infty}^{\infty}\Pi(\mathsf{BA}_\omega \mid \omega)e^{-i\omega\tau}\,\mathrm{d}\omega, \tag{12.1.22}$$

where K is defined only for positive τ.

The physical reason for considering only $\tau>0$ (causality) is now seen to be associated mathematically with the limiting process in (12.1.17). The summand in (12.1.17) has poles on the real axis at $\pm\omega_{0n}$, displaced downwards by $i\epsilon$ before going to the limit $\epsilon=0$; the integrand in (12.1.22) is analytic, except at these poles, throughout the lower half of the complex plane *provided that* $\tau>0$ but not otherwise. To evaluate the integral (12.1.22), along the real axis, we can therefore use the contour shown in Fig. 12.1, noting that the integrand vanishes over the infinite semicircle, and use Cauchy's residue theorem. Thus

$$-I=2\pi i\times(\text{sum of residues at } \omega=\pm\omega_{0n}-i\epsilon),$$

and this leads immediately to (12.1.9) with $t-t'=\tau$. If $\tau<0$, on the other hand, the integrand is analytic only in the *upper* half-plane—which

‡ Note that $K(\mathsf{BA} \mid t-t')$ is defined for $\mathsf{A}=\frac{1}{2}(\mathsf{A}_\omega+\mathsf{A}_{-\omega})$, and hence $K(\mathsf{BA} \mid \tau)=\frac{1}{2}[K(\mathsf{BA}_\omega \mid \tau)+K(\mathsf{BA}_{-\omega} \mid \tau)]$: the sign of ω may be reversed in the second integral since $f(-\omega)=f(\omega)$.

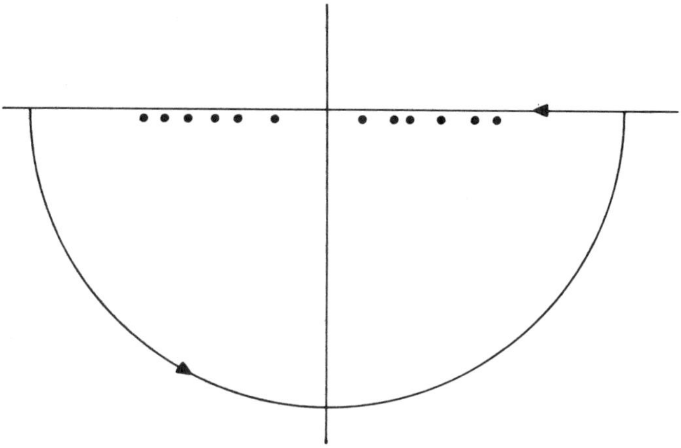

Fig. 12.1 Infinite semicircular contour for evaluating the integral (12.1.22) with τ positive.

contains no poles; on using the same argument with the contour in the upper half-plane, the result is zero, justifying the presence of the factor $\theta(\tau)$ in the inverse transform (12.1.22).

The FDP defined in (12.1.7) exhibits certain symmetry properties. For any pair of operators A and B, not necessarily Hermitian, it is easily verified that

$$\Pi(\mathsf{BA} \mid -\omega)^* = \Pi(\mathsf{B}^\dagger\mathsf{A}^\dagger \mid \omega), \tag{12.1.23}$$

where the new FDP (that of $\mathsf{B}^\dagger$ with respect to $\mathsf{A}^\dagger$) is defined exactly as in (12.1.17) and therefore also has poles just below the real axis for $\epsilon > 0$. It is also evident that in a certain sense the roles of A and B may be reversed, since

$$\Pi(\mathsf{BA} \mid -\omega) = \bar{\Pi}(\mathsf{AB} \mid \omega), \tag{12.1.24}$$

where the quantity on the right is an FDP of A with respect to B but for one small difference—the term $i\epsilon$ in the definition (12.1.17) is replaced by $-i\epsilon$. The FDP $\bar{\Pi}(\mathsf{AB} \mid \omega)$ therefore has poles just *above* the real axis and is analytic in the *upper* half-plane except at these poles. The apparent lack of symmetry in (12.1.24) may be removed by conventionally extending the definition of an FDP to admit all values of ω in the whole plane *except* (for $\epsilon \to 0$) on the real axis: in the lower half-plane Π defined in (12.1.17) is analytic (except at the poles), while in the *upper* half-plane we may define the FDP by adopting $\bar{\Pi}$ in place of Π, i.e. we *redefine the* FDP by reversing the sign of the positive infinitesimal ϵ in (12.1.17). Thus Π becomes analytic throughout the complex plane except along a "cut" containing the real axis. On crossing the cut, it is necessary to jump from one sheet to the other by making the sign change. With this wider definition, (12.1.24) becomes

$$\Pi(\mathsf{BA} \mid -\omega) = \Pi(\mathsf{AB} \mid \omega), \tag{12.1.25}$$

where the bar is no longer needed. The significance of the FDPs will be examined in more detail in the next chapter; for the moment we note only that these quantities generalize the static polarizabilities encountered in Chapter 11, and that if A and B are taken to be electric-moment operators then the zero-frequency limit of the FDP is a polarizability-tensor component, while (12.1.24) corresponds to the symmetry of the tensor. Away from the poles discussion of the limit $\epsilon \to 0$ is unnecessary.

12.2 TIME-DEPENDENT VARIATION THEORY

In practice, we cannot solve the time-dependent Schrödinger equation

$$\mathsf{H}\Psi = i\hbar \frac{\partial \Psi}{\partial t} \tag{12.2.1}$$

by expanding Ψ in terms of exact stationary-state eigenfunctions, and the theory of Section 12.1 therefore remains formal. To avoid this difficulty, however, we may start from the Frenkel principle (Frenkel, 1934):

$$\boxed{\left\langle \delta\Psi \middle| \left(\mathsf{H} - i\hbar\frac{\partial}{\partial t}\right) \middle| \Psi \right\rangle = 0,} \tag{12.2.2}$$

where $\delta\Psi$ is an arbitrary variation of the time-dependent approximate wavefunction Ψ. The most satisfactory formulation of this principle has been discussed by many authors (e.g. McLachlan and Ball, 1964; Löwdin and Mukherjee, 1972; Langhoff *et al.*, 1972). Here we shall follow the treatment due to Moccia (1973b), which confirms the validity of (12.2.2) subject to certain conditions.

To allow for the presence of a rapidly oscillating phase factor, the normalized wavefunction (which we now denote by Ψ_t) may be written as

$$\Psi_t = \frac{g(t)\Psi}{\langle \Psi | \Psi \rangle^{1/2}}, \tag{12.2.3}$$

where $g(t)$ is a unimodular complex number, depending on the time but not on electronic variables. On using standard calculus-of-variation arguments (see e.g. Margenau and Murphy, 1956), with an appropriate Lagrangian density and a lower time boundary at $t = t_0$, the factor $g(t)$ is uniquely determined as

$$g(t) = \exp\left[-\frac{i}{\hbar} \int_{t_0}^{t} \alpha(t')\, dt' \right], \tag{12.2.4a}$$

where

$$\alpha(t) = \langle \Psi \mid \Psi \rangle^{-1} [\langle \Psi | \mathsf{H} | \Psi \rangle - \tfrac{1}{2} i\hbar (\langle \Psi \mid \dot{\Psi} \rangle - \langle \dot{\Psi} \mid \Psi \rangle)]. \tag{12.2.4b}$$

The variation principle for Ψ itself takes the form

$$\left\langle \delta\Psi \middle| \mathsf{Q}\left(\mathsf{H} - i\hbar \frac{\partial}{\partial t}\right) \middle| \Psi \right\rangle + \text{c.c.} = 0, \tag{12.2.5a}$$

where c.c. as usual indicates complex conjugate and

$$\mathsf{Q} = \langle \Psi \mid \Psi \rangle^{-1} (1 - |\Psi\rangle \langle \Psi \mid \Psi \rangle^{-1} \langle \Psi |) \tag{12.2.5b}$$

is a projection operator onto the orthogonal complement of $|\Psi\rangle$, to which a convenient normalizing factor has been attached. Provided that the two terms in (12.2.5a) are *linearly independent*, each may separately be equated to zero, and Frenkel's statement (12.2.2) is then valid, with $\delta\Psi$ interpreted (without loss of generality) as a variation orthogonal to Ψ.

When H is time-independent, (12.2.5a) reduces to the usual stationary-state variation principle, while (12.2.4a) yields the corresponding phase factor $e^{-iEt/\hbar}$, where E is the stationary value of the usual energy functional $\langle \Psi | \mathsf{H} | \Psi \rangle / \langle \Psi \mid \Psi \rangle$.

In discussing time evolution, we assume as in (12.1.6) that

$$\mathsf{H}(t) = \mathsf{H}_0 + \mathsf{H}'(t) = \mathsf{H}_0 + F(t)\mathsf{A}, \tag{12.2.6}$$

where H_0 is time-independent while the second term represents a time variation with strength parameter F. The variational wavefunction is

$$\Psi = \Psi(p_1, p_2, \ldots; \boldsymbol{x}_1, \boldsymbol{x}_2, \ldots, \boldsymbol{x}_N) = \Psi(\mathbf{p}; \boldsymbol{x}), \tag{12.2.7}$$

as in (2.4.7), and it will be assumed that the parameter values $\mathbf{p}_0$ apply to the case $F = 0$, i.e. in the absence of the perturbation. As in Section 2.4, we take $p_j = p_j^0 + d_j$ and describe the variation in terms of the d_j, noting that both $\langle \delta\Psi |$ and $|\Psi\rangle$ in (12.2.3) will acquire a time dependence through the time-dependent changes $d_j(t)$. Thus, working to first order in the d_j,

$$\Psi = \Psi_0 + \sum_j \left(\frac{\partial \Psi}{\partial p_j} \right)_0 d_j(t) + \ldots,$$

and the variation of Ψ due both to first-order variation of parameters and to their time-dependent changes will be

$$\delta\Psi = \sum_j \frac{\partial\Psi}{\partial p_j}\,\delta p_j = \sum_j \left[\left(\frac{\partial\Psi}{\partial p_j}\right)_0 + \sum_k \left(\frac{\partial^2\Psi}{\partial p_k\,\partial p_j}\right)_0 d_k(t)\right]\delta p_j.$$

On adopting the tensor notation of Section 2.4, these last two expressions yield

$$|\Psi\rangle = |\Psi_0\rangle + |(\boldsymbol{\nabla}\Psi)_0\rangle\,\mathbf{d} + \ldots, \tag{12.2.8}$$

$$|\delta\Psi\rangle = [|(\boldsymbol{\nabla}\Psi)_0\rangle + |(\boldsymbol{\nabla}\boldsymbol{\nabla}\Psi)_0\rangle\,\mathbf{d} + \ldots]\delta\mathbf{p} \tag{12.2.9}$$

With the same notation, the components $\langle(\partial\Psi/\partial p_j)_0|$ are written as $\langle(\Psi\boldsymbol{\nabla})_j^0|$, and $\langle\delta\Psi|$ becomes

$$\langle\delta\Psi| = \delta\mathbf{p}^*\langle(\Psi\boldsymbol{\nabla})_0| = \delta\mathbf{p}^*[\langle(\Psi\boldsymbol{\nabla})_0| + \mathbf{d}^*\langle(\Psi\boldsymbol{\nabla}\boldsymbol{\nabla})_0|]. \tag{12.2.10}$$

On substituting (12.2.10) and (12.2.8) in (12.2.5), and remembering that Ψ_0 is the time-independent variation function corresponding to $\mathsf{H}\rightarrow\mathsf{H}_0$, we obtain

$$\begin{aligned}\delta\mathbf{p}^*[\langle(\Psi\boldsymbol{\nabla})_0|\,\mathsf{QH}\,|\Psi_0\rangle + \mathbf{d}^*\langle(\Psi\boldsymbol{\nabla}\boldsymbol{\nabla})_0|\,\mathsf{QH}\,|\Psi_0\rangle]\\ + \delta\mathbf{p}^*[\langle(\Psi\boldsymbol{\nabla})_0|\,\mathsf{QH}\,|(\boldsymbol{\nabla}\Psi)_0\rangle\mathbf{d} - i\hbar\langle(\Psi\boldsymbol{\nabla})_0|\,\mathsf{Q}\,|(\boldsymbol{\nabla}\Psi)_0\rangle\dot{\mathbf{d}}] = 0,\end{aligned} \tag{12.2.11}$$

in which we work to first order in small quantities. The various terms in this equation are easily identified in terms of the tensors defined in Section 2.4. Thus, in general,

$$\langle(\Psi\boldsymbol{\nabla})_j|\,\mathsf{QH}\,|\Psi\rangle = \frac{\langle(\partial\Psi/\partial p_j)|\,\mathsf{H}\,|\Psi\rangle - \langle(\partial\Psi/\partial p_j)\,|\,\Psi\rangle\langle\Psi\,|\,\Psi\rangle^{-1}\langle\Psi|\,\mathsf{H}\,|\Psi\rangle}{\langle\Psi\,|\,\Psi\rangle},$$

which is the quantity $(\nabla H)_j$ defined in (2.4.11b). To summarize,

$$\left.\begin{aligned}\langle(\Psi\boldsymbol{\nabla})|\,\mathsf{QH}\,|\Psi\rangle &= \boldsymbol{\nabla}H, & \langle(\Psi\boldsymbol{\nabla})|\,\mathsf{QH}\,|(\boldsymbol{\nabla}\Psi)\rangle &= \boldsymbol{\nabla}H\boldsymbol{\nabla},\\ \langle(\Psi\boldsymbol{\nabla}\boldsymbol{\nabla})|\,\mathsf{QH}\,|\Psi\rangle &= \boldsymbol{\nabla}\boldsymbol{\nabla}H, & \langle(\Psi\boldsymbol{\nabla})|\,\mathsf{Q}\,|(\boldsymbol{\nabla}\Psi)\rangle &= \boldsymbol{\nabla}Q\boldsymbol{\nabla},\end{aligned}\right\} \tag{12.2.12}$$

where the only new term is $\boldsymbol{\nabla}Q\boldsymbol{\nabla}$, with components

$$(\boldsymbol{\nabla}Q\boldsymbol{\nabla})_{jk} = \left\langle\frac{\partial\Psi}{\partial p_j}\right|\mathsf{Q}\left|\frac{\partial\Psi}{\partial p_k}\right\rangle = V_{jk}. \tag{12.2.13}$$

The same symmetry relations apply as in (2.4.15).

In (12.2.11) the various tensors are all evaluated for $\mathbf{p} = \mathbf{p}_0$, corresponding to absence of the perturbation, though from now on we drop the subscript zero. The equation may therefore be written as

$$\delta\mathbf{p}^*\{\boldsymbol{\nabla}H + \mathbf{d}^*(\boldsymbol{\nabla}\boldsymbol{\nabla}H) + (\boldsymbol{\nabla}H\boldsymbol{\nabla})\mathbf{d} - i\hbar(\boldsymbol{\nabla}Q\boldsymbol{\nabla})\dot{\mathbf{d}}\} = 0, \tag{12.2.14}$$

and as this must hold for arbitrary parameter variations the quantity in braces must vanish.

In Section 2.4 we found that the tensors that now appear in (12.2.14) were of importance in defining variational "stability conditions". The 2-index quantities $(\nabla\nabla H)_{jk}$ and $(\nabla H\nabla)_{jk}$ are elements of the matrices $\mathbf{Q}$ and $\mathbf{M}$; and if we collect the elements $\boldsymbol{\nabla} Q \boldsymbol{\nabla}$ into a matrix $\mathbf{V}$ then the basic equation for time evolution of the parameter values around $\mathbf{p} = \mathbf{p}_0$ (i.e. $\mathbf{d} = \mathbf{0}$) is

$$\boldsymbol{\nabla} H + \mathbf{Md} + \mathbf{Qd}^* = i\hbar \mathbf{V}\dot{\mathbf{d}}. \tag{12.2.15}$$

This linear equation in the "parameter displacements" and their derivatives is somewhat analogous to a classical equation of motion. If the Hamiltonian contains an *oscillating* perturbation $\mathsf{H}'(t)$ then the solution will describe "forced oscillations". Such oscillations may persist, however, even for $\mathsf{H}' = 0$, for certain characteristic frequencies; these "free oscillations" will describe possible time variations of Ψ around the Ψ_0 corresponding to fully optimized parameter values ($\mathbf{p} = \mathbf{p}_0$). The two cases are conveniently treated together; both are considered in the next section.

12.3 FREE AND FORCED OSCILLATIONS

Let us now introduce (12.2.6) in the definition of the tensors, noting that the energy functional $H(p_1, p_2, \ldots)$ will then become a sum of two parts, H_0 and H', where (with $\mathsf{H}' = F(t)\mathsf{A}$)

$$H'(p_1, p_2, \ldots) = \frac{\langle \Psi | \mathsf{H}' | \Psi \rangle}{\langle \Psi \mid \Psi \rangle}. \tag{12.3.1}$$

$\boldsymbol{\nabla} H$, $\boldsymbol{\nabla\nabla} H$ etc. will then contain terms $\boldsymbol{\nabla} H'$, $\boldsymbol{\nabla\nabla} H'$ etc., defined as in (2.4.11), (2.4.13), etc., and the quantities in (12.2.15) may be expanded in powers of F around the point $\mathbf{p} = \mathbf{p}_0$. Thus we take

$$\mathbf{d} = \mathbf{d}_0 + F\mathbf{d}' + \ldots, \tag{12.3.2}$$

and make the replacements‡

$$\left.\begin{aligned} &\boldsymbol{\nabla} H \to \boldsymbol{\nabla} H + F(\boldsymbol{\nabla} H'), \\ &\mathbf{M} \to \boldsymbol{\nabla} H \boldsymbol{\nabla} + F(\boldsymbol{\nabla} H' \boldsymbol{\nabla}) = \mathbf{M} + F\mathbf{M}', \\ &\mathbf{Q} \to \boldsymbol{\nabla\nabla} H + F(\boldsymbol{\nabla\nabla} H') = \mathbf{Q} + F\mathbf{Q}', \end{aligned}\right\} \tag{12.3.3}$$

‡ Note that the leading terms on the right in (12.3.3) contain the unperturbed *Hamiltonian*; they correspond both to $\mathbf{p} = \mathbf{p}_0$ *and* to $F = 0$. This is consistent with our definition of the unperturbed state.

where all derivatives are evaluated using the "unperturbed" variation function. On substituting in (12.2.15) and separating zero- and first-order terms, we obtain (noting that the stationary condition in the absence of the perturbation is $(\nabla H)_0 = 0$, which we assume satisfied)

$$\mathbf{M}\mathbf{d}_0 + \mathbf{Q}\mathbf{d}_0^* = i\hbar\mathbf{V}\dot{\mathbf{d}}_0, \qquad (12.3.4)$$

$$\nabla H' + \mathbf{M}'\mathbf{d}_0 + \mathbf{M}\mathbf{d}' + \mathbf{Q}'\mathbf{d}_0^* + \mathbf{Q}\mathbf{d}'^* = i\hbar\mathbf{V}\dot{\mathbf{d}}'. \qquad (12.3.5)$$

Let us now consider each equation in turn, using atomic units from now on to eliminate the repeated appearance of $\hbar$.

(i) Free oscillations

If we look for a solution in which $\mathbf{d}$ contains a time-factor $e^{-i\omega t}$ (ω real) then the presence of $\mathbf{d}_0^*$ in (12.3.4) will lead also to terms in $e^{i\omega t}$. We therefore seek a solution with

$$\mathbf{d}_0 = \mathbf{X}e^{-i\omega t} + \mathbf{Y}^* e^{i\omega t}, \qquad (12.3.6)$$

where $\mathbf{X}$ and $\mathbf{Y}$ are time-independent "amplitude" vectors and the use of the complex conjugate in the second term simply preserves a convenient symmetry in what follows. Substitution in (12.3.4) then yields

$$\left.\begin{aligned} \mathbf{M}\mathbf{X} + \mathbf{Q}\mathbf{Y} &= \omega\mathbf{V}\mathbf{X}, \\ \mathbf{Q}\mathbf{X}^* + \mathbf{M}\mathbf{Y}^* &= -\omega\mathbf{V}\mathbf{Y}^*. \end{aligned}\right\} \qquad (12.3.7)$$

On taking the complex conjugate of the second equation, both may be presented together as

$$\begin{pmatrix} \mathbf{M} & \mathbf{Q} \\ \mathbf{Q}^* & \mathbf{M}^* \end{pmatrix}\begin{pmatrix} \mathbf{X} \\ \mathbf{Y} \end{pmatrix} = \omega\begin{pmatrix} \mathbf{V} & \mathbf{0} \\ \mathbf{0} & -\mathbf{V} \end{pmatrix}\begin{pmatrix} \mathbf{X} \\ \mathbf{Y} \end{pmatrix}. \qquad (12.3.8a)$$

This equation, which may be abbreviated to

$$\mathbf{\Omega}\mathbf{C} = \omega\mathbf{\Delta}\mathbf{C}, \qquad (12.3.8b)$$

has solutions with interesting properties, which we now examine.

First we note that, although $\mathbf{\Omega}^\dagger = \mathbf{\Omega}$ (from the symmetry properties of the tensor components), (12.3.8) is not an ordinary eigenvalue equation owing to the unusual form of the matrix on the right, which is not positive-definite. Nevertheless, this linear system will possess solutions for certain discrete eigenvalues $\omega_1, \omega_2, \ldots, \omega_p, \ldots,$ and these will occur in pairs, since the changes

$$\omega \to -\omega, \qquad \begin{pmatrix} \mathbf{X}_\omega \\ \mathbf{Y}_\omega \end{pmatrix} \to \begin{pmatrix} \mathbf{Y}_\omega^* \\ \mathbf{X}_\omega^* \end{pmatrix} = \begin{pmatrix} \mathbf{X}_{-\omega} \\ \mathbf{Y}_{-\omega} \end{pmatrix} \qquad (12.3.9)$$

will simply interchange the two equations in (12.3.7). It is therefore sufficient to confine attention to *positive values of* ω, obtaining the second family of solutions by use of (12.3.9). Moreover, solutions for different ω values are orthogonal with the metric described by the matrix $\mathbf{\Delta}$ in (12.3.8b), since if $(\mathbf{C}_\omega, \omega)$ and $(\mathbf{C}_{\omega'}, \omega')$ are solutions then we obtain at once

$$\mathbf{C}_{\omega'}^\dagger \mathbf{\Omega} \mathbf{C}_\omega = \omega \mathbf{C}_{\omega'}^\dagger \mathbf{\Delta} \mathbf{C}_\omega,$$

$$\mathbf{C}_{\omega}^\dagger \mathbf{\Omega} \mathbf{C}_{\omega'} = \omega' \mathbf{C}_{\omega}^\dagger \mathbf{\Delta} \mathbf{C}_{\omega'},$$

and on taking the Hermitian conjugate of the second equation and subtracting from the first we obtain (remembering that $\mathbf{\Omega}^\dagger = \mathbf{\Omega}$)

$$(\omega - \omega')\mathbf{C}_{\omega'}^\dagger \mathbf{\Delta} \mathbf{C}_\omega = 0. \qquad (12.3.10)$$

The vectors are therefore orthogonal with metric $\mathbf{\Delta}$ provided that $\omega' \neq \omega$, and when $\omega' = \omega$ they can be made so in the usual way. The vectors cannot all be normalized to unity, however, since if we use an arbitrary factor to ensure that

$$\mathbf{C}_\omega^\dagger \mathbf{\Delta} \mathbf{C}_\omega = \mathbf{X}_\omega^\dagger \mathbf{V} \mathbf{X}_\omega - \mathbf{Y}_\omega^\dagger \mathbf{V} \mathbf{Y}_\omega = 1$$

then (12.3.9) requires that

$$\mathbf{C}_{-\omega}^\dagger \mathbf{\Delta} \mathbf{C}_{-\omega} = \mathbf{Y}_\omega^{*\dagger} \mathbf{V} \mathbf{Y}_\omega^* - \mathbf{X}_\omega^{*\dagger} \mathbf{V} \mathbf{X}_\omega^* = \mathbf{Y}_\omega^\dagger \mathbf{V} \mathbf{Y}_\omega - \mathbf{X}_\omega^\dagger \mathbf{V} \mathbf{X}_\omega = -1,$$

where the second step involves transposing each (single-number) matrix product and noting that $\mathbf{V}$ is real-symmetric. It is therefore convenient to adopt the normalization

$$\mathbf{C}_\omega^\dagger \mathbf{\Delta} \mathbf{C}_\omega = \pm 1, \qquad (12.3.11)$$

where the upper and lower signs apply for positive and negative values of ω respectively.

In one important special case, arising when the matrices $\mathbf{M}$ and $\mathbf{Q}$ are real, (12.3.8) may be reduced to a normal matrix-eigenvalue equation of half the dimension. In this case, taking the complex conjugate of the second equation in (12.3.7), addition and subtraction of the first yields

$$(\mathbf{M} + \mathbf{Q})(\mathbf{X} + \mathbf{Y}) = \omega \mathbf{V}(\mathbf{X} - \mathbf{Y}),$$

$$(\mathbf{M} - \mathbf{Q})(\mathbf{X} - \mathbf{Y}) = \omega (\mathbf{V}(\mathbf{X} + \mathbf{Y}),$$

whence

$$(\mathbf{M} + \mathbf{Q})(\mathbf{X} + \mathbf{Y}) = \omega^2 \mathbf{V}(\mathbf{M} - \mathbf{Q})^{-1} \mathbf{V}(\mathbf{X} + \mathbf{Y}). \qquad (12.3.12)$$

Thus the sums of the $\mathbf{X}$ and $\mathbf{Y}$ components of $\mathbf{C}_\omega$ in (12.3.8) appear as eigenvectors of the matrix $\mathbf{M} + \mathbf{Q}$, with metric $\mathbf{V}(\mathbf{M} - \mathbf{Q})^{-1}\mathbf{V}$, each eigenvalue being the square of a natural frequency. The difference $\mathbf{X} - \mathbf{Y}$ may then be obtained using

$$(\mathbf{X} - \mathbf{Y}) = \omega (\mathbf{M} - \mathbf{Q})^{-1} \mathbf{V}(\mathbf{X} + \mathbf{Y}), \qquad (12.3.13)$$

thus completing the solution. The existence of positive eigenvalues in (12.3.12), and of a positive-definite metric, is guaranteed (cf. (2.4.17) *et seq.*) when the matrices $\mathbf{M}+\mathbf{Q}$ and $\mathbf{M}-\mathbf{Q}$ are both positive-definite—a situation that we normally assume.

(ii) Forced oscillations

When the time-dependent perturbation $\mathsf{H}'(t)$ is absent, and the parameters $\mathbf{p}$ have their "equilibrium" value $\mathbf{p}_0$ (i.e. when $\mathbf{d}_0=\mathbf{0}$), (12.3.4) ensures that $\mathbf{d}_0$ remain zero and the variation function Ψ_0 be time-independent (remembering that the trivial phase factor $\exp(-iE_0t/\hbar)$ has been discarded). When the perturbation is applied, starting we suppose at $t=-\infty$, the initial conditions are $\mathbf{d}_0=\mathbf{0}$ and $\boldsymbol{\nabla}H=\mathbf{0}$, and the time evolution is determined to first order by solving (12.3.5) for $\mathbf{d}'$. We imagine the perturbation to be turned on infinitely slowly, corresponding to an adiabatic process, and seek the best variational approximation to Ψ at time t. On putting $\mathbf{d}_0=\mathbf{0}$ in (12.3.5), we have

$$\boldsymbol{\nabla}H'+\mathbf{Md}'+\mathbf{Qd}'^*=i\mathbf{V}\dot{\mathbf{d}}', \tag{12.3.14}$$

where the perturbation operator enters only through $\boldsymbol{\nabla}H'$, which has components derived from (12.3.1), namely

$$(\boldsymbol{\nabla}H')_j=\frac{\langle(\partial\Psi/\partial p_j)|\,(\mathsf{H}'-H')\,|\Psi\rangle}{\langle\Psi\,|\,\Psi\rangle}, \tag{12.3.15}$$

evaluated in the limit $\mathbf{p}\rightarrow\mathbf{p}_0$.

Let us now follow the approach of Section 12.1, using a single Fourier component of $\mathsf{H}'(t)$, or more conveniently the Hermitian combination (12.1.14), namely

$$\mathsf{H}'(\omega)=\tfrac{1}{2}(\mathsf{A}_\omega e^{-i\omega t}+\mathsf{A}_{-\omega}e^{i\omega t}), \tag{12.3.16}$$

where as usual $\mathsf{A}_{-\omega}=\mathsf{A}_\omega^\dagger$. After finding a solution of (12.3.14), for a perturbation (12.3.16) at given frequency ω, the solution corresponding to an arbitrary perturbation (12.1.11) will then follow on using the same procedure—multiply by $f(\omega)/2\pi$ and integrate over ω.

We now look for a solution of (12.3.14), at frequency ω, in a form analogous to (12.3.16):

$$\mathbf{d}'=\tfrac{1}{2}(\mathbf{X}'e^{-i\omega t}+\mathbf{Y}'^*e^{i\omega t}). \tag{12.3.17}$$

On substituting (12.3.16) and (12.3.17) in (12.3.14), we obtain a pair of equations similar to (12.3.7):

$$\left.\begin{aligned}\boldsymbol{\nabla}A_\omega+\mathbf{MX}'+\mathbf{QY}'&=\omega\mathbf{VX}',\\ \boldsymbol{\nabla}A_{-\omega}+\mathbf{QX}'^*+\mathbf{MY}'^*&=-\omega\mathbf{VY}'^*,\end{aligned}\right\} \tag{12.3.18}$$

and, on taking the complex conjugate of the second equation, the results may be written as

$$\boxed{\begin{pmatrix} \nabla A \\ A\nabla \end{pmatrix} + \begin{pmatrix} \mathbf{M} & \mathbf{Q} \\ \mathbf{Q}^* & \mathbf{M}^* \end{pmatrix}\begin{pmatrix} \mathbf{X}' \\ \mathbf{Y}' \end{pmatrix} = \omega\begin{pmatrix} \mathbf{V} & \mathbf{0} \\ \mathbf{0} & -\mathbf{V} \end{pmatrix}\begin{pmatrix} \mathbf{X}' \\ \mathbf{Y}' \end{pmatrix}.} \qquad (12.3.19)$$

Here ∇A and $A\nabla$ are defined in the same way as ∇H and $H\nabla$, in (2.4.10) and (2.4.11), but in terms of the expectation value of A; and we have noted that in general (with $A^\dagger$ indicating the expectation value of $\mathsf{A}^\dagger$)

$$\nabla A^\dagger = (A\nabla)^*, \qquad (12.3.20)$$

which reduces to the first result in (2.4.15) when A is Hermitian.

Formal solution of (12.3.19) involves the inverse of the matrix defined in the free-oscillation equation (12.3.8), namely

$$\mathbf{\Omega} - \omega\mathbf{\Delta} = \begin{pmatrix} \mathbf{M} & \mathbf{Q} \\ \mathbf{Q}^* & \mathbf{M}^* \end{pmatrix} - \omega\begin{pmatrix} \mathbf{V} & \mathbf{0} \\ \mathbf{0} & -\mathbf{V} \end{pmatrix}, \qquad (12.3.21)$$

which may be written in terms of its eigenvectors in "spectral form". Thus, if we use $\mathbf{U}$ to denote the full matrix whose pth column $\mathbf{C}_p$ contains $\mathbf{X}_p$, $\mathbf{Y}_p$, then both $\mathbf{\Omega}$ and $\mathbf{\Delta}$ are brought to diagonal form by the corresponding unitary transformation, and we obtain

$$\mathbf{U}^\dagger(\mathbf{\Omega} - \omega\mathbf{\Delta})\mathbf{U} = \mathbf{D} \quad \text{(diagonal)}.$$

The required inverse is then

$$(\mathbf{\Omega} - \omega\mathbf{\Delta})^{-1} = \mathbf{U}\mathbf{D}^{-1}\mathbf{U}^\dagger.$$

Now, with the normalization (12.3.10) and ω_p taken positive,

$$\mathbf{C}_p^\dagger(\mathbf{\Omega} - \omega\mathbf{\Delta})\mathbf{C}_p = \omega_p - \omega,$$
$$\mathbf{C}_{-p}^\dagger(\mathbf{\Omega} - \omega\mathbf{\Delta})\mathbf{C}_{-p} = \omega_p + \omega,$$

and the inverse therefore becomes

$$(\mathbf{\Omega} - \omega\mathbf{\Delta})^{-1} = \sum_{\omega_p > 0}\left(\frac{\mathbf{C}_p\mathbf{C}_p^\dagger}{\omega_p - \omega} + \frac{\mathbf{C}_{-p}\mathbf{C}_{-p}^\dagger}{\omega_p + \omega}\right). \qquad (12.3.22)$$

When ω has a value coinciding with any free-oscillation eigenvalue ω_p there is a singularity; at these poles we encounter a resonance, and the amplitude of the forced oscillation becomes infinite.

The solution of (12.3.19) may now be written down for any value of ω; on using the pairing property (12.3.9), we find

$$\begin{pmatrix} \mathbf{X}' \\ \mathbf{Y}' \end{pmatrix} = \sum_{\omega_p > 0}\left\{\frac{1}{\omega - \omega_p}\begin{pmatrix} \mathbf{X}_p \\ \mathbf{Y}_p \end{pmatrix}(\mathbf{X}_p^\dagger \quad \mathbf{Y}_p^\dagger) - \frac{1}{\omega + \omega_p}\begin{pmatrix} \mathbf{Y}_p^* \\ \mathbf{X}_p^* \end{pmatrix}(\mathbf{Y}_p^{*\dagger} \quad \mathbf{X}_p^{*\dagger})\right\}\begin{pmatrix} \nabla A \\ A\nabla \end{pmatrix}, \qquad (12.3.23)$$

which may be written more neatly as

$$\begin{pmatrix} \mathbf{X}' \\ \mathbf{Y}' \end{pmatrix} = \begin{pmatrix} \mathbf{\Pi}_{XX}(\omega) & \mathbf{\Pi}_{XY}(\omega) \\ \mathbf{\Pi}_{YX}(\omega) & \mathbf{\Pi}_{YY}(\omega) \end{pmatrix} \begin{pmatrix} \mathbf{\nabla} A \\ A\mathbf{\nabla} \end{pmatrix}. \tag{12.3.24}$$

The square matrix on the right completely determines the linear response of the system, at any frequency, to the "applied force" corresponding to any perturbation operator A. This *response matrix* is an intrinsic characteristic of the system itself and depends in no way on the applied perturbation: its determination (using an approximate wavefunction of any suitable form) provides a practicable route to the study of all linear response properties.

12.4 SIGNIFICANCE OF THE RESPONSE MATRIX

The response matrix $\mathbf{\Pi}$, defined in (12.3.23) and (12.3.24), enables us to find the time-dependent change in the expectation value of *any* quantity, B say, as the result of a perturbation (12.3.16) containing the operator A. It therefore yields an approximation to the FDP $\Pi(\mathsf{BA} \mid \omega)$ defined formally in Section 12.1.

The variation $\delta\langle \mathsf{B} \rangle$ of $\langle \mathsf{B} \rangle$ is given by an expression similar to (2.4.16) but with B in place of H. Thus

$$\langle \mathsf{B} \rangle = \langle \mathsf{B} \rangle_0 + [\mathbf{d}^*(\mathbf{\nabla} B) + (B\mathbf{\nabla})\mathbf{d}] + \dots,$$

and, on using (12.3.17), we obtain a time-dependent fluctuation that reduces to

$$\begin{aligned} \langle \mathsf{B} \rangle = {} & \tfrac{1}{2} e^{-i\omega t}[(B\mathbf{\nabla})\mathbf{X}' + (\mathbf{\nabla} B)\mathbf{Y}'] \\ & + \tfrac{1}{2} e^{i\omega t}[\mathbf{X}'^*(\mathbf{\nabla} B) + \mathbf{Y}'^* B\mathbf{\nabla})]. \end{aligned} \tag{12.4.1}$$

On making use of (12.3.24) and (12.3.20), this becomes

$$\delta\langle \mathsf{B} \rangle = \frac{1}{2}\left\{ \begin{pmatrix} \mathbf{\nabla} B \\ B\mathbf{\nabla} \end{pmatrix}^{\dagger} \begin{pmatrix} \mathbf{\Pi}_{XX} & \mathbf{\Pi}_{XY} \\ \mathbf{\Pi}_{YX} & \mathbf{\Pi}_{YY} \end{pmatrix} \begin{pmatrix} \mathbf{\nabla} A \\ A\mathbf{\nabla} \end{pmatrix} e^{-i\omega t} + (\omega \to -\omega) \right\}, \tag{12.4.2}$$

where $(\omega \to -\omega)$ is a similar term with ω replaced by $-\omega$ and (with the usual convention) containing $\mathsf{A}^{\dagger}$ in place of A, and $\mathsf{B}^{\dagger}$ in place of B. The coefficient

$$\Pi(\mathsf{BA} \mid \omega) = \begin{pmatrix} \mathbf{\nabla} B \\ B\mathbf{\nabla} \end{pmatrix}^{\dagger} \begin{pmatrix} \mathbf{\Pi}_{XX} & \mathbf{\Pi}_{XY} \\ \mathbf{\Pi}_{YX} & \mathbf{\Pi}_{YY} \end{pmatrix} \begin{pmatrix} \mathbf{\nabla} A \\ A\mathbf{\nabla} \end{pmatrix} \tag{12.4.3}$$

is thus the variational approximation to the FDP of B with respect to A, at frequency ω, as defined in (12.1.16).

The symmetry properties of the FDPs, noted in Section 12.1, are reflected in those of the response matrix. Thus the $\mathbf{\Pi}$ matrix that appears in (12.3.23) is unchanged if we replace ω by $-\omega$, exchange $\mathbf{X}$ and $\mathbf{Y}$, and take the complex conjugate. For the individual blocks, we have

$$\left.\begin{aligned}\mathbf{\Pi}_{XX}(-\omega)^* &= \mathbf{\Pi}_{YY}(\omega), \qquad & \mathbf{\Pi}_{XY}(-\omega)^* &= \mathbf{\Pi}_{YX}(\omega),\\ \mathbf{\Pi}_{YX}(-\omega)^* &= \mathbf{\Pi}_{XY}(\omega), \qquad & \mathbf{\Pi}_{YY}(-\omega)^* &= \mathbf{\Pi}_{XX}(\omega).\end{aligned}\right\} \tag{12.4.4}$$

On putting the row and column submatrices of (12.4.3) in reverse order and taking the complex conjugate (noting the property (12.3.20)), we then obtain, from (12.4.4),

$$\Pi(\mathsf{BA} \mid \omega)^* = \begin{pmatrix}\nabla B^\dagger \\ B^\dagger \nabla\end{pmatrix}^\dagger \begin{pmatrix}\mathbf{\Pi}_{XX}(-\omega) & \mathbf{\Pi}_{XY}(-\omega)\\ \mathbf{\Pi}_{YX}(-\omega) & \mathbf{\Pi}_{YY}(-\omega)\end{pmatrix}\begin{pmatrix}\nabla A^\dagger \\ A^\dagger \nabla\end{pmatrix} = \Pi(\mathsf{B}^\dagger\mathsf{A}^\dagger \mid -\omega). \tag{12.4.5}$$

The two terms in (12.4.2) are therefore complex-conjugate, and the fluctuation $\delta\langle \mathsf{B}\rangle$ is real.

It is of some importance to examine the behaviour of the response matrix at the poles, where it becomes singular and the expression (12.3.23) is no longer useful. To consider this case, we should go back to (12.3.16), add a convergence factor $e^{\epsilon t}$, so that the perturbation builds up from zero at $t = -\infty$, and then take the limit $\epsilon \to 0$. It is clear, however, that, since (12.4.3) is the variational analogue of (12.1.17), the correct behaviour at the poles will follow on putting $\omega + i\epsilon$ in place of ω in the two denominators of (12.3.23). Again using $\mathbf{C}_p$ to denote the column containing $\mathbf{X}_p$, $\mathbf{Y}_p$, it follows that

$$\mathbf{\Pi}(\omega) = \lim_{\epsilon\to 0} \sum_{\omega_p > 0} \left(\frac{\mathbf{C}_p \mathbf{C}_p^\dagger}{\omega + i\epsilon - \omega_p} - \frac{\mathbf{C}_{-p}\mathbf{C}_{-p}^\dagger}{\omega + i\epsilon + \omega_p}\right), \tag{12.4.6}$$

which falls into blocks as in (12.3.24) and may be used in (12.4.3) to obtain any desired FDP.

The effect of the $i\epsilon$ term is revealed when we consider the time variation $\delta\langle \mathsf{B}\rangle$ resulting from the perturbation

$$\mathsf{H}'(\mathrm{t}) = \tfrac{1}{2}F(t)(\mathsf{A}_\omega + \mathsf{A}_{-\omega}) \tag{12.4.7}$$

with an arbitrary time factor $F(t)$. In this case we must perform an ω integration, according to (12.1.11), with $f(\omega)$ chosen to be the Fourier transform of $F(t)$ as in (12.1.10). In other words, when dealing with an arbitrary time-dependent perturbation, the appropriate FDP will occur as a factor in any integration over the frequency domain $(-\infty, \infty)$—as we have seen in deriving (12.1.18). On keeping in mind this frequency integration, in which ω is a real continuous variable, we may use the symbolic identity

$$\frac{1}{\omega + i\epsilon - \omega_p} = \mathrm{P}\left(\frac{1}{\omega - \omega_p}\right) - i\pi\delta(\omega - \omega_p) \tag{12.4.8}$$

to deal with the two terms in (12.4.6). Any frequency integration of a quantity containing $\mathbf{\Pi}(\omega)$ will then become a sum of two parts: (i) the *principal value* of the integral, obtained by integrating along the real axis with the poles $\omega = \pm\omega_p$ excluded; and (ii) the pole contributions, which arise from the delta-function terms.

The pole contributions may be evaluated explicitly by separating $\mathbf{\Pi}(\omega)$, and consequently any FDP obtained as in (12.4.3), into two parts using (12.4.8). Thus we put

$$\mathbf{\Pi}(\omega) = \mathbf{\Pi}'(\omega) + i\mathbf{\Pi}''(\omega), \tag{12.4.9}$$

where $\mathbf{\Pi}'(\omega)$ is the quantity used previously (with $\epsilon = 0$) and it is assumed that in any integration a principal value will be taken, while $i\mathbf{\Pi}''(\omega)$ is the new term that arises from the poles. The expression for $\mathbf{\Pi}''(\omega)$ is evidently

$$\mathbf{\Pi}''(\omega) = -\pi \sum_{\omega_p > 0} [\delta(\omega - \omega_p)\mathbf{C}_p\mathbf{C}_p^\dagger - \delta(\omega + \omega_p)\mathbf{C}_{-p}\mathbf{C}_{-p}^\dagger]. \tag{12.4.10}$$

When the response matrix is written in the partitioned form (12.3.24) there will of course be corresponding contributions to every block. Thus, for example,

$$\mathbf{\Pi}_{XX}(\omega) = \mathbf{\Pi}_{XX}(\omega) + i\mathbf{\Pi}_{XX}(\omega), \tag{12.4.11}$$

where, with the proviso following (12.4.9),

$$\mathbf{\Pi}'_{XX}(\omega) = \sum_{\omega_p > 0} \left(\frac{\mathbf{X}_p\mathbf{X}_p^\dagger}{\omega - \omega_p} - \frac{\mathbf{Y}_p^*\mathbf{Y}_p^{*\dagger}}{\omega + \omega_p}\right), \tag{12.4.12a}$$

$$\mathbf{\Pi}''_{XX}(\omega) = -\pi \sum_{\omega_p > 0} [\delta(\omega - \omega_p)\mathbf{X}_p\mathbf{X}_p^\dagger - \delta(\omega + \omega_p)\mathbf{Y}_p^*\mathbf{Y}_p^{*\dagger}], \tag{12.4.12b}$$

12.5 TIME-DEPENDENT HARTREE–FOCK THEORY

A simple application of the very general approach used in earlier sections leads to the time-dependent generalization of Hartree–Fock theory. The time-dependent Hartree–Fock (TDHF) equations (Dirac, 1929) were first formulated variationally by Frenkel (1934); they are also widely used in nuclear physics (see e.g. Thouless, 1961) under the name "random-phase approximation" (RPA). Since the equations describe response to a perturbation, as in Section 11.9 but now time-dependent, they will obviously resemble those for the static case—to which they must reduce in the zero-frequency limit—and it will thus be sufficient merely to sketch the derivations.

We start from a one-determinant wavefunction Ψ_0 of spin-orbitals $\{\psi_i\}$, assumed to be orthonormal, and consider the fluctuation of these orbitals under the influence of a perturbation operator (12.1.8) or, equivalently, a single Fourier component (12.1.14). The variation is conveniently introduced by considering $\boldsymbol{\psi} \to \boldsymbol{\psi} e^{\boldsymbol{\Delta}}$, as in Section 8.2, where $\boldsymbol{\Delta}$ is an anti-Hermitian matrix that describes the admixture of virtual orbitals $\{\psi_m\}$ into the occupied set $\{\psi_i\}$. On expanding the exponential and taking the distinct elements Δ_{rs} $(r > s)$ as the parameters p_{rs}, the variation yields, up to second order,

$$\Psi = \Psi_0 + \sum_{i,m} p_{mi} \Psi(i \to m) + \sum_{\substack{i \neq j \\ m \neq n}} p_{mi} p_{nj} \Psi(i \to m, j \to n) + \ldots, \qquad (12.5.1)$$

which is exactly like (11.9.6) except that now the parameters are time-dependent. Again the only parameters that appear are of the type p_{mi} (lower triangle of the matrix $\boldsymbol{\Delta}$).

The derivatives needed in evaluating the tensor components (12.2.12) *et seq.* have been given already in (11.9.7), while expressions for the tensor components themselves are available in (11.9.8)–(11.9.10). In particular, the gradient components are

$$(\nabla H)_{mi} = \langle m | \mathsf{F} | i \rangle, \qquad (12.5.2)$$

where F is the usual Fock operator. The other components define explicitly the elements of the matrices $\mathbf{M}$ and $\mathbf{Q}$ in the basic time-development equation (12.2.15). The only other matrix required is $\mathbf{V}$, with elements given in (12.2.13), which in this case becomes the unit matrix.

Again, in the absence of the perturbation, $\mathsf{H} = \mathsf{H}_0$ and we assume the parameters p_{mi} to be chosen so that the energy functional is stationary for $\Psi = \Psi_0$, $p_{mi}^{(0)} = 0$; in other words, the Brillouin condition is satisfied, from (12.5.2), and Ψ_0 must satisfy the usual SCF equations.

We now include in the Hamiltonian the time-dependent term (12.1.14) and write down explicitly the equations for the free and forced oscillations. The parameter displacements $d_{mi} = p_{mi} - p_{mi}^{(0)} = p_{mi}$ ($p_{mi}^{(0)} = 0$ for the unperturbed stationary state) will be written as

$$\mathbf{d} = \mathbf{d}_0 + \mathbf{d}' + \ldots \qquad (12.5.3)$$

where $\mathbf{d}_0$ and $\mathbf{d}'$ are defined as in (12.3.6) and (12.3.17). Free oscillations $\mathbf{d}_0$ are obtained when $\mathbf{X}$ and $\mathbf{Y}$ satisfy (12.3.8), and forced oscillations $\mathbf{d}'$ when the corresponding $\mathbf{X}'$ and $\mathbf{Y}'$ satisfy (12.3.19). Thus, for the free oscillations, (12.3.8) takes the form

$$\boxed{\begin{aligned}\sum_{n} [\delta_{mi,nj}(\epsilon_m - \epsilon_i) + \langle mj \| in \rangle] X_{nj} + \sum_{nj} \langle mn \| ij \rangle Y_{nj} = \omega X_{mi},\\ \sum_{nj} [\delta_{mi,nj}(\epsilon_m - \epsilon_i) + \langle in \| mj \rangle] Y_{nj} + \sum_{nj} \langle ij \| mn) X_{nj} = -\omega Y_{mi},\end{aligned}} \tag{12.5.4}$$

which are the standard TDHF equations (see e.g. Thouless, 1961).

The forced-oscillation equations are similar, except for the replacement of X, Y by X', Y' and the addition of extra terms on the left (cf. (12.3.19))—these latter being simply matrix elements of the perturbation operators. Explicitly,

$$(\nabla A)_{mi} = \langle m | \mathsf{A} | i \rangle, \qquad (A\nabla)_{mi} = \langle m | \mathsf{A}^\dagger | i \rangle^*, \tag{12.5.5}$$

and, on inserting these terms in the first and second equations of (12.5.4), the resultant equations determine the forced oscillations $\mathbf{d}'$ through (12.3.17).

Finally, we note that for a closed-shell ground state (cf. Section 6.2) spin may be eliminated at once. Every orbital appears twice in Ψ_0, once in $\phi_i\alpha$ $(=\psi_i)$ and once in $\phi_i\beta$ $(=\bar{\psi}_i)$, and the determinants $\Psi(i \to m)$ and $\Psi(\bar{i} \to \bar{m})$ both appear in (12.5.1) with the same coefficient (Ψ being a singlet state): these coefficients p_{mi} $(=d_{mi})$ describe *orbital*, rather than spin-orbital variations. The equations for free and forced oscillations of these parameters are then exactly like those given above, except that the 1- and 2-electron integrals are defined over orbital factors alone and the coulomb-exchange combination that occurs is

$$\langle rs \| tu \rangle = \langle rs | g | tu \rangle - \tfrac{1}{2} \langle rs | g | ut \rangle. \tag{12.5.6}$$

It should of course be remembered that all orbitals are here assumed to be solutions of the usual SCF equations (Section 6.2) for the unperturbed system and ϵ_i, ϵ_m are assumed to the corresponding occupied and virtual orbital energies. It is not in fact essential that *canonical* SCF orbitals be used in any of the preceding derivations, and it is sometimes useful to employ, for example, more localized linear combinations: but the modified equations are easily obtained (Problem 12.9).

The above TDHF equations are linear, like those in Section 11.9, because we work only to first order in the parameter displacements (from their unperturbed SCF values). Again, however, there are some advantages in an alternative formulation, more obviously related to the usual finite-basis form of HF theory, in which solutions are obtained by an iterative procedure akin to that used in the stationary-state SCF method. This approach makes use of TDHF equations in the original density-matrix form due to Dirac (1931).

12.6 DENSITY-MATRIX FORM OF TDHF THEORY

The variational derivation of TDHF equations (Frenkel, 1934) relates immediately to the stationary state approach and the SCF method. The quantities needed for substitution in the variation principle (12.2.2) are $\langle \delta\Psi \mid \Psi \rangle$ and $\langle \delta\Psi | \mathsf{H} | \Psi \rangle$: the first follows from first principles (for example from Slater's rules, noting that $\partial/\partial t$ behaves like a 1-electron operator) as

$$\langle \delta\Psi \mid \Psi \rangle = \sum_i \left(\langle \delta\psi_i \mid \dot{\psi}_i \rangle + \sum_{j(\neq i)} \langle \delta\psi_i \mid \psi_i \rangle \langle \psi_j \mid \dot{\psi}_j \rangle \right), \quad (12.6.1)$$

while the second is already familiar (cf. (6.1.29)), being

$$\langle \delta\Psi | \mathsf{H} | \Psi \rangle = \sum_i \langle \delta\psi_i | \mathsf{F} | \psi_i \rangle. \quad (12.6.2)$$

Substitution in (12.2.2) then gives

$$\sum_i \left[\left\langle \delta\psi_i \right| \left(\mathsf{F} - i\hbar \frac{\partial}{\partial t} \right) \left| \psi_i \right\rangle - i\hbar \sum_{j(\neq i)} \langle \delta\psi_i \mid \psi_i \rangle \langle \psi_j \mid \dot{\psi}_j \rangle \right] = 0. \quad (12.6.3)$$

As usual, however, there is an orthonormality constraint, which gives

$$\langle \delta\psi_i \mid \psi_i \rangle + \langle \psi_i \mid \delta\psi_i \rangle = 0, \qquad \langle \delta\psi_i \mid \psi_j \rangle = 0 \quad (i \neq j), \quad (12.6.4)$$

and, on using Lagrange multipliers as in Section 6.1, we easily find

$$\left(\mathsf{F} - i\hbar \frac{\partial}{\partial t} - i\hbar \sum_{j(\neq i)} \langle \psi_j \mid \dot{\psi}_j \rangle \right) | \psi_i \rangle - \sum_j | \psi_j \rangle b_{ji} = 0. \quad (12.6.5)$$

The multipliers are identified by taking scalar products: thus with $\langle \psi_k |$ on the left $(k \neq i)$

$$\langle \psi_k | \mathsf{F} | \psi_i \rangle - i\hbar \langle \psi_k \mid \dot{\psi}_i \rangle = b_{ki} \quad (k \neq i),$$

while with $\langle \psi_i |$ on the left

$$\langle \psi_i | \mathsf{F} | \psi_i \rangle - i\hbar \langle \psi_i \mid \dot{\psi}_i \rangle - i\hbar \sum_{j(\neq i)} \langle \psi_j \mid \dot{\psi}_j \rangle = b_{ii}.$$

On defining

$$\epsilon_{ki} = \langle \psi_k | \mathsf{F} | \psi_i \rangle - i\hbar \langle \psi_k \mid \dot{\psi}_i \rangle \quad (\text{all } i, k), \quad (12.6.6)$$

the stationary condition (12.6.5) then becomes

$$\left(\mathsf{F} - i\hbar\frac{\partial}{\partial t}\right)|\psi_i\rangle = \sum_j |\psi_j\rangle\epsilon_{ji}. \tag{12.6.7}$$

It is also easily established (Problem 12.10) that the matrix of Lagrange multipliers is Hermitian, and (12.6.7) is thus exactly like the usual stationary-state HF equation (6.1.18) except for the presence of the time derivative.

To obtain the finite-basis form of (12.6.7), the same argument may be followed in matrix form. With the basis-set expansion

$$\boldsymbol{\psi} = \boldsymbol{\chi}\mathbf{T} \tag{12.6.8}$$

($\boldsymbol{\chi}$ containing, say, m orthonormal spin-orbitals), the variations (12.6.1) and (12.6.2) may be rewritten as

$$\langle \delta\Psi \mid \Psi\rangle = \sum_i \left[\delta\mathbf{c}_i^\dagger\dot{\mathbf{c}}_i + \sum_{j(\neq i)} (\delta\mathbf{c}_i^\dagger\mathbf{c}_i)(\mathbf{c}_j^\dagger\dot{\mathbf{c}}_j)\right], \tag{12.6.9}$$

$$\langle \delta\Psi \mid \mathsf{H} \mid \Psi\rangle = \sum_i \delta\mathbf{c}_i^\dagger\mathbf{F}\mathbf{c}_i, \tag{12.6.10}$$

where $\mathbf{c}_i$ is the ith column of $\mathbf{T}$, representing ψ_i. The Fock *matrix* $\mathbf{F}$ is defined as in Section 6.2; thus

$$\mathbf{F} = \mathbf{h} + \mathbf{G}(\boldsymbol{\rho}), \qquad \boldsymbol{\rho} = \sum_i \mathbf{c}_i\mathbf{c}_i^\dagger = \mathbf{T}\mathbf{T}^\dagger, \tag{12.6.11}$$

$\mathbf{G}$ being the usual electron-interaction matrix. The same argument as before then yields, instead of (12.6.7),

$$\left(\mathbf{F} - i\hbar\frac{\partial}{\partial t}\right)\mathbf{c}_i = \sum_j \mathbf{c}_j\epsilon_{ji}, \tag{12.6.12}$$

or, collecting the columns into the matrix $\mathbf{T}$,

$$\left(\mathbf{F} - i\hbar\frac{\partial}{\partial t}\right)\mathbf{T} = \mathbf{T}\boldsymbol{\epsilon}. \tag{12.6.13}$$

This reduces, of course, to the ordinary SCF equation $\mathbf{FT} = \mathbf{T}\boldsymbol{\epsilon}$ when $\mathbf{F}$ is time-independent; otherwise $\mathbf{T}$ will acquire a time dependence and will introduce (through the time-dependent density in the $\mathbf{G}$ matrix) a coupling between the density fluctuations and the effective field. When the time-dependent part of $\mathbf{h}$ is specified it is therefore necessary to solve (12.6.13) iteratively until the fluctuations of $\mathbf{F}$ and $\boldsymbol{\rho}$ are self-consistent. There is thus a fully coupled *time-dependent* analogue of the SCF perturbation theory in Section 11.9.

The Dirac (1931) form of the equations follows on multiplying (12.6.13) from the right by $\mathbf{T}^\dagger$ and then subtracting from the resultant equation its Hermitian transpose to obtain

$$\mathbf{F}\boldsymbol{\rho} - \boldsymbol{\rho}\mathbf{F} = i\hbar\dot{\boldsymbol{\rho}}, \tag{12.6.14}$$

which is the time-dependent generalization of (6.2.21).

We now turn to the case $\mathsf{H} = \mathsf{H}_0 + \mathsf{H}'$, where H' is a single Fourier component (12.1.14) and the operator A is of one-body form, describing for example an oscillating applied field. This corresponds to taking

$$\mathbf{h} = \mathbf{h}_0 + \mathbf{h}', \qquad \mathbf{h}' = \tfrac{1}{2}(\mathbf{A}e^{-i\omega t} + \mathbf{A}^\dagger e^{i\omega t}), \tag{12.6.15}$$

where $\mathbf{h}_0$ refers to the unperturbed system in a stationary state. It is assumed that a time-*in*dependent solution is known, satisfying the unperturbed SCF equation $\mathbf{F}_0\boldsymbol{\rho}_0 - \boldsymbol{\rho}_0\mathbf{F}_0 = \mathbf{0}$, and we take

$$\boldsymbol{\rho} = \boldsymbol{\rho}_0 + \boldsymbol{\rho}', \qquad \boldsymbol{\rho}' = \tfrac{1}{2}(\mathbf{d}e^{-i\omega t} + \mathbf{d}^\dagger e^{i\omega t}) \tag{12.6.16}$$

to determine the first-order (i.e. linear) response.

To proceed, we substitute these expressions in (12.6.14) and separate the orders, obtaining a first-order equation to be satisfied by $\mathbf{d}$:

$$\mathbf{F}_0\mathbf{d} - \mathbf{d}\mathbf{F}_0 + [\mathbf{A} + \mathbf{G}(\mathbf{d})]\boldsymbol{\rho}_0 - \boldsymbol{\rho}_0[\mathbf{A} + \mathbf{G}(\mathbf{d})] = \hbar\omega\dot{\mathbf{d}}. \tag{12.6.17}$$

This equation arises from the $e^{-i\omega t}$ terms, a Hermitian-conjugate equation coming from terms in $e^{i\omega t}$. From an assumed first approximation (e.g. $\mathbf{d} = \mathbf{0}$), the equations may be solved for $\mathbf{d}$, and the revised approximation will define $\mathbf{G}(\mathbf{d})$ for starting the next iteration.

A solution of (12.6.17) may be obtained by the method used in Section 11.9. Idempotency of the matrix $\boldsymbol{\rho}$ requires that (to first order) $\mathbf{d}$ in (12.6.16) can only have projections of the form

$$\mathbf{x} = (\mathbf{1} - \boldsymbol{\rho}_0)\mathbf{d}\boldsymbol{\rho}_0, \qquad \mathbf{y} = \boldsymbol{\rho}_0\mathbf{d}(\mathbf{1} - \boldsymbol{\rho}_0) \tag{12.6.18}$$

On putting $\mathbf{d} = \mathbf{x} + \mathbf{y}$ in (12.6.17), taking appropriate projections, and noting that $\boldsymbol{\rho}_0$ commutes with $\mathbf{F}_0$, we obtain two equations:

$$\left.\begin{aligned} (\mathbf{F}_0 - \hbar\omega\mathbf{1})\mathbf{x} - \mathbf{x}\mathbf{F}_0 + (\mathbf{1} - \boldsymbol{\rho}_0)[\mathbf{A} + \mathbf{G}(\mathbf{x} + \mathbf{y})]\boldsymbol{\rho}_0 = \mathbf{0}, \\ (\mathbf{F}_0 - \hbar\omega\mathbf{1})\mathbf{y} - \mathbf{y}\mathbf{F}_0 - \rho_0[\mathbf{A} + \mathbf{G}(\mathbf{x} + \mathbf{y})](\mathbf{1} - \boldsymbol{\rho}_0) = \mathbf{0}. \end{aligned}\right\} \tag{12.6.19}$$

The most general forms of $\mathbf{x}$ and $\mathbf{y}$, in terms of the "occupied' and "virtual" eigenvectors of $\mathbf{F}_0$ ($\mathbf{c}_i$ and $\mathbf{c}_m$ respectively) are

$$\mathbf{x} = \sum_{i,m} X_i^m \mathbf{c}_m \mathbf{c}_i^\dagger, \qquad \mathbf{y} = \sum_{i,m} Y_i^m \mathbf{c}_i \mathbf{c}_m^\dagger, \tag{12.6.20}$$

and the coefficients are easily identified by multiplying from the left with $\mathbf{c}_n^\dagger$ and $\mathbf{c}_j^\dagger$ respectively, and using the orthonormality property. With the definitions

$$\mathbf{\Delta} = \mathbf{A} + \mathbf{G}(\mathbf{x} + \mathbf{y}), \qquad \Delta_{rs} = \mathbf{c}_r^\dagger \mathbf{\Delta} \mathbf{c}_s, \tag{12.6.21}$$

it then follows that $\mathbf{d}$ $(=\mathbf{x}+\mathbf{y})$, and hence $\boldsymbol{\rho}$, may be obtained from

$$\mathbf{x} = -\sum_{i,m} \frac{\Delta_{mi}}{\epsilon_m - \epsilon_i - \hbar\omega} \mathbf{c}_m \mathbf{c}_i^\dagger, \tag{12.6.22a}$$

$$\mathbf{y} = -\sum_{i,m} \frac{\Delta_{im}}{\epsilon_m - \epsilon_i + \hbar\omega)} \mathbf{c}_i \mathbf{c}_m^\dagger, \tag{12.6.22b}$$

in terms of the unperturbed eigenvectors of $\mathbf{F}_0$.

The above equations may be solved iteratively until the perturbation matrix $\mathbf{\Delta}$ in (12.6.21) is self-consistent with the density fluctuation $\mathbf{d}$ $(=\mathbf{x}+\mathbf{y})$ to which it leads. This is a standard SCF procedure, and, starting from $\mathbf{x}=\mathbf{y}=\mathbf{0}$, often converges well except in the region of the poles—where one or more of the denominators is small.

Once $\mathbf{d}$ has been determined, it is easy to find any FDP $\Pi(\mathsf{BA} \mid \omega)$, B and A being arbitrary 1-electron operators, on remembering (p. 126) that

$$\langle \mathsf{B} \rangle = \mathrm{tr}\, \mathbf{B}\boldsymbol{\rho},$$

where $\mathbf{B}$ is the matrix associated with B in the χ basis. When $\boldsymbol{\rho} = \boldsymbol{\rho}_0 + \boldsymbol{\rho}'$ we obtain for the fluctuation $\delta\langle \mathsf{B} \rangle = \mathrm{tr}\, \mathbf{B}\boldsymbol{\rho}'$

$$\delta\langle \mathsf{B} \rangle = \tfrac{1}{2}[\mathrm{tr}\,(\mathbf{B}\mathbf{d})e^{-i\omega t} + \mathrm{tr}\,(\mathbf{B}\mathbf{d}^\dagger)e^{i\omega t}], \tag{12.6.23}$$

and hence

$$\Pi(\mathsf{BA} \mid \omega) = \mathrm{tr}\,[\mathbf{B}(\mathbf{x}_A + \mathbf{y}_A)], \tag{12.6.24}$$

where a subscript A has been added to the $\mathbf{x}$ and $\mathbf{y}$ matrices to indicate their dependence on the perturbation A.

Evidently, if the response is needed for only a few values of ω (none of them near a resonance) the iterative procedure for solving the time-dependent equations, conveniently referred to as a TDSCF method to distinguish it from the non-iterative (TDHF) method introduced earlier, can provide an efficient route to the calculation of FDPs. In particular, orthogonality of the basis may be relaxed provided that an overlap matrix is included in the SCF equations in the usual way: the working equations are unchanged. Consequently "raw" basis functions may be used, no integral transformations being necessary, and the dimension of all matrix equations is that of the basis. On the other hand, if the response is required for all values of ω, covering the range of the resonances, it is necessary to use the TDHF approach, obtaining all eigenvalues and eigenvectors of a very large matrix. Each method thus has its own advantages and disadvantages.

Finally, it is easy to demonstrate the complete equivalence of the

TDHF and TDSCF procedures. The density matrix follows from the TDHF wavefunction (12.5.1) on integrating $\Psi\Psi^*$ over all variables except $\mathbf{x}_1$, and the *change* $\boldsymbol{\rho}'$ will thus arise from integration of

$$\sum_{i,m} [p_{mi}^* \Psi_0 \Psi(i \to m)^* + p_{mi} \Psi(i \to m) \Psi_0^*].$$

From Slater's rules, this yields

$$\rho' = \sum_{i,m} (p_{mi}^* \psi_i \psi_m^* + p_{mi} \psi_m \psi_i^*),$$

and, on expressing the spin-orbitals in LCAO form, substituting for p_{mi} ($= d_{mi}$) from (12.5.3), and making a comparison with (12.6.16) (with $\mathbf{d} = \mathbf{x} + \mathbf{y}$), it follows (after some rearrangement) that

$$\mathbf{x} = \sum_{i,m} X_{mi} \mathbf{c}_m \mathbf{c}_i^\dagger, \qquad \mathbf{y} = \sum_{i,m} Y_{mi} \mathbf{c}_i \mathbf{c}_m^\dagger. \tag{12.6.25}$$

In other words, the X_i^m and Y_i^m coefficients in (12.6.20) coincide with the amplitudes X_{mi} and Y_{mi} in the conventional TDHF equations. The equivalence of the equations themselves is easily demonstrated by expressing the 2-electron integrals in the **G**-matrix notation appropriate to the finite-basis approximation. Thus, for example,

$$\langle rs \parallel tu \rangle = \mathbf{c}_r^\dagger \mathbf{G}(\mathbf{c}_u \mathbf{c}_s^\dagger) \mathbf{c}_t, \tag{12.6.26}$$

and, on using this result in (12.5.4), multiplying from left and right by $\mathbf{c}_m$ and $\mathbf{c}_i^\dagger$ respectively, and summing over i and m, we obtain (noting (12.6.25)) the first equation of (12.6.19). The TDHF (or RPA) and TDSCF equations are thus mathematically equivalent, although their appearance and the solution methods they suggest are so different.

12.7 MULTICONFIGURATION (MC) TDHF THEORY

The variational equations formulated in Section 12.2 may be used to obtain a *multi*configuration form of the TDHF equations, the generalization being similar to that made in Section 11.9 for a static perturbation. One or two new features appear, however, in the time-dependent case: in particular, special care is needed in using the Frenkel principle (McWeeny, 1983). We therefore proceed cautiously, writing the principle in the fuller form (12.2.5a), namely

$$\left\langle \delta\Psi \middle| \mathsf{Q}\left(\mathsf{H} - i\hbar\frac{\partial}{\partial t}\right) \middle| \Psi \right\rangle + \left\langle \Psi \middle| \left(\mathsf{H} + i\hbar\frac{\partial}{\partial t}\right)\mathsf{Q} \middle| \delta\Psi \right\rangle = 0. \tag{12.7.1}$$

It is only when the two terms are linearly independent that it is sufficient to equate each separately to zero, obtaining Frenkel's original form.

Orbital variation may be introduced as in Section 8.2 by using

$$\Psi = e^{\mathsf{R}}\Psi_0, \tag{12.7.2}$$

where R has the second-quantization form

$$\mathsf{R} = \sum_{r,s} \Delta_{rs}\mathsf{a}_r^\dagger \mathsf{a}_s = \sum_{r>s} (\Delta_{rs}\mathsf{E}_{rs} + \Delta_{sr}\mathsf{E}_{sr})$$

and thus contains two types of parameter:

$$p_\mu = \Delta_{rs}, \qquad p_{\bar{\mu}} = \Delta_{sr} \quad (\mu = rs,\ \bar{\mu} = sr;\ r > s). \tag{12.7.3}$$

Changes in Ψ will involve derivatives with respect to both p_μ and $p_{\bar{\mu}}$: for variations preserving orthonormality, these are not independent, since $\mathbf{\Delta}$ must then be anti-Hermitian and hence $p_{\bar{\mu}} = -p_\mu^*$; but this constraint may be introduced later.‡

The first term in (12.7.1) has already been reduced in deriving (12.2.14): it may now be written in full as

$$\begin{aligned}\left\langle \delta\Psi \middle| \mathsf{Q}\left(H - i\hbar\frac{\partial}{\partial t}\right) \middle| \Psi \right\rangle = {} & \sum_\mu \delta p_\mu^* (\nabla H)_\mu \\ & + \sum_{\mu,\nu} \delta p_\mu^* d_\nu^* (\nabla\nabla H)_{\mu\nu} + \sum_{\mu,\nu} \delta p_\mu^* (\nabla H \nabla)_{\mu\nu} d_\nu \\ & - i \sum_{\mu,\nu} \delta p_\mu^* (\nabla Q \nabla)_{\mu\nu} \dot{d}_\nu + \sum_{\mu,\bar{\nu}} \delta p_\mu^* s_{\bar{\nu}}^* (\nabla\nabla H)_{\mu\bar{\nu}} \\ & + \sum_{\mu,\bar{\nu}} \delta p_\mu^* (\nabla H \nabla)_{\mu\bar{\nu}} d_{\bar{\nu}} - i \sum_{\mu,\bar{\nu}} \delta p_\mu^* (\nabla Q \nabla)_{\mu\bar{\nu}} \dot{d}_{\bar{\nu}} \\ & + (\text{similar terms with } \mu \to \bar{\mu}),\end{aligned} \tag{12.7.4}$$

where δp_μ, $\delta p_{\bar{\mu}}$ are arbitrary variations, and d_ν, $d_{\bar{\nu}}$ are first-order time displacements of the parameters.

On introducing the constraints $\delta p_{\bar{\mu}} = -\delta p_\mu^*$, $d_{\bar{\nu}} = -d_\nu^*$, the variation (12.7.4) takes the form

$$\sum_\mu \delta p_\mu^* [\ldots] + \sum_\mu [\ldots] \delta p_\mu,$$

which contains *both* the arbitrary variations δp_μ and their complex conjugates δp_μ^*. The second part of (12.7.1) reduces in a similar way, to

‡ The derivation of the TDHF equations (Section 12.5) was simplified by the fact that with a 1-determinant reference function only the p_μ appeared ($\mu = mi$ for an excitation $i \to m$, no de-excitations $m \to i$ being possible).

the complex conjugate of (12.7.4), and it is now clear that there are terms in δp_μ (and terms in δp_μ^*) in both expressions; in other words, the two parts of (12.7.1) are *not* linearly independent and it is no longer correct to formulate the variation principle by equating either of them separately to zero. Instead, it is necessary to collect terms of the same type. The resultant expression is of the above form, the two terms still being complex-conjugate; and the coefficients of δp_μ and δp_μ^* must then separately vanish. The resultant variation condition, analogous to (12.2.14), is

$$\overline{\boldsymbol{\nabla} H} + \mathbf{d}^*(\overline{\boldsymbol{\nabla\nabla} H}) + (\overline{\boldsymbol{\nabla} H \boldsymbol{\nabla}})\mathbf{d} - i\mathbf{V}\dot{\mathbf{d}} - i\mathbf{W}\dot{\mathbf{d}}^* = \mathbf{0},$$

where $\mathbf{d}$ contains only the independent elements ($d_\mu = d_{rs}$, $r > s$) and the constrained derivatives of the energy functional are

$$\left.\begin{aligned}(\overline{\nabla H})_\mu &= [(\nabla H)_\mu - (\nabla H)_{\bar{\mu}}],\\ (\overline{\nabla\nabla H})_{\mu\nu} &= [(\nabla\nabla H)_{\mu\nu} - (\nabla H \nabla)_{\mu\bar{\nu}} - (\nabla H \nabla)_{\nu\bar{\mu}} + (H\nabla\nabla)_{\bar{\mu}\bar{\nu}}],\\ (\overline{\nabla H \nabla})_{\mu\nu} &= [(\nabla H \nabla)_{\mu\nu} - (\nabla\nabla H)_{\mu\bar{\nu}} - (H\nabla\nabla)_{\nu\bar{\mu}} + (\nabla H \nabla)_{\bar{\nu}\bar{\mu}}],\end{aligned}\right\} \quad (12.7.5)$$

while the remaining derivatives are

$$\left.\begin{aligned}V_{\mu\nu} &= [(\nabla Q \nabla)_{\mu\nu} - (\nabla Q \nabla)^*_{\bar{\mu}\bar{\nu}}],\\ W_{\mu\nu} &= [(\nabla Q \nabla)_{\mu\bar{\nu}} - (\nabla Q \nabla)^*_{\bar{\mu}\nu}).\end{aligned}\right\} \quad (12.7.6)$$

With a matrix notation (cf. (12.2.15)), the equation for the parameter displacements becomes, dropping the bars,

$$\boldsymbol{\nabla} H + \mathbf{Md} + \mathbf{Qd}^* = i(\mathbf{V}\dot{\mathbf{d}} + \mathbf{W}\dot{\mathbf{d}}^*), \quad (12.7.7)$$

and this forms the starting point for the discussion of free and forced oscillations, closely following the treatment in Section 12.3.

With the assumption that $\boldsymbol{\nabla} H = (\boldsymbol{\nabla} H)_0 = \mathbf{0}$ in the absence of the perturbation (i.e. that Ψ_0 is optimized) the free-oscillation solutions are obtained by substituting (12.3.6) in (12.7.7) and separating the $e^{\pm i\omega t}$ components. The result is

$$\mathbf{MX} + \mathbf{QY} = \omega(\mathbf{VX} + \mathbf{WY}),$$
$$\mathbf{MY}^* + \mathbf{QX}^* = -\omega(\mathbf{VY}^* + \mathbf{WX}^*),$$

and, on taking the complex conjugate of the second equation, both may be displayed together as

$$\begin{pmatrix}\mathbf{M} & \mathbf{Q}\\ \mathbf{Q}^* & \mathbf{M}^*\end{pmatrix}\begin{pmatrix}\mathbf{X}\\ \mathbf{Y}\end{pmatrix} = \omega\begin{pmatrix}\mathbf{V} & \mathbf{W}\\ -\mathbf{W}^* & -\mathbf{V}^*\end{pmatrix}\begin{pmatrix}\mathbf{X}\\ \mathbf{Y}\end{pmatrix}, \quad (12.7.8a)$$

which resembles (12.3.8a) and may likewise be abbreviated to

$$\mathbf{\Omega C} = \omega \mathbf{\Delta C}. \tag{12.7.8b}$$

The solutions follow as on p. 429 and have similar properties.

The forced-oscillation solutions at frequency ω (away from a resonance) are obtained by putting $\mathbf{d}_0 = \mathbf{0}$ and looking for a *first*-order term $\mathbf{d}'$ of the form (12.3.17). The resultant equations are just like (12.7.8) except for a perturbation term on the left: instead of (12.3.9), we find (dropping the primes since these are obviously the first-order equations)

$$\boxed{\begin{pmatrix} \mathbf{\nabla} A \\ A\mathbf{\nabla} \end{pmatrix} + \begin{pmatrix} \mathbf{M} & \mathbf{Q} \\ \mathbf{Q}^* & \mathbf{M}^* \end{pmatrix}\begin{pmatrix} \mathbf{X} \\ \mathbf{Y} \end{pmatrix} = \omega \begin{pmatrix} \mathbf{V} & \mathbf{W} \\ -\mathbf{W}^* & -\mathbf{V}^* \end{pmatrix}\begin{pmatrix} \mathbf{X} \\ \mathbf{Y} \end{pmatrix},} \tag{12.7.9a}$$

or, more briefly, collecting the gradient terms into a single column $\mathbf{\nabla}_A$,

$$\boxed{\mathbf{\nabla}_A + \mathbf{\Omega C} = \omega \mathbf{\Delta C}.} \tag{12.7.9b}$$

The solutions may be obtained, as in Section 12.3, in terms of the free-oscillation solutions (ω_p, $\mathbf{C}_p$) of (12.7.8). Formally,

$$\mathbf{C} = -(\mathbf{\Omega} - \omega \mathbf{\Delta})^{-1} \mathbf{\nabla}_A. \tag{12.7.10}$$

The inverse in this equation is essentially the response matrix introduced in (12.3.24) and discussed in Section 12.4: in terms of the eigenvectors $\mathbf{C}_p$ it takes the form

$$\mathbf{\Pi} = -\sum_{\omega_p > 0} \left(\frac{\mathbf{C}_p \mathbf{C}_p^\dagger}{\omega_p - \omega} + \frac{\mathbf{C}_{-p} \mathbf{C}_{-p}^\dagger}{\omega_p + \omega} \right). \tag{12.7.11}$$

Since $\mathbf{C}_p$ and $\mathbf{C}_{-p}$ (for frequencies $\pm\omega_p$) are related by the pairing property (12.3.9), the response matrix takes more explicitly the block form indicated in (12.3.23) and (12.3.24). All the properties discussed in Section 12.4 follow unchanged.

The MC TDHF equations in the form given above were first derived, using a different approach, by Yeager and Jørgensen (1979b) and Dalgaard (1980). To see the connection with their work, it is sufficient to use (12.7.3) in (12.7.2) *et seq.* to obtain

$$\Psi = \Big[1 + \sum_{\mu} (d_\mu \mathsf{E}_\mu + d_{\bar{\mu}} \mathsf{E}_\mu^\dagger) + \tfrac{1}{2} \sum_{\mu,\nu} (d_\mu \mathsf{E}_\mu + d_{\bar{\mu}} \mathsf{E}_\mu^\dagger)(d_\nu \mathsf{E}_\nu + d_{\bar{\nu}} \mathsf{E}_\nu^\dagger) + \dots \Big] \Psi_0$$

and to identify the derivatives Ψ^μ, $\Psi^{\bar{\mu}}$, $\Psi^{\mu\nu}$, etc. required in evaluating (see Section 2.4) the tensor components $(\nabla H)_\mu$, $(\nabla H)_{\bar{\mu}}$, $(\nabla\nabla H)_{\mu\nu}$, etc.

that appear in (12.7.5). In this way we find

$$\left.\begin{aligned}(\nabla H)_\mu &= \langle \Psi_0 | [\mathsf{E}_\mu^\dagger, \mathsf{H}] | \Psi_0 \rangle, \\ M_{\mu\nu} &= (\nabla H \nabla)_{\mu\nu} = \langle \Psi_0 | [\mathsf{E}_\mu^\dagger, \mathsf{H}, \mathsf{E}_\nu] | \Psi_0 \rangle, \\ Q_{\mu\nu} &= (\nabla\nabla H)_{\mu\nu} = -\langle \Psi_0 | [\mathsf{E}_\mu^\dagger, \mathsf{H}, \mathsf{E}_\nu^\dagger] | \Psi_0 \rangle, \\ V_{\mu\nu} &= \langle \Psi_0 | [\mathsf{E}_\mu^\dagger, \mathsf{E}_\nu] | \Psi_0 \rangle, \\ W_{\mu\nu} &= -\langle \Psi_0 | [\mathsf{E}_\mu^\dagger, \mathsf{E}_\nu^\dagger] | \Psi_0 \rangle,\end{aligned}\right\} \qquad (12.7.12)$$

where the symmetric double commutator is defined by

$$2[A, B, C] = [A, [B, C]] + [[A, B], C]. \qquad (12.7.13)$$

In the derivation of (12.7.12), terms involving "overlaps" of the form $\langle \Psi_0 | \mathsf{E}_\mu | \Psi_0 \rangle$ and $\langle \Psi_0 | \mathsf{E}_\mu^\dagger | \Psi_0 \rangle$ cancel in all the above expressions as a result of the constraints (whether or not Ψ_0 is fully optimized). Again (cf. Section 8.6), these results lead directly to working equations for the numerical calculation of response properties, given any form of multiconfiguration reference function Ψ_0; and again it is not difficult to extend the equations to include variation of the CI coefficients along with the orbitals (see e.g. Dalgaard, 1980; Jaszunski and McWeeny, 1982; McWeeny, 1983) as in Sect. 8.6. Many applications (see e.g. Albertson *et al.*, 1980) confirm the validity of these powerful methods.

PROBLEMS 12

12.1 Fill in the details in the derivation of the first-order expression (12.1.7) for the coefficients $c_n(t)$ in the wavefunction (12.1.3) when the perturbation $\mathsf{H}'(t)$ is turned on infinitely slowly, starting at $t = -\infty$. Hence derive the fundamental results (12.1.8) and (12.1.9) for the linear response.

12.2 Repeat the derivation in Problem 12.1 to obtain the frequency-dependent response to an oscillatory perturbation, as defined in (12.1.16) and (12.1.17). Then verify the relationships (12.1.21) and (12.1.22) between the time-dependent and frequency-dependent quantities.

12.3 Discuss, along the lines of Section 12.1, the effect of an exponentially growing perturbation $\mathsf{H}'(t) = \mathsf{A}e^{\omega t}$, switched on at $t = -\infty$, and show that the time-dependent response of an expectation value $\langle \mathsf{B} \rangle$ is given by $\delta\langle \mathsf{B} \rangle = \Pi(\mathsf{BA} \mid i\omega)e^{\omega t}$, where $\Pi(\mathsf{BA} \mid i\omega)$ is obtained from (12.1.17) by changing ω to $i\omega$ and putting $\epsilon = 0$.

12.4 Let $\Psi = \sum_\kappa c_\kappa \Phi_\kappa$ be a finite-basis approximation to a ground-state wavefunction, satisfying $\mathbf{H}\mathbf{c}_1 = E_1\mathbf{c}_1$. Show that a parameter displacement $\mathbf{d}$, orthogonal to $\mathbf{c}_1$, will evolve *approximately* according to $(\mathbf{H} - E_1\mathbf{1})\mathbf{d} = i\hbar\dot{\mathbf{d}}$—thus giving a "best" approximation (in the sense of satisfying the Frenkel principle) to a solution of the time-dependent Schrödinger equation. [*Hint:* Use (12.2.15) for the time evolution of the parameters.]

12.5 Use the result obtained in Problem 12.4 to show that the natural frequencies of the free oscillations of the ground-state parameters are given by $\hbar\omega = E_n - E_1$, where E_n is the nth eigenvalue of the usual secular equations—and hence that in this simple approximation the resonance frequencies (which correspond to excitation energies) coincide with the usual stationary-state approximations.

12.6 Work through the derivation of the "pairing property" of the solutions of (12.3.8), verifying the possibility of adopting the normalization (12.3.11).

12.7 A hydrogen atom experiences an oscillating electric field along the z axis. Use a variation function $\psi = \exp\{-a[x^2 + y^2 + (z - R)^2]^{1/2}\}$ (i.e. a 1s orbital whose centre is able to "float" away from the nucleus, along the field direction) to discuss the linear response. Obtain approximations (poor ones) to the frequency-dependent electric dipole polarizability and the first two resonant frequencies. [*Hint:* Evaluate the derivatives needed in setting up the 2×2 matrices **M**, **Q**, **V** in (12.3.7); they are conveniently expressed in terms of unnormalized Slater orbitals (Appendix 1) of ns, np, nd, . . . type, as in Problem 2.25. Obtain the response matrix in (12.3.24) from the solutions of the free-oscillation equations, and hence the required FDP from (12.4.3) with $\mathsf{A} = \mathsf{B} = \mathsf{d}_z$ (the electric-dipole operator).]

12.8 Extend the analysis in Problem 12.7 to a variation function containing an additional polarization factor $1 + \lambda z$, showing that the static limit of $\alpha(\omega)$ is improved from 0.1875 to 4.15 (in units of $e^2a_0^2/E_\text{h}$) compared with the exact value 4.5. Comment also on the 1s→2p resonance. [*Hint:* The matrices **M**, **Q** and **V** are now 3×3, but break into 1×1 and 2×2 blocks—for "breathing" and "shaking" oscillations respectively.]

12.9 Obtain the TDHF (RPA) equations, analogous to (12.5.4), for a system described by a closed-shell HF function, eliminating spin at the start. What modifications will be needed if non-canonical orbitals are used? [*Hint:* Look back at Problem 11.18; once again the derivation runs parallel to that sketched in the text.]

12.10 In what sense is it possible to say that "$\partial/\partial t$ behaves like a 1-electron operator" in the derivation of (12.6.1)? Continue the derivation of the time-dependent analogue (12.6.7) of the HF equations, giving a proof that the matrix of Lagrangian multipliers is Hermitian. [*Hint:* Write down a complex-conjugate equation, take appropriate scalar products, and eliminate the time.]

12.11 Follow through in detail the derivation of a matrix form of the coupled SCF equations, analogous to (12.6.13), using a closed-shell form of the wavefunction and eliminating spin at the beginning. [*Hint:* Note that the density matrix $\boldsymbol{\rho}$ is replaced throughout by the projection matrix **R** defined in (6.4.18), while **h** and **G** are replaced by their orbital analogues (cf. p. 171).]

12.12 Show how the equations obtained in Problem 12.11 might be used to discuss the response of a closed-shell system to an oscillating electric field. Make a numerical application to the system used in Problem 6.5. [*Hint:* Put the field in any convenient direction, roughly estimating the perturbation matrix in terms of the molecular geometry. Keep ω small (or very large) compared with orbital energy differences (—why?).]

12.13 Obtain a modified form of the equations in Problem 12.12, appropriate to the case of a perturbation at purely imaginary frequency (i.e. exponentially rising, as in Problem 12.3). Show that the convergence difficulties that can arise at real frequencies will in this case be absent; and test this conclusion numerically.

12.14 Introduce the constraints $\delta p_{\bar{\mu}} = -\delta p_{\mu}^*$ etc. in (12.7.4) and in the second term of (12.7.1) to obtain an explicit form of the variational condition. Then regroup the terms so as to separate those that contain δp_{μ} from those that contain δp_{μ}^*, thus confirming the statements in the text and verifying the expressions (12.7.5) for the various derivatives.

12.15 The multiconfigurational (MC) TDHF equations (12.7.8a) and (12.7.9a) look very much like their counterparts in the simple TDHF approach. Take a simple example and clarify the differences between the two cases, indicating carefully the dimensions of all the blocks in the matrices, the significance of their elements, and the origin of the **W** blocks in the MC TDHF equations. [*Hint:* You might take 4 electrons, 6 "occupied" spin-orbitals and 10 virtual. Remember that there are de-excitations as well as excitations.]

12.16 Show that with the second-quantization form of the exponential operator (12.7.2) and the parametrization (12.7.3) the tensor components in (12.7.6) take the forms given in (12.7.12)—in which only *commutators* (or double commutators) appear.

12.17 Generalize the MC TDHF equation (12.7.8a) to include simultaneous variation of both spin-orbitals and CI coefficients. Show that the matrices all assume block forms in which, for example, $\mathbf{M}_{oo}$ is an orbital–orbital block, while $\mathbf{M}_{oc}$ and $\mathbf{M}_{cc}$ are orbital–coefficient and coefficient–coefficient blocks respectively. Obtain explicit forms for all the new elements. [*Hint:* Use $\Psi = \sum_{\kappa} C_{\kappa}\Phi_{\kappa}$ (keeping μ, ν to label orbital parameters) and note that with $C_{\kappa} = C_{\kappa}^0 + d_{\kappa}$ the derivatives with respect to the d_{κ} follow from $\Psi^{(\kappa)} = \partial\Psi/\partial C_{\kappa} = \Phi_{\kappa}$. A typical mixed derivative is $\Psi^{\mu\kappa} = \mathsf{E}_{\mu}\Phi_{\kappa}$. Evaluate the tensor components as in Section 8.6]

12.18 Show how the matrix elements containing the operators E_{μ}, $\mathsf{E}_{\mu}^{\dagger}$ may be expressed in terms of density-matrix elements (ρ_{rs}, $\pi_{rs,tu}$).

13 Propagator and Equation-of-Motion Methods

13.1 PRELIMINARY NOTIONS

In the last chapter, starting from the time-dependent Schrödinger equation (which is the "equation of motion" for the wavefunction), a number of new ideas were introduced. In particular, the linear response of a system to a time-dependent perturbation was characterized by (i) a time-correlation function $K(\mathsf{BA} \mid \tau)$ and (ii) a frequency-dependent polarizability (FDP), $\Pi(\mathsf{BA} \mid \omega)$, each being a Fourier transform of the other.

To summarize: we consider the response to a perturbation of the form

$$\mathsf{H}'(t) = \mathsf{A}(t) = F(t)\mathsf{A} \tag{13.1.1}$$

(A being time-*in*dependent), applied during the interval $(-\infty, t)$. The linear response of an expectation value $\langle \mathsf{B} \rangle$ is given by

$$\delta\langle \mathsf{B} \rangle_t = \int_{-\infty}^{\infty} \theta(t - t')K(\mathsf{BA} \mid t - t')\mathsf{A}(t')\, dt', \tag{13.1.2}$$

and may thus be regarded as the result of a train of "impulses" $\mathsf{A}(t')$, applied at all times from $t' = -\infty$ to $t' = t$ (the θ factor ensuring the absence of contributions from later times). The proportionality factor $\theta(t - t')K(\mathsf{BA} \mid t - t')$ describes how these impulses *propagate* in time to produce their effect on $\langle \mathsf{B} \rangle$; this factor is an example of a *propagator.*

Alternatively, by introducing the Fourier transform $f(\omega) = \int_{-\infty}^{+\infty} F(t)e^{i\omega t}\, dt$, of the time factor in (13.1.1), the perturbation may be expressed in terms of Fourier components‡

$$\mathsf{H}'(\omega) = \mathsf{A}\,\mathrm{Re}\,(e^{-i\omega t}), \tag{13.1.3}$$

‡ In what follows, it is convenient to regard A and B as Hermitian operators representing real observables. In this case, from (12.1.23), we have $\Pi(\mathsf{BA} \mid -\omega) = \Pi(\mathsf{BA} \mid \omega)^*$.

to which the response of $\langle \mathsf{B} \rangle$ at frequency ω is given by

$$\delta\langle \mathsf{B} \rangle_\omega = \mathrm{Re}\,[\Pi(\mathsf{BA} \mid \omega)e^{-i\omega t}]. \tag{13.1.4}$$

On multiplying (13.1.3) by the weight factor $(2\pi)^{-1}f(\omega)$ and integrating over all frequencies, we may transform back to obtain (13.1.1); and by doing likewise with the response (13.1.4) we obtain (13.1.2) with

$$\theta(\tau)K(\mathsf{BA} \mid \tau) = (2\pi)^{-1} \int_{-\infty}^{+\infty} \Pi(\mathsf{BA} \mid \omega)e^{-i\omega t}\,\mathrm{d}\omega. \tag{13.1.5}$$

The Fourier transform $\Pi(\mathsf{BA} \mid \omega)$ of the propagator that appears in (13.1.2) is often called the "frequency form" of the propagator; it is the form that will be used most often.

In the next section we give a more general and formal definition of the propagator associated with two operators, but we are already in a position to anticipate the essential nature of propagators. First, it is convenient to introduce a common notation by writing

$$\theta(t-t')K(\mathsf{BA} \mid t-t') = \langle\!\langle \mathsf{B}(t); \mathsf{A}(t') \rangle\!\rangle \tag{13.1.6}$$

for the "time" form and

$$\Pi(\mathsf{BA} \mid \omega) = \langle\!\langle \mathsf{B}; \mathsf{A} \rangle\!\rangle_\omega \tag{13.1.7}$$

for the frequency form, noting that so far the former has been defined only for $t > t'$.

The frequency form (13.1.7) has been investigated already and written out explicitly in (12.1.17), which becomes

$$\langle\!\langle \mathsf{B}; \mathsf{A} \rangle\!\rangle_\omega = \lim_{\epsilon \to 0} \frac{1}{\hbar} \sum_{n(\neq 0)} \left\{ \frac{\langle 0|\,\mathsf{B}\,|n\rangle\langle n|\,\mathsf{A}\,|0\rangle}{\omega + i\epsilon - \omega_{0n}} - \frac{\langle 0|\,\mathsf{A}\,|n\rangle\langle n|\,\mathsf{B}\,|0\rangle}{\omega + i\epsilon + \omega_{0n}} \right\}, \tag{13.1.8}$$

and contains summations over the complete set of exact stationary states $|n\rangle$ of the unperturbed system. The infinitesimals $i\epsilon$ in the denominators are associated with the causality requirement $(t > t')$, the "first" operator $\mathsf{A}(t')$ always being turned on (starting at $t' = -\infty$) *before* its effect is observed by measuring $\delta\langle \mathsf{B} \rangle_t$. Except at the poles, however, when ω approaches a natural frequency $\pm\omega_{0n}$ (for excitation or de-excitation), the infinitesimals may be discarded.

Equation (13.1.8) expresses the propagator in "spectral form", i.e. in terms of the complete spectrum (eigenvalues and eigenfunctions) of the unperturbed time-independent Hamiltonian; and the excitation energies

$$E_n - E_0 = \hbar\omega_{0n} \tag{13.1.9}$$

that connect *stationary* states of the system may thus be looked at as the *poles of the propagator*—for any pair of operators A, B. In other words, by applying an oscillating perturbation (13.1.3), with arbitrary A, and using the change $\delta\langle \mathsf{B}\rangle_\omega$ of any observable to "probe" its effect, we can discover what *transitions* the system is capable of making—without ever calculating any excited states! The propagator approach is thus of great potential value. In principle, it could bypass one of the greatest difficulties of computational quantum chemistry—that of calculating small energy *changes* (as usually observed in experiments) by taking the difference of exceedingly large, almost equal, and independently computed *total* energies. If it were possible to calculate a propagator, as a function of frequency, then such energy differences could be obtained *directly*, simply by locating the poles. Moreover, by finding the *residues* at the poles, we should obtain directly the quantities $\langle 0|\,\mathsf{B}\,|n\rangle\langle n|\,\mathsf{A}\,|0\rangle$, without ever calculating the individual ground and excited states $|0\rangle$ and $|n\rangle$; and it is such quantities, for particular choices of the operators, that give information about all possible transition processes. Thus, with $\mathsf{A}=\mathsf{B}=d_x$, the x component of the electric-moment operator (as used in Section 11.5), we should obtain $|\langle n|\,d_x\,|0\rangle|^2$—which determines the probability of an electric-dipole-induced transition ($n\leftarrow 0$) due to a radiation field polarized in the x direction.

The propagator in (13.1.8), although it seems to require knowledge of an infinite set of exact N-electron wavefunctions, is not a remote "dream object": practical methods of calculating *approximate* propagators have been devised, and, indeed, such methods have been developed in Section 12.4. The worst is over! All that remains is to tie up a few loose ends, to explore the formal properties of propagators, to indicate principal applications, and to examine the connection between propagator and other "equation-of-motion" (EOM) approaches.

13.2 PROPAGATORS. THE POLARIZATION PROPAGATOR

In the general treatment of propagators (see e.g. Zubarev 1960, 1964) it is usual to define all operators in the so-called "Heisenberg picture"; this picture (implicit in the elementary treatment (Section 12.1) of time-dependent perturbations) must therefore be made explicit. To do this, we momentarily add a superscript S to distinguish wavefunctions and operators in the usual Schrödinger representation, and start from the time-dependent Schrödinger equation, which now becomes

$$i\hbar\frac{\partial\Psi^{\mathrm{S}}}{\partial t}=\mathsf{H}^{\mathrm{S}}\Psi^{\mathrm{S}}. \tag{13.2.1}$$

When H^{S} is the usual time-dependent Hamiltonian of the (unperturbed) system, a formal solution may be written:

$$\Psi^{\mathrm{S}}(t) = \exp\left(-\frac{i\mathsf{H}^{\mathrm{S}}t}{\hbar}\right)\Psi^{\mathrm{S}}(0), \tag{13.2.2}$$

as may be verified by differentiation. The operator

$$\mathsf{U}(t) = \exp\left(-\frac{i\mathsf{H}^{\mathrm{S}}t}{\hbar}\right) \tag{13.2.3}$$

is the *evolution operator* in the Schrödinger picture, and carries an initial state $\Psi^{\mathrm{S}}(0)$ into a final state $\Psi^{\mathrm{S}}(t)$; it is clearly unitary, satisfying

$$\mathsf{U}^{\dagger}(t)\mathsf{U}(t) = \mathsf{U}(t)\mathsf{U}^{\dagger}(t) = 1.$$

When $\Psi^{\mathrm{S}}(0)$ is a (time-independent) eigenfunction of H^{S}, the exponential in (13.2.2) becomes the trivial time factor $e^{-iEt/\hbar}$ for a stationary state of given energy E.

In the Heisenberg picture we simply make a unitary transformation in which wavefunctions and operators are replaced (cf. Section 2.2) according to

$$\Psi^{\mathrm{S}}(t) \rightarrow \Psi = \mathsf{U}^{\dagger}(t)\Psi^{\mathrm{S}}(t), \qquad \mathsf{A}^{\mathrm{S}}(t) \rightarrow \mathsf{A}(t) = \mathsf{U}^{\dagger}(t)\mathsf{A}^{\mathrm{S}}(t)\mathsf{U}^{\mathrm{S}}(t).$$

The resultant wavefunction in the Heisenberg picture is thus "reduced to rest", becoming $\Psi = \mathsf{U}^{\dagger}(t)\mathsf{U}(t)\Psi^{\mathrm{S}}(0) = \Psi^{\mathrm{S}}(0)$; but *all* operators (even those, like the Hamiltonian, that do not depend on time in the Schrödinger picture) in general acquire a time dependence. Thus a general Heisenberg operator is

$$\mathsf{A}(t) = \exp\left(\frac{i\mathsf{H}^{\mathrm{S}}t}{\hbar}\right)\mathsf{A}^{\mathrm{S}}(t)\exp\left(-\frac{i\mathsf{H}^{\mathrm{S}}t}{\hbar}\right), \tag{13.2.4}$$

and all relationships among operators and state functions are preserved in the usual way (for example, $\mathsf{A}^{\mathrm{S}}\mathsf{B}^{\mathrm{S}} = \mathsf{C}^{\mathrm{S}}$ implies $\mathsf{AB} = \mathsf{C}$ and vice versa). When A^{S} is time-*in*dependent it follows easily from (13.2.1) that

$$i\hbar\frac{\mathrm{d}\mathsf{A}}{\mathrm{d}t} = [\mathsf{A}, \mathsf{H}], \tag{13.2.5}$$

where the superscript on the Hamiltonian has been dropped since, by (13.2.4), $\mathsf{H} = \mathsf{H}^{\mathrm{S}}$. To determine the time development of any observable, with operator A, it is therefore necessary to integrate the equation of motion (13.2.5).

As a trivial example of the use of (13.2.5), we note that in the Heisenberg picture the creation and annihilation operators themselves acquire a time dependence and that *for an IPM Hamiltonian*, $\mathsf{H} = \sum_r \epsilon_r \mathsf{a}_r^\dagger \mathsf{a}_r$, the equation of motion becomes (using the anticommutation rules)

$$i\hbar \dot{\mathsf{a}}_r = \mathsf{a}_r \mathsf{H} - \mathsf{H}\mathsf{a}_r = \sum_s \epsilon_s (\mathsf{a}_r \mathsf{a}_s^\dagger \mathsf{a}_s - \mathsf{a}_s^\dagger \mathsf{a}_s \mathsf{a}_r) = \epsilon_r \mathsf{a}_r,$$

which may be integrated immediately: initially, the operators are those of the original time-independent Hamiltonian ($\mathsf{a}_r(0) = \mathsf{a}_r$), but subsequently they evolve in time. The solutions for a_r, and similarly for $\mathsf{a}_r^\dagger$, are clearly

$$\mathsf{a}_r(t) = e^{-i\epsilon_r t/\hbar} \mathsf{a}_r, \qquad \mathsf{a}_r^\dagger(t) = e^{i\epsilon_r t/\hbar} \mathsf{a}_r^\dagger.$$

The Hamiltonian remains time-independent; but in the Heisenberg picture the constituent operators a_r, $\mathsf{a}_r^\dagger$ are replaced by $\mathsf{a}_r(t)$, $\mathsf{a}_r^\dagger(t)$.

For a *real* system, described by the full Hamiltonian (3.6.9), the equations of motion for a_r and $\mathsf{a}_r^\dagger$ cannot be solved explicitly, and we must be content to use the formal solutions (13.2.4); the operators in the (second-quantization) Hamiltonian used in earlier chapters will thus be replaced by

$$\mathsf{a}_r(t) = e^{i\mathsf{H}t/\hbar} \mathsf{a}_r e^{-i\mathsf{H}t/\hbar}, \qquad \mathsf{a}_r^\dagger(t) = e^{i\mathsf{H}t/\hbar} \mathsf{a}_r^\dagger e^{-i\mathsf{H}t/\hbar}. \tag{13.2.6}$$

In the rest of this chapter all creation and annihilation operators will be assumed to have this form.

Let us now study the general "two-time" propagator, defined by Zubarev (1960) as

$$\langle\langle \mathsf{B}(t); \mathsf{A}(t') \rangle\rangle = (i\hbar)^{-1} \langle \Psi | \mathsf{T}[\mathsf{B}(t)\mathsf{A}(t')] | \Psi_0 \rangle. \tag{13.2.7}$$

in which Ψ_0 is the state under consideration, normally assumed to be the ground state‡, while T puts the operators in chronological order, whichever operates *first* (lower time value) standing on the right, and adds a factor $\eta = \pm 1$ according to the nature of the operators. Thus

$$\mathsf{T}[\mathsf{B}(t)\mathsf{A}(t')] = \begin{cases} \mathsf{B}(t)\mathsf{A}(t') & (t > t'), \\ \eta \mathsf{A}(t')\mathsf{B}(t) & (t < t'). \end{cases} \tag{13.2.8}$$

It is customary to take $\eta = +1$ for operators that contain only creation–annihilation *pairs* (e.g. typical 1- and 2-body operators in the Hamiltonian); such number-conserving operators are said to be of Bose type. For operators that contain odd numbers of creation/annihilation factors, thus changing particle numbers, η is instead given the value -1 and the operators are said to be of Fermi type.

‡ Zubarev also introduced *ensemble* expectation values, in which *all* states are present; but such developments are mainly of interest in statistical mechanics and are not required here.

It follows from the definition, using the Heisenberg forms (13.2.4), that the propagator (13.2.7) does not depend separately on t and t' but only on their difference. Thus, noting that the operator H standing next to $\langle \Psi_0|$ or $|\Psi_0\rangle$ is equivalent to the *number* E_0 (ground-state energy),

$$\langle\langle B(t); \mathsf{A}(t')\rangle\rangle = \begin{cases} (i\hbar)^{-1}\langle \Psi_0|\,\mathsf{B}\exp[i(E_0-\mathsf{H})(t-t')/\hbar]\mathsf{A}\,|\Psi_0\rangle & (t-t'>0), \\ \eta(i\hbar)^{-1}\langle \Psi_0|\,\mathsf{A}\exp[-i(E_0-\mathsf{H})(t-t')/\hbar]\mathsf{B}\,|\Psi_0\rangle & (t-t'<0), \end{cases}$$

and this may be written more compactly in the form, with $\tau = t - t'$,

$$\langle\langle \mathsf{B}(\tau); \mathsf{A}\rangle\rangle = (i\hbar)^{-1}\langle \Psi_0|\,\{\theta(\tau)\mathsf{B}e^{i(E_0-\mathsf{H})\tau/\hbar}\mathsf{A} + \eta\theta(-\tau)\mathsf{A}e^{-i(E_0-\mathsf{H})\tau/\hbar}\mathsf{B}\}\,|\Psi_0\rangle, \quad (13.2.9)$$

where $\theta(\tau)$ is the Heaviside function (12.1.20) and A, B (without time arguments) are the usual time-independent Schrödinger operators.

Let us now pass to the frequency form of the propagator, by multiplying (13.2.9) by $e^{i\omega\tau}$ and integrating over τ. More cautiously, a convergence factor will be included in this Fourier transformation by adding a negative term $-\epsilon\,|\tau|$ to the exponent; and, in order to reveal the structure of the result, a unit operator $\sum_n |\Psi_n\rangle\langle \Psi_n|$ will be inserted between the exponentials and the other operators A, B. The operator exponential, between say $\langle \Psi_n|$ and $|\Psi_{\mathrm{m}}\rangle$, then becomes a numerical factor $\delta_{nm}\exp[i(E_0-E_n)]$ and, on performing the integration, we find

$$\langle\langle \mathsf{B}; \mathsf{A}\rangle\rangle_\omega = \lim_{\epsilon\to 0+}\frac{1}{\hbar}\sum_n\left(\frac{\langle \Psi_0|\,\mathsf{B}\,|\Psi_n\rangle\langle \Psi_n|\,\mathsf{A}\,|\Psi_0\rangle}{\omega + i\epsilon - \omega_{0n}} - \eta\frac{\langle \Psi_0|\,\mathsf{A}\,|\Psi_n\rangle\langle \Psi_n|\,\mathsf{B}\,|\Psi_0\rangle}{\omega - i\epsilon + \omega_{0n}}\right), \quad (13.2.10)$$

where, for the number-conserving operators used so far, $\eta = 1$. The propagator in the form (13.2.10) is identical with the quantity defined previously, in (12.1.17) and (13.1.8), except for one small detail—the second term contains $\omega - i\epsilon$ instead of $\omega + i\epsilon$ and therefore has a pole at $\omega = -\omega_{0n} + i\epsilon$ (just above the real axis, as in Fig. 13.1, instead of just below). Away from the poles, we may of course go to the limit simply by putting $\epsilon = 0$, and the two expressions give exactly the same propagator.

The propagator defined in (13.2.10), as the Fourier transform of (13.2.7), is usually called the "causal" propagator (left-hand poles above the real axis, right-hand poles below); that used earlier (with $\epsilon \to -\epsilon$ in the second term of (13.2.10) is called the "retarded" propagator and has all poles *below* the real axis; and by reversing the sign of ϵ in the *first*

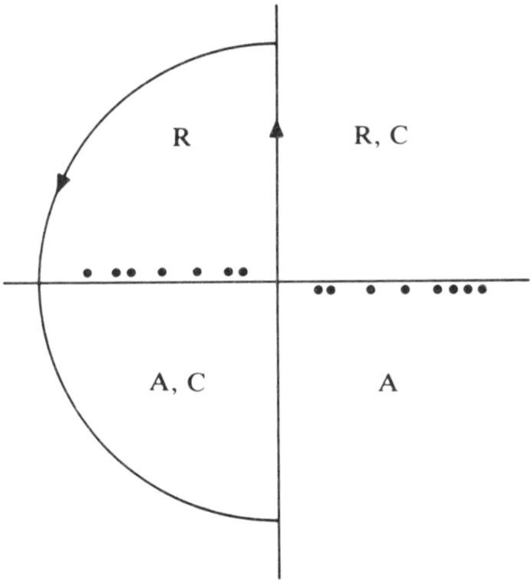

Fig. 13.1 Poles of the causal propagator. Quadrants in which the retarded, advanced, and causal propagators are analytic are marked with the letters R, A and C respectively. (The contour shown is needed in a later section.)

term of (13.2.10) we obtain the "advanced" propagator, with all poles *above* the real axis.‡ The retarded propagator refers to a real physical process: $\langle\langle \mathsf{B}(t); \mathsf{A}(t')\rangle\rangle$ with $t > t'$ determines the response of $\langle \mathsf{B}\rangle$ only *after* the application of a perturbation, and the positive infinitesimal in the "switching factor" $e^{\epsilon t'}$ (zero for $t' = -\infty$) is dictated by causality requirements. The advanced propagator, denoted by $\bar{\Pi}(\mathsf{BA} \mid \omega)$ in (12.1.24), corresponds to an *imaginary* physical process in which t' runs from $+\infty$ down to $-t$ (i.e. "backwards") and $\tau = t - t'$ is always negative; the required switching factor is then $e^{-\epsilon t'}$ and, as we have noted (p. 424), the propagator has poles at $i\epsilon$ *above* the real axis—thus leading to a time-correlation function $\theta(\tau)\bar{K}(\mathsf{BA} \mid \tau)$ that is zero unless τ is *negative* (i.e. t earlier than t'). The retarded propagator $\langle\langle \mathsf{B}; \mathsf{A}\rangle\rangle_{\omega}^{\mathrm{R}}$ is analytic within the entire upper half-plane, and the advanced propagator $\langle\langle \mathsf{B}, \mathsf{A}\rangle\rangle_{\omega}^{\mathrm{A}}$ within the entire lower half-plane. The causal propagator is analytic throughout *neither*; this does not, however, detract from its great value as a mathematical object, and it has a slight advantage in so far as its Fourier (time) transform exists for all values of τ except $\tau = 0$. When the behaviour at the poles is of no special interest all of these fine distinctions disappear, it is sufficient to put $\epsilon = 0$ at once, and the resultant propagator is analytic everywhere except in a cut containing the real axis.

‡ Sometimes (see e.g. Mattuck 1976) the propagators thus defined are said to be "of the second kind", the separate terms in (13.2.10) being used to define retarded and advanced propagators "of the first kind"; but we shall make no use of the latter.

The time forms of the retarded and advanced propagators are most neatly defined on introducing the generalized commutator

$$[\mathsf{A}, \mathsf{B}]_{-\eta} = \mathsf{AB} - \eta \mathsf{BA}. \tag{13.2.11}$$

The propagators then become, in terms of the time difference τ,

$$\langle\langle \mathsf{B}(\tau); \mathsf{A} \rangle\rangle^{\mathrm{R}} = (i\hbar)^{-1}\theta(\tau)\langle \Psi_0 | [\mathsf{B}(\tau), \mathsf{A}]_{-\eta} | \Psi_0 \rangle \quad \text{(retarded)}, \tag{13.2.12a}$$

$$\langle\langle \mathsf{B}(\tau); \mathsf{A} \rangle\rangle^{\mathrm{A}} = (i\hbar)^{-1}\theta(-\tau)\langle \Psi_0 | [\mathsf{B}(\tau), \mathsf{A}]_{-\eta} | \Psi_0 \rangle \quad \text{(advanced)}. \tag{13.2.12b}$$

The causal propagator is expressed similarly, directly from (13.2.7) and (13.2.8), as

$$\langle\langle \mathsf{B}(\tau); \mathsf{A} \rangle\rangle = (i\hbar)^{-1}\langle \Psi_0 | \{\theta(\tau)\mathsf{B}(\tau)\mathsf{A} + \eta\theta(-\tau)\mathsf{AB}(\tau)\} | \Psi_0 \rangle \quad \text{(causal)}. \tag{13.2.13}$$

The three propagators have many properties in common, and it is often unnecessary to distinguish them, provided that the pole behaviour in their frequency forms is correctly recognized when necessary.

Equations of motion

How can we actually calculate a propagator? The only method explored so far is the one developed in Chapter 12, which is based on an approximate variational solution of the time-dependent Schrödinger equation for the wavefunction of the system in the presence of an oscillating perturbation, i.e. variational solution of the equations of motion for the *wavefunction*. This is a powerful and fairly general approach, but is not without its limitations: in the first place it involves explicit optimization of a time-dependent N-electron wavefunction; and secondly it requires consideration of a real physical system in which the operators A and B are associated with actual *observables*. It is well suited to the determination of *polarization propagators* in which A and B are typical 1-electron operators and the propagator is essentially a dynamic polarizability. But the propagator (13.2.7) is itself a well-defined mathematical quantity, whatever the nature of the operators A and B, and it is natural to expect that its time development will conform to equations of motion—which might be solved directly, *without the need to seek a variational approximation to the many-electron wavefunction and its time evolution*. These equations, which we now derive, form the starting point for most approximation schemes for calculating propagators.

Let us take the time-dependent form (13.2.13) and differentiate with respect to τ, remembering that the derivative of the step function $\theta(\tau)$ is simply the Dirac delta function $\delta(\tau)$, which vanishes except at $\tau = 0$. The result is easily found to be‡

$$i\hbar \frac{d}{d\tau} \langle\langle \mathsf{B}(\tau); \mathsf{A} \rangle\rangle = \delta(t) \langle \Psi_0 | [\mathsf{B}(\tau), \mathsf{A}]_{\pm} | \Psi_0 \rangle - \langle\langle [\mathsf{H}, \mathsf{B}(\tau)]; \mathsf{A} \rangle\rangle \tag{13.2.14}$$

which determines the time development of the given propagator in terms of a more complicated propagator. It is easily confirmed that the retarded and advanced propagators satisfy the same differential equation, and superscripts are thus unnecessary.

The equation of motion (13.2.14) is not very useful as it stands, and to make progress it is necessary to make specific choices of the operators A and B, relating the propagators that occur to physically meaningful quantities that might suggest solution methods and approximations. In any case, however, it is usually preferable to use a frequency form of (13.2.14), and this is obtained in the usual way by Fourier transformation. Thus we obtain

$$\hbar\omega \langle\langle \mathsf{B}; \mathsf{A} \rangle\rangle_{\omega} = \langle \Psi_0 | [\mathsf{B}, \mathsf{A}]_{\pm} | \Psi_0 \rangle - \langle\langle [\mathsf{H}, \mathsf{B}]; \mathsf{A} \rangle\rangle_{\omega}, \tag{13.2.15}$$

where the left-hand side is the result of integration by parts and, with inclusion of the usual convergence factor ($\omega \to \omega \pm i\epsilon$), the term $e^{i\omega\tau} \langle\langle \mathsf{B}(\tau); \mathsf{A} \rangle\rangle$ vanishes at the limits $\pm\infty$. This "frequency form" of the equations of motion is a valuable starting point for the development of approximation schemes; its use is illustrated in later sections.

13.3 PROPAGATORS, GREEN'S FUNCTIONS AND DENSITY MATRICES

One choice of the operators A and B in the propagator equations has been explored already: that in which both are of 1-body form, so that

$$\mathsf{A} = \sum_{r,s} \langle r | \mathsf{A} | s \rangle \mathsf{a}_r^{\dagger} \mathsf{a}_s, \tag{13.3.1}$$

where $\langle r | \mathsf{A} | s \rangle$ is a matrix element in the usual (Schrödinger) sense, and thus merely a numerical coefficient. In this case $\langle\langle \mathsf{B}; \mathsf{A} \rangle\rangle_{\omega}$ is a type of

‡ Henceforth it will usually be convenient to denote the generalized commutator (13.2.11) by $[\mathsf{A}, \mathsf{B}]_{\pm}$, choosing the sign according to operator type. When no sign is shown an ordinary commutator is implied ($\eta = +1$).

polarization propagator, whose poles determine excitation frequencies of the N-electron system, and, since the number of creation/annihilation operators is even, η in the definition (13.2.10) has the value +1.

Another very important choice occurs when $\mathsf{A} = a_s^\dagger(t')$, $\mathsf{B} = a_r(t)$, and, since these are *not* number-conserving operators, $\eta = -1$. The time form of the corresponding *electron propagator* is then, with $\tau = t - t'$ as in (13.2.9),

$$\langle\langle \mathsf{a}_r(\tau); \mathsf{a}_s^\dagger \rangle\rangle = (i\hbar)^{-1} \langle \Psi_0 | \{\theta(\tau)\mathsf{a}_r(\tau)\mathsf{a}_s^\dagger - \theta(-\tau)\mathsf{a}_s^\dagger \mathsf{a}_r(\tau)\} | \Psi_0 \rangle. \quad (13.3.2)$$

This equation reveals an important connection between propagators and density matrices, since in Chapter 5 (p. 132) we noted that ρ_{rs}, a density-matrix element in the discrete representation provided by a set of spin-orbitals $\{\psi_r(\mathbf{x})\}$, could be expressed quite generally as an expectation value of $\mathsf{a}_s^\dagger \mathsf{a}_r$; and if we let $\tau \to 0$ from below, the first term in the braces, in (13.3.2), will be zero while the second will yield $\langle \mathsf{a}_s^\dagger \mathsf{a}_r \rangle$. Thus we find

$$\rho_{rs} = -i\hbar \lim_{\tau \to 0-} \langle\langle \mathsf{a}_r(\tau); \mathsf{a}_s^\dagger \rangle\rangle. \quad (13.3.3)$$

In other words, the electron propagator is a generalization of the 1-electron density matrix; and if we can find an approximation to the propagator then we can determine $\boldsymbol{\rho}$ (and hence all 1-electron expectation values) simply by taking a suitable limit.

It is often convenient to express the above result in a slightly different way; using the form (5.3.5), the density matrix $\rho(\boldsymbol{x}; \boldsymbol{x}')$ may be expressed as

$$\rho(\boldsymbol{x}; \boldsymbol{x}') = -i\hbar \lim_{\tau \to 0-} \langle\langle \psi(\boldsymbol{x}, \tau); \psi^\dagger(\boldsymbol{x}', 0) \rangle\rangle, \quad (13.3.4)$$

where

$$\psi(\boldsymbol{x}, \tau) = \sum_r \psi_r(\boldsymbol{x})\mathsf{a}_r(\tau), \qquad \psi^\dagger(\boldsymbol{x}', 0) = \sum_s \psi_s^*(\boldsymbol{x}')\mathsf{a}_s^\dagger(0). \quad (13.3.5)$$

These quantities are called *field operators*; they are simply linear combinations of the (time-dependent) creation and annihilation operators, with numerical coefficients that are defined in terms of the variables $\boldsymbol{x}$, $\boldsymbol{x}'$—which are to be regarded as *parameters*. The propagator in (13.3.4), which results when $\mathsf{A} = \psi^\dagger(\boldsymbol{x}', t')$, $\mathsf{B} = \psi(\boldsymbol{x}, t)$ in (13.2.7), is often called the 1-electron *Green's function* (having properties similar to the Green's functions encountered in solving differential equations—see Problem 13.6); it is often denoted by

$$G(\boldsymbol{x}, \tau; \boldsymbol{x}', 0) = \langle\langle \psi(\boldsymbol{x}, \tau); \psi^\dagger(\boldsymbol{x}', 0) \rangle\rangle, \quad (13.3.6)$$

where $\tau = t - t'$ and we have remembered the invariance property under time displacements.

It is also clearly possible to define a 2-electron Green's function $G_2(\boldsymbol{x}_1, t_1, \boldsymbol{x}'_1, t'_1; \boldsymbol{x}_2, t_2, \boldsymbol{x}'_2, t'_2)$ using *pairs* of field operators for A and B (e.g. $\mathsf{B} = \psi(\boldsymbol{x}_1, t_1)\psi^\dagger(\boldsymbol{x}'_1, t'_1)$); but the properties of this general many-time Green's function (see e.g.Csanak *et al.*, 1971; Jørgenson, 1975) are more complicated and will not be required here, although such quantities are important in scattering theory.

For most purposes it is preferable to use the frequency forms of the previous equations. Thus the Fourier transforms of (13.3.2) and (13.3.6) are respectively (inserting suitable unit operators, as in (13.2.10))

$$\boxed{G_{rs}(\omega) = \lim_{\epsilon \to 0+} \frac{1}{\hbar}\left(\sum_m \frac{\langle \Psi_0 | \mathsf{a}_r | \Psi_m^- \rangle \langle \Psi_m^- | \mathsf{a}_s^\dagger | \Psi_0 \rangle}{\omega + i\epsilon - \omega_{0m}} + \sum_n \frac{\langle \Psi_0 | \mathsf{a}_s^\dagger | \Psi_n^+ \rangle \langle \Psi_n^+ | \mathsf{a}_r | \Psi_0 \rangle}{\omega - i\epsilon + \omega_{0n}}\right)} \tag{13.3.7}$$

and

$$G(\omega \mid \boldsymbol{x}; \boldsymbol{x}') = \lim_{\epsilon \to 0+} \frac{1}{\hbar}\left(\sum_m \frac{\langle \Psi_0 | \psi(\boldsymbol{x}) | \Psi_m^- \rangle \langle \Psi_m^- | \psi^\dagger(\boldsymbol{x}') | \Psi_0 \rangle}{\omega + i\epsilon - \omega_{0m}} \sum_n \frac{\langle \Psi_0 | \psi^\dagger(\boldsymbol{x}') | \Psi_n^+ \rangle \langle \Psi_n^+ | \psi(\boldsymbol{x}) | \Psi_0 \rangle}{\omega - i\epsilon + \omega_{0n}}\right), \tag{13.3.8}$$

where we have used the spectral form (13.2.10). It is important to note that these expressions differ radically from anything used previously: for example, the single operators a_r and $\mathsf{a}_s^\dagger$ connect states of systems containing *different numbers of electrons*. Thus Ψ_m^- must refer to the system after addition of an $(N+1)$th electron; it is the mth state of the *negative ion*. Similarly, Ψ_n^+ must be an $(N-1)$-electron function: the nth state of the *positive ion*. It is for this reason that the summation in (13.2.10) must turn into two separate summations, the first corresponding to attachment processes ($\Psi_0 \to \Psi_m^-$, the superscript indicating explicitly a state of the negative ion) and the second to ionization ($\Psi_0 \to \Psi_n^+$).

The analogue of the result (13.3.3) is obtained on noting that the residue at the pole $\omega = -\omega_{0n} + i\epsilon$ is $\langle \Psi_0 | \mathsf{a}_s^\dagger | \Psi_n \rangle \langle \Psi_n | \mathsf{a}_r | \Psi_0 \rangle$ and that the *sum* of these residues (which is given by $(2\pi i)^{-1}$ times an integral over any contour enclosing them) will be

$$\sum_n \langle \Psi_0 | \mathsf{a}_s^\dagger | \Psi_n^+ \rangle \langle \Psi_n^+ | \mathsf{a}_r | \Psi_0 \rangle = \langle \Psi_0 | \mathsf{a}_s^\dagger \mathsf{a}_r | \Psi_0 \rangle = \rho_{rs}.$$

A convenient contour (apparently first used in quantum chemistry by Coulson, 1937) is the one shown in Fig. 13.1. The required result is then

$$\rho_{rs} = \frac{\hbar}{2\pi i} \int_C \langle\!\langle \mathsf{a}_r; \mathsf{a}_s^\dagger \rangle\!\rangle_\omega \, d\omega = \frac{\hbar}{2\pi i} \int_C G_{rs}(\omega) \, d\omega, \tag{13.3.9}$$

where G_{rs} is a discrete *matrix element* of the Green's function in the spin-orbital basis. A similar treatment of (13.3.8) yields

$$\rho(\boldsymbol{x};\boldsymbol{x}')=\frac{\hbar}{2\pi i}\int_C \langle\langle \psi(\boldsymbol{x});\,\psi^\dagger(\boldsymbol{x}')\rangle\rangle_\omega\,\mathrm{d}\omega=\frac{\hbar}{2\pi i}\int_C G(\omega\mid\boldsymbol{x};\boldsymbol{x}')\,\mathrm{d}\omega \tag{13.3.10}$$

—which relates the density matrix to the Green's function itself.

Since any kind of polarization propagator can clearly be handled along similar lines, it is possible to obtain expressions for $\pi_{rs,tu}$ and $\pi(\boldsymbol{x}_1, \boldsymbol{x}_2; \boldsymbol{x}_1', \boldsymbol{x}_2')$ in essentially the same way. Thus (Problem 13.5) it may be shown that

$$\delta_{st}\rho_{ru}+\pi_{rs,tu}=-\frac{\hbar}{2\pi i}\int_C \langle\langle \mathsf{a}_t^\dagger\mathsf{a}_r;\,\mathsf{a}_u^\dagger\mathsf{a}_s\rangle\rangle_\omega\,\mathrm{d}\omega, \tag{13.3.11}$$

or, in terms of the field operators,

$$\delta(\boldsymbol{x}_2-\boldsymbol{x}_1')\rho(\boldsymbol{x}_1;\boldsymbol{x}_2')+\pi(\boldsymbol{x}_1,\boldsymbol{x}_2;\boldsymbol{x}_1',\boldsymbol{x}_2')$$
$$=-\frac{\hbar}{2\pi i}\int_C \langle\langle \psi^\dagger(\mathbf{x}_1')\psi(\mathbf{x}_1);\,\psi^\dagger(\mathbf{x}_2')\psi(\mathbf{x}_2)\rangle\rangle_\omega\,\mathrm{d}\omega. \tag{13.3.12}$$

The propagator on the right in (13.3.12) is a special case of the 2-electron Green's function referred to above, corresponding to a limit in which $t_1-t_1'\to 0+$, $t_2-t_2'\to 0+$ and the field operators are ordered accordingly. In practice, we seldom need the general many-time quantities, and the polarization propagators that we do need can usually be calculated by the methods already indicated. More remarkably, however, although the ground-state energy depends on the *two*-electron density matrix, it may be determined entirely in terms of the *one*-electron Green's function, as we shall discover in the next section.

13.4 SOME PROPERTIES OF THE ELECTRON PROPAGATOR

In the last section two special propagators (the electron propagator and the polarization propagator) were introduced and related to the 1- and 2-electron density matrices. The polarization propagator is by now a familiar object, and can indeed often be calculated rather efficiently by the variational methods of the previous chapter. The same cannot be said, however, of the electron propagator (or 1-electron Green's function), and its calculation will require the development of various new techniques. First of all, however, we shall want to know more about its potential value.

Let us therefore examine $G(\omega \mid \boldsymbol{x};\boldsymbol{x}')$, the electron propagator in frequency form, introduced in (13.3.8) *et seq.*; and consider some of the reasons for its great importance. The matrix elements in the numerators involve wavefunctions for the N-electron system, and for its *negative and positive ions* (with $N \pm 1$ electrons), and are of two types

$$f_m(\boldsymbol{x}) = \langle \Psi_0 |\, \psi(\boldsymbol{x}) \,| \Psi_m^- \rangle, \qquad g_n(\boldsymbol{x}) = \langle \Psi_n^+ |\, \psi(\boldsymbol{x}) \,| \Psi_0 \rangle, \quad (13.4.1)$$

the other elements involving the corresponding adjoint operators. For example,

$$\langle \Psi_m^- |\, \psi^\dagger(\boldsymbol{x}') \,| \Psi_0 \rangle = \langle \psi(\boldsymbol{x}') \Psi_m^- \mid \Psi_0 \rangle = \langle \Psi_0 |\, \psi(\boldsymbol{x}') \,| \Psi_m^- \rangle^* = f_m(\boldsymbol{x}')^*.$$

The result is thus

$$G(\omega \mid \boldsymbol{x};\boldsymbol{x}') = \lim_{\epsilon \to 0+} \frac{1}{\hbar} \left(\sum_m \frac{f_m(\boldsymbol{x}) f_m^*(\boldsymbol{x}')}{\omega + i\epsilon - \omega_{0m}^-} + \sum_n \frac{g_n(\boldsymbol{x}) g_n^*(\boldsymbol{x}')}{\omega - i\epsilon + \omega_{0n}^+} \right), \quad (13.4.2)$$

where $f_m(\boldsymbol{x})$ and $g_n(\boldsymbol{x})$ are frequently called "Dyson orbitals" or "overlap amplitudes": the only question at this point is how to get these quantities, which involve vectors and operators in Fock space, from any (Schrödinger) wavefunctions that we may possess. In other words: what is the Schrödinger operator equivalent to the *single* destruction operator $\psi(\boldsymbol{x})$ or, in a discrete basis to a_r?

When the operators a_r, $\mathsf{a}_r^\dagger$ were introduced in Section 3.6, it was postulated that they carried a normalized N-electron ket into a normalized ket *either* with the Nth electron missing, or with an $(N+1)$th electron added, respectively. It was then noted that the properties of number-conserving operators, containing equal numbers of a and $\mathsf{a}^\dagger$ factors, were independent of the number of particles in the system—an important asset in using Fock-space methods. In dealing with individual operators, however, the Schrödinger equivalent of the Fock operator a_r, which removes the Nth electron to leave a system containing electrons $1, 2, \ldots, N-1$, does depend on N. Thus if $\mathsf{a}_r |ij \ldots pr\rangle = |ij \ldots p\rangle$, with N occupied states in the first ket and $N-1$ in the second, then the Schrödinger equivalent which destroys electron N in spin orbital ψ_r must have the property

$$\mathsf{d}_r(N) \left(\frac{1}{N!} \right)^{1/2} \det |\psi_i(\boldsymbol{x}_1)\ \psi_j(\boldsymbol{x}_2)\ \ldots\ \psi_p(\boldsymbol{x}_{N-1})\ \psi_r(\boldsymbol{x}_N)|$$
$$= \left[\frac{1}{(N-1)!} \right]^{1/2} \det |\psi_i(\boldsymbol{x}_1)\ \psi_j(\boldsymbol{x}_2)\ \ldots\ \psi_p(\boldsymbol{x}_{N-1})|.$$

In other words, $\mathsf{d}_r(N)$ must strike out the spin-orbital ψ_r wherever it appears (in any expansion over a complete set of determinants), destroying any term not containing ψ_r, and then multiply by $(N!)^{1/2}$ in order to renormalize the determinant. The explicit form of the association $\mathsf{a}_r \to \mathsf{d}_r(N)$, within the subspace of N-electron kets, is thus found to be

$$\mathsf{a}_r \to \mathsf{d}_r(N) = N^{1/2} \int \mathrm{d}\boldsymbol{x}_N \, \psi_r^*(\boldsymbol{x}_N) \ldots, \tag{13.4.3}$$

while for the field operator $\psi(\boldsymbol{x}) = \sum_r \psi_r(\boldsymbol{x})\mathsf{a}_r$ we shall have to make the association

$$\left.\begin{aligned} \psi(\boldsymbol{x}) \to \psi^{\mathrm{S}}(N) &= N^{1/2} \int \mathrm{d}\boldsymbol{x}_N \sum_r \psi_r(\boldsymbol{x}) \psi_r^*(\boldsymbol{x}_N) \ldots, \\ &= N^{1/2} \int \mathrm{d}\boldsymbol{x}_N \, \delta(\boldsymbol{x} - \boldsymbol{x}_N) \ldots \quad . \end{aligned}\right\} \tag{13.4.4}$$

This rule is really very simple:

The effect of the field operator $\psi(\boldsymbol{x})$ on any N-electron Schrödinger wavefunction is to replace the Nth variable $\boldsymbol{x}_N$ by the parameter $\boldsymbol{x}$, multiplying the result by $N^{1/2}$

(13.4.5)

Similar rules may be derived for creation operators (Problem 13.7), but are less simple; and, since a creation operator $\mathsf{d}_r^\dagger(N)$ working on a ket function is equivalent to its adjoint (destruction operator), transferred to the bra, it is sufficient to know (13.4.5). For number-conserving operators, which contain equal numbers of d and $\mathsf{d}^\dagger$ factors, the result is always N-independent; Slater's rules are thus valid whatever the number of particles, and no N-dependence appears in the second-quantization Hamiltonian (3.6.9).

The overlap amplitudes in (13.4.1) are now seen to be expressible as

$$f_m(\boldsymbol{x}) = (N+1)^{1/2} \int \Psi_0(\boldsymbol{x}_1, \boldsymbol{x}_2, \ldots, \boldsymbol{x}_N)^* \Psi_m^-(\boldsymbol{x}_1, \boldsymbol{x}_2, \ldots, \boldsymbol{x}_N, \boldsymbol{x}) \, \mathrm{d}\boldsymbol{x}_1 \ldots \mathrm{d}\boldsymbol{x}_N, \tag{13.4.6}$$

where the $(N+1)$th variable in the negative-ion wavefunction has been replaced by $\boldsymbol{x}$, and

$$g_n(\boldsymbol{x}) = N^{1/2} \int \Psi_n^+(\boldsymbol{x}_1, \boldsymbol{x}_2, \ldots, \boldsymbol{x}_{N-1})^* \Psi_0(\boldsymbol{x}_1, \boldsymbol{x}_2, \ldots, \boldsymbol{x}_{N-1}, \boldsymbol{x}) \, \mathrm{d}\boldsymbol{x}_1 \ldots \mathrm{d}\boldsymbol{x}_{N-1}, \tag{13.4.7}$$

where the Nth variable in the neutral-system wavefunction has been replaced by $\boldsymbol{x}$. Such functions resemble the transition density matrices defined in Section 5.4, but connect states that differ by the removal or addition of one electron; they play an important role in the discussion of ionization and attachment processes (see e.g. Pickup 1977). From (13.4.2), the products $f_m(\boldsymbol{x})f_m^*(\boldsymbol{x}')$ and $g_n(\boldsymbol{x})g_n^*(\boldsymbol{x}')$ arise as *residues* at the poles ($\omega = \omega_{0m}^-$ or $-\omega_{0n}^+$ respectively) of the Green's function. The poles themselves determine, in principle, the *exact* attachment and ionization energies (for *all* attachment and ionization processes), and the great importance of the electron propagator, particularly in fields such as photoelectron spectroscopy, rests upon this property.

We now show how the Green's function (13.4.2) can also give further information on the *ground* state of the neutral system—even the total electronic energy! This is remarkable, since E_0 depends on the 2-electron density matrix and hence, apparently, on the 2-electron Green's function. The fact that knowledge of $G(\omega\,|\,\boldsymbol{x};\boldsymbol{x}')$ (or its general matrix element $G_{pq}(\omega)$) is sufficient for this purpose emerges when we consider the equation of motion (13.2.14). Thus, putting $\mathsf{A} = \mathsf{a}_q^\dagger$, $\mathsf{B} = \mathsf{a}_p$, we obtain (with $\eta = -1$)

$$i\hbar\frac{\mathrm{d}}{\mathrm{d}\tau}\langle\!\langle \mathsf{a}_p(\tau); \mathsf{a}_q^\dagger(0)\rangle\!\rangle = \delta(\tau)\langle\Psi_0|\,[\mathsf{a}_p(\tau), \mathsf{a}_q^\dagger(0)]_+\,|\Psi_0\rangle + \langle\!\langle [\mathsf{a}_p(\tau), \mathsf{H}]; \mathsf{a}_q^\dagger(0)\rangle\!\rangle. \tag{13.4.8}$$

A more explicit form follows on inserting the Hamiltonian (3.6.9), most conveniently expressed in terms of operators at time τ. The commutator in the last term then reduces as in (8.2.28), to give for $\tau \neq 0$

$$(i\hbar)^2\frac{\mathrm{d}}{\mathrm{d}\tau}G_{pq}(\tau;0) = \left\langle\Psi_0\right|\mathsf{T}\Big\{\sum_s \langle p|\,\mathsf{h}\,|s\rangle \mathsf{a}_s(\tau)\mathsf{a}_q^\dagger(0) + \sum_{s,t,u}\langle ps|\,g\,|tu\rangle \mathsf{a}_s^\dagger(\tau)\mathsf{a}_u(\tau)\mathsf{a}_t(\tau)\mathsf{a}_q^\dagger(0)\Big\}\left|\Psi_0\right\rangle. \tag{13.4.9}$$

If now we let τ be a negative infinitesimal, the expectation value in (13.4.9) can be written, remembering that $\eta = -1$ for Bose-type operators,

$$\lim_{\tau\to 0-}\left\langle\Psi_0\right|\Big\{-\sum_s \langle p|\,\mathsf{h}\,|s\rangle \mathsf{a}_q^\dagger(0)\mathsf{a}_s(\tau) - \sum_{s,t,u}\langle ps|\,g\,|tu\rangle \mathsf{a}_q^\dagger(0)\mathsf{a}_s^\dagger(\tau)\mathsf{a}_u(\tau)\mathsf{a}_t(\tau)\Big\}\left|\Psi_0\right\rangle$$
$$= -\sum_s \langle p|\,\mathsf{h}\,|s\rangle\langle \mathsf{a}_q^\dagger\mathsf{a}_s\rangle - \sum_{s,t,u}\langle ps|\,g\,|tu\rangle\langle \mathsf{a}_q^\dagger\mathsf{a}_s^\dagger\mathsf{a}_u\mathsf{a}_t\rangle.$$

But the 1- and 2-electron expectation values are known from (5.4.20) to be ρ_{sq} and $\pi_{tu,qs}$ respectively. If, then, we put $q = p$ in (13.4.9) and take τ

to be a negative infinitesimal, we shall obtain on the right-hand side *almost* the ground-state total energy. The result is not quite right, since the 2-electron term lacks the usual factor $\frac{1}{2}$, but this may be corrected by adding $-\sum_{s,p} \langle p|\,\mathrm{h}\,|s\rangle \rho_{sp}$ to both sides of the equation, and, using (13.3.3), this gives

$$-2E_0 = (i\hbar)^2 \left[\frac{\mathrm{d}}{\mathrm{d}\tau} \sum_p G_{pp}(\tau; 0) + i\hbar \sum_{s,p} \langle p|\,\mathrm{h}\,|s\rangle G_{sp}(\tau; 0)\right]_{\tau \to 0-}. \quad (13.4.10)$$

The ground-state total energy can thus be expressed in terms of the 1-body Green's function alone—or, in a discrete basis, in terms of its matrix elements $G_{pq}(t; t')$.

This remarkable result is expressed more conveniently in frequency form by using

$$G_{pq}(\tau; 0) = \frac{1}{2\pi} \int G_{pq}(\omega) e^{-i\omega\tau}\,\mathrm{d}\omega,$$

$$\frac{\mathrm{d}}{\mathrm{d}\tau} G_{pq}(\tau; 0) = \frac{1}{2\pi} \int (-i\omega) G_{pq}(\omega) e^{-i\omega\tau}\,\mathrm{d}\omega.$$

On substituting these transforms in (13.4.10) and taking the limit $\tau \to 0-$,

$$\boxed{E_0 = \frac{\hbar}{4\pi i} \int_C \sum_{p,q} (\hbar\omega\delta_{pq} + \langle p|\,\mathrm{h}\,|q\rangle) G_{pq}(\omega)\,\mathrm{d}\omega.} \quad (13.4.11)$$

As usual, the contour C is arbitrary as long as it encloses the (ionization) poles in the second quadrant of the complex ω plane; a suitable contour is again the one shown in Fig. 13.1.

Although, generally speaking, the power of the Green's function approach is most clearly revealed in the discussion of *processes* (e.g. ionization, attachment, excitation) that involve more than one state, the result in (13.4.11) has been used as the basis of numerical calculations of the energy of a molecular ground state (Moccia *et al.*, 1976; Caravetta and Moccia, 1978). Before any kind of application can be made, however, it is of course necessary to develop methods of calculating $G(\omega \mid \boldsymbol{x}; \boldsymbol{x}')$ or its matrix elements $G_{qp}(\omega)$. Several families of methods are available; in the next section we consider only one, but this will be sufficiently general to expose most of the methodology currently in use.

13.5 THE CALCULATION OF PROPAGATORS

The starting point for most methods of calculating the electron propagator is the equation of motion (13.2.15). With minor changes, the same methods apply to the polarization (or any other) propagator, and the operators A, B will therefore be left unspecified at this point. By using the basic equation

$$\hbar\omega \langle\langle \mathsf{B};\mathsf{A}\rangle\rangle_\omega = \langle \Psi_0 | [\mathsf{B},\mathsf{A}]_\pm | \Psi_0\rangle - \langle\langle [\mathsf{H},\mathsf{B}];\mathsf{A}\rangle\rangle_\omega \qquad (13.5.1)$$

to express the propagator on the extreme right in terms of similar terms with B replaced by [H, B], we obtain a 3-term expression for $\langle\langle \mathsf{B};\mathsf{A}\rangle\rangle_\omega$; and by repetition we generate the so-called "moment expansion"

$$\langle\langle \mathsf{B};\mathsf{A}\rangle\rangle_\omega = \frac{1}{\hbar\omega}\left\{\langle \Psi_0 | [\mathsf{B},\mathsf{A}]_\pm | \Psi_0\rangle + \left(\frac{-1}{\hbar\omega}\right)\langle \Psi_0 | [[\mathsf{H},\mathsf{B}],\mathsf{A}]_\pm | \Psi_0\rangle + \left(\frac{-1}{\hbar\omega}\right)^2 \langle \Psi_0 | [[\mathsf{H},[\mathsf{H},\mathsf{B}]],\mathsf{A}]_\pm | \Psi_0\rangle + \ldots\right\}, \qquad (13.5.2)$$

in which all the terms in the expansion are simply expectation values, with respect to Ψ_0 (which in principle will be an exact ground state function), of increasingly complicated operators. Apparently, this expansion could be used to generate approximate propagators; but in practice truncation of the series at any reasonable level has been found unsatisfactory (see e.g. Linderberg and Öhrn, 1973), and more efficient approaches have therefore been sought. The ones that we shall discuss are essentially *algebraic* and depend neither on the separation of H into an IPM H_0 plus a perturbation term nor on the use of an IPM reference function as Ψ_0. Perturbation methods have been devised, and implemented using diagrammatic techniques akin to those used in Chapter 9, but lie outside the scope of the present exposition, although they have given impressive results (see e.g. the reviews by Cederbaum and Domcke, 1977, Öhrn and Born, 1981; von Niessen *et al.*, 1984). The algebraic approach has certain advantages, especially in its generality and flexibility.

As a first step, the appearance of (13.5.2) can be improved by introducing the concept of a "superoperator" (Banwell and Primas, 1963; Goscinski and Lukman, 1980): the Hamiltonian *super*operator‡, denoted by $\hat{\mathsf{H}}$, works on an arbitrary *operator* to generate a commutator. Thus§

$$\hat{\mathsf{H}}\mathsf{B} = [\mathsf{H},\mathsf{B}], \qquad (13.5.3)$$

and by repetition

$$\hat{\mathsf{H}}^2\mathsf{B} = [\mathsf{H},[\mathsf{H},\mathsf{B}]], \qquad \hat{\mathsf{H}}^3\mathsf{B} = [\mathsf{H},[\mathsf{H},[\mathsf{H},\mathsf{B}]]], \ldots,$$

‡ This is essentially the Liouville operator commonly used in statistical mechanics.

§ Some authors use the definition $\hat{\mathsf{H}}\mathsf{B} = \mathsf{BH} - \mathsf{HB}$: troublesome sign difference often occur and the conventions used must be carefully observed.

so that a "nest" of commutators can be represented simply as $\hat{\mathsf{H}}^n\mathsf{B}$. The expression (13.5.2) thus contains a power series and may be written as

$$\langle\!\langle \mathsf{B}; \mathsf{A} \rangle\!\rangle_\omega = \frac{1}{\hbar\omega} \sum_{n=0}^{\infty} \left(\frac{-1}{\hbar\omega}\right)^n \langle \Psi_0 | [\hat{\mathsf{H}}^n \mathsf{B}, \mathsf{A}]_\pm | \Psi_0 \rangle. \qquad (13.5.4)$$

Before summing the series, we note that, as long as we are proceeding formally with Ψ_0 regarded as an exact ground-state function, the superoperator may be passed from the B to the A, since it is easily verified that

$$\langle \Psi_0 | [\hat{\mathsf{H}}\mathsf{X}, \mathsf{A}]_\pm | \Psi_0 \rangle = -\langle \Psi_0 | [\mathsf{X}, \hat{\mathsf{H}}\mathsf{A}]_\pm | \Psi_0 \rangle, \qquad (13.5.5)$$

and $\hat{\mathsf{H}}^n$ may thus be applied to A instead of B by introducing a factor $(-1)^n$. The sum in (13.5.4) is immediately recognized as the expansion of an inverse. With the conventions

$$\hat{1}\mathsf{X} = \mathsf{X}, \qquad \hat{\mathsf{H}}^0 = \hat{1}, \qquad (13.5.6)$$

in which $\hat{1}$ is the "unit superoperator", (13.5.4) then yields

$$\langle\!\langle \mathsf{B}; \mathsf{A} \rangle\!\rangle_\omega = \langle \Psi_0 | [\mathsf{B}, (\hbar\omega\hat{1} - \hat{\mathsf{H}})^{-1}\mathsf{A}]_\pm | \Psi_0 \rangle, \qquad (13.5.7)$$

where the so-called "superoperator resolvent" $(\hbar\omega\hat{1} - \hat{\mathsf{H}})^{-1}$ is defined by the power series. This is a compact form of (13.5.2), although the improvement is at this stage purely cosmetic.

The next stage is to develop the concept of "operator space" ($\hat{\mathsf{H}}$ being defined through its effect on arbitrary *operators*, rather than vectors) by introducing a metric as in Section 2.2. This means looking for a complex number to associate with every pair of elements in the space (i.e. operators), and a rather natural choice is to adopt

$$(\mathsf{X} | \mathsf{Y}) = \langle \Psi_0 | [\mathsf{X}^\dagger, \mathsf{Y}]_\pm | \Psi_0 \rangle, \qquad (13.5.8)$$

where the dagger on the operator coming from the bra factor is like the star in the vector-space definition, and its presence ensures that the usual metric axiom

$$(\mathsf{Y} | \mathsf{X}) = (\mathsf{X} | \mathsf{Y})^*$$

is satisfied. With this definition, (13.5.7) can evidently be expressed as

$$\langle\!\langle \mathsf{B}; \mathsf{A} \rangle\!\rangle_\omega = (\mathsf{B}^\dagger | (\hbar\omega\hat{1} - \hat{\mathsf{H}})^{-1} | \mathsf{A}), \qquad (13.5.9)$$

with a Dirac-type notation in which the operator works on the "ket" $|\mathsf{A})$, considered as an element of the operator space. This result is formally exact, but is of course useless until we can find some way of generating approximations. The value of introducing superoperators and scalar products in that it allows us to draw on experience and intuition gained in using the ordinary vector spaces of Chapter 2.

The kind of approximation to be employed is analogous to the *truncation* of a basis in a linear vector space $\mathcal{V}$; truncation to n elements produces a *subspace* $\mathcal{V}_n$, and the truncation of a general vector corresponds to finding its *projection* onto the subspace. We are therefore concerned with the projection operator, ρ say, associated with a truncated basis; and if the basis ($\mathbf{e}$, say) is orthonormal we may conveniently write

$$\rho = \sum_i \mathsf{e}_i \mathsf{e}_i^* = \mathbf{e}\mathbf{e}^\dagger$$

in the symbolic form (p. 177) where the "dyad" $\mathsf{e}_i\mathsf{e}_i^*$ operates on an arbitrary element v to produce‡ $\mathsf{v}' = (\mathsf{e}_i\mathsf{e}_i^*)\mathsf{v} = \mathsf{e}_i(\mathsf{e}_i^*\mathsf{v}) = \mathsf{e}_i\langle \mathsf{e}_i \mid \mathsf{v}\rangle$. For a *complete* space, ρ becomes the unit operator and we recognize the familiar "resolution of the identity"

$$\rho = \sum_i |\mathsf{e}_i\rangle\langle \mathsf{e}_i| \rightarrow 1$$

employed frequently in previous chapters. When the basis is *not* orthonormal, the metric matrix must be included: this is $\mathbf{S} = \mathbf{e}^\dagger\mathbf{e}$, and by using in place of $\mathbf{e}$ the Löwdin basis $\bar{\mathbf{e}} = \mathbf{e}\mathbf{S}^{-1/2}$ it follows at once that§

$$\rho = \mathbf{e}\mathbf{S}^{-1}\mathbf{e}^\dagger = \sum_{i,j} \mathsf{e}_i(S^{-1})_{ij}\mathsf{e}_j^*, \tag{13.5.10}$$

which again gives a resolution of the identity when the basis is complete.

When using a truncated basis in *operator* space, two kinds of projection are useful; these are (Löwdin, 1977, 1982)

$$\mathsf{A}' = \rho\mathsf{A}\rho, \qquad \mathsf{A}'' = \mathsf{A}^{1/2}\rho\mathsf{A}^{1/2} \tag{13.5.11}$$

and are called respectively the *outer projection* and the *inner projection* of the operator A onto the space $\mathcal{V}_n$ defined by ρ. The outer projection has been used already, in matrix form, in Chapter 6 (p. 181). The inner projection is particularly useful in propagator theory (see e.g. Öhrn and Born, 1981); and such projections are also useful in establishing various theorems (Löwdin, 1977) concerning upper and lower bounds on the eigenvalues of an operator. It should be emphasized that neither

‡ With Dirac-type notation each dyad is a ket–bra product $|\mathsf{e}_i\rangle\langle \mathsf{e}_i|$ and the same result is written $|\mathsf{v}'\rangle = (|\mathsf{e}_i\rangle\langle \mathsf{e}_i|)\,|\mathsf{v}\rangle = |\mathsf{e}_i\rangle\langle \mathsf{e}_i|\,\mathsf{v}\rangle$.

§ Note that any operator can be written in the form $\mathsf{A} = \sum_{rs} A_{rs}\mathsf{e}_r\mathsf{e}_s^* = \mathbf{e}\mathbf{A}\mathbf{e}^\dagger$ where the row $\mathbf{e}$ and column $\mathbf{e}^\dagger$ appear on the left and the right respectively of the *matrix* $\mathbf{A}$.

projection preserves the "multiplication table" for a set of operators: the relationship $\mathsf{AB} = \mathsf{C}$ implies neither that $\mathsf{A'B'} = \mathsf{C'}$, nor that $\mathsf{A''B''} = \mathsf{C''}$; nevertheless, the projected forms may be regarded as *approximations* that become increasingly precise as the basis is extended, and this is the key to the formulation of convergent approximation methods. The nature of the convergence involved has been discussed at length by Löwdin (1982).

The inner projection in (13.5.11) may be put into a useful form by inserting (13.5.10) to give

$$\mathsf{A}'' = \mathsf{A}^{1/2}\mathbf{e}\mathbf{S}^{-1}\mathbf{e}^{\dagger}\mathsf{A}^{1/2},$$

where for instance $\mathsf{A}^{1/2}\mathbf{e} = (\mathsf{A}^{1/2}\mathsf{e}_1 \; \mathsf{A}^{1/2}\mathsf{e}_2 \; \ldots) = \mathbf{f}$, say. Here it is assumed that A is Hermitian and positive-definite‡ (so that $\mathsf{A}^{1/2}$ can be defined, A having only positive eigenvalues). On introducing $\mathbf{f}$ in the last equation, and noting that $\mathbf{S} = \mathbf{e}^{\dagger}\mathbf{e} = (\mathsf{A}^{-1/2}\mathbf{f})^{\dagger}(\mathsf{A}^{-1/2}\mathbf{f}) = \mathbf{f}^{\dagger}\mathsf{A}^{-1}\mathbf{f}$, we obtain

$$\mathsf{A}'' = \mathbf{f}(\mathbf{f}^{\dagger}\mathsf{A}^{-1}\mathbf{f})^{-1}\mathbf{f}^{\dagger}.$$

But since A is an arbitrary operator, it may be replaced by A^{-1}; and since $\mathbf{f}^{\dagger}\mathsf{A}\mathbf{f} = \mathbf{A}$, the metrically defined matrix (see p. 32) associated with A (with elements $A_{ij} = \mathsf{f}_i^*\mathsf{A}\mathsf{f}_j = \langle \mathsf{f}_i | \, \mathsf{A} \, | \mathsf{f}_j \rangle$), the result may be written as

$$(\mathsf{A}^{-1})'' = \mathbf{f}(\mathbf{f}^{\dagger}\mathsf{A}\mathbf{f})^{-1}\mathbf{f}^{\dagger} = \mathbf{f}\mathbf{A}^{-1}\mathbf{f}^{\dagger}. \qquad (13.5.12)$$

When the basis approaches completeness ($\mathcal{V}_n \to \mathcal{V}$) the inner projection approaches the operator A^{-1} itself; otherwise it represents a finite-basis approximation. In this way it is possible to express an *operator* inverse, approximately, in terms of a *matrix* inverse.

Since $\mathbf{e}$ was an arbitrary basis, defining $\mathcal{V}_n$, $\mathbf{f}$ will also define an n-dimensional subspace $\mathcal{V}'_n$. We may therefore use the vectors of $\mathbf{f}$ to define the inner projection and, reverting to the original notation ($\mathbf{e}$ in place of $\mathbf{f}$), write

$$\mathsf{A}^{-1} \simeq \mathbf{e}\mathbf{A}^{-1}\mathbf{e}^{\dagger} = \sum_{i,j} \mathsf{e}_i(\mathbf{e}^{\dagger}\mathsf{A}\mathbf{e})^{-1}_{ij}\mathsf{e}_j^*. \qquad (13.5.13)$$

The basis is arbitrary, but its choice, to be discussed later, will determine the goodness of the approximation.

The inner-projection concept now leads to an approximation for the propagator defined in (13.5.9). The Dirac scalar product $\langle \mathsf{x} \,|\, \mathsf{y} \rangle = \mathsf{x}^*\mathsf{y}$ is replaced by the binary product (13.5.8), but everything else is formally similar. To preserve the parallel, we define an *operator basis*

$$\boldsymbol{\eta} = (\eta_1 \; \eta_2 \; \ldots \; \eta_r \; \ldots) \qquad (13.5.14)$$

‡ In some applications this requirement may be waived.

(like the row of basis vectors $\mathbf{e}$) and its conjugate column $\boldsymbol{\eta}^\dagger$, whose elements are the adjoint operators $\eta_1^\dagger, \eta_2^\dagger, \ldots$, and write (13.5.8) as the symbolic scalar product

$$\mathsf{X}^\dagger \cdot \mathsf{Y} = (\mathsf{X} \mid \mathsf{Y}) = \langle \Psi_0 | [\mathsf{X}^\dagger, \mathsf{Y}]_\pm | \Psi_0 \rangle, \tag{13.5.15}$$

resembling $\mathsf{x}^*\mathsf{y} = \langle \mathsf{x} \mid \mathsf{y} \rangle$ for vectors. The *resolvent superoperator*

$$\hat{\mathsf{R}}(\omega) = (\hbar\omega\hat{1} - \hat{\mathsf{H}})^{-1} \tag{13.5.16}$$

may then be approximated, using (13.5.13), in the form

$$\hat{\mathsf{R}}(\omega) \simeq \boldsymbol{\eta}\mathbf{R}(\omega)\boldsymbol{\eta}^\dagger = \sum_{r,s} \eta_r [R(\omega)]_{rs} \eta_s^\dagger. \tag{13.5.17}$$

Here η_r and $\eta_s^\dagger$ are the analogues of e_i and e_j^* in (13.5.13), but are elements of the *operator* space, while the resolvent *matrix* is

$$\mathbf{R}(\omega) = \mathbf{M}(\omega)^{-1}, \qquad \mathbf{M}(\omega) = \boldsymbol{\eta}^\dagger(\hbar\omega\hat{1} - \hat{\mathsf{H}})\boldsymbol{\eta}. \tag{13.5.18}$$

The matrix elements of $\mathbf{M}(\omega)$ are easily calculated once we have chosen an operator basis, as we shall see presently. Finally, from (13.5.9),

$$\boxed{\langle\!\langle \mathsf{B}; \mathsf{A} \rangle\!\rangle_\omega = \mathsf{B} \cdot \mathsf{R}(\omega)\mathsf{A} = \mathsf{B} \cdot \boldsymbol{\eta}\mathbf{R}(\omega)\boldsymbol{\eta}^\dagger \cdot \mathsf{A},} \tag{13.5.19}$$

which is the key equation for calculating approximations to any kind of propagator. The approximation depends only on the choice of basis $\boldsymbol{\eta}$ and of the reference function used in defining the scalar product.

The parallel between operator and vector spaces, used above, depends only on the existence of a scalar product. The metric defined in (13.5.8) is not the only possible choice (see e.g. Löwdin 1982, 1985), but it is the one most commonly employed in propagator theory. The same matrix notation applies to both vector and operator spaces, provided that the basic conventions are carefully observed; $\boldsymbol{\eta}$ and $\boldsymbol{\eta}^\dagger$ (playing the part of *basis* elements and their duals) are always of row and column form respectively, and all equations are then internally consistent.

At this point it is clear that two types of approximation will be involved in calculating a propagator: (i) the truncation of the complete set of operators $\{\eta_r\}$; and (ii) the use of a reference state $|\Psi_0\rangle$ that is not the *exact* ground state. To illustrate the possibilities available, we consider a reference state of IPM form, in which spin-orbitals $\{\psi_i\}$ are occupied and $\{\psi_m\}$ are empty, and discuss two important examples.

The electron propagator

The 1-body Green's function (13.3.7) arises on taking $\mathsf{B} = \mathsf{a}_r$, $\mathsf{A} = \mathsf{a}_s^\dagger$ in (13.5.19), and may be "blocked" accordingly as $r, s = i, j, \ldots$ or $r, s =$

$m, n, \ldots$. The "occupied" block for example, will have elements

$$\langle\langle \mathsf{a}_i; \mathsf{a}_j^\dagger \rangle\rangle_\omega = \sum_{t,u} \mathsf{a}_i \cdot \eta_t R(\omega)_{tu} \eta_u^\dagger \cdot \mathsf{a}_j^\dagger, \tag{13.5.20}$$

where $R(\omega)_{tu}$ is an element of the inverse $\mathbf{M}(\omega)^{-1}$, with elements†

$$M(\omega)_{rs} = \eta_r^\dagger \cdot (\hbar\omega\hat{1} - \hat{\mathsf{H}})\eta_s, \tag{13.5.21}$$

and we must first decide what operators to include in the $\{\eta_r\}$. The scalar products in (13.5.20) are

$$\mathsf{a}_i \cdot \eta_t = \langle \Psi_0 | (\mathsf{a}_i \eta_t + \eta_t \mathsf{a}_i) | \Psi_0 \rangle,$$
$$\eta_u^\dagger \cdot \mathsf{a}_j^\dagger = \langle \Psi_0 | (\eta_u^\dagger \mathsf{a}_j^\dagger + \mathsf{a}_j^\dagger \eta_u^\dagger) | \Psi_0 \rangle,$$

and will vanish unless η_t contains a creation operator of the type $\mathsf{a}_k^\dagger$, possibly multiplied by some number-conserving product such as $\mathsf{a}_p^\dagger \mathsf{a}_q$. If we restrict the set to *single* operators then the only choices giving non-zero results will be $\eta_t = \mathsf{a}_k^\dagger$, $\eta_u = \mathsf{a}_l^\dagger$—for which (from the anticommutation rules)

$$\mathsf{a}_i \cdot \eta_t \to \mathsf{a}_i \cdot \mathsf{a}_k^\dagger = \delta_{ik},$$
$$\eta_u^\dagger \cdot \mathsf{a}_j^\dagger \to \mathsf{a}_l \cdot \mathsf{a}_j^\dagger = \delta_{lj}.$$

Only the occupied block of $\mathbf{R}(\omega)$ will therefore be required in evaluating (13.5.20). Other blocks of the propagator matrix will be determined similarly, and to calculate the inverse we shall need to consider the matrix $\mathbf{M}(\omega)$, in (13.5.21) with a similar partitioning.

The explicit form of (13.5.21) follows on noting that (taking for example the occupied block)

$$\mathsf{a}_k \cdot \mathsf{a}_l^\dagger = \langle \Psi_0 | [\mathsf{a}_k, \mathsf{a}_l^\dagger]_+ | \Psi_0 \rangle = \delta_{kl}, \tag{13.5.22a}$$

$$\mathsf{a}_k \cdot \hat{\mathsf{H}} \mathsf{a}_l^\dagger = \langle \Psi_0 | [\mathsf{a}_k, [\mathsf{H}, \mathsf{a}_l^\dagger]]_+ | \Psi_0 \rangle. \tag{13.5.22b}$$

The double commutator reduces easily (cf. p. 261), giving first

$$[\hat{\mathsf{H}}, \mathsf{a}_l^\dagger] = \sum_r \langle r | \mathsf{h} | l \rangle \mathsf{a}_r^\dagger + \tfrac{1}{2} \sum_{r,s,u} \langle rs \| lu \rangle \mathsf{a}_r^\dagger \mathsf{a}_s^\dagger \mathsf{a}_u, \tag{13.5.23}$$

(using the antisymmetrized 2-electron integral) and then

$$[\mathsf{a}_k, [\hat{\mathsf{H}}, \mathsf{a}_l^\dagger]]_+ = \langle k | \mathsf{h} | l \rangle + \sum_{s,u} \langle ks \| lu \rangle \mathsf{a}_s^\dagger \mathsf{a}_u. \tag{13.5.24}$$

Since, however, $\langle \Psi_0 | \mathsf{a}_s^\dagger \mathsf{a}_u | \Psi_0 \rangle = \rho_{us}$, and vanishes except for s, u running over *occupied* spin-orbitals, the right-hand side of (13.5.22b) is immediately seen to be

$$F_{kl} = \langle k | \mathsf{h} | l \rangle + \sum_{i,j} \rho_{ji} \langle ki \| lj \rangle, \tag{13.5.25}$$

†From now on, = will be used also for an *approximate* equality.

a matrix element of the Fock operator (Section 6.1). Without loss of generality, in the usual way, we may use canonical orbitals that diagonalize the matrix $\mathbf{F}$, and in that case

$$F_{kl} = \epsilon_k \delta_{kl}, \tag{13.5.26}$$

where ϵ_k is an orbital energy in the usual sense.

The same argument applies to the other blocks of $\mathbf{M}(\omega)$, which then takes wholly diagonal form, with elements

$$M(\omega)_{ij} = (\hbar\omega - \epsilon_i)\delta_{ij},$$
$$M(\omega)_{mn} = (\hbar\omega - \epsilon_m)\delta_{mn}.$$

Inversion, and substitution in (13.5.20), yields a diagonal form of the Green's matrix $\mathbf{G}(\omega)$, which gives on passing to the field-operator form with $G(\omega \mid \boldsymbol{x};\boldsymbol{x}') = \sum_{r,s} G(\omega)_{rs}\psi_r(\boldsymbol{x})\psi_s^*(\boldsymbol{x}')$,

$$G(\omega \mid \boldsymbol{x};\boldsymbol{x}') = \sum_m \frac{\psi_m(\boldsymbol{x})\psi_m^*(\boldsymbol{x}')}{\hbar\omega - \epsilon_m} + \sum_i \frac{\psi_i(\boldsymbol{x})\psi_i^*(\boldsymbol{x}')}{\hbar\omega - \epsilon_i}. \tag{13.5.27}$$

On comparing this with the exact result (13.4.2), it is evident that the poles of the approximate propagator correspond to $\hbar\omega = -\hbar\omega_{0i}^+ = -(E_i^+ - E_0) = \epsilon_i$, for ionization processes, and $\hbar\omega = \hbar\omega_{0m}^- = E_m^- - E_0 = \varepsilon_m$ for attachment processes; the numerators give the corresponding residues at the two types of pole, and it is clear that in this approximation the Dyson orbitals are represented by the canonical Hartree–Fock spin-orbitals.

The polarization propagator

When A and B in (13.5.19) are (1-electron) number-conserving operators of the type

$$\mathsf{A} = \sum_{r,s} A_{rs}\mathsf{a}_r^\dagger \mathsf{a}_s, \qquad \mathsf{B} = \sum_{r,s} B_{rs}\mathsf{a}_r^\dagger \mathsf{a}_s \tag{13.5.28}$$

we shall obtain a general *polarization propagator*, and it will clearly be sufficient to consider typical elements, $\mathsf{A} \to \mathsf{a}_r^\dagger \mathsf{a}_s$ and $\mathsf{B} \to \mathsf{a}_t^\dagger \mathsf{a}_u$.

Again various types of approximation may be distinguished. If the spin-orbitals of the reference function are $\psi_i, \psi_j \ldots$, those of a complementary set being $\psi_m, \psi_n, \ldots$, then a "particle–hole" basis would include only "excitation" and "de-excitation" operators, $\mathsf{a}_m^\dagger \mathsf{a}_i$ and $\mathsf{a}_i^\dagger \mathsf{a}_m$ respectively. In this case the basis and its dual could be written as

$$\boldsymbol{\eta} = (\mathbf{e} \quad \mathbf{d}), \qquad \boldsymbol{\eta}^\dagger = \begin{pmatrix} \mathbf{e}^\dagger \\ \mathbf{d}^\dagger \end{pmatrix}, \tag{13.5.29}$$

where $\mathbf{e}$ is the row of excitation operators $\mathsf{e}_{mi} = \mathsf{a}_m^\dagger \mathsf{a}_i = \mathsf{E}_{mi}$ (the operator used in earlier chapters); similarly, $\mathbf{d}$ is the row of de-excitation operators $\mathsf{d}_{im} = \mathsf{a}_i^\dagger \mathsf{a}_m = \mathsf{E}_{mi}^\dagger$. The resolvent matrix is then, from (13.5.18),

$$\mathbf{R}(\omega) = [\boldsymbol{\eta}^\dagger \cdot (\hbar\omega\hat{1} - \hat{\mathsf{H}})\boldsymbol{\eta}]^{-1} = \left(\begin{array}{c|c} \mathbf{e}^\dagger \cdot (\hbar\omega 1 - \hat{\mathsf{H}})\mathbf{e} & \mathbf{e}^\dagger \cdot (\hbar\omega\hat{1} - \hat{\mathsf{H}})\mathbf{d} \\ \mathbf{d}^\dagger \cdot (\hbar\omega 1 - \hat{\mathsf{H}})\mathbf{e} & \mathbf{d}^\dagger \cdot (\hbar\omega\hat{1} - \hat{\mathsf{H}})\mathbf{d} \end{array}\right)^{-1}. \tag{13.5.30}$$

The elements in the first block are

$$\mathsf{E}_{mi}^\dagger \cdot (\hbar\omega\hat{1} - \hat{\mathsf{H}})\mathsf{E}_{nj} = \hbar\omega \mathsf{E}_{mi}^\dagger \cdot \mathsf{E}_{nj} - \mathsf{E}_{mi}^\dagger \cdot [\mathsf{H}, \mathsf{E}_{nj}]$$

and the scalar products are (with $\eta = +1$ for Bose-type operators)

$$\left.\begin{aligned} \mathsf{E}_{mi}^\dagger \cdot \mathsf{E}_{nj} &= \langle \Psi_0 | [\mathsf{E}_{mi}^\dagger, \mathsf{E}_{ni}] | \Psi_0 \rangle, \\ \mathsf{E}_{mi}^\dagger \cdot [\mathsf{H}, \mathsf{E}_{nj}] &= \langle \Psi_0 | [\mathsf{E}_{mi}^\dagger, [\mathsf{H}, \mathsf{E}_{nj}]] | \Psi_0 \rangle. \end{aligned}\right\} \tag{13.5.31a}$$

The first scalar product reduces easily to $\delta_{mn}\delta_{ij}$; giving a contribution $\hbar\omega\mathbf{1}$ in the first diagonal block of $\mathbf{R}(\omega)^{-1}$; the second is similar to those encountered in MC SCF theory (p. 279), where they determined the elements of the M block of the Hessian matrix, but appears to be less symmetrical. It is easily verified that the symmetric double-commutator (12.7.13) may be used *provided that* Ψ_0 is an exact eigenstate of H; in this case, using μ and ν for the double indices (mi, nj),

$$\left.\begin{aligned} \mathsf{E}_\mu^\dagger \cdot \mathsf{E}_\nu &= \langle \Psi_0 | [\mathsf{E}_\mu^\dagger, \mathsf{E}_\nu] | \Psi_0 \rangle = V_{\mu\nu}, \\ \mathsf{E}_\mu^\dagger \cdot [\mathsf{H}, \mathsf{E}_\nu] &= \langle \Psi_0 | [\mathsf{E}_\mu^\dagger, \mathsf{H}, \mathsf{E}_\nu] | \Psi_0 \rangle = M_{\mu\nu}, \end{aligned}\right\} \tag{13.5.31b}$$

where $V_{\mu\nu}$ and $M_{\mu\nu}$ are matrix elements already defined (p. 446) in the time-dependent form of MC SCF theory. Since the symmetric form becomes appropriate as the precision of the reference function increases, and arises naturally in the variational discussion of time-dependent phenomena, it might be argued that (13.5.31b) provides the more "correct" foundation for propagator calculations. It is also noteworthy that the reference state does not necessarily correspond to a single determinant; "occupied" and "unoccupied" spin-orbitals have been distinguished, and we are calculating a "particle–hole" propagator in which μ, ν are of (mi)-type—but a given ψ_i is not necessarily occupied in *all* determinants of a multideterminant reference function.

The other blocks of the matrix on the right in (13.5.30) are determined similarly and yield a resolvent matrix (and from (13.5.19) the propagator)

$$\mathbf{R}(\omega)^{-1} = \begin{pmatrix} \mathbf{M} & \mathbf{Q} \\ \mathbf{Q}^* & \mathbf{M}^* \end{pmatrix} - \hbar\omega \begin{pmatrix} \mathbf{V} & \mathbf{W} \\ -\mathbf{W}^* & -\mathbf{V}^* \end{pmatrix}, \tag{13.5.32}$$

where the off-diagonal blocks are also exactly as defined in Section 12.7.

It is now clear that the determination of the poles of the matrix $\mathbf{R}(\omega)$, and hence of the approximate polarization propagator, is equivalent to finding the "free-oscillation" solutions of the MC TDHF equation (12.7.8a): for $\mathbf{R}(\omega)$ will become infinite at the ω-values that make the determinant of the matrix on the right in (13.5.30) vanish—and these are simply the eigenvalues of

$$\begin{pmatrix} \mathbf{M} & \mathbf{Q} \\ \mathbf{Q}^* & \mathbf{M}^* \end{pmatrix}\begin{pmatrix} \mathbf{X} \\ \mathbf{Y} \end{pmatrix} = \hbar\omega\begin{pmatrix} \mathbf{V} & \mathbf{W} \\ -\mathbf{W}^* & -\mathbf{V}^* \end{pmatrix}\begin{pmatrix} \mathbf{X} \\ \mathbf{Y} \end{pmatrix}. \tag{13.5.33}$$

The eigencolumns determine linear combinations of the excitation and de-excitation operators that produce corresponding approximate excited states when working on $|\Psi_0\rangle$.

The Dyson equation

The previous example shows that in practice it is not necessary to perform a large matrix inversion to obtain the Green's function itself; it will usually be sufficient to look for the appropriate solutions of a familiar matrix eigenvalue equation. Historically, however, a common procedure was to relate the required Green's function (the electron propagator or the polarization propagator) to the corresponding quantity for a "model" system by using perturbation theory. For completeness, we indicate the essential features of this alternative approach to the calculation of the electron propagator.

For an IPM (Hartree–Fock) Hamiltonian H_0, the propagator in (13.5.27) would be exact, since extending the operator basis, for example to include terms of type $\mathsf{a}_m^\dagger\mathsf{a}_p^\dagger\mathsf{a}_i$, would correspond to admitting negative-ion states with an added electron in ψ_m *and* with an excitation $i \to p$, and such states would have *zero* overlap amplitudes with $\Psi_0\rangle$. The effect of using the full Hamiltonian $\mathsf{H} = \mathsf{H}_0 + \mathsf{H}'$ (see p. 293) is then (i) to introduce modified superoperator matrix elements, $\boldsymbol{\eta}^\dagger \cdot (\hbar\omega\hat{1} - \hat{\mathsf{H}})\boldsymbol{\eta}$, and (ii) to necessitate the expansion of the basis $\boldsymbol{\eta}$ to include $\mathsf{a}_m^\dagger\mathsf{a}_p^\dagger\mathsf{a}_i$ and higher terms. Let us therefore partition the basis by writing $\boldsymbol{\eta} = (\boldsymbol{\eta}_1 \quad \boldsymbol{\eta}_3)$, where $\boldsymbol{\eta}_1$ contains the *single* creation operators $\mathsf{a}_r^\dagger$ while $\boldsymbol{\eta}_3$ contains triple (and possibly higher) products. The matrix $\mathbf{M}(\omega)$ in (13.5.18) then assumes the form

$$\mathbf{M}(\omega) = (\boldsymbol{\eta}_1 \quad \boldsymbol{\eta}_3)^\dagger \cdot (\hbar\omega\hat{1} - \hat{\mathsf{H}})(\boldsymbol{\eta}_1 \quad \boldsymbol{\eta}_3) = \begin{pmatrix} \mathbf{M}_{11}(\omega) & \mathbf{M}_{12}(\omega) \\ \mathbf{M}_{21}(\omega) & \mathbf{M}_{22}(\omega) \end{pmatrix}, \tag{13.5.34}$$

where, for example,

$$\mathbf{M}_{11}(\omega) = \hbar\omega\mathbf{V}_{11} - \mathbf{H}_{11} = \hbar\omega\boldsymbol{\eta}_1^\dagger \cdot \boldsymbol{\eta}_1 - \boldsymbol{\eta}_1^\dagger \cdot \hat{\mathsf{H}}\boldsymbol{\eta}_1. \tag{13.5.35}$$

The elements of $\mathbf{M}_{11}(\omega)$ have been calculated already (p. 470)—including the terms from H', since a general Hamiltonian was assumed; and this block, on inversion, would therefore lead to a Green's matrix of IPM *form,* which may be denoted by $\mathbf{G}_0(\omega)$. The partitioned form of $\mathbf{M}(\omega)$ in (13.5.34) then suggests that we expand the inverse of the whole matrix in such a way that $\mathbf{M}_{11}^{-1}$ $(=\mathbf{G}_0)$ becomes the leading term.

To do this, we first note an important simplication; in any metric space, the Schmidt process (p. 33) can be used to ensure that two subspaces (in this case $\boldsymbol{\eta}_1$ and $\boldsymbol{\eta}_3$) are *orthogonal.* We therefore assume that (by adding suitable multiples of the $\boldsymbol{\eta}_1$ operators) the new $\boldsymbol{\eta}_3$ operators yield scalar-product matrices $\boldsymbol{\eta}_1^\dagger \cdot \boldsymbol{\eta}_3$ and $\boldsymbol{\eta}_3^\dagger \cdot \boldsymbol{\eta}_1$ that are identically zero. This means, since $\boldsymbol{\eta}_1 = (\mathsf{a}_1^\dagger \quad \mathsf{a}_2^\dagger \quad \ldots \quad \mathsf{a}_r^\dagger \quad \ldots)$, that $\mathsf{a}_r \cdot \eta_t^{(3)} = \eta_u^{(3)\dagger} \cdot \mathsf{a}_s^\dagger = 0$ for all values of the indices; consequently the scalar products in (13.5.22) will vanish unless η_t and η_u belong to the $\boldsymbol{\eta}_1$ set, and only the 11-block of the resolvent matrix $\mathbf{R}(\omega)$ will be required. Simple matrix algebra then shows (cf. Problem 2.19) that

$$\mathbf{R}_{11} = [\mathbf{M}^{-1}]_{11} = (\mathbf{M}_{11} - \mathbf{M}_{12}\mathbf{M}_{22}^{-1}\mathbf{M}_{21})^{-1}, \qquad (13.5.36)$$

in which the main term in any expansion of the inverse will indeed be $\mathbf{M}_{11}^{-1}$. To generate the expansion explicitly, we use the matrix identity

$$(\mathbf{A} - \mathbf{B})^{-1} = \mathbf{A}^{-1} + \mathbf{A}^{-1}\mathbf{B}(\mathbf{A} - \mathbf{B})^{-1},$$

and obtain

$$\mathbf{R}_{11} = \mathbf{M}_{11}^{-1}(\mathbf{1} + \mathbf{M}_{12}\mathbf{M}_{22}^{-1}\mathbf{M}_{21}\mathbf{R}_{11}) \qquad (13.5.37)$$

from which $\mathbf{R}_{11}(\omega)$ may be obtained by iteration. Since, however, $\boldsymbol{\eta}_1 = \mathbf{a}^\dagger$ and $\mathsf{a}_r \cdot \mathsf{a}_s^\dagger = \delta_{rs}$, (13.5.19) becomes $G_{rs}(\omega) = \langle\!\langle \mathsf{a}_r; \mathsf{a}_s^\dagger \rangle\!\rangle_\omega = (\mathbf{R}_{11}(\omega))_{rs}$; and then substitution in (13.5.35) gives

$$\mathbf{G}(\omega) = \mathbf{G}_0(\omega) + \mathbf{G}_0(\omega)\boldsymbol{\Sigma}(\omega)\mathbf{G}(\omega), \qquad (13.5.38)$$

where

$$\boldsymbol{\Sigma}(\omega) = \mathbf{M}_{12}(\omega)\mathbf{M}_{22}(\omega)^{-1}\mathbf{M}_{21}(\omega) \qquad (13.5.39)$$

and is usually referred to as a "self-energy" term. The result (13.5.38) is a *Dyson equation* for the Green's matrix; since $\mathbf{G}(\omega)$ itself appears in the "correction" term on the right, it must clearly be solved iteratively. Methods of solution, which usually involve perturbation expansions and partial summations by diagrammatic methods (cf. Chapter 9) have been discussed in great detail elsewhere (e.g. von Niessen *et al.*, 1984).

The three topics treated above must serve as an introduction to a vast and rapidly growing field: for further details and applications reference may be made to reviews by Csanak *et al.* (1971), Jørgensen (1975), Cederbaum and Domcke (1977), Öhrn and Born (1981), von Niessen *et al.* (1984) and Oddershede *et al.* (1984), all with full bibliographies.

13.6 EQUATION-OF-MOTION (EOM) METHODS

In the last section we introduced the concept of an operator space, with a well-defined metric, in which a particular type of mapping was produced by the Hamiltonian superoperator (or Liouvillian) $\hat{\mathsf{H}}$ according to the rule (13.5.3). This led to the possibility of calculating approximate propagators, and hence obtaining information on excitation, ionization and attachment processes, by purely algebraic methods. The same concepts may be used, however, in developing the so-called "equation-of-motion" (EOM) methods, which in some ways appear to be less sophisticated and more direct. The EOM approach, first proposed by Rowe (1968) and later extended by many others (for a review see McCurdy *et al.*, 1975) makes no explicit reference to propagators or response theory, but leads to working equations strikingly similar (sometimes identical) to those derived in previous sections. We therefore conclude this chapter with a brief exposition, emphasizing the links with other approaches.

The operator sets $\boldsymbol{\eta} = (\boldsymbol{\eta}_1 \quad \boldsymbol{\eta}_2 \quad \boldsymbol{\eta}_3 \quad \ldots)$ used in the last section contain products of $1, 2, 3, \ldots$ creation/annihilation operators, which act on an N-electron reference state $|0\rangle$, say, to produce other elements of Fock space, which describe the system (or its ions) with $N, N \pm 1, \ldots$ electrons. These elements are not eigenstates of the Hamiltonian, but of course may be combined to give approximations to actual state vectors, for both the neutral system and its ions. In the EOM approach generalized operators are introduced, which work on the ground state $|0\rangle$ to produce any desired excited state $|n\rangle$, neutral or ionic, and an attempt is made to determine these operators directly.

A typical operator, which creates an excitation, may be defined formally by

$$\mathsf{O}_n^\dagger |0\rangle = |n\rangle, \tag{13.6.1}$$

and it is evident that such operators exist, since they may be represented by ket–bra products

$$\mathsf{O}_n^\dagger = |n\rangle\langle 0|, \tag{13.6.2}$$

which clearly produce a multiple of $|n\rangle$ from any reference vector containing a non-zero component of $|0\rangle$. There is a corresponding de-excitation operator,

$$\mathsf{O}_n = |0\rangle\langle n|, \tag{13.6.3}$$

which must be the adjoint of $\mathsf{O}_n^\dagger$.

The effect of the superoperator $\hat{\mathsf{H}}$, acting on $\mathsf{O}_n^\dagger$, is to produce $[\mathsf{H}, \mathsf{O}_n^\dagger]$, and for exact state vectors it follows at once that

$$\hat{\mathsf{H}}\mathsf{O}_n^\dagger |0\rangle = \mathsf{H}\mathsf{O}_n^\dagger |0\rangle - \mathsf{O}_n^\dagger \mathsf{H} |0\rangle = (E_n - E_0)\mathsf{O}_n^\dagger |0\rangle. \tag{13.6.4}$$

In other words, when acting on the exact ground state,

$$\hat{\mathsf{H}}\mathsf{O}_n^\dagger = \Delta E_{0n}\mathsf{O}_n^\dagger \qquad (\Delta E_{0n} = E_n - E_0). \tag{13.6.5}$$

Thus $\mathsf{O}_n^\dagger$ is an *eigenoperator* of $\hat{\mathsf{H}}$, with the exact *excitation energy* as its eigenvalue.

To determine approximate solutions of (13.6.5), we may proceed in the usual way, expanding $\mathsf{O}_n^\dagger$ in terms of an operator basis as in Section 13.5. and seeking equations to determine the expansion coefficients. One possibility is to take a scalar product of (13.6.4) with $\mathsf{O}_n^\dagger |0\rangle$ from the left; it follows from the turnover rule that

$$\Delta E_{0n} = \frac{\langle 0|\, \mathsf{O}_n[\mathsf{H}, \mathsf{O}_n^\dagger]\, |0\rangle}{\langle 0|\, \mathsf{O}_n\mathsf{O}_n^\dagger\, |0\rangle}, \tag{13.6.6}$$

and this provides a useful functional, whose stationary values will (at a complete-set limit) coincide with the eigenvalues in (13.6.5). A second possibility is to use the commutator metric (13.5.8), in which case (13.6.6) is replaced by

$$\Delta E_{0n} = \frac{\langle 0|\, [\mathsf{O}_n, [\mathsf{H}, \mathsf{O}_n^\dagger]]_\pm\, |0\rangle}{\langle 0|\, [\mathsf{O}_n, \mathsf{O}_n^\dagger]_\pm\, |0\rangle}. \tag{13.6.7}$$

As usual, the choice of $\pm$ is open, but is usually made according to the type of operator (Bose or Fermi). The two functionals are clearly equivalent when $|0\rangle$ is an exact ground state, because the double-commutator term in (13.6.7) with O_n on the right will contain $\mathsf{O}_n\,|0\rangle = |0\rangle\langle n \mid 0\rangle = 0$.

It is instructive to make an immediate application (Smith and Day, 1975) to the calculation of ionization potentials. Thus, on putting ‡

$$\mathsf{O}_n^\dagger = \sum_r c_r^* \mathsf{a}_r = \mathsf{O}^+$$

and using the first form (13.6.6), the "excitation energy" (ΔE^+, say) for producing a *positive ion* (i.e. destroying an electron) becomes

$$\Delta E^+ = \frac{\sum\limits_{r,s} c_r^* c_s \langle 0|\, \mathsf{a}_s^\dagger[\mathsf{H}, \mathsf{a}_r]\, |0\rangle}{\sum\limits_{r,s} c_r^* c_s \langle 0|\, \mathsf{a}_s^\dagger \mathsf{a}_r\, |0\rangle} = \frac{\mathbf{c}^\dagger\mathbf{K}\mathbf{c}}{\mathbf{c}^\dagger\boldsymbol{\rho}\mathbf{c}}, \tag{13.6.8}$$

‡ The complex-conjugate coefficients are used only for notational consistency: according to (13.4.3), a destruction operator that removes an electron from a *linear combination* of basis orbitals will contain complex conjugates of the expansion coefficients.

where **K** is the Koopmans *matrix*, whose general element is

$$K_{rs} = \langle 0| \mathsf{a}_s^\dagger[\mathsf{H}, \mathsf{a}_r] |0\rangle = \langle 0| \mathsf{K}_{rs} |0\rangle, \tag{13.6.9}$$

being an expectation value of the Koopmans *operator* defined in (8.2.27). The denominator in (13.6.8) contains the 1-body density matrix, with elements $\rho_{rs} = \langle 0| \mathsf{a}_s^\dagger \mathsf{a}_r |0\rangle$ as noted in (5.4.20), and all expectation values are for the reference state $|0\rangle$.

The significance of the Koopmans operator, whose expectation value is given explicitly in (8.2.29) for any kind of wavefunction (exact or approximate), is now clear. According to (8.2.30), the elements of **K** coincide with those of $-\boldsymbol{\epsilon}$, the matrix of Lagrangian multipliers in the MC SCF equations; and they in turn are matrix elements of a 1-electron Hamiltonian containing the *effective* potential "felt" by any electron in the presence of the others. In full,

$$K_{rs} = -\left[\sum_{t} \langle r| \mathsf{h} |t\rangle \rho_{ts} + \sum_{t,u,v} \langle rt| g |uv\rangle \pi_{uv,st}\right]. \tag{13.6.10}$$

The condition for a stationary value of ΔE^+ in (13.6.8) would appear to be

$$\mathbf{Kc} = \Delta E^+ \boldsymbol{\rho}\mathbf{c}, \tag{13.6.11}$$

and the eigenvalues of the Koopmans matrix, with metric $\boldsymbol{\rho}$, should thus give the ionization potentials (exact or approximate) of the system described by $|0\rangle$. It is easy to verify that this "extended Koopmans theorem", first stated by Smith and Day (1975), gives the usual results (p. 164) for a single-determinant reference function, the matrices in (13.6.11) then referring to the occupied spin-orbitals. For then $-\mathbf{K}$ is the diagonal matrix of orbital energies obtained from the usual Fock operator, while $\boldsymbol{\rho}$ becomes the unit matrix; and consequently the ith eigenvalue is $\Delta E_i^+ = -\epsilon_i$, while the eigenoperator is a_i—which destroys the electron in spin-orbital ψ_i.

Although excellent numerical results are often obtained from multi-configuration reference functions (see e.g. Day *et al.*, 1975), the matrix **K** is not automatically Hermitian and may thus possess both right- and left-hand eigenvectors (cf. p. 250) with different eigenvalues. Fortunately, Hermitian symmetry can be obtained by completely *optimizing* the reference function, since this is a necessary and sufficient condition (p. 261) for optimal orbitals, but this may not always be feasible. Lack of Hermitian symmetry is indeed a recurrent problem in methods of EOM (and propagator) type in which approximate reference functions are used. The basic difficulty is to decide at what point in the formal theory to stop assuming $|0\rangle$ exact and to start making approximations.

Use of the alternative functional (13.6.7), which is the more conventional starting point for EOM methods (Rowe, 1968), does not remove such problems, but does offer certain advantages: these become clear in the application to ionization processes, where (13.6.8) is replaced by

$$\Delta E^{+} = \frac{\sum_{r,s} c_r^* c_s \langle 0|\,[\mathrm{a}_s^\dagger, [\mathrm{H}, \mathrm{a}_r]]_+\,|0\rangle}{\sum_{r,s} c_r^* c_s \langle 0|\,[\mathrm{a}_s^\dagger, \mathrm{a}_r]_+\,|0\rangle} = \frac{\mathbf{c}^\dagger \mathbf{K}' \mathbf{c}}{\mathbf{c}^\dagger \mathbf{c}}. \tag{13.6.12}$$

The eigenvalue equation corresponding to (13.6.11) is then

$$\mathbf{K}'\mathbf{c} = \Delta E^{+}\mathbf{c} \tag{13.6.13}$$

and is simpler than (13.6.11)—not only on account of the unit metric (i.e. orthonormal operator basis) but also because

$$K'_{rs} = \langle 0|\,[\mathrm{a}_s^\dagger, [\mathrm{H}, \mathrm{a}_r]]_+\,|0\rangle$$
$$= -\Big[\langle r|\,\mathrm{h}\,|s\rangle + \sum_{t,u} \langle rt \,\|\, su\rangle \rho_{ut}\Big], \tag{13.6.14}$$

which (cf. (13.6.10)) involves only the *one*-electron density matrix, being Fock-like even for a multiconfiguration reference function. Both simplifications arise from the anticommutators in (13.6.12), which lead to cancellations, but they are not peculiar to the case of Fermi-type operators—the outer commutators in (13.6.7) usually lead to equations containing densities of lower rank than those resulting from (13.6.6).

Let us return to the general EOM functional, recalling ((13.6.7) *et seq.*) that the alternative forms will be equivalent, even for an *approximate* reference function, provided that

$$\mathrm{O}_n\,|0\rangle = 0 \tag{13.6.15}$$

—often referred to as the "killer condition". The condition may in principle be satisfied without disturbing the validity of the expressions for ΔE_{on} by noting that

$$\tilde{\mathrm{O}}_n^\dagger = |n\rangle\langle 0| + \sum_{m(\neq 0)} c_m\,|n\rangle\langle m| = \mathrm{O}_n^\dagger + \Delta\mathrm{O}_n^\dagger \tag{13.6.16}$$

may be used in (13.6.6) in place of $\mathrm{O}_n^\dagger$, without changing the value of ΔE_{on}, when the coefficients are chosen appropriately. The arbitrariness in the definition of an approximate $\mathrm{O}_n^\dagger$ suggests that (13.6.7) may provide a satisfactory variational principle; but the operator in the numerator may still not be Hermitian, and ΔE_{0n} is therefore not necessarily even real! It is therefore usual to make a further modification of (13.6.7), noting that *for an exact reference function* the operator in the numerator may be replaced by $[[\mathsf{O}_n, \mathsf{H}], \mathsf{O}_n^\dagger]_-$ without disturbing the identity, and that by taking half the sum of the alternative forms we obtain an operator that is Hermitian-symmetric. The result is

$$\boxed{\Delta E_{0n} = \frac{\langle 0|\, [\mathsf{O}_n, \mathsf{H}, \mathsf{O}_n^\dagger]_\pm \,|0\rangle}{\langle 0|\, [\mathsf{O}_n, \mathsf{O}_n^\dagger]_\pm \,|0\rangle},} \tag{13.6.17}$$

which is real for an *arbitrary* reference state, and for any approximation to $\mathsf{O}_n^\dagger$, and gives the correct limit when $|0\rangle$ and $\mathsf{O}_n^\dagger$ are exact. The symmetric double commutator has been met already (p. 446), being

$$[\mathsf{A}, \mathsf{B}, \mathsf{C}]_\pm = \tfrac{1}{2}\{[[\mathsf{A}, \mathsf{B}], \mathsf{C}]_\pm + [\mathsf{A}, [\mathsf{B}, \mathsf{C}]]_\pm\}. \tag{13.6.18}$$

The excitation-energy expression (13.6.17) provides the general basis for variational EOM methods.

To proceed, we express the operator $\mathsf{O}_n^\dagger$ in terms of excitation and de-excitation operators ($\eta_\mu^\dagger$ and η_μ) of a general operator basis; they will be products of elementary creation and annihilation operators whose forms need not be made explicit at this stage. On putting

$$\mathsf{O}_n^\dagger = \sum_\mu (a_\mu \eta_\mu + b_\mu \eta_\mu^\dagger) \tag{13.6.19}$$

and varying the coefficients, the stationary-value condition $\delta(\Delta E_{0n}) = 0$ leads from (13.6.17) to a set of secular equations that may be written in a form very similar to that encountered (p. 444) in the discussion of time-dependent response, starting from the Frenkel principle.

The most common application of EOM procedures is to the determination of excitation energies and related transition quantities. In this case the lower sign is chosen in (13.6.18), since O_n and $\mathsf{O}_n^\dagger$ must be number-conserving operators, and the simplest operator basis to consider is that in which η_μ and $\eta_\mu^\dagger$ are elementary hole–particle pairs, $\eta_\mu = \mathsf{a}_m^\dagger \mathsf{a}_i = \mathsf{E}_\mu$ and $\eta_\mu^\dagger = \mathsf{a}_i^\dagger \mathsf{a}_m = \mathsf{E}_\mu^\dagger$, where i labels a spin-orbital that appears in the reference function while m refers to a complementary virtual set. To preserve a parallel with earlier equations, it is then convenient to write (13.6.19) in the form

$$\mathsf{O}_n^\dagger = \sum_\mu (X_\mu \mathsf{E}_\mu - Y_\mu \mathsf{E}_\mu^\dagger) \qquad (\mu = mi), \tag{13.6.20}$$

which resembles an infinitesimal rotation operator such as (8.2.20), except that in general $Y_\mu \neq X_\mu^*$ since the operator is no longer necessarily anti-Hermitian; indeed, the operator $\mathsf{O}_n^\dagger$ does not describe an infinitesimal fluctuation of the reference function, due to an oscillating perturbation, but rather a *finite* change that carries the system from the ground state to an excited state. On inserting (13.6.20) in (13.6.17), varying the numerical coefficients and requiring that $\delta(\Delta E_{0n})=0$, the resultant secular equations are found to be

$$\begin{pmatrix} \mathbf{M} & \mathbf{Q} \\ \mathbf{Q}^* & \mathbf{M}^* \end{pmatrix}\begin{pmatrix} \mathbf{X} \\ \mathbf{Y} \end{pmatrix} = \hbar\omega \begin{pmatrix} \mathbf{V} & \mathbf{W} \\ -\mathbf{W}^* & -\mathbf{V}^* \end{pmatrix}\begin{pmatrix} \mathbf{X} \\ \mathbf{Y} \end{pmatrix}, \qquad (13.6.21)$$

where

$$\left.\begin{aligned} M_{\mu\nu} &= \langle 0|\,[\mathsf{E}_{\mu}^{\dagger}, \mathsf{H}, \mathsf{E}_{\nu}]\,|0\rangle, & Q_{\mu\nu} &= -\langle 0|\,[\mathsf{E}_{\mu}^{\dagger}, \mathsf{H}, \mathsf{E}_{\nu}^{\dagger}]\,|0\rangle, \\ V_{\mu\nu} &= \langle 0|\,[\mathsf{E}_{\mu}^{\dagger}, \mathsf{E}_{\nu}]\,|0\rangle, & W_{\mu\nu} &= -\langle 0|\,[\mathsf{E}_{\mu}^{\dagger}, \mathsf{E}_{\nu}^{\dagger}]\,|0\rangle. \end{aligned}\right\} \qquad (13.6.22)$$

Equation (13.6.21) is identical with (12.7.8a), derived by considering "free oscillations" of the system in its reference state, and the matrix elements defined in (13.6.22) agree exactly with those obtained by variational methods, given in (12.7.12). The equations are also in complete agreement with those derived in the last section by propagator methods. In other words, at the level of approximation considered, EOM, propagator and MC TDHF methods all lead to the same working equations for discussing electronic excitation processes, and will therefore (given identical basis sets, reference functions and operator manifolds) give the same numerical predictions. For example, with a 1-determinant closed-shell reference function and a single-excitation manifold all methods give the standard RPA equations, while various types of correlated reference function, with the same excitation manifold, lead to various kinds of "higher" RPA equations (see e.g. Szabo and Ostlund, 1977). Some interesting numerical comparisons have been made by Chojnacki and Styrcz (1981), using a reference function $e^{\mathsf{U}}|0\rangle$ with various truncations of the exponential operator and classifying different types of HRPA accordingly; but such a form is not essential, and the expressions (13.6.22) provide a convenient and general interface between RPA-type equations and any correlated reference function that may be available.

Much more work remains to be done in this area and many points of detail require further discussion (for example the symmetrization of commutator expressions, the choice of operator manifolds, and the definition of the metric in operator space) to say nothing of the obvious need for efficient solution of very-high-order matrix equations. But the theoretical equivalence of various approaches, under well-defined assumptions and approximations, is a reassuring feature of the methods developed in this chapter. Some fundamental aspects of the methodology have been discussed by Löwdin (1985).

PROBLEMS 13

13.1 Verify that the evolution operator defined in (13.2.3) yields a wavefunction $\Psi^S(t) = \mathsf{U}(t)\Psi^S(0)$ that does indeed satisfy the time-dependent Schrödinger equation. Confirm also that all Schrödinger-operator relationships (e.g. $\mathsf{A}^S\mathsf{B}^S = \mathsf{C}^S$) are echoed by the Heisenberg operators $\mathsf{A} = \mathsf{U}^\dagger(t)\mathsf{A}^S\mathsf{U}(t)$ and that the latter develop in time according to (13.2.5).

13.2 Perform the necessary differentiations to show that the advanced, retarded and causal propagators all satisfy (13.2.14).

13.3 Obtain a "frequency form" of the equation discussed in Problem 13.2 by making a Fourier transformation. [*Hint:* Multiply by $e^{i\omega t}$ and integrate by parts with respect to τ in the interval $(-\infty, +\infty)$. Use a factor $e^{-\epsilon|\tau|}$ to take care of limits.]

13.4 Use the Schrödinger interpretation (p. 461) of the effect of the operator a_r on a many-electron state to show how the numerators in the Green's function (13.3.7) might be calculated in practice for a 2-electron system. [*Hint:* The negative-ion function Ψ_m^- will be a linear combination of 3-electron Slater determinants (e.g. $c_1|\psi_1\psi_2\psi_3| + c_2|\psi_1\psi_2\psi_6| + c_3|\psi_1\psi_3\psi_5| + \ldots$. What will be the effect of, say, a_2?]

13.5 Start from the result $\pi_{rs,tu} = \langle \mathsf{a}_t^\dagger \mathsf{a}_u^\dagger \mathsf{a}_s \mathsf{a}_r \rangle$ in (5.4.20) and use an argument parallel to that which leads to (13.3.10) to express the 2-electron density matrix in terms of an appropriate polarization propagator. [*Hint:* How should you choose the operators in (13.2.10) in order to obtain poles whose residues (numerators) will yield the desired elements? Use the anticommutation rules and remember that $\sum_n |n\rangle\langle n|$ is the unit operator.]

13.6 Green's functions originated in the study of differential equations; they are essentially kernels (cf. 175) representing operators G, such that (for one variable) $\mathsf{G}\phi(x) = \int G(x;x')\phi(x')\,dx'$. Show the following.

(i) An integral kernel can be associated with a (1-variable) Hamiltonian operator in the form $H(x;x') = \sum_n E_n\phi_n(x)\phi_n(x')$.

(ii) The operator that is defined through its kernel as $G(\omega\,|\,x;x') = \sum_n (\omega - E_n)^{-1}\phi_n(x)\phi_n^*(x')$ represents the *inverse operator* $(\omega\mathsf{I} - \mathsf{H})^{-1}$ in the sense that

$$(\omega\mathsf{I} - \mathsf{H})\mathsf{G}(\omega) = \mathsf{G}(\omega)(\omega\mathsf{I} - \mathsf{H}) = \mathsf{I},$$

where I is the unit operator with integral kernel $I(x';x'') = \sum_n \phi_n(x')\phi_n^*(x'')$ and is evidently a representation of the Dirac delta function $\delta(x' - x'')$.

(iii) The solutions of the equation $(\omega\mathsf{I} - \mathsf{H})\phi(x) = \rho(x)$ can be found, once $G(\omega\,|\,x;x')$ is known, in the form

$$\phi(x) = (\omega\mathsf{I} - \mathsf{H})^{-1}\rho(x) = \int G(\omega\,|\,x;x')\rho(x')\,dx'.$$

(iv) The poles of $G(\omega \mid x; x')$, the function satisfying $(\omega \mathsf{I} - \mathsf{H})G(\omega \mid x; x') = \delta(x - x')$, are the eigenvalues of H.

(v) The solution of Poisson's equation $\nabla^2\phi = \rho$ (for the potential due to a distribution of charge of density ρ) can be written down immediately in terms of the solution of Laplace's equation $\nabla^2\phi = 0$ for the field outside a single point charge. What is the equation for the Green's function and what is its solution?

13.7 Find a Schrödinger operator c_s equivalent to the Fock-space creation operator $\mathsf{a}_s^\dagger$. Hence show, using the form of the destruction operator d_r found on p. 462, that any creation/annihilation pair such as $\mathsf{a}_s^\dagger \mathsf{a}_r$ will give a result that is independent of the number of electrons in the ket upon which it acts (in conformity with the properties established and used in various chapters). [*Hint:* The operator will act on an N-electron function, adding an $(N+1)$th electron, and will be N-dependent. It must involve an antisymmetrizer for the $(N+1)$-electron function; and this may be "factorized"—see (14.1.4).]

13.8 Fill in the steps leading to the frequency form (13.4.11) of the ground-state energy formula. Indicate (cf. Fig. 13.1) various contours in the complex plane that could be used in the integration.

13.9 Show that the Hamiltonian superoperator has the property $\langle[\hat{\mathsf{H}}\mathsf{X}, \mathsf{Y}]\rangle = -\langle[\mathsf{X}, \hat{\mathsf{H}}\mathsf{Y}]\rangle$ when the expectation values refer to the exact ground state. Show also, on adopting the metric defined in (13.5.8), that $\hat{\mathsf{H}}$ is Hermitian in the sense that $(\hat{\mathsf{H}}\mathsf{B} \mid \mathsf{A}) = (\mathsf{B} \mid \hat{\mathsf{H}}\mathsf{A})$. Use these results to pass from (13.5.4) to (13.5.7).

13.10 Evaluate the commutator $[\mathsf{H}, \mathsf{a}_l^\dagger]$ and then the double commutator $[\mathsf{a}_k, [\mathsf{H}, \mathsf{a}_l^\dagger]]_+$. Obtain the expectation values that appear in (13.5.22) and complete the derivation of the Green's function (13.5.27). [*Hint:* Use the Hamiltonian (3.6.9) and the anticommutation rules to reduce all terms. Note how the "double-bar" quantities can be introduced and how the factor $\frac{1}{2}$ disappears in going to (13.5.24).]

13.11 The Green's function in Problem 13.10 is a first approximation. Show how a better approximation may be obtained by extending the operator basis to include terms such as $(\mathsf{a}_p^\dagger \mathsf{a}_q)\mathsf{a}_k^\dagger$, formulating equations analogous to those in Problem 13.10.

13.12 Evaluate explicitly the operator scalar products in (13.5.31a), in terms of density-matrix elements for the reference function. Verify that the symmetric double-commutator form (13.5.31b), would be exactly equivalent provided that the reference function were *exact*.

13.13 Extend the analysis in Problem 13.12 to the scalar products that determine the off-diagonal blocks in (13.5.30), and thus verify that the full matrices in (13.5.32) coincide with those obtained in the TD MC SCF approach (Section 12.7).

13.14 Verify the steps leading up to the Dyson equation (13.5.38), first establishing the expression (13.5.36), for the 11-block of the resolvent matrix, and the matrix identity that follows it. [*Hint:* Similar matrix manipulations were involved in Problem 2.19.]

13.15 In the formal derivation of EOM and propagator equations, the operator basis has normally contained creation/annihilation operators for electrons in *spin*-orbitals. Consider how spin might eventually be eliminated, taking as an example the EOM equation (13.6.8) for the determination of ionization potentials. Hence obtain a spin-free form of (13.6.11). [*Hint:* Use a more explicit notation $a^{\dagger}_{r\sigma}$, $a_{s\sigma}$ $(\sigma = \alpha, \beta)$ for the operators, referring back to Section 8.2 (p. 262), where "spin-traced" combinations were introduced.]

13.16 Use a 1-determinant closed-shell reference function to obtain a corresponding approximation to the spin-free Koopmans matrix obtained in Problem 13.15 and show that the eigenvalues of **K** coincide with the usual "Koopmans theorem" IPs. Then consider an all-pair-excitation reference function (Problem 6.19) to obtain a simple multiconfiguration method for improving the results. [*Hint:* In each case substitute appropriate density-matrix elements into the basic equations.]

13.17 Repeat the derivations in Problem 13.16, using instead the double-commutator definition (13.6.14) of the Koopmans matrix. Show that the predicted IPs will be unchanged in the 1-determinant approximation: then obtain new results for the all-pair-excitation reference function.

13.18 Expand the operator basis used in defining the ionization operator O_n^+ and derive new equations, parallel to (13.6.8)–(13.6.11), (13.6.12) and (13.6.13) for a 1-determinant reference function. Confirm that the double-commutator forms are simpler. [*Hint:* Look back at Problem 13.11, where a similar extension of the basis was made.]

13.19 Re-examine the equations occurring in Problems 13.16–13.18, to see what changes will be necessary in using the *symmetric* double-commutator expressions, which follow from (13.6.17), instead of the earlier forms.

13.20 Fill in the details of the demonstration, sketched in the text, that use of the EOM excitation operator (13.6.20) leads back to the same equation for the excitation energies (namely (13.6.21)) as that obtained in Section 12.7 from the linear-response approach.

14 Intermolecular Forces

14.1 WEAK INTERACTIONS. GENERALIZED PRODUCT FUNCTIONS

So far, we have been concerned only with single molecules and their properties. But most of chemistry is concerned with *interacting* molecules, usually in the fluid phase, and it is therefore fitting that, before reaching the end of this book, some attention be given to the subject of interactions. The interactions that we shall consider are "weak" in the sense that the interacting systems retain their individuality, at least in good approximation. Thus water molecules in the liquid phase are still water molecules; naphthalene molecules in solid naphthalene are still naphthalene molecules; and so on: and even, say, different CH_3 groups in the same molecule are still recognizable as CH_3 groups. Indeed, the definition of "weakly interacting systems" may be extended to include, with varying degrees of precision, atomic inner shells, lone pairs, bond pairs, and all similar entities that may be recognized by certain characteristic properties that are not too sensitive to molecular environment. The great importance of such interactions, which are sometimes called collectively "non-bonded interactions", cannot be overemphasized; their dominant role in molecular biology, for example, is well known (Monod, 1971; see also Hobdza and Zahradnik, 1980). The problem we approach in this chapter is how to set up a mathematical formalism for dealing *not* with covalent bonding but rather with *non*-covalent bonding.

At the quantum-mechanical level, two systems A and B will be recognizably distinct if each is described by "its own" wavefunction; thus

$$\Phi(\boldsymbol{x}_1, \boldsymbol{x}_2, \ldots, \boldsymbol{x}_N) = \Phi^{\mathrm{A}}(\boldsymbol{x}_1, \ldots, \boldsymbol{x}_{N_{\mathrm{A}}})\Phi^{\mathrm{B}}(\boldsymbol{x}_{N_{\mathrm{A}}+1}, \ldots, \boldsymbol{x}_{N_A+N_{\mathrm{B}}})$$

would describe a system containing $N_{\mathrm{A}} + N_{\mathrm{B}}$ electrons, with system A containing the first "group" of N_{A} electrons in a state represented by Φ^{A}, and system B containing the remaining N_{B} electrons in a state represented by Φ^{B}. Obviously, such a product would not be appropriate, except at very large distances (such that electrons $1, 2, \ldots, N_{\mathrm{A}}$ could be regarded as belonging unequivocally to system A), because it violates the

Pauli principle; but it may be "symmetry-adapted" by applying the antisymmetrizer (3.1.7). On adding a normalizing factor, and generalizing to the case of any number of interacting electron groups, we may write

$$\Phi(\boldsymbol{x}_1, \boldsymbol{x}_2, \ldots, \boldsymbol{x}_N) = M\mathsf{A}[\Phi^{\mathrm{A}}(\boldsymbol{x}_1, \ldots, \boldsymbol{x}_{N_\mathrm{A}})\Phi^{\mathrm{B}}(\boldsymbol{x}_{N_\mathrm{A}+1}, \ldots, \boldsymbol{x}_{N_\mathrm{A}+N_\mathrm{B}}) \ldots] \tag{14.1.1}$$

as a properly antisymmetric wavefunction for the composite system ABC . . . , capable of yielding an exact energy

$$E = E^{\mathrm{A}} + E^{\mathrm{B}} + E^{\mathrm{C}} + \ldots$$

for a Hamiltonian

$$\mathsf{H} = \mathsf{H}^{\mathrm{A}} + \mathsf{H}^{\mathrm{B}} + \mathsf{H}^{\mathrm{C}} + \ldots$$

in which the groups are so well separated that interaction terms are negligible.

It will be recognized that a 1-determinant wavefunction may be regarded as a special case of (14.1.1) in which each "group" consists of a single electron in a corresponding spin-orbital. Thus

$$\Phi(\boldsymbol{x}_1, \boldsymbol{x}_2, \ldots, \boldsymbol{x}_N) = M\mathsf{A}[\psi_{\mathrm{A}}(\boldsymbol{x}_1)\psi_{\mathrm{B}}(\boldsymbol{x}_2) \ldots]$$

is equivalent to

$$\Phi(\boldsymbol{x}_1, \boldsymbol{x}_2, \ldots, \boldsymbol{x}_N) = MN!^{-1} \det |\psi_{\mathrm{A}}(\boldsymbol{x}_1) \;\; \psi_{\mathrm{B}}(\boldsymbol{x}_2) \;\; \ldots|$$

Evidently, the antisymmetrized product of "group functions" may be regarded as a generalization of the antisymmetrized spin-orbital product, the individual factors now describing whole groups of electrons instead of a single electron. If the electron groups are in fact weakly interacting (for example as a result of their physical separation in space) then a *generalized product* of the form (14.1.1) may be an exceedingly accurate wavefunction because each factor may, in principle, describe correlation within the group of electrons to which it refers. A further refinement may be added, however, by allowing the *mixing* of a number of generalized product functions of the form (14.1.1), each product corresponding to a particular selection of "states" for the individual groups. This mixing is entirely analogous to configuration interaction in the usual sense, where each Slater determinant corresponds to a particular selection of spin-orbitals or *one*-electron states. We shall therefore consider more generally wavefunctions of the form

$$\Psi = \sum_{\kappa} c_\kappa \Phi_\kappa, \tag{14.1.2}$$

where Φ_κ is a generalized product function corresponding to a particular assignment of states to the various electron groups,

$$\Phi_\kappa = M_\kappa \mathsf{A}[\Phi_a^{\mathrm{A}}\Phi_b^{\mathrm{B}}\ldots] \qquad (\kappa = \mathrm{A}a,\ \mathrm{B}b, \ldots). \tag{14.1.3}$$

Here the group function Φ_r^{R} describes group R in state r, and κ denotes a particular selection of group states, in exactly the same way that κ in (3.1.4) and (3.1.5) denoted a particular selection of spin-orbitals.

It will normally be assumed that the group functions Φ_r^{R} are individually antisymmetric, and in this case it is useful to write A in terms of *partial* antisymmetrizers $\mathsf{A}_\mathrm{A}, \mathsf{A}_\mathrm{B}, \ldots, \mathsf{A}_\mathrm{R}, \ldots$ for the separate groups. If we write (cf. (3.1.7)) $\mathsf{A}_\mathrm{R} = N_\mathrm{R}^{-1}\sum \epsilon_\mathsf{P}\mathsf{P}$, where P is a permutation of variables in Φ_r^{R} only, it is well known from permutation theory that

$$\mathsf{A} = \frac{N_\mathrm{A}!N_\mathrm{B}!\ldots}{N!}\mathsf{A}'\mathsf{A}_\mathrm{A}\mathsf{A}_\mathrm{B}\ldots, \tag{14.1.4}$$

where $\mathsf{A}' = \sum \epsilon_\mathsf{T}\mathsf{T}$ is a sum of transpositions that exchange variables *between different groups*. With antisymmetric group functions the partial antisymmetrizers simply multiply by unity, and may be discarded, and it is then only necessary in forming (14.1.3), to replace A by A', which contains a relative small number of distinct transpositions.

Generalized products can be handled with the same facility as spin-orbital products only if we impose certain orthogonality requirements; these are more stringent than those imposed on ordinary spin-orbitals and take the form

$$\int \Phi_r^{\mathrm{R}}(\boldsymbol{x}_1, \boldsymbol{x}_i, \boldsymbol{x}_j, \ldots)^* \Phi_s^{\mathrm{S}}(\boldsymbol{x}_1, \boldsymbol{x}_k, \boldsymbol{x}_l, \ldots)\,\mathrm{d}\boldsymbol{x}_1 \equiv 0 \qquad (\mathrm{R} \neq \mathrm{S}). \tag{14.1.5a}$$

This means that the result of integrating over any one variable $\boldsymbol{x}_1$, common to two different group functions, must vanish identically *for all values of the other variables* $\boldsymbol{x}_i, \boldsymbol{x}_j, \ldots$, and is evidently a much stronger condition than the usual orthonormality requirement

$$\int \Phi_r^{\mathrm{R}}(\boldsymbol{x}_1, \boldsymbol{x}_2, \ldots, \boldsymbol{x}_{N_\mathrm{R}})^* \Phi_{r'}^{\mathrm{R}}(\boldsymbol{x}_1, \boldsymbol{x}_2, \ldots, \boldsymbol{x}_{N_\mathrm{R}})\,\mathrm{d}\boldsymbol{x}_1\,\mathrm{d}\boldsymbol{x}_2\ldots\mathrm{d}\boldsymbol{x}_{N_\mathrm{R}} = \delta_{rr'}, \tag{14.1.5b}$$

which will normally be assumed for the different state functions permitted *within* each group. The *strong-orthogonality* condition (14.1.5a), first discussed by Parr *et al.* (1956), is not as strange as may at first appear; if we have $m_\mathrm{R} + m_\mathrm{S}$ orthonormal spin-orbitals, and we make any determinant Φ_r^{R} from the first m_R of them, and any Φ_s^{S} from the remaining m_S, then Φ_r^{R} and Φ_s^{S} are stong-orthogonal. In general, if a spin-orbital set is partitioned into disjoint subsets (i.e. no two subsets having a spin-orbital in common) then functions built from any two of the subsets A, B, . . . , R, . . . will satisfy the strong orthogonality requirement. It is now apparent that the expansion (14.1.2) cannot be complete; since if every group function were expanded in terms of its own spin-orbitals then each function Φ_κ (and hence Ψ itself) would consist of determinants of a particular type, always containing m_A spin-orbitals from set A, m_B from set B, etc.; but in a *complete* set expansion of Ψ there would be determinants with, for example, $m_A + 1$ spin-orbitals from set A and only $m_B - 1$ from set B. Clearly, functions of the form (14.1.2), with a *fixed* partitioning of the electrons into groups, disregard the possibility of "electron transfer" between groups; on the other hand, when such effects are important the groups between which electron transfer must be admitted have effectively lost their "individuality", and the group-function description would be physically inappropriate. The value of the group-function approach, with the strong-orthogonality restriction, is therefore largely dependent upon making an intelligent choice of the electron groups in terms of which to describe the system.

There are other analogies between group-function expansions (14.1.2) and spin-orbital expansions. A single generalized product may be a good wavefunction if the individual group functions are chosen variationally, just as a single determinant may be a good approximation if the spin-orbitals satisfy the Hartree–Fock equations; however, sometimes, owing to (space or spin) symmetry, a number of generalized products may be equally eligible and must be grouped into linear combinations with symmetry-determined coefficients, just as the determinants of an open-shell configuration must be suitably "vector-coupled" before they can properly represent actual states of the system.

The further development of group-function theory clearly requires the determination of matrix-element expressions, the procedure and results again running entirely parallel to those of the Slater method. The results are expressed most succinctly in terms of the density (and transition density) matrices, those for the whole system now being expressible in terms of those for the component electron groups. As in Section 5.4, we write $\rho(\kappa\kappa' \mid \mathbf{x}_1;\mathbf{x}_1')$, $\pi(\kappa\kappa' \mid \mathbf{x}_1, \mathbf{x}_2;\mathbf{x}_1', \mathbf{x}_2')$ for the transition densities connecting Φ_κ and $\Phi_{\kappa'}$ while those for the individual groups are defined analogously, so that $\rho_R(rr' \mid \mathbf{x}_1:\mathbf{x}_1')$, for example, is a transition density for group R alone, connecting states Φ_r^R and $\Phi_{r'}^R$. As in Slater's rules (Section 3.3), there are only three cases to consider: (i) κ' not differing from κ in any group; (ii) κ' different from κ in one group R, say; and (iii) κ' different from κ in two groups R and S, say. If there are differences in three or more groups, the 1- and 2-electron transition densities vanish as a result of the strong orthogonality (14.1.5a). The other results are

(McWeeny, 1959, 1960):

$$
\left.
\begin{aligned}
&\text{(i)}\ \kappa=\kappa'=(a,b,\ldots,r,\ldots,s,\ldots)\\
&\qquad \rho(\kappa\kappa\mid \boldsymbol{x}_1;\boldsymbol{x}_1')=\sum_{\mathrm{R}}\rho_{\mathrm{R}}(rr\mid \boldsymbol{x}_1;\boldsymbol{x}_1'),\\
&\quad \pi(\kappa\kappa\mid \boldsymbol{x}_1,\boldsymbol{x}_2;\boldsymbol{x}_1',\boldsymbol{x}_2')=\sum_{\mathrm{R}}\pi_{\mathrm{R}}(rr\mid \boldsymbol{x}_1,\boldsymbol{x}_2;\boldsymbol{x}_1',\boldsymbol{x}_2')\\
&\qquad\qquad +(1-\mathsf{P}_{12})\sum_{\mathrm{R,S}}{}'\rho_{\mathrm{R}}(rr\mid \boldsymbol{x}_1;\boldsymbol{x}_1')\rho_{\mathrm{S}}(ss\mid \boldsymbol{x}_2;\boldsymbol{x}_2');\\
&\text{(ii)}\ \kappa=(a,b,\ldots,r,\ldots),\ \kappa'=(a,b,\ldots,r',\ldots)\\
&\qquad \rho(\kappa\kappa'\mid \boldsymbol{x}_1;\boldsymbol{x}_1')=\rho_{\mathrm{R}}(rr'\mid \boldsymbol{x}_1;\boldsymbol{x}_1'),\\
&\quad \pi(\kappa\kappa'\mid \boldsymbol{x}_1,\boldsymbol{x}_2;\boldsymbol{x}_1',\boldsymbol{x}_2')=\pi_{\mathrm{R}}(rr'\mid \boldsymbol{x}_1,\boldsymbol{x}_2;\boldsymbol{x}_1',\boldsymbol{x}_2')\\
&\qquad\qquad +(1-\mathsf{P}_{12})\sum_{\mathrm{S}(\neq\mathrm{R})}[\rho_{\mathrm{R}}(rr'\mid \boldsymbol{x}_1;\boldsymbol{x}_1')\rho_{\mathrm{S}}(ss\mid \boldsymbol{x}_2;\boldsymbol{x}_2')\\
&\qquad\qquad +\rho_{\mathrm{S}}(ss\mid \boldsymbol{x}_1;\boldsymbol{x}_1')\rho_{\mathrm{R}}(rr'\mid \boldsymbol{x}_2;\boldsymbol{x}_2')];\\
&\text{(iii)}\ \kappa=(a,b,\ldots,r,\ldots,s,\ldots),\ \kappa''=(a,b,\ldots,r',\ldots,s',\ldots)\\
&\quad \pi(\kappa\kappa''\mid \boldsymbol{x}_1,\boldsymbol{x}_2;\boldsymbol{x}_1',\boldsymbol{x}_2')=(1-\mathsf{P}_{12})[\rho_{\mathrm{R}}(rr'\mid \boldsymbol{x}_1:\boldsymbol{x}_1')\rho_{\mathrm{S}}(ss'\mid \boldsymbol{x}_2;\boldsymbol{x}_2')\\
&\qquad\qquad +\rho_{\mathrm{S}}(ss'\mid \boldsymbol{x}_1;\boldsymbol{x}_1')\rho_{\mathrm{R}}(rr'\mid \boldsymbol{x}_2;\boldsymbol{x}_2')];
\end{aligned}
\right\}
\tag{14.1.6}
$$

where P_{12} interchanges the unprimed variables $\boldsymbol{x}_1$ and $\boldsymbol{x}_2$. Here the density matrices for the separate groups are defined exactly as in Section 5.4, and it must be stressed again (cf. p. 117) that the variables $\boldsymbol{x}_1$, $\boldsymbol{x}_2$, $\boldsymbol{x}_1'$, $\boldsymbol{x}_2'$, label points in configuration space, not the coordinates of specific particles.

It is readily verified that in the special case where each group consists of only *one* electron, in a spin-orbital $\psi_{\mathrm{R}r}$ $(=\Phi_r^{\mathrm{R}})$ these results reduce to the expressions that were derived from Slater's rules (Section 5.4). Thus for $\kappa'=\kappa$ we note that $\pi_{\mathrm{R}}(rr\mid \boldsymbol{x}_1,\boldsymbol{x}_2;\boldsymbol{x}_1',\boldsymbol{x}_2')$ does not occur, as there is only one electron in each group, and that

$$\rho_{\mathrm{R}}(rr\mid \boldsymbol{x}_1;\boldsymbol{x}_1')=\psi_{\mathrm{R}r}(\boldsymbol{x}_1)\psi_{\mathrm{R}r}^*(\boldsymbol{x}_1'),$$

and consequently

$$
\begin{aligned}
\pi(\kappa\kappa\mid \boldsymbol{x}_1,\boldsymbol{x}_2;\boldsymbol{x}_1',\boldsymbol{x}_2')=\sum_{\mathrm{R,S}}{}'\,&[\psi_{\mathrm{R}r}(\boldsymbol{x}_1)\psi_{\mathrm{R}r}^*(\boldsymbol{x}_1')\psi_{\mathrm{S}s}(\boldsymbol{x}_2)\psi_{\mathrm{S}s}^*(\boldsymbol{x}_2')\\
&-\psi_{\mathrm{R}r}(\boldsymbol{x}_2)\psi_{\mathrm{R}r}^*(\boldsymbol{x}_1')\psi_{\mathrm{S}s}(\boldsymbol{x}_1)\psi_{\mathrm{S}s}^*(\boldsymbol{x}_2')],
\end{aligned}
$$

which (apart from the slight difference of notation) is identical with case (i) of (5.4.14).

The correspondence with Slater's results is seen most clearly in the expressions for the matrix elements of the Hamiltonian H, the only non-zero results being as follows:

$$\left.\begin{aligned}
&\text{(i)} \quad H_{\kappa\kappa} = \sum_{\mathrm{R}} H^{\mathrm{R}}(rr) + \tfrac{1}{2}\sum_{\mathrm{R,S}}{}' \, [J^{\mathrm{RS}}(rr, ss) - K^{\mathrm{RS}}(rr, ss)],\\
&\text{(ii)} \quad H_{\kappa'\kappa} = H^{\mathrm{R}}(rr') + \sum_{S(\neq R)} [J^{\mathrm{RS}}(rr', ss) - K^{\mathrm{RS}}(rr', ss)],\\
&\text{(iii)} \quad H_{\kappa''\kappa} = J^{\mathrm{RS}}(rr', ss') - K^{\mathrm{RS}}(rr', ss'),
\end{aligned}\right\} \qquad (14.1.7)$$

where κ' and κ'' refer to configurations differing from κ in one group (R, with $r \rightarrow r'$) and two groups (R, S, with $r \rightarrow r'$ and $s \rightarrow s'$) respectively, and, with the usual brief notation (5.2.8),

$$H^{\mathrm{R}}(rr') = \int_{\boldsymbol{x}_1' \rightarrow \boldsymbol{x}_1} \mathsf{h}(1)\rho_{\mathrm{R}}(rr' \mid \boldsymbol{x}_1; \boldsymbol{x}_1')\, \mathrm{d}\boldsymbol{x}_1$$

$$+ \tfrac{1}{2}\int g(1,2)\pi_{\mathrm{R}}(rr' \mid \boldsymbol{x}_1, \boldsymbol{x}_2)\, \mathrm{d}\boldsymbol{x}_1\, \mathrm{d}\boldsymbol{x}_2, \qquad (14.1.8)$$

$$J^{\mathrm{RS}}(rr', ss') = \int g(1,2)\rho_{\mathrm{R}}(rr' \mid \boldsymbol{x}_1)\rho_{\mathrm{S}}(ss' \mid \boldsymbol{x}_2)\, \mathrm{d}\boldsymbol{x}_1\, \mathrm{d}\boldsymbol{x}_2, \qquad (14.1.9)$$

$$K^{\mathrm{RS}}(rr', ss') = \int g(1,2)\rho_{\mathrm{R}}(rr' \mid \boldsymbol{x}_2; \boldsymbol{x}_1)\rho_{\mathrm{S}}(ss' \mid \boldsymbol{x}_1; \boldsymbol{x}_2)\, \mathrm{d}\boldsymbol{x}_1\, \mathrm{d}\boldsymbol{x}_2. \qquad (14.1.10)$$

It is evident that $H^{\mathrm{R}}(rr')$ has the usual form of a matrix-element expression of a Hamiltonian *for* N_{R} *electrons of group* R *alone*, taken between states r and r': for if we write

$$\mathsf{H}^{\mathrm{R}} = \sum_{i=1}^{N_{\mathrm{R}}} \mathsf{h}(i) + \tfrac{1}{2}\sum_{i,j=1}^{N_{\mathrm{R}}}{}' \, g(i,j) \qquad (14.1.11)$$

then comparison of (14.1.8) with (5.4.3) and (5.4.4) shows that

$$H^{\mathrm{R}}(rr') = \langle \Phi_{\mathrm{R}r'} | \, \mathsf{H}^{\mathrm{R}} \, | \Psi_{\mathrm{R}r} \rangle. \qquad (14.1.12)$$

This reduces simply to $\langle \psi_{\mathrm{R}r'} | \, \mathsf{h} \, | \psi_{\mathrm{R}r} \rangle$ when $N_{\mathrm{R}} = 1$, and the first terms in (14.1.7) (i) and (ii) therefore correspond to the matrix elements of the 1-electron Hamiltonian in Slater's rules. On the other hand, $J^{\mathrm{RS}}(rr', ss')$ and $K^{\mathrm{RS}}(rr', ss')$ correspond to generalizations of the 2-electron coulomb and exchange integrals. Thus $J^{\mathrm{RS}}(rr', ss')$ is the electrostatic interaction energy of two charge distributions whose densities are $\rho_R(rr' \mid \boldsymbol{x}_1)$ and $\rho_S(ss' \mid \boldsymbol{x}_2)$ respectively, namely transition densities for $r \to r'$ in group R and $s \to s'$ in group S. Since $g(1, 2)$ $(=1/r_{12})$ is spinless, and J^{RS} depends only on diagonal elements $(\boldsymbol{x}_1' = \boldsymbol{x}_1, \boldsymbol{x}_2' = \boldsymbol{x}_2)$, the spin integrations in (14.1.9) may be completed immediately to yield

$$J^{RS}(rr', ss') = \int g(1, 2) P_R(rr' \mid \boldsymbol{r}_1) P_S(ss' \mid \boldsymbol{r}_2)\, d\boldsymbol{r}_1\, d\boldsymbol{r}_2, \quad (14.1.13)$$

in which the densities are in ordinary three-dimensional space. The exchange integral does not reduce immediately, since it involves *off-diagonal* elements of the density matrices, and the effect of spin integration then depends on the spin states of the two groups; if these are both singlet states, the density matrices take the two-component form (5.3.12), and reduction then gives

$$K^{RS}(rr', ss') = \tfrac{1}{2} \int g(1, 2) P_R(rr' \mid \boldsymbol{r}_2; \boldsymbol{r}_1) P_S(ss' \mid \boldsymbol{r}_1; \boldsymbol{r}_2)\, d\boldsymbol{r}_1\, d\boldsymbol{r}_2. \quad (14.1.14)$$

For electron groups that do not overlap appreciably the exchange integrals are usually small; their significance is discussed in the next section.

The matrix-element expressions given above provide the basis for group-function calculations completely parallel to those based on Slater determinants. Accordingly, we first consider in more detail the variational determination of an optimum *single* generalized product, corresponding to the one determinant of Hartree–Fock theory, and then the nature of the refinements introduced by "configuration interaction" (where the term "configuration" now refers to a specification of *group states*).

14.2 VARIATION THEORY FOR SUBSYSTEMS

We expect to find a group-function analogue of the Hartree–Fock method, by starting with a single generalized product (14.1.3) and seeking a minimum value of the corresponding variational energy expression: from (14.1.7)(i) this is seen to be

$$E = \sum_{R} H^R(rr) + \sum_{R<S} [J^{RS}(rr, ss) - K^{RS}(rr, ss)]. \quad (14.2.1)$$

To simplify matters, we take first a two-group system AB, with wavefunction

$$\Psi = M\mathsf{A}'[\Phi_a^A \Phi_b^B]$$

and energy

$$E = H^{A}(aa) + H^{B}(bb) + J^{AB}(aa, bb) - K^{AB}(aa, bb),$$

and require that E be stationary for first-order variation of Φ_a^A and Φ_b^B subject to orthonormality of the two functions. Since each variation makes its own first-order change in E, we need consider only one; let us take $\Phi_a^A \to \Phi_a^A + \delta\Phi_a^A$, where $\delta\Phi_a^A$ is a variation constructed from A-group orbitals alone, so that Φ_a^A will remain automatically strong-orthogonal to Φ_b^B. It is then useful to write

$$E = E_{\text{eff}}^{A} + E^{B},$$

where

$$E_{\text{eff}}^{A} = H^{A}(aa) + J^{AB}(aa, bb) - K^{AB}(aa, bb),$$
$$E^{B} = H^{B}(bb).$$

Thus E^B is the energy the B-group electrons would have *by themselves* in the field of the nuclei, while E_{eff}^{A} is the energy of the A-group electrons *in the presence of those belonging to* group B. Variation of the A-group wavefunction affects all three parts of E_{eff}^{A}, but leaves E^B quite unchanged. The stationary-value condition is thus

$$E_{\text{eff}}^{A} = H^{A}(aa) + J^{AB}(aa, bb) - K^{AB}(aa, bb) = \text{stationary value},$$

subject to

$$\langle \Phi_a^A \mid \Phi_a^A \rangle = 1.$$

An exactly similar criterion, obtained by interchanging A, B and a, b, applies to the optimum B-group wavefunction, and the same argument applies to the general energy expression (14.2.1) to give a variation condition *for each group in the presence of all other groups*. Thus, in general, the energy to be minimized to optimize group R is

$$E_{\text{eff}}^{R} = H^{R}(rr) + \sum_{S(\neq R)} [J^{RS}(rr, ss) - K^{RS}(rr, ss)]. \tag{14.2.2}$$

This means that the whole wavefunction may be optimized by considering just one group at a time, replacing an N-electron calculation effectively by a succession of smaller calculations on systems of $N_A, N_B, \ldots$ electrons. To complete this reduction, we express the interaction terms in (14.2.2) as matrix elements of actual *one*-electron operators describing the potential due to the electrons outside group R. Let us introduce for any group S the following operators, defined by their effect on any spin-orbital ψ:

$$\left.\begin{aligned} \mathsf{J}^{S}(1)\psi(\boldsymbol{x}_1) &= \left[\int g(1, 2)\rho_S(ss \mid \boldsymbol{x}_2; \boldsymbol{x}_2)\, d\boldsymbol{x}_2\right]\psi(\boldsymbol{x}_1), \\ \mathsf{K}^{S}(1)\psi(\boldsymbol{x}_1) &= \int g(1, 2)\rho_S(ss \mid \boldsymbol{x}_1; \boldsymbol{x}_2)\psi(\boldsymbol{x}_2)\, d\boldsymbol{x}_2. \end{aligned}\right\} \tag{14.2.3}$$

Clearly, $\mathsf{J}^{\mathrm{S}}(1)$ just multiplies $\psi(\boldsymbol{x}_1)$ by the value of the potential at point $\boldsymbol{x}_1$ due to electrons distributed according to the density function for group S in state s. On the other hand $\mathsf{K}^{\mathrm{S}}(1)$ is an integral operator, of the kind used in Hartree–Fock theory (Sections 6.1 and 6.4). These two operators are the "coulomb" and "exchange" operators for an electron in the effective field due to the electrons of group S. It is now possible to write the interaction terms in (14.2.2) in the form

$$\left.\begin{aligned} J^{\mathrm{RS}}(rr, ss) &= \left\langle \Phi_{\mathrm{R}r} \middle| \sum_{i=1}^{N_{\mathrm{R}}} \mathsf{J}^{\mathrm{S}}(i) \middle| \Phi_{\mathrm{R}r} \right\rangle, \\ K^{\mathrm{RS}}(rr, ss) &= \left\langle \Phi_{\mathrm{R}r} \middle| \sum_{i=1}^{N_{\mathrm{R}}} \mathsf{K}^{\mathrm{S}}(i) \middle| \Phi_{\mathrm{R}r} \right\rangle; \end{aligned}\right\} \tag{14.2.4}$$

namely as expectation values of that part of the energy of the N_{R} electrons of group R that arise from interaction (coulomb and exchange respectively) with the electrons of group S. The R-group electrons thus behave as if they were quite alone, but each in a field described by the (1-electron) Hamiltonian

$$\mathsf{h}_{\mathrm{eff}}^{\mathrm{R}} = \mathsf{h} + \sum_{\mathrm{S}(\neq \mathrm{R})} (\mathsf{J}^{\mathrm{S}} - \mathsf{K}^{\mathrm{S}}) \tag{14.2.5}$$

instead of h. The complete R-group energy expression (14.2.2) may then be written as

$$E_{\mathrm{eff}}^{\mathrm{R}} = \langle \Phi_r^{\mathrm{R}} | \, \mathsf{H}_{\mathrm{eff}}^{\mathrm{R}} \, | \Phi_r^{\mathrm{R}} \rangle, \tag{14.2.6}$$

where

$$\mathsf{H}_{\mathrm{eff}}^{\mathrm{R}} = \sum_{i=1}^{N_{\mathrm{R}}} \mathsf{h}_{\mathrm{eff}}^{\mathrm{R}}(i) + \tfrac{1}{2} \sum_{i,j=1}^{N_{\mathrm{R}}}{}' \, g(i,j). \tag{14.2.7}$$

the stationary-value condition for $E_{\mathrm{eff}}^{\mathrm{R}}$, subject to Φ_r^{R} remaining normalized, is of course equivalent to the usual unconstrained stationary-value problem

$$E_{\mathrm{eff}}^{\mathrm{R}} = \frac{\langle \Phi_r^{\mathrm{R}} | \, \mathsf{H}_{\mathrm{eff}}^{\mathrm{R}} \, | \Phi_r^{\mathrm{R}} \rangle}{\langle \Phi_t^{\mathrm{R}} \, | \, \Phi_r^{\mathrm{R}} \rangle} = \text{stationary value}. \tag{14.2.8}$$

All groups other than R have been formally eliminated, their presence being absorbed into the effective one-electron Hamiltonian (14.2.5), and the optimization of the R-group wavefunction is essentially an N_{R}-electron problem.

The preceding result is extremely powerful, and is not at all difficult to apply. Suppose, for example, that we construct Φ_r^{R} from spin-orbitals ψ_i^{R}: then in working with group R we may simply forget about all the other electrons, provided that we replace matrix elements $\langle \psi_i^{\mathrm{R}} | \, \mathsf{h} \, | \psi_j^{\mathrm{R}} \rangle$

between the basic R-group spin-orbitals by

$$\langle \psi_i^{\rm R} |\, \mathsf{h}_{\rm eff}^{\rm R} \,| \psi_j^{\rm R} \rangle = \langle \psi_i^{\rm R} |\, \mathsf{h} \,| \psi_j^{\rm R} \rangle + \left\langle \psi_i^{\rm R} \middle| \sum_{\rm S(\neq R)} (\mathsf{J}^{\rm S} - \mathsf{K}^{\rm S}) \middle| \psi_j^{\rm R} \right\rangle. \quad (14.2.9)$$

When the S-group density matrix is expressed in the form

$$\rho_{\rm S}(ss \mid \boldsymbol{x};\boldsymbol{x}') = \sum_{k,l} \rho_{kl}^{\rm S} \psi_k^{\rm S}(\boldsymbol{x}) \psi_l^{\rm S*}(\boldsymbol{x}')$$

the matrix elements of $\mathsf{J}^{\rm S}$ and $\mathsf{K}^{\rm S}$ then reduce to

$$\left.\begin{aligned} \langle \psi_i^{\rm R} |\, \mathsf{J}^{\rm S} \,| \psi_j^{\rm R} \rangle &= \sum_{k,l} \rho_{kl}^{\rm S} \langle \psi_i^{\rm R} \psi_l^{\rm S} |\, g \,| \psi_j^{\rm R} \psi_k^{\rm S} \rangle, \\ \langle \psi_i^{\rm R} |\, \mathsf{K}^{\rm S} \,| \psi_j^{\rm R} \rangle &= \sum_{k,l} \rho_{kl}^{\rm S} \langle \psi_i^{\rm R} \psi_l^{\rm S} |\, g \,| \psi_k^{\rm S} \psi_j^{\rm R} \rangle, \end{aligned}\right\} \quad (14.2.10)$$

and are thus simply linear combinations of 2-electron integrals connecting the two groups, the coefficients being elements of the charge-density matrix for group S. If we consider a non-degenerate ground state (cf. the closed-shell Hartree–Fock case), and a spinless Hamiltonian, these results may be further reduced: the matrix elements between basis *orbitals* $\phi_j^{\rm R}$ (all spins eliminated) are

$$\boxed{\langle \phi_i^{\rm R} |\, \mathsf{h}_{\rm eff}^{\rm R} \,| \phi_j^{\rm R} \rangle = \langle \phi_i^{\rm R} |\, \mathsf{h} \,| \phi_j^{\rm R} \rangle + \left\langle \phi_i^{\rm R} \middle| \sum_{\rm S(\neq R)} (\mathsf{J}^{\rm S} - \mathsf{K}^{\rm S}) \middle| \phi_j^{\rm R} \right\rangle,} \quad (14.2.11)$$

where

$$\left.\begin{aligned} \langle \phi_i^{\rm R} |\, \mathsf{J}^{\rm S} \,| \phi_j^{\rm R} \rangle &= \sum_{k,l} P_{kl}^{\rm S} \langle \phi_i^{\rm R} \phi_l^{\rm S} |\, g \,| \phi_j^{\rm R} \phi_k^{\rm S} \rangle, \\ \langle \phi_i^{\rm R} |\, \mathsf{K}^{\rm S} \,| \phi_j^{\rm R} \rangle &= \tfrac{1}{2} \sum_{k,l} P_{kl}^{\rm S} \langle \phi_i^{\rm R} \phi_l^{\rm S} |\, g \,| \phi_k^{\rm S} \phi_j^{\rm R} \rangle, \end{aligned}\right\} \quad (14.2.12)$$

and the elements of $\mathbf{P}^{\rm S}$ are simply the "charges and bond orders" for group S in state s, possibly calculated with extensive CI.

The variational determination of optimum group functions is a well-defined problem, which may solved iteratively in much the same way as the SCF problem. From plausible initial approximations to the group wavefunctions $\Phi_a^{\rm A}, \Phi_b^{\rm B}, \ldots, \Phi_r^{\rm R}, \ldots,$ the group density matrices are constructed and, using (14.2.11) and (14.2.12), the matrix elements of the effective 1-electron Hamiltonian $\mathsf{h}_{\rm eff}^{\rm R}$ are formed for each group; each group function is then optimized, so as to satisfy (14.2.8), and, after improving all functions, the cycle is repeated. Iteration is continued until the computed density matrices agree, to any prescribed accuracy, with those used in the preceding cycle. Convergence of this process is normally quite rapid (see e.g. Klessinger and McWeeny, 1965).

The potential superiority of a single generalized product, over the single determinant of Hartree–Fock theory, rests upon the fact that individual groups may in principle be described by wavefunctions of high accuracy, giving due recognition to electron correlation within each group. To see why this is so, we need only consider the pair function $\Pi(\boldsymbol{r}_1, \boldsymbol{r}_2)$, obtained by integrating over spins in (14.1.6)(i). On assuming for simplicity a singlet state, and dropping the state label κ, we find

$$\Pi(\boldsymbol{r}_1, \boldsymbol{r}_2) = \sum_{\mathrm{R}} \Pi_{\mathrm{R}}(rr \mid \boldsymbol{r}_1, \boldsymbol{r}_2) + \sum_{\mathrm{R,S}}{}' [P_{\mathrm{R}}(rr \mid \boldsymbol{r}_1) P_{\mathrm{S}}(ss \mid \boldsymbol{r}_2) - \tfrac{1}{2} P_{\mathrm{R}}(rr \mid \boldsymbol{r}_2; \boldsymbol{r}_1) P_{\mathrm{S}}(ss \mid \boldsymbol{r}_1; \boldsymbol{r}_2)]. \qquad (14.2.13)$$

Now if the different electron groups are localized (cf. p. 204) in different regions of space (corresponding to not too strongly interacting groups), it may be assumed that a density function for any group becomes small when the variables refer to points "outside" that group. Suppose then that $\boldsymbol{r}_1$ and $\boldsymbol{r}_2$ both refer to points within the R-group region; in this case

$$\Pi(\boldsymbol{r}_1, \boldsymbol{r}_2) \simeq \Pi_{\mathrm{R}}(rr \mid \boldsymbol{r}_1, \boldsymbol{r}_2) \quad \text{(pair function for R-group electrons).}$$

On the other hand, if $\boldsymbol{r}_1$ is a point in R but $\boldsymbol{r}_2$ is in S then

$$\Pi(\boldsymbol{r}_1, \boldsymbol{r}_2) \simeq P_{\mathrm{R}}(rr \mid \boldsymbol{r}_1) P_{\mathrm{S}}(ss \mid \boldsymbol{r}_2) \quad \begin{pmatrix} \text{pair function for one} \\ \text{electron in R, the other in S} \end{pmatrix}.$$

In other words, when two electrons are close together in the same group they are described by a pair function appropriate to that group alone; when they are far apart, in different groups, the pair function is just the *product* of the probabilities of finding them at their separate places. This is exactly the kind of flexibility that we need in setting up wavefunctions that can correctly describe electron correlation in a molecule with physically separate groups of electrons. Even when the groups are not completely separate, the situation is still fairly satisfactory since in the region of overlap between R and S the last term in (14.2.13) becomes significant and introduces at least the Fermi correlation (Section 5.8) between electrons of like spin.

Quantum chemistry contains countless examples of situations in which the group-function approach is applicable. In particular, it provides a natural and mathematically rigorous basis for the various "models" originally proposed on intuitive grounds, in which theoretical discussion of a system is simplified by confining attention to only a few electrons (for example the valence electrons, or the π-electrons of a conjugated molecule).

14.3 CONFIGURATION INTERACTION. INTERPRETATION OF THE TERMS

We now examine the possibility of transcending the single-term approximation considered in the last section by admitting other terms in the expansion (14.1.2); and, in view of the close analogy with the orbital approximations used in earlier chapters, we then refer to the admission of "configuration interaction", where now the "excited" configurations are associated with local excitations in which one, two or several groups may be described as "excited-state" functions $\Phi_{a'}^{A}, \Phi_{b'}^{B}, \ldots$ $(a', b', \ldots \neq a, b, \ldots)$. It will also be possible to estimate the energy corrections due to admixture of such configurations by the partitioning method of Section 2.5. Thus from (2.5.6) and (14.1.7) it is clear that, to first order, only the functions with excitations in *one* or *two* groups will mix with the one-term approximation $\Psi = \Phi_\kappa$. We use κ' and κ'' to denote singly and doubly excited products, and thus require the matrix elements evaluated in (14.1.7). It is convenient to express the second of these in terms of the effective Hamiltonian introduced in (14.2.7), and the two results required are then

(*i*) κ, κ' *differing in group* R

$$\langle \Phi_{\kappa'} | \mathsf{H} | \Phi_\kappa \rangle = H^{\mathrm{R}}(rr') + \sum_{\mathrm{S}(\neq \mathrm{R})} [J^{\mathrm{RS}}(rr', ss) - K^{\mathrm{RS}}(rr', ss)] = H^{\mathrm{R}}_{\mathrm{eff}}(rr'), \tag{14.3.1}$$

where

$$H^{\mathrm{R}}_{\mathrm{eff}}(rr') = \langle \Phi_{Rr'} | \mathsf{H}^{\mathrm{R}}_{\mathrm{eff}} | \Phi_{\mathrm{R}r} \rangle \tag{14.3.2}$$

and formally involves only the group in which the single excitation occurs; and

(*ii*) κ, κ'' *differing in groups* R, S

$$\langle \Phi_{\kappa''} | \mathsf{H} | \Phi_\kappa \rangle = J^{\mathrm{RS}}(rr', ss') - K^{\mathrm{RS}}(rr', ss'). \tag{14.3.3}$$

The energy after configuration interaction follows from (2.5.4) as

$$E = E^{(0)} + E^{(1)} + E^{(2)} + \ldots, \tag{14.3.4}$$

where $E^{(0)}$ is the initial approximation given by (14.2.1) while $E^{(1)}$ and $E^{(2)}$ arise from singly and doubly excited configurations respectively:

$$E^{(1)} = -\sum_{\mathrm{R}} \sum_{r'(\neq r)} \frac{|H^{\mathrm{R}}_{\mathrm{eff}}(rr')|^2}{\Delta E(\mathrm{R}r \to \mathrm{R}r')}, \tag{14.3.5}$$

$$E^{(2)} = -\sum_{\mathrm{R<S}} \sum_{\substack{r'(\neq r) \\ s'(\neq s)}} \frac{|J^{\mathrm{RS}}(rr', ss') - K^{\mathrm{RS}}(rr', ss')|^2}{\Delta E(\mathrm{R}r \to \mathrm{R}r', \mathrm{S}s \to \mathrm{S}s')}. \tag{14.3.6}$$

The "excitation energies" in the denominators are

$$\left.\begin{aligned} &\Delta E(\mathrm{R}r \to \mathrm{R}r') = H_{\kappa'\kappa'} - H_{\kappa\kappa} \quad (\kappa' = \mathrm{A}a, \mathrm{B}b, \ldots, \mathrm{R}r', \ldots), \\ &\Delta E(\mathrm{R}r \to \mathrm{R}r', \mathrm{S}s \to \mathrm{S}s') = H_{\kappa''\kappa''} - H_{\kappa\kappa} \\ &\qquad\qquad (\kappa'' = \mathrm{A}a, \mathrm{B}b, \ldots, \mathrm{R}r', \ldots, \mathrm{S}s', \ldots), \end{aligned}\right\} \tag{14.3.7}$$

and may easily be deduced from the diagonal matrix-element expression (i) in (14.1.7).

These results have an interesting significance. If the group functions have been determined variationally, so that (14.2.8) is satisfied for arbitrary variations $\Phi_r^{\mathrm{R}} \to \Phi_r^{\mathrm{R}} + \delta\Phi_r^{\mathrm{R}}$ within the subspace of functions belonging to group R, then it follows easily (expressing $\delta\Phi_r^{\mathrm{R}}$ in terms of the $\Phi_{r'}^{\mathrm{R}}$) that

$$H_{\mathrm{eff}}^{\mathrm{R}}(rr') = \langle \Phi_{r'}^{\mathrm{R}} | \mathsf{H}_{\mathrm{eff}}^{\mathrm{R}} | \Phi_r^{\mathrm{R}} \rangle = 0, \tag{14.3.8}$$

and hence that the single-excitation contribution $E^{(1)}$ must vanish. This is the group-function analogue of Brillouin's theorem (6.1.29) *et seq.* Since admixture of only the *singly* excited functions modifies (i.e. "polarizes") the *charge density*, as follows from the presence of the term $\rho_{\mathrm{R}}(rr' \mid \boldsymbol{x}_1; \boldsymbol{x}_1')$ in (14.1.6)(ii), it is clear that the charge density cannot be further improved to this order of perturbation theory: in other words, *polarization* of each group by the presence of the others has been fully accounted for by optimizing the one-configuration approximation. On the other hand, the double-excitation contribution $E^{(2)}$ is normally non-zero and is associated with modification of the pair function (as may be seen from (iii) in (14.1.6)); it derives essentially from coulomb interactions (with exchange corrections) between *transition charge densities* located in the various electron groups and is formally similar to the "dispersion energy" introduced by London (1930). Usually, in the theory of intermolecular forces, such interactions are discussed using a transition *dipole* approximation, but the present results are rigorously valid, without the use of multipole expansions and without neglect of exchange effects, *as long as the strong-orthogonality condition is satisfied.*

The most attractive features of the approach developed so far are (i) the possibility of separating the total electronic energy into "exact" *molecular* energies together with terms representing intermolecular effects, and (ii) the identification of the interactions in terms of quantities with a more immediate physical significance. Thus for two systems A and

B the one-configuration energy (14.2.1) may be written as ‡

$$E = E_a^{\mathrm{A}} + E_b^{\mathrm{B}} + E_{ab}^{\mathrm{AB}}(\mathrm{elec}) + E_{ab}^{\mathrm{AB}}(\mathrm{ex}), \tag{14.3.9}$$

where, for example, E_a^{A} is the energy of molecule A in state *a in the absence of molecule B,* while the third term is a classically computed *electrostatic* interaction and the final term is a small exchange correction—which rapidly goes to zero as the overlap of the two systems diminishes. The expressions for the last two terms in (14.3.9) are

$$E_{ab}^{\mathrm{AB}}(\mathrm{elec}) = E^{\mathrm{AB}}(\mathrm{nuc}) + \int V_{\mathrm{B}}(1)P_{\mathrm{A}}(aa \mid \boldsymbol{r}_1)\,\mathrm{d}\boldsymbol{r}_1 + \int V_{\mathrm{A}}(1)P_{\mathrm{B}}(bb \mid \boldsymbol{r}_1)\,\mathrm{d}\boldsymbol{r}_1 + \int g(1,2)P_{\mathrm{A}}(aa \mid \boldsymbol{r}_1)P_{\mathrm{B}}(bb \mid \boldsymbol{r}_2)\,\mathrm{d}\boldsymbol{r}_1\,\mathrm{d}\boldsymbol{r}_2, \tag{14.3.10}$$

$$E_{ab}^{\mathrm{AB}}(\mathrm{ex}) = -\tfrac{1}{2}\int g(1,2)P_{\mathrm{A}}(aa \mid \boldsymbol{r}_2;\boldsymbol{r}_1)P_{\mathrm{B}}(bb \mid \boldsymbol{r}_1;\boldsymbol{r}_2)\,\mathrm{d}\boldsymbol{r}_1\,\mathrm{d}\boldsymbol{r}_2, \tag{14.3.11}$$

and the reason for the designation "electrostatic" is evident: the terms in $E_{ab}^{\mathrm{AB}}(\mathrm{elec})$ refer respectively to nuclear–nuclear (A and B) repulsions, nuclear B and charge-cloud A attractions and vice versa, and charge-cloud (A and B) repulsions.

When the one-configuration result is modified by adding single- and double-excitation terms, as in (14.3.4), the *interaction* energy, which is $E - E_a^{\mathrm{A}} - E_b^{\mathrm{B}}$, may be expressed as

$$E_{\mathrm{int}} = E_{ab}^{\mathrm{AB}}(\mathrm{elec}) + E_{ab}^{\mathrm{AB}}(\mathrm{ex}) + E_{ab}^{\mathrm{AB}}(\mathrm{pol}) + E_{ab}^{\mathrm{AB}}(\mathrm{disp}), \tag{14.3.12}$$

where $E_{ab}^{\mathrm{AB}}(\mathrm{pol})$ and $E_{ab}^{\mathrm{AB}}(\mathrm{disp})$ are the terms defined in (14.3.5) and (14.3.6) respectively. In principle, E_{int} in (4.3.12) could be evaluated with *exact* molecular wavefunctions, and would lead to a correspondingly "exact" second-order interaction (for non-overlapping systems). In practice, of course, approximate wavefunctions must be employed; but at least they can now be used directly in calculating the quantity of interest (E_{int}), which is usually an infinitesimal part of the total electronic energy, thus avoiding the familiar difficulty of estimating a small interaction as a difference of two exceedingly large energies.

There are two main difficulties in exploiting this approach: (i) extended sets of excited-state functions, needed in the second-order sums, are not

‡ The electron–nuclear potential energy is $V = V_{\mathrm{A}} + V_{\mathrm{B}}$ (a sum of terms for the nuclei of A and B respectively), and the reduction follows on using (14.2.8)–(14.2.10) with integration over spins. It is convenient to also include the nuclear interactions (a constant term) at this point. The systems are assumed to be in non-degenerate ground states.

usually available even in approximate form; and (ii) the strong-orthogonality condition is violated as soon as the different systems begin to overlap, however slightly. In the next section we deal with the second-order sum in a different way, using response theory, and in the following section (14.5) with non-orthogonality effects.

14.4 LONG-RANGE EFFECTS. THE DISPERSION ENERGY

For charged or highly polar species the electrostatic and polarization terms in (14.3.12) are dominant: the former is easy to calculate, given good ground-state electron densities, but the latter is of second order and is consequently more difficult as noted already. For uncharged and non-polar species the long-range interaction is dominated by the dispersion energy, and this again is difficult to calculate. In this section we first establish the general approach, by obtaining an expression for the dispersion energy in terms of the FDPs of Section 12.1; and then show that the polarization energy may be written in terms of the same FDPs.

The dispersion energy

With only two groups, A and B, the dispersion term (14.3.6) becomes

$$E_{ab}^{\mathrm{AB}}(\mathrm{disp}) = -\sum_{a'b'} \frac{|\langle \mathrm{A}a'\mathrm{B}b'|\,\mathsf{H}\,|\mathrm{A}a\mathrm{B}b\rangle|^2}{\Delta E(\mathrm{A}a \to \mathrm{A}a';\, \mathrm{B}b \to \mathrm{B}b')}, \tag{14.4.1}$$

with numerator given by (14.3.3). For negligible overlap,

$$\langle \mathrm{A}a'\mathrm{B}b'|\,\mathsf{H}\,|\mathrm{A}a\mathrm{B}b\rangle = \frac{e^2}{4\pi\epsilon_0}\int \frac{P_{\mathrm{A}}(aa' \mid \boldsymbol{r}_1)(P_{\mathrm{B}}(bb' \mid \boldsymbol{r}_2)}{r_{12}}\, \mathrm{d}\boldsymbol{r}_1\, \mathrm{d}\boldsymbol{r}_2 \tag{14.4.2}$$

—the exchange term in (14.3.6) being discarded. The Hamiltonian H refers to the *whole system* AB, and neither "unperturbed part" nor perturbation need be defined; nor is it necessary to develop r_{12}^{-1} to obtain the familiar multipolar expansion—although this is very commonly done.

Instead of using the transition densities explicitly, we revert to a matrix-element form of (14.4.2), noting that in second quantization (Problem 14.4)

$$P_{\mathrm{A}}(aa' \mid \boldsymbol{r}_1) = \sum_{r,s} \langle \Phi_{a'}^{\mathrm{A}}|\, \mathsf{E}_{rs}^{\mathrm{A}}\, |\Phi_{s}^{\mathrm{A}}\rangle\, \phi_s^{\mathrm{A}}(\boldsymbol{r}_1)\phi_r^{\mathrm{A}}(\boldsymbol{r}_1)^*, \tag{14.4.3}$$

where E_{rs} is the spin-traced operator defined in (10.4.9), whose

Schrödinger form (10.4.11) is an integral operator:

$$\mathsf{E}^{\mathrm{A}}_{rs}(i) \rightarrow \mathsf{E}^{\mathrm{A}}_{rs}(\boldsymbol{r}_i; \boldsymbol{r}'_i) = \phi^{\mathrm{A}}_r(\boldsymbol{r}_i)\phi^{\mathrm{A}}_s(\boldsymbol{r}'_i)^*. \tag{14.4.4}$$

This operator, as usual, replaces orbital ϕ^{A}_s by ϕ^{A}_r wherever it may appear in Φ^{A}_a (destroying any terms in which ϕ^{A}_s is absent). Similar equations apply to system B; and the orbital sets $\{\phi^{\mathrm{A}}_r\}$, $\{\phi^{\mathrm{B}}_r\}$ used in describing the two systems are disjoint (the systems being assumed to be non-overlapping) but otherwise arbitrary. Evidently, (14.4.3) may be written as

$$\left.\begin{aligned} P_{\mathrm{A}}(aa' \mid \boldsymbol{r}_1) &= \langle \Phi^{\mathrm{A}}_{a'} \mid \mathsf{d}^{\mathrm{A}}_{\boldsymbol{r}_1} \mid \Phi^{\mathrm{A}}_a \rangle, \\ \mathsf{d}^{\mathrm{A}}_{\boldsymbol{r}_1} &= \sum_{r,s} \mathsf{E}^{\mathrm{A}}_{rs}\phi^{\mathrm{A}}_s(\boldsymbol{r}_1)\phi^{\mathrm{A}}_r(\boldsymbol{r}_1)^* \end{aligned}\right\} \tag{14.4.5}$$

and $\mathsf{d}^{\mathrm{A}}_{\boldsymbol{r}_1}$ is the (spinless) density *operator* whose expectation value in a given state yields the electron density at point $\boldsymbol{r}_1$, while for a pair of states the matrix element yields the *transition* density at point $\boldsymbol{r}_1$.

The density operators may now be used in (14.4.2) to give

$$\langle \mathrm{A}a'\mathrm{B}b' | \mathsf{H} | \mathrm{A}a\mathrm{B}b \rangle = \frac{e^2}{4\pi\epsilon_0} \int \langle \Phi^{\mathrm{A}}_{a'} | \mathsf{d}^{\mathrm{A}}_{\boldsymbol{r}_1} | \Phi^{\mathrm{A}}_a \rangle \langle \Phi^{\mathrm{B}}_{b'} | \mathsf{d}^{\mathrm{B}}_{\boldsymbol{r}_2} | \Phi_b \rangle r_{12}^{-1}\, \mathrm{d}\boldsymbol{r}_1\, \mathrm{d}\boldsymbol{r}_2, \tag{14.4.6}$$

and on using this, and a similar result for system B, in (14.4.1), we obtain

$$E^{\mathrm{AB}}_{ab}(\mathrm{disp}) = -\left(\frac{e^2}{4\pi\epsilon_0}\right)^2 \sum_{a'b'} \times \int \frac{\langle \Phi^{\mathrm{A}}_a | \mathsf{d}^{\mathrm{A}}_{\boldsymbol{r}'_1} | \Phi^{\mathrm{A}}_{a'} \rangle \langle \Phi^{\mathrm{A}}_{a'} | \mathsf{d}^{\mathrm{A}}_{\boldsymbol{r}_1} | \Phi^{\mathrm{A}}_a \rangle \langle \Phi^{\mathrm{B}}_b | \mathsf{d}^{\mathrm{B}}_{\boldsymbol{r}'_2} | \Phi^{\mathrm{B}}_{b'} \rangle \langle \Phi^{\mathrm{B}}_b | \mathsf{d}^{\mathrm{B}}_{\boldsymbol{r}_2} | \Phi^{\mathrm{B}}_b \rangle}{\hbar(\omega^{\mathrm{A}}_{aa'} + \omega^{\mathrm{B}}_{bb'}) r_{12} r'_{12}} \times \mathrm{d}\boldsymbol{r}_1\, \mathrm{d}\boldsymbol{r}_2\, \mathrm{d}\boldsymbol{r}'_1\, \mathrm{d}\boldsymbol{r}'_2, \tag{14.4.7}$$

where we have written

$$\hbar\omega^{\mathrm{A}}_{aa'} = E^{\mathrm{A}}_{a'} - E^{\mathrm{A}}_a, \qquad \hbar\omega^{\mathrm{B}}_{bb'} = E^{\mathrm{B}}_{b'} - E^{\mathrm{B}}_b \tag{14.4.8}$$

for the excitation energies. It should be noted that, although, for example, $\hbar\omega^{\mathrm{A}}_{aa'}$ is an excitation energy for molecule A *in the presence of* molecule B, such energies are typically of the order of several electron volts, and therefore virtually no error is incurred by using free-molecule excitation energies (cf. Fig. 14.1).

The products in the numerator of (14.4.7) are reminiscent of those that occur in the polarization propagator used in (13.3.12). There is one product for molecule A, which would determine how a perturbation at point $\boldsymbol{r}_1$ propagated to point $\boldsymbol{r}'_1$ and a similar product for molecule B. But

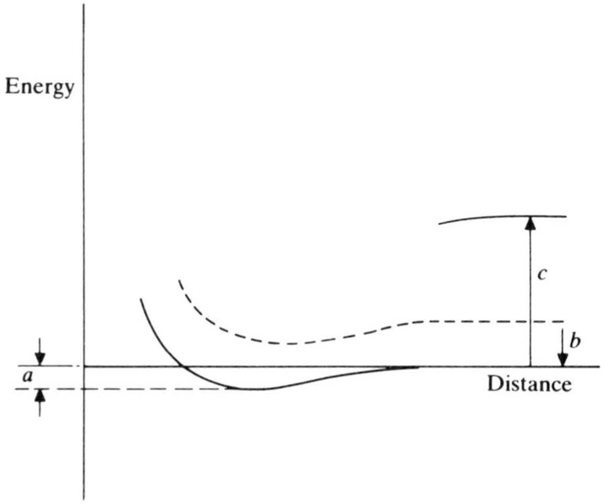

Fig. 14.1 Energy relationships for weakly interacting molecules. Solid curves indicate a ground-state potential-energy surface and part of an excited-state surface; the broken line being an approximation (e.g. Hartree–Fock) to the former. Typically, the required interaction energy (a) is less than 10^{-4} of the correlation error (b), while the correlation and excitation energies (b, c) are of the same order of magnitude (several eV).

the sum over states cannot be performed for each molecule separately because the denominator involves *both* molecules and does not factorize into a product of an A part and a B part: if factorization were possible then the summations could be performed and would lead to a product of two quantities recognizable as FDPs for the separate molecules. In fact, however, a product form can be obtained at no great expense by using the integral transform

$$\frac{1}{x+y} = \frac{2}{\pi}\int_0^\infty \frac{x}{x^2+\omega^2}\frac{y}{y^2+\omega^2}\,\mathrm{d}\omega, \qquad (14.4.9)$$

in which x- and y-*factors* appear in the integrand. By applying this transformation to (14.4.7), we obtain (McWeeny, 1984)

$$E_{ab}^{\mathrm{AB}}(\mathrm{disp}) = -\frac{\hbar}{2\pi}\left(\frac{e^2}{4\pi\epsilon_0}\right)^2 \int \frac{\mathrm{d}\boldsymbol{r}_1\,\mathrm{d}\boldsymbol{r}_2\,\mathrm{d}\boldsymbol{r}_1'\,\mathrm{d}\boldsymbol{r}_2'}{r_{12}r_{12}'}$$

$$\times \int_0^\infty \Pi_A(\mathrm{d}_{r_1'}^{\mathrm{A}}\mathrm{d}_{r_1}^{\mathrm{A}} \mid i\omega)\Pi_B(\mathrm{d}_{r_2'}^{\mathrm{B}}\mathrm{d}_{r_2}^{\mathrm{B}} \mid i\omega)\,\mathrm{d}\omega. \qquad (14.4.10)$$

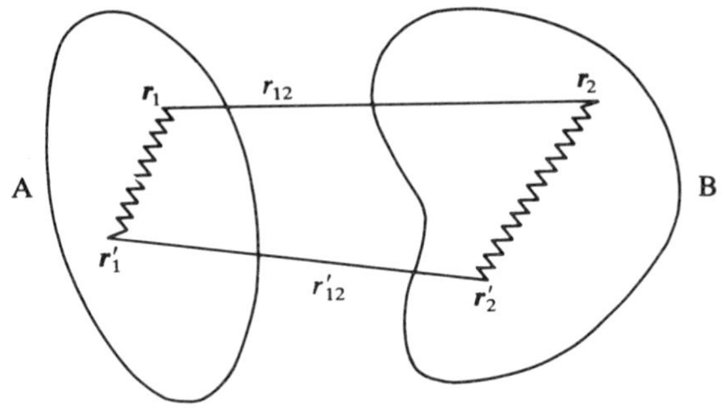

Fig. 14.2 Pictorial interpretation of the dispersion energy. The energy formula (14.4.10) involves two coulomb interactions (solid lines), multiplied by two FDPs (wavy lines), integrated over all positions of the volume elements. Each FDP describes the propagation of a density fluctuation within the molecule.

The Π factors in the integrand are in fact dynamic polarizabilities (FDPs) of the kind discussed in Chapter 12) evaluated at purely imaginary frequency (i.e. with ω replaced by $i\omega$); thus by continuing (12.1.17) along the imaginary axis‡ we obtain

$$\Pi(\mathsf{GF} \mid i\omega) = -\frac{1}{\hbar}\sum_{n>0}\frac{2\omega_{0n}\langle 0|\,\mathsf{G}\,|n\rangle\langle n|\,\mathsf{F}\,|0\rangle}{\omega_{0n}^2+\omega^2}, \qquad (14.4.11)$$

provided that the operators are Hermitian and the matrix elements are real (an imaginary term then vanishing), this being the most common situation. The FDPs in (14.4.10) arise when F and G are taken to be the density operators and are evidently special forms of *polarization propagator*. Thus, for example,

$$\Pi_{\mathrm{A}}(\mathsf{d}^{\mathrm{A}}_{r'_1}\mathsf{d}^{\mathrm{A}}_{r_1} \mid i\omega) = -\frac{1}{\hbar}\sum_{n>0}\frac{2\omega_{0n}\langle 0|\,\mathsf{d}^{\mathrm{A}}_{r'_1}\,|n\rangle\langle n|\,\mathsf{d}^{\mathrm{A}}_{r_1}\,|0\rangle}{\omega_{0n}^2+\omega^2}, \qquad (14.4.12)$$

and such quantities can be calculated by any one of the methods developed in Chapters 12 and 13 (for example from the RPA equations or their multiconfiguration generalization; or even more directly from the methods of Section 12.6 with an exponentially growing perturbation; or from the EOM equations).

The expression (14.4.10), while it may appear computationally intractable, has an interesting physical interpretation (Fig. 14.2). There are two electrostatic interaction terms, that for electron density in volume elements $d\boldsymbol{r}_1$ (in A) and $d\boldsymbol{r}_2$ (in B), bringing in the factor r_{12}^{-1}, and that for

‡ The same result is obtained by using an exponentially switched perturbation $\mathsf{F}e^{\omega t}$ instead of the oscillating term (12.1.11).

density in $d\boldsymbol{r}_1'$ and $d\boldsymbol{r}_2'$ supplying the factor $r_{12}'^{-1}$: these interactions are multiplied by FDPs, which are properties of the *individual molecules* and determine how density fluctuations propagate from point $\boldsymbol{r}_1$ to point $\boldsymbol{r}_1'$ (in A) or $\boldsymbol{r}_2$ to $\boldsymbol{r}_2'$ (in B), and (after a frequency integration) the results are summed over all positions of the volume elements. If the FDPs for the individual molecules were accurately known then (14.4.10) would provide in principle a very precise value of the dispersion energy, for arbitrary geometry and with none of the usual approximations (multipolar expansions, summation of perturbation series, etc). It may seem curious that a dependence on only an inverse *second* power of intermolecular distance is suggested by (14.4.10), whereas London's form contains an inverse *sixth* power: an r^{-6} dependence arises when the integrations are performed—but only on making a bipolar expansion of r_{12}^{-1} and $r_{12}'^{-1}$ and admitting only leading terms in the series, namely those leading to dipole–dipole interactions. In fact, however, the dipolar expansion is only conditionally convergent and, even when it does not diverge, leads to rather poor results at the dipole–dipole level.

To introduce orbital approximations, it is only necessary to write the operators in each FDP explicitly as in (14.4.5): the integrations then lead to sums of ordinary 2-electron integrals multiplied by FDPs for pairs of the E operators. In the most commonly occurring case, where the orbitals are real, the permutation symmetry of the integrals may be exploited by introducing the Hermitian combinations

$$\mathsf{D}_{rs} = \tfrac{1}{2}(2-\delta_{rs})(\mathsf{E}_{rs}+\mathsf{E}_{sr}) \quad (r \geqslant s) \qquad (14.4.13)$$

and restricting the summations accordingly. The final expression (with a "charge-cloud" notation for the integrals) is then

$$E_{ab}^{AB}(\text{disp}) = -\frac{\hbar}{2\pi}\sum_{\substack{p\geqslant q\\ r\geqslant s}}\sum_{\substack{t\geqslant u\\ v\geqslant w}} (\phi_p^A\phi_q^A \mid \phi_t^B\phi_u^B)(\phi_r^A\phi_s^A \mid \phi_v^B\phi_w^B) \times \int_0^\infty \Pi_A(\mathsf{D}_{rs}^A\mathsf{D}_{pq}^A \mid i\omega)\Pi_B(\mathsf{D}_{vw}^B\mathsf{D}_{tu}^B \mid i\omega)\,d\omega. \qquad (14.4.14)$$

This expression is now in a computationally manageable form, and contains no approximations except the usual truncation of a complete set to a large but finite basis, and the assumption that this may be divided into disjoint subsets for the two molecules.

To proceed, it is necessary to specify more closely the forms of the orbitals employed: they may, for example, be the Löwdin orthogonalized AOs, on the various atoms of each molecule; or perhaps localized bond orbitals, obtained by a unitary transformation (p. 203) from the MOs of an SCF calculation. Let us consider two concrete examples (McWeeny, 1984).

When the orbitals are not only orthonormal but are also strongly *localized*, the energy formula will be dominated by terms of a particular type and may be approximated accordingly. On using $\{\chi_r^A\}$ and $\{\chi_t^B\}$ to denote the sets on A and B respectively, an integral such as $(\chi_r^A\chi_s^A, \chi_t^B\chi_u^B)$ represents the coulomb repulsion between two fragments of charge, described by density functions

$$d_{rs}^A(\boldsymbol{r}) = \chi_r^A(\boldsymbol{r})\chi_s^A(\boldsymbol{r}), \qquad d_{tu}^B(\boldsymbol{r}) = \chi_t^B(\boldsymbol{r})\chi_u^B(\boldsymbol{r}), \tag{14.4.15}$$

and (cf. p. 134) such densities contain zero total charge (i.e. integrate to zero) unless $r = s$ and $t = u$—in which case each contains *unit* charge (normalized orbitals). Since the densities (14.4.5) are located in different molecules, and are thus well separated, the expression (14.4.14) will evidently be dominated by charge-charge interactions between densities d_{pp}^A and d_{tt}^B on the atoms of A and B. In the extreme approximation in which all other terms are neglected, the summation will then reduce to give

$$E_{ab}^{AB}(\text{disp}) = -\frac{\hbar}{2\pi}\sum_{p,r,t,v} \gamma_{pt}^{AB}\gamma_{rv}^{AB} \int_0^\infty \pi_{rr,pp}^A(i\omega)\pi_{vv,tt}^B(i\omega)\,d\omega, \tag{14.4.16}$$

where γ_{rt}^{AB} denotes the coulomb integral $(\chi_r^A\chi_r^A, \chi_t^B\chi_t^B)$, while

$$\pi_{rr,pp}^A(i\omega) = \Pi_A(\mathsf{D}_{rr}^A\mathsf{D}_{pp}^A \mid i\omega) \tag{14.4.17}$$

is the frequency-dependent analogue of the "atom–atom polarizability" introduced many years ago in Hückel theory by Coulson and Longuet-Higgins (1947). When the charge densities are written in the usual form $P(\boldsymbol{r}) = \sum_{r,s} P_{rs}d_{rs}$ and the 1-electron matrix element $\langle\chi_p|\,\mathsf{h}\,|\chi_q\rangle$ is denoted by α_p for $q = p$ (β_{pq} for $q \neq p$), $\pi_{rr,pp}$ describes the response of P_{rr} to unit change in the diagonal element α_p applied at purely imaginary frequency (i.e. growing exponentially in time); and for $\omega \to 0$ it reduces to the static polarizability $\pi_{rr,pp} = \partial P_{rr}/\partial\alpha_{pp}$. Such quantities can be calculated very easily and directly by the methods of Section 12.5, for molecules of any kind, with the usual approximations of SCF or MC SCF theory.

The fact that (14.4.16) contains only "atom–atom" coulomb interactions, with suitable coefficients, suggests a theoretical basis for various empirical procedures (see Amos and Crispin, 1970) in which the dispersion energy (as well as the electrostatic and polarization terms) is represented as a sum of empirically determined interactions between certain "sites", usually atoms, in the two molecules. If the coulomb integrals were suitably approximated then there would be some basis for the use of localized *multipoles* on the various atomic sites, and indeed such procedures are in use (Stone and Alderton, 1985). On the other hand, (14.4.14) reduces even further when the coulomb integrals γ_{pt}^{AB} are given average values for the atoms (A_m and $A_{m'}$, say) to which p and t refer, since then the summation over the orbitals on the various atoms may be completed, yielding "atom–atom" FDPs $\pi(A_m A_{m'} \mid i\omega)$ and a very simple form of the interaction energy. Thus, with an inverse-distance approximation,

$$E_{ab}^{AB}(\text{disp}) = -\frac{\hbar}{2\pi}\left(\frac{e^2}{4\pi\epsilon_0}\right)^2 \sum_{m,m',n,n'} \frac{F(A_m A_{m'}; B_n B_{n'})}{R_{mn}^{AB} R_{m'n'}^{AB}}, \quad (14.4.18)$$

which involves only atom–atom distances, in the denominator, and a "strength factor"

$$F(A_m A_{m'}; B_n B_{n'}) = \int_0^\infty \pi(A_m A_{m'} \mid i\omega)\pi(B_n B_{n'} \mid i\omega)\,d\omega, \quad (14.4.19)$$

which is completely independent of intermolecular geometry.

The atom–atom FDPs (which are sums of orbital-orbital contributions) describe the flow of charge from one atom to another that occurs in a density fluctuation: this provides an interpretation of the charge–charge interactions between atomic sites in the two molecules, an effect that will evidently be important for extended systems, but which has no counterpart in the interaction between free atoms.

The second extreme situation occurs when the orbitals of each system are completely *de*localized, extending over the whole molecule, and takes a particularly simple form when the FDPs are calculated at the TDHF (RPA) level. The free-oscillation solutions are then of the form (12.5.3), in which $i, j, \ldots$ are occupied MOs and $m, n, \ldots$ are virtual; and the summation ranges in (14.4.12) will be similarly restricted.

To perform the reduction explicitly, we note that the free-oscillation solutions of (12.3.8) will possess elements only of the type $X_{p,mi}$, $Y_{p,mi}$; and that $\mathbf{X}_p$ and $\mathbf{Y}_p$ will be obtained via the reduced equation (12.3.12) as the sums and differences $\mathbf{X}_p \pm \mathbf{Y}_p$. On introducing $\mathbf{U}_p = \mathbf{X}_p + \mathbf{Y}_p$, it follows easily that the response matrix in (12.3.24) and the general FDP (12.4.3) may be written in the reduced forms‡

$$\mathbf{\Pi}(i\omega) = -2 \sum_{p(\omega_p>0)} \frac{2\omega_p}{\omega^2 + \omega_p^2} \mathbf{U}_p \mathbf{U}_p^\dagger, \quad (14.4.20)$$

$$\Pi(\mathsf{GF} \mid i\omega) = (\nabla\mathsf{G})^\dagger \mathbf{\Pi}(i\omega)(\nabla\mathsf{F}), \quad (14.4.21)$$

‡ We assume a closed-shell ground state and that spin has been eliminated in the usual way (see Problem 12.9); this leads to the extra factor of 2 in (14.4.20).

where, for example, $(\nabla \mathsf{G})_{mi} = \langle m | \mathsf{G} | i \rangle$. The only non-zero FDPs in (14.4.14) then appear in the form

$$\Pi_{\mathrm{A}}(\mathsf{D}^{\mathrm{A}}_{mi}\mathsf{D}^{\mathrm{A}}_{nj} \mid i\omega) = -\frac{2}{\hbar} \sum_{p(\omega_p > 0)} \frac{2\omega_p}{\omega^2 + \omega_p^2} U^{\mathrm{A}}_{p,mi} U^{\mathrm{A}}_{p,nj}, \qquad (14.4.22)$$

with a similar result for molecule B. Since the FDPs are thus expressed as simple functions of ω, it is possible to perform the integration in (14.4.14) analytically, which amounts to using (14.4.9) in reverse. The result is (Jaszunski and McWeeny, 1985)

$$E^{\mathrm{AB}}_{ab}(\mathrm{disp}) = -\frac{4}{\hbar} \sum_{\substack{i,i',j,j'\\ m,m',n,n'}} \sum_{p,q} \frac{U^{\mathrm{A}}_{p,mi} U^{\mathrm{A}}_{p,m'i'} U^{\mathrm{B}}_{q,nj} U^{\mathrm{B}}_{q,n'j'}}{\omega^{\mathrm{A}}_p + \omega^{\mathrm{B}}_q}$$

$$\times (\phi^{\mathrm{A}}_m \phi^{\mathrm{A}}_i, \phi^{\mathrm{B}}_n \phi^{\mathrm{B}}_j)(\phi^{\mathrm{A}}_{m'} \phi^{\mathrm{A}}_{i'}, \phi^{\mathrm{B}}_{n'} \phi^{\mathrm{B}}_{j'}), \qquad (14.4.23)$$

and is similar to a formula first derived by Szabo and Ostlund (1977), in a somewhat different context, but does *not* contain multipole approximations.

The polarization energy

To show that the polarization energy is expressible in terms of FDPs evaluated at the static limit $\omega \to 0$, we start from the expression (14.3.5), namely

$$E^{\mathrm{AB}}_{ab}(\mathrm{pol}) = -\sum_{a'} \frac{|\langle \Phi^{\mathrm{A}}_{a'} | \mathsf{H}^{\mathrm{A}}_{\mathrm{eff}} | \Phi^{\mathrm{A}}_a \rangle|^2}{\Delta E(Aa \to Aa')} - (\mathrm{A} \to \mathrm{B}), \qquad (14.4.24)$$

where $(\mathrm{A} \to \mathrm{B})$ means a similar sum for B instead of A. Here $\mathsf{H}^{\mathrm{A}}_{\mathrm{eff}}$, as defined in Section 14.2, is an N_{A}-electron Hamiltonian for the electrons of A in the "effective field" due to B (in its ground state); and the denominators are transition energies, computed as differences of energy expectation values for the ground- and excited-state functions.

The terms in (14.4.24) reduce on writing, for example,

$$\mathsf{H}^{\mathrm{A}}_{\mathrm{eff}} = \mathsf{H}^{\mathrm{A}} + \sum_{i=1}^{N_{\mathrm{A}}} [V_{\mathrm{B}}(\boldsymbol{r}_i) + eJ_{\mathrm{B}}(\boldsymbol{r}_i)] = \mathsf{H}^{\mathrm{A}} + \sum_{i=1}^{N_{\mathrm{A}}} \delta V^{(\mathrm{B})}_{\mathrm{A}}(\boldsymbol{r}_i), \quad (14.4.25)$$

where H^{A} is the free-molecule Hamiltonian for A, while the sum gives the change due to the electron distribution of B and the nuclei it contains (i.e. the usual coulomb interactions). The off-diagonal elements of H^{A} vanish, and hence

$$\langle \Phi^{\mathrm{A}}_{a'} | \mathsf{H}^{\mathrm{A}}_{\mathrm{eff}} | \Phi^{\mathrm{A}}_a \rangle = \int \delta V^{(\mathrm{B})}_{\mathrm{A}}(\boldsymbol{r}) P_{\mathrm{A}}(aa' \mid \boldsymbol{r})\, \mathrm{d}\boldsymbol{r} \qquad (14.4.26)$$

—again with a classical interpretation.

To obtain the polarization energy, we now write the first term of (14.4.24) in the form

$$\sum_{a'} \frac{\langle \Phi_a^{\mathrm{A}} |\, \delta V_{\mathrm{A}}^{(\mathrm{B})} \,| \Phi_{a'}^{\mathrm{A}} \rangle \langle \Phi_{a'}^{\mathrm{A}} |\, \delta V_{\mathrm{A}}^{(\mathrm{B})} \,| \Phi_a^{\mathrm{A}} \rangle}{E_a^{\mathrm{A}} - E_{a'}^{\mathrm{A}}} = \tfrac{1}{2} \int \delta V_{\mathrm{A}}^{(\mathrm{B})}(\boldsymbol{r}) \delta P_{\mathrm{A}}(\boldsymbol{r})\, \mathrm{d}\boldsymbol{r}. \tag{14.4.27}$$

Here we have noted that the second-order sum may be written as either $\langle \Phi_a^{\mathrm{A}} |\, \delta V_{\mathrm{A}}^{(\mathrm{B})} \,| \delta \Phi_a^{(\mathrm{A})} \rangle$ or $\langle \delta \Phi_a^{\mathrm{A}} |\, \delta V_{\mathrm{A}}^{(\mathrm{B})} \,| \Phi_a^{\mathrm{A}} \rangle$, where $\delta \Phi_a^{\mathrm{A}}$ is the first-order change in Φ_a^{A} due to the perturbation $\delta V_{\mathrm{A}}^{(\mathrm{B})}$; and half the sum of the two forms gives half the *change* in the expectation value of $\delta V_{\mathrm{A}}^{(\mathrm{B})}$ arising from the density *change* δP_{A}. But $\delta V_{\mathrm{A}}^{(\mathrm{B})}$ may be written (cf. (3.6.8)) as a weighted sum of E operators in which E_{pq} (switched on exponentially) makes a change $\delta P_{sr} = \Pi(\mathsf{E}_{rs}^{\mathrm{A}} \mathsf{E}_{pq}^{\mathrm{A}} \mid i\omega) e^{\omega t}$ in P_{sr}^{A}; and (14.4.27) then easily reduces to give (for $\omega \to 0$)

$$\int \delta V_{\mathrm{A}}^{(\mathrm{B})}(\boldsymbol{r}) \delta P_{\mathrm{A}}(\boldsymbol{r})\, \mathrm{d}\boldsymbol{r}$$

$$= \sum_{r,s,p,q} \langle \phi_r^{\mathrm{A}} |\, \delta V_{\mathrm{A}}^{(\mathrm{B})} \,| \phi_s^{\mathrm{A}} \rangle \Pi_{\mathrm{A}}(\mathsf{E}_{rs}^{\mathrm{A}} \mathsf{E}_{pq}^{\mathrm{A}} \mid 0) \langle \phi_p^{\mathrm{a}} |\, \delta V_{\mathrm{A}}^{(\mathrm{B})} \,| \phi_q^{\mathrm{A}} \rangle. \tag{14.4.28}$$

On introducing the Hermitian combinations (14.4.13), the polarization energy (14.4.24) becomes

$$E_{ab}^{\mathrm{AB}}(\mathrm{pol}) = \tfrac{1}{2} \sum_{\substack{r \geqslant s \\ p \geqslant q}} \langle \phi_r^{\mathrm{A}} |\, \delta V_{\mathrm{A}}^{(\mathrm{B})} \,| \phi_s^{\mathrm{A}} \rangle \Pi_{\mathrm{A}}(\mathsf{D}_{rs}^{\mathrm{A}} \mathsf{D}_{pq}^{\mathrm{A}} \mid 0) \langle \phi_p^{\mathrm{A}} |\, \delta V_{\mathrm{A}}^{(\mathrm{B})} \,| \phi_q^{\mathrm{A}} \rangle + (\mathrm{A} \leftrightarrow \mathrm{B}), \tag{14.4.29}$$

where $(\mathrm{A} \leftrightarrow \mathrm{B})$ denotes a similar term with A, B interchanged. The polarization energy is thus determined by the same FDPs, defined in (14.4.22), as the dispersion energy.

14.5 NON-ORTHOGONALITY AND SHORT-RANGE EFFECTS

The formalism developed so far owes its simplicity to the assumption of strong orthogonality between the separated systems and, as already stressed, is valid only in the absence of any interpenetration of their electron distributions. At intermediate and short range, strong orthogonality is incompatible with the use of "exact" wavefunctions (assumed in separating the interaction energy from the total energy), and it is essential to take full account of non-orthogonality. Indeed, as soon as overlap occurs, certain "penetration terms" become dominant, leading either to covalency effects (for systems with unpaired electrons) or to strong repulsive interactions (for systems in closed-shell states). In van der Waals molecules, for example, the minimum-energy conformation results from the balance between long-range attractions and short-range repulsions; and, without the repulsions, colliding molecules would always coalesce! It is therefore necessary to treat non-orthogonality effects on the same footing as the long-range interactions discussed in Section 14.4. This is a matter of some difficulty, since we wish to retain (in principle) exact free-molecule functions and must therefore work in terms of density functions rather than orbitals. Such an approach has been investigated already (McWeeny and Sutcliffe, 1963; Dacre and McWeeny, 1970) at the single-configuration level; here we include multiconfigurational effects.

First we note that the partitioning method used in obtaining the energy expression (14.3.12) for a wavefunction (14.1.2) is substantially unchanged: for functions that are neither orthogonal nor normalized (Problem 2.27) the result becomes

$$E = \frac{H_{00}}{M_{00}} + \sum_{\kappa} \frac{[H_{\kappa 0} - M_{\kappa 0}(H_{00}/M_{00})]^2}{[(H_{00}/M_{00}) - (H_{\kappa\kappa}/M_{\kappa\kappa})]}, \tag{14.5.1}$$

which is good to second order in off-diagonal elements. It is also clear that, since addition of a constant term to the Hamiltonian and to the energy E will leave invariant all the elements $H_{\kappa\kappa'} - EM_{\kappa\kappa'}$, we may make the replacements

$$\left.\begin{aligned} \mathsf{H} &\to \mathsf{H} - (E_a^{\mathrm{A}} + E_b^{\mathrm{B}})\mathsf{1}, \\ E &\to E - (E_a^{\mathrm{A}} + E_b^{\mathrm{B}}) = E_{\mathrm{int}} \end{aligned}\right\} \tag{14.5.2}$$

in all that follows. On introducing the one-configuration estimates of the excited-state energies, namely (with $\kappa = a'b'$)

$$E_\kappa = \frac{H_{\kappa\kappa}}{M_{\kappa\kappa}} = \frac{\langle \Phi_{a'b'}^{\mathrm{AB}} | \mathsf{H} | \Phi_{a'b'}^{\mathrm{AB}} \rangle}{\langle \Phi_{a'b'}^{AB} | \Phi_{a'b'}^{AB} \rangle}, \tag{14.5.3}$$

we obtain

$$E_{\mathrm{int}} = E_0 + \sum_{\kappa} \frac{(H_{\kappa 0} - E_0 M_{\kappa 0})^2}{M_{\kappa\kappa}(E_0 - E_\kappa)}, \tag{14.5.4}$$

as follows from (14.5.1).

It should be noted that, since E_0 and E_κ are energies of AB relative to the reference level $E^{\mathrm{A}} + E^{\mathrm{B}}$, E_0 approaches zero at long range while E_κ becomes an electronic excitation energy for $a \to a'$, $b \to b'$. Since excitation energies are usually of the order of a few electron volts, the overlap corrections in the denominator (which are of the order of van der Waals interactions) can safely be neglected; while in the numerators

E_0 ($\simeq E_{\text{int}}$) appears with an overlap factor and is small compared with the term $H_{\kappa 0}$, which contains the interaction of two charge densities in close proximity. Consequently, the polarization and dispersion energies associated with the second-order sum in (14.5.4) cannot be greatly influenced by non-orthogonality effects. There is thus a certain formal justification for the common practice of estimating the neglected (repulsive) terms in (14.3.12), whether empirically or by calculation, as if they were quite independent of those already discussed.

To proceed, we must evaluate the matrix elements. We use the functions

$$\Phi_{ab}^{\text{AB}} = K\mathsf{A}[\Phi_a^{\text{A}}\Phi_b^{\text{B}}], \tag{14.5.5}$$

with the antisymmetrizer expressed as in (14.1.4), and normalize to unity at infinite separation by choosing

$$K = \left(\frac{N!}{N_{\text{A}}!\,N_{\text{B}}!}\right)^{1/2} \tag{14.5.6}$$

—a value common to all functions.

To indicate the reduction, it will be sufficient to consider the non-orthogonality integral $M_{\kappa 0}$ in (14.5.4). Thus

$$\begin{aligned}\langle \Phi_{a'b'}^{\text{AB}} \mid \Phi_{ab}^{\text{AB}} \rangle &= K^2 \langle \mathsf{A}[\Phi_{a'}^{\text{A}}\Phi_{b'}^{\text{B}}] \mid \mathsf{A}[\Phi_a^{\text{A}}\Phi_b^{\text{B}}] \rangle \\ &= \langle \Phi_{a'}^{\text{A}}\Phi_{b'}^{\text{B}} \mid \mathsf{A}'[\Phi_a^{\text{A}}\Phi_b^{\text{B}}] \rangle \\ &= \left\langle \Phi_{a'}^{\text{A}}\Phi_{b'}^{\text{B}} \middle| \left[1 - \sum_{i,\bar{r}} (i,\bar{r}) + \sum_{i,j,\bar{r},\bar{s}} (i,\bar{r})(j,\bar{s}) \ldots \right] \middle| \Phi_a^{\text{A}}\Phi_b^{\text{B}} \right\rangle\end{aligned}$$

where Φ_a^{A} etc. are assumed to be normalized, while indices $i, j, \ldots$ occur in the A functions and $\bar{r}, \bar{s}, \ldots$, with $\bar{p} = p + N_{\text{A}}$, in the B functions. The result is thus a sum of zero-, single- and multiple-interchange contributions. The first term is obviously $\delta_{a'a}\delta_{b'b}$ (orthogonal states within the separate molecules), while the second is

$$\begin{aligned}-\left\langle \Phi_{a'}^{\text{A}}\Phi_{b'}^{\text{B}} \middle| \sum_{i,r} (i,\bar{r}) \middle| \Phi_a^{\text{A}}\Phi_b^{\text{B}} \right\rangle \\ = -N_{\text{A}}N_{\text{B}} \int \Phi_{a'}^{\text{A}}(\boldsymbol{x}_1, \boldsymbol{x}_2, \ldots, \boldsymbol{x}_{N_{\text{A}}})^* \Phi_{b'}^{\text{B}}(\boldsymbol{x}_{\bar{1}}, \boldsymbol{x}_{\bar{2}}, \ldots, \boldsymbol{x}_{\bar{N}_{\text{B}}})^* \\ \times (1,\bar{1})\Phi_a^{\text{A}}(\boldsymbol{x}_1, \boldsymbol{x}_2, \ldots, \boldsymbol{x}_{N_{\text{A}}})\Phi_b^{\text{B}}(\boldsymbol{x}_{\bar{1}}, \boldsymbol{x}_{\bar{2}}, \ldots, \boldsymbol{x}_{\bar{N}_{\text{B}}})\, \mathrm{d}\boldsymbol{x}_1 \ldots \mathrm{d}\boldsymbol{x}_{N_{\text{A}}}\, \mathrm{d}\boldsymbol{x}_{\bar{1}} \ldots \mathrm{d}\boldsymbol{x}_{\bar{N}_{\text{B}}},\end{aligned}$$

since there are N_{A} ways of choosing i and N_{B} ways of choosing $\bar{r}$. The single transposition $(1,\bar{1})$ interchanges $\boldsymbol{x}_1$ and $\boldsymbol{x}_{\bar{1}}$, and, from the definition of the one-body transition densities, the result is (the variables being dummies)

$$-\int \rho_{\mathrm{A}}(aa' \mid \boldsymbol{x};\boldsymbol{x}')\rho_{B}(bb' \mid \boldsymbol{x}';\boldsymbol{x})\,\mathrm{d}\boldsymbol{x}\,\mathrm{d}\boldsymbol{x}'.$$

Subsequent terms follow in the same way to give

$$\begin{aligned}\langle \Phi^{\mathrm{AB}}_{a'b'} \mid \Phi^{\mathrm{AB}}_{ab}\rangle &= \delta_{aa'}\delta_{bb'} - \int \rho_{\mathrm{A}}(aa' \mid \boldsymbol{x};\boldsymbol{x}')\rho_{\mathrm{B}}(bb' \mid \boldsymbol{x}';\boldsymbol{x})\,\mathrm{d}\boldsymbol{x}\,\mathrm{d}\boldsymbol{x}'\\ &+ \int \pi_{\mathrm{A}}(aa' \mid \boldsymbol{x}_1, \boldsymbol{x}_2;\boldsymbol{x}_1', \boldsymbol{x}_2')\pi_{\mathrm{B}}(bb' \mid \boldsymbol{x}_1', \boldsymbol{x}_2';\boldsymbol{x}_1, \boldsymbol{x}_2)\,\mathrm{d}\boldsymbol{x}_1\,\mathrm{d}\boldsymbol{x}_2\,\mathrm{d}\boldsymbol{x}_1'\,\mathrm{d}\boldsymbol{x}_2'\\ &+ \text{etc.,}\end{aligned} \tag{14.5.7}$$

where the n-interchange term contains n-electron densities analogous to those defined in (5.4.1) and (5.4.2).

The matrix element $\langle \Phi^{\mathrm{AB}}_{a'b'}| \mathsf{H} |\Phi^{\mathrm{AB}}_{ab}\rangle$ is determined according to (5.4.3) and (5.4.4) in terms of 1- and 2-electron transition densities, and may be obtained in a precisely similar manner (McWeeny and Sutcliffe, 1963). Since, however, it is usual to employ a spinless Hamiltonian, further reductions will be necessary, and these are less straightforward. Two new factors must be recognized: (i) even when the functions Φ^{A}_a, Φ^{B}_b are spatially non-degenerate, they may appear in spin-degenerate sets (e.g. triplets, quintuplets, . . .), and the excited functions for AB will then be obtained by *vector coupling*; and (ii) when non-singlet states occur the further reduction will also lead to spin densities.

Here we shall simply indicate the general results, up to single-interchange terms (which are sufficient for most purposes). The spin-coupled functions formed from the $(2S_a + 1) \times (2S_b + 1)$ antisymmetrized products of $\Phi^{\mathrm{A}}_{a,M}$ and $\Phi^{\mathrm{B}}_{b,M}$ will be denoted by $\Phi^{\mathrm{AB}}_{ab,SM}$, where S, M indicate the resultant spin eigenvalues. The overlap matrix elements and the transition densities between the spin-coupled functions (with common spin eigenvalues S, M) may then be denoted as follows:

$$\left.\begin{array}{ll}\text{overlap (for } \Phi^{\mathrm{AB}}_{a'b',SM}, \Phi^{\mathrm{AB}}_{ab,SM}) & \tilde{M}^{SM}_{\mathrm{AB}}(ab, a'b'),\\ \text{1-electron transition density} & \tilde{P}^{SM}_{\mathrm{AB}}(ab, a'b' \mid \boldsymbol{r};\boldsymbol{r}'),\\ \text{2-electron transition density} & \tilde{\Pi}^{SM}_{\mathrm{AB}}(ab, a'b' \mid \boldsymbol{r}_1, \boldsymbol{r}_2;\boldsymbol{r}_1', \boldsymbol{r}_2'),\end{array}\right\} \tag{14.5.8}$$

where the tilde has been added to indicate that these quantities (with non-orthogonality admitted) do not derive from *normalized* functions. All may be developed in orders of interchange, taking the forms

$$\left.\begin{aligned}\tilde{M}^{SM}_{AB}(\ldots) &= \tilde{M}^{SM}_{AB}(\ldots)^{(0)} - \tilde{M}^{SM}_{AB}(\ldots)^{(1)} + \ldots,\\ \tilde{P}^{SM}_{AB}(\ldots) &= \tilde{P}^{SM}_{AB}(\ldots)^{(0)} - \tilde{P}^{SM}_{AB}(\ldots)^{(1)} + \ldots,\\ \tilde{\Pi}^{SM}_{AB}(\ldots) &= \tilde{\Pi}^{SM}_{AB}(\ldots)^{(0)} - \tilde{\Pi}^{SM}_{AB}(\ldots)^{(1)} + \ldots,\end{aligned}\right\} \tag{14.5.9}$$

and the leading terms (with superscript zero) are exactly as for strong-orthogonal functions, except for omission of the exchange part of $\Pi^{SM}_{AB}(\ldots)^{(0)}$ (which goes into the single-interchange term). The results are independent of M, and each term consists of a functional, involving transition densities and spin densities for the separate systems (calculated using the $M = S$ "standard" states as in Section 5.9), multiplied by a "geometric" factor that depends only on the spin coupling. Thus we find, for example,

$$\tilde{M}^{SM}_{AB}(ab, a'b')^{(1)} = \tfrac{1}{2}\int P_A(\bar{a}\bar{a}' \mid \boldsymbol{r};\boldsymbol{r}')P_B(\bar{b}\bar{b}' \mid \boldsymbol{r}';\boldsymbol{r})\,\mathrm{d}\boldsymbol{r}\,\mathrm{d}\boldsymbol{r};$$

$$+2f(S_a, S_{a'}, S_b, S_b, S_{b'}, S)\int Q_A(\bar{a}\bar{a}' \mid \boldsymbol{r};\boldsymbol{r}')Q_B(\bar{b}\bar{b}' \mid \boldsymbol{r}';\boldsymbol{r})\,\mathrm{d}\boldsymbol{r}\,\mathrm{d}\boldsymbol{r}', \tag{14.5.10}$$

where the transition densities are defined for $M_a = S_a$, $M_{a'} = S_{a'}$, $M_b = S_b$, $M_{b'} = S_{b'}$. The f factor is

$$f(S_a, S_{a'}, S_b, S_{b'}, S)$$

$$= (-1)^{(S_a+S_{b'}+S)}\begin{pmatrix} S_{a'} & 1 & S_a \\ -S_{a'} & \bar{m}_A & S_a\end{pmatrix}^{-1}\begin{pmatrix} S_{b'} & 1 & S_b \\ -S_{b'} & \bar{m}_B & S_b\end{pmatrix}^{-1}\begin{Bmatrix} S & S_{b'} & S_{a'} \\ 1 & S_a & S_b\end{Bmatrix}, \tag{14.5.11}$$

where $\bar{m}_A = S_{a'} - S_a$, $\bar{m}_B = S_{b'} - S_b$, and the round- and curly-bracket terms are the Wigner $3j$ and $6j$ symbols respectively. In cases of practical importance such factors are easily evaluated from available tables (e.g. Brink and Satchler, 1968). With a singlet ground state, for example, the first interacting excited state will often be $\Phi^{AB}_{a'b',SM}$, with a', b' both triplet states, coupled to a singlet. In this case the single-interchange contributions in (14.5.9) all contain the factor

$$f(0, 1, 0, 1, 0) = (-1)^1\begin{pmatrix} 1 & 1 & 0 \\ -1 & 1 & 0\end{pmatrix}^{-2}\begin{Bmatrix} 0 & 1 & 1 \\ 1 & 0 & 0\end{Bmatrix}$$

$$= -\left(\frac{1}{\sqrt{3}}\right)^{-2}\left(\frac{1}{\sqrt{3}}\right) = -\sqrt{3}.$$

It is only for higher-order interchanges (Dacre and McWeeny, 1970) that more complicated factors (involving $9j$, $12j$, ... symbols) are encountered.

The single-interchange terms in the 1- and 2-electron densities take a similar form. Thus

$$\tilde{P}^{SM}_{AB}(ab, a'b' \mid \boldsymbol{r};\boldsymbol{r}')^{(1)} = \tfrac{1}{2}\mathscr{F}_{AB}(\bar{a}\bar{b}, \bar{a}'\bar{b}' \mid \boldsymbol{r};\boldsymbol{r}') + 2f(S_a, S_{a'}, S_b, S_{b'}, S)\mathscr{F}^{s}_{AB}(\bar{a}\bar{b}, \bar{a}'\bar{b}' \mid \boldsymbol{r};\boldsymbol{r}'), \tag{14.5.12}$$

$$\tilde{\Pi}^{SM}_{AB}(ab, a'b' \mid \boldsymbol{r}_1, \boldsymbol{r}_2;\boldsymbol{r}_1', \boldsymbol{r}_2')^{(1)} = \tfrac{1}{2}\mathscr{G}_{AB}(\bar{a}\bar{b}, \bar{a}'\bar{b}' \mid \boldsymbol{r}_1, \boldsymbol{r}_2;\boldsymbol{r}_1', \boldsymbol{r}_2') + 2f(S_a, S_{a'}, S_b, S_{b'}, S)\mathscr{G}^{s}_{AB}(\bar{a}\bar{b}, \bar{a}'\bar{b}' \mid \boldsymbol{r}_1, \boldsymbol{r}_2;\boldsymbol{r}_1', \boldsymbol{r}_2'). \tag{14.5.13}$$

$\mathscr{F}_{AB}$ and $\mathscr{G}_{AB}$ are functionals of the transition densities (P_A, P_B), transition pair functions (Π_A, Π_B), and transition "triplet" functions (T_A, T_B) of the separate systems (between standard states $\bar{a}$, $\bar{a}'$ and $\bar{b}$, $\bar{b}'$ respectively: the functionals with superscript s are formed similarly but using the analogous *spin*-density functions. For example (using ξs for integration variables),

$$\mathscr{F}_{AB}(\bar{a}\bar{b}, \bar{a}'\bar{b}' \mid \boldsymbol{r};\boldsymbol{r}) = \int P_A(\bar{a}\bar{a}' \mid \xi;\boldsymbol{r}')P_B(\bar{b}\bar{b}' \mid \boldsymbol{r};\xi)\,d\xi + (A \leftrightarrow B) + \int \Pi_A(\bar{a}\bar{a}' \mid \xi, \boldsymbol{r};\xi', \boldsymbol{r}')P_B(\bar{b}\bar{b}' \mid \xi';\xi)\,d\xi\,d\xi' + (A \leftrightarrow B), \tag{14.5.14}$$

where $(A \leftrightarrow B)$ is a term formed by exchanging the roles of A, B. The functional $\mathscr{F}^{s}_{AB}(\bar{a}\bar{b}, \bar{a}'\bar{b}' \mid \boldsymbol{r};\boldsymbol{r}')$ is obtained by putting Q_S in place of P and Q_{SL} (see p. 151) in place of Π. The functional $\mathscr{G}_{AB}(\bar{a}\bar{b}, \bar{a}'\bar{b}' \mid \boldsymbol{r}_1, \boldsymbol{r}_2;\boldsymbol{r}_1', \boldsymbol{r}_2')$ is more unwieldy, besides involving the 3-electron functions T_A, T_B, and will not be given here. Fortunately, the convergence of the series (14.5.9) is rapid for not too large overlap (Dacre and McWeeny, 1970) and the cumbersome multiple-interchange analogues of these functionals are not required.

To illustrate the general nature of the results, let us consider the case of perhaps the greatest practical importance—that of two systems in non-degenerate (singlet) states. From (14.5.3) *et seq.*, the interaction energy may be written, including up to single-interchange terms, as

$$E_{\text{int}} = \frac{\tilde{H}_{AB}(ab, ab)^{(0)} - \tilde{H}_{AB}(ab, ab)^{(1)}}{1 - \tilde{M}_{AB}(ab, ab)^{(1)}} + \Sigma_{AB}, \tag{14.5.15}$$

where the superscripts indicate the zero- and single-interchange terms, evaluated in the usual way from the density functions in (14.5.9), while Σ_{AB} stands for the second-order sum in (14.5.3). Since the zero-interchange terms give the same densities as for strong-orthogonal functions (minus the exchange part) $\tilde{H}^{(0)}_{AB}$ yields a first approximation $E^{(0)}_{\text{int}}$

to the interaction energy in agreement with (14.3.12), namely

$$E_{\text{int}}^{(0)} = \tilde{H}_{\text{AB}}(ab, ab)^{(0)} = E_{ab}^{\text{AB}}(\text{elec}). \tag{14.5.16}$$

The full expression (14.5.15) may then be rearranged to give

$$E_{\text{int}} = E_{\text{int}}^{(0)} - [\tilde{H}_{\text{AB}}(ab, ab)^{(1)} - E_{\text{int}}^{(0)}\tilde{M}_{\text{AB}}(ab, ab)^{(1)}] + \Sigma_{\text{AB}}. \tag{14.5.17}$$

The middle term in this result contains all the single-interchange corrections to $E_{\text{int}}^{(0)}$: it may conveniently be separated into an "exchange energy", as defined for strong-orthogonal systems in (14.3.11), and a "penetration energy", which vanishes when the overlap goes to zero. The final term in (14.5.17) contains the polarization and dispersion terms (in good approximation without overlap corrections), as discussed in the previous section. The interaction energy thus becomes

$$\boxed{E_{\text{int}} = E_{ab}^{\text{AB}}(\text{elec}) + E_{ab}^{\text{AB}}(\text{ex}) + E_{ab}^{\text{AB}}(\text{pen}) + E_{ab}^{\text{AB}}(\text{pol}) + E_{ab}^{\text{AB}}(\text{disp}).} \tag{14.5.18}$$

The exchange term is

$$E_{ab}^{\text{AB}}(\text{ex}) = -\tfrac{1}{2}(1 - \tfrac{1}{2}m_{ab}^{\text{AB}})^{-1} \int r_{12}^{-1} P_{\text{A}}(aa \mid \boldsymbol{r}_2; \boldsymbol{r}_1) P_{\text{B}}(bb \mid \boldsymbol{r}_1; \boldsymbol{r}_2)\, \mathrm{d}\boldsymbol{r}_1\, \mathrm{d}\boldsymbol{r}_2, \tag{14.5.19}$$

where

$$m_{ab}^{\text{AB}} = \int P_{\text{A}}(aa \mid \boldsymbol{r}_2; \boldsymbol{r}_1) P_{\text{B}}(bb \mid \boldsymbol{r}_1; \boldsymbol{r}_2)\, \mathrm{d}\boldsymbol{r}_1\, \mathrm{d}\boldsymbol{r}_2 \tag{14.5.20}$$

is a convenient measure of the degree of overlap of the two systems. The remaining terms from the functionals $\mathscr{F}_{\text{AB}}$ and $\mathscr{G}_{\text{AB}}$ are collected in $E_{ab}^{\text{AB}}(\text{pen})$.

It is important to note that at short range $E_{ab}^{\text{AB}}(\text{pen})$ rapidly becomes the dominant term in (14.5.18) and is solely responsible for the strong repulsions that arise when closed-shell systems begin to interpenetrate: its detailed form depends on how the molecular wavefunctions, so far assumed to be exact, are approximated. Since, however, the *interaction energy* is being calculated directly, it is not unreasonable to expect that simple orbital approximations will yield useful explicit formulae. In conclusion, therefore, it seems worth putting on record the expression for the penetration energy corresponding to the use of 1-determinant Hartree–Fock functions for the separate molecules. In this case the many-electron density matrices assume the simple form (5.3.11), and the required functionals can be reduced easily. After considerable cancellation and rearrangement, it follows, using $a, a', \ldots$ and $b, b', \ldots$ for the doubly-occupied MOs in Φ_a^{A} and Φ_b^{B} respectively, that

$$
\begin{aligned}
E_{ab}^{\mathrm{AB}}(\mathrm{pen}) = -(1-\tfrac{1}{2}m)^{-1}\Big\{ & 2\sum_{a,b} h_{ab}S_{ba} - 2\sum_{a,a',b} h_{aa'}S_{a'b}S_{ba} \\
& + 4\sum_{a,b,b'} \langle ab' \parallel bb' \rangle S_{ba} + 4\sum_{a,a'b} \langle aa' \parallel ba' \rangle S_{ba} \\
& - 4\sum_{a,a',b,b'} \langle a'b' | g | ab' \rangle S_{ab}S_{ba'} - 4\sum_{a,a',a'',b} \langle a'a'' \parallel aa'' \rangle S_{ab}S_{ab'} \\
& + \sum_{a,a',b,b'} \langle a'b | g | ab' \rangle S_{ab}S_{b'a'} \Big\} + (\mathrm{A} \leftrightarrow \mathrm{B}),
\end{aligned}
\tag{14.5.21}
$$

where the double-bar quantities $\langle \ \parallel \ \rangle$ are defined in (12.5.6) and m ($= m_{ab}^{\mathrm{AB}}$ in (14.5.20)) reduces to a sum of squares of overlap integrals:

$$m = 4\sum_{a,b} S_{ab}S_{ba}. \tag{14.5.22}$$

Such expressions are easy to evaluate, and are in principle capable of giving a good account of the repulsive part of the interaction energy. They have been used successfully, for example, in calculations (Amovilli and McWeeny, 1986) of the conformations of simple dimers. It should be noted, however, that there is a heavy dependence on overlap integrals, and the results are therefore sensitive to the quality of the wavefunctions in the region of the "tails" of the orbitals employed.

4.6 EXCITED STATES AND RESONANCE INTERACTIONS

So far the molecules A and B have been assumed to be in their ground states. When one or both of them are in excited states new effects must be considered. In that case the reference state (exact in the long-range limit) is already excited relative to the *ground* state: if we use $\kappa = (\mathrm{A}a'\mathrm{B}b)$ to label the corresponding reference function then the long-range limit of the energy will be

$$E_\kappa = E_0 + \Delta E(\mathrm{A}a \rightarrow \mathrm{A}a'), \tag{14.6.1}$$

E_0 being the energy limit on the *ground*-state surface.

To exploit the formalism used so far, the wavefunction will be written as

$$\Psi_\kappa = \Phi_\kappa + \sum_{\kappa'} c_{\kappa'} \Phi_{\kappa'} + \sum_{\kappa''} c_{\kappa''} \Phi_{\kappa''}, \tag{14.6.2}$$

where $\Phi_{\kappa'}$ differs from the *reference* function Φ_κ in one factor (A or B), and $\Phi_{\kappa''}$ differs in both factors. Clearly, however, κ' will now include (AaBb), corresponding to the *de*-excitation of molecule A; but if Ψ_κ is to be orthogonal to the true ground state then Φ_{ab}^{AB} must enter with zero coefficient. We may therefore write, more explicitly, for the function deriving from $\Phi_{a'b}^{AB}$,

$$\Psi = \Phi_{a'b}^{AB} + \sum_{b'} c_{a'b'} \Phi_{a'b'}^{AB} + \sum_{a''(\neq a, a')} c_{a''b} \Phi_{a''b}^{AB} + \sum_{a''(\neq a'), b'} c_{a''b'} \Phi_{a''b'}^{AB}, \tag{14.6.3}$$

and again use the partitioning method to estimate the interaction energy in a form similar to (14.3.12).

The leading term in E_{int} is an electrostatic interaction similar to (14.3.10), except that A is now in state a' instead of a; the next two sums in (14.6.3) give rise to polarization energies for A (in state a'), in the presence of B (in state b), and for B (in state b) in the presence of A (in state a'). The final term introduces interesting new possibilities: the energy denominators in the second-order sum are now, instead of those that appear in (14.4.1),

$$\Delta E(\text{A}a' \to \text{A}a''; \text{B}b \to \text{B}b') \simeq \Delta E_{\text{A}}(a' \to a'') + \Delta E_{\text{B}}(b \to b'), \tag{14.6.4}$$

where the interaction energy is neglected (cf. (14.4.8) *et seq.*) in comparison with the excitation energy. These denominators may thus be of *either sign*, giving attractive or repulsive contributions to E_{int}, or they may indeed by close to (or exactly) zero, in which case the corresponding terms will dominate the second-order sum and the convergence of the series will need special attention.

One special case, of great importance, arises when the two molecules A and B are identical. For then the last sum in (14.6.4) will contain a term with $a'' = a$ whose energy denominator will vanish exactly when a', b' refer to the *same* excited state; formally, a de-excitation of A ($a' \to a$) may be accompanied by a corresponding excitation ($b \to b'$) of B. The expansion (14.6.3) then needs modification to admit the *degenerate* functions $\Phi_{a'b}^{AB}$ and $\Phi_{ab'}^{AB}$ on an equal footing. As always, the degeneracy will be resolved in first order by solving an appropriate 2×2 secular equation to obtain (overlap neglected)

$$\Psi_\pm = (\Phi_{a'b}^{AB} \pm \Phi_{ab'}^{AB})/\sqrt{2}, \tag{14.6.5}$$

$$E_\pm = \langle \Phi_{a'b}^{AB} | \mathsf{H} | \Phi_{ab'}^{AB} \rangle \pm \langle \Phi_{a'b}^{AB} | \mathsf{H} | \Phi_{ab'}^{AB} \rangle. \tag{14.6.6}$$

Such an effect is well known in the theory of molecular crystals (see e.g. Davydov, 1962), where it is usually referred to as 'Davydov splitting'.

The matrix elements in (14.6.6) are evaluated as in earlier sections, and give immediately a *first*-order interaction energy

$$E_{\text{int}} = E^{\text{AB}}_{a'b}(\text{elec}) \pm E^{\text{AB}}_{a'b}(\text{res}), \tag{14.6.7}$$

where the so-called *resonance interaction* is

$$E^{\text{AB}}_{a'b}(\text{res}) = \frac{e^2}{4\pi\epsilon_0}\int \frac{P_{\text{A}}(aa' \mid \boldsymbol{r}_1)P_{\text{B}}(b'b \mid \boldsymbol{r}_2)}{r_{12}}\, \mathrm{d}\boldsymbol{r}_1\, \mathrm{d}\boldsymbol{r}_2. \tag{14.6.8}$$

The excited-state energy surface is thus split into two sheets by the resonance interaction; and this interaction is relatively strong, representing the coulomb energy of two transition charge densities. At distances where a multipole expansion is permissible, the leading term would evidently give a distance dependence of the form R^{-3} instead of the R^{-6} in London's approximation to the dispersion energy.

Higher-order effects and overlap corrections may be treated as in earlier sections, using the "correct zeroth-order functions" defined in (14.6.5). The energy formula (14.6.6) will then assume the general form (14.5.18), except for the presence of an extra term, $\pm E^{\text{AB}}_{a'b}(\text{res})$, obtained above.

The study of interactions between molecules in excited states, which has received comparatively little attention, is clearly of great potential importance in photochemistry, and raises many new problems. Throughout this chapter, for example, it has been assumed that the distance between A and B be small enough to permit a study of the complex AB *as if it were in a stationary state.* But at a sufficiently large distance this assumption must become untenable: it will eventually be possible to distinguish physically *which* of the two molecules is in an excited state, and the delocalization of the excitation (by transfer from one molecule to the other) will take place after a measurable period of time. In such cases it will be necessary to employ time-dependent theory (using, for example, the techniques of Chapter 12) to discuss the *propagation* of the excitation; the stationary-state approach used so far will evidently be appropriate in the limit where the propagation time turns out to be brief compared with the shortest attainable times of observation. Such studies, which would involve the calculation of time-correlation functions (Section 12.1), are certainly feasible (see e.g. McWeeny, 1985), but the detailed discussion of dynamical processes involving the intermolecular transfer of electronic energy belongs to the future.

PROBLEMS 14

14.1 Establish the factorization property (14.1.4) for the antisymmetrizer A in the case of a two-group system. [*Hint:* Divide the electron indices into two sets $1, 2, \ldots, i, \ldots, N_A$ and $\bar{1}, \bar{2}, \ldots, \bar{j}, \ldots, \bar{N}_B$, where $\bar{j} = j + N_A$. The permutations of S_N can be generated from those of the *sub*group of permutations that permute indices only *within* the two sets by combination with a *coset* of elements $(T_1, T_2, \ldots,)$ that do not belong to the subgroup. Express a general permutation in the form $P = TP_AP_B$; verify that T (a multiple transposition) can be chosen in a sufficient number of ways to give all $N!$ elements; make the simplest choice, and finally obtain the antisymmetrizer by appropriate summation and normalization.]

14.2 Derive the matrix elements $\langle \Phi_{\kappa'} | \mathrm{H} | \Phi_\kappa \rangle$ for a two-group system with wavefunctions of the form $\Phi_\kappa = M\mathrm{A}[\Phi_a^A \Phi_b^B]$, where the antisymmetric A and B factors (assumed to be strong-orthogonal) may describe various "local" states $(a, a', \ldots$ and $b, b', \ldots)$. [*Hint:* Use the antisymmetrizer of Problem 14.1, noting that A_A, A_B are redundant and that A′ need only contain transpositions such as $(i, \bar{j})$, $(i, \bar{j})(k, \bar{l})$, etc. Use the "turnover rule" and the property $\mathrm{A}^2 = \mathrm{A}$. Note that H will contain terms such as $\mathrm{h}(i)$, $\mathrm{h}(\bar{j})$, $g(i, j)$, $g(i, \bar{j}), \ldots$ and that the integrations will give large numbers of equal terms or zeros.]

14.3 Use the result obtained in Problem 14.2 to verify the forms of the 1- and 2-electron densities and transition densities (ρ, π) as given (14.1.6). Show that when Φ_b^A, Φ_b^B are singlet functions the corresponding spinless densities (P, Π) follow easily. [*Hint:* Compare your expressions with (5.4.3) and (5.4.4), which give the general matrix elements in terms of density functions. Remember that ρ has only two components when $S = 0$.]

14.4 Use a second-quantization approach to obtain $\rho(\boldsymbol{x}_1; \boldsymbol{x}_1')$ and $\pi(\boldsymbol{x}_1, \boldsymbol{x}_2; \boldsymbol{x}_1', \boldsymbol{x}_2')$ for a state $|\Phi_{AB}\rangle = G_A^\dagger G_B^\dagger |\mathrm{vac}\rangle$, where $G_A^\dagger$, $G_B^\dagger$ are generally linear combinations of N_A, N_B creation operators respectively selected from two disjoint sets $\{a_i^\dagger\}$ and $\{\bar{a}_j^\dagger\}$—this being the Fock-space analogue of a group function. You should reproduce the results in Problem 14.3. [*Hint:* Introduce field operators $\psi(\boldsymbol{x})$, $\psi^\dagger(\boldsymbol{x}')$ as in Section 13.3 (but now time-independent), where for example $\psi(\boldsymbol{x}) = \psi_a(\boldsymbol{x}) + \psi_b(\boldsymbol{x})$ and the two parts contain operators of the (orthogonal) disjoint sets. Then find $\langle \Phi_{AB} | \psi^\dagger(\boldsymbol{x}_1')\psi(\boldsymbol{x}_1) | \Phi_{AB} \rangle$ and $\langle \Phi_{AB} | \psi^\dagger(\boldsymbol{x}_1')\psi^\dagger(\boldsymbol{x}_2')\psi(\boldsymbol{x}_2)\psi(\boldsymbol{x}_1) | \Phi_{AB} \rangle$. Note that the anticommutation rules permit the free left–right passage of operators built from different sets.]

14.5 Justify the core–valence separation used in Section 7.4, assuming the core to be in a singlet state and eliminating spin.

14.6 Consider two neutral atoms A, B, at a large distance R, and expand the inverse distance factor in (14.4.2) in terms of R and the "local" coordinates (x_1, y_1, z_1) and (x_2, y_2, z_2). Hence obtain a dipole–dipole approximation to the integral, showing how the terms in the sum (14.4.1) may be associated with spectroscopic information. [*Hint:* Put atom A at the origin, atom B at $z = R$, and expand r_{12}^{-1} up to terms quadratic in $x_1, y_1, \ldots, z_2$. Look back at Section 11.5, where transition dipoles also appeared.]

14.7 Use the integral-transform method to obtain a dipole–dipole approximation to the dispersion energy (cf. Problem 14.6), clearly defining the nature of the dynamic polarizabilities involved.

14.8 Formulate in detail, using the methods discussed in Sections 12.5 and 12.6, the equations required to calculate the dynamic polarizabilities that appear in the dipole–dipole approximation to the dispersion energy of two diatomic molecules.

14.9 Generalize the results obtained in Problem 14.2 to admit slight non-orthogonality between the two groups, using a single antisymmetrized product with both groups in singlet states. Then separate the energy into the terms $E_a^{\mathrm{A}} + E_b^{\mathrm{B}} + E_{\mathrm{int}}^{\mathrm{AB}}$, obtaining the component parts of $E_{\mathrm{int}}^{\mathrm{AB}}$ as defined in the text. [*Hint:* Admit only single transpositions in the antisymmetrizer and consider first the normalization integral $\langle \Phi_{ab}^{\mathrm{AB}} \mid \Phi_{ab}^{\mathrm{AB}} \rangle$, remembering the definition of the 1-electron density matrix.]

14.10 Suppose that two molecules A, B, whose closed-shell ground states are described using MOs $\{\phi_i^{\mathrm{A}}\}$, $\{\phi_j^{\mathrm{B}}\}$, overlap weakly (so that the approximation in Problem 14.9 is adequate). Show how the energy terms already obtained reduce to the MO forms given in (14.5.21). [*Hint:* Insert orbital approximations to the density matrices and note how the integrations introduce products of MO overlap integrals.]

14.11 Make a calculation from first principles of the interaction energy of two hydrogen molecules, in simple MO approximation, obtaining exchange and penetration energies in terms of the 1-electron, 2-electron and overlap integrals over MOs. Check that the results in Problem 14.10 are in agreement up to terms in S^2. Identify the origin of the strong repulsive interaction.

14.12 From the results obtained in Problem 14.9, investigate how the electron density in the region of overlap of two systems (in singlet states) is modified by non-orthogonality effects. On denoting the generalized overlap integral in (14.5.20) by M_{AB}^P, you should find that $P(\boldsymbol{r})$ differs from the density sum $P_{\mathrm{A}}(\boldsymbol{r}) + P_{\mathrm{B}}(\boldsymbol{r})$ by an "interference term"

$$\Delta P(\boldsymbol{r}) = -\frac{[P_{\mathrm{B}}P_{\mathrm{A}}(\boldsymbol{r}) + P_{\mathrm{A}}P_{\mathrm{B}}(\mathbf{r})]}{2(1 - M_{\mathrm{AB}}^P)} + \frac{[P_{\mathrm{A}}P_{\mathrm{B}}P_{\mathrm{A}}(\boldsymbol{r}) + P_{\mathrm{B}}P_{\mathrm{A}}P_{\mathrm{B}}(\mathbf{r})]}{4(1 - M_{\mathrm{AB}}^P)},$$

where, for example, $P_{\mathrm{A}}P_{\mathrm{B}}(\boldsymbol{r})$ means $\int P_{\mathrm{A}}(\boldsymbol{r};\boldsymbol{r}')P_{\mathrm{B}}(\boldsymbol{r}';\boldsymbol{r})\,d\boldsymbol{r}'$.

14.13 Assume an orbital approximation to the density matrices used in Problem 14.12 to show how the total charge ($N_{\mathrm{A}} + N_{\mathrm{B}}$ electrons) is divided between the separate systems and the region in which their orbitals overlap. Evaluate M_{AB}^P, showing it to be positive, and prove that (with an appropriate convention) the total charge migration from the systems into the overlap region is $\Delta = -M_{\mathrm{AB}}^P/2(1 - M_{\mathrm{AB}}^P)$—so that charge is "pushed out" from the overlap region. How do your results relate to collision-induced dipole moments? [*Hint:* In an orbital approximation a density function $P_{\mathrm{A}}(\boldsymbol{r};\boldsymbol{r}')$ is always a linear combination of products $\phi_i^{\mathrm{A}}(\boldsymbol{r})\phi_j^{\mathrm{A}}(\boldsymbol{r}')^*$; so the integral $P_{\mathrm{A}}P_{\mathrm{B}}(\boldsymbol{r})$ will contain products $\phi_i^{\mathrm{A}}(\boldsymbol{r})\phi_j^{\mathrm{B}}(\boldsymbol{r})$, while $P_{\mathrm{A}}P_{\mathrm{B}}P_{\mathrm{A}}(\boldsymbol{r})$ will contain only A orbitals. Density in the overlap region is associated with the "mixed" products (cf. "population analysis" in Section 5.5).]

14.14 Use the functionals given in (14.5.12) and (14.5.13) to obtain results analogous to those in Problem 14.13, but now for systems in ground states of maximum multiplicity (Hund's rule), S_A, $S_B \neq 0$, approaching along a reaction path with spins coupled to a resultant S ($S = S_A + S_B$, $S_A + S_B - 1, \ldots, |S_A - S_B|$). Show that in this case the electron density in the overlap region may be reduced or enhanced, depending on spin coupling, and obtain orbital approximations to the density changes by using simple 1-determinant wavefunctions. How do your results relate to the fact that atoms adsorbed on surfaces, or in cavities, may acquire an electric moment? [*Hint:* States belonging to the same multiplet ($a, a', \ldots$) may be labelled by quantum numbers M_a, $M'_a, \ldots$; and the factor (14.5.11) becomes the expectation value of a spin scalar product. M^P_{AB} in Problem 14.13 is replaced by $M^P_{AB} + \theta M^Q_{AB}$, where $\theta = \langle \mathbf{S}_A \cdot \mathbf{S}_B \rangle / S_A S_B$ and M^Q_{AB} contains standard-state spin densities instead of the charge densities.]

Appendix 1: Atomic Orbitals

A bound-state wavefunction for a single electron in a central field, corresponding either to an actual or hypothetical potential (for example the "effective" field used in an independent-particle model), is usually referred to as an *atomic orbital*. We denote the potential energy by $V(r)$ (angle-independent), and the eigenvalue equation is thus

$$\mathsf{h}\chi = \epsilon\chi,$$

with Hamiltonian

$$\mathsf{h} = -\tfrac{1}{2}\nabla^2 + V(r).$$

When spin terms are omitted, the operators that commute with h, and thus permit a full classification of each stationary state in terms of corresponding constants of the motion, are

$$\mathsf{L}_z = \frac{1}{i}\left(x\frac{\partial}{\partial y} - y\frac{\partial}{\partial x}\right), \qquad \mathsf{L}^2 = \mathsf{L}_x^2 + \mathsf{L}_y^2 + \mathsf{L}_z^2, \tag{A1.1}$$

where L_z is associated with a (cartesian) component of *orbital angular momentum* (conventionally taken as the z component), L_x and L_y are defined similarly, and L^2 denotes the operator associated with the square of the total angular momentum.

Simultaneous eigenfunctions of h, L^2 and L_z can be found (see e.g. Eyring *et al.*, 1944) in the form

$$\chi_{nlm}(r, \theta, \phi) = F_{nl}(r)Y_{lm}(\theta, \phi) \tag{A1.2}$$

where r, θ, ϕ are polar coordinates and n, l, m, are integers.‡ The function $Y_{lm}(\theta, \phi)$ is a spherical harmonic. The eigenvalues of the squared angular momentum and its z component (in units of $\hbar^2$ and $\hbar$) are $l(l+1)$ and m, the wavefunction satisfying

$$\left.\begin{aligned} \mathsf{L}^2\chi_{nlm} &= l(l+1)\chi_{nlm} \quad (l = 0, 1, 2, \ldots), \\ \mathsf{L}_z\chi_{nlm} &= m\chi_{nlm} \quad (m = l, l-1, \ldots, -l). \end{aligned}\right\} \tag{A1.3}$$

‡ For *one*-electron functions it is customary to use small letters (l, m) for the angular-momentum quantum numbers. In this Appendix atomic units are also used.

These results are independent of the form of $V(r)$. Moreover, $F_{nl}(r)$ is in general the same for all m values (given l) and there is a $(2l+1)$-fold degeneracy in the energy:

$$\mathsf{h}\chi_{nlm} = \epsilon_{nl}\chi_{nlm}. \tag{A1.4}$$

The integer n is needed generally to establish the energy order of solutions with the same l value, but may have a more special significance for certain potential functions.

The spherical harmonics are conveniently normalized so that on squaring, multiplying by $\sin\theta\, d\theta\, d\phi$ (an element of solid angle) and integrating over θ and ϕ the result is unity. The first few spherical harmonics are listed in Table A.1.1.

A standard "phase factor" has been incorporated in Table A1.1 by taking

$$Y_{lm}(\theta, \phi) = (-1)^{(m+|m|)/2}\left[\left(\frac{2l+1}{4\pi}\right)\frac{(l-|m|)!}{(l+|m|)!}\right]^{1/2} P_l^{|m|}(\cos\theta)e^{im\phi}, \tag{A1.5}$$

where $P_l^{|m|}$ denotes the associated Legendre polynomial. This represents the phase choice of Condon and Shortley (1935)‡ and should be observed in relating different solutions through their symmetry and angular momentum properties: for most other purposes the resultant minus signs may be discarded. Atomic orbitals are designated s, p, d, f, g, ... type according as $l = 0, 1, 2, 3, \ldots$ respectively, while m values are frequently indicated by a subscript: thus a $2p_{-1}$ orbital has $l = 1$, $m = -1$.

It is often convenient to introduce the *real* combinations

$$Y_{lm}^C = \frac{(-1)^m Y_{lm} + Y_{l,-m}}{\sqrt{2}}, \quad Y_{lm}^S = -\frac{i[(-1)^m Y_{lm} - Y_{l,-m}]}{\sqrt{2}}, \tag{A1.6}$$

in which the $e^{\pm im\phi}$ (with its phase factor) is replaced by $\sqrt{2}\cos m\phi$ or $\sqrt{2}\sin m\phi$ respectively. For s, p and d functions the real combinations coincide with the "cubic harmonics" (Table A1.2).

The degeneracy of functions with different m values may be exploited in various ways to obtain atomic orbitals that are still eigenfunctions of h and L^2, though no longer of L_z. Thus if we are expanding a wavefunction for a molecule of tetrahedral symmetry then the orbitals with the same l values but different m will always appear in certain specific combinations, giving *real* functions that are expressed more naturally in terms of cartesian coordinates. These "cubic harmonics" are so frequently pre-

‡ Other phase conventions are discussed by Edmonds (1957).

Table A1.1 Normalized spherical harmonics $Y_{lm}(\theta, \phi)$.

s functions ($l = 0$)	
$m = 0$	$\frac{1}{2}\left(\frac{1}{\pi}\right)^{1/2}$

p functions ($l = 1$)	
$m = 0$	$\frac{1}{2}\left(\frac{3}{\pi}\right)^{1/2} \cos\theta$
$m = \pm 1$	$\mp\frac{1}{2}\left(\frac{3}{2\pi}\right)^{1/2} \sin\theta\, e^{\pm i\phi}$

d functions ($l = 2$)	
$m = 0$	$\frac{1}{4}\left(\frac{5}{\pi}\right)^{1/2} (3\cos^2\theta - 1)$
$m = \pm 1$	$\mp\frac{1}{2}\left(\frac{15}{2\pi}\right)^{1/2} \sin\theta\cos\theta\, e^{\pm i\phi}$
$m = \pm 2$	$\frac{1}{4}\left(\frac{15}{2\pi}\right)^{1/2} \sin^2\theta\, e^{\pm 2i\phi}$

f functions ($l = 3$)	
$m = 0$	$\frac{1}{4}\left(\frac{7}{\pi}\right)^{1/2} (5\cos^2\theta - 1)\cos\theta$
$m = \pm 1$	$\mp\frac{1}{8}\left(\frac{21}{\pi}\right)^{1/2} (5\cos^2\theta - 1)\sin\theta\, e^{\pm i\phi}$
$m = \pm 2$	$\frac{1}{4}\left(\frac{105}{2\pi}\right)^{1/2} \sin^2\theta\cos\theta\, e^{\pm 2i\phi}$
$m = \pm 3$	$\mp\frac{1}{8}\left(\frac{35}{\pi}\right)^{1/2} \sin^3\theta\, e^{\pm 3i\phi}$

ferable in quantum chemistry that we list the first few in Table A1.2, indicating the modified notation for the orbitals in which they appear. The symmetry properties of such functions have been discussed elsewhere (e.g. McWeeny, 1963).

To complete the specification of an atomic orbital it is only necessary to define the radial factor. This is usually normalized so that on squaring, multiplying by r^2 and integrating over r, the result is unity: the product (A1.2) is then normalized to unity in the usual sense. Three common choices of potential function are as follows:

$$V(r) = -Z/r \qquad \text{(hydrogen-like orbitals)}, \tag{A1.7}$$

$$V(r) = \frac{-\zeta n}{r} + \frac{n(n-1) - l(l+1)}{2r^2} \qquad \text{(Slater orbitals)}, \tag{A1.8}$$

$$V(r) = 2\zeta^2 r^2 + \frac{n(n-1) - l(l+1)}{2r^2} \qquad \text{(Gaussian orbitals)}. \tag{A1.9}$$

Table A1.2 Cubic harmonics in cartesian form.

s functions $N_s \times 1$

p functions

	p_x	p_y	p_z
$(N_p/r)\times$	x	y	z

d functions

	d_{z^2}	$d_{x^2-y^2}$	
$(N_d/r_2)\times$	$\frac{1}{2}(3z^2 - r^2)$	$\frac{1}{2}\sqrt{3}\,(x^2 - y^2)$	
	d_{yz}	d_{xz}	d_{xy}
$(N_d/r^2)\times$	$\sqrt{3}\,yz$	$\sqrt{3}\,zx$	$\sqrt{3}\,xy$

f functions

	f_x	f_y	f_z
$(N_f/r^3)\times$	$\frac{1}{2}(5x^2 - 3r^2)x$	$\frac{1}{2}(5y^2 - 3r^2)y$	$\frac{1}{2}(5z^2 - 3r^2)z$
	f'_x	f'_y	f'_z
$(N_f/r^3)\times$	$\frac{1}{2}\sqrt{15}\,(y^2 - z^2)x$	$\frac{1}{2}\sqrt{15}\,(z^2 - x^2)y$	$\frac{1}{2}\sqrt{15}\,(x^2 - y^2)z$
	f_{xyz}		
$(N_f/r^3)\times$	$\sqrt{15}\,xyz$		

Normalizing factors

$$N_s = \frac{1}{2}\left(\frac{1}{\pi}\right)^{1/2} \quad N_p = \frac{1}{2}\left(\frac{3}{\pi}\right)^{1/2} \quad N_d = \frac{1}{2}\left(\frac{5}{\pi}\right)^{1/2} \quad N_f = \frac{1}{2}\left(\frac{7}{\pi}\right)^{1/2}$$

Notes. Each function is an lth degree polynomial in x, y, z divided by r^l, and is therefore equivalent to a function of θ and ϕ only (combination of spherical harmonics); each is normalized in the corresponding sense (for integration over solid angle) and may therefore be combined with a normalized radial factor (see below) to give a normalized atomic orbital. The subdivisions indicate sets of functions carrying distinct irreducible representations of the cubic point groups (for example, the sets p_x, p_y, p_z and f_x, f_y, f_z carry the T_{1u} representation of the symmetry group of the cube).

The radial factors that correspond to the various choices of potential are as follows.

Hydrogen-like orbitals

These orbitals arise when the potential (A1.7) is chosen. The standard form is

$$F_{nl}(r) = C_{nl}\left(\frac{2Z}{n}\right)^{l} e^{-Zr/n} L_{n+l}^{2l+1}\left(\frac{2Zr}{n}\right),$$
$$C_{nl} = -\left[\frac{(n-l+1)!}{2n([n+l]!)^{3}}\left(\frac{2Z}{n}\right)^{3}\right]^{1/2}, \qquad \text{(A1.10)}$$

where the L_{n+l}^{2l+1} are *associated Laguerre polynomials*. Their definition can be found in most books on quantum mechanics (e.g. Eyring *et al.*, 1944, p. 63). The $F_{nl}(r)$ form an orthonormal set, and the first few members of this set are tabulated, for example, in Eyring *et al.* (1944, p. 85). The set is not complete, however, unless *continuum* functions (describing a free electron scattered by the nucleus) are admitted; without the continuum, the convergence of wavefunction expansions based on hydrogen-like AOs is poor and they are consequently seldom used.

Slater orbitals

The standard form is

$$F_n(r) = C_n r^{n-1} e^{-\zeta r}, \qquad C_n = \frac{(2\zeta)^{n+1/2}}{[(2n)!]^{1/2}}, \qquad \text{(A1.11)}$$

where n corresponds to the principal quantum number in the hydrogen-like orbitals. This radial factor may be associated with any harmonic (Tables A.1 and A.2) with $l = 0, 1, \ldots, n-1$: for $l = n-1$ the result coincides with a hydrogen-like orbital for $\zeta = Z/n$, but for lower l-values the orbital satisfies an eigenvalue equation with the "effective" potential (A1.8), which eliminates the radial nodes. The nodeless functions are not orthogonal, and different exponents ζ may be used for different l-values (within the same "shell").

The flexibility of these functions lends them well to the construction of approximate wavefunctions. Unfortunately, their use leads to difficulties in the calculation of 1- and 2-electron many-centre integrals. In molecular calculations it is usually preferable to *fit* such functions in terms of others (usually of Gaussian form) for which the evaluation of the integrals is less demanding. Many methods of fitting have been employed: the first (McWeeny, 1952) made use of the Schrödinger equation with potential

(A1.8), satisfied by the Slater orbitals, while other methods (see e.g. Huzinaga, 1965) minimize the deviation between the Slater function and its approximant.

Gaussian orbitals

There are many types of Gaussian function, the two most commonly used being "spherical" and "cartesian". Spherical Gaussian orbitals have the standard form (A1.2) but with

$$F_n(r) = C_n r^{n-1} e^{-\zeta r^2}, \qquad C_n = \left[\frac{2^{2n}(n-1)!}{(2n-1)}\left(\frac{(2\zeta)^{2n+1}}{\pi}\right)^{1/2}\right]^{1/2}. \qquad \text{(A1.12)}$$

These orbitals are again central-field eigenfunctions, corresponding to the potential (A1.9), and are again non-orthogonal unless differing in angular-momentum quantum numbers. The cartesian type, instead, consists of functions with cubic symmetry (cf. Table A1.2), in which the spherical-harmonic separation in (A1.2) is not introduced. The normalized orbitals are instead

$$\chi_{abc}(x, y, z) = N_a N_b N_c x^a y^b z^c e^{-\zeta r^2}, \qquad N_a = \left[\frac{(2a-1)!!}{\zeta^a}\left(\frac{\pi}{\zeta}\right)^{-1/2}\right]^{1/2} \qquad \text{(A1.13)}$$

(with similar expressions for N_b, N_c), a, b, c being non-negative integers. Such functions may of course be combined to give orbitals in which the angle-dependent factor is a spherical or cubic harmonic; but they are often used in the single-term form given above.

Very efficient routines exist for the evaluation of many-centre integrals over Gaussian orbitals, and the use of such functions in molecular calculations has become almost standard. Tables have also been prepared (for a recent compilation see Poirer *et al.*, 1985), which contain Gaussian approximations, of various qualities, to Slater functions.

Appendix 2: Angular Momentum

The orbital angular-momentum operators are of great importance in any central-field system: other operators with formally similar properties are the spin operators (used extensively in Chapters 4 and 11), and it is therefore useful to collect the properties that characterize *any* kind of angular momentum. If the operators associated with the components are denoted by $\hbar\mathsf{K}_x$, $\hbar\mathsf{K}_y$ and $\hbar\mathsf{K}_z$ then

$$[\mathsf{K}_x, \mathsf{K}_y] = i\mathsf{K}_z, \qquad [\mathsf{K}_y, \mathsf{K}_z] = i\mathsf{K}_x, \qquad [\mathsf{K}_z, \mathsf{K}_x] = i\mathsf{K}_y, \tag{A2.1}$$

where, for example, the commutator $[\mathsf{K}_x, \mathsf{K}_y] = \mathsf{K}_x\mathsf{K}_y - \mathsf{K}_y\mathsf{K}_x$. These are the "commutation relations". It is easy to verify that the orbital angular momentum $\mathbf{L}$ (i.e. the set of operators L_x, L_y, L_z) exhibits these properties; this is also true for the total orbital angular momentum, in which each component is a sum of one-particle contributions; and it is true for spin and for spin plus orbital angular momenta. The interpretation is therefore very wide: $\mathbf{K}$ may stand for $\mathbf{L}$, $\mathbf{S}$, $\mathbf{J}$ $(= \mathbf{L} + \mathbf{S})$, etc.

Instead of K_x and K_y, it is often convenient to use the step-up and step-down operators

$$\mathsf{K}^+ = \mathsf{K}_x + i\mathsf{K}_y, \qquad \mathsf{K}^- = \mathsf{K}_x - i\mathsf{K}_y, \tag{A2.2}$$

in terms of which $\mathsf{K}^2 = \mathsf{K}_x^2 + \mathsf{K}_y^2 + \mathsf{K}_z^2$ may be written in the alternative forms

$$\mathsf{K}^2 = \mathsf{K}^-\mathsf{K}^+ + \mathsf{K}_z + \mathsf{K}_z^2 = \mathsf{K}^+\mathsf{K}^- - \mathsf{K}_z + \mathsf{K}_z^2. \tag{A2.3}$$

It is then possible to show that the simultaneous eigenfunctions of K^2 and K_z satisfy

$$\left.\begin{aligned} \mathsf{K}^2\Psi_{KM_K} &= K(K+1)\Psi_{KM_K}, \\ \mathsf{K}_z\Psi_{KM_K} &= M_K\Psi_{KM_K} \quad (M_K = K, K-1, \ldots, -K), \end{aligned}\right\} \tag{A2.4}$$

where K is a positive half-integer, and that

$$\mathsf{K}^{\pm}\Psi_{K,M_K} = [(K \mp M_K)(K \pm M_K + 1)]^{1/2}\Psi_{K,M_K\pm 1}. \tag{A2.5}$$

If $\mathbf{K}$ is an *orbital* angular-momentum operator $(= \mathbf{L}$, say) then K is integral; but for spin and for total angular momenta (**S** or **J**) K may be half an odd integer.

Since most of the applications treated in this book refer to *spin* angular momentum we shall henceforth use S^2, S_z, etc. for the operators and S, M for the corresponding quantum numbers; but exactly parallel results apply to all kinds of angular momentum.

Let us consider first (cf. p. 92) a set of product functions $\Theta^{(1)}_{S_1M_1}\Theta^{(2)}_{S_2M_2}$ whose factors refer to two systems with mutually commuting operators S_1^2, S_{1z}, S_2^2, S_{2z}. The first and second factors are assumed to be normalized eigenfunctions of the operators with subscripts 1 and 2 respectively, with quantum numbers S_1, M_1 and S_2, M_2; and for given values of S_1 and S_2 there will be $(2S_1+1)(2S_2+1)$ possible products. The problem is how to combine these products in order to obtain eigenfunctions of the total spin operators S^2 and S_z defined in (4.1.1). For present purposes, noting that S_α ($\alpha = x, y, z$) is a sum of two parts $\mathsf{S}_{1\alpha}$, $\mathsf{S}_{2\alpha}$, it is useful to write

$$\mathsf{S}^2 = \mathsf{S}_1^2 + \mathsf{S}_2^2 + 2\mathbf{S}_1 \cdot \mathbf{S}_2, \tag{A2.6}$$

where S_1 and S_2 refer to the separate systems while (as follows by using (A2.3))

$$\mathbf{S}_1 \cdot \mathbf{S}_2 = \tfrac{1}{2}(\mathsf{S}_1^+\mathsf{S}_2^- + \mathsf{S}_1^-\mathsf{S}_2^+) + \mathsf{S}_{1z}\mathsf{S}_{2z} \tag{A2.7}$$

gives the spin scalar product.

The key to the solution is the recognition that the single product $\Theta^{(1)}_{S_1S_1}\Theta^{(2)}_{S_2S_2}$ (with $M_1 = S_1$, $M_2 = S_2$) is already an eigenfunction with $S = M = S_1 + S_2$. This follows at once from (A2.6) and (A2.7) since S_1^+ and S_2^+ each annihilate one factor (the "top" state cannot be stepped up) while S_{1z} and S_{2z} multiply the product by M_1 $(= S_1)$ and M_2 $(= S_2)$ respectively; from the assumed properties of $\mathbf{S}_1$ and $\mathbf{S}_2$ it follows that S^2 in (A2.6) will multiply the product by $S_1(S_1+1) + S_2(S_2+1) + 2S_1S_2$ (which is $S(S+1)$ with $S = S_1 + S_2$), while S_z multiplies by $M_1 + M_2 = S_1 + S_2$ (which is $M = S$), and this completes the proof.

The next step is to apply the operator $\mathsf{S}^- = \mathsf{S}_1^- + \mathsf{S}_2^-$, using (A2.5) for the separate factors. The result will be a linear combination of $\Theta^{(1)}_{S_1,S_1-1}\Theta^{(2)}_{S_2S_2}$ and $\Theta^{(1)}_{S_1S_1}\Theta^{(2)}_{S_2,S_2-1}$, which must be a function with $S = S_1 + S_2$ but $M = S_1 + S_2 - 1$. However, a second combination can be found, orthogonal to the first, also corresponding to $M = S_1 + S_2 - 1$; this must correspond to the "top" state (S, S) of a new series with $S = S_1 + S_2 - 1$, no other products contributing, because all those that remain correspond to lower values of M. The process may be continued, applications of S^- to the two eigenfunctions already found generating two new functions with $S = S_1 + S_2$ and $S = S_1 + S_2 - 1$ respectively, both with $M = S_1 + S_2 - 2$; there will then be a third linear combination, orthogonal to the first two, which will be the top state $(M = S)$ of a new series with

$S = S_1 + S_2 - 2$. In this way, by a straightforward but monotonous construction, we can find all possible "vector-coupled" states with $S = S_1 + S_2, S_1 + S_2 - 1, \ldots, |S_1 - S_2|$ and $2S + 1$ possible values for each resultant spin S.

The "Clebsch–Gordan" (CG) coefficients that occur in the linear combinations obtained above depend in a very complicated way on the quantum numbers (S, M), (S_1, M_1) and (S_2, M_2); and, although a general formula can be obtained (see e.g. Wigner, 1959) the coefficients are usually tabulated for the most frequently occurring situations. Various notations are in use; here we adopt $\left(\begin{matrix} S_1 & S_2 \\ M_1 & M_2 \end{matrix}\middle|\begin{matrix} S \\ M \end{matrix}\right)$ as the coefficient of the normalized product $\Theta^{(1)}_{S_1M_1}\Theta^{(2)}_{S_2M_2}$ in the S, M-eigenfunction. Thus, with a more suggestive Dirac-type notation,

$$\left|\begin{matrix} S \\ M \end{matrix}\right\rangle = \sum_{M_1, M_2} \left|\begin{matrix} S_1 & S_2 \\ M_1 & M_2 \end{matrix}\right\rangle \left(\begin{matrix} S_1 & S_2 \\ M_1 & M_2 \end{matrix}\middle|\begin{matrix} S \\ M \end{matrix}\right), \tag{A2.8}$$

which makes it clear that, since the product functions represented by $\left|\begin{matrix} S_1 & S_2 \\ M_1 & M_2 \end{matrix}\right\rangle$ are orthogonal, the CG coefficient on the right may be identified with a scalar product $\left\langle\begin{matrix} S_1 & S_2 \\ M_1 & M_2 \end{matrix}\middle|\begin{matrix} S \\ M \end{matrix}\right\rangle$. It is easily confirmed that the number of possible states Θ_{SM} is $(2S_1 + 1)(2S_2 + 1)$, and the full set of CG coefficients therefore forms a square matrix, which is unitary (Section 2.2), connecting two alternative orthonormal bases.

Although many tables of CG coefficients are available (e.g. Brink and Satchler, 1968), it is nowadays more common to use the closely related Wigner $3j$ symbols (Wigner, 1959), which better display various symmetry properties and have been more extensively tabulated (e.g. Rotenberg *et al.*, 1959).

The connection between the alternative coupling coefficients is

$$\left(\begin{matrix} S_1 & S_2 \\ M_1 & M_2 \end{matrix}\middle|\begin{matrix} S \\ M \end{matrix}\right) = (-1)^{(S_1 - S_2 + M)}(2S + 1)^{1/2}\begin{pmatrix} S_1 & S_2 & S \\ M_1 & M_2 & -M \end{pmatrix}, \tag{A2.9}$$

where the Wigner $3j$ symbol (in this case a "$3S$" symbol) appears in the parentheses on the right. The phase conventions adopted for the angular momentum eigenfunctions are such that all coupling coefficients are real; and the $3j$ symbols are multiplied by ± 1 under even or odd permutations (respectively) of the three columns. It should also be noted that the coefficients in (A2.8) are zero unless $M = M_1 + M_2$; the $3j$ symbols in (A2.9) correspondingly vanish unless $M_1 + M_2 + (-M) = 0$. There are also limitations on the S values; if any one is greater than the sum of the other two, or less than their difference, then the $3j$ symbol is zero; for obvious reasons, this is usually called a "triangle condition".

An important special case of angular-momentum coupling occurs in the construction of spin eigenfunctions, following the branching-diagram procedure (p. 94). In this case the only coefficients needed are those that describe the parallel or antiparallel coupling of an $(N+1)$th electron spin to an N-electron system in a state $\Theta_N^{(S,M)}$, to give a state with $S \rightarrow S \pm 1$. On using these coefficients, a very simple algorithm is obtained (Kotani *et al.*, 1955). It is sufficient to consider the case $M = S$, since eigenfunctions with other M-values are easily obtained by applying a step-down operator, and the results needed are then simply (showing only S-values)

$$\left.\begin{aligned} \Theta_{N+1}^{S+1/2} &= \Theta_N^S \alpha, \\ \Theta_{N+1}^{S-1/2} &= \frac{(-\mathrm{S}^- \Theta_N^S)\alpha + (2S)^{1/2}\Theta_N^S \beta}{(2S+1)}. \end{aligned}\right\} \qquad \text{(A2.10)}$$

Here the step-down operator (S^-) works on $\Theta_N^S(s_1, s_2, \ldots . s_N)$, a spin eigenfunction with $M = S$, and the α, β factors refer to spin s_{N+1}. It is a simple matter to construct the branching-diagram functions directly by repeated use of (A2.10).

Appendix 3: Symmetry and Group Concepts

We make fairly frequent reference in the text to functions of "given symmetry" and to results that depend on the theory of groups. Even though a deep understanding of group theory is not absolutely essential, some familiarity with basic concepts is necessary in many parts of quantum chemistry.

The form of a wavefunction is often largely determined by the *symmetry* of the potential provided by the fixed nuclei. For example, the characteristic angular dependence of the atomic orbitals of the hydrogen atom (Appendix 1) is determined completely by the fact that the electrostatic field due to the nucleus is spherically symmetric. In molecules the fixed nuclei often define some regular geometrical figure (for example, H_2O defines an isosceles triangle, NH_3 a trigonal pyramid). We say that the molecule has the symmetry of that figure, and it is intuitively clear that the molecular wavefunctions must reflect this symmetry.

When we discuss a physical system from a quantum-mechanical point of view we imagine that system is constructed in a certain frame of reference, for example, a rectangular cartesian coordinate system. Thus, the hydrogen molecule in Fig. A3.1a is constructed by placing two protons at $\pm\frac{1}{2}R$ along the x axis. Let us next consider the system constructed by the same rules but in a reference frame obtained from the original frame (the *fixed* frame) by a rotation through angle θ about the origin (Fig. A3.1b). In general, there will be an observable difference between the system in the rotated frame and in the fixed frame, but for some special rotations (Fig. A3.1c) the system in the rotated frame will be indistinguishable from that in the fixed frame. Such a rotation is called a *symmetry* operation.

For brevity we shall call the "system constructed in the rotated frame" the "rotated system". We note that the term "rotation" usually includes *reflections* ("improper" rotations) and that these need not be mechanically feasible: for example, the reflection that interchanges two of the

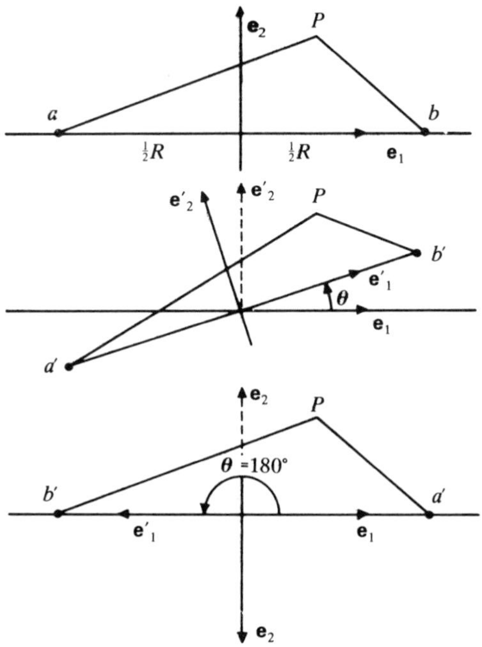

Fig. A3.1 Invariance of the Hamiltonian. (a) The potential at an arbitrary field point P depends on the positions of the identical nuclei a, b. (b) After rotation ($\theta \neq 180°$), a new potential function is obtained. (c) Rotation through 180° is a "symmetry" or "invariance" operation.

hydrogen atoms in the pyramidal molecule NH_3 is a symmetry operation, although it could not be performed without pulling the molecule apart.

Since free space is isotropic, quantities referring to the rotated system will always satisfy equations formally identical with those for the fixed system. Thus the Schrödinger equation

$$\mathsf{H}\Psi = E\Psi \tag{A3.1}$$

for the original system will be replaced by

$$\mathsf{H}'\Psi' = E\Psi'. \tag{A3.2}$$

Here H' and Ψ' are defined in the rotated frame just as H and Ψ were defined in the fixed frame, while E is again simply the numerical parameter whose eigenvalues are to be found by solving the equation. Equation (A3.2) tells us nothing new, although H' and Ψ' generally differ from H and Ψ when expressed in a common language (which we take to be that appropriate to the original "fixed" frame). If, however, the rotation were a symmetry operation, there would be no observable difference between the two operators:

$$\mathsf{H}' = \mathsf{H}. \tag{A3.3}$$

A symmetry operation is thus characterized mathematically as one that leaves the Hamiltonian invariant. This invariance property is not necessarily shared by the eigenfunctions, however, and symmetry plays a large part in determining their forms and transformation properties.

Example: The hydrogen molecule

The only term in the Hamiltonian that refers to the nuclear positions, and hence to orientation in space, is the electron–nuclear potential-energy function (atomic units)

$$V_{\text{en}} = -\left(\frac{1}{r_{a1}} + \frac{1}{r_{a2}} + \frac{1}{r_{b1}} + \frac{1}{r_{b2}}\right), \tag{A3.4}$$

where 1 and 2 are arbitrary "field points" at which we suppose the electrons are located. The rotated Hamiltonian is defined by putting the nuclei in their new positions a', b' (Fig. A3.1b), the new potential-energy function being

$$V'_{en} = -\left(\frac{1}{r_{a'1}} + \frac{1}{r_{a'2}} + \frac{1}{r_{b'1}} + \frac{1}{r_{b'2}}\right), \tag{A3.5}$$

where, for example, $r_{a'1}$ is the distance between field point 1 and nucleus *a in its rotated position* a'. In general (Fig. A3.1b), V'_{en} will differ from V_{en}: but if a' coincides with b and b' with a (Fig. A3.1c) the two potentials will be identical functions of the electronic variables, the terms in (A3.4) merely appearing in (A3.5) in a different order. Thus rotation through $\theta = \pi$ is a symmetry operation.

The basic implication of the existence of a symmetry operation, which we shall denote by R, is that the rotated function $\Psi' = \mathsf{R}\Psi$ must satisfy the same equation as the unrotated function Ψ; since when $\mathsf{H}' = \mathsf{H}$ it is clear that (A3.2) may be written as

$$\mathsf{H}(\mathsf{R}\Psi) = E(\mathsf{R}\Psi). \tag{A3.6}$$

Now rotation of the functions on each side of (A3.1) yields

$$\mathsf{R}(\mathsf{H}\Psi) = E(\mathsf{R}\Psi). \tag{A3.7}$$

The last two statements are generally compatible (for all Ψ) only if the operators on their left-hand sides are equivalent:

$$\mathsf{HR} = \mathsf{RH}. \tag{A3.8}$$

Symmetry operations therefore commute with the Hamiltonian, and we expect that their eigenvalues will be useful in classifying electronic states. Wavefunctions that are symmetric or antisymmetric under some given operation provide obvious examples.

The operations that describe the spatial symmetry of molecules may be specified in terms of rotations of a set of basis vectors‡, just in Section 2.3. Here the basis (e_1 e_2 e_3) may comprise the three unit vectors along the x, y, z axes. Any rotation about the origin sends these into a new set, the relationship between the two being (cf. (2.2.14))

$$(\mathsf{e}'_1\ \mathsf{e}'_2\ \mathsf{e}'_3) = (\mathsf{e}_1\ \mathsf{e}_2\ \mathsf{e}_3)\begin{pmatrix} R_{11} & R_{12} & R_{13} \\ R_{21} & R_{22} & R_{23} \\ R_{31} & R_{32} & R_{33} \end{pmatrix},$$

or, more briefly,

$$\mathbf{e}' = \mathbf{eR}, \tag{A3.9}$$

where it is understood that the basis vectors are collected into *row* matrices. The position vector of any point, fixed in the system, may be written (cf. (2.2.1)) as

$$\mathsf{r} = \sum_i \mathsf{e}_i r_i = \mathbf{er}, \tag{A3.10}$$

and in the rotation (A3.9) this vector is sent into the new vector

$$\mathsf{r}' = \sum_i \mathsf{e}'_i r_i = \mathbf{e}'\mathbf{r} = \mathbf{eRr}.$$

The components of the rotated vector are thus collected in the column

$$\mathbf{r}' = \mathbf{Rr}, \tag{A3.11}$$

which should be contrasted with (A3.9). The symmetry operations with which we are concerned, besides leaving (at least) one point fixed, also restrict the form of the matrix $\mathbf{R}$. This matrix is "real-orthogonal", which means that

$$\mathbf{R}^{-1} = \tilde{\mathbf{R}} \tag{A3.12}$$

($\tilde{\mathbf{R}}$ being the transpose of $\mathbf{R}$), and the determinant of $\mathbf{R}$ has the value ± 1 for proper and improper rotations, respectively. This restriction ensures that all distances and angles are preserved in the rotation, and hence that the rotated molecule *can* be exactly superimposed on the original.

When we collect all the (g) distinct symmetry operations that bring a

‡ In this Appendix we preserve, typographically, the distinction between a vector (r) and the set of components ($\mathbf{r}$)—a distinction that is relaxed elsewhere (for example in referring to the electric field $\boldsymbol{E}$) whenever no misunderstanding is likely to occur.

molecule into self-coincidence, including the "identity" or "unit" operation (which does nothing), we obtain a *group* G of order g. The symmetry operations are *elements* of the group, and for any two (A, B) there is a *product* $\mathsf{C} = \mathsf{AB}$, which is the element equivalent to B *followed by* A (this convention is important, for very often $\mathsf{BA} \neq \mathsf{AB}$); the group must contain the *identity* element (usually denoted by E) such that $\mathsf{EA} = \mathsf{A}$ for every operation A; and for every operation A there must be an *inverse* operation, denoted by A^{-1}, which entirely annuls the effect of A, $\mathsf{A}^{-1}\mathsf{A} = \mathsf{E}$. In the present example, we are concerned with the *point groups,* in which the elements are spatial symmetry operations that leave at least one point in the molecule (e.g. centre of rotation) unmoved. But we have also met other types of operation (for example *permutations* of indistinguishable particles in Chapter 4), which may also form groups. Generally, then, group theory can be applied in quantum mechanics whenever the Hamiltonian possesses at least one *invariance group*.

In quantum chemistry it is customary to use the Schoenflies notation for the molecular point groups and symmetry operations. It there is an n-fold principal axis then this is chosen as the "vertical" direction, and rotation through $2\pi/n$ is denoted by an operator C_n. Reflection in a plane containing this (vertical) axis is denoted by σ_v, while reflection in the perpendicular (i.e. horizontal) plane is σ_h. The *inversion* operation, which sends every point through the origin into a corresponding point on the opposite side, is denoted by i. If there are axes of symmetry perpendicular to the principal axis then these are denoted by C_n', and various extensions and refinements of the notation may be found in the literature (e.g. Herzberg, 1950). Point groups often contain many elements (e.g. 48 for a cube) but can then be defined in terms of a few typical operations that serve as *generators*; thus all the operations of the group of the cube may be built up by repeated application of (i) C_4 about the z axis, (ii) C_3 about the diagonal connecting two opposite corners, and (iii) the inversion i. For this reason, it is usually possible to concentrate entirely upon two or three basic symmetry operations in discussing, for example, the symmetry classification of wavefunctions.

SYMMETRY OF WAVEFUNCTIONS

In defining the rotated function,

$$\Psi' = \mathsf{R}\Psi, \tag{A3.13}$$

we rotated the *basis* in which the nuclear framework of the molecule, and the molecular wavefunctions, were defined. Nothing else changes; the point P, for example, is an arbitrary reference point in space, with coordinates x, y, z relative to the original basis **e**—which defines the "laboratory frame" to which everything will be referred. With this "active" interpretation of a symmetry operation, a rotated function is constructed just like the original function, but with respect to the rotated or "molecule-fixed" frame. As a simple example, we may consider orbitals (ϕ); thus a set of p-type AOs, expressed in terms of the laboratory-fixed basis will be

$$p_x = \mathrm{e}_1 \cdot \mathrm{r} f(r), \qquad p_y = \mathrm{e}_2 \cdot \mathrm{r} f(r), \qquad p_z = \mathrm{e}_3 \cdot \mathrm{r} f(r)$$

The rotated basis is obtained (remembering that each column of the orthogonal matrix **R** contains direction cosines of a rotated basis vector relative to the fixed basis) as

$$(\mathrm{e}_1' \ \mathrm{e}_2' \ \mathrm{e}_3') = (\mathrm{e}_1 \ \mathrm{e}_2 \ \mathrm{e}_3)\begin{pmatrix} l_1 & l_2 & l_3 \\ m_1 & m_2 & m_3 \\ n_1 & n_2 & n_3 \end{pmatrix}, \tag{A3.14}$$

and the p functions constructed relative to the new basis are then immediately found to be

$$(p_x' \ p_y' \ p_z') = (p_x \ p_y \ p_z)\begin{pmatrix} l_1 & l_2 & l_3 \\ m_1 & m_2 & m_3 \\ n_1 & n_2 & n_3 \end{pmatrix}. \tag{A3.15}$$

A set of p functions transforms under rotations exactly like a set of cartesian unit vectors; the p functions have "vector properties".

It is not always convenient, with more complicated functions, to work explicitly in terms of the basis vectors; since normally the value of a function is expressed in terms of the *coordinates*. An alternative prescription is therefore useful and reference to the example just considered shows this to be as follows: When $\mathbf{e} \rightarrow \mathbf{e}' = \mathbf{eR}$, the transformation of any function of x, y, z ($= \mathbf{r}$) is obtained by substituting the coordinates x', y', z' (contained in $\mathbf{r}' = \mathbf{R}^{-1}\mathbf{r}$) of the same reference point with respect to "molecule-fixed" axes. With cartesian coordinates, $\mathbf{R}^{-1} = \tilde{\mathbf{R}}$ (real orthogonal matrix) and no inversion is necessary. This prescription may be written formally as

$$\phi'(\mathbf{r}) = \mathsf{R}\phi(\mathbf{r}) = \phi(\mathbf{r}') = \phi(\mathbf{R}^{-1}\mathbf{r}) = \phi(\tilde{\mathbf{R}}\mathbf{r}), \tag{A3.16}$$

and obviously makes good sense because, replacing the arbitrary set of coordinates $\mathbf{r}$ by $\mathbf{Rr}$, it gives $\mathsf{R}\phi(\mathbf{Rr}) = \phi(\mathbf{r})$. In words: the numerical value of the function ϕ at point $\mathbf{r}$ is unchanged if we rotate both the function (i.e. $\phi \rightarrow \mathsf{R}\phi$) and the point (i.e. $\mathbf{r} \rightarrow \mathbf{Rr}$) at which it is evaluated. By using (A3.16), it is a simple matter to examine the transformation properties of all the functions in Table A1.2. The rotation of the set of five d functions, for example, is described by a 5×5 matrix whose columns contain the expansion coefficients when each rotated function is expressed in terms of the original set. Many-electron functions (Ψ), built from orbitals (ϕ), may evidently be handled in a similar way.

TRANSFORMATION OF OPERATORS

Suppose that we wish to calculate the components of the electric moment of a molecule, referred to molecule-fixed axes. To do this we evaluate the expectation values of the operators $\mathsf{d}_x = \sum_i \mathsf{d}_x(i) = \sum_i (-ex_i)$ etc. But what operators should we use to calculate the same quantities for the rotated molecule? The new operators will be defined in the same way as the old—but relative to the basis $\mathbf{e}' = \mathbf{eR}$. Thus d'_z will contain x'_i (measured along the new x direction) instead of x_i, and, since $x'_i = l_1 x_i + m_1 y_i + n_1 z_i$, it follows readily that the rotated operators will be $(\mathsf{d}'_x \ \mathsf{d}'_y \ \mathsf{d}'_z) = (\mathsf{d}_x \ \mathsf{d}_y \ \mathsf{d}_z)\mathbf{R}$, where $\mathbf{R}$ is given explicitly in (A3.14).

More complicated operators may be dealt with similarly. Components of linear momentum, for example, are defined for the rotated system in terms of derivatives $\partial/\partial x'$ etc. and, since $\partial/\partial x' = (\partial x/\partial x')\partial/\partial x + (\partial y/\partial x')\partial/\partial y + (\partial z/\partial z')\partial/\partial x'$ and $\boldsymbol{r} = \mathbf{R}\boldsymbol{r}'$, it follows that $\mathsf{p}'_x = l_1\mathsf{p}_x + m_1\mathsf{p}_y + n_1\mathsf{p}_z$. In general, all such *vector operators* follows the transformation rule

$$(\mathsf{V}'_x \ \mathsf{V}'_y \ \mathsf{V}'_z) = (\mathsf{V}_x \ \mathsf{V}_y \ \mathsf{V}_z)\mathbf{R}. \tag{A3.17}$$

On the other hand, the *angular*-momentum operators do not behave in quite the same way: they are *pseudo*vector operators and transform according to

$$(\mathsf{L}'_x \ \mathsf{L}'_y \ \mathsf{L}'_z) = \pm(\mathsf{L}_x \ \mathsf{L}_y \ \mathsf{L}_z)\mathbf{R}, \tag{A3.18}$$

where the negative sign appears when $\mathbf{R}$ describes an *improper* rotation (i.e. a rotation combined with reflection or inversion).

Very frequently an operator is *invariant* under the operations of some group (for example the Hamiltonian of a molecule with point-group symmetry); and in many cases such operators are constructed from sets of operators that are not invariant. Thus, if $\mathbf{V}_1$ and $\mathbf{V}_2$ are two sets, with elements $\mathsf{V}_i^{(1)}$, $\mathsf{V}_j^{(2)}$ that behave like (A3.17) but with matrices $\mathbf{R}_1$, $\mathbf{R}_2$ respectively, it is sometimes possible to obtain an invariant operator in the form $\mathsf{U} = \sum_i \mathsf{V}_i^{(1)}\mathsf{V}_i^{(2)}$. The condition that this be so is that $\mathsf{U}' = \sum_i \sum_{k,l} \mathsf{V}_k^{(1)} R_{ki}^{(1)} \mathsf{V}_l^{(2)} R_{li}^{(2)} = \mathsf{U}$; and this implies that $\sum_i R_{ki}^{(1)} R_{li}^{(2)} = \delta_{kl}$. In matrix notation this means that $\tilde{\mathbf{R}}_2 = \mathbf{R}_1^{-1}$, i.e. that one matrix is the transposed inverse of the other. Two transformations related in this way are said to be contragredient (a property exploited in Chapter 10) and when the two sets carry representations (D and $\check{D}$, say) of a group then $\check{D}$, with matrices $\tilde{\mathbf{D}}(\mathsf{R}^{-1})$, is the *contragredient representation*. For quantities whose transformation matrices are real-orthogonal, $\mathbf{R}^{-1} = \mathbf{R}$ and then the representations coincide, $\check{D} = D$. Thus, for example, the linear-momentum and the angular-momentum operators lead to operators T (kinetic energy) and L^2 that are both invariant with respect to all rotations (proper or improper).

DEGENERACY OF ENERGY LEVELS

In discussing "rotated functions", using atomic orbitals for illustration, we noted that it was the *degenerate* eigenfunctions of the corresponding Hamiltonian that were mixed under the rotations comprising the symmetry group of the system (in that case the rotation group R_3). This is a general result, applying to the eigenfunctions of any Hamiltonian with an invariance group. A given set of degenerate functions thus yields, on applying a symmetry operation R, the new set

$$\Psi_i' = \mathsf{R}\Psi_i = \sum_j \Psi_j R_{ji}, \tag{A3.19}$$

or, in matrix notation,

$$\boldsymbol{\Psi}' = \mathsf{R}\boldsymbol{\Psi} = \boldsymbol{\Psi}\mathbf{R}, \tag{A3.20}$$

where we revert to capital letters (indicating many-electron wavefunctions) to emphasize the generality of the result—the orbitals considered earlier being a very special case. The matrix $\mathbf{R}$ is thus characteristic of the symmetry operation R. Exactly as in Section 2.2, the matrices must behave like the rotations with which they are associated. If we denote the symmetry operations by A, B, C, . . . and have a relationship

$$\mathsf{AB} = \mathsf{C} \qquad \text{(combined by sequential performance)}$$

then the associated matrices must have a similar property

$$\mathbf{AB} = \mathbf{C} \qquad \text{(combined by matrix multiplication)},$$

and the full set $\{\mathbf{A}, \mathbf{B}, \mathbf{C}, \ldots\}$ thus forms a *representation* of the group $\{\mathsf{A}, \mathsf{B}, \mathsf{C}, \ldots\}$.

The representations introduced in the preceding examples are said to be *carried* by sets of eigenfunctions, these taking the place of the basis vectors in (A3.9) *et seq.* It must be noted, however, that representations are not necessarily *faithful,* i.e. the matrices associated with distinct elements are not necessarily distinct in a given representation. Thus, the sets of p and d orbitals carry representations in 3 and 5 dimensions respectively of the group of rotations about the origin; but any s orbital carries a *one*-dimensional representation in which the "matrix" 1 is associated with *every* rotation (rotation leaves an s orbital invariant, i.e. multiplies it by 1). Every point group possesses this trivial *identity* or *totally symmetric* representation, together with a number of other *inequivalent irreducible representations,* such as those carried by the p and d orbitals in the examples above. These basic representations are characteristic of the group considered and are tabulated in various textbooks (e.g. McWeeny, 1963).

The number of representations that need to be tabulated is limited for two reasons. First, there is no essential difference between two representations that differ only through choice of basis: if we use $\bar{\mathbf{e}} = \mathbf{eU}$ instead of $\mathbf{e}$ then it turns out (cf. 2.2.19)) that we must use $\bar{\mathbf{R}} = \mathbf{U}^{-1}\mathbf{RU}$ instead of $\mathbf{R}$, and the alternative representations, $\{\mathbf{A}, \mathbf{B}, \ldots\}$ and $\bar{\mathbf{A}}, \bar{\mathbf{B}}, \ldots\}$ are counted *equivalent.* Secondly, if by a change of basis it is possible to bring all the matrices of a representation to a common "block form"

$$\mathbf{A} = \left(\begin{array}{c|c} \mathbf{A}_1 & \mathbf{0} \\ \hline \mathbf{0} & \mathbf{A}_2 \end{array}\right), \qquad \mathbf{B} = \left(\begin{array}{c|c} \mathbf{B}_1 & \mathbf{0} \\ \hline \mathbf{0} & \mathbf{B}_2 \end{array}\right) \quad \text{etc.}$$

then each set of submatrices, $\{\mathbf{A}_1, \mathbf{B}_1, \ldots\}$ and $\{\mathbf{A}_2, \mathbf{B}_2, \ldots\}$ forms a representation by itself: the original representation, which is called their *direct sum,* is said to be *reduced* by the basis change and need not be given separate consideration. The *irreducible representations* are those that remain when no further reduction can be achieved: the forms and dimensions of these "*irreps*" are fundamental characteristics of the group.

We may summarize the preceding ideas as follows. Given a molecule whose electronic Hamiltonian H possesses an invariance group $\mathsf{G} = (\mathsf{A}, \mathsf{B}, \mathsf{C}, \ldots)$, the eigenfunctions will fall into degenerate sets; and the functions in any such set, $\Psi_1^{(\alpha)}$, $\Psi_2^{(\alpha)}, \ldots, \Psi_g^{(\alpha)}$, say, transform only among themselves under the operations of the symmetry group. Each set carries a representation of the group, normally irreducible, which may be denoted by D_α. It is customary in group theory to use a "functional" notation in which $\mathbf{D}_\alpha(\mathsf{R})$, rather than $\mathbf{R}$, denotes the matrix associated with group element R in representation D_α; and (A3.19) is then written as

$$\Psi_i'^{(\alpha)} = \mathsf{R}\Psi_i^{(\alpha)} = \sum_j \Psi_j^{(\alpha)} D_\alpha(\mathsf{R})_{ji}. \tag{A3.21}$$

In this case we say that $\Psi_i^{(\alpha)}$ "belongs to the representation D_α", the different members of the set being "partners" in the representation.

Whenever we come across another set, $\{\Omega_i^{(\alpha)}\}$, say, in which $\Omega_i^{(\alpha)}$ and its partners are mixed, by the group operations, by the same prescription (A3.21), then we say $\Omega_i^{(\alpha)}$ "transforms like the ith basis function of representation D_α" or is "of symmetry species (α, i)". It is of course possible to set up functions $\Phi_i^{(\alpha)}$, of any given symmetry species (α, i), that are *not* eigenfunctions of the Hamiltonian; and the nomenclature "symmetry functions" or "symmetry-adapted functions" is usually employed in this wider sense.

The special property of the eigenfunctions that carry an irrep D_α of a group under which H is invariant is, of course, their degeneracy; and indeed one of the main uses of group theory in quantum chemistry is in the discussion of degeneracies and their resolution. To appreciate the connection, we return to (A3.6) and note that if Ψ is an eigenfunction, with energy E, then so is $\mathsf{R}\Psi$; but this requires that $\mathsf{R}\Psi$ be a linear combination of the full set of eigenfunctions with this eigenvalue (this being the most general solution with energy E), and it follows that every eigenfunction, on rotation, turns into a linear combination containing only itself and the eigenfunctions degenerate with it. In other words, any set of eigenfunctions that carries a representation (D_α) of the group, as in (A3.21), is necessarily degenerate. Moreover, there is no group-theoretical reason why functions that are *not* mixed under symmetry operations (for example those from *different* irreducible representations) should be degenerate. A degenerate set is therefore normally assumed to carry an *irreducible* representation. Thus, any three p functions mix under operations of the group of all rotations in 3-dimensional space, and all must have the same energy; the same is true for a set of d functions, but there is no reason to expect a degeneracy between p and d states.

For an electron in a central field, the existence of s, p, d, f, . . . states, with degeneracies 1, 3, 5, 7, . . . is a consequence of the fact that the rotation group possesses irreducible representations of 1, 3, 5, 7, . . . dimensions. These degeneracies are due to symmetry alone, independent of the form of the central field so long as it has spherical symmetry: when, in rare cases, there is a further degeneracy (e.g. in the case of the hydrogen atom, where the coulombic form of the field results in $E_{2s} = E_{2p}$, $E_{3s} = E_{3p} = E_{3d}$, etc.) it is termed an "accidental degeneracy".

SPECIAL THEOREMS

Familiarity with the preceding notions is sufficient for an understanding of most of the text references to symmetry. The only fundamental results that we actually use are the following.

(i) The functions of any complete set may be grouped into linear combinations that are symmetry functions with respect to some group G. If $\Phi_{i\mu}^{(\alpha)}$ is the μth function of symmetry species (α, i) then only functions of this species are required in the expansion of an arbitrary function $\Psi_i^{(\alpha)}$ of the same given symmetry

$$\Psi_i^{(\alpha)} = \sum_\mu c_\mu \Phi_{i,\mu}^{(\alpha)}. \tag{A3.22}$$

A familiar example is the expansion of a function that is symmetric (or antisymmetric) under some operation in terms of basis functions with the same behaviour (see e.g. p. 73).

(ii) If the matrices of an irrep D_α (of dimension d_α) are known then symmetry functions may be projected from an *arbitrary* function by the prescription

$$\Phi_i^{(\alpha)} = \frac{d_\alpha}{g} \sum_R \breve{D}_\alpha(\mathsf{R})_{ij} \mathsf{R}\Phi = \rho_{ij}^{(\alpha)} \Phi \tag{A3.23}$$

—operate on Φ with each element R, attach the numerical coefficients shown, and sum over all g elements of the group. The operator $\rho_{ij}^{(\alpha)}$, defined‡ generally in this way, is usually referred to as a "Wigner operator". In fact, the quantities $\rho_{ii}^{(\alpha)}$ are projection operators, with the characteristic property of idempotency (p. 57), while the $\rho_{ij}^{(\alpha)}$ $(i \neq j)$ are "shift operators" that change a function of species (α, j) into a partner of species (α, i). In general,

$$\rho_{ij}^{(\alpha)} \rho_{kl}^{(\beta)} = \delta_{\alpha\beta} \delta_{jk} \rho_{il}^{(\alpha)}, \tag{A3.24}$$

from which the shift-operator property $\rho_{ij}^{(\alpha)} \Phi_j^{(\alpha)} = \Phi_i^{(\alpha)}$ easily follows. The Wigner operators are of fundamental importance in, for example, the theory of the unitary group (Chapter 10) and their properties are echoed in those of the second-quantization operators E_{ij}. If the matrices $\mathbf{D}_\alpha(\mathsf{R})$ are not available, we may use their *traces* $\chi_\alpha(\mathsf{R}) = \mathrm{tr}\,\mathbf{D}_\alpha(\mathsf{R})$ ($\chi_\alpha(\mathsf{R})$ is called the *character* of element R in representation D_α) to obtain a weaker form of (A3.23), namely that

$$\Phi^{(\alpha)} = \frac{1}{g} \sum_{\mathsf{R}} \breve{\chi}_\alpha(\mathsf{R}) \mathsf{R}\Phi \tag{A3.25}$$

belongs to the representation D_α. Consequently, $\Phi^{(\alpha)}$ is a linear combination of the $\Phi_i^{(\alpha)}$ given above in (A3.23). An example is

‡The definition and properties of such operators are presented here in general form. In the text they have been used mainly for *unitary* representations, in which case the matrix elements in the contragredient representation (p. 538) are $\breve{D}_\alpha(\mathsf{R})_{ij} = D_\alpha(\mathsf{R})_{ij}^*$.

the antisymmetrizer (p. 57), in which the operations are permutations and the characters in the totally antisymmetric one-dimensional representation are ±1 according to parity. Equation (A3.23) is also valid and useful when D_α is *reducible* (see McWeeny, 1963, p. 199).

(iii) In evaluating elements of an operator T (such as the Hamiltonian) that is invariant under all group operations, it turns out that

$$\langle \Phi_{i\mu}^{(\alpha)} | \mathsf{T} | \Phi_{j\nu}^{(\beta)} \rangle = \delta_{\alpha\beta}\delta_{ij} T_{\mu\nu} \qquad (T_{\mu\nu} \text{ independent of } i, j). \qquad \text{(A3.26)}$$

This result is closely related to (i): the matrix element vanishes unless the functions are of identical symmetry species, and consequently, in setting up a secular problem, there will be no mixing between functions of different species. Furthermore, since the result is independent of i, the same secular problem will appear for each species of given α but with $i = 1, 2, \ldots, d_\alpha$, corresponding to the d_α-fold degeneracy of eigenfunctions belonging to the representation D_α.

A generalization of the last result is needed when T is not invariant but is instead one of a family of operators (e.g. angular-momentum operators) carrying a representation. And when such an operator set is chosen so as to carry an irrep D_α, thus behaving under the group operations according to the rule (A3.21), it is usually described as an "irreducible tensorial set" (see e.g. Fano and Racah, 1959; Griffith, 1961). In the case of the vector operators (i.e. "rank-1" tensor operators) we have frequently used (for example in Chapter 11) "spherical" components (for example the spin operators S_{+1}, S_0, S_{-1} instead of the usual S_x, S_y, S_x), which behave under rotations exactly like the p-type AOs p_{+1}, p_0, p_{-1}, with spherical-harmonic angular factors and Condon–Shortley phase conventions; and both sets in this case carry a common standard irrep of the 3-dimensional rotation group. For this, and many of the other most commonly encountered groups, the required generalization of (iii) is as follows.

(iv) The matrix elements of a tensor operator of symmetry species (γ, k), between functions of species (α, i) and (β, j), where i, j, k take all possible values, are related by (Wigner–Eckart theorem: Wigner, 1959; Eckart, 1930).

$$\langle \Phi_{i\mu}^{(\alpha)} | \mathsf{T}_k^\gamma | \Phi_{j\nu}^{(\beta)} \rangle = \begin{pmatrix} \gamma & \beta & | & \alpha \\ k & j & | & i \end{pmatrix}^* \times \text{constant}, \qquad \text{(A3.27)}$$

where the "constant" is independent of i, j, k, and therefore depends on the nature of the functions and the representations‡ to which they (and the operators) belong, while the bracketed factor is a numerical quantity, a "coupling coefficient", taking values characteristic of the group and independent of the particular physical problem. For the rotation group the species symbols (α, i), (γ, k), (β, j) become pairs of angular-momentum quantum numbers, for example (L_1, M_1), (L_2, M_2), (L_3, M_3), and the coupling coefficient is the quantity appearing when two angular momenta $\boldsymbol{L}_2$ and $\boldsymbol{L}_3$ are coupled to a resultant $\boldsymbol{L}_1$ according to the formula

$$\Psi(L_1M_1) = \sum_{M_2M_3} \Psi(L_2M_2)\Psi(L_3M_3)\begin{pmatrix} L_2 & L_3 & | & L_1 \\ M_2 & M_3 & | & M_1 \end{pmatrix}. \quad \text{(A3.28)}$$

In this case the bracketed factor is a Clebsch–Gordan coefficient (Appendix 2) and is real so that the star in (A3.27) may be omitted.

The last result is used in two ways: (i) we may evaluate the constant in (A3.27) by working out the matrix element for one choice of (i, j, k), the simplest, and a knowledge of the coupling constants then gives the whole family of matrix elements corresponding to other choices; or (ii) we may relate matrix elements of one tensor operator $\mathsf{X}_k^{(\gamma)}$ to those of another $\mathsf{Y}_k^{(\gamma)}$, of the same species, by applying (A3.27) to each and taking the ratio to obtain

$$\langle \Phi_{i\mu}^{(\alpha)} | \mathsf{X}_k^{(\gamma)} | \Phi_{j\nu}^{(\beta)} \rangle = \text{constant} \times \langle \Phi_{i\mu}^{(\alpha)} | \mathsf{Y}_k^{(\gamma)} | \Phi_{j\nu}^{(\beta)} \rangle. \quad \text{(A3.29)}$$

This result, in which the coupling coefficients no longer appear explicitly, is often called the "replacement theorem": it is of limited value (although sometimes useful in the applications in Chapter 11). For further discussion of (A3.27), particularly for the point groups, the reader is referred elsewhere (e.g. McWeeny, 1963; Griffith, 1961, 1962).

‡ Here assumed to be unitary.

Appendix 4: Relativistic Terms in the Hamiltonian

During the last 20 years much progress has been made in the actual implementation of relativistic calculations; but there remains no satisfactory foundation for a "fully relativistic" many-electron Hamiltonian. The most striking progress (see e.g. Grant, 1970; Pyykö, 1978; Grant and Quinney, 1987; and, for a rather comprehensive bibliography, Pyykö, 1986) has been in the use of 4-component wavefunctions in the approximate solution of the Dirac equation (11.2.9), usually in stationary-state form, and its somewhat *ad hoc* N-electron generalization—in which each electron has its own Dirac Hamiltonian but electron–electron interactions are represented classically by the usual $g(i, j)$ terms. Even at this level, practical difficulties have come to light in attempts to make accurate finite-basis calculations. For example, the Dirac equation has negative-energy solutions, representing positron states, and any variational approximation is therefore liable to "collapse" towards the positron solutions. Other difficulties are also recognized in the reduction of the Dirac equation to 2-component form: the spin terms appear to be satisfactory, at least to order $1/c^2$, but the so-called "mass-variation" (π^4) term, which arises from an expansion whose convergence is open to discussion, may be incorrect (Farazdel and Smith, 1986).

For many-particle systems, in spite of much progress, the status of some of the small terms of relativistic origin is still obscure. All that can be said is that there remains a certain consensus about the forms of the many spin-dependent terms whose effects are *experimentally observable*—however dubious their parentage may be—and that these forms (usually presented in a Pauli-type wavefunction) are just the ones that would be expected on the basis of simple semiclassical arguments. It seems likely that for many years to come these terms will continue to be accepted, and their effects to be discussed using Pauli-type spin operators and wavefunctions built from familiar (2-component) spin-orbitals.

In this admittedly unsatisfactory situation we merely sketch the origin of the Pauli-type equations and list the most significant terms that occur in the many-electron generalization (originating in the work of Breit (1929, 1930, 1932) and further discussed by Itoh (1965)).

PAULI FORM OF THE DIRAC EQUATION

There are several ways of reducing the one-electron relativistic equation (11.2.7) to a form involving the Pauli spin operators. The method we use here is due to Löwdin (1964) and utilizes the matrix partitioning technique of Section 2.5.

It is convenient to adopt the Dirac representation of the operators α_ν and β, in which

$$\beta \to \begin{pmatrix} \mathbf{I} & \mathbf{0} \\ \mathbf{0} & -\mathbf{I} \end{pmatrix}, \qquad \alpha_\nu \to \begin{pmatrix} \mathbf{0} & \boldsymbol{\sigma}_\nu \\ \boldsymbol{\sigma}_\nu & \mathbf{0} \end{pmatrix} \qquad (\nu = 1, 2, 3). \tag{A4.1}$$

Here $\mathbf{I}$ is a 2×2-unit matrix and the $\boldsymbol{\sigma}_\nu$ are the Pauli spin matrices

$$\boldsymbol{\sigma}_1 = \begin{pmatrix} 0 & 1 \\ 1 & 0 \end{pmatrix}, \qquad \boldsymbol{\sigma}_2 = \begin{pmatrix} 0 & -i \\ i & 0 \end{pmatrix}, \qquad \boldsymbol{\sigma}_3 = \begin{pmatrix} 1 & 0 \\ 0 & -1 \end{pmatrix}. \tag{A4.2}$$

It is readily confirmed that these matrices have the required commutation properties given by (11.2.5), and we now work entirely in the 4×4 representation, making no typographical distinction between operators and the corresponding matrices. The four components of the wavefunction may then be collected in a column

$$\boldsymbol{\psi} = \begin{pmatrix} \psi_1 \\ \psi_2 \\ \psi_3 \\ \psi_4 \end{pmatrix}. \tag{A4.3}$$

Such a wavefunction is called a *four-component spinor,* the components being functions of space variables only.

The equation (11.2.7) is exactly soluble in a number of simple cases, for example the free particle and the particle moving in a fixed coulomb potential (Dirac, 1958; Slater, 1960). In these situations it is found that there are always four associated solutions; two in which the total energy (i.e. energy including the rest-mass energy mc^2) is positive and two in which it is negative. For the solutions of positive total energy it is found that at low enough energies ($\simeq mc^2$) the components ψ_1 and ψ_2 are very much larger than the components ψ_3 and ψ_4 and that in this non-relativistic limit the functions ψ_1 and ψ_2 become the solutions of the corresponding Schrödinger equation. These results suggest that we might be able to find an equivalent 2×2 equation to determine the 2-component spinor composed of ψ_1 and ψ_2, and that the corresponding effective Hamiltonian may be expressible in terms of the Pauli operators alone.

The form of the effective Hamiltonian has been given already in equation (11.2.10). This equation may be derived in the following way.‡ First we use the freedom of phase allowed in any wavefunction to eliminate the rest-mass energy (for positive-energy states) by writing

$$\boldsymbol{\psi} = \exp(-imc^2t/\hbar)\bar{\boldsymbol{\psi}}. \tag{A4.4}$$

Equation (11.2.7) then becomes §

$$[\boldsymbol{\pi}_0 - \boldsymbol{\alpha} \cdot \boldsymbol{\pi} - (\boldsymbol{\beta} - \mathbf{1})mc]\bar{\boldsymbol{\psi}} = \mathbf{0}. \tag{A4.5}$$

Next we partition $\bar{\boldsymbol{\psi}}$ as

$$\bar{\boldsymbol{\psi}} = \begin{pmatrix} \boldsymbol{\psi}_A \\ \boldsymbol{\psi}_B \end{pmatrix}, \tag{A4.6}$$

where $\boldsymbol{\psi}_A$ contains the transformed large components and $\boldsymbol{\psi}_B$ the small components. With the partitioning of $\boldsymbol{\alpha}$ and $\boldsymbol{\beta}$ given in (A4.1), (11.2.7) is equivalent to

$$\pi_0 \boldsymbol{\psi}_A - (\boldsymbol{\alpha} \cdot \boldsymbol{\pi})\boldsymbol{\psi}_B = \mathbf{0}, \tag{A4.7a}$$

$$\boldsymbol{\pi}_0 \boldsymbol{\psi}_B - (\boldsymbol{\sigma} \cdot \boldsymbol{\pi})\boldsymbol{\psi}_A + 2mc\boldsymbol{\psi}_B = \mathbf{0}. \tag{A4.7b}$$

Here, as in (11.2.7), $\boldsymbol{\sigma} \cdot \boldsymbol{\pi}$ denotes the "scalar product" $\sum_\nu \boldsymbol{\sigma}_\nu \boldsymbol{\pi}_\nu$, the $\boldsymbol{\pi}_\nu$ now being regarded as 2×2 diagonal matrices. From (A4.7b) we obtain, proceeding as in Section 2.5,

$$\boldsymbol{\psi}_B = (\pi_0 + 2mc)^{-1}(\boldsymbol{\sigma} \cdot \boldsymbol{\pi})\boldsymbol{\psi}_A, \tag{A4.8}$$

and hence the equation formally determining $\boldsymbol{\psi}_A$ is

$$[\boldsymbol{\pi}_0 - \boldsymbol{\sigma} \cdot \boldsymbol{\pi}(\pi_0 + 2mc)^{-1}\boldsymbol{\sigma} \cdot \boldsymbol{\pi}]\boldsymbol{\psi}_A = \mathbf{0}, \tag{A4.9}$$

which is entirely of 2-component form.

Let us now introduce the operator

$$\mathsf{k} = \left(1 + \frac{\pi_0}{2mc}\right)^{-1}, \tag{A4.10}$$

‡ Although the approach presented here is due to Löwdin (1964), the idea of division into large and small components is a very old one. For a more traditional approach see Slater (1960), or Akhiezer and Berestetskii (1965). For a different approach see Messiah (1962).

§ Note here that each $\boldsymbol{\pi}_\nu$ is regarded as a diagonal matrix, with the operator π_ν along the diagonal.

and use the fact that for any vectors **a** and **b** that commute with $\boldsymbol{\sigma}$

$$(\boldsymbol{\sigma}\cdot\mathbf{a})(\boldsymbol{\sigma}\cdot\mathbf{b}) = \mathbf{a}\cdot\mathbf{b} + i\boldsymbol{\sigma}\cdot(\mathbf{a}\times\mathbf{b}). \qquad \text{(A4.11)}$$

With the aid of this result, (A4.9) may be rewritten as

$$\left(i\hbar\frac{\partial}{\partial t} - q\phi - \frac{1}{2m}\boldsymbol{\pi}\cdot\mathsf{k}\boldsymbol{\pi} - \frac{i}{2m}\boldsymbol{\sigma}\cdot(\boldsymbol{\pi}\times\mathsf{k}\boldsymbol{\pi})\right)\boldsymbol{\psi}_A = \mathbf{0}. \qquad \text{(A4.12)}$$

To introduce the usual spin operators, we simply note that there is a one-to-one correspondence between the effect of the matrices $\frac{1}{2}\boldsymbol{\sigma}_v$ on the matrices $\begin{pmatrix}1\\0\end{pmatrix}$ and $\begin{pmatrix}0\\1\end{pmatrix}$ and the effect of the spin operators S_v on the spin functions $\alpha(s)$ and $\beta(s)$ respectively. Thus we may reinterpret the 2-component spinor

$$\boldsymbol{\psi}_A = \psi_1\begin{pmatrix}1\\0\end{pmatrix} + \psi_2\begin{pmatrix}0\\1\end{pmatrix} \qquad \text{(A4.13)}$$

as the usual spin-orbital

$$\psi_A = \psi_1\alpha(s) + \psi_2\beta(s) \qquad \text{(A4.14)}$$

if, at the same time, we replace $\frac{1}{2}\boldsymbol{\sigma}$ by **S** in (A4.12). This procedure yields equation (11.2.10).

To pass from (A4.12), or (11.2.10), to a Dirac–Pauli equation containing the Hamiltonian (11.2.17), it is customary to expand the operator **k** in powers of $1/c$; and it is also necessary to replace ψ_A (which is only a truncated part of the 4-component quantity (A4.3)) by a modified 2-component function, "corrected" to admit the small components in (A4.8). It is this step that leads to the "mass-variation" term and that has certain dubious features (Farazdel and Smith, 1986); there is, however, general agreement that the final Hamiltonian (11.2.17) is satisfactory apart from the π^4 term and that its 2-component eigenfunctions—the usual spin-orbitals—will correctly allow for relativistic effects up to order $1/c^2$.

MANY-ELECTRON SYSTEMS

The extension to many-particle systems (Breit, 1929, 1932, 1939) is based on, first, a generalization of the Dirac Hamiltonian to two (possibly unlike) particles, with the "retarded" interactions (see Chapter 11: (11.3.2)) to allow for the finite velocity of light, and then its application to a many-particle system in which all pairwise interactions are treated as additive.

Here we simply put on record the results obtained (see e.g. Moss, 1973) for a many-electron molecule in the fixed-nucleus approximation. The Hamiltonian will be written (cf. (1.1.13)) in the form

$$\mathsf{H} = \mathsf{H}_e + \mathsf{H}_n + \mathsf{H}_{en}. \tag{A4.15}$$

The pure electronic term is then

$$\mathsf{H}_e = \sum_{m=1}^{12} \mathsf{H}_m^e, \tag{A4.16}$$

where (using κ_0, as usual, to denote $4\pi\epsilon_0$)

$$\mathsf{H}_1^e = \sum_i \pi^2(i)/2m, \tag{A4.17a}$$

$$\mathsf{H}_2^e = -e \sum_i \phi(i), \tag{A4.17b}$$

$$\mathsf{H}_3^e = g\beta \sum_i \mathbf{S}(i) \cdot \boldsymbol{B}(i), \tag{A4.17c}$$

$$\mathsf{H}_4^e = -\sum_i \boldsymbol{\pi}^4(i)/8m^3c^2, \tag{A4.17d}$$

$$\mathsf{H}_5^e = -(g\beta/4mc^2) \sum_i [\mathbf{S}(i) \cdot \boldsymbol{\pi}(i) \times \boldsymbol{E}(i) - \mathbf{S}(i) \cdot \boldsymbol{E}(i) \times \boldsymbol{\pi}(i)], \tag{A4.17e}$$

$$\mathsf{H}_6^e = (\hbar\beta/4mc^2) \sum_i \operatorname{div} \boldsymbol{E}(i), \tag{A4.17f}$$

$$\mathsf{H}_7^e = (e^2/2\kappa_0) \sum_{i,j}{}' (1/r_{ij}), \tag{A4.17g}$$

$$\mathsf{H}_8^e = -(2\pi\beta^2/\kappa_0c^2) \sum_{i,j}{}' \delta(\boldsymbol{r}_i - \boldsymbol{r}_j), \tag{A4.17h}$$

$$\mathsf{H}_9^e = -(g\beta^2/\hbar\kappa_0c^2) \sum_{i,j}{}' r_{ij}^{-3}[2\mathbf{S}(i) \cdot \boldsymbol{r}_{ij} \times \boldsymbol{\pi}(j) + \mathbf{S}(j) \cdot \boldsymbol{r}_{ij} \times \boldsymbol{\pi}(j)], \tag{A4.17i}$$

$$\mathsf{H}_{10}^e = -(\beta^2/\hbar^2\kappa_0c^2) \sum_{i,j}{}' r_{ij}^{-3}[r_{ij}^2\boldsymbol{\pi}(i) \cdot \boldsymbol{\pi}(j) + \boldsymbol{r}_{ij} \cdot (\boldsymbol{r}_{ij} \cdot \boldsymbol{\pi}(i))\boldsymbol{\pi}(j)], \tag{A4.17j}$$

$$\mathsf{H}_{11}^e = -(g^2\beta^2/2\kappa_0c^2) \sum_{i,j}{}' r_{ij}^{-5}[3(\mathbf{S}(j) \cdot \boldsymbol{r}_{ij})(\mathbf{S}(i) \cdot \boldsymbol{r}_{ij}) - r_{ij}^2\mathbf{S}(i) \cdot \mathbf{S}(j)], \tag{A4.17k}$$

$$\mathsf{H}_{12}^e = -(4\pi g^2\beta^2/3\kappa_0c^2) \sum_{i,j}{}' \mathbf{S}(i) \cdot \mathbf{S}(j)\delta(\boldsymbol{r}_i - \boldsymbol{r}_j). \tag{A4.17l}$$

The pure nuclear term is

$$\mathsf{H}_n = \sum_{m=1}^{4} \mathsf{H}_m^n, \tag{A4.18}$$

where (with the fixed-nucleus approximation)

$$\mathsf{H}_1^{\mathrm{n}} = e \sum_n Z_n \phi(n), \tag{A4.19a}$$

$$\mathsf{H}_2^{\mathrm{n}} = -\beta_{\mathrm{p}} \sum_n g_n \mathbf{I}(n) \cdot \boldsymbol{B}(n), \tag{A4.19b}$$

$$\mathsf{H}_3^{\mathrm{n}} = (e^2/2\kappa_0) \sum_{n,n'}{}' Z_n Z_{n'}/R_{nn'}, \tag{A4.19c}$$

$$\mathsf{H}_4^{\mathrm{n}} = -(\beta_{\mathrm{p}}^2/2\kappa_0 c^2) \sum_{n,n'}{}' g_n g_{n'} R_{nn'}^{-5}[3(\mathbf{I}(n) \cdot \boldsymbol{R}_{nn'})(\mathbf{I}(n') \cdot \boldsymbol{R}_{nn'}) - R_{nn'}^2 \mathbf{I}(n) \cdot \mathbf{I}(n')]. \tag{A4.19d}$$

Finally, the electron–nuclear term is

$$\mathsf{H}_{\mathrm{en}} = \sum_{m=1}^{6} \mathsf{H}_m^{\mathrm{en}}, \tag{A4.20}$$

where

$$\mathsf{H}_1^{\mathrm{en}} = -(e^2/\kappa_0) \sum_{i,n} Z_n/r_{ni}, \tag{A4.21a}$$

$$\mathsf{H}_2^{\mathrm{en}} = (g\beta\beta_{\mathrm{p}}/\kappa_0 c^2) \sum_{i,n} g_n r_{ni}^{-5}[3(\mathbf{S}(i) \cdot \mathbf{r}_{ni})(\mathbf{I}(n) \cdot \mathbf{r}_{ni}) - r_{ni}^2 \mathbf{S}(i) \cdot \mathbf{I}(n)], \tag{A4.21b}$$

$$\mathsf{H}_3^{\mathrm{en}} = (8\pi g\beta\beta_{\mathrm{p}}/3\kappa_0 c^2) \sum_{i,n} g_n \mathbf{S}(i) \cdot \mathbf{I}(n) \delta(\boldsymbol{r}_i - \boldsymbol{r}_n), \tag{A4.21c}$$

$$\mathsf{H}_4^{\mathrm{en}} = (2\beta\beta_p/\hbar\kappa_0 c^2) \sum_{i,n} g_n r_{ni}^{-3}[\mathbf{I}(n) \cdot \boldsymbol{r}_{ni} \times \boldsymbol{\pi}(i)], \tag{A4.21d}$$

$$\mathsf{H}_5^{\mathrm{en}} = (g\beta^2/\hbar\kappa_0 c^2) \sum_{i,n} Z_n r_{ni}^{-3}[\mathbf{S}(i) \cdot \boldsymbol{r}_{ni} \times \boldsymbol{\pi}(i)], \tag{A4.21e}$$

$$\mathsf{H}_6^{\mathrm{en}} = (2\pi\beta^2/\kappa_0 c^2) \sum_{i,n} Z_n \delta(\boldsymbol{r}_i - \boldsymbol{r}_n). \tag{A4.21f}$$

In the formulae given above all the signs are explicitly taken care of, so e and Z_n should be entered in them as positive numbers. We have used the Bohr magneton β and the nuclear magneton $\beta_{\mathrm{p}} = e\hbar/2M$ (where M is the proton mass), and the electron and nuclear moments are thus

$$\boldsymbol{\mu}_i = -g\beta \mathbf{S}(i) \quad \text{(electron)}, \tag{A4.22a}$$

$$\boldsymbol{\mu}_n = g_n \beta_{\mathrm{p}} \mathbf{I}(n) \quad \text{(nucleus)}. \tag{A4.22b}$$

There is of course no justification, in the theory as we have presented it, for the inclusion of the observed free-electron g factor ($g \neq 2$), and its inclusion is required simply on phenomenological grounds. The nuclear factors g_n, must be found experimentally and tables of them may be found, for example, in Pople *et al.* (1959). Sometimes nuclear magnetic moments are given in terms of the nuclear magnetogyric ratios γ_n defined so that $\gamma_n\hbar = g_n\beta_p$.

It is also noteworthy that the dipole–dipole interaction terms (A4.17k), (A4.19d) and (A4.21b) may appear to give divergent expectation values. Only the last of these, however, needs special consideration since in (A4.19d) $R_{nn'}$ is normally finite, and since the spin–spin coupling function (p. 150) behaves in such a way as to eliminate any possible divergence arising from (A4.17k). Any divergence in dealing with (A4.21b) can be avoided either by evaluating the integrals in momentum space (see e.g. Bethe and Saltpeter, 1957, p. 180) or by using coordinate space but excluding a small spherical volume element near the nucleus and then letting that element tend to zero.

It should be noticed that there are no nuclear quadrupole terms in the above formulae, as these arise only when the nucleus is considered as a structured particle. Such terms are important in, for example, the consideration of relaxation phenomena in solids. Their forms are discussed elsewhere (e.g. Abragam, 1961; Slichter, 1964). Briefly, the extra term required is

$$\mathsf{H}_7^{\mathrm{en}} = \sum_{i,n} \tfrac{1}{6}\mathbf{Q}(n) \cdot \mathbf{V}_n(i), \qquad \text{(A4.21g)}$$

where $\mathbf{Q}(n) \cdot \mathbf{V}_n(i)$ is a tensor scalar product between the nuclear quadrupole tensor $\mathbf{Q}(n)$ and the electric-field-gradient tensor $\mathbf{V}_n(i)$ evaluated at the position of nucleus n. In cartesian coordinates $\mathbf{V}_n(i)$ has components

$$V_n(i)_{xx} = -\frac{e}{\kappa_0 r_{ni}^5}(3x_{ni}^2 - r_{ni}^2), \qquad V_n(i)_{xy} = -\frac{e}{\kappa_0 r_{ni}^5}(3x_{ni}y_{ni}) \qquad \text{(A4.23)}$$

etc. It is usually more convenient to introduce the spherical tensor components (cf. p. 149) and then

$$\begin{aligned} V_{\pm 2} &= \tfrac{1}{2}\sqrt{3}\,(V_{xx} \pm 2iV_{xy} - V_{yy}), \\ V_{\pm 1} &= \mp\sqrt{6}\,(V_{zx} \pm iV_{yz}), \\ V_0 &= 2V_{zz} - V_{xx} - V_{yy}. \end{aligned} \qquad \text{(A4.24)}$$

In this case the scalar product in (A4.21g) takes the form

$$\mathbf{Q}(n) \cdot \mathbf{V}_n(i) = \sum_m (-1)^m Q(n)_{-m} V_n(i)_m \quad (m = 0, \pm 1, \pm 2). \quad \text{(A4.25)}$$

The numerical factor in the definition (A4.21g) may be removed if the components in (A4.24) are conventionally normalized by adding a factor $1/\sqrt{6}$, but usually this is not done.

References

Abragam, A. (1961). *The Principles of Nuclear Magnetism.* Oxford University Press.

Ahlrichs, R. (1979). *Comput. Phys. Commun.* **17,** 31.

Ahlrichs, R. (1983). In *Methods in Computational Molecular Physics* (ed. G. H. F. Diercksen and S. Wilson), p. 209. Reidel, Dordrecht.

Ahlrichs, R. and Scharf, P. (1987). *Adv. Chem. Phys.* **67,** 501.

Akhiezer, A. and Berestetskii, V. (1965). *Quantum Electrodynamics,* Chap. 8. Wiley-Interscience, New York.

Albertson, P., Jørgensen, P. and Yeager, D. L. (1980). *Mol. Phys.* **41,** 409.

Amos, A. T. and Crispin, R. J. (1970). In *Theoretical Chemistry: Advances and Perspectives* (ed. H. Eyring and D. Henderson), p. 1. Academic Press, New York.

Amos, A. T. and Hall, G. G. (1961). *Proc. R. Soc. Lond.* **A263,** 483.

Amovilli, C. and McWeeny, R. (1986). *Chem. Phys. Lett.* **128,** 11.

Archbold, J. W. (1961). *Algebra,* p. 338. Pitman, London.

Arrighimi, G.-P., Biondi, F. and Guidotti, C. (1971). *J. Chem. Phys.* **55,** 4090.

Atkins, P. W. and Seymour, P. A. (1973). *Mol. Phys.* **25,** 113.

Balasumbramanian, K. and Pitzer, K. S. (1987). *Adv. Chem. Phys.* **67,** 287.

Balint-Kurti, G. and Karplus, M. (1968). *J. Chem. Phys.* **50,** 478.

Balint-Kurti, G. and Karplus, M. (1974). In *Orbital Theories of Molecules and Solids* (ed. N. H. March), p. 250. Oxford University Press.

Bangudu, E. A., Jankowski, K. and Dion, D. R. (1974). *Chem. Phys. Lett.* **19,** 418.

Banwell, C. N. and Primas, H. (1963). *Mol. Phys.* **6,** 225.

Bartlett, R. J., Wilson, S. and Silver, D. M. (1977). *Int. J. Quantum Chem.* **12,** 737.

Bethe, H. and Salpeter, E. (1957). *Quantum Mechanics of One- and Two-Electron Atoms,* Sects. 38 and 39. Springer-Verlag, Berlin.

Born, M. (1964). *Natural Philosophy of Cause and Chance.* Dover, New York.

Born, M. and Green, H. S. (1947). *Proc. R. Soc. Lond.* **A191,** 168.

Born, M. and Huang, K. (1954). *Dynamical Theory of Crystal Lattices,* Appendix 8. Oxford University Press.

Born, M. and Oppenheimer, J. R. (1927). *Annln Phys.* **84,** 457.

Boys, S. F. (1955). *Svensk Kem. Tidskr.* **67,** 367. (See Reeves (1957) for the theorem referred to here.)

Boys, S. F. (1969). *Proc. R. Soc. Lond.* **A309,** 195

Boys, S. F. and Foster, J. (1960). *Rev. Mod. Phys.* **32,** 305.

Boys, S. F. and Handy, N. C. (1969). *Proc. R. Soc. Lond.* **A309,** 209.

Brandow, B. H. (1967). *Rev. Mod. Phys.* **39,** 771.

Brandow, B. H. (1977). *Adv. Quantum Chem.* **10,** 187.

Breit, G. (1929). *Phys. Rev.* **34,** 553.
Breit, G. (1930). *Phys. Rev.* **36,** 383.
Breit, G. (1932). *Phys. Rev.* **39,** 616.
Brink, D. M. and Satchler, G. R. (1968). *Angular Momentum,* 2nd edn. Clarendon Press, Oxford.
Burrau, O. (1927). *Kgl. Danske. Videnskab. Selskab.* **7,** 1.
Calais, J.-L. (1985). *Adv. Quantum Chem.* **17,** 225.
Campion, W. J. and Karplus, M. (1973). *Mol. Phys.* **25,** 921.
Cantu, A. A., Klein, D. J., Matsen, F. A. and Seligman, T. H. (1975). *Theor. Chim. Acta* **38,** 341.
Caravetta, V. and Moccia, R. (1978). *Mol. Phys.* **35,** 129.
Cederbaum, L. S. and Domcke, W. (1977). *Adv. Chem. Phys.* **36,** 205.
Chojnacki, H. and Styrcz, S. (1981). *Acta Phys. Polon.* **A60,** 687.
Chojnacki, H., Laskowski, Z. and Andre, J.-M. (1984). *Acta Phys. Polon.* **A65,** 183.
Čįžek, J. (1966). *J. Chem. Phys.* **45,** 4256.
Čįžek, J. and Paldus, J. (1967). *J. Chem. Phys.* **47,** 3976.
Clementi, E. and Roetti, C. (1974). *At. Nucl. Data Tables* **14,** 177.
Clementi, E. and Veillard, A. (1967). *Theor. Chim. Acta* **7,** 133.
Cohen, E. R. and DuMond, J. W. (1965). *Rev. Mod. Phys.* **37,** 537.
Cohen, L. and Frishberg, C. (1976a). *J. Chem. Phys.* **65,** 4234.
Cohen, L. and Frishberg, C. (1976b). *Phys. Rev.* **A13,** 927.
Coleman, A. J. (1963). *Rev. Mod. Phys.* **35,** 668.
Coleman, A. J. (1968). *Adv. Quantum Chem.* **4,** 83.
Colle, R. and Salvetti, O. (1975). *Theor. Chim. Acta* **37,** 329.
Colle, R. and Salvetti, O. (1979). *Theor. Chim. Acta* **53,** 55.
Condon, E. U. (1930). *Phys. Rev.* **36,** 1121.
Condon, E. U. and Shortley, E. (1935). *The Theory of Atomic Spectra.* Cambridge University Press.
Cook, D. B. (1975). *Mol. Phys.* **30,** 733.
Cook, D. B. (1981). *Theor. Chim. Acta* **58,** 155.
Cook, D. B. (1984). *Mol. Phys.* **53,** 631.
Cooper, D. L., Gerratt, J. and Raimondi, M. (1987). *Adv. Chem. Phys.* **69,** 319.
Cooper, I. L. and McWeeny, R. (1966a). *J. Chem. Phys.* **45,** 226.
Cooper, I. L. and McWeeny, R. (1966b). *J. Chem. Phys.* **45,** 3484.
Cooper, I. L. and Pounder, C. (1980). *Int. J. Quantum Chem.* **17,** 759.
Cooper, J. L. B. (1948). *Q. Appl. Maths* **6,** 179.
Coulson, C. A. (1937). *Trans. Faraday Soc.* **33,** 388.
Coulson, C. A. (1940). *Proc. Camb. Phil. Soc.* **36,** 201.
Coulson, C. A. (1953). *Electricity,* Chap. 14. Oliver & Boyd, Edinburgh.
Coulson, C. A. and Longuet-Higgins, H. C. (1947). *Proc. R. Soc. Lond.* **A191,** 39.
Craig, D. P. and Thirunamachandran, T. (1984). *Molecular Quantum Electrodynamics.* Academic Press, London.
Csanak, G., Taylor, H. S. and Yaris, R. (1971). *Adv. At. Mol. Phys.* **7,** 287.
Daborn, G. T. and Handy, N. C. (1983). *Mol. Phys.* **49,** 1277.
Dacre, P. D. and McWeeny, R. (1970). *Proc. R. Soc. Lond.* **A317,** 435.
Dahl, P. J. and Avery, J. (eds.) (1982). *Local Density Approximations in Quantum Chemistry.* Plenum, New York.
Dalgaard, E. (1979). *Int. J. Quantum Chem.* **15,** 197 (see also *Chem. Phys. Lett.* **65,** 559).

Dalgaard, E. (1980). *J. Chem. Phys.* **72,** 816.
Dalgaard, E. and Jørgensen, P. (1978). *J. Chem. Phys.* **69,** 3833.
Dalgarno, A. (1961). In *Quantum Theory* (ed. D. R. Bates), Part I, Chap. 5. Academic Press, London.
Dalton, B. J. (ed.) (1980). *Theory and Appliction of Moment Methods in Many-Fermion Systems.* Plenum, New York.
Darwin, C. G. (1920). *Phil. Mag.* **39,** 537.
Das, G. and Wahl, A. C. (1967). *J. Chem. Phys.* **47,** 2934.
Davidson, E. R. (1962). *J. Chem. Phys.* **37,** 577.
Davidson, E. R. (1983). In *Methods in Computational Molecular Physics* (ed. G. H. F. Diercksen and S. Wilson), p. 95. Reidel, Dordrecht.
Davies, D. W. (1960). *J. Chem. Phys.* **33,** 781.
Davydov, A. S. (1962). *Theory of Molecular Exitons.* McGraw-Hill, New York.
Davydov, S. A. (1976). *Quantum Mechanics.* Pergamon, Oxford.
Day, O. W., Smith, D. W. and Morrison, R. C. (1975). *J. Chem. Phys.* **62,** 115.
Deb, B. M. (ed.) (1981). *The Force Concept in Chemistry.* Van Nostrand, New York.
Diercksen, G. H. F. and McWeeny, R. (1966). *J. Chem. Phys.* **44,** 3554.
Diercksen, G. H. F. and Sutcliffe, B. T. (1974). *Theor. Chim. Acta* **34,** 105.
Diercksen, G. H. F. and Sadlej, A. J. (1985). *Chem. Phys.* **96,** 17.
Dirac, P. A. M. (1929). *Proc. Camb. Phil. Soc.* **25,** 62.
Dirac, P. A. M. (1930). *Proc. Camb. Phil. Soc.* **26,** 376.
Dirac, P. A. M. (1931). *Proc. Camb. Phil. Soc.* **27,** 240.
Dirac, P. A. M. (1958). *The Principles of Quantum Mechanics,* 4th edn. Oxford University Press.
do Ameral, O. A. and McWeeny, R. (1983). *Theor. Chim. Acta (Berl.)* **64,** 171.
Dodds, J. and McWeeny, R. (1972). *Chem. Phys. Lett.* **13,** 9.
Dodds, J., McWeeny, R., Raynes, W. T. and Riley, J. P. (1977a). *Mol. Phys.* **33,** 611.
Dodds, J., McWeeny, R., Raynes, W. T. and Riley, J. P. (1977b). *Mol. Phys.* **34,** 1779.
Duch, W. and Karwowski, J. (1985). *Comput. Phys. Rep.* **2,** 93.
Dykstra, C. E. (1977a). *J. Chem. Phys.* **67,** 4716.
Dykstra, C. E. (1977b). *J. Chem. Phys.* **68,** 1829.
Eckart, C. (1930). *Rev. Mod. Phys.* **2,** 305.
Edmiston, C. and Ruedenberg, K. (1963). *Rev. Mod. Phys.* **35,** 457.
Edmonds, A. R. (1957). *Angular Momentum in Quantum Mechanics.* Princeton University Press.
Epstein, P. S. (1926). *Phys. Rev.* **28,** 695.
Epstein, S. T. (1974). *The Variation Method in Quantum Chemistry.* Academic Press, New York.
Epstein, S. T., Hurley, A. C., Parr, R. G. and Wyatt, R. E. (1967). *J. Chem. Phys.* **47,** 1275.
Eyring, H., Walter, J. and Kimball, G. (1944). *Quantum Chemistry.* Wiley, New York.
Fano, U. and Racah, G. (1959). *Irreducible Tensorial Sets.* Academic Press, New York.
Farazdel, A. and Smith, V. H. (1986). *Int. J. Quantum Chem.* **29,** 311.
Feynman, R. P. (1939). *Phys. Rev.* **56,** 340.
Feynman, R. P. and Hibbs, A. R. (1965). *Quantum Mechanics and Path Integrals.* McGraw-Hill, New York.

Feynman, R. P., Leighton, R. B. and Sands, M. (1964). *The Feynman Lectures on Physics*, Vol. II, 21-5. Addison-Wesley, Reading, Massachusetts.
Firsht, D. and McWeeny, R. (1976). *Mol. Phys.* **32,** 1673.
Firsht, D. and Pickup, B. T. (1977). *Int. J. Quantum Chem.* **12,** 765.
Fischer, G. (1984). *Vibronic Coupling*. Academic Press, London.
Fock, V. A. (1930). *Z. Phys.* **15,** 126.
Franz, L. M. and Mills, R. L. (1960). *Nucl. Phys.* **15,** 16.
Frenkel, J. (1934). *Wave Mechanics: Advanced General Theory*. Clarendon Press, Oxford.
Fueno, T., Nagase, S. and Rorokum, A. K. (1979). *J. Am. Chem. Soc.* **101,** 5849.
Fukutome, H. (1981). *Int. J. Quantum Chem.* **20,** 955.
Gallup, G. A. (1973). *Adv. Quantum Chem.* **16,** 229.
Gallup, G. A., Vance, R. L., Collins, J. R. and Norbeck, J. M. (1982). *Adv. Quantum Chem.* **16,** 229.
Gallup, G. A. and Norbeck, J. M. (1976). *J. Chem. Phys.* **64,** 2179.
Garton, D. and Sutcliffe, B. T. (1974). *Chem. Soc. Spec. Periodical Report: Theoretical Chemistry*, Vol. 1, p. 34.
Gerratt, J. (1967). PhD Thesis, University of Reading, England.
Gerratt, J. (1971). *Adv. Atom. Mol. Phys.* **7,** 141.
Gerratt, J. and Mills, I. M. (1968). *J. Chem. Phys.* **49,** 1719.
Gerratt, J. and Raimondi, M. (1980). *Proc. R. Soc. Lond.* **A371,** 525.
Gilbert, T. L. (1964). In *Molecular Orbitals in Chemistry, Physics, and Biology* (ed. P. O. Löwdin and B. Pullman), p. 465. Academic Press, London.
Gilbert, T. L. (1965). *J. Chem. Phys.* **43,** S245.
Gillespie, R. J. (1973). *Molecular Geometry*. Van Nostrand Reinhold, New York
Goddard, W. A. (1967a). *Phys. Rev.* **157,** 73.
Goddard, W. A. (1967b). *Phys. Rev.* **157,** 81.
Goddard, W. A. (1968a). *J. Chem. Phys.* **48,** 450.
Goddard, W. A. (1968b). *J. Chem. Phys.* **48,** 5337.
Goldstone, J. (1957). *Proc. R. Soc. Lond.* **A239,** 267.
Golebiewski, A., Hinze, J. and Yurtsever, E. (1979). *J. Chem. Phys.* **70,** 1101.
Goscinki, O. and Lukman, B. (1980). *Chem. Phys. Lett.* **7,** 573.
Gouyet, J. F. (1973). *J. Chem. Phys.* **59,** 4637.
Gouyet, J. F. (1981). In *The Unitary Group* (Lecture Notes in Chemistry, Vol. 22, ed. J. Hinze), p. 177. Springer-Verlag, Berlin.
Grant, I. (1970). *Adv. Phys.* **19,** 747.
Grant, I. and Quinney, H. M. (1988). *Adv. At. Mol. Phys.* **23,** 37.
Grein, F. and Banerjee, A. (1975). *Int. J. Quantum Chem. Symp.* **9,** 147.
Grein, F. and Chang, T. C. (1971). *Chem. Phys. Lett.* **12,** 44.
Griffith, J. S. (1961). *The Theory of Transition Metal Ions*, p. 396. Cambridge University Press.
Griffith, J. S. (1962). *The Irreducible Tensor Method for Molecular Symmetry Groups*. Prentice-Hall, Englewood Cliffs, New Jersey.
Gritsenko, O. V., Bagaturjants, A. A. and Kazansky, V. B. (1986). *Int. J. Quantum Chem.* **29,** 1799.
Guest, M. F. and Saunders, V. R. (1974). *Mol. Phys.* **28,** 819.
Hall, G. G. (1951). *Proc. R. Soc. Lond.* **A205,** 541.
Hall, G. G. (1961). *Phil. Mag.* **6,** 249.
Hall, G. G. (1964). *Adv. Quantum Chem.* **1,**241.

Hamermesh, M. (1962). *Group Theory and its Applications to Physical Problems.* Addison-Wesley, Reading, Massachusetts.
Handy, N. C., Knowles, P. J. and Somasundran, K. (1985a). *Theor. Chim. Acta* **68,** 87.
Handy, N. C., Knowles, P. J. and Somasundran, K. (1985b). *Chem. Phys. Lett.* **113,** 8.
Harriman, J. (1980). *Int. J. Quantum Chem.* **17,** 689.
Hartree, D. R. (1928). *Proc. Camb. Phil. Soc.* **24,** 328.
Heitler, W. and London, F. (1927). *Z. Phys.* **44,** 455.
Hellmann, H. (1937). *Einführung in die Quantenchemie.* Franz Deutiche, Leipzig.
Herzberg, G. (1945). *Molecular Spectra and Molecular Structure,* Vol. 1. Van Nostrand, New York.
Herzberg, G. (1950). *Molecular Spectra and Molecular Structure,* Vol. 2. Van Nostrand, New York.
Herzberg, G. (1966). *Molecular Spectra and Molecular Structure,* Vol. 3. Van Nostrand, New York.
Hestenes, M. R. (1980). *Conjugate Direction Methods in Optimization.* Springer-Verlag, New York.
Hibbert, A. (1975). *Rep. Prog. Phys.* **38,** 1217.
Hinze, J. (1973). *J. Chem. Phys.* **59,** 6424.
Hinze, J. (ed.) (1981). *The Unitary Group* (Lecture Notes in Chemistry, Vol. 22). Springer-Verlag, Berlin.
Hinze, J. and Yurtsever, E. (1979). *J. Chem. Phys.* **70,** 3188.
Hirao, K. (1974a) *J. Chem. Phys.* **60,** 3215.
Hirao, K. (1974b). *J. Chem. Phys.* **61,** 3247.
Hirschfelder, J. O. (1977). *J. Chem. Phys.* **67,** 5477.
Hirschfelder, J. O., Byers-Brown, W. and Epstein, S. T. (1964). *Adv. Quantum Chem.* **1,** 256.
Hobdza, P. and Zahradnik, R. (1980). *Weak Intermolecular Interactions in Chemistry and Biochemistry.* Pergamon, Oxford.
Hückel, E. (1931). *Z. Phys.* **70,** 204.
Hurley, A. C. (1954). *Proc. R. Soc. Lond.* **A226,** 179.
Hurley, A. C. (1976). *Electron Correlation in Small Molecules.* Academic Press, London.
Husimi, K. (1940). *Proc. Phys.-Math. Soc. Jpn* **22,** 264.
Hutchinson, C. and Mangum, B. (1958). *J. Chem. Phys.* **29,** 952.
Huzinaga, S. (1965). *J. Chem. Phys.* **42,** 1293.
Hylleraas, E. A. and Undheim, B. (1930). *Z. Phys.* **65,** 759.
Ishida, K. (1985). *Int. J. Quantum Chem.* **28,** 349.
Itoh, T. (1965). *Rev. Mod. Phys.* **37,** 159.
Jankowski, K. (1987). In *Methods in Computational Chemistry,* Vol. 1 (ed. S. Wilson), p. 1. Plenum, New York.
Jaszunski, M. (1976). *Acta Phys. Polon.* **A49,** 785.
Jaszunski, M. (1978). *Theor. Chim. Acta* **48,** 323.
Jaszunski, M. and McWeeny, R. (1982). *Mol. Phys.* **46,** 863.
Jaszunski, M. and McWeeny, R. (1985). *Mol. Phys.* **55,** 1275.
Jaszunski, M. and Sadlej, A. J. (1975). *Theor. Chim. Acta* **40,** 157.
Jaszunski, M. and Sadlej, A. J. (1977). *Int. J. Quantum Chem.* **11,** 233.
Jørgen, H., Jensen, A., Jørgensen, P. and Ågren, H. (1987). *J. Chem. Phys.* **87,** 451.

Jørgensen, P. (1975). *Am. Rev. Phys. Chem.* **26,** 359.
Jørgensen, P. and Simons, J. (1981). *Second-Quantization-Based Methods in Quantum Chemistry*. Academic Press, New York.
Jørgensen, P. and Simons, J. (1983). *J. Chem. Phys.* **79,** 334.
Kaldor, U. (1975a). *J. Chem. Phys.* **62,** 4634.
Kaldor, U. (1975b). *J. Chem. Phys.* **63,** 2199.
Kaplan, I. G. (1975). *Symmetry of Many-Electron Systems*. Academic Press, New York.
Kato, T. (1951). *Commun. Pure Appl. Maths* **10,** 151.
Kelly, H. P. (1969). *Adv. Chem. Phys.* **14,** 129.
Kelly, H. P. (1979). *Comput. Phys. Commun.* **17,** 99.
Kemble, E. (1958). *The Fundamental Principles of Quantum Mechanics*. Dover, New York.
King, H. F., Stanton, R. E., Kim, H., Wyatt, R. E. and Parr, R. G. (1967). *J. Chem. Phys.* **47,** 1936.
Klessinger, M. and McWeeny, R. (1965). *J. Chem. Phys.* **42,** 3343.
Knowles, P. and Werner, H.-J. (1985). *Chem. Phys. Lett.* **115,** 259.
Koopmans, T. (1933). *Physica,'s Grav.* **1,** 104.
Kotani, M., Ameniya, A., Ishiguro, E. and Kimura, T. (1955). *Tables of Molecular Integrals*. Maruzen, Tokyo.
Kramers, H. A. (1930). *Proc. Amsterdam Acad.* **33,** 959.
Kvasnička, V. (1974). *Czech. J. Phys.* **B24,** 605.
Kvasnička, V. (1977a). *Czech. J. Phys.* **B27,** 599.
Kvasnička, V. (1977b). *Adv. Chem. Phys.* **36,** 345.
Kutzelnigg, W. (1973). *Top. Curr. Chem.* **41,** 31.
Kutzelnigg, W. (1977). In *Modern Theoretical Chemistry,* Vol. 3: *Methods of Electronic Structure Theory* (ed. H. F. Schaeffer), p. 129. Plenum, New York.
Kutzelnigg, W. (1979). *Chem. Phys. Lett.* **64,** 383.
Kutzelnigg, W., Del Re, G. and Berthier, G. (1968). *Phys. Rev.* **172,** 49.
Langhoff, P. W., Epstein, S. T. and Karplus, M. (1972). *Rev. Mod. Phys.* **44,** 602.
Lefebvre-Brion, H. and Field, R. W. (1986). *Perturbations in the Spectra of Diatomic Molecules*. Academic Press, New York.
Lennard-Jones, J. (1931). *Proc. Camb. Phil. Soc.* **27,** 469.
Lennard-Jones, J. (1949a). *Proc. R. Soc. Lond.* **A198,** 1.
Lennard-Jones, J. (1949b). *Proc. R. Soc. Lond.* **A198,** 14.
Levy, B. and Berthier, G. (1968). *Int. J. Quantum Chem.* **2,** 307.
Linderberg, J. and Öhrn, Y. (1972). *Propagators in Quantum Chemistry*. Academic Press, London.
Lindgren, I. (1974). *J. Phys.* **B7,** 2441.
Lindgren, I. (1978). *Int. J. Quantum Chem. Symp.* **12,** 33.
Lindgren, I. and Morrison, J. (1982). *Atomic Many-Body Theory*. Springer-Verlag, Berlin.
Liu, B. (1978). In *Numerical Algorithms in Chemistry: Algebraic Methods,* p. 49. Lawrence Berkeley Lab. Rep. NRCC LBL-8158, Berkeley.
London, F. (1930). *Z. Phys.* **63,** 245.
London, F. (1937). *J. Phys. Radium* **8,** 397.
Longuet-Higgins, H. C. (1961). *Adv. Spectrosc.* **2,** 429.
Louck, J. D. (1970). *Am. J. Phys.* **38,** 3.

Löwdin, P.-O. (1950). *J. Chem. Phys.* **18,** 365.
Löwdin, P.-O. (1951). *J. Chem. Phys.* **19,** 1396.
Löwdin, P.-O. (1955). *Phys. Rev.* **97,** 1474.
Löwdin, P.-O. (1959). *Adv. Chem. Phys.* **2,** 207.
Löwdin, P.-O. (1962a). *J. Math. Phys.* **3,** 969.
Löwdin, P.-O. (1962b). *J. Math. Phys.* **3,** 1171.
Löwdin, P.-O. (1964). *J. Mol. Spectrosc.* **14,** 131.
Löwdin, P.-O. (1977). *Int. J. Quantum Chem. Suppl.* 1, **12,** 197.
Löwdin, P.-O. (1982). *Int. J. Quantum Chem. Symp.* **16,** 485.
Löwdin, P.-O. (1985). *Adv. Quantum Chem.* **17,** 285.
Löwdin, P.-O. and Mukherjee, P. K. (1972). *Chem. Phys. Lett.* **14,** 1.
Löwdin, P.-O. and Shull, H. (1956). *J. Chem. Phys.* **25,** 1035.
McConnell, H. M. (1958). *J. Chem. Phys.* **28,** 1188.
McCurdy, C. W., Rescigno, T. N., Yeager, D. L. and McCoy, V. (1977). In *Modern Theoretical Chemistry,* Vol. 3: *Methods of Electronic Structure Theory* (ed. H. F. Schaeffer), p. 339. Plenum, New York.
MacDonald, J. K. L. (1933). *Phys. Rev.* **43,** 830.
McLachlan, A. D. (1962). *Mol. Phys.* **5,** 51.
McLachlan, A. D. (1963). *Mol. Phys.* **6,** 441.
McLachlan, A. D. and Ball, M. A. (1964). *Rev. Mod. Phys.* **36,** 844.
McWeeny, R. (1951a). *J. Chem. Phys.* **19,** 1614.
McWeeny, R. (1951b). *J. Chem. Phys.* **20,** 920.
McWeeny, R. (1952). *Acta Crystallogr.* **5,** 463.
McWeeny, R. (1954a). *Proc. R. Soc. Lond.* **A223,** 63.
McWeeny, R. (1954b). *Proc. R. Soc. Lond.* **A233,** 306.
McWeeny, R. (1955a). *Proc. R. Soc. Lond.* **A227,** 288.
McWeeny, R. (1955b). *Proc. R. Soc. Lond.* **A232,** 114.
McWeeny, R. (1956). *Proc. R. Soc. Lond.* **A235,** 496.
McWeeny, R. (1957). *Proc. R. Soc. Lond.* **A241,** 243.
McWeeny, R. (1959). *Proc. R. Soc. Lond.* **A253,** 242.
McWeeny, R. (1960). *Rev. Mod. Phys.* **32,** 335.
McWeeny, R. (1961a). *Phys. Rev.* **126,** 1028.
McWeeny, R. (1961b). *J. Chem. Phys.* **34,** 399 (see also *ibid.* **34,** 1065).
McWeeny, R. (1963). *Symmetry*. Pergamon Press, Oxford.
McWeeny, R. (1965). *J. Chem. Phys.* **42,** 1717.
McWeeny, R. (1967). *Int. J. Quantum. Chem.* **15,** 351.
McWeeny, R. (1973). *Quantum Mechanics: Methods and Basic Applications.* Pergamon, Oxford.
McWeeny, R. (1974). *Mol. Phys.* **28,** 1273.
McWeeny, R. (1975). *Chem. Phys. Lett.* **35,** 13.
McWeeny, R. (1976). In *The New World of Quantum Chemistry* (ed. B. Pullman and R. Parr), p. 3. Reidel, Dordrecht.
McWeeny, R. (1983). *Int. J. Quantum Chemistry* **23,** 405.
McWeeny, R. (1984). *Croat. Chem. Acta* **57,** 865.
McWeeny, R. (1985). *J. Mol. Struct.* (*Theochem*) **123,** 231.
McWeeny, R. (1986). *Proc. Ind. Acad. Sci.* **96,** 263.
McWeeny, R. (1988). *Int. J. Quantum Chem.* **34,** 25.
McWeeny, R. and Del Re, G. (1968). *Theor. Chim. Acta* **10,** 13.
McWeeny, R. and Kutzelnigg, W. (1968). *Int. J. Quantum Chem.* **2,** 187.
McWeeny, R. and Mizuno, Y. (1961). *Proc. R. Soc. Lond.* **A259,** 554.

McWeeny, R. and Steiner, E. (1965). *Adv. Quantum. Chem.* **2,** 93.
McWeeny, R. and Sutcliffe, B. T. (1963). *Proc. R. Soc. Lond.* **A273,** 103.
Magnasco, V. and Figaro, G. (1985). *Mol. Phys.* **55,** 319.
Magnasco, V. and Perico, A. (1967). *J. Chem. Phys.* **48,** 800.
Manne, R. (1977). *Int. J. Quantum. Chem. Symp.* **11,** 175.
Manne, R. and Zerner, M. F. (1985). *Int. J. Quantum Chem. Symp.* **19,** 165.
March, N. H., Young, W. H. and Sampanther, S. (1967). *The Many-Body Problem in Quantum Mechanics.* Cambridge University Press.
Margenau, H. and Murphy, G. (1956). *The Mathematics of Physics and Chemistry.* Van Nostrand, New York.
Matsen, F. A. (1964). *Adv. Quantum Chem.* **1,** 59.
Matsen, F. A. (1976). *Int. J. Quantum Chem.* **10,** 525.
Matsen, F. A. and Pauncz, R. (1986). *The Unitary Group in Quantum Chemistry.* Elsevier, Amsterdam.
Mattuck, R. D. (1976). *A Guide to Feynman Diagrams.* McGraw-Hill, New York.
Meyer, W. (1976). *J. Chem. Phys.* **64,** 2901.
Meyer, W. (1977). In *Modern Theoretical Chemistry*, Vol. 3: *Methods of Electronic Structure Theory* (ed. H. F. Schaeffer), p. 413. Plenum, New York.
Moccia, R. (1973a). Notes on Diagrammatic Perturbation Theory (unpublished).
Moccia, R. (1973b). *Int. J. Quantum. Chem.* **7,** 779.
Moccia, R. (1974). *Int. J. Quantum. Chem.* **8,** 293.
Moccia, R., Resta, R. and Zandomeneghi, M. (1976). *Chem. Phys. Lett.* **37,** 556.
Møller, C. and Plesset, M. S. (1934). *Phys. Rev.* **46,** 618.
Monod, J. (1971). *Chance and Necessity.* Random House, New York.
Moores, W. H. and McWeeny, R. (1973). *Proc. R. Soc. Lond.* **A332,** 365.
Morse, P. M. and Feshbach, H. (1953). *Methods of Theoretical Physics,* Vols. 1 and 2. McGraw-Hill, New York.
Moshinsky, M. (1963). *J. Math. Phys.* **4,** 1128.
Moshinsky, M. (1966). *J. Math. Phys.* **7,** 691.
Moshinsky, M. (1968). *Group Theory and the Many-Body Problem.* Gordon and Breach, New York.
Moshinsky, M. and Seligman, T. H. (1971). *Ann. Phys.* (*N.Y.*) **66,** 311.
Moss, R. E. (1973). *Advanced Molecular Quantum Mechanics.* Chapman & Hall, London.
Mukherjee, D. (1978). *Proc. Ind. Acad. Sci.* **87A,** 37.
Mukherjee, N. G. and McWeeny, R. (1970). *Int. J. Quantum. Chem.* **4,** 97.
Mulder, J. J. C. (1966). *Mol. Phys.* **10,** 479.
Mulliken, R. (1955a). *J. Chem. Phys.* **23,** 1833.
Mulliken, R. (1955b). *J. Chem. Phys.* **23,** 2343.
Murrell, J. N. and Harget, A. J. (1972). *Semi-Empirical Self-Consistent-Field Molecular-Orbital Theory of Molecules.* Wiley-Interscience, London.
Nesbet, R. K. (1955a). *Proc. R. Soc. Lond.* **A230,** 312.
Nesbet, R. K. (1955b). *Proc. R. Soc. Lond.* **A230,** 322.
Nesbet, R. K. (1965a). *Adv. Chem. Phys.* **9,** 321.
Nesbet, R. K. (1965b). *J. Chem. Phys.* **43,** 311.
Norbeck, J. M. and McWeeny, R. (1975). *Chem. Phys. Lett.* **34,** 206.
Nyholm, R. S. and Gillespie, R. J. (1957). *Q. Rev. Chem. Soc.* **11,** 339.
Oddershede, J., Jørgensen, P. and Yeager, D. L. (1984). *Comput. Phys. Rep.* **2,** 33.

Öhrn, Y. and Born, G. (1981). *Adv. Quantum Chem.* **13,** 1.
Olsen, J., Yeager, D. and Jørgensen, P. (1983). *Adv. Chem. Phys.* **54,** 1.
Paldus, J. (1975). *Int. J. Quantum Chem. Symp.* **9,** 165.
Paldus, J. and Čížek, J. (1969). *Prog. Theor. Phys.* **42,** 769.
Paldus, J. and Čížek, J. (1970a). *J. Chem. Phys.* **52,** 2919.
Paldus, J. and Čížek, J. (1970b). *Phys. Rev.* **A2,** 2268.
Paldus, J. and Čížek, J. (1975). *Adv. Quantum Chem.* **7,** 105.
Paldus, J. and Wormer, P. (1989). *Group Theory of Many-Electron Systems.* Academic Press, London.
Parr, R. (1983). *Ann. Rev. Phys. Chem.* **34,** 631.
Parr, R. G., Ellison, F. O. and Lykos, P. G. (1956). *J. Chem. Phys.* **24,** 1106.
Pauling, L. (1933). *J. Chem. Phys.* **1,** 280.
Pauling, L. and Wilson, E. B. (1935). *Introduction to Quantum Mechanics.* McGraw-Hill, New York.
Pauncz, R. (1967). *The Alternant Molecular Orbital Method.* Saunders, Philadelphia.
Pauncz, R. (1975). *Chem. Phys. Lett.* **31,** 443.
Pauncz, R. (1977). *Int. J. Quantum Chem.* **12,** 369.
Pauncz, R. (1979). *Spin Eigenfunctions: Construction and Use.* Plenum, New York.
Peng, M. (1941). *Proc. R. Soc. Lond.* **A178,** 449.
Pickup, B. T. (1977). *Chem. Phys.* **19,** 193.
Poirier, P., Kari, R. and Csizmadia, I. G. (1985). *Handbook of Gaussian Basis Sets.* Elsevier, Amsterdam.
Polezzo, S. (1975). *Theor. Chim. Acta* **38,** 211.
Polezzo, S. and Fantucci, P. (1978). *Mol. Phys.* **36,** 1835.
Pople, J. A. (1953). *Trans. Faraday Soc.* **49,** 1375.
Pople, J. A. and Beveridge, D. L. (1970). *Approximate Molecular Orbital Theory.* McGraw-Hill, New York.
Pople, J. A. and Nesbet, R. K. (1954). *J. Chem. Phys.* **22,** 521.
Pople, J. A. and Nesbet, R. K. (1954). *J. Chem. Phys.* **22,** 521.
Pople, J. A., Schneider, W. and Bernstein, H. (1959). *High Resolution Nuclear Magnetic Resonance,* Appendix A. McGraw-Hill, New York.
Pople, J. A., Santry, D. and Segal, G. (1965). *J. Chem. Phys.* **43,** 5129.
Pople, J. A., McIver, J. W. and Ostlund, N. S. (1968). *J. Chem. Phys.* **49,** 2960.
Pople, J., Binkley, J. and Seeger, R. (1976). *Int. J. Quantum Chem. Symp.* **10,** 1.
Pople, J. A., Krishnan, R., Schlegel, H. B. and Binkley, J. S. (1979). *Int. J. Quantum Chem. Symp.* **13,** 225.
Primas, H. (1965). In *Modern Theoretical Chemistry* (ed. O. Sinanoglu), Vol. 2, p. 45. Academic Press, New York.
Prosser, F., and Hagstrom, S. (1968). *Int. J. Quantum Chem.* **2,** 89.
Pulay, P. (1977). In *Modern Theoretical Chemistry,* Vol. 4: *Applications of Electronic Structure Theory* (ed. H. F. Schaeffer), p. 153. Plenum, New York.
Purvis, G. D. and Bartlett, R. J. (1981). *J. Chem. Phys.* **75,** 1284.
Pyykö, P. (1978). *Adv. Quantum Chem.* **11,** 354.
Pyykö, P. (1986). *Relativistic Theory of Atoms and Molecules* (Lecture Notes in Chemistry, Vol. 41). Springer-Verlag, Berlin.
Pyper, N. C. and Gerratt, J. (1977). *Proc. R. Soc. Lond.* **A355,** 402.
Raffinetti, R. C. (1979). *J. Comput. Phys.* **32,** 403.

Raimondi, M. and Simonetta, M. (1976). *Int. J. Quantum Chem.* **10,** 1057.
Raimondi, M., Simonetta, M. and Tantardini, G. F. (1975). *Mol. Phys.* **30,** 703.
Raimondi, M., Simonetta, M. and Tantardini, G. F. (1985). *Comput. Phys. Rep.* **2,** 171.
Rajagopal, A. K. (1980). *Adv. Chem. Phys.* **41,** 59.
Ratajczak, H. and Orville-Thomas, W. J. (eds.) (1982). *Molecular Interactions,* Vols. 1–3. Wiley, Chichester.
Reeves, C. (1957). PhD Thesis, University of Cambridge, England.
Reeves, C. (1966). *Commun. ACM* **9,** 276.
Rettrup, S. (1982). *J. Comput. Phys.* **45,** 100.
Rettrup, S. (1986). *Int. J. Quantum Chem.* **29,** 119.
Robb, M. A. and Niazi, U. (1984). *Comput. Phys. Rep.* **1,** 127.
Roos, B. O. (1972). *Chem. Phys. Lett.* **15,** 153.
Roos, B. O. (1983). In *Methods in Computational Molecular Physics* (ed. G. H. F. Diercksen and S. Wilson), p. 161. Reidel, Dordredt.
Roos, B. O. and Siegbahn, P. E. M. (1977). In *Modern Theoretical Chemistry,* Vol. 3: *Methods of Electronic Structure Theory* (ed. H. F. Schaeffer), p. 277. Plenum, New York.
Roos, B. O., Taylor, P. and Siegbahn, P. (1980). *Chem. Phys.* **48,** 157.
Roothaan, C. C. J. (1951). *Rev. Mod. Phys.* **23,** 69.
Roothaan, C. C. J. (1960). *Rev. Mod. Phys.* **32,** 179.
Roothaan, C. C. J. and Bagus, P. (1964). *Methods Comput. Phys.* **2,** 47.
Rotenberg, M., Bivins, R., Metropolis, N. C. and Wooten, J. K. (1959). *The 3-j and 6-j Symbols.* The Technology Press, Cambridge, Massachusetts.
Rowe, D. J. (1968). *Rev. Mod. Phys.* **40,** 153.
Ruedenberg, K. (1962). *Rev. Mod. Phys.* **34,** 326.
Ruedenberg, K., Cheung, L. M. and Elbert, S. T. (1979). *Int. J. Quantum Chem.* **16,** 1069.
Rumer, G. (1932). *Göttingen Nachr.* 377.
Rutherford, D. E. (1948). *Substitutional Analysis.* Edinburgh University Press.
Sadlej, A. J. *Chem. Phys. Lett.* **47,** 50.
Saxe, P., Fox, P. J., Schaeffer, H. F. and Handy, N. C. (1982). *J. Chem. Phys.* **77,** 5584.
Schmidt, E. (1907). *Math. Annln* **63,** 433.
Seeger, R. and Pople, J. A. (1977). *J. Chem. Phys.* **66,** 3045.
Seligman, T. H. (1981). In *The Unitary Group* (Lecture Notes in Chemistry, Vol. 22, ed. J. Hinze), p. 362. Springer-Verlag, Berlin.
Serber, R. (1934a) *Phys. Rev.* **45,** 461.
Serber, R. (1934b). *J. Chem. Phys.* **2,** 697.
Shavitt, I., Bender, C. F., Pipano, A. and Hosteney, R. P. (1973). *J. Comput. Phys.* **11,** 90.
Shepard, R. (1987). *Adv. Chem. Phys.* **69,** 63.
Sidwick, N. and Powell, H. (1940). *Proc. R. Soc. Lond.* **A176,** 153.
Silver, D. M. and Wilson, S. (1978). *J. Chem. Phys.* **69,** 3837.
Simonetta, M., Gianinetti, E. and Vandoni, I. (1968). *J. Chem. Phys.* **48,** 1579.
Sinanoglu, O. (1964). *Adv. Chem. Phys.* **6,** 315.
Slater, J. C. (1929). *Phys. Rev.* **34,** 1293.
Slater, J. C. (1930). *Phys. Rev.* **35,** 210.
Slater, J. C. (1931). *Phys. Rev.* **38,** 1109.
Slater, J. C. (1951). *Phys. Rev.* **81,** 385.

Slater, J. C. (1960). *Quantum Theory of Atomic Structure,* Vols. 1 and 2, McGraw-Hill, New York.
Slichter, C. P. (1964). *Principles of Magnetic Resonance.* Harper and Row, New York.
Smith, D. W. and Day, O. W. (1975). *J. Chem. Phys.* **62,** 113.
Sneddon. I. N. (1951). *Fourier Transforms.* McGraw-Hill, New York.
Stewart, G. W. (1973). *Introduction to Matrix Computations.* Academic Press, New York.
Stone, A. J. and Alderton, M. (1985). *Mol. Phys.* **56,** 1047.
Sutcliffe, B. T. (1966). *J. Chem. Phys.* **45,** 235.
Szabo, A. and Ostlund, N. S. (1977). *J. Chem. Phys.* **67,** 4251.
Talmi, I. and de Shalit, A. (1963). *Nuclear Shell Theory.* Academic Press, London.
ter Haar, D. (1954). *Elements of Statistical Mechanics.* Holt, Rinehart, Winston, New York.
ter Haar, D. (1961). *Rep. Prog. Phys.* **24,** 304.
Thouless, D. J. (1961). *The Quantum Mechanics of Many-Body Systems.* Academic Press, New York.
Tolman, R. C. (1938). *The Principles of Statistical Mechanics.* Oxford University Press.
Urban, M., Černušák, I., Kellö, V. and Noga, J. (1987). In *Methods in Computational Chemistry,* Vol. 1: *Electron Correlation in Atoms and Molecules* (ed. S. Wilson), p. 117. Plenum, New York.
van Vleck, J. H. and Sherman, A. (1935). *Rev. Mod. Phys.* **7,** 167.
von Neumann, J. (1927). *Göttingen Nachr.* 245.
von Niessen, W., Schirmer, J. and Cederbaum, L. S. (1984). *Comput. Phys. Rep.* **1,** 57.
Weissman, S. I. (1956). *J. Chem. Phys.* **25,** 890.
Werner, H.-J. (1987). *Adv. Chem. Phys.* **69,** 1.
Werner, H.-J. and Knowles, P. (1985). *J. Chem. Phys.* **82,** 5053.
Werner, H.-J. and Meyer, W. (1980). *J. Chem. Phys.* **73,** 2342.
Weyl, H. (1956). *The Theory of Groups and Quantum Mechanics.* Dover, New York.
Wigner, E. P. (1959). *Group Theory and its Application to the Quantum Mechanics of Atomic Spectra,* pp. 131, 257. Academic Press, New York.
Wilson, S. (1977). *Int. J. Quantum Chem.* **12,** 604.
Wilson, S. (1978). *Mol. Phys.* **35,** 1.
Wilson, S. (1983). In *Methods in Computational Molecular Physics* (ed. G. H. F. Diercksen and S. Wilson), p. 273. Reidel, Dordrecht.
Wilson, S. (1984). *Electron Correlation in Molecules.* Oxford University Press.
Wilson, S. (1986). *Chemistry by Computer—An Overview of the Applications of Computers in Chemistry.* Plenum, New York.
Wilson, S. and Silver, D. M. (1979a). *Int. J. Quantum Chem.* **15,** 683.
Wilson, S. and Silver, D. M. (1979b). *Chem. Phys. Lett.* **63,** 367.
Wilson, E. B., Decius, J. and Cross, P. (1955). *Molecular Vibrations.* McGraw-Hill, New York.
Wilson, S., Jankowski, K. and Paldus, J. (1983). *Int. J. Quantum Chem.* **23,** 1781.
Wooley, G. and Sutcliffe, B. T. (1977). *Chem. Phys. Lett.* **45,** 393.
Wormer, P. (1981). In *The Unitary Group* (Lecture Notes in Chemistry, Vol. 22, ed. J. Hinze), p. 286. Springer-Verlag, Berlin.

Wormer, P., Visser, F. and Paldus, J. (1982). *J. Comp. Phys.* **48,** 23.
Yang, C. N. (1962). *Rev. Mod. Phys.* **34,** 694.
Yeager, D. L. and Jørgensen, P. (1979a). *J. Chem. Phys.* **71,** 755.
Yeager, D. L. and Jørgensen, P. (1979b). *Chem. Phys. Lett.* **65,** 77.
Zubarev, D. N. (1960). *Sov. Phys. Usp.* **3,** 320.
Zubarev, D. N. (1964). *Non-Equilibrium Statistical Mechanics.* Consultants Bureau, New York.

Index